P_k	kth percentile, where k is an integer from 1 to 99
$P(A)$	Probability that event A will occur
$P(A\|B)$	Probability that event A will occur given event B has occurred
q	$1 - p$, probability of failure for the binomial experiments
$\hat{q}$	$1 - \hat{p}$
$\bar{q}$	$1 - \bar{p}$
Q_1, Q_2, Q_3	First, second, and third quartiles, respectively
R	Number of rows in a contingency table
r	Linear correlation coefficient
r^2	Coefficient of determination
S	Sample space
s	Sample standard deviation
s^2	Sample variance
s_b	Estimator of σ_b
s_d	Standard deviation of the paired differences for a sample
$s_{\bar{d}}$	Estimator of $\sigma_{\bar{d}}$
s_e	Standard deviation of errors for the sample regression model
s_p	Pooled standard deviation
$s_{\hat{p}}$	Estimator of $\sigma_{\hat{p}}$
$s_{\bar{x}}$	Estimator of $\sigma_{\bar{x}}$
$s_{\hat{y}_m}$	Estimator of $\sigma_{\hat{y}_m}$
$s_{\hat{y}_p}$	Estimator of $\sigma_{\hat{y}_p}$
Σ	(Greek letter capital sigma) Summation
σ	(Greek letter lowercase sigma) Population standard deviation
σ^2	Population variance
σ_b	Standard deviation of the sampling distribution of b
σ_d	Standard deviation of the paired differences for the population
$\sigma_{\bar{d}}$	Standard deviation of the sampling distribution $\bar{d}$
σ_ϵ	Standard deviation of errors for the population regression model
$\sigma_{\hat{p}}$	Standard deviation of the sampling distribution of $\hat{p}$
$\sigma_{\bar{x}}$	Standard deviation of the sampling distribution of $\bar{x}$
$\sigma_{\hat{y}_m}$	Standard deviation of $\hat{y}$ when estimating $\mu_{y\|x}$
$\sigma_{\hat{y}_p}$	Standard deviation of $\hat{y}$ when predicting y_p
t	t distribution
T_i	Sum of the values included in sample i in ANOVA
x	(1) Variables; (2) a random variable; (3) independent variable in a regression model
$\bar{x}$	Sample mean
y	(1) Variable; (2) dependent variable in a regression model
$\hat{y}$	Estimated or predicted value of y using a regression model
z	Units of the standard normal distribution

LIST OF KEY FORMULAS

Prem S. Mann • Introductory Statistics

CHAPTER 3 • NUMERICAL DESCRIPTIVE MEASURES

- Mean for ungrouped data: $\mu = \Sigma x/N$ and $\bar{x} = \Sigma x/n$ where μ is the population mean and $\bar{x}$ is the sample mean.
- Mean for grouped data: $\mu = \Sigma mf/N$ and $\bar{x} = \Sigma mf/n$ where m is the midpoint and f is the frequency of a class.
- Variance for ungrouped data:

$$\sigma^2 = \frac{\Sigma x^2 - \frac{(\Sigma x)^2}{N}}{N} \quad \text{and} \quad s^2 = \frac{\Sigma x^2 - \frac{(\Sigma x)^2}{n}}{n-1}$$

where σ^2 is the population variance and s^2 is the sample variance.

- Standard deviation for ungrouped data:

$$\sigma = \sqrt{\frac{\Sigma x^2 - \frac{(\Sigma x)^2}{N}}{N}} \quad \text{and} \quad s = \sqrt{\frac{\Sigma x^2 - \frac{(\Sigma x)^2}{n}}{n-1}}$$

where σ and s are the population and sample standard deviations, respectively.

- Variance for grouped data:

$$\sigma^2 = \frac{\Sigma m^2 f - \frac{(\Sigma mf)^2}{N}}{N} \quad \text{and} \quad s^2 = \frac{\Sigma m^2 f - \frac{(\Sigma mf)^2}{n}}{n-1}$$

- Standard deviation for grouped data:

$$\sigma = \sqrt{\frac{\Sigma m^2 f - \frac{(\Sigma mf)^2}{N}}{N}} \quad \text{and} \quad s = \sqrt{\frac{\Sigma m^2 f - \frac{(\Sigma mf)^2}{n}}{n-1}}$$

CHAPTER 4 • PROBABILITY

- Relative frequency as an approximation of probability:
$$P(A) = f/n$$
- Conditional probability of an event:
$$P(A|B) = P(A \text{ and } B)/P(B)$$
- Condition for independence of events:
$$P(A) = P(A|B) \quad \text{and/or} \quad P(B) = P(B|A)$$
- For complementary events: $P(A) + P(\bar{A}) = 1$
- Multiplication rule for dependent events:
$$P(A \text{ and } B) = P(A)\, P(B|A)$$
- Multiplication rule for independent events:
$$P(A \text{ and } B) = P(A)\, P(B)$$
- Joint probability of two mutually exclusive events:
$$P(A \text{ and } B) = 0$$
- Addition rule for nonmutually exclusive events:
$$P(A \text{ or } B) = P(A) + P(B) - P(A \text{ and } B)$$
- Addition rule for mutually exclusive events:
$$P(A \text{ or } B) = P(A) + P(B)$$

CHAPTER 5 • DISCRETE RANDOM VARIABLES AND THEIR PROBABILITY DISTRIBUTIONS

- Mean of a discrete random variable x: $\mu = \Sigma x P(x)$
- Standard deviation of a discrete random variable x:
$$\sigma = \sqrt{\Sigma x^2 P(x) - \mu^2}$$
- n factorial: $n! = n(n-1)(n-2)\ldots 3 \cdot 2 \cdot 1$
- Number of combinations of n items selected x at a time:
$$\binom{n}{x} = \frac{n!}{(n-x)!\,x!}$$
- Binomial probability formula: $P(x) = \binom{n}{x} p^x q^{n-x}$
- Mean and standard deviation of the binomial distribution:
$$\mu = np \quad \text{and} \quad \sigma = \sqrt{npq}$$
- Poisson probability formula: $P(x) = \lambda^x e^{-\lambda}/x!$

CHAPTER 6 • CONTINUOUS RANDOM VARIABLES AND THE NORMAL DISTRIBUTION

- z value for an x value: $z = (x - \mu)/\sigma$

CHAPTER 7 • SAMPLING DISTRIBUTIONS

- z value for $\bar{x}$: $z = (\bar{x} - \mu)/\sigma_{\bar{x}}$ where $\sigma_{\bar{x}} = \sigma/\sqrt{n}$
- z value for $\hat{p}$: $z = (\hat{p} - p)/\sigma_{\hat{p}}$ where $\sigma_{\hat{p}} = \sqrt{pq/n}$

CHAPTER 8 • ESTIMATION OF THE MEAN AND PROPORTION

- Confidence interval for μ for a large sample:

$\bar{x} \pm z\,\sigma_{\bar{x}}$ if σ is known

$\bar{x} \pm z\,s_{\bar{x}}$ if σ is not known

where $\sigma_{\bar{x}} = \sigma/\sqrt{n}$ and $s_{\bar{x}} = s/\sqrt{n}$

- Confidence interval for μ for a small sample:
$\bar{x} \pm t\,s_{\bar{x}}$ where $s_{\bar{x}} = s/\sqrt{n}$
- Confidence interval for p for a large sample:
$\hat{p} \pm z\,s_{\hat{p}}$ where $s_{\hat{p}} = \sqrt{\hat{p}\hat{q}/n}$
- Determining sample size for estimating μ: $n = z^2\sigma^2/E^2$
- Determining sample size for estimating p: $n = \dfrac{z^2pq}{E^2}$

CHAPTER 9 • HYPOTHESIS TESTS ABOUT THE MEAN AND PROPORTION

- Test statistic for a test of hypothesis about μ (large sample):

$z = (\bar{x} - \mu)/\sigma_{\bar{x}}$ if σ is known, where $\sigma_{\bar{x}} = \sigma/\sqrt{n}$

or $z = (\bar{x} - \mu)/s_{\bar{x}}$ if σ is not known, where $s_{\bar{x}} = s/\sqrt{n}$

- Test statistic for a test of hypothesis about μ (small sample):
$t = (\bar{x} - \mu)/s_{\bar{x}}$ where $s_{\bar{x}} = s/\sqrt{n}$
- Test statistic for a test of hypothesis about p (large sample):
$z = (\hat{p} - p)/\sigma_{\hat{p}}$ where $\sigma_{\hat{p}} = \sqrt{pq/n}$

CHAPTER 10 • ESTIMATION AND HYPOTHESIS TESTING: TWO POPULATIONS

- Confidence interval for $\mu_1 - \mu_2$ for large and independent samples:

$(\bar{x}_1 - \bar{x}_2) \pm z\,\sigma_{\bar{x}_1 - \bar{x}_2}$ if σ_1 and σ_2 are known

or $(\bar{x}_1 - \bar{x}_2) \pm z\,s_{\bar{x}_1 - \bar{x}_2}$ if σ_1 and σ_2 are not known

where $\sigma_{\bar{x}_1 - \bar{x}_2} = \sqrt{\dfrac{\sigma_1^2}{n_1} + \dfrac{\sigma_2^2}{n_2}}$ and $s_{\bar{x}_1 - \bar{x}_2} = \sqrt{\dfrac{s_1^2}{n_1} + \dfrac{s_2^2}{n_2}}$

- Test statistic for a test of hypothesis about $\mu_1 - \mu_2$ (large and independent samples):

$$z = \frac{(\bar{x}_1 - \bar{x}_2) - (\mu_1 - \mu_2)}{\sigma_{\bar{x}_1 - \bar{x}_2}}$$

If σ_1 and σ_2 are not known, then replace $\sigma_{\bar{x}_1 - \bar{x}_2}$ by its point estimator $s_{\bar{x}_1 - \bar{x}_2}$.

- For two small and independent samples taken from two populations with equal standard deviations:

Pooled standard deviation:

$$s_p = \sqrt{\frac{(n_1 - 1)s_1^2 + (n_2 - 1)s_2^2}{n_1 + n_2 - 2}}$$

Estimate of the standard deviation of $\bar{x}_1 - \bar{x}_2$:

$$s_{\bar{x}_1-\bar{x}_2} = s_p\sqrt{(1/n_1) + (1/n_2)}$$

Confidence interval for $\mu_1 - \mu_2$: $(\bar{x}_1 - \bar{x}_2) \pm t\, s_{\bar{x}_1-\bar{x}_2}$

Test statistic: $t = \dfrac{(\bar{x}_1 - \bar{x}_2) - (\mu_1 - \mu_2)}{s_{\bar{x}_1-\bar{x}_2}}$

- For two small and independent samples taken from two populations with unequal standard deviations:

Degrees of freedom: $$df = \frac{\left(\dfrac{s_1^2}{n_1} + \dfrac{s_2^2}{n_2}\right)^2}{\dfrac{\left(\dfrac{s_1^2}{n_1}\right)^2}{n_1 - 1} + \dfrac{\left(\dfrac{s_2^2}{n_2}\right)^2}{n_2 - 1}}$$

Estimate of the standard deviation of $\bar{x}_1 - \bar{x}_2$:

$$s_{\bar{x}_1-\bar{x}_2} = \sqrt{(s_1^2/n_1) + (s_2^2/n_2)}$$

Confidence interval for $\mu_1 - \mu_2$: $(\bar{x}_1 - \bar{x}_2) \pm t\, s_{\bar{x}_1-\bar{x}_2}$

Test statistic: $t = \dfrac{(\bar{x}_1 - \bar{x}_2) - (\mu_1 - \mu_2)}{s_{\bar{x}_1-\bar{x}_2}}$

- For two paired or matched samples:

Sample mean for paired differences: $\bar{d} = \Sigma d/n$

Sample standard deviation for paired differences:

$$s_d = \sqrt{\frac{\Sigma d^2 - \dfrac{(\Sigma d)^2}{n}}{n - 1}}$$

Confidence interval for μ_d: $\bar{d} \pm t\, s_{\bar{d}}$

where $s_{\bar{d}} = s_d/\sqrt{n}$

Test statistic for a test of hypothesis about μ_d:

$$t = (\bar{d} - \mu_d)/s_{\bar{d}}$$

- For two large and independent samples:

Confidence interval for $p_1 - p_2$: $(\hat{p}_1 - \hat{p}_2) \pm z\, s_{\hat{p}_1-\hat{p}_2}$

where $s_{\hat{p}_1-\hat{p}_2} = \sqrt{(\hat{p}_1\hat{q}_1/n_1) + (\hat{p}_2\hat{q}_2/n_2)}$

For a test of hypothesis about $p_1 - p_2$ with H_0: $p_1 - p_2 = 0$, pooled sample proportion:

$\bar{p} = (x_1 + x_2)/(n_1 + n_2)$ or $(n_1\hat{p}_1 + n_2\hat{p}_2)/(n_1 + n_2)$

Estimate of the standard deviation of $\hat{p}_1 - \hat{p}_2$:

$$s_{\hat{p}_1-\hat{p}_2} = \sqrt{\bar{p}\bar{q}\left(\frac{1}{n_1} + \frac{1}{n_2}\right)}$$

Test statistic: $z = \dfrac{(\hat{p}_1 - \hat{p}_2) - (p_1 - p_2)}{s_{\hat{p}_1-\hat{p}_2}}$

CHAPTER 11 ▪ CHI-SQUARE TESTS

- Expected frequency for a category for a goodness-of-fit test:

$$E = np$$

- Expected frequency for a cell for an independence or homogeneity test:

$$E = \frac{\text{(Row total)(Column total)}}{n}$$

- Test statistic for a goodness-of-fit test and a test of independence or homogeneity:

$$\chi^2 = \Sigma(O - E)^2/E$$

- Confidence interval for the population variance σ^2:

$$(n - 1)\, s^2/\chi^2_{\alpha/2} \quad \text{to} \quad (n - 1)\, s^2/\chi^2_{1-\alpha/2}$$

- Test statistic for a test of hypothesis about σ^2:

$$\chi^2 = (n - 1)\, s^2/\sigma^2$$

CHAPTER 12 ▪ SIMPLE LINEAR REGRESSION

$$SS_{xy} = \Sigma xy - \frac{(\Sigma x)(\Sigma y)}{n}$$

$$SS_{xx} = \Sigma x^2 - \frac{(\Sigma x)^2}{n} \qquad SS_{yy} = \Sigma y^2 - \frac{(\Sigma y)^2}{n}$$

- Least squares estimates of A and B:

$$b = SS_{xy}/SS_{xx} \quad \text{and} \quad a = \bar{y} - b\bar{x}$$

- Standard deviation of the sample errors:

$$s_e = \sqrt{\frac{SS_{yy} - b\, SS_{xy}}{n - 2}}$$

- Error sum of squares: $SSE = \Sigma e^2 = \Sigma(y - \hat{y})^2$
- Total sum of squares: $SST = \Sigma y^2 - \dfrac{(\Sigma y)^2}{n}$
- Regression sum of squares: $SSR = SST - SSE$
- Coefficient of determination: $r^2 = bSS_{xy}/SS_{yy}$
- Confidence interval for B: $b \pm t\, s_b$ where $s_b = s_e/\sqrt{SS_{xx}}$
- Test statistic for a test of hypothesis about B: $t = (b - B)/s_b$
- Linear correlation coefficient: $r = SS_{xy}/\sqrt{SS_{xx}SS_{yy}}$
- Confidence interval for $\mu_{y|x}$:

$$\hat{y} \pm t\, s_{\hat{y}_m} \quad \text{where} \quad s_{\hat{y}_m} = s_e\sqrt{\frac{1}{n} + \frac{(x_0 - \bar{x})^2}{SS_{xx}}}$$

- Prediction interval for y_p:

$$\hat{y} \pm t\, s_{\hat{y}_p} \quad \text{where} \quad s_{\hat{y}_p} = s_e\sqrt{1 + \frac{1}{n} + \frac{(x_0 - \bar{x})^2}{SS_{xx}}}$$

CHAPTER 13 ▪ ANALYSIS OF VARIANCE

Let

k = the number of different samples (or treatments)

n_i = the size of sample i

T_i = the sum of the values in sample i

N = the number of values in all samples $= n_1 + n_2 + n_3 + \cdots$

Σx = the sum of the values in all samples $= T_1 + T_2 + T_3 + \cdots$

Σx^2 = the sum of the squares of values in all samples

- For the F distribution:

Degrees of freedom for the numerator $= k - 1$

Degrees of freedom for the denominator $= N - k$

- Between-samples sum of squares:

$$SSB = \left(\frac{T_1^2}{n_1} + \frac{T_2^2}{n_2} + \frac{T_3^2}{n_3} + \cdots\right) - \frac{(\Sigma x)^2}{N}$$

- Within-samples sum of squares:

$$SSW = \Sigma x^2 - \left(\frac{T_1^2}{n_1} + \frac{T_2^2}{n_2} + \frac{T_3^2}{n_3} + \cdots\right)$$

- Total sum of squares: $SST = \Sigma x^2 - \dfrac{(\Sigma x)^2}{N}$
- Variance between samples: $MSB = SSB/(k - 1)$
- Variance within samples: $MSW = SSW/(N - k)$
- Test statistic for an ANOVA test: $F = MSB/MSW$

INTRODUCTORY STATISTICS

INTRODUCTORY STATISTICS

PREM S. MANN
EASTERN CONNECTICUT STATE UNIVERSITY

JOHN WILEY & SONS, INC.
NEW YORK • CHICHESTER • BRISBANE • TORONTO • SINGAPORE

Cover Photo *Tom Drake*

ACQUISITIONS EDITOR	Brad Wiley, II
DEVELOPMENTAL EDITOR	Joan Carrafiello
PRODUCTION MANAGER	Katharine Rubin
DESIGNER	Ann Marie Renzi
PRODUCTION SUPERVISOR	Sandra Russell
MANUFACTURING MANAGER	Lorraine Fumoso
COPY EDITOR	Richard Blander
ILLUSTRATION	John Balbalis

Recognizing the importance of preserving what has been written, it is a policy of John Wiley & Sons, Inc. to have books of enduring value published in the United States printed on acid-free paper, and we exert our best efforts to that end.

Library of Congress Cataloging in Publication Data:

Mann, Prem.
Introductory statistics / Prem Mann.
Includes index.
ISBN 0-471-52733-5
1. Statistics I. Title.
QA276.12.M29 1991 91-15043
519.5--dc20 CIP

Printed in the United States of America

10 9 8 7 6 5 4 3 2

To my parents

A Note From the Publisher

John Wiley and Sons, Inc., has an abiding commitment to publishing quality textbooks. To this end, several measures have been systematically undertaken to ensure that *Introductory Statistics* is pedagogically sound and accurate.

While still in manuscript form, *Introductory Statistics* was class tested for two years by Professor Mann. His classroom experiences resulted in refinements to accuracy and instructional techniques. Suggestions from three rounds of reviewing by statistics professors from two- and four-year colleges were analyzed and incorporated into the text. In addition, while revising manuscript, Professor Mann worked closely with a developmental editor and a statistics professor who reviewed all changes.

To sustain accuracy through the book production process, a professor of statistics read the text in galley form.

Finally, to ensure accuracy of answers while, at the same time, monitoring the appropriateness of exercises, Professor Mann and yet another expert in statistics prepared the answers and solutions in the text and Solutions Manual.

PREFACE

Introductory Statistics is written for a first course in applied statistics. The book is intended for students who do not have a strong background in mathematics. The only prerequisite for this text is a knowledge of elementary algebra. Today, college students from several fields of study are required to take at least one course in statistics. It is the goal of this text to make statistics both interesting and accessible to such a wide and varied audience. Three major characteristics of this text support this goal: the realistic content of its examples and exercises, the clarity and brevity of its presentation, and the soundness of its pedagogical approach. These characteristics are exhibited through the interplay of a variety of significant text features. The following is a description of these features and their purposes.

MAIN FEATURES OF THE TEXT

Style and Pedagogy

Clear and Concise Exposition The explanation of statistical methods and concepts is clear and concise. Moreover, the style is informal and readable. In chapter introductions and in transition from section to section, new ideas are related to those discussed earlier. These pedagogical accomplishments are due, in part, to the classroom use of this text prior to its publication.

Abundant Examples

Examples The text contains a wealth of examples, a total of 209 for 13 chapters. Moreover, a large number of these examples contain several parts. The examples are usually given in a format showing a problem and its solution. Examples are well sequenced and thorough, displaying all facets of concepts. The examples capture student interest because they cover a wide variety of relevant topics; they are based on typical situations practicing statisticians encounter, and on real data that are taken from sources such as books, government and private data sources and reports, magazines, newspapers, and professional journals.

Realistic Settings

Solutions A clear, concise solution follows each problem presented in an example. When the solution to an example involves many steps, it is presented in a step-by-step format. For instance, examples related to tests of hypotheses contain five steps that are consistently used to solve such examples in all chapters. Thus, procedures are presented in the concrete setting of an application rather than as isolated abstractions. Frequently, solutions contain highlighting remarks that recall and reinforce ideas critical to the solution of the problem. Such remarks add to the clarity of presentation.

Guideposts

Margin Notes for Examples Appearing in the margin notes beside each example is a brief description of what is being done in that example. Students can use these margin notes to assist them as they read through sections and to quickly locate appropriate model problems as they work through exercises.

Frequent Use of Diagrams Concepts can often be made less compex by describing them visually, with diagrams. This text has used the diagrams frequently to help students understand concepts and problems. For example, tree diagrams are extensively used in Chapters 4 and 5 to assist in explaining the probability concepts and in computing probabilities. Similarly, solutions to all examples about tests of hypotheses contain diagrams showing the rejection and nonrejection regions and the critical values.

Highlighting Definitions of important terms, formulas, and key concepts are enclosed within color boxes so that students can easily locate them. A similar use of color is found in Using MINITAB sections where a color tint highlights MINITAB commands, their explanations, and their usage along with MINITAB solutions. Important terms appear in the text either in boldface or italic type. Glossary terms are printed boldface in the text.

Cautions Certain items need special attention. Such items may deal with potential trouble spots that commonly are the source of errors. Or, they may deal with ideas that students often overlook. Special emphasis is placed on such items through the headings: "Remember," "An Observation," or "Warning."

Realistic Applications

Case Studies Case studies, which appear in most chapters, provide additional illustrations of the application of statistics in research and statistical analysis. Most of these case studies are based on articles published in journals, magazines, and newspapers. All case studies are based on real data.

Abundant Exercises

Exercises and Supplementary Exercises The text contains an abundance of exercises, 1010 in total for the 13 chapters and Appendix A (excluding Computer Assignments). Moreover, a large number of these exercises contain several parts. Exercise sets appearing at the end of each section include problems on the topics of that section. Supplementary exercises appear at the end of each chapter and contain exercises on all sections and topics discussed in that chapter. Most of the exercises are based on real data, which are taken from a large number of data sources such as books, government and private data sources and reports, magazines, newspapers, and professional journals. Exercises given in the text do not only provide a source of practice but their real data provide interesting information and insight into social, economic, political, and psychological aspects of life and in many cases provide entertaining facts about our popular culture. The exercise sets also contain many conceptual problems that demand critical thinking skills. The answers to odd-numbered exercises appear in the answers section toward the end of the book. Optional exercises are indicated by the use of an asterisk (*).

Summary and Review

Key Formulas Each chapter contains a list of key formulas used in that chapter.

Glossary Each chapter has a glossary that lists the key terms introduced in that chapter along with their brief explanations. Almost all the terms that are boldfaced in the text appear in the glossary.

Self-Review Tests Each chapter contains a Self-Review Test, which appears immediately after the Supplementary Exercises. These problems can help students to test their grasp of the concepts and skills presented in the corresponding chapters and to monitor their understanding of statistical methods. The problems marked by an asterisk (*) in the self-review tests are optional. The answers to all problems of the self-review tests appear in the answer section toward the end of the book.

Technology

Computer Usage Another feature of this text is the detailed instruction on the usage of MINITAB.† Eleven of the 13 chapters contain a Using MINITAB section at the end of each chapter. Each section contains a detailed description of the MINITAB commands that are used to perform the statistical analysis presented in that chapter. In addition, each section contains several illustrations that demonstrate how MINITAB can be used to solve statistical problems—a total of 28 such illustrations. These MINITAB instructions and illustrations are so complete that students will not need to purchase any other MINITAB supplement. Computer assignments are also given at the end of each MINITAB section so that students can further practice MINITAB. A total of 46 computer assignments are contained in these sections.

Calculator Usage The text contains many footnotes that explain how a calculator can be used to evaluate complex mathematical expressions.

Data Sets Four data sets appear in Appendix B. These data sets, collected from different sources, contain information on many variables. They can be used to perform statistical analysis with statistical computer software such as MINITAB. **These data sets are available from the publisher on a diskette in MINITAB format and in ASCII format.**

OPTIONAL SECTIONS

Because each instructor has different preferences, the text does not indicate optional sections. This decision has been left to the instructor. The instructors may cover the sections or chapters that they think are important. However, the following sections may be considered optional: Sections 1.6, 2.5, 2.6, 2.7, 3.5, 3.6, 5.7, 6.7, Sections 7.5, 7.8, 8.6, 8.7, 9.3, 10.3, 10.4, 11.5, and 12.8; Chapter 13; and Appendix A.

Complete Learning System

SUPPLEMENTS

The following supplements are available to accompany this text.

Instructor's Manual This manual contains complete solutions to all exercises and the self-review test problems.

Student's Solutions Manual This manual contains complete solutions to most of the odd-numbered exercises and to almost all the self-review test problems.

Study Guide and Review This guide contains review material about studying and learning patterns for a first course in statistics. Special attention is given to the critical material of each chapter. Review of mathematical notation, formulas, and table reading is also included.

Printed Test Bank The printed copy of the test bank contains a large number of multiple choice questions and quantitative problems, which are categorized by chapter.

Computerized Test Bank All the questions that are in the printed test bank are available on a diskette. This diskette can be obtained from the publisher.

†MINITAB is a registered trademark of Minitab, Inc., 3081 Enterprise Drive, State College, PA 16801. Phone: 814-238-3280; fax: 814-238-4383; telex: 881612. The author would take this opportunity to thank Minitab, Inc., for their help.

Data Diskette A diskette that contains all data sets given in Appendix B is available to the adopters of the text. This diskette is available in MINITAB format and in ASCII format.

ACKNOWLEDGMENTS

I thank the following reviewers whose comments and suggestions were invaluable in improving the manuscript.

James Curl	Modesto Community College
William D. Ergle	Roanoke College
Ronald Ferguson	San Antonio College
Larry Griffey	Florida Community College, Jacksonville
Gary S. Itzkowitz	Glassboro State College
Michael Karelius	American River College
Dix J. Kelly	Central Connecticut State University
Martin Kotler	Pace University, Pleasantville
Marlene Kovaly	Florida Community College, Jacksonville
Jeffrey Mock	Diablo Valley College
Luis Moreno	Broome Community College
Mary Parker	Austin Community College
Roger Peck	University of Rhode Island
Joseph Pigeon	Villanova University
Richard Quindley	Bridgewater State College
Gerald Rogers	New Mexico State University, Las Cruces
Kathryn Schwarz	Scottsdale Community College
Bruce Trumbo	California State University, Hayward
Terry Wilson	San Jacinto College

My special thanks to Professor James Curl, who not only read the whole manuscript twice and made many suggestions but was also of great help on many other occasions. I am thankful to Professor Gerald Rogers for checking all the examples and solutions for mathematical accuracy, and to my colleague Dr. Chandrasekhar Krishnamurti for checking all answers given in the answer section. I apologize to all those students at Eastern on whom I tested this manuscript and thank those from whom I received feedback. I am thankful to Eastern Connecticut State University for partially releasing me from my teaching duties. I am also thankful to many of my students, especially Kashif Faruqui, Michael Magala, James Nassiff, Rickie Pero, and Alwyn White, who were of immense help during the writing of this text. I thank my colleagues, friends, and family whose support was a source of encouragement during this project.

It is my pleasure to thank all the professionals at John Wiley, with whom I enjoyed working. Among them are Bradford Wiley, II, acquisitions; Joan Carrafiello, development; Richard Blander, copy editing; Ernie Kohlmetz, copy editing; Ann Marie Renzi, design; Sandra Russell, and Katharine Rubin, production; and John Balbalis, art and illustration. My special thanks to Joan Carrafiello who supervised this manuscript at all stages and was available during all times of need. I extend my special thanks to Richard Blander also for his careful review of this manuscript and helpful suggestions.

Prem S. Mann

CONTENTS

Chapter 1 **INTRODUCTION** 2

1.1 WHAT IS STATISTICS? 4
1.2 TYPES OF STATISTICS 4
1.2.1 DESCRIPTIVE STATISTICS 4
1.2.2 INFERENTIAL STATISTICS 5

CASE STUDY 1-1 NATION TRAVELED 3.651 TRILLION MILES IN 1986 6

1.3 POPULATION VERSUS SAMPLE 7

CASE STUDY 1-2 SURVEY FINDS SHARP DROP IN TOOTH DECAY IN YOUNG 8
CASE STUDY 1-3 ASPIRIN REDUCES THE RISK OF HEART ATTACK 9
CASE STUDY 1-4 ASPIRIN: YES, NO, MAYBE? 10

1.4 BASIC TERMS 11
1.5 TYPES OF VARIABLES 14
1.5.1 QUANTITATIVE VARIABLE 14
1.5.2 QUALITATIVE OR CATEGORICAL VARIABLE 15
1.6 CROSS-SECTION VERSUS TIME-SERIES DATA 16
1.6.1 CROSS-SECTION DATA 16
1.6.2 TIME-SERIES DATA 17
1.7 SOURCES OF DATA 17
1.8 SUMMATION NOTATION 18

GLOSSARY 22 SUPPLEMENTARY EXERCISES 23
SELF-REVIEW TEST 24 USING MINITAB: AN INTRODUCTION 26
COMPUTER ASSIGNMENTS 33

Chapter 2 **ORGANIZING DATA** 36

2.1 RAW DATA 38
2.2 ORGANIZING AND GRAPHING QUALITATIVE DATA 38
2.2.1 FREQUENCY DISTRIBUTIONS 39
2.2.2 RELATIVE FREQUENCY AND PERCENTAGE DISTRIBUTIONS 40
2.2.3 GRAPHICAL PRESENTATION OF QUALITATIVE DATA 41

CASE STUDY 2-1 NUMBER OF TELEPHONES OWNED BY HOUSEHOLDS? 42
CASE STUDY 2-2 HOW STUDENTS GRADE THEIR SCHOOLS? 43
CASE STUDY 2-3 HOW MUCH CHILDREN KNOW ABOUT GRANDPARENTS! 44

CASE STUDY 2-4 SERVING WITH DISTINCTION 45

2.3 ORGANIZING AND GRAPHING QUANTITATIVE DATA 47
2.3.1 FREQUENCY DISTRIBUTIONS 47
2.3.2 CONSTRUCTING FREQUENCY DISTRIBUTION TABLES 49
2.3.3 RELATIVE FREQUENCY AND PERCENTAGE DISTRIBUTIONS 52
2.3.4 GRAPHING GROUPED DATA 52

CASE STUDY 2-5 APPENDECTOMY CASES BY AGE AND SEX, 1986 54

2.3.5 MORE ON CLASSES AND FREQUENCY DISTRIBUTIONS 56
2.4 SHAPES OF HISTOGRAMS 58

CASE STUDY 2-6 NUMBER OF FIRE UNITS FOR FIVE BUSIEST FIRE ENGINE COMPANIES 60

2.5 CUMULATIVE FREQUENCY DISTRIBUTIONS 66
2.5.1 OGIVES 67
2.6 STEM-AND-LEAF DISPLAYS 70
2.7 GRAPHING TIME-SERIES DATA 74

CASE STUDY 2-7 WHAT IS WRONG WITH THIS PICTURE? 76

GLOSSARY 79 KEY FORMULAS 80
SUPPLEMENTARY EXERCISES 80 SELF-REVIEW TEST 85
USING MINITAB 88 COMPUTER ASSIGNMENTS 91

Chapter 3 NUMERICAL DESCRIPTIVE MEASURES 94

3.1 MEASURES OF CENTRAL TENDENCY FOR UNGROUPED DATA 96
3.1.1 MEAN 96

CASE STUDY 3-1 HOOPSTERS ARE MAKING THE MOST 99

3.1.2 MEDIAN 100

CASE STUDY 3-2 MEDIAN SALE PRICES OF EXISTING SINGLE-FAMILY HOMES FOR METROPOLITAN AREAS 102

3.1.3 MODE 103
3.1.4 RELATIONSHIP BETWEEN THE MEAN, MEDIAN, AND MODE 105
3.2 MEASURES OF DISPERSION FOR UNGROUPED DATA 109
3.2.1 RANGE 110
3.2.2 VARIANCE AND STANDARD DEVIATION 110

CASE STUDY 3-3 DOES NECKWEAR TIGHTNESS AFFECT VISUAL PERFORMANCE? 114

3.2.3 POPULATION PARAMETERS AND SAMPLE STATISTICS 115
3.3 MEAN, VARIANCE, AND STANDARD DEVIATION FOR GROUPED DATA 117
3.3.1 MEAN FOR GROUPED DATA 117
3.3.2 VARIANCE AND STANDARD DEVIATION FOR GROUPED DATA 120

3.4 USE OF STANDARD DEVIATION 126
3.4.1 CHEBYSHEV'S THEOREM 126
3.4.2 EMPIRICAL RULE 128
3.5 MEASURES OF POSITION 130
3.5.1 QUARTILES 130
3.5.2 PERCENTILES 132
3.6 BOX-AND-WHISKER PLOT 136

GLOSSARY 140 KEY FORMULAS 141
SUPPLEMENTARY EXERCISES 143 APPENDIX 3.1 147
SELF-REVIEW TEST 150 USING MINITAB 153
COMPUTER ASSIGNMENTS 156

Chapter 4 PROBABILITY 158

OCT/15/93

4.1 EXPERIMENT, OUTCOMES, AND SAMPLE SPACE 160
4.1.1 SIMPLE AND COMPOUND EVENTS 162
4.2 CALCULATING PROBABILITY 167
4.2.1 THREE CONCEPTUAL APPROACHES TO PROBABILITY 168

CASE STUDY 4-1 PROBABILITY AND ODDS 173

4.3 MARGINAL AND CONDITIONAL PROBABILITIES 176

CASE STUDY 4-2 DOES DRINKING INCREASE THE PROBABILITY OF BREAST CANCER? 180
CASE STUDY 4-3 DO PARENTS' PSYCHOLOGICAL PROBLEMS INCREASE THE PROBABILITIES OF CHILDREN'S PSYCHOLOGICAL PROBLEMS? 181

4.4 MUTUALLY EXCLUSIVE EVENTS 182
4.5 INDEPENDENT VERSUS DEPENDENT EVENTS 183
4.6 COMPLEMENTARY EVENTS 186
4.7 INTERSECTION OF EVENTS AND THE MULTIPLICATION RULE 190
4.7.1 INTERSECTION OF EVENTS 190
4.7.2 MULTIPLICATION RULE 191

CASE STUDY 4-4 BASEBALL PLAYERS HAVE "SLUMPS" AND "STREAKS" 198

4.8 COUNTING RULE 200
4.9 UNION OF EVENTS AND THE ADDITION RULE 204
4.9.1 UNION OF EVENTS 204
4.9.2 ADDITION RULE 206

GLOSSARY 213 KEY FORMULAS 214
SUPPLEMENTARY EXERCISES 215 SELF-REVIEW TEST 219

Chapter 5 DISCRETE RANDOM VARIABLES AND THEIR PROBABILITY DISTRIBUTIONS 222

5.1 RANDOM VARIABLES 224
5.1.1 DISCRETE RANDOM VARIABLE 224
5.1.2 CONTINUOUS RANDOM VARIABLE 225

5.2 THE PROBABILITY DISTRIBUTION OF A DISCRETE RANDOM VARIABLE 226
5.3 THE MEAN OF A DISCRETE RANDOM VARIABLE 233

CASE STUDY 5-1 INSTANT BASEBALL LOTTERY 235

5.4 THE STANDARD DEVIATION OF A DISCRETE RANDOM VARIABLE 238
5.5 FACTORIALS AND COMBINATIONS 242
5.5.1 FACTORIALS 242
5.5.2 COMBINATIONS 244

CASE STUDY 5-2 PLAYING LOTTO 246

5.5.3 THE TABLE OF COMBINATIONS 246
5.6 THE BINOMIAL PROBABILITY DISTRIBUTION 248
5.6.1 THE BINOMIAL EXPERIMENT 249
5.6.2 THE BINOMIAL PROBABILITY DISTRIBUTION AND BINOMIAL FORMULA 250

CASE STUDY 5-3 MISSING WOMEN 256

5.6.3 THE TABLE OF BINOMIAL PROBABILITIES 257
5.6.4 THE PROBABILITY OF SUCCESS AND THE SHAPE OF THE BINOMIAL DISTRIBUTION 259
5.6.5 THE MEAN AND STANDARD DEVIATION OF THE BINOMIAL DISTRIBUTION 260
5.7 THE POISSON PROBABILITY DISTRIBUTION 264
5.7.1 THE TABLE OF POISSON PROBABILITIES 268

GLOSSARY 271 KEY FORMULAS 272
SUPPLEMENTARY EXERCISES 273 SELF-REVIEW TEST 277
USING MINITAB 279 COMPUTER ASSIGNMENTS 282

Chapter 6 CONTINUOUS RANDOM VARIABLES AND THE NORMAL DISTRIBUTION 284

6.1 A CONTINUOUS PROBABILITY DISTRIBUTION 286

CASE STUDY 6-1 DISTRIBUTION OF TIME TAKEN TO RUN A ROAD RACE 289

6.2 THE NORMAL DISTRIBUTION 291
6.3 THE STANDARD NORMAL DISTRIBUTION 294
6.4 STANDARDIZING A NORMAL DISTRIBUTION 303
6.5 APPLICATIONS OF THE NORMAL DISTRIBUTION 311
6.6 DETERMINING THE z VALUE WHEN AN AREA UNDER THE STANDARD NORMAL CURVE IS KNOWN 318
6.7 THE NORMAL APPROXIMATION TO THE BINOMIAL DISTRIBUTION 321

GLOSSARY 327 KEY FORMULAS 328
SUPPLEMENTARY EXERCISES 328 SELF-REVIEW TEST 332
USING MINITAB 334 COMPUTER ASSIGNMENTS 336

Chapter 7 **SAMPLING DISTRIBUTIONS** 338

7.1 THE POPULATION AND SAMPLING DISTRIBUTIONS 340
7.1.1 THE POPULATION DISTRIBUTION **340**
7.1.2 THE SAMPLING DISTRIBUTION **341**
7.2 THE SAMPLING ERROR 343

CASE STUDY 7-1 SILBER'S VICTORY RENEWS QUESTIONS ON POLL ANSWERS 344

7.3 THE MEAN AND STANDARD DEVIATION OF $\bar{x}$ 345
7.4 THE SHAPE OF THE SAMPLING DISTRIBUTION OF $\bar{x}$ 349
7.4.1 SAMPLING FROM A NORMALLY DISTRIBUTED POPULATION **349**
7.4.2 SAMPLING FROM A NONNORMALLY DISTRIBUTED POPULATION **352**
7.5 CALCULATING THE PROBABILITY OF $\bar{x}$ 356
7.6 POPULATION AND SAMPLE PROPORTIONS 360
7.7 THE MEAN, STANDARD DEVIATION, AND SHAPE OF THE SAMPLING DISTRIBUTION OF $\hat{p}$ 361
7.7.1 THE SAMPLING DISTRIBUTION OF $\hat{p}$ **361**
7.7.2 THE MEAN AND STANDARD DEVIATION OF $\hat{p}$ **363**
7.7.3 THE SHAPE OF THE SAMPLING DISTRIBUTION OF $\hat{p}$ **364**
7.8 CALCULATING THE PROBABILITY OF $\hat{p}$ 366

GLOSSARY 370 KEY FORMULAS 370
SUPPLEMENTARY EXERCISES 371 SELF-REVIEW TEST 373

Chapter 8 **ESTIMATION OF THE MEAN AND PROPORTION** 376

8.1 ESTIMATION: AN INTRODUCTION 378
8.2 POINT AND INTERVAL ESTIMATES 379
8.2.1 A POINT ESTIMATE **379**
8.2.2 AN INTERVAL ESTIMATE **380**
8.3 INTERVAL ESTIMATION OF A POPULATION MEAN: LARGE SAMPLES 381

CASE STUDY 8-1 CRYING BEHAVIOR IN THE HUMAN ADULT 387

8.4 INTERVAL ESTIMATION OF A POPULATION MEAN: SMALL SAMPLES 390
8.4.1 THE t DISTRIBUTION **391**
8.4.2 THE CONFIDENCE INTERVAL FOR μ USING THE t DISTRIBUTION **393**

CASE STUDY 8-2 OPENING MEDICATION CONTAINERS 396

8.5 INTERVAL ESTIMATION OF A POPULATION PROPORTION: LARGE SAMPLES 400

CASE STUDY 8-3 WHAT ENHANCES THE PROFESSIONAL IMAGE OF CAREER WOMEN 403

8.6 SAMPLE SIZE DETERMINATION FOR THE ESTIMATION OF MEAN 407
8.7 SAMPLE SIZE DETERMINATION FOR THE ESTIMATION OF PROPORTION 409

GLOSSARY 411 KEY FORMULAS 412
SUPPLEMENTARY EXERCISES 413 APPENDIX 8.1 415
SELF-REVIEW TEST 416 USING MINITAB 418
COMPUTER ASSIGNMENTS 421

Chapter 9 HYPOTHESIS TESTS ABOUT THE MEAN AND PROPORTION 424

9.1 HYPOTHESIS TESTS: AN INTRODUCTION 426
9.1.1 TWO HYPOTHESES 426
9.1.2 REJECTION AND NONREJECTION REGIONS 427
9.1.3 TWO TYPES OF ERRORS 428
9.1.4 TAILS OF THE TEST 430
9.2 HYPOTHESIS TESTS ABOUT A POPULATION MEAN: LARGE SAMPLES 434
9.3 HYPOTHESIS TESTS USING THE p-VALUE APPROACH 442
9.4 HYPOTHESIS TESTS ABOUT A POPULATION MEAN: SMALL SAMPLES 446
9.5 HYPOTHESIS TESTS ABOUT A POPULATION PROPORTION: LARGE SAMPLES 453

GLOSSARY 459 KEY FORMULAS 460
SUPPLEMENTARY EXERCISES 461 SELF-REVIEW TEST 463
USING MINITAB 466 COMPUTER ASSIGNMENTS 470

Chapter 10 ESTIMATION AND HYPOTHESIS TESTING: TWO POPULATIONS 472

10.1 INFERENCES ABOUT THE DIFFERENCE BETWEEN TWO POPULATION MEANS FOR LARGE AND INDEPENDENT SAMPLES 474
10.1.1 INDEPENDENT VERSUS DEPENDENT SAMPLES 474
10.1.2 THE MEAN, STANDARD DEVIATION, AND SAMPLING DISTRIBUTION OF $\bar{x}_1 - \bar{x}_2$ 475
10.1.3 INTERVAL ESTIMATION OF $\mu_1 - \mu_2$ 476

CASE STUDY 10-1 INFERENCES MADE BY THE CENSUS BUREAU 479

10.1.4 HYPOTHESIS TESTING ABOUT $\mu_1 - \mu_2$ 479

CASE STUDY 10-2 TIME TOGETHER AMONG DUAL-EARNER COUPLES 483

10.2 INFERENCES ABOUT THE DIFFERENCE BETWEEN TWO POPULATION MEANS FOR SMALL AND INDEPENDENT SAMPLES: EQUAL STANDARD DEVIATIONS 488
10.2.1 INTERVAL ESTIMATION OF $\mu_1 - \mu_2$ 489
10.2.2 HYPOTHESIS TESTING ABOUT $\mu_1 - \mu_2$ 491
10.3 INFERENCES ABOUT THE DIFFERENCE BETWEEN TWO POPULATION MEANS FOR SMALL AND INDEPENDENT SAMPLES: UNEQUAL STANDARD DEVIATIONS 497
10.3.1 INTERVAL ESTIMATION OF $\mu_1 - \mu_2$ 498
10.3.2 HYPOTHESIS TESTING ABOUT $\mu_1 - \mu_2$ 499

10.4 INFERENCES ABOUT THE DIFFERENCE BETWEEN TWO POPULATION MEANS FOR PAIRED SAMPLES 503
10.4.1 INTERVAL ESTIMATION OF μ_d 506
10.4.2 HYPOTHESIS TESTING ABOUT μ_d 508
10.5 INFERENCES ABOUT THE DIFFERENCE BETWEEN TWO POPULATION PROPORTIONS FOR LARGE AND INDEPENDENT SAMPLES 514
10.5.1 THE MEAN, STANDARD DEVIATION, AND SAMPLING DISTRIBUTION OF $\hat{p}_1 - \hat{p}_2$ 514
10.5.2 INTERVAL ESTIMATION OF $p_1 - p_2$ 515
10.5.3 HYPOTHESIS TESTING ABOUT $p_1 - p_2$ 516

CASE STUDY 10-3 MORE ON THE INFERENCES MADE BY THE CENSUS BUREAU 521

GLOSSARY 524 KEY FORMULAS 524
SUPPLEMENTARY EXERCISES 527 SELF-REVIEW TEST 531
USING MINITAB 534 COMPUTER ASSIGNMENTS 542

Chapter 11 CHI-SQUARE TESTS 544

11.1 THE CHI-SQUARE DISTRIBUTION 546
11.2 A GOODNESS-OF-FIT TEST 548

CASE STUDY 11-1 ARE LEADERSHIP STYLES EVENLY DISTRIBUTED? 555

11.3 CONTINGENCY TABLES 558
11.4 A TEST OF INDEPENDENCE OR HOMOGENEITY 558
11.4.1 A TEST OF INDEPENDENCE 558

CASE STUDY 11-2 PREVALENCE OF SUICIDAL BEHAVIOR AMONG HIGH SCHOOL STUDENTS 564

11.4.2 A TEST OF HOMOGENEITY 566
11.5 INFERENCES ABOUT THE POPULATION VARIANCE 571
11.5.1 ESTIMATION OF THE POPULATION VARIANCE 572
11.5.2 HYPOTHESIS TESTS ABOUT THE POPULATION VARIANCE 574

GLOSSARY 578 KEY FORMULAS 579
SUPPLEMENTARY EXERCISES 580 SELF-REVIEW TEST 583
USING MINITAB 586 COMPUTER ASSIGNMENTS 589

Chapter 12 SIMPLE LINEAR REGRESSION 592

12.1 THE SIMPLE LINEAR REGRESSION MODEL 594
12.1.1 SIMPLE REGRESSION 594
12.1.2 LINEAR REGRESSION 594
12.2 THE SIMPLE LINEAR REGRESSION ANALYSIS 596
12.2.1 SCATTER DIAGRAM 598
12.2.2 LEAST SQUARES LINE 599
12.2.3 INTERPRETATION OF a AND b 603

CASE STUDY 12-1 HEIGHTS AND WEIGHTS OF NBA PLAYERS 605

12.2.4 ASSUMPTIONS OF THE REGRESSION MODEL 607
12.2.5 A NOTE ON THE USE OF SIMPLE LINEAR REGRESSION 609

12.3 THE STANDARD DEVIATION OF THE RANDOM ERROR 612
12.4 THE COEFFICIENT OF DETERMINATION 614
12.5 INFERENCES ABOUT *B* 619

12.5.1 SAMPLING DISTRIBUTION OF *b* 619
12.5.2 ESTIMATION OF *B* 620
12.5.3 HYPOTHESIS TESTING ABOUT *B* 621

12.6 LINEAR CORRELATION 625
12.7 REGRESSION ANALYSIS: A COMPLETE EXAMPLE 630
12.8 USING THE REGRESSION MODEL 638

12.8.1 USING THE REGRESSION MODEL FOR ESTIMATING THE MEAN VALUE OF *y* 638
12.8.2 USING THE REGRESSION MODEL FOR PREDICTING A PARTICULAR VALUE OF *y* 641

12.9 CAUTIONS IN USING REGRESSION 643

GLOSSARY 644 KEY FORMULAS 645
SUPPLEMENTARY EXERCISES 647 SELF-REVIEW TEST 653
USING MINITAB 655 COMPUTER ASSIGNMENTS 660

Chapter 13 ANALYSIS OF VARIANCE 662

13.1 THE *F* DISTRIBUTION 664
13.2 ONE-WAY ANALYSIS OF VARIANCE 666

13.2.1 CALCULATING THE VALUE OF THE TEST STATISTIC 668
13.2.2 ONE-WAY ANOVA TEST 671

GLOSSARY 679 KEY FORMULAS 679
SUPPLEMENTARY EXERCISES 680 SELF-REVIEW TEST 683
USING MINITAB 685 COMPUTER ASSIGNMENTS 687

Appendix A DATA SOURCES, SAMPLING, AND THE USE OF RANDOM NUMBERS 688

A.1 SOURCES OF DATA 689

1. INTERNAL SOURCES 689
2. EXTERNAL SOURCES 689
3. SURVEYS AND EXPERIMENTS 689

CASE STUDY A-1 IS IT A SIMPLE QUESTION? 691
CASE STUDY A-2 HOW TO SKEW A POLL: LOADED QUESTIONS AND OTHER TRICKS 691

A.2 WHY SAMPLE? 692
A.3 A REPRESENTATIVE SAMPLE 693

A.4 SAMPLING AND NONSAMPLING ERRORS 694
A.4.1 SAMPLING ERROR 694
A.4.2 NONSAMPLING ERRORS 694
A.5 RANDOM SAMPLING TECHNIQUES 695
1. SIMPLE RANDOM SAMPLING 695
2. SYSTEMATIC RANDOM SAMPLING 696
3. STRATIFIED RANDOM SAMPLING 697
4. CLUSTER SAMPLING 698

GLOSSARY 698 USING MINITAB 700
COMPUTER ASSIGNMENTS 701

Appendix B DATA SETS 702

DATA SET I CITY DATA 703
DATA SET II DATA ON STATES 709
DATA SET III NBA DATA 711
DATA SET IV SAMPLE OF 500 OBSERVATIONS FROM MANCHESTER ROAD RACE DATA 718

Appendix C STATISTICAL TABLES 719

TABLE I RANDOM NUMBERS 720
TABLE II FACTORIALS 724
TABLE III VALUES OF $\binom{n}{x}$ (COMBINATION) 725
TABLE IV TABLE OF BINOMIAL PROBABILITIES 726
TABLE V VALUES OF $e^{-\lambda}$ 734
TABLE VI TABLE OF POISSON PROBABILITIES 735
TABLE VII STANDARD NORMAL DISTRIBUTION TABLE 741
TABLE VIII THE t DISTRIBUTION TABLE 742
TABLE IX CHI-SQUARE DISTRIBUTION TABLE 744
TABLE X THE F DISTRIBUTION TABLE 745

ANSWERS TO EXERCISES 749

INDEX 767

CASE STUDIES

NATION TRAVELED 3.651 TRILLION MILES IN 1986
SURVEY FINDS SHARP DROP IN TOOTH DECAY IN YOUNG
ASPIRIN REDUCES THE RISK OF HEART ATTACK
ASPIRIN: YES, NO, MAYBE?
NUMBER OF TELEPHONES OWNED BY HOUSEHOLDS
HOW STUDENTS GRADE THEIR SCHOOLS?
HOW MUCH CHILDREN KNOW ABOUT GRANDPARENTS!
SERVING WITH DISTINCTION
APPENDECTOMY CASES BY AGE AND SEX, 1986
NUMBER OF FIRE UNITS FOR FIVE BUSIEST FIRE ENGINE COMPANIES
WHAT IS WRONG WITH THIS PICTURE?
HOOPSTERS ARE MAKING THE MOST
MEDIAN SALE PRICES OF EXISTING SINGLE-FAMILY HOMES FOR METROPOLITAN AREAS
DOES NECKWEAR TIGHTNESS AFFECT VISUAL PERFORMANCE?
PROBABILITY AND ODDS
DOES DRINKING INCREASE THE PROBABILITY OF BREAST CANCER?
DO PARENTS' PSYCHOLOGICAL PROBLEMS INCREASE THE PROBABILITIES OF CHILDREN'S PSYCHOLOGICAL PROBLEMS?
BASEBALL PLAYERS HAVE "SLUMPS" AND "STREAKS"
INSTANT BASEBALL LOTTERY
PLAYING LOTTO
MISSING WOMEN
DISTRIBUTION OF TIME TAKEN TO RUN A ROAD RACE
SILBER'S VICTORY RENEWS QUESTIONS ON POLL ANSWERS
CRYING BEHAVIOR IN THE HUMAN ADULT
OPENING MEDICATION CONTAINERS
WHAT ENHANCES THE PROFESSIONAL IMAGE OF CAREER WOMEN
INFERENCES MADE BY THE CENSUS BUREAU
TIME TOGETHER AMONG DUAL-EARNER COUPLES
MORE ON THE INFERENCES MADE BY THE CENSUS BUREAU
ARE LEADERSHIP STYLES EVENLY DISTRIBUTED?
PREVALENCE OF SUICIDAL BEHAVIOR AMONG HIGH SCHOOL STUDENTS
HEIGHTS AND WEIGHTS OF NBA PLAYERS

1 INTRODUCTION

1.1 WHAT IS STATISTICS?
1.2 TYPES OF STATISTICS
1.3 POPULATION VERSUS SAMPLE
1.4 BASIC TERMS
1.5 TYPES OF VARIABLES
1.6 CROSS-SECTION VERSUS TIME-SERIES DATA
1.7 SOURCES OF DATA
1.8 SUMMATION NOTATION
SELF-REVIEW TEST
USING MINITAB

The study of statistics has become more popular during the past two decades. The increasing availability of computers and statistical software packages has enlarged the role of statistics as a tool for empirical research. Statistics is used for research in almost all professions, from medicine to sports. Today, college students in almost all disciplines are required to take at least one statistics course.

Every field of study has its own terminology. Statistics is no exception. This introductory chapter explains the basic terms of statistics. These terms will bridge our understanding of the concepts and techniques presented in subsequent chapters.

1.1 WHAT IS STATISTICS?

The word **statistics** has two meanings. In the more common usage, statistics refers to numerical facts. The numbers that represent the percentage of passes completed by the quarterback of a football team, the ages of all students in a mathematics class, the number of cars registered in each state of the United States, and the starting salary of a typical college graduate are examples of statistics in this sense of the word. Recently an article in *U.S. News & World Report* declared, "Statistics are an American obsession."† During the 1988 baseball World Series between the Los Angeles Dodgers and Oakland A's, NBC commentator Joe Garagiola reported to the viewers numerical data about the players' performances. In response, fellow commentator Vin Scully said, "I love it when you talk statistics." In all these examples, the word "statistics" refers to numbers.

The second meaning of statistics refers to the field or discipline of study. In this sense of the word, statistics is defined as follows.

STATISTICS

Statistics is a group of methods used to collect, analyze, present, and interpret data and to make decisions.

Everyday we make decisions that may be personal, business related, or of some other kind. Usually these decisions are made under conditions of uncertainty. Many times, the situations or problems we face in the real world have no precise or definite solution. Statistical methods help us to make scientific and intelligent decisions in such situations. Decisions made using statistical methods are called *educated guesses*. Decisions made without using statistical (or scientific) methods may prove to be unreliable.

Like almost all fields of study, statistics has two aspects: theoretical and applied. *Theoretical* or *mathematical statistics* deals with the development, derivation, and proof of statistical theorems, formulas, rules, and laws. *Applied statistics* describes the application of those theorems, formulas, rules, and laws to solve real-world problems. This text is concerned with applied statistics and not with theoretical statistics. By the time you finish studying this book, you will learn how to think statistically and how to make educated guesses.

1.2 TYPES OF STATISTICS

Broadly speaking, applied statistics can be divided into two areas: descriptive statistics and inferential statistics.

1.2.1 DESCRIPTIVE STATISTICS

Suppose we collect information on the ages of all students in a statistics class. In statistical terminology, the whole set of numbers that represent the ages of students

†"The Numbers Racket: How Polls and Statistics Lie," *U.S. News & World Report*, July 11, 1988, pp. 44–47.

is called a **data set,** and the age of each student is called an **element** of that data set. (These terms are defined in detail in Section 1.4.)

A data set in its original form is usually very large. Consequently, such a data set is not very helpful in drawing conclusions or making decisions. We reduce data into manageable size by constructing tables, by displaying data with diagrams, or by calculating summary measures such as averages. It is easier to draw conclusions from tables and diagrams than from the original version of a data set. The portion of statistics that helps us to conduct this type of statistical analysis is called **descriptive statistics.**

DESCRIPTIVE STATISTICS

Descriptive statistics consists of methods for organizing, displaying, and describing data by using tables, graphs, and summary measures.

Chapters 2 and 3 of this text cover descriptive statistical methods. Example 1-1 illustrates one technique that is used to organize data.

Illustrating descriptive statistics.

EXAMPLE 1-1 According to Metropolitan Life Insurance Co., 275,192 appendectomies (surgical removal of the appendix) were performed in the United States in 1986. If the age of each of the 275,192 persons is recorded individually, the whole set of information will be spread over many pages. It is essential to reduce the size of this data set by constructing a table similar to Table 1.1.

Table 1.1 Appendectomies by Age: United States, 1986

Age (years)	Total Number
Under 15	58,600
15–19	47,850
20–24	40,539
25–34	50,605
35–44	34,788
45–54	15,982
55–64	12,041
65 and over	14,787
Total	275,192

Source: Statistical Bulletin, Metropolitan Life Insurance Co., 69(2), April–June 1988, p. 21.

It is easier to draw conclusions based on a table such as Table 1.1 than based on the original data set in which the ages of all 275,192 persons are recorded individually. The construction of such tables is discussed in Chapter 2. ■

1.2.2 INFERENTIAL STATISTICS

The collection of all elements of interest is called a **population,** and the selection of a few elements from this population is called a **sample.** (Population and sample are discussed in more detail in Section 1.3.)

A major portion of statistics deals with making decisions, inferences, predictions, and forecasts about populations based on results obtained from samples. For example, we may make some decisions about the political views of all college and university students based on the political views of 1000 students selected from a few colleges and universities. The area of statistics that deals with such decision-making procedures is referred to as **inferential statistics.** This branch of statistics is also called *inductive reasoning* or *inductive statistics.*

INFERENTIAL STATISTICS

Inferential statistics consists of methods that use sample results to help make decisions or predictions about the population.

Chapters 8 through 13 and parts of Chapter 7 deal with inferential statistics. Case Study 1-1 is an example of inferential statistics. In this study, the Hertz Corporation uses the information collected from a poll (consisting of a sample of Americans interviewed) to make estimates about the population of all Americans.

CASE STUDY 1-1 NATION TRAVELED 3.651 TRILLION MILES IN 1986

According to a poll conducted by the Hertz Corporation, each American traveled an estimated average of 41.8 miles a day in 1986. This calculation is based on commutation, business, vacation, and personal trips. Based on this estimate, the total miles traveled by all Americans comes to 3.651 trillion. The total money spent on these journeys was approximately $1.256 trillion or $5251 per person for the whole year. The travelers also spent an additional $193.6 billion (or $1055 per person) on travel-related items in 1986.

The study includes "nearly all forms of transportation—trains, trucks, taxis, autos, airplanes, ships, buses . . . and walking." The study does not include personal boating and private flying. The estimates of expenses include "food and lodging away from home" and "family shopping and other incidental local trips."

Source: *Press Information,* The Hertz Corporation, September 28, 1987.

Probability, which gives a measurement of the likelihood that a certain outcome will occur, acts as a link between descriptive and inferential statistics. We use probability to make statements about the occurrence or nonoccurrence of a certain event under uncertain conditions. Probability and probability distributions are discussed in Chapters 4 through 6 and parts of Chapter 7.

EXERCISES

1.1 Briefly describe the two meanings of the word "statistics."

1.2 Briefly explain the types of statistics.

1.3 POPULATION VERSUS SAMPLE

We will encounter the terms "population" and "sample" on almost every page of this text.† Consequently, understanding the definition and meaning of these two terms is crucial. Suppose a statistician is interested in knowing

1. The percentage of all voters who will vote for a particular candidate in an election
2. The 1989 gross sales of all firms in New York City
3. The prices of all statistics books published during the last five years

In these examples, the statistician is interested in all voters, all firms, and all statistics books. Each of these groups is called the population for the respective example. In statistics, a population does not always mean a collection of people. It can, in fact, be a collection of people or of any kind of items such as books, television sets, or cars. A population of interest is usually called the **target population.**

POPULATION

A population consists of all elements—individuals, items, or objects—whose characteristics are being studied. The population being studied is also called the *target population*.

Most of the time, decisions are made based on portions of populations. For example, the various election polls conducted in the United States to estimate the percentage of voters favoring various candidates in any presidential election are based on only a few hundred or thousand voters selected from across the country. The selection of a few elements from the population is called a **sample.**

SAMPLE

A portion of the population selected for study is referred to as a sample.

Figure 1.1 illustrates the selection of a sample from a population.

The collection of information from the elements of a population or a sample is called a **survey.** A survey that includes every element of the population is called a **census.** Often the size of the target population is large. Hence, in practice, a census is rarely taken because it is very expensive and time consuming. In many cases it is even impossible to identify each element of the target population. Usually, to conduct

†To learn more about sampling and sampling techniques, refer to Appendix A.

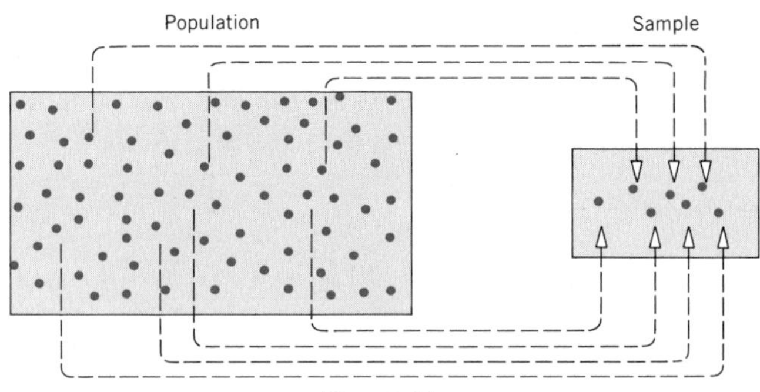

Figure 1.1

a survey, we select a sample and collect the required information from the elements included in that sample. We then make decisions based on this sample information. Such a survey conducted on a sample is called a **sample survey.**

Case Study 1-2, which is about tooth decay in children, is based on a sample survey of about 40,000 children selected from 970 schools around the country.

CASE STUDY 1-2 SURVEY FINDS SHARP DROP IN TOOTH DECAY IN YOUNG

Half the nation's schoolchildren have no cavities or other tooth decay in a continuation of gains that health officials say could mean the virtual end of dental disease as a major public health problem, a new Federal survey released today shows.

Experts credit widespread use of fluoride and high levels of dental care for the improvements seen in the survey, which was conducted in the 1986–1987 school year. . . .

The survey of almost 40,000 children at 970 schools around the country was conducted by the National Institute of Dental Research. The findings indicate that tooth decay and cavities have declined at a dramatic rate over the last 15 years and that the decline is generally uniform throughout the nation. . . .

The new survey showed that 49.9 percent of all children had no decay in their permanent teeth, as against 36.6 percent in a similar 1979–1980 study and an estimated 28 percent in the early 1970's.

Most of the children who were entirely free of tooth decay were in the under-10 age group. The majority of older students surveyed, up through the age of 17, had some cavities or other decay problems. By the age of 17, only 16 percent of those surveyed were entirely free of cavities. But experts at the institute said that older children were also getting fewer cavities than in the past. . . .

In the newest survey, dentists performed oral examinations on subjects ranging in age from 5 to 17 who were chosen to represent the nation's 43 million schoolchildren. . . .

Source: Warren E. Leary, "Survey Finds Sharp Drop in Tooth Decay in Young," *The New York Times,* June 22, 1988.

The purpose of conducting a sample survey is to make decisions about the corresponding population. Hence, it is important that the results (e.g., average family income; percentage of voters favoring a candidate) obtained from a sample survey closely match the results that we would obtain by conducting a census. Otherwise, any decision made based on a sample survey will not apply to the corresponding population. As an example, to find the average income of families living in New York City by conducting a sample survey, the sample must contain families that belong to different income groups in almost the same proportion as they exist in the population. Such a sample is called a **representative sample.** Inferences derived from a representative sample will be more reliable.

REPRESENTATIVE SAMPLE

A representative sample contains the characteristics of the population as closely as possible.

Case Study 1-3, based on a sample of 22,000 male U.S. physicians, describes whether taking aspirin reduces the risk of heart attack.

CASE STUDY 1-3 ASPIRIN REDUCES THE RISK OF HEART ATTACK

The Steering Committee of the Physicians' Health Study Research Group conducted an experiment to investigate if taking one adult-size aspirin every other day reduces the risk of heart attack.

The researchers selected a group of 22,000 male U.S. physicians, aged 40 to 84, from 59,000 volunteers. The men were randomly assigned to receive either aspirin or placebo (a dummy pill). The experiment was conducted over a period of five years. According to the editorial by Arnold S. Relman, published in *The New England Journal of Medicine,* "the total number of myocardial infarctions among the physicians taking aspirin had been reduced by nearly half. Strokes, on the other hand, were slightly, although not significantly, more numerous among the aspirin takers. Aspirin had no effect on the total number of vascular deaths or deaths from all causes. When the numbers of important vascular events were combined (nonfatal myocardial infarctions plus nonfatal strokes plus vascular deaths from all causes), those receiving aspirin still had a significant 23 percent reduction in risk."

Source: Arnold S. Relman, "Aspirin for the Primary Prevention of Myocardial Infarctions," *The New England Journal of Medicine,* 318(4), January 28, 1988, pp. 245–246; The Steering Committee of the Physicians' Health Study Research Group, "Preliminary Report: Findings from the Aspirin Component of the Ongoing Physicians' Health Study," pp. 262–264.

From Case Study 1-3, can we deduce that aspirin always works for all people to reduce the risk of heart attacks? In other words, was the sample selected for this study representative of the general population? Case Study 1-4 provides an answer to this question.

CASE STUDY 1-4 ASPIRIN: YES, NO, MAYBE?

. . . Why were only doctors chosen for the study? Doctors are an accessible, fairly uniform group of skilled medical observers who could be tracked accurately by questionnaire. But it's worth thinking about the selection process. Of the 59,000 doctors who originally volunteered, 26,000 were excluded because of disqualifying medical conditions, such as a previous heart attack or aspirin intolerance. Of the remaining 33,000, a third dropped out for various reasons. It's legitimate to wonder at what rate the dropouts had heart attacks, since the answer could have bearing on the heart attack rate among the men studied. Are these doctors typical of the general population? Clearly they are not. . . .

Does what applies to white males apply to everyone? The report raised this question and concluded that "there seems little reason to suspect that the biologic effects of aspirin would be materially different in other populations with comparable or higher risks of cardiovascular disease." But though the biologic effect may be the same, its relative importance may well depend on what other risk factors are present.

For example, women are a special group when it comes to heart attacks. Presumably, also, the majority of the doctors were white. Yet we know that the risk factors for black men are different from those of white men. Blacks have a 33% greater chance of having hypertension—a significant risk factor for heart attack—than whites. So the answer to this question is tentative at best. . . .

Source: Excerpted from "Aspirin: Yes, No, Maybe?" *University of California, Berkeley Wellness Letter,* 4(7), April 1988, p. 1. Reprinted by permission.

A sample may be a random or a nonrandom sample. In a **random sample,** each element of the population has some chance of being included in the sample. However, in a nonrandom sample this may not be the case.

RANDOM SAMPLE

A sample drawn in such a way that each element of the population has some chance of being selected is called a random sample. If the chance of being selected is the same for each element of the population, it is called a **simple random sample.**

A random sample may or may not be a simple random sample. One way to select a simple random sample is by lottery or draw. For example, if we are to select 5 students from a class of 50, we write each of the 50 names on a separate piece of paper. Then we place all 50 slips in a box and mix them thoroughly. Finally, we randomly draw five slips from the box. The five names drawn will give a simple random sample. On the other hand, if we arrange all 50 names alphabetically and then select the first 5 names on the list to include in a sample, it would be a nonrandom sample because the students listed sixth to fiftieth have no chance of being included in the sample.

A sample may be selected with or without replacement. In sampling **with replacement,** each time we select an element from the population, we put it back in the population before we select the next element. Thus, in sampling with replacement, the population will contain the same number of items each time a selection is made. As a result, we may select the same item more than once in such a sample. Consider an example of a box that contains 25 balls of different colors. Suppose we draw a ball, record its color, and put it back in the box before drawing the next ball. Every time we draw a ball from this box, the box contains 25 balls. This is an example of sampling with replacement.

Sampling **without replacement** occurs when the selected element is not replaced in the population. Each time we select an item, the size of the population is reduced by one element. Thus, we cannot select the same item more than once in this type of sampling. Most of the time, samples taken in statistics are without replacement. Consider the example of an opinion poll based on a certain number of voters selected from the population of all eligible voters. In this case, the same voter would not be selected more than once. Therefore, this is an example of sampling without replacement.

EXERCISES

1.3 Briefly explain the following terms: population, sample, representative sample, random sample, sampling with replacement, sampling without replacement.

1.4 Explain which of the following constitute a population and which constitute a sample.

a. Scores of all students in a statistics class
b. Yield of potatoes per acre for 10 pieces of land
c. Cattle owned by 100 farmers in Iowa
d. Ages of all members of a family
e. Number of days missed by all employees of a company during the past month
f. Marital status of 50 persons selected from a large city

1.5 Give one example each of sampling with and sampling without replacement.

1.6 Why is conducting a sample survey preferable to conducting a census?

1.4 BASIC TERMS

It is important to understand the meaning of some basic terms that will be used frequently in this text. This section explains the meaning of an element (or member), a variable, an observation, and a data set. An element and a data set were briefly defined earlier in Section 1.2. This section defines these terms formally and illustrates them with the help of an example.

Table 1.2 gives information on cotton production in thousands of bales (a bale weighs 480 pounds) for a sample of 10 states. Each state in Table 1.2 is called an **element** or **member** of the sample. Table 1.2 contains information on 10 elements.

Table 1.2 Production of Cotton, 1988

State	Production of Cotton (thousands of bales) ← Variable
Alabama	380
Arizona	1120
Arkansas	1050
Element or member → California	2853 ← An observation or measurement
Georgia	370
Louisiana	950
Mississippi	1830
Missouri	310
Tennessee	590
Texas	5260

Source: Statistical Abstract of the United States 1990.

ELEMENT OR MEMBER

An element or member of a sample or population is a specific subject or object (for example, a person, firm, item, state, or country) about which the information is collected.

The "production of cotton" in Table 1.2 is called a **variable.** The production of cotton is a characteristic of states that we are studying or investigating.

VARIABLE

A variable is a characteristic under study that assumes different values for different elements.

A few other examples of variables are incomes of households, number of children per family, marital status of persons, gross sales of firms, and number of crimes committed in a city per day.

In general, a variable assumes different values for different elements, as does the cotton production for 10 states in Table 1.2. For some elements, however, the value of the variable may be the same. For example, if we are collecting information on incomes of households, all households are expected to have different incomes, although some of them may have the same income. In contrast to a variable, the value of a *constant* is fixed.

A variable is often denoted by x, y, or z. For instance, in Table 1.2 the production of cotton may be denoted by any one of these letters. Starting with Section 1.8, we will begin to use these letters to denote variables.

Each of the values representing the production of cotton for 10 states in Table 1.2 is called an **observation** or **measurement.**

OBSERVATION OR MEASUREMENT

The value of a variable for a single element is called an observation or measurement.

According to Table 1.2, the production of cotton for Alabama was 380 thousand bales in 1988. The value 380 is an observation or measurement. Table 1.2 contains 10 observations, one for each of the 10 states.

The information given in Table 1.2 on the production of cotton is called the **data** or a **data set.**

DATA SET

A data set is a collection of observations on a variable.

Another example of a data set would be a list of prices of 25 recently sold homes.

EXERCISES

1.7 Explain the meaning of the following terms: element, variable, observation, data set.

1.8 The following table gives the scores of five students on a statistics test.

Student	Score
Bill	83
Susan	91
Allison	78
Jeff	69
Neil	87

Briefly explain the meaning of member, variable, measurement, and data set with reference to this table.

1.9 The following table lists the 1989 profits for the seven U.S. companies that made maximum profits in that year.

Company	1989 Profits (billions of dollars)
General Motors	4.2
General Electric	3.9
Ford Motor	3.8
IBM	3.8
Exxon	3.0
Philip Morris	2.9
AT&T	2.7

Briefly explain the meaning of member, variable, observation, and data set with reference to this table.

1.5 TYPES OF VARIABLES

In Section 1.4, we learned that a variable is a characteristic under study or investigation that assumes different values for different elements. The incomes of families, heights of persons, gross sales of firms, prices of college textbooks, number of cars owned by each family, number of accidents, and status (freshman, sophomore, junior, or senior) of each student at a university are a few examples of variables.

A variable may be classified as quantitative or qualitative.

1.5.1 QUANTITATIVE VARIABLE

Some variables can be measured numerically whereas others cannot. A variable that can assume numerical values is called a **quantitative variable.**

QUANTITATIVE VARIABLE

A variable that can be measured numerically is called a quantitative variable. The data collected on a quantitative variable are called **quantitative data.**

Income, height, gross sales, price, number of cars owned, and accidents are examples of quantitative variables since each of them can be expressed numerically. For instance, the income of a family may be $31,520.75 per year and a family may own two cars. Such quantitative variables can either be classified as *discrete variables* or *continuous variables*.

Discrete Variable

The values that a certain quantitative variable can assume may be countable or not. For example, we can count the number of students in a class but we cannot count the weight of a student. The variable whose values are countable is called a **discrete variable**. There is a gap between any two consecutive values of a discrete variable.

DISCRETE VARIABLE

A variable whose values are countable is called a discrete variable. In other words, a discrete variable can assume only certain values with no intermediate values.

For example, the number of cars sold on any day at a car dealership is a discrete variable because the number of cars sold must be 0, 1, 2, 3, ... The number of cars sold cannot be between 0 and 1, or between 1 and 2. A few other examples of discrete variables are the number of people visiting Disneyland on any day, the number of auto accidents on a highway on a given day, the number of cattle owned by a farmer, and the number of students in a class.

Continuous Variable

A **continuous variable** can assume any value over a certain range. Moreover, we cannot count these values.

> **CONTINUOUS VARIABLE**
>
> A variable that can assume any numerical value over a certain interval or intervals is called a continuous variable.

For instance, the time taken to complete an examination can assume any value, let us say, between 30 and 60 minutes. The time taken may be 42.6 minutes or 42.67 minutes or 42.674 minutes. (Theoretically, we can measure time as precisely as we want.) Similarly, the height of a person can be measured to a tenth of an inch or to a hundredth of an inch. However, neither time nor height can be counted in a discrete fashion. A few other examples of continuous variables are weights of people, amount of soda in a 12-ounce can (note that a can will not contain exactly 12 ounces of soda), and the yield of potatoes per acre.

1.5.2 QUALITATIVE OR CATEGORICAL VARIABLE

Certain variables cannot be measured numerically, but they can be divided into different categories. Such variables are called **qualitative** or **categorical variables.**

> **QUALITATIVE OR CATEGORICAL VARIABLE**
>
> A variable that cannot assume a numerical value but can be divided into two or more nonnumeric categories is called a qualitative or categorical variable. The data collected on such a variable are called **qualitative data.**

For example, the status of an undergraduate college student is a qualitative variable because a student can fall into any one of four categories: freshman, sophomore, junior, or senior. A few other examples of qualitative variables are the sex of a person, the color of hair, and the make of a car. Figure 1.2 illustrates the types of variables.

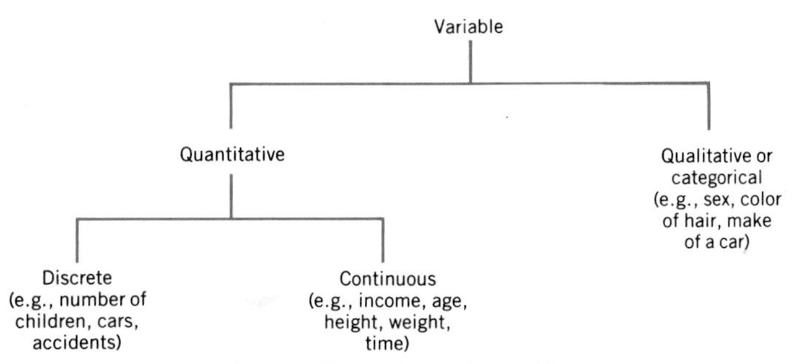

Figure 1.2 Types of variables.

EXERCISES

1.10 Explain the meaning of the following terms.

a. Quantitative variable
b. Qualitative variable
c. Discrete variable
d. Continuous variable
e. Quantitative data
f. Qualitative data

1.11 Explain which of the following variables are quantitative and which are qualitative.

a. Number of persons in a family
b. Rent paid by tenants
c. Marital status of people
d. Length of a frog's jump
e. Number of students in a class
f. Color of eyes

1.12 Classify the quantitative variables in Exercise 1.11 as discrete or continuous.

1.6 CROSS-SECTION VERSUS TIME-SERIES DATA

Based on the time over which they are collected, the data can be classified as either cross-section or time-series data.

1.6.1 CROSS-SECTION DATA

Cross-section data contain information on different elements of a population or sample for the *same* period of time. The information on incomes of 100 families for the year 1990 is an example of cross-section data.

CROSS-SECTION DATA

Cross-section data are data collected on different elements at the same point in time or for the same period of time.

Table 1.3 gives the number of live births for each of the six New England states for the same period of time, February 1990. This is an example of cross-section data.

Table 1.3 Number of Live Births

State	Live Births, February 1990
Connecticut	3,200
Maine	1,200
Massachusetts	10,998
New Hampshire	1,228
Rhode Island	1,176
Vermont	668

Source: Monthly Vital Statistics Report, U.S. Department of Health and Human Services, 39(2), June 7, 1990.

1.6.2 TIME-SERIES DATA

Time-series data contain information on the same element for *different* periods of time. Information on U.S. exports for the years 1975 to 1990 is an example of time-series data.

> **TIME-SERIES DATA**
>
> Time-series data are data collected on the same element for the same variable at different points in time or for different periods of time.

Data given in Table 1.4 are an example of time-series data. This table gives the total number of doctorates awarded by all U.S. universities for each year from 1962 through 1988.

Table 1.4 Doctorates Awarded by U.S. Universities, 1962–1988

Year	Number	Year	Number	Year	Number
1962	11,500	1971	31,867	1980	31,020
1963	12,728	1972	33,041	1981	31,357
1964	14,325	1973	33,755	1982	31,106
1965	16,340	1974	33,047	1983	31,280
1966	17,949	1975	32,951	1984	31,332
1967	20,403	1976	32,946	1985	31,291
1968	22,936	1977	31,716	1986	31,896
1969	25,743	1978	30,875	1987	32,367
1970	29,498	1979	31,239	1988	33,456

Source: Summary Report 1988, Doctorate Recipients from United States Universities, National Research Council, Washington, D.C., 1989.

1.7 SOURCES OF DATA

The availability of accurate and appropriate data is essential for deriving reliable results.† Data may be obtained from internal sources, external sources, or surveys and experiments.

Many times data come from *internal sources*, such as a company's own personnel files or accounting records. For example, a company that wants to forecast the future sales of its product may use the data of past periods from its own records. However, for most studies, all the needed data are not usually available from internal sources. In such cases, one may have to depend on outside sources to obtain data. These sources are called *external sources*. For instance, the *Statistical Abstract of the United States* (published annually), which contains various kinds of data on the United States, is an external source of data.

A large number of government and private publications can be used as external sources of data. The following is a list of some of the government publications.‡

1. *Statistical Abstract of the United States*
2. *Employment and Earnings*

†Sources of data are discussed in more detail in Appendix A.

‡All of these books, and the catalog of U.S. government publications, can be obtained from the Superintendent of Documents, U.S. Government Printing Office, Washington, D.C. 20402.

3. *Handbook of Labor Statistics*
4. *Source Book of Criminal Justice Statistics*
5. *Economic Report of the President*
6. *County & City Data Book*
7. *State & Metropolitan Area Data Book*
8. *Digest of Education Statistics*
9. *Health United States*
10. *Agricultural Statistics*

Besides these government publications, a large number of private publications (e.g., *Standard and Poors' Security Owner's Stock Guide* and *World Almanac and Book of Facts*) and periodicals (e.g., *The Wall Street Journal, U.S.A. Today, Fortune,* and *Business Week*) can be used as external data sources.

Sometimes the needed data may not be available from either internal or external sources. In such cases, the investigator may have to conduct a survey or experiment to obtain the required data.

EXERCISES

1.13 Explain the difference between cross-section and time-series data. Give an example of each of these two types of data.

1.14 Classify the following as cross-section or time-series data.

a. Gross sales of IBM for the period 1970 to 1989
b. Weights of 20 chickens
c. Poverty rates in the United States for 1975 to 1989
d. Auto insurance premiums paid by 100 persons

1.15 Describe briefly the internal and external sources of data.

1.8 SUMMATION NOTATION

Sometimes mathematical notation helps to express a mathematical relationship concisely. This section explains the *summation notation* that is used to denote the sum of values.

Suppose we have five books in a sample and their prices are \$25, \$40, \$37, \$33, and \$16. The variable *price of a book* can be denoted by x. The prices of five books can be written as follows.

$$\text{Price of the first book} = x_1 = \$25$$

↑ Subscript of x denotes the number of the book

Similarly

$$\text{Price of the second book} = x_2 = \$40$$
$$\text{Price of the third book} = x_3 = \$37$$
$$\text{Price of the fourth book} = x_4 = \$33$$
$$\text{Price of the fifth book} = x_5 = \$16$$

In this display, x represents price and the subscript denotes a particular book. Now, suppose we want to add the prices of all five books. Then,

$$x_1 + x_2 + x_3 + x_4 + x_5 = 25 + 40 + 37 + 33 + 16 = \$151$$

The uppercase Greek letter Σ (pronounced sigma) is used to denote the sum of all values. Using Σ notation, we can write the foregoing sum as follows.

$$\Sigma x = x_1 + x_2 + x_3 + x_4 + x_5 = \$151$$

The notation Σx in this expression represents the sum of all the values of x.

Using summation notation: one variable.

EXAMPLE 1-2 Suppose the ages of four persons are 35, 47, 28, and 60. Find
(a) Σx (b) $\Sigma(x - 6)$ (c) $(\Sigma x)^2$ and (d) Σx^2

Solution Let x_1 be the age of the first person, x_2 that of the second person, x_3 of the third person, and x_4 of the fourth person. Then,

$$x_1 = 35, \quad x_2 = 47, \quad x_3 = 28, \quad \text{and} \quad x_4 = 60$$

(a) $\Sigma x = x_1 + x_2 + x_3 + x_4 = 35 + 47 + 28 + 60 = 170$

(b) To calculate $\Sigma(x - 6)$, first we subtract 6 from each value of x and then add the resulting values. Thus,

$$\begin{aligned}\Sigma(x - 6) &= (x_1 - 6) + (x_2 - 6) + (x_3 - 6) + (x_4 - 6)\\ &= (35 - 6) + (47 - 6) + (28 - 6) + (60 - 6)\\ &= 29 + 41 + 22 + 54 = 146\end{aligned}$$

(c) Note that $(\Sigma x)^2$ is the square of the sum of all x values. Hence,

$$(\Sigma x)^2 = (170)^2 = 28{,}900$$

(d) The expression Σx^2 is the sum of the squares of x values. To calculate Σx^2, we first square each of the x values and then add the squared values. Thus,

$$\Sigma x^2 = (35)^2 + (47)^2 + (28)^2 + (60)^2 = 1225 + 2209 + 784 + 3600 = 7818$$

■

Using summation notation: one and two variables.

EXAMPLE 1-3 The following table lists four pairs of m and f values.

m	12	15	20	30
f	5	9	10	16

Compute the following.

(a) Σm (b) Σf^2 (c) Σmf (d) $\Sigma m^2 f$ (e) $\Sigma(m - 5)^2 f$

Solution We can write

$$m_1 = 12, \quad m_2 = 15, \quad m_3 = 20, \quad m_4 = 30,$$
$$f_1 = 5, \quad f_2 = 9, \quad f_3 = 10, \quad f_4 = 16.$$

(a) $\Sigma m = 12 + 15 + 20 + 30 = 77$

(b) $\Sigma f^2 = (5)^2 + (9)^2 + (10)^2 + (16)^2 = 462$

(c) To compute Σmf, we multiply the corresponding values of m and f and add the products as follows.

$$\Sigma mf = m_1f_1 + m_2f_2 + m_3f_3 + m_4f_4$$
$$= 12(5) + 15(9) + 20(10) + 30(16) = 875$$

(d) To calculate Σm^2f, we square each m value, then multiply the corresponding m^2 and f values, and add the products.

$$\Sigma m^2f = (m_1)^2f_1 + (m_2)^2f_2 + (m_3)^2f_3 + (m_4)^2f_4$$
$$= (12)^2(5) + (15)^2(9) + (20)^2(10) + (30)^2(16) = 21{,}145$$

The calculations done in parts (a) through (d) to find the values of Σm, Σf^2, Σmf, and Σm^2f can be performed in a tabular form, as shown in Table 1.5.

Table 1.5

m	f	f^2	mf	m^2f
12	5	25	60	720
15	9	81	135	2,025
20	10	100	200	4,000
30	16	256	480	14,400
$\Sigma m = 77$	$\Sigma f = 40$	$\Sigma f^2 = 462$	$\Sigma mf = 875$	$\Sigma m^2f = 21{,}145$

The columns of Table 1.5 can be explained as follows.

1. The first column lists the values of m. The sum of these values gives $\Sigma m = 77$.
2. The second column lists the values of f. The sum of this column gives $\Sigma f = 40$.
3. The third column lists the squares of the f values. The sum of the values in this column gives $\Sigma f^2 = 462$.
4. The fourth column records the products of the corresponding m and f values. The sum of this column gives $\Sigma mf = 875$.
5. Next, the m values are squared and multiplied by the corresponding f values. The resulting products, denoted by m^2f, are recorded in the fifth column of the table. The sum of this column gives $\Sigma m^2f = 21{,}145$.

(e) The value of $\Sigma(m - 5)^2 f$ is computed as follows.

$$\begin{aligned}\Sigma(m - 5)^2 f &= (m_1 - 5)^2 f_1 + (m_2 - 5)^2 f_2 + (m_3 - 5)^2 f_3 + (m_4 - 5)^2 f_4 \\ &= (12 - 5)^2(5) + (15 - 5)^2(9) + (20 - 5)^2(10) + (30 - 5)^2(16) \\ &= 13{,}395\end{aligned}$$

We can do these computations to calculate $\Sigma(m - 5)^2 f$ in a tabular form, as shown in Table 1.6.

Table 1.6

m	f	$(m - 5)$	$(m - 5)^2$	$(m - 5)^2 f$
12	5	$12 - 5 = 7$	49	$49 \times 5 = 245$
15	9	$15 - 5 = 10$	100	$100 \times 9 = 900$
20	10	$20 - 5 = 15$	225	$225 \times 10 = 2{,}250$
30	16	$30 - 5 = 25$	625	$625 \times 16 = 10{,}000$
$\Sigma m = 77$	$\Sigma f = 40$			$\Sigma(m - 5)^2 f = 13{,}395$

The third column of Table 1.6 lists the $(m - 5)$ values, which are obtained by subtracting 5 from the values listed in the column labeled m. The fourth column lists the squares of $(m - 5)$ values recorded in the third column. The fifth column contains the products of the corresponding values of $(m - 5)^2$ and f. The sum of the values listed in the fifth column gives $\Sigma(m - 5)^2 f$. ■

EXERCISES

1.16 The scores of five students in a statistics class are 75, 80, 97, 91, and 63. Find

a. Σx b. $\Sigma(x - 12)$ c. $(\Sigma x)^2$ d. Σx^2

1.17 The number of cars owned by six families are 3, 2, 1, 4, 1, and 2. Find

a. Σx b. $\Sigma(x - 1)$ c. $(\Sigma x)^2$ d. Σx^2

1.18 The heating bills for February 1990 for four families are $122, 72, 96, and 110. Find

a. Σx b. $\Sigma(x - 25)$ c. $(\Sigma x)^2$ d. Σx^2

1.19 The weights of seven newborn babies are 7, 9, 6, 12, 10, 9, and 8 pounds. Find

a. Σx b. $\Sigma(x - 4)$ c. $(\Sigma x)^2$ d. Σx^2

1.20 The following table lists five pairs of m and f values.

m	5	10	15	20	25
f	12	8	5	16	4

Find the value of each of the following.

a. Σm b. Σf^2 c. Σmf d. $\Sigma m^2 f$ e. $\Sigma(m - 15)^2 f$

1.21 The following table lists six pairs of m and f values.

m	3	6	9	12	15	18
f	16	11	6	8	4	14

Calculate the value of each of the following.

a. Σf b. Σm^2 c. Σmf d. $\Sigma m^2 f$ e. $\Sigma(m - 10)^2 f$

1.22 The following table lists five pairs of x and y values.

x	15	20	11	8	5
y	10	7	14	9	18

Compute a. Σx b. Σy c. Σxy d. Σx^2 e. Σy^2

1.23 The following table lists six pairs of x and y values.

x	4	18	25	9	12	20
y	12	5	14	7	12	8

Compute a. Σx b. Σy c. Σxy d. Σx^2 e. Σy^2

GLOSSARY

Census A survey that includes every member of the population.

Continuous variable A (quantitative) variable that can assume any numerical value over a certain interval or intervals.

Cross-section data Data collected on different elements at the same point in time or for the same period of time.

Data or data set Collection of observations or measurements on a variable.

Descriptive statistics Methods for organizing, displaying, and describing data using tables, graphs, and summary measures.

Discrete variable A quantitative variable whose values are countable.

Element or **member** A specific subject or object included in a sample or population.

Inferential statistics Methods for helping to make decisions about a population based on sample results.

Observation or **measurement** The value of a variable for a single element.

Population or **target population** The collection of all elements whose characteristics are being studied.

Qualitative or **categorical data** Data generated by a qualitative variable.

Qualitative or **categorical variable** A variable that cannot assume numerical values but is classified into two or more categories.

Quantitative data Data generated by a quantitative variable.

Quantitative variable A variable that can be measured numerically.

Random sample A sample drawn in such a way that each element of the population has some chance of being selected in the sample.

Representative sample A sample that contains the characteristics of the corresponding population.

Sample A portion of the population of interest.

Sample survey A survey that includes the elements of a sample.

Simple random sample A sample drawn in such a way that each element of the population has the same chance of being selected in the sample.

Statistics Methods used to collect, analyze, present, and interpret data and to make decisions.

Survey Collection of data on the elements of a population or a sample.

Time-series data Data that give the values for the same variable for the same element at different points in time or for different periods of time.

Variable A characteristic under study or investigation that assumes different values for different elements.

SUPPLEMENTARY EXERCISES

1.24 The following table gives the average hourly earnings (in current dollars) of production or nonsupervisory workers on private nonagricultural payrolls in the United States for 9 months, from May 1989 through January 1990.

Month and Year	Average Hourly Earnings
May 1989	$9.60
June 1989	9.62
July 1989	9.69
August 1989	9.69
September 1989	9.74
October 1989	9.78
November 1989	9.78
December 1989	9.83
January 1990	9.84

Source: U.S. Bureau of Labor Statistics.

Describe the meaning of a variable, a measurement, and a data set with reference to this table.

1.25 The following table gives the total number of divorces for April 1990 for the four East South Central states.

State	Number of Divorces
Kentucky	1544
Tennessee	2564
Alabama	2242
Mississippi	896

Source: Monthly Vital Statistics Report, August 2, 1990.

Describe the meaning of a variable, a measurement, a data set, and an element with reference to this table.

1.26 Refer to Exercises 1.24 and 1.25. Classify these data sets as either cross-section or time-series.

1.27 Indicate which of the following examples refer to a population and which refer to a sample.

a. A group of 25 patients selected to test a new drug
b. Total items produced on a machine during one week's period
c. Yearly expenditure on clothes for 50 persons
d. Number of houses sold by each of the 10 employees of a real estate agency
e. Salaries of all employees of a bank

1.28 State which of the following are examples of sampling with replacement and which are of sampling without replacement.

a. Selecting 10 patients out of 100 to test a new drug
b. Selecting five students to form a committee
c. Selecting one professor to be a member of the university senate and then selecting one professor from the same group to be a member of the curriculum committee

1.29 The number of ties owned by six persons are 10, 9, 14, 12, 7, and 4. Find

a. Σx **b.** $\Sigma(x - 6)$ **c.** $(\Sigma x)^2$ **d.** Σx^2

1.30 The following table lists five pairs of m and f values.

m	3	16	11	9	20
f	7	32	17	12	34

Calculate the value of each of the following.

a. Σm **b.** Σf^2 **c.** Σmf **d.** $\Sigma m^2 f$ **e.** $\Sigma(m - 5)^2 f$

SELF-REVIEW TEST

1. A population in statistics means

a. A collection of all people of interest
b. A collection of all subjects or objects of interest
c. A collection of all people living in a country

2. A sample in statistics means

a. A portion of the people selected from the population of a country
b. A portion of the people selected from the population of an area
c. A portion of the population of interest

3. Which of the following is an example of a representative sample, a sample with replacement, or a sample without replacement?

a. Ten students are selected from a statistics class in such a way that as soon as a student is selected his or her name is deleted from the list before the next student is selected.
b. A sample of 100 families is taken from a town in such a way that it contains the characteristics of the population.
c. A box contains five balls of different colors. A ball is drawn from this box, its color is recorded, and it is put back in the box before the next ball is drawn. This experiment is repeated 12 times.

4. Explain which of the following variables are quantitative and which are qualitative. Classify the quantitative variables as discrete or continuous.

a. Type of blood
b. Number of physicians in a city
c. Weekly earnings of employees

5. The following table gives the 1989 populations for five European countries.

Country	Population (millions)
Austria	7.6
France	56.1
West Germany	61.5
Netherlands	14.9
United Kingdom	57.3

Explain the meaning of a member, a variable, a measurement, and a data set with reference to this table.

6. Classify the following as cross-section or time-series data.

a. Blood types of 20 persons
b. Average summer temperature in New York from 1980 to 1990
c. 1990 population of 50 states

7. The stress scores (on a scale of 1 to 10) of six students before a test are found to be 5, 7, 3, 8, 7, and 6. Calculate

a. Σx **b.** Σx^2 **c.** $(\Sigma x)^2$ **d.** $\Sigma(x - 2)$ **e.** $\Sigma(x - 2)^2$

8. The following table lists five pairs of m and f values.

m	3	6	9	12	15
f	15	25	40	20	12

Calculate the value of each of the following.

a. Σm **b.** Σf **c.** Σmf **d.** $\Sigma m^2 f$

USING MINITAB†

AN INTRODUCTION

In recent years the use of computers has significantly increased in almost every aspect of life. The use of computers has reduced the computation time for quantitative work to almost negligible fractions. Calculations that may take weeks or even months for a human to perform can be done in seconds by computers.

In the real world, when doing a statistical analysis, we usually do not deal with 20 or 30 observations but with many hundreds or thousands of observations. For this reason, it is either very time consuming or almost impossible to make all the required calculations manually. The use of computers is of invaluable assistance in such situations. Consequently, learning to use a statistical software package has become as important as learning statistics.

A large number of statistical software packages, both for mainframe computers and for microcomputers, have been developed in recent years. Most of these software packages are very user friendly and self-explanatory.

Four of the major statistical software packages are BMDP, MINITAB, SAS, and SPSS.‡ All four of these packages are available for mainframe and for microcomputers. Besides these four packages, a large number of software packages have been developed for personal computers. In this text, we will provide brief instructions on how to use MINITAB to solve statistical problems. Two formal manuals that explain MINITAB commands in detail are

1. *MINITAB Reference Manual,* MINITAB Inc., State College, 1989.
2. Barbara F. Ryan, Brian L. Joiner, and Thomas A. Ryan, Jr., *MINITAB Handbook,* 2nd ed., PWS-KENT Publishing Company, Boston, 1985.

You may consult these manuals for a detailed discussion of MINITAB.

MINITAB is available on both mainframe computers and microcomputers. The MINITAB commands are the same whether you are working on a mainframe computer or on a microcomputer. However, procedures to start a system and to access MINITAB are different for these two systems. If you are using a mainframe computer system, the first step is to *log on* to the system using an account number and a password.

†The MINITAB commands illustrated in this textbook are for the PC version of MINITAB. Some of these commands may not work for other versions of MINITAB, especially the student version of MINITAB.

‡BMDP is a registered trademark of BMDP Statistical Software, Inc. MINITAB is a registered trademark of Minitab, Inc. SAS is a registered trademark of SAS Institute, Inc. SPSS is a registered trademark of SPSS, Inc.

Your instructor or an assistant in the computer center can explain how to use your school's mainframe computer system. Also, remember that you must *log off* the system before you leave. You must also ask either your instructor or the computer center how to obtain a hard (printed) copy of the MINITAB *output*. After you log on to the mainframe system, the next step is to enter the MINITAB environment. Again, your instructor or the computer center can show you how to enter the MINITAB *environment*.

If you are working on a microcomputer, you need to know how to *load* MINITAB and how to enter the MINITAB environment. Your instructor or the computer lab assistant can explain this to you.

Once you have completed these formalities and entered the MINITAB environment, your computer terminal will display the following message.

```
MTB >
```

This message is called the *MINITAB prompt*. At this point, the computer is ready to receive MINITAB commands. The MINITAB commands are entered on the computer next to the MINITAB prompt and the *enter/return* key is pressed after completing each command. The MINITAB commands can be entered in uppercase letters, lowercase letters, or in any combination of the two. To end a MINITAB session, type STOP next to the MINITAB prompt and hit the enter/return key.

Any information that you enter on the computer terminal is called *input*, and the solution provided by MINTAB in response to your commands is called *output*.

If you are working on a microcomputer (an IBM PC or a compatible), the following command will send all your subsequent MINITAB commands and MINITAB output to the printer.

```
MTB > PAPER
```

If you want to print everything from the very beginning, type PAPER at the very first MINITAB prompt. When you want to stop printing, type NOPAPER. Note that the printer must be turned on before you type the PAPER command. The current screen can also be printed by pressing the key marked 'Print Screen'.

Usually you need to enter some data before you do any statistical analysis. Illustration M1-1 demonstrates how you can enter data using MINITAB commands.

Illustration M1-1 Following are the scores of seven students in a statistics test.

86 91 74 80 65 97 79

To enter data into MINITAB, you can use the SET or READ command. The SET command is used to enter data on one variable only. The READ command can be used to enter data on one or more variables. The data are always entered into a column that is denoted by C followed by a number (e.g., C1, C2, C3, . . .). The notation used to enter a constant is K followed by a number (e.g., K1, K2, K3, . . .). A MINITAB command that begins with NOTE is not processed by MINITAB. This command is only for the information of the user. Using the NOTE command, you can enter any comments for your information. The data on scores of seven students are entered into MINITAB as shown in the following set of commands.

```
MTB  > NOTE: DATA ON SCORES OF SEVEN STUDENTS
MTB  > SET C1   ← This command instructs MINITAB that you are to enter the data in
                  column C1
DATA > 86 91 74 80   ← When using the SET command, you can enter these data values
                       in as many rows as you want
DATA > 65 97 79
DATA > END   ← This command indicates the end of data entry
```

In the foregoing MINITAB display, "DATA >" prompt indicates that MINITAB is ready for data entry. When SET command is used to enter data on one variable, as shown in the display, all values can be entered in one row or in any number of rows. In the foregoing display, four values were entered in the first row and three in the second row. However, if the READ command is used to enter data on one variable, only one value should be entered in each row. After entering all data values, type END at the next "DATA >" prompt to instruct MINITAB that all data have been entered.

You can give a column a name using the NAME command. The following command will name the data entered in column C1 as 'SCORES'.

```
MTB > NAME C1 'SCORES'
```

The PRINT C1 command will show the data on scores on the computer screen as follows.

```
MTB > PRINT C1
C1

86   91   74   80   65   97   79
```

PRINT 'SCORES' command (after naming C1 as 'SCORES') will also print these data. Note that when you use a proper name for a column, as SCORES, this name must always be enclosed within single quotes.

If you plan to use the data you entered into MINITAB again at some later time, you need to save it as a file before you end the current MINITAB session. Consult your instructor or the computer center at your school to find out how to save a data file if you are using a mainframe computer system. To save a data file, you must give it a name. Suppose you call the current data file 'SCORES'. If you are using a microcomputer, any of the following commands will save this file. Use only one of these commands, depending on whether you want to save your file on the hard disk, disk drive A, or disk drive B. Note that the file name is enclosed within single quotes.

```
MTB > SAVE 'SCORES'     ← This command will save the data file as SCORES.MTW on
                          the hard disk
MTB > SAVE 'A:SCORES'   ← This command will save the data file on the floppy disk
                          in drive A
MTB > SAVE 'B:SCORES'   ← This command will save the data file on the floppy disk
                          in drive B
```

MINITAB will attach the extension ".MTW" (which stands for MINITAB Worksheet) at the end of the file name when you save a file, unless you use another extension.

To work on SCORES file at a later date, the RETRIEVE command will bring it back into the current worksheet.

```
MTB > RETRIEVE 'SCORES'     ← This command is used if the saved file is on the hard
                              disk
MTB > RETRIEVE 'A:SCORES'   ← This command is used if the saved file is on a
                              floppy disk in drive A
MTB > RETRIEVE 'B:SCORES'   ← This command is used if the saved file is on a
                              floppy disk in drive B
```

MINITAB recognizes a command from the first four letters. Hence, if you use the command RETR 'SCORES' instead of RETRIEVE 'SCORES', the computer will respond with the same answer. This is true of all MINITAB commands. Remember that whenever you use the file name—either to save it or to retrieve it or to work on it—you must enclose the file name within single quotes.

The following INFORMATION command will help you to know what is in a retrieved file.

```
MTB > INFO
```

MINITAB SAMPLE command can be used to take a sample from a population. Assume that the data on seven scores entered in column C1 in MINITAB display of Illustration M1-1 belong to a population. Using the following procedure, a sample of three scores can be selected from this population.

```
MTB > NOTE: SELECTING A SAMPLE OF 3 SCORES
MTB > SAMPLE 3 FROM C1 PUT IN C2  ← This command instructs MINITAB to
                                     take a sample of three observations
                                     (without replacement) from the data
                                     of column C1 and put those in column
                                     C2
MTB > PRINT C2  ← This command will print the sample data of column C2 on the
                  computer terminal screen
C2
91   65   97   ← These values give the required sample
```

Remember that if you already have entered some data in column C1 and you want to enter some new data into MINITAB, the new data must be entered in a new column, say, C2. If you enter the new data in C1, the data entered in C1 previously will be lost.

Illustration M1-2 explains how data on more than one variable can be entered into MINITAB using the READ command.

Illustration M1-2 Suppose you need to enter the following data on heights and weights of six persons into MINITAB.

Height (inches)	Weight (pounds)
69	178
67	135
65	121
71	210
68	149
66	142

To enter these data using the READ command, enter both the height and weight of each person in one row. This illustration contains information on six persons. Therefore, enter these data in six rows—each row containing information on the height and weight of one person.

```
MTB    > NOTE: DATA ON HEIGHTS AND WEIGHTS
MTB    > READ C1 C2  ← This command instructs MINITAB that you are to enter data
                       on two variables in two columns C1 and C2
DATA > 69 178
DATA > 67 135
DATA > 65 121
DATA > 71 210
DATA > 68 149
DATA > 66 142
DATA > END
```

You should repeat the READ command, try the following commands for MINITAB display of Illustration M1-2, and analyze the output for each command.

```
MTB > SAMPLE 3 FROM C1-C2 PUT IN C3-C4
MTB > PRINT C3-C4
MTB > NAME C1 'HEIGHT'
MTB > NAME C2 'WEIGHT'
MTB > PRINT C1 C2
MTB > PRINT 'HEIGHT' 'WEIGHT'
MTB > PRINT C1
MTB > PRINT 'WEIGHT'
MTB > SAVE 'A:PERSONAL'   ← You have named this file 'PERSONAL'
MTB > RETR 'A:PERSONAL'
MTB > INFO
MTB > PRINT C1-C2
MTB > PRINT 'HEIGHT'
MTB > PRINT C2
MTB > LET K1 = 5
MTB > LET C5 = C1 - K1
MTB > LET C6 = C2 + K1
MTB > MULTIPLY C1 BY K1 PUT IN C7
MTB > DIVIDE C2 BY K1 PUT IN C8
MTB > PRINT C5 C6 C7 C8
```

The SAVE command in this set of commands will save the file on a floppy disk in disk drive A on a microcomputer. However, if you are saving on a hard disk, or on disk drive B, or on a mainframe computer system, follow the earlier instructions.

Following are some of the additional MINITAB commands and their explanations. (Note that these commands will not be used in the sequence they are presented here.)

```
MTB > HELP HELP   ← This command can be used to seek help about MINITAB commands
MTB > HELP COMMANDS   ← This command also provides help about MINITAB commands
MTB > HELP OVERVIEW   ← This command can be used to have an overview of MINITAB
MTB > STOP   ← This command will end the current MINITAB session
MTB > COPY C1 TO C2   ← This command will copy all data values from column C1 to column C2
MTB > ERASE C2   ← This command will delete all data values entered in column C2
MTB > DELETE ROW 2 C1   ← This command will delete the second value entered in column C1
MTB > DELETE ROW 2 C1-C3   ← This command will delete the data values entered in the second row of columns C1, C2, and C3
MTB > INSERT BETWEEN 2 AND 3 C1-C3   ← This command will insert a new row between second and third rows for columns C1, C2, and C3
MTB > LET C1(4) = 10   ← This command will replace the fourth entry in column C1 by 10
```

```
MTB > SORT C1 PUT IN C4  ← This command will sort the data of column C1 in
                           increasing order and put those in column C4
MTB > ADD C1 C2 PUT IN C5  ← This command will add the corresponding values
                             of columns C1 and C2 and put the new data in
                             column C5
MTB > SUBTRACT C2 FROM C1 PUT IN C6  ← This command will subtract each
                                       value of column C2 from the cor-
                                       responding value of column C1
                                       and put the new data in column
                                       C6
MTB > MULTIPLY C1 BY C2 PUT IN C7  ← This command will multiply the
                                     corresponding values of columns
                                     C1 and C2 and put the new data
                                     in column C7
MTB > DIVIDE C1 BY C2 PUT IN C8  ← This command will divide each value of
                                   column C1 by the corresponding value of
                                   column C2 and put the new data in col-
                                   umn C8
MTB > LET C9 = C1 * C2  ← This command will multiply the corresponding values of
                          columns C1 and C2 and put the new data in column C9
MTB > LET C10 = C1**2  ← This command will square each value entered in column
                         C1 and put the new data in column C10
MTB > ADD 5 TO C1 PUT IN C11  ← This command will add 5 to each value of
                                column C1 and put the new data in column
                                C11
MTB > SUBTRACT 8 FROM C1 PUT IN C12  ← This command will subtract 8
                                       from each value of column C1
                                       and put the new data in column
                                       C12
MTB > MULTIPLY C1 BY 2 PUT IN C13  ← This command will multiply each value
                                     of column C1 by 2 and put the new data
                                     in column C13
MTB > DIVIDE C1 BY 3 PUT IN C14  ← This command will divide each value of
                                   column C1 by 3 and put the new data in
                                   column C14
```

MINITAB HELP command followed by a specific command provides information about that command. For example, the command HELP SET can be used to find information about the SET command and its usage.

In MINITAB, data can also be entered using the MINITAB *worksheet,* which is like a spreadsheet. MINITAB worksheet lets you enter the information on a variable that contains alpha characters. In other words, using MINITAB worksheet, you can enter the names of persons, places, and so forth. Illustration M1-3 describes this procedure. (Note: This procedure cannot be used in the student version of MINITAB.)

Illustration M1-3 The following data give the 1989 profits (in billions of dollars) for the top five U.S. companies.

Company	1989 Profits (billions of dollars)
General Motors	4.2
General Electric	3.9
Ford Motor	3.8
IBM	3.8
Exxon	3.0

Suppose you need to enter these data into MINITAB. Use the following procedure to enter these data.

At MTB prompt, press the key marked ESC. The computer terminal will show a spreadsheet with column numbers at the top and observation numbers on the left. The cursor will be positioned next to observation number 1 below C1. Type General Motors and press the enter/return key. The cursor will move to the second column (below C2). Type 4.2 and press enter/return key. The cursor will move to the position below C3. Because you have information on only two variables, move the cursor to the beginning of the second row using the keys marked [↓] and [←]. Type General Electric and press the enter/return key. Type 3.9 and press the enter/return key. Move the cursor to the beginning of the third row using the keys [↓] and [←]. Continue this procedure until you have entered information on all five companies. When you have entered all data, the computer screen will look as follows.

```
                     C1  C2  C3  C4  C5  C6  C7  C8  C9  C10  C11  C12
 1 General Motors        4.2
 2 General Electric      3.9
 3 Ford Motor            3.8
 4 IBM                   3.8
 5 Exxon                 3.0
 6
 7
 8
 9
10
11
12
13
14
15
16
17
18
19
20
21
22
                                      Last Column: C2        Last Row: 5
```

At this point, press the ESC key again. You will be back at the MTB prompt. Now you can save this file or do any statistical analysis.

This procedure can also be used to modify any information in any file.

COMPUTER ASSIGNMENTS

M1.1 Refer to Data Set III of Appendix B (page 711) on the heights and weights of all NBA players. Enter the data on the heights and weights of all players in columns C1 and C2 using MINITAB READ command. (If you are using the data disk that contains the data on names also, delete the column of names using ERASE C1 command. Then copy the data of C2 into C1 and those of C3 into C2 using the COPY command.) Take a sample of heights and weights of 15 players using the MINITAB SAMPLE command and put the sample data in columns C3 and C4. (Use the command: MTB > SAMPLE 15 C1 C2 PUT IN C3 C4.) Using the sample data, do the following.

a. Name C3 heights and C4 weights
b. Print heights and weights
c. Delete the fifth row of columns C3 and C4
d. Insert a new row between fourth and fifth rows with values 81 198
e. Replace the fifth entry in column C3 by 78
f. Add 5 to C3 and put in C5
g. Subtract 10 from C4 and put in C6
h. Multiply C3 by 3 and put in C7
i. Divide C4 by 2 and put in C8
j. Print columns C5 to C8

M1.2 The following table gives the 1989 assets and sales of five U.S. companies.

Company	1989 Assets (billions of dollars)	1989 Sales (billions of dollars)
General Motors	173.3	126.9
Ford Motor	160.9	96.1
Exxon	83.2	88.1
IBM	77.7	62.7
General Electric	128.3	53.9

a. Enter these data into MINITAB using MINITAB worksheet
b. Print C1, C2, and C3
c. Name C1 'Company', C2 'Assets', and C3 'Sales'
d. Print 'Company', 'Assets', and 'Sales'
e. Take a sample of three companies
f. Print the sample data

2 ORGANIZING DATA

2.1 RAW DATA

2.2 ORGANIZING AND GRAPHING QUALITATIVE DATA

2.3 ORGANIZING AND GRAPHING QUANTITATIVE DATA

2.4 SHAPES OF HISTOGRAMS

2.5 CUMULATIVE FREQUENCY DISTRIBUTIONS

2.6 STEM-AND-LEAF DISPLAYS

2.7 GRAPHING TIME-SERIES DATA

SELF-REVIEW TEST

USING MINITAB

In addition to thousands of private organizations and individuals, a large number of U.S. government agencies such as the Bureau of the Census, the Bureau of Labor Statistics, the National Agricultural Statistics Service, the National Center for Education Statistics, the National Center for Health Statistics, and the Bureau of Justice Statistics conduct hundreds of surveys every year. The data collected from each of these surveys comprise hundreds of thousands of pages. In their original form, these data do not make sense to most of us. Descriptive statistics, however, supplies the techniques that help to condense large data sets using tables, graphs, and summary measures. We see such tables, graphs, and summary measures in newspapers and magazines every day. At a glance, these data displays present information on every aspect of life. Consequently, descriptive statistics is of immense importance because it provides efficient and effective methods for analyzing and summarizing information.

This chapter explains how to organize and display data using tables and graphs. We will learn how to prepare frequency tables for qualitative and quantitative data; how to construct bar graphs, pie charts, histograms, and polygons for such data; and how to prepare stem-and-leaf displays.

2.1 RAW DATA

When data are collected, the information obtained from each member of a population or sample is recorded in the sequence in which it becomes available. This sequence of data recording is random and unordered. Such data, before they are grouped or ranked, are called **raw data.**

RAW DATA

Data recorded in the sequence in which they are collected and before they are processed are called raw data.

Suppose we collect information on the ages (in years) of 50 students selected from a university. The data values, in the order they are collected, are recorded in Table 2.1. For instance, the first student's age is 21 years, the second student's age is 19 years (second number in the first row), and so forth. Data given in Table 2.1 are quantitative raw data.

Table 2.1 Ages of 50 Students

21	19	24	25	29	34	26	27	37	33
18	20	19	22	19	19	25	22	25	23
25	19	31	19	23	18	23	19	23	26
22	28	21	20	22	22	21	20	19	21
25	23	18	37	27	23	21	25	21	24

Suppose we ask the same 50 students about their student status. The responses of the students are recorded in Table 2.2. In this table, F, SO, J, and SE are the abbreviations for freshman, sophomore, junior, and senior, respectively. This is an example of qualitative (or categorical) raw data.

Table 2.2 Status of 50 Students

J	F	SO	SE	J	J	SE	J	J	J
F	F	J	F	F	F	SE	SO	SE	J
J	F	SE	SO	SO	F	J	F	SE	SE
SO	SE	J	SO	SO	J	J	SO	F	SO
SE	SE	F	SE	J	SO	F	J	SO	SO

The data presented in Tables 2.1 and 2.2 are also called *ungrouped data*. An ungrouped data set contains information on each member of a sample or population individually.

2.2 ORGANIZING AND GRAPHING QUALITATIVE DATA

This section discusses how to organize and display qualitative (or categorical) data. A data set is organized by constructing tables and is displayed by making graphs.

2.2.1 FREQUENCY DISTRIBUTIONS

According to the U.S. Bureau of the Census, there was a total of 92,901 thousand women, aged 18 and over, living in the United States in 1988. Table 2.3 gives the breakdown of these women according to marital status. In this table, the variable is the *marital status,* which is a qualitative variable. The categories (representing the marital status) listed in the first column of the table are mutually exclusive. In other words, each of the 92,901 thousand women belongs to one and only one of these categories. The number of women who belong to a certain category is called the *frequency* of that category. A **frequency distribution** exhibits how the frequencies are distributed over various categories. Table 2.3 is called a *frequency distribution table* or simply a *frequency table.*

Table 2.3 Marital Status of Women Aged 18 and Over, 1988

Marital Status (← Variable)	Number of Women (thousands) (← Frequency column)
Single	17,364
Married (← Category)	56,128 (← Frequency)
Widowed	11,239
Divorced	8,170
Total	92,901

Source: Statistical Abstract of the United States 1990.

> **FREQUENCY DISTRIBUTION FOR QUALITATIVE DATA**
>
> A frequency distribution for qualitative data lists all categories and the number of elements that belong to each of the categories.

Example 2-1 illustrates how to construct a frequency distribution table for qualitative data.

Constructing frequency distribution table for qualitative data.

EXAMPLE 2-1 A sample was taken of 25 high school seniors who are planning to go to college. Each of the students was asked which of the following majors he or she intended to choose: business, economics, management information systems (MIS), behavioral sciences (BS), other. The responses of these students are listed as follows.

Economics	MIS	Economics	Business	Business
Business	Business	Other	Other	Other
BS	BS	MIS	Other	MIS
Other	Business	MIS	Business	Other
Economics	MIS	Other	Other	MIS

Construct a frequency distribution table for these data.

Solution Note that the *major a student intends to choose* is the variable in this example. This variable is classified into five categories: business, economics, MIS, BS, and other. We record these categories in the first column of Table 2.4. Then we

Table 2.4 Frequency Distribution of Majors

Major	Tally	Frequency (f)
Business	卌 |	6
Economics	|||	3
MIS	卌 |	6
BS	||	2
Other	卌 |||	8
		Sum = 25

read each student's response from the given data and mark a *tally,* denoted by the symbol |, in the second column of Table 2.4 next to the corresponding category. For example, the first student intends to major in economics. We show this in the frequency distribution table by marking a tally in the second column next to the category *Economics*. Note that the tallies are marked in blocks of fives for counting convenience. Finally, we record the total of tallies for each category in the third column of the table. This column is called the column of frequencies and is usually denoted by f. The sum of the entries in the frequency column gives the sample size or total frequency. In Table 2.4, this total is 25, which is the sample size. ■

2.2.2 RELATIVE FREQUENCY AND PERCENTAGE DISTRIBUTIONS

The **relative frequency** of a category is obtained by dividing the frequency of that category by the sum of all frequencies. Thus, the relative frequency shows what fractional part or proportion of the total frequency belongs to the corresponding category. A *relative frequency distribution* lists the relative frequencies for all categories.

RELATIVE FREQUENCY OF A CATEGORY

$$\text{Relative frequency of a category} = \frac{\text{Frequency of that category}}{\text{Sum of all frequencies}}$$

The **percentage** for a category is obtained by multiplying the relative frequency of that category by 100. A *percentage distribution* lists the percentages for all categories.

PERCENTAGE

$$\text{Percentage} = (\text{Relative frequency}) \cdot 100$$

Constructing relative frequency and percentage distributions.

EXAMPLE 2-2 Construct the relative frequency and percentage columns for Table 2.4.

Solution We have calculated the relative frequencies and percentages for Table 2.4 and listed them in Table 2.5.

Table 2.5 Relative Frequency and Percentage Distributions Table

Major	Relative Frequency	Percentage
Business	6/25 = .24	.24 (100) = 24
Economics	3/25 = .12	.12 (100) = 12
MIS	6/25 = .24	.24 (100) = 24
BS	2/25 = .08	.08 (100) = 8
Other	8/25 = .32	.32 (100) = 32
	Sum = 1.00	Sum = 100

Based on Table 2.5, we can state that .24 or 24 percent of the students in the sample said that they intend to major in business. By adding the percentages for the first two categories, we can state that 36% of the students said they intend to major in business or economics. The other numbers in Table 2.5 (and also in Table 2.4) can be interpreted in the same way. ■

Notice that the sum of the relative frequencies is always 1.00 (or approximately 1.00 if the relative frequencies are rounded), and the sum of the percentages is always 100 (or approximately 100 if the percentages are rounded).

2.2.3 GRAPHICAL PRESENTATION OF QUALITATIVE DATA

All of us have heard the saying "a picture is worth a thousand words." A graphic display can reveal at a glance the main characteristics of a data set. The bar graph and the pie chart are two types of graphs used to display qualitative data.

Bar Graphs

To construct a **bar graph** (also called a *bar chart*), we mark the various categories on the horizontal axis as in Figure 2.1. Note that all categories are represented by intervals of equal width. We mark the frequencies on the vertical axis. Then we draw one bar for each category such that the height of the bar represents the frequency of the corresponding category. We leave a small gap between adjacent bars. Figure 2.1 gives the bar graph for the frequency distribution of Table 2.4.

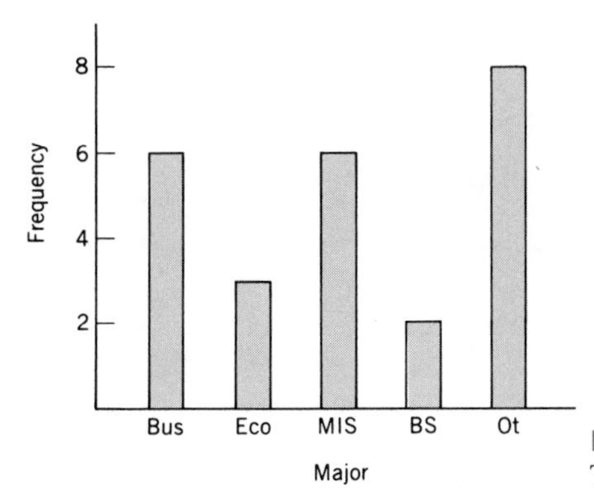

Figure 2.1 Bar graph for the frequency distribution of Table 2.4.

BAR GRAPH

A graph made of bars, whose heights represent the frequencies of respective categories, is called a bar graph.

We can draw the bar graphs for relative frequency and percentage distributions simply by marking the relative frequencies or percentages, instead of the class frequencies, on the vertical axis.

A bar graph can also be constructed by marking the categories on the vertical axis and the frequencies on the horizontal axis. Case Study 2-2 shows a bar graph with percentages on the horizontal axis and categories on the vertical axis.

CASE STUDY 2-1 NUMBER OF TELEPHONES OWNED BY HOUSEHOLDS

The following bar graph, reprinted from *USA Today*, shows the number of telephones owned by households. Data used to construct this bar graph are taken from a national poll of 1000 households conducted by Maritz Ameripoll. According to this poll, 27% of the households in the sample owned one telephone, 35% owned two, 22% owned three, 10% owned four, and 6% owned five or more telephones.

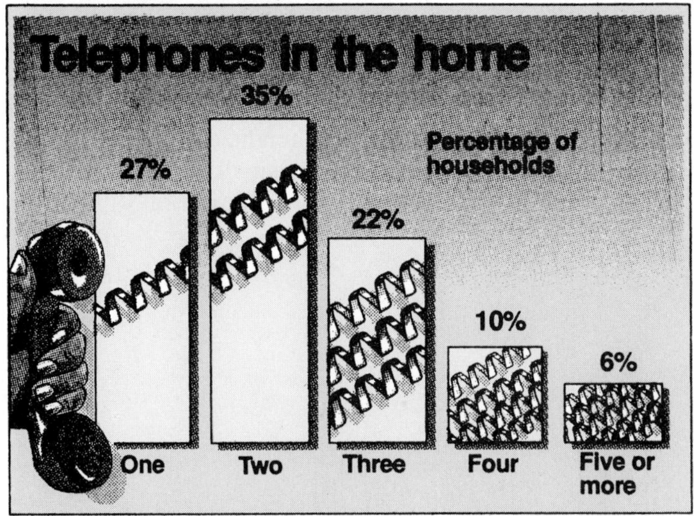

Source: *USA Today*, October 15, 1990. Copyright 1990, *USA Today*. Reprinted with permission.

CASE STUDY 2-2 HOW STUDENTS GRADE THEIR SCHOOLS

The Roper Organization, Inc., conducted a nationwide survey of 1000 American young people, aged 8 to 17, who were attending school. The interviews were conducted in 1987 for the American Chicle Youth Poll commissioned by The American Chicle Group, Warner-Lambert Company. "In order to assess the perceived quality of today's schools, the respondents were asked to assign a grade to their school on a scale of A to F." The bar graph displays the responses to the survey. The numbers in the bar graph represent the percentages of students that assigned different grades.

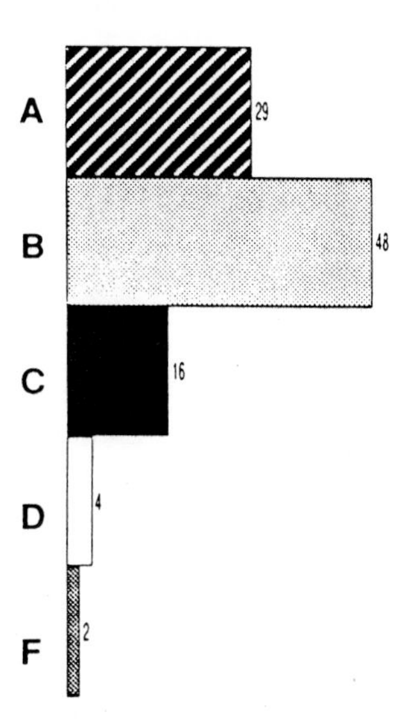

Source: The American Chicle Youth Poll, March 1987. Copyright © 1987 by Warner-Lambert Company. Reprinted with permission.

Pie Charts

A **pie chart** is more commonly used to display percentages, although it can be used to display frequencies or relative frequencies. The whole pie (or circle) represents the total sample or population. The pie is divided into different portions that represent the percentages belonging to different categories of the population or sample.

PIE CHART

A circle divided into portions that represent the percentages of a population or a sample that belong to different categories is called a pie chart.

Table 2.6 Calculation of Angle Sizes for the Pie Chart of Figure 2.2

Major	Relative Frequency	Angle Size
Business	.24	360 (.24) = 86.4
Economics	.12	360 (.12) = 43.2
MIS	.24	360 (.24) = 86.4
BS	.08	360 (.08) = 28.8
Other	.32	360 (.32) = 115.2
	Sum = 1.00	Sum = 360

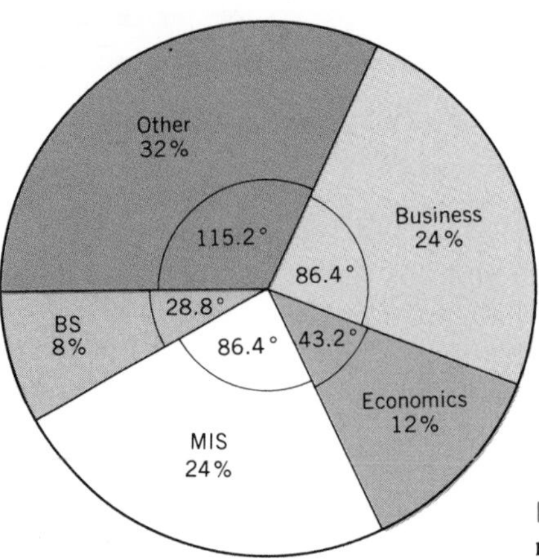

Figure 2.2 Pie chart for the percentage distribution of majors.

As we know, a circle contains 360°. To construct a pie chart, we multiply 360 by the relative frequency for each category to obtain the degree measure or size of the angle for the corresponding category. Table 2.6 shows the calculation of angle sizes for various categories of Table 2.5.

Figure 2.2 shows the pie chart for the percentage distribution of Table 2.5, which uses the angle sizes calculated in Table 2.6.

Case Studies 2-3 and 2-4 show examples of applications of pie charts.

CASE STUDY 2-3 HOW MUCH CHILDREN KNOW ABOUT GRANDPARENTS!

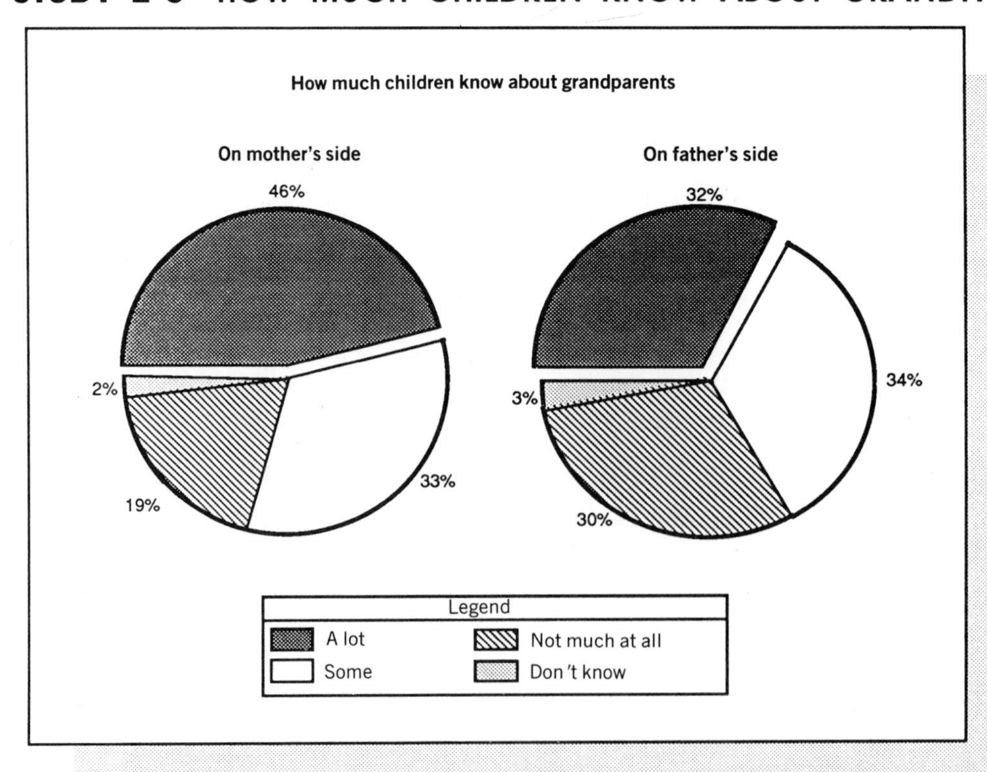

In 1987 the Roper Organization, Inc., interviewed for the American Chicle Youth Poll 1000 American young people, aged 8 to 17, who were attending school. The students were asked, "How much do you feel you know about your grandparents" on "your" mother's and father's sides. The two pie charts show the percentage distributions of responses of the children. Notice that because of rounding, the sum of the percentages for the pie chart titled "On Father's Side" is not 100%.

Source: *The American Chicle Youth Poll,* March 1987. Copyright © 1987 by Warner-Lambert Company. The pie charts reprinted with permission.

CASE STUDY 2-4 SERVING WITH DISTINCTION

The following pie charts show the results of a poll conducted of 2000 adults. The respondents were asked how satisfied they were with various institutions. The pie charts show the percentages of people who said they were either very satisfied, somewhat satisfied, or dissatisfied. Because of "no opinion" responses, percentages do not add up to 100 in all cases. The gaps shown in some of the pie charts represent the "no opinion" percentages.

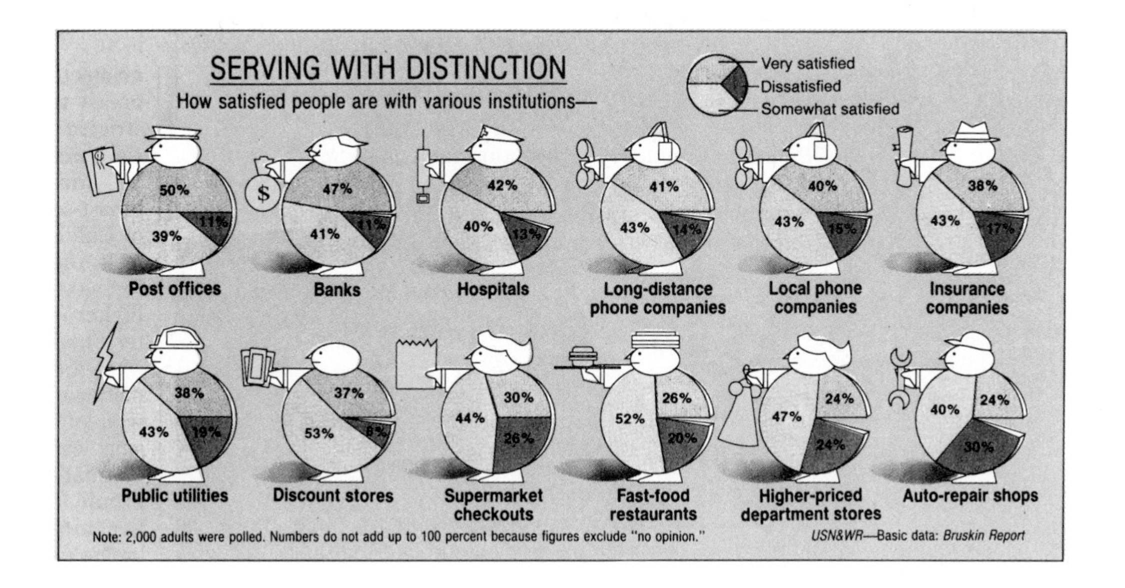

Source: *U.S. News & World Report,* July 11, 1988. Copyright © 1988 by U.S. News & World Report, Inc. Reprinted with permission.

EXERCISES

2.1 Data on the status of 50 students given in Table 2.2 of Section 2.1 are reproduced below.

J	F	SO	SE	J	J	SE	J	J	J
F	F	J	F	F	F	SE	SO	SE	J
J	F	SE	SO	SO	F	J	F	SE	SE
SO	SE	J	SO	SO	J	J	SO	F	SO
SE	SE	F	SE	J	SO	F	J	SO	SO

a. Construct a frequency distribution table.
b. Calculate the relative frequencies and percentages for all categories.
c. What percentage of these students are juniors or seniors?
d. Draw a bar graph for the frequency distribution.

2.2 The following are the responses of 20 students of a statistics class who were asked to evaluate their instructor. The students were asked to choose one of the five answers: Excellent (E), Above average (AA), Average (A), Below average (B), and Poor (P).

AA	B	A	A	E	AA	P	E	AA	B
E	AA	E	B	E	A	B	P	AA	E

a. Construct a frequency distribution table.
b. Calculate the relative frequencies and percentages for all categories.
c. What percentage of these students ranked this instructor as excellent or above average?
d. Draw a bar graph for the relative frequency distribution.

2.3 Fifteen workers of a company were asked about their opinions on an issue. The responses of the workers are listed here. (F, A, and N indicate that a worker is in favor, against, or has no opinion, respectively.)

F A A F N N F F F A A N F F A

a. Prepare a frequency distribution table.
b. Calculate the relative frequencies and percentages for all categories.
c. What percentage of the workers in this sample are in favor of this issue?
d. Draw a pie chart for the percentage distribution.

2.4 Twelve persons were asked to taste two types of soft drinks, *A* and *B*, and indicate if the taste of *A* was superior (S), same (M), or inferior (I) to that of *B*. Their responses are listed below.

S I I M M S M M S I S S

a. Prepare a frequency distribution table.
b. Calculate the relative frequencies and percentages for all categories.
c. Draw a pie chart for the percentage distribution.

2.5 According to the U.S. Bureau of the Census, 48% of preschool children in families with mothers in the work force are *minded* by their relatives, 6% by sitters, 22% by family day care, and 23% by day-care centers. (Figures do not add up to 100 because of rounding.) Draw a bar graph to display these data.

2.6 According to the U.S. Bureau of Labor Statistics, in 1988 consumers spent 31.2% of their incomes on housing, 19.7% on transportation, 14.4% on food, 5.0% on health care, 5.8% on apparel and services, 8.7% on personal insurance and pensions, and 15.2% on other things. Draw a bar graph to display these data.

2.7 A survey of 216,362 full-time freshmen, enrolled at 403 colleges and universities in fall 1989, showed that politically 1.9% of them were of the far left, 21.7% were liberal, 53.6% were middle of the road, 21.3% were conservative, and 1.5% were of the far right. (Alexander W. Astin et al., *The American Freshman: National Norms for Fall 1989,* American Council on Education and UCLA.) Draw a pie chart to display these data.

2.8 The following table gives the major sources of individual income for the tax year 1986. (The percentages do not add up to 100 because of rounding.)

Source	Percentage of Individual Income
Salaries and wages	78.0
Interest	6.4
Business	3.2
Dividends	2.4
Capital gains	5.1
Other	4.7

Source: Statistics of Income Bulletin, IRS, 8(2), Fall 1988.

Display these data by constructing a bar graph.

2.3 ORGANIZING AND GRAPHING QUANTITATIVE DATA

This section explains how to group and display quantitative data.

2.3.1 FREQUENCY DISTRIBUTIONS

Table 2.7 (based on the U.S. Bureau of the Census estimates) gives the number of men (in thousands), aged 15 to 44, who never married. The data are for 1989. The first column of this table lists the *classes,* which represent the (quantitative) variable, the age. Note that the classes always represent a variable. As we can observe, the classes are nonoverlapping; that is, each age belongs to one and only one class. The second column in the table gives the number of (never married) men for each class. For example, in 1989, there were 8760 thousand men, aged 15 to 19, who had never

Table 2.7 Men (in Thousands), Aged 15 to 44, Who Never Married, 1989

Age	Men who Never Married f
15–19	8760
20–24	6915
25–29	4890
30–34	2789
35–39	1461
40–44	673

Variable → Age; Frequency column ← Men who Never Married f; Third class → 25–29; Frequency of the third class ← 4890; Lower limit of the sixth class; Upper limit of the sixth class married. These numbers, listed in the second column of this table, are called the *frequencies* of the respective classes. The frequencies are denoted by f.

For quantitative data, the frequency of a class represents the number of values in the data set that fall in that class. Table 2.7 contains six classes. Each class has a *lower limit* and an *upper limit*. The values 15, 20, 25, 30, 35, and 40 give the lower limits and 19, 24, 29, 34, 39, and 44 give the upper limits of the six classes, respectively. The data presented in Table 2.7 are an illustration of a *frequency distribution table* for quantitative data. Data presented in the form of a frequency table are called **grouped data.** As mentioned earlier in this chapter, a data set that contains information on each observation is called an ungrouped data set.

FREQUENCY DISTRIBUTION FOR QUANTITATIVE DATA

A frequency distribution for quantitative data lists all the classes and the number of values that belong to each class. Data presented in the form of a frequency distribution is called *grouped data.*

To find the midpoint of the upper limit of the first class and the lower limit of the second class in Table 2.7, we divide the sum of these two limits by 2. Thus, this midpoint is

$$\frac{19 + 20}{2} = 19.5$$

The value 19.5 is called the *upper boundary* of the first class and the *lower boundary* of the second class. By use of this technique, we can convert the class limits of Table 2.7 to **class boundaries** (also called *real class limits*). The second column of Table 2.8 lists the boundaries for Table 2.7.

CLASS BOUNDARY

The class boundary is given by the midpoint of the upper limit of one class and the lower limit of the next class.

By taking the difference between two boundaries of a class we obtain the **class width.** The class width is also called the **class size.**

WIDTH OF A CLASS

$$\text{Width of a class} = \text{Upper boundary} - \text{Lower boundary}$$

Thus

$$\text{Width of the first class} = 19.5 - 14.5 = 5$$

Table 2.8 Class Boundaries, Class Widths, and Class Midpoints for Table 2.7

Class Limits	Class Boundaries	Class Width	Class Midpoint
15–19	14.5 to less than 19.5	5	17
20–24	19.5 to less than 24.5	5	22
25–29	24.5 to less than 29.5	5	27
30–34	29.5 to less than 34.5	5	32
35–39	34.5 to less than 39.5	5	37
40–44	39.5 to less than 44.5	5	42

The class widths for the frequency distribution of Table 2.7 are listed in the third column of Table 2.8. Each class in Table 2.8 (and Table 2.7) has the same width of 5.

The **class midpoint** or **mark** is obtained by dividing the sum of the two limits (or the two boundaries) of a class by 2.

CLASS MIDPOINT OR MARK

$$\text{Class midpoint or mark} = \frac{\text{Lower limit} + \text{Upper limit}}{2}$$

Thus, the midpoint of the first class is calculated as follows.

$$\text{Midpoint of the first class} = \frac{15 + 19}{2} = 17$$

The class midpoints for the frequency distribution of Table 2.7 are listed in the fourth column of Table 2.8.

Note that in Table 2.8, when we write classes using class boundaries, we write "to less than" in order to ensure that each value belongs to one and only one class. As we can see, the upper boundary of the preceding class and the lower boundary of the succeeding class are the same.

2.3.2 CONSTRUCTING FREQUENCY DISTRIBUTION TABLES

While constructing a frequency table, we make the following three major decisions.

Number of Classes

Usually the number of classes for a frequency table varies from 5 to 20, depending mainly on the number of observations in the data set. It is preferable to have more classes as the size of a data set increases. The decision about the number of classes is arbitrarily made by the data organizer.

Class Width

Although it is not uncommon to have classes of different sizes, most of the time it is preferable to have the same width for all classes. To determine the class width, when all classes are of the same size, we first find the difference between the largest and the smallest values in the data. Then we obtain the approximate width of a class by dividing this difference by the number of desired classes.

CALCULATION OF CLASS WIDTH

$$\text{Approximate class width} = \frac{\text{Largest value} - \text{Smallest value}}{\text{Number of classes}}$$

Usually this approximate class width is rounded to a convenient number, which is then used as the class width. Rounding this number may slightly change the number of classes initially intended.

Lower Limit of the First Class or the Starting Point

Any convenient number, which is equal to or less than the smallest value in the data set, can be used as the lower limit of the first class.

Example 2-3 illustrates the procedure for constructing a frequency table for quantitative data.

Constructing frequency distribution for quantitative data.

EXAMPLE 2-3 The following data give the number of dentists per 100,000 people in the 50 states for the year 1987.† (*Statistical Abstract of the United States 1990*. Based on American Dental Association data.)

41	60	43	40	61	66	77	47	43	43
65	56	62	49	62	53	55	45	50	68
76	65	71	36	56	67	73	44	53	69
48	75	43	58	57	47	75	61	53	43
55	55	44	68	64	54	70	48	71	58

Construct a frequency distribution table.

Solution In these data, the minimum value is 36 and the maximum value is 77. Suppose we decide to group these data using five classes of equal width. Then

$$\text{Approximate width of a class} = \frac{77 - 36}{5} = 8.2$$

Suppose we round this approximate width to a convenient number, say, 10. The lower limit of the first class can be taken as 36 or any number less than 36. Suppose we take 31 as the lower limit of the first class. Then our classes will be

31–40, 41–50, 51–60, 61–70, and 71–80

We record these five classes in the first column of Table 2.9.

†The data for the 50 states are entered (by row) in the following order: Alabama, Alaska, Arizona, Arkansas, California, Colorado, Connecticut, Delaware, Florida, Georgia, Hawaii, Idaho, Illinois, Indiana, Iowa, Kansas, Kentucky, Louisiana, Maine, Maryland, Massachusetts, Michigan, Minnesota, Mississippi, Missouri, Montana, Nebraska, Nevada, New Hampshire, New Jersey, New Mexico, New York, North Carolina, North Dakota, Ohio, Oklahoma, Oregon, Pennsylvania, Rhode Island, South Carolina, South Dakota, Tennessee, Texas, Utah, Vermont, Virginia, Washington, West Virginia, Wisconsin, Wyoming.

Now we read each value from the given data and mark a tally in the second column of Table 2.9 next to the corresponding class. The first value in our original data is 41, which belongs to the 41–50 class. To record it, we mark a tally in the second column of Table 2.9 next to the 41–50 class. We continue this process until all the data values have been read and entered in the tally column. Note that tallies are marked in blocks of fives for counting convenience. After the tally column is completed, we count the tally marks for each class and write those numbers in the third column. This is the *column of frequencies* and is denoted by f. These frequencies represent the number of states that belong to each of the five different classes. For example, two states have 41–50 dentists per 100,000 people.

Table 2.9 Frequency Distribution of Number of Dentists per 100,000 People for 50 States, 1987

Number of Dentists per 100,000 People	Tally	f
31–40	||	2
41–50	~~||||~~ ~~||||~~ ~~||||~~	15
51–60	~~||||~~ ~~||||~~ |||	13
61–70	~~||||~~ ~~||||~~ |||	13
71–80	~~||||~~ ||	7
		$\Sigma f = 50$

In Table 2.9, we can denote the frequencies of the five classes by f_1, f_2, f_3, f_4, and f_5, respectively. Therefore,

$$f_1 = \text{Frequency of the first class} = 2$$

Similarly

$$f_2 = 15, \quad f_3 = 13, \quad f_4 = 13, \quad \text{and} \quad f_5 = 7$$

Using the Σ notation (see Section 1.8 of Chapter 1), we can denote the sum of the frequencies of all classes by Σf. Hence,

$$\Sigma f = f_1 + f_2 + f_3 + f_4 + f_5 = 2 + 15 + 13 + 13 + 7 = 50$$

The number of observations in a sample is usually denoted by n. Thus, for the sample data, Σf is equal to n. The number of observations in a population is denoted by N. Consequently, Σf is equal to N for population data. Because the data set on dentists in Table 2.9 is for all 50 states, it represents the population. Therefore, in Table 2.9 we can denote the sum of frequencies by N instead of Σf. ■

Note that when we present data in the form of a frequency table, as in Table 2.9, we lose the information on individual observations. We cannot know the exact number of dentists for any particular state from Table 2.9. All we know is that for two states the number of dentists is somewhere in the interval 31 to 40, and so forth.

2.3.3 RELATIVE FREQUENCY AND PERCENTAGE DISTRIBUTIONS

Using Table 2.9, we can compute the *relative frequency* and *percentage* columns the same way we did in Section 2.2.2 for qualitative data. The relative frequencies and percentages for a quantitative data set are obtained as follows.

RELATIVE FREQUENCY AND PERCENTAGE

$$\text{Relative frequency of a class} = \frac{\text{Frequency of that class}}{\text{Sum of all frequencies}} = \frac{f}{\Sigma f}$$

$$\text{Percentage} = (\text{Relative frequency}) \cdot 100$$

Example 2-4 illustrates how to construct relative frequency and percentage distributions.

Constructing relative frequency and percentage distributions.

EXAMPLE 2-4 Calculate the relative frequencies and percentages for Table 2.9.

Solution We calculate the relative frequencies and percentages for Table 2.9 and list them in the third and fourth columns of Table 2.10, respectively. We have listed the class boundaries in the second column of Table 2.10.

Table 2.10 Relative Frequency and Percentage Distributions for Table 2.9

Dentists	Class Boundaries	Relative Frequency	Percentage
31–40	30.5 to less than 40.5	2/50 = .04	4
41–50	40.5 to less than 50.5	15/50 = .30	30
51–60	50.5 to less than 60.5	13/50 = .26	26
61–70	60.5 to less than 70.5	13/50 = .26	26
71–80	70.5 to less than 80.5	7/50 = .14	14
		Sum = 1.00	Sum = 100

From Table 2.10, we can make statements about the percentage of states with a certain number of dentists per 100,000 people. For example, 4% of the states had 31 to 40 dentists per 100,000 people. By adding the percentages for the first two classes, we can state that 34% of the states had 50 or fewer dentists per 100,000 people. By adding the percentages of the last two classes, we can state that 40% of the states had 61 or more dentists per 100,000 people. ■

2.3.4 GRAPHING GROUPED DATA

Grouped (quantitative) data can be displayed by using a *histogram* or a *polygon*. This section describes how to construct such graphs. We can also draw a pie chart to display the percentage distribution for a quantitative data set. The procedure to construct a pie chart is similar to the one explained in Section 2.2.3 for qualitative data; it will not be repeated in this section.

Histograms

A histogram is a certain kind of graph that can be drawn for a frequency distribution, a relative frequency distribution, or a percentage distribution. To draw a histogram, we first mark classes on the horizontal axis and frequencies (or relative frequencies or percentages) on the vertical axis. Notice that either class limits or class boundaries can be used to mark classes on the horizontal axis. Next, we draw a bar for each class so that its height represents the frequency of that class. The bars in a histogram are drawn adjacent to each other without leaving any gap between them. If a histogram is drawn by marking the frequencies on the vertical axis, it is called a **frequency histogram.** However, if the relative frequencies are marked on the vertical axis it is called a **relative frequency histogram.** The one drawn by using the percentages is called a **percentage histogram.**

> **HISTOGRAM**
>
> A histogram is a graph in which classes are marked on the horizontal axis and either the frequencies, relative frequencies, or percentages are marked on the vertical axis. The frequencies, relative frequencies, or percentages are represented by the heights of the bars. In a histogram, the bars are drawn adjacent to each other.

Figures 2.3 and 2.4 show the frequency and relative frequency histograms, respectively, for the data of Tables 2.9 and 2.10 given in the two previous sections. The two histograms look alike because they represent the same data. We can draw a percentage histogram for the percentage distribution of Tables 2.9 and 2.10 by marking the percentages on the vertical axis.

The symbol "-//-" used in the horizontal axis of both Figures 2.3 and 2.4 represents a break in the horizontal axis, also called the **truncation.** It indicates that the entire horizontal axis is not shown in these figures. As can be noticed, the zero to 30.5 portion of the horizontal axis has been omitted in each of these figures.

In Figures 2.3 and 2.4, we have used class boundaries to mark classes on the horizontal axis. However, we can show the intervals on the horizontal axis by marking the class limits instead of the class boundaries, as in Case Study 2-5.

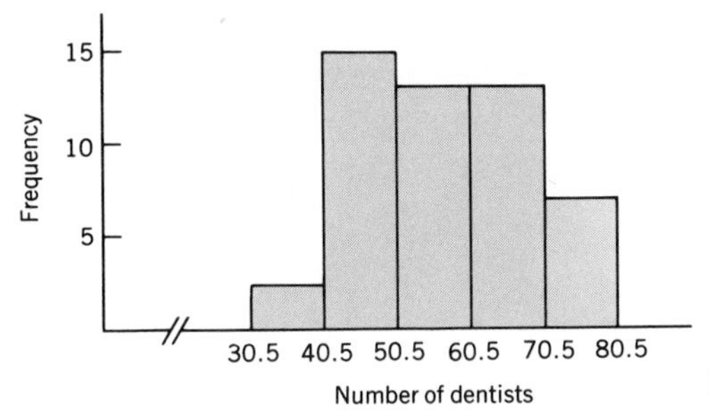

Figure 2.3 Frequency histogram for Table 2.9.

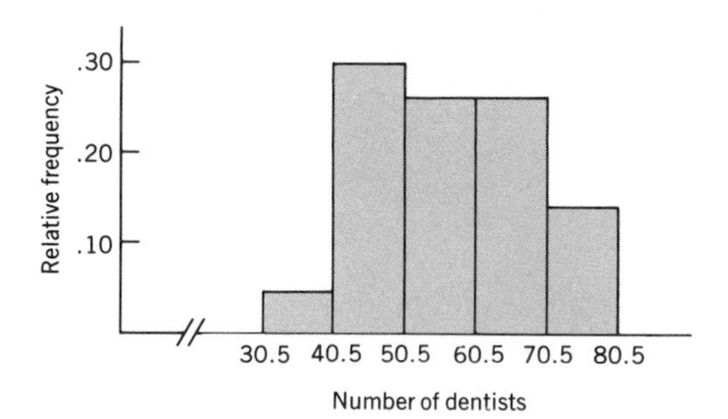

Figure 2.4 Relative frequency histogram for Table 2.10.

CASE STUDY 2-5 APPENDECTOMY CASES BY AGE AND SEX, 1986

A total of 275,192 appendectomies (surgical removal of the appendix) were performed in the United States in 1986. The following graph gives the histograms for the percentage distributions of the appendectomies performed by age and sex. For example, 7.3 percent of the males and 6.4 percent of the females who had such surgeries in 1986 were zero to nine years of age.

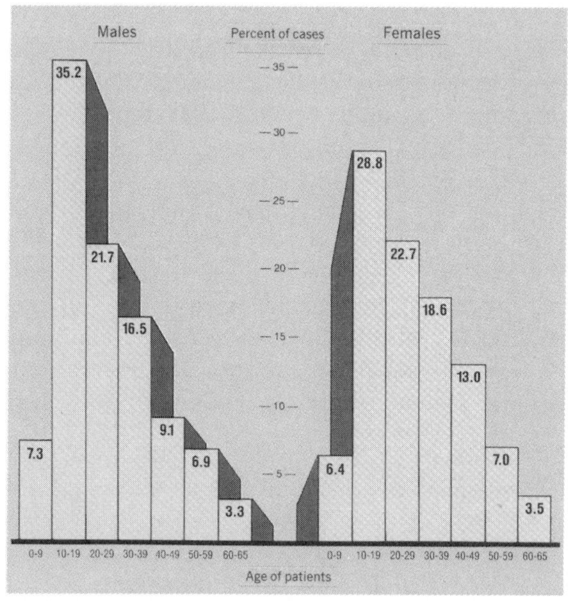

*Surgeries performed on employees and dependents of group health customers of Metropolitan Life Insurance Company
Source of data: Corporate Health Strategies, Inc.

Source: *Statistical Bulletin,* 69(2), April–June 1988, p. 25. Copyright © 1988 by Metropolitan Life Insurance Company. Reprinted with permission.

Polygons

A polygon is another device that can be used to present quantitative data in graphic form. To draw a frequency polygon, we first mark a dot above the midpoint of each class at a height equal to the frequency of that class. This is the same as marking the midpoint at the top of each bar in a histogram. Next, we mark two more classes, one at each end, with zero frequencies and mark their midpoints. In the last step, we join the adjacent dots with straight lines. The resulting line graph is called a **frequency polygon** or simply a *polygon*.

A polygon with relative frequencies marked on the vertical axis is called a *relative frequency polygon*. Similarly, a polygon with percentages marked on the vertical axis is called a *percentage polygon*.

> **POLYGON**
>
> A graph formed by joining the midpoints of the tops of successive bars in a histogram by straight lines is called a polygon.

Figure 2.5 shows the frequency polygon for the frequency distribution of Table 2.9.

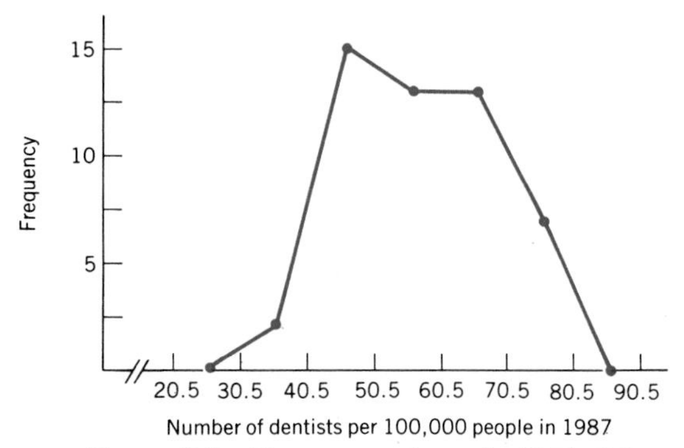

Figure 2.5 Frequency polygon for Table 2.9.

For a very large data set, as the number of classes is increased (and the width of classes is decreased), the frequency polygon eventually becomes a smooth curve. Such a curve is called a *frequency distribution curve* or simply a *frequency curve*. Figure 2.6 shows the frequency curve for a large data set with a large number of classes.

Figure 2.6 Frequency distribution curve.

2.3.5 MORE ON CLASSES AND FREQUENCY DISTRIBUTIONS

This section presents two alternative methods for writing classes to construct a frequency distribution for quantitative data.

Less Than Method for Writing Classes

The classes in frequency Table 2.9 for the data on dentists in 50 states were written as 31 to 40, 41 to 50, etc. Alternatively, we can write the classes in a frequency table using the *less than* method. The technique of Table 2.9 for writing classes is more commonly used for data sets that do not contain fractional values. The *less than* method is more appropriate when a data set contains fractional values. Example 2-5 illustrates the *less than* method.

Constructing frequency distribution using less than *method.*

EXAMPLE 2-5 The following data give the normal annual precipitation (in inches) for 29 cities of the United States.† (U.S. National Oceanic and Atmospheric Administration. The data, taken at the airports, are based on the 30-year period from 1951 through 1980.)

8.12	48.61	41.84	15.36	11.71	43.81
33.34	35.40	29.46	15.31	30.97	7.82
23.47	44.76	12.08	43.56	57.55	26.36
48.49	59.74	44.12	30.34	7.11	37.39
7.49	33.91	15.31	19.71	28.61	

Construct a frequency distribution table. Calculate the relative frequencies and percentages for all classes.

Solution The minimum value in this data set is 7.11 and the maximum value is 59.74. Because the data set contains only 29 values, five classes (of equal width) may be enough to group these data. Then,

$$\text{Approximate width of a class} = \frac{59.74 - 7.11}{5} = 10.53$$

We round this number to a more convenient number, say, 12. Then, we take 12 as the width of each class. If we start the first class at zero, the classes will be written as *0 to less than 12*, *12 to less than 24*, and so on. We record the five classes, which cover all the data, in the first column of Table 2.11. The second column shows the tallies. The third column of this table lists the frequencies of classes. A value in the data set that is 0 or larger but less than 12 belongs to the first class, and a value that is 12 or larger but less than 24 falls in the second class, and so on. We have recorded the relative frequencies and percentages for classes in the fourth and fifth columns of Table 2.11, respectively.

†The data, entered by row, are for the following cities: Albuquerque, Atlanta, Baltimore, Bismarck, Boise, Boston, Chicago, Cleveland, Dallas, Denver, Detroit, El Paso, Honolulu, Houston, Los Angeles, Louisville, Miami, Minneapolis, Nashville, New Orleans, New York, Omaha, Phoenix, Portland (Oregon), Reno, St. Louis, Salt Lake City, San Francisco, Wichita.

Table 2.11 Normal Annual Precipitation (inches) in 29 U.S. Cities

Normal Annual Precipitation	Tallies	f	Relative Frequency	Percentage
0 to less than 12	~~||||~~	5	.172	17.2
12 to less than 24	~~||||~~ |	6	.207	20.7
24 to less than 36	~~||||~~ |||	8	.276	27.6
36 to less than 48	~~||||~~ |	6	.207	20.7
48 to less than 60	||||	4	.138	13.8
		$\Sigma f = 29$	Sum = 1.0	Sum = 100

■

A histogram and a polygon for the data of Table 2.11 can be drawn in the same way as for the data of Tables 2.9 and 2.10.

Single-Valued Classes

If the observations in a data set assume only a few distinct values, it may be appropriate to prepare a frequency distribution table using *single-valued classes*, that is, classes that are made of single values and not of intervals. This technique is especially useful in cases of discrete data with only a few possible values. Example 2-6 exhibits such a situation.

Constructing frequency distribution using single-valued classes.

EXAMPLE 2-6 The U.S. Bureau of Labor Statistics conducts a survey called the Diary Survey every quarter of the year.† The following data give the number of members in each of 40 randomly selected households from the 1987 Diary Survey public-use tape, which in total contains information on 13,098 households.

2	2	4	4	3	3	4	2	5	4
4	1	2	1	1	2	1	5	3	5
1	1	3	2	2	2	4	3	1	1
3	1	6	4	2	2	1	1	3	3

Construct a frequency distribution table for these data using single-valued classes.

Solution The observations in this data set assume only six distinct values: 1, 2, 3, 4, 5, and 6. We use each of these six values as a class in the frequency distribution Table 2.12. We have listed the six classes in the first column of Table 2.12. To find the frequencies of these classes, we count the observations in the data that belong to each class. These results are recorded in the second column of Table 2.12. Thus, in these data 11 households have one member each, 10 have two members each, and so on.

†In this and subsequent chapters, the data that refer to Interview and Diary Surveys are taken from the public-use tapes containing data from the following sources: *Interview Survey:* Bureau of Labor Statistics 1987 Interview Survey based on 23,536 consumer units containing information on families, households, and individuals; *Diary Survey:* Bureau of Labor Statistics 1987 Diary Survey based on 13,098 consumer units containing information on families, households, and individuals. The latest year for which these tapes are available at this time is 1987.

Table 2.12 Frequency Distribution for Household Size

Household Size	Number of Households (f)
1	11
2	10
3	8
4	7
5	3
6	1
	$\Sigma f = 40$

■

The data of Table 2.12 can be displayed by drawing a bar graph as shown in Figure 2.7.

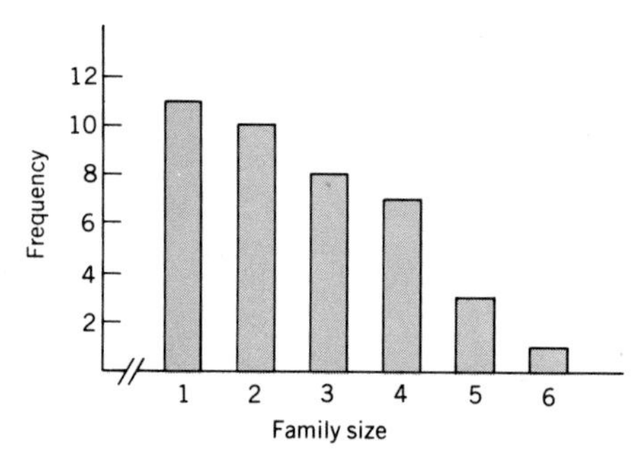

Figure 2.7 Bar graph for household size.

To construct a bar graph, we mark the classes as intervals on the horizontal axis with a little gap between consecutive intervals. The bars represent the frequencies of respective classes. We can convert the frequencies of Table 2.12 to relative frequencies and percentages in the same way as we did in Table 2.10. Then we can construct a bar graph to display the relative frequency or percentage distribution by marking the relative frequencies or percentages, respectively, on the vertical axis.

2.4 SHAPES OF HISTOGRAMS

A histogram can assume any one of a large number of shapes. The most common of these shapes are

1. Symmetric
2. Skewed
3. Uniform or rectangular

A **symmetric histogram** is identical on both sides of its central point. The histograms shown in Figure 2.8 are symmetric around the dashed lines that represent their central points.

A **skewed histogram** is nonsymmetric. For a skewed histogram, the tail on one side is longer than the tail on the other side. A **skewed to the right histogram** has a longer tail on the right side (see Figure 2.9*a*). A **skewed to the left histogram** has a longer tail on the left side (see Figure 2.9*b*).

A **uniform** or **rectangular histogram** has the same frequency for each class. Figure 2.10 is an illustration of such a case.

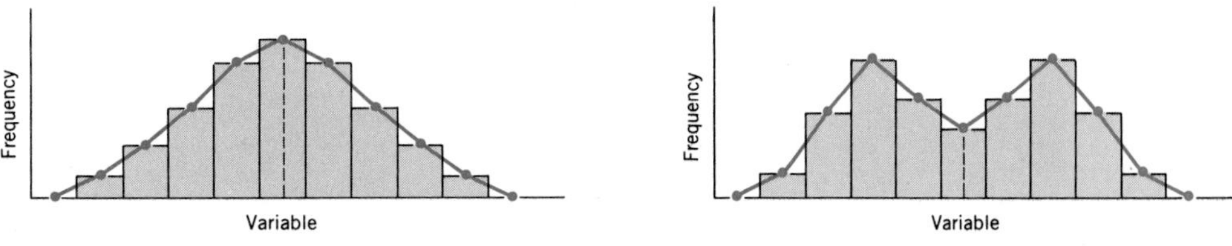

Figure 2.8 Symmetric histograms.

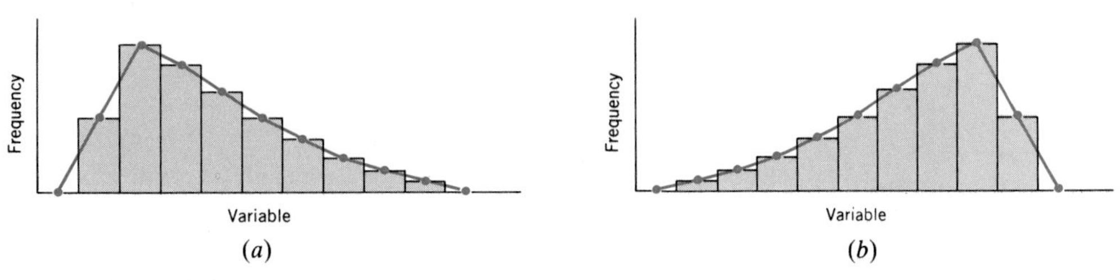

Figure 2.9 (*a*) A histogram skewed to the right. (*b*) a histogram skewed to the left.

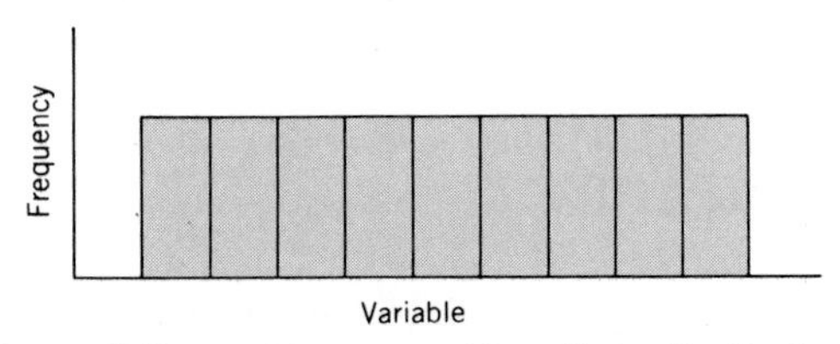

Figure 2.10 A histogram with uniform distribution.

Figures 2.11*a* and *b* display symmetric frequency curves. Figures 2.11*c* and *d* show frequency curves skewed to the right and skewed to the left, respectively.

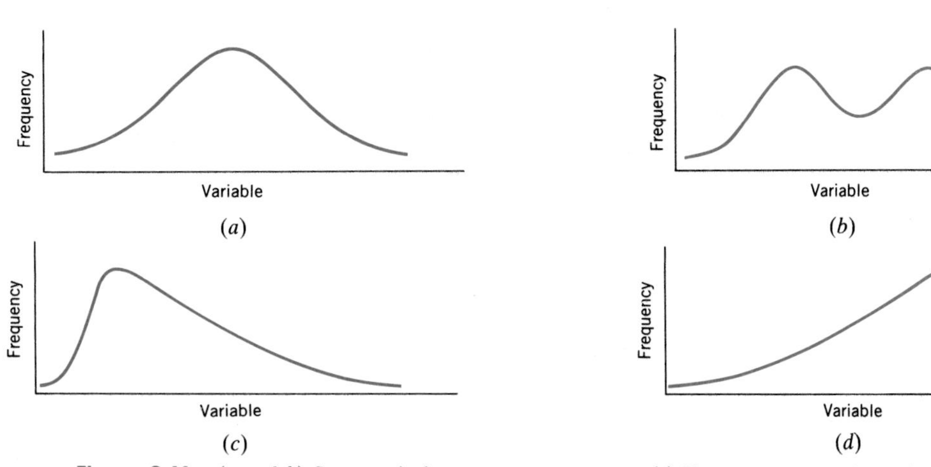

Figure 2.11 (*a* and *b*) Symmetric frequency curves. (*c*) Frequency curve skewed to the right. (*d*) Frequency curve skewed to the left.

WARNING

Describing data using graphs helps to give us insights into the main characteristics of the data. But graphs, unfortunately, can also be used, intentionally or unintentionally, to distort the facts and to deceive the readers. Following are two ways to manipulate graphs to convey a particular opinion or impression.

1. *Changing the scale* either on one or both axes. That is, shortening or stretching one or both of the axes.
2. *Truncating the frequency axis*. That is, starting the frequency axis at a number greater than zero.

When interpreting a graph, we should be very cautious. We should observe carefully whether the frequency axis has been truncated or whether any axis has been unnecessarily shortened or stretched.

In Case Study 2-6, the bar graph gives the number of fire units for the five busiest fire engine companies in the United States. Note that the horizontal (frequency) axis is truncated because it starts at 3500. If we ignore the truncation, we may get the impression that the number of fire units in Sacramento Co. #6 is about one third of the number of fire units in New York City Co. #48. However, as indicated by the frequencies, the actual difference between these two companies is much smaller.

CASE STUDY 2-6 NUMBER OF FIRE UNITS FOR FIVE BUSIEST FIRE ENGINE COMPANIES

Source: *USA Today*, July 1, 1987. Copyright 1987, *USA Today*. Reprinted with permission.

Case Study 2-7 of Section 2.7 shows how time-series data can be manipulated to distort facts.

EXERCISES

2.9 The following table gives the number of live births (in thousands) to mothers aged 15 to 49 in the United States for the year 1985.

Age of Mother	Live Births (thousands)
15 to 19	467
20 to 24	1141
25 to 29	1201
30 to 34	696
35 to 39	214
40 to 44	28
45 to 49	1

Source: Statistical Bulletin, Metropolitan Life Insurance Company, January–March 1988.

a. Find the class boundaries and class midpoints.
b. Do all classes have the same width? If yes, what is that width?
c. Prepare the relative frequency and percentage distribution columns.

2.10 The following table gives the frequency distribution of persons, aged 25 to 64, who possess a doctorate degree.

Age	Persons with a Doctorate (thousands)
25 to 34	147
35 to 44	255
45 to 54	123
55 to 64	125

Source: U.S. Bureau of the Census.

a. Find the class boundaries and class midpoints.
b. Do all classes have the same width? If yes, what is that width?
c. Prepare the relative frequency and percentage distribution columns.

2.11 A data set on the political contributions made during 1990 by 200 households has the lowest value of $1 and the highest value of $683. Suppose we want to group these data into seven classes of equal widths.

a. Assuming we take the lower limit of the first class as $1 and the width of each class equal to $100, find the class limits for all seven classes.
b. Calculate the class boundaries and class midpoints.

2.12 A data set on weekly expenditures on bakery products for a sample of 500 households has a minimum value of $1 and a maximum value of $18. Suppose we want to group these data into five classes of equal widths.

a. Assuming we take the lower limit of the first class as $1 and the upper limit of the fifth class as $20, find the class limits for all five classes.
b. Determine the class boundaries and class widths.
c. Find the class midpoints.

2.13 The following data give the number of hospital beds per 100,000 people for 1987 for 50 states. (*Source:* U.S. National Center for Health Statistics.)

593	381	394	544	389	428	476	525	507	564
351	400	538	515	631	675	521	598	514	469
639	468	610	663	603	606	696	408	454	513
448	617	463	878	522	496	378	592	517	441
803	622	482	351	457	510	363	595	531	633

a. Construct a frequency distribution table with the classes 300–399, 400–499, . . . , 800–899.

b. Prepare the relative frequency and percentage columns for the table of part a.

c. Based on the frequency distribution, can you say whether the data are symmetric or skewed?

d. What percentage of the states had 700 or more beds per 100,000 people?

2.14 Following are the heights (in inches) of a sample of 30 NBA players taken from Data Set III of Appendix B.

82	78	83	85	80	81	86	78	85	82
80	81	81	76	79	81	76	81	82	82
82	84	79	84	80	76	81	76	82	82

a. Construct a frequency distribution table. Take the classes as 76–77, 78–79, 80–81, 82–83, 84–85, and 86–87.

b. Calculate the relative frequencies and percentages for all six classes.

c. Draw a histogram and a polygon for the relative frequency distribution.

d. What percentage of the players in this sample are 81 inches tall or below?

2.15 Following are the weights (in pounds) of a sample of 30 NBA players taken from Data Set III of Appendix B.

225	232	232	245	235	245	270	225	240	240
217	195	225	185	200	220	200	210	271	240
220	230	215	252	225	220	206	185	227	236

a. Construct a frequency distribution table. Take the classes as 180–199, 200–219, 220–239, 240–259, 260–279.

b. Calculate the relative frequencies and percentages for all classes.

c. Construct a histogram and a polygon for the percentage distribution.

Exercises 2.16 through 2.20 are based on the following data.

The table given on the next page, based on the American Chamber of Commerce Researchers Association Survey, gives the prices of five items in 25 urban areas across the United States. (See Data Set I of Appendix B.)

EXPLANATION OF VARIABLES

Apartment rent Monthly rent of an unfurnished two-bedroom apartment (excluding all utilities except water), 1-½ or 2 baths, approximately 950 square feet

Phone bill Monthly telephone charges for a private residential line (customer owns instruments)

Beauty salon Price for woman's shampoo, trim, and blow dry

Dry cleaning Price of dry cleaning, men's two-piece suit

Price of wine Price of Paul Masson Chablis, 1.5-liter bottle

City	Apartment Rent	Phone Bill	Beauty Salon	Dry Cleaning	Price of Wine
Huntsville (AL)	422	25.81	15.29	5.20	6.62
Phoenix (AZ)	512	15.65	18.81	7.02	4.41
Little Rock (AR)	400	22.79	25.20	5.44	5.10
San Diego (CA)	899	14.26	23.48	5.98	3.83
Denver (CO)	471	17.24	20.18	5.76	4.69
Orlando (FL)	511	17.48	18.10	5.80	5.05
Bloomington (IN)	486	23.24	15.00	5.29	5.00
Ames (IA)	450	21.44	13.19	5.88	5.20
New Orleans (LA)	339	22.37	22.60	5.03	4.10
Baltimore (MD)	524	23.19	16.20	6.26	5.65
Minneapolis (MN)	542	19.24	11.80	6.18	4.19
Lincoln (NE)	429	16.46	16.80	6.04	4.04
Las Vegas (NV)	513	10.98	23.20	6.80	3.81
Albuquerque (NM)	480	20.49	17.47	5.47	4.39
Albany (NY)	523	27.30	16.60	5.29	5.91
Charlotte (NC)	426	17.98	20.60	5.52	4.03
Cincinnati (OH)	522	20.91	14.00	5.56	4.99
Salem (OR)	411	17.82	17.40	6.34	5.03
Rapid City (SD)	522	19.90	15.50	5.29	5.68
Nashville (TN)	519	19.46	15.00	5.40	4.96
Houston (TX)	443	17.84	22.12	5.05	4.97
Salt Lake City (UT)	377	17.12	15.00	6.74	4.79
Seattle (WA)	608	18.50	14.38	7.12	5.51
Charleston (WV)	396	29.43	16.42	5.50	5.92
Green Bay (WI)	427	21.68	14.45	5.74	4.63

2.16 **a.** Construct a frequency distribution table for apartment rents with the classes 300–399, 400–499, . . . , 800–899.

b. Prepare the relative frequency and percentage distribution columns.

c. Find the class midpoints.

d. From the frequency distribution, tell how many cities have a rent of \$700 or more in this sample.

2.17 **a.** Construct a frequency distribution table for the price of dry cleaning with the classes 5.00 to less than 5.50, 5.50 to less than 6.00, . . . , 7.00 to less than 7.50.

b. Calculate the relative frequencies and percentages for all classes.

c. What percentage of the cities have a dry cleaning price of less than \$6.00?

2.18 **a.** Construct a frequency distribution table for the price of wine with the classes 3.70 to less than 4.50, 4.50 to less than 5.30, 5.30 to less than 6.10, 6.10 to less than 6.90.

b. Prepare the relative frequency and percentage distribution columns.

c. What percentage of the cities in this sample have the price of wine in the interval \$4.50 to less than \$6.10?

2.19 **a.** Construct a frequency distribution table for phone bills with the classes 10 to less than 14, 14 to less than 18, . . . , 26 to less than 30.

b. Prepare the relative frequency and percentage distribution columns.

c. Draw a histogram and a polygon for the relative frequency distribution.

2.20 **a.** Prepare a frequency distribution table for the price of beauty salon with the classes 11 to less than 14, 14 to less than 17, . . . , 23 to less than 26.

b. Calculate the relative frequencies and percentages for all classes.

c. Draw a histogram and a polygon for the percentage distribution.

2.21 The following data give the quarterly housing expenditures (in dollars) for 30 households randomly selected from the 1987 Diary Survey.

872	705	2679	475	1099	2465
126	1528	1903	456	825	1042
529	719	401	1322	563	1675
2532	2393	1421	685	209	647
1151	975	1675	876	672	1078

a. Construct a frequency distribution table with the classes 1–600, 601–1200, . . . , and 2401–3000.

b. Prepare the relative frequency and percentage columns.

c. Draw a histogram and a polygon for the relative frequency distribution.

2.22 The following data give the annual value (in dollars) of food stamps received by 40 households randomly selected from the 1987 Diary Survey. (These households were selected only from those who received food stamps.)

1308	2160	348	624	1488	2628	240	2784
1020	1344	252	576	1560	1272	1224	3048
1260	2868	2160	768	120	1920	3000	804
1068	2172	1464	2956	1440	1116	600	2532
564	2112	3648	1104	1944	480	1332	720

a. Construct a frequency distribution table with the classes 1–800, 801–1600, . . . , and 3201–4000.

b. Prepare the relative frequency and percentage columns for the table of part a.

c. Draw a histogram for the percentage distribution.

2.23 The following data give the annual earnings (in thousands of dollars) before taxes for 36 families randomly selected from the 1987 Interview Survey. (These households were selected only from those who had positive earnings before taxes.)

11.5	10.9	23.3	5.0	40.0	20.0	47.4	16.2	48.6
55.0	18.8	8.0	35.0	22.7	8.9	19.9	8.0	58.0
27.7	9.0	12.4	34.8	26.0	16.6	3.9	14.0	25.0
21.6	29.0	43.0	34.3	26.1	14.7	43.0	28.0	16.8

a. Construct a frequency distribution table with the classes 0 to less than 12, 12 to less than 24, . . . , and 48 to less than 60.

b. Calculate the relative frequencies and percentages for all classes.

c. Construct a histogram and a polygon for the relative frequency distribution.

2.24 The following data give the number of new cars sold at a dealership during a 20-day period.

8	5	12	3	9	10	6	3	8	8
4	6	10	11	7	7	3	5	9	11

a. Prepare a frequency distribution table with the classes 3–4, 5–6, 7–8, 9–10, 11–12.

b. Calculate the relative frequencies and percentages for all classes.

c. Draw a histogram and a polygon for the percentage distribution.

2.25 The following data give the infant mortality rate for 1987 for all 50 states based on the U.S. National Center for Health Statistics estimates. (The data give the deaths of infants under 1 year of age per 1000 live births.)

12.2	10.4	9.5	10.3	9.0	9.8	8.8	11.7	10.6	12.7
8.9	10.4	11.6	10.1	9.1	9.5	9.7	11.8	8.3	11.5
7.2	10.7	8.7	13.7	10.2	10.0	8.6	9.6	7.8	9.4
8.1	10.7	11.9	8.7	9.3	9.6	10.4	10.4	8.4	12.7
9.9	11.7	9.1	8.8	8.5	10.2	9.7	9.8	8.6	9.2

a. Construct a frequency distribution table with the classes 7 to less than 8.4, 8.4 to less than 9.8, . . . , 12.6 to less than 14.0.

b. Prepare the relative frequency and percentage distribution columns for the table.

c. From the frequency distribution, can you tell whether the data are symmetric or skewed? If skewed, are they skewed to the left or right?

2.26 The following data give the number of children less than 18 years of age for 40 families randomly selected from the 1987 Diary Survey.

0	0	2	1	0	1	2	0	0	2
3	2	1	0	0	3	2	0	1	0
1	0	2	2	0	1	0	0	2	3
0	1	1	0	3	0	0	2	1	2

a. Construct a frequency distribution table for these data using single-valued classes.

b. Calculate the relative frequencies and percentages for all classes.

c. How many families in this sample have two or three children under 18 years of age?

d. Draw a bar graph for the frequency distribution.

2.27 The following data give the number of automobiles owned by 30 families randomly selected from the 1987 Interview Survey.

2	0	1	2	2	1	0	2	2	1
1	2	1	3	2	3	2	2	1	1
1	3	1	1	0	2	1	3	1	1

a. Construct a frequency distribution table for these data using single-valued classes.

b. Calculate the relative frequencies and percentages for all classes.

c. What percentage of the families in this sample own one or two automobiles?

d. Draw a bar graph for the relative frequency distribution.

2.28 The following data give the number of bedrooms in homes of 50 families randomly selected from the 1987 Interview Survey.

2	3	2	1	3	3	2	2	6	1
5	3	5	4	2	2	3	2	3	2
4	3	3	2	2	2	4	1	3	3
2	3	4	2	3	3	4	2	3	4
1	5	3	1	3	4	2	2	3	2

a. Prepare a frequency distribution table for these data using single-valued classes.

b. Calculate the relative frequencies and percentages for all classes.

c. What percentage of the families in this sample own homes with four or more bedrooms?

2.5 CUMULATIVE FREQUENCY DISTRIBUTIONS

Consider again Example 2-3 of Section 2.3.2 about the number of dentists per 100,000 people in the 50 states. Suppose we want to know how many states have fewer than a certain number of dentists per 100,000 people. Such a question can be answered using a **cumulative frequency distribution.** Each class in a cumulative frequency distribution table gives the total number of values that fall below a certain value. A cumulative frequency distribution is constructed for quantitative data only.

CUMULATIVE FREQUENCY DISTRIBUTION

A cumulative frequency distribution gives the total number of values that fall below the upper boundary of each class.

In a *less than* cumulative frequency table, each class has the same lower limit but a different upper limit. Example 2-7 illustrates the procedure for preparing a cumulative frequency distribution.

Constructing cumulative frequency distribution table.

EXAMPLE 2-7 Using the frequency distribution of Table 2.9, reproduced here, prepare a cumulative frequency distribution for the data on the number of dentists per 100,000 people in 50 states.

Number of Dentists per 100,000 People	f
31–40	2
41–50	15
51–60	13
61–70	13
71–80	7

Solution Table 2.13 gives the cumulative frequency distribution for the number of dentists per 100,000 people in 50 states. As we can observe, 31 (which is the lower limit of the first class in Table 2.9) is taken as the lower limit of each class in Table 2.13. The upper limits of all classes in Table 2.13 are the same as those in Table 2.9.

Table 2.13 Cumulative Frequency Distribution of Number of Dentists per 100,000 People in 50 States, 1987

Class Limits	Class Boundaries	Cumulative Frequency
31–40	30.5 to less than 40.5	2
31–50	30.5 to less than 50.5	2 + 15 = 17
31–60	30.5 to less than 60.5	2 + 15 + 13 = 30
31–70	30.5 to less than 70.5	2 + 15 + 13 + 13 = 43
31–80	30.5 to less than 80.5	2 + 15 + 13 + 13 + 7 = 50

To obtain the cumulative frequency of a class, we have added the frequency of that class in Table 2.9 to the frequencies of all the preceding classes. The cumulative frequencies are recorded in the third column of Table 2.13. The second column of this table lists the boundaries. ■

From Table 2.13, we can determine the number of observations that fall below the upper boundary of a class. For example, from Table 2.13, 30 states had 60 or fewer dentists per 100,000 people in 1987.

The **cumulative relative frequencies** are obtained by dividing the cumulative frequencies by the total number of observations in the data. The **cumulative percentages** are obtained by multiplying the cumulative relative frequencies by 100.

CUMULATIVE RELATIVE FREQUENCY AND CUMULATIVE PERCENTAGE

$$\text{Cumulative relative frequency} = \frac{\text{Cumulative frequency}}{\text{Total observations in the data}}$$

$$\text{Cumulative percentage} = (\text{Cumulative relative frequency}) \cdot 100$$

Table 2.14 contains both the cumulative relative frequencies and the cumulative percentages for Table 2.13. We can observe from this table, for example, that 34% of the states had 50 or fewer dentists per 100,000 people.

Table 2.14 Cumulative Relative Frequency and Cumulative Percentage Distributions of Dentists for 50 States, 1987

Class Limits	Cumulative Relative Frequency	Cumulative Percentage
31–40	2/50 = .04	4
31–50	17/50 = .34	34
31–60	30/50 = .60	60
31–70	43/50 = .86	86
31–80	50/50 = 1.00	100

2.5.1 OGIVES

When plotted on a diagram, the cumulative frequencies give a curve that is called an **ogive** (pronounced o-jive). Figure 2.12 gives an ogive for the cumulative frequency distribution of Table 2.13. We mark the variable, the number of dentists per 100,000 people, on the horizontal axis and the cumulative frequencies on the vertical axis. Then we mark the points above the upper boundaries of the various classes at the heights equal to the corresponding cumulative frequencies. We obtain the ogive by joining consecutive points with straight lines. Note that the ogive starts at the lower boundary of the first class and ends at the upper boundary of the last class.

> **OGIVE**
>
> An ogive is a curve drawn for the cumulative frequency distribution by joining the points marked above the upper boundaries of classes at heights equal to the cumulative frequencies of respective classes.

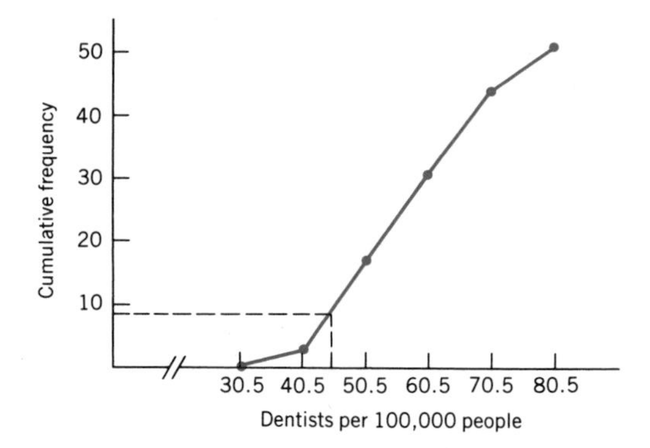

Figure 2.12 Ogive for Table 2.13.

One advantage of an ogive is that we can use it to approximate the cumulative frequency for any interval. For example, to find the number of states with 45 or fewer dentists per 100,000 people, first we draw a vertical line from 45 on the horizontal axis up to the ogive. Then we draw a horizontal line from the point where this line intersects the ogive to the vertical axis. This point gives the cumulative frequency of the class 31 to 45. In Figure 2.12, this cumulative frequency is approximately 9. Therefore, approximately 9 states had 45 or fewer dentists per 100,000 people in 1987.

We can draw an ogive for cumulative relative frequency or cumulative percentage distribution in the same way as we did for the cumulative frequency distribution.

EXERCISES

2.29 The following table gives the age distribution of householders aged 15 to 44 for 1987.

Age	Number of Householders (thousands)
15–19	398
20–24	4,799
25–29	9,652
30–34	10,850
35–39	10,155
40–44	8,549

Source: U.S. Bureau of the Census

a. Construct a cumulative frequency distribution table.

b. Calculate the cumulative relative frequencies and cumulative percentages for all classes.

c. What percentage of the householders were 15 to 34 years of age in 1987?

d. Draw an ogive for the cumulative percentage distribution.

e. Using the ogive, find the percentage of householders who were 15 to 27 years of age in 1987.

2.30 The following table gives the number of persons (in thousands) aged 20 to 69 who had four years of college education in 1985.

Age	Number of Persons (thousands)
20–29	4939
30–39	5648
40–49	2770
50–59	2010
60–69	1366

Source: U.S. Department of Education

a. Construct a cumulative frequency distribution table.

b. Calculate the cumulative relative frequencies and cumulative percentages for all classes.

c. What percentage of the persons with four years of college education were 40 to 69 years of age in 1985?

d. Draw an ogive for the cumulative percentage distribution.

e. Using the ogive, find the percentage of persons with four years of college education who were 20 to 45 years of age in 1985.

2.31 Using the frequency distribution table constructed in Exercise 2.13, prepare the cumulative frequency, cumulative relative frequency, and cumulative percentage distributions.

2.32 Using the frequency distribution table constructed in Exercise 2.14, prepare the cumulative frequency, cumulative relative frequency, and cumulative percentage distributions.

2.33 Using the frequency distribution table constructed in Exercise 2.15, prepare the cumulative frequency, cumulative relative frequency, and cumulative percentage distributions.

2.34 Construct the cumulative frequency, cumulative relative frequency, and cumulative percentage distributions using the frequency distribution table of Exercise 2.19.

2.35 Refer to Exercise 2.20. Construct the cumulative frequency, cumulative relative frequency, and cumulative percentage distributions by using the frequency distribution table of that exercise.

2.36 Refer to Exercise 2.21. Prepare the cumulative frequency, cumulative relative frequency, and cumulative percentage distributions by using the frequency distribution table. Draw an ogive for the cumulative percentage distribution. What percentage of the households in that sample spent $1200 or less on housing?

2.37 Refer to Exercise 2.22. Prepare the cumulative frequency, cumulative relative frequency, and cumulative percentage distributions by using the frequency distribution table. Draw an ogive for the cumulative frequency distribution. What percentage of the households received food stamps worth $2400 or less?

2.38 Prepare the cumulative frequency, cumulative relative frequency, and cumulative percentage distributions by using the frequency distribution constructed in Exercise 2.23. Draw an ogive for the cumulative frequency distribution. Using the ogive, find the approximate percentage of families who earned less than $30 thousand.

2.39 Using the frequency distribution table constructed for the data of Exercise 2.24, prepare the cumulative frequency, cumulative relative frequency, and cumulative percentage distributions.

2.40 Using the frequency distribution table constructed for the data of Exercise 2.25, prepare the cumulative frequency, cumulative relative frequency, and cumulative percentage distributions. Draw an ogive for the cumulative percentage distribution. Using the ogive, find the approximate percentage of states with a 1987 infant mortality rate of 10.5 or less.

2.6 STEM-AND-LEAF DISPLAYS

Another technique used to present quantitative data in condensed form is the **stem-and-leaf display.** An advantage of a stem-and-leaf display over a frequency distribution is that by preparing a stem-and-leaf display we do not lose information on individual observations. A stem-and-leaf display is constructed only for quantitative data.

> **STEM-AND-LEAF DISPLAY**
>
> In a stem-and-leaf display of quantitative data, each value is divided into two portions—a stem and a leaf. Then the leaves for each stem are shown separately in a display.

Example 2-8 describes the procedure for constructing a stem-and-leaf display.

Constructing stem-and-leaf display for two-digit numbers.

EXAMPLE 2-8 The following are the scores of 30 college students on a statistics test.

75	52	80	96	65	79	71	87	93	95
69	72	81	61	76	86	79	68	50	92
83	84	77	64	71	87	72	92	57	98

Construct a stem-and-leaf display.

Solution To construct a stem-and-leaf display of these scores, we split each score into two parts. The first part contains the first digit, which is called the *stem*. The second part contains the second digit, which is called the *leaf*. Thus, for the score of the first student, which is 75, 7 is the stem and 5 is the leaf. For the score of the second student, which is 52, the stem is 5 and the leaf is 2. We observe from the data that stems for all scores are 5, 6, 7, 8, and 9 as all the scores lie in the range 50 to 98. To obtain a stem-and-leaf display, we draw a vertical line and write the stems on the left side of it arranged in increasing order, as shown in Figure 2.13.

After we have listed the stems, we read the leaves for all scores and record them next to the corresponding stems on the right side of the vertical line. For example, for the first score we write the leaf 5 next to the stem 7; for the second score we write the leaf 2 next to the stem 5. The recording of these two scores in a stem-and-leaf display is shown in Figure 2.13.

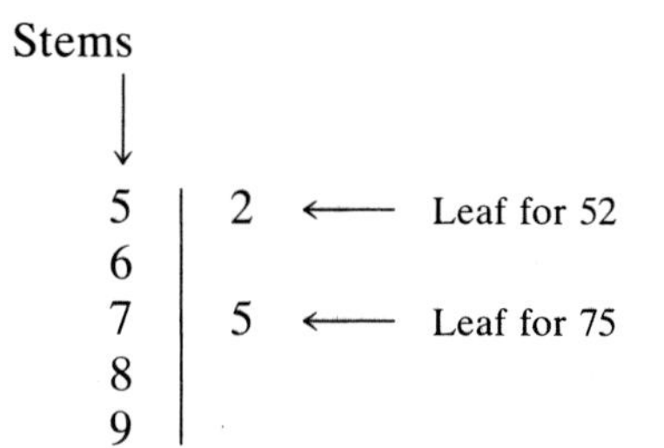

Figure 2.13 Stem-and-leaf display.

Now we read all the scores and write the leaves on the right side of the vertical line in the rows of corresponding stems. The complete stem-and-leaf display for scores is shown in Figure 2.14.

```
5 | 2 0 7
6 | 5 9 1 8 4
7 | 5 9 1 2 6 9 7 1 2
8 | 0 7 1 6 3 4 7
9 | 6 3 5 2 2 8
```

Figure 2.14 Stem-and-leaf display of test scores.

By looking at the stem-and-leaf display of Figure 2.14, we can observe how the data values are distributed. For example, stem 7 has the highest frequency, followed by stems 8, 9, 6, and 5.

The leaves of the stem-and-leaf display of Figure 2.14 are *ranked* (in increasing order) and presented in Figure 2.15.

```
5 | 0 2 7
6 | 1 4 5 8 9
7 | 1 1 2 2 5 6 7 9 9
8 | 0 1 3 4 6 7 7
9 | 2 2 3 5 6 8
```

Figure 2.15 Ranked stem-and-leaf display of test scores.

■

As mentioned earlier, one advantage of a stem-and-leaf display is that we do not lose information on individual observations. We can rewrite the individual scores of 30 college students from the stem-and-leaf display of Figure 2.14 or 2.15. By contrast, the information on individual observations is lost when data are grouped into a frequency table.

Constructing stem-and-leaf display for three- and four-digit numbers.

EXAMPLE 2-9 The following data give the monthly rents paid by a sample of 30 households selected from a city.

429	585	732	675	550	989	1020	620	750	660
540	578	956	1030	1070	930	871	765	880	975
650	1020	950	840	780	870	900	800	750	820

Prepare a stem-and-leaf display.

Solution Each of the values in the given data set contains either three or four digits. We will take the first digit for three-digit numbers and the first two digits for four-digit numbers as stems. Then, we will use the last two digits of each number as a leaf. Thus, for the first value, which is 429, the stem is 4 and the leaf is 29. The stems for the entire data are 4, 5, 6, 7, 8, 9, and 10. They are recorded on the left side of the vertical line in Figure 2.16. The leaves for various numbers are recorded on the right side.

4	29
5	85 50 40 78
6	75 20 60 50
7	32 50 65 80 50
8	71 80 40 70 00 20
9	89 56 30 75 50 00
10	20 30 70 20

Figure 2.16 Stem-and-leaf display of rents.

■

Sometimes a data set may contain too many stems, with each stem containing only a few leaves. In such cases, we may want to condense the stem-and-leaf display by *grouping the stems*. The following example describes this procedure.

Preparing a grouped stem-and-leaf display.

EXAMPLE 2-10 Prepare a new stem-and-leaf display by grouping the stems for the following stem-and-leaf display.

1	3 5
2	2 5 6
3	0 1
4	2 3 6
5	0
6	5 6
7	0 3 9
8	1 5 7
9	2 6

Solution To condense the given stem-and-leaf display, we can combine the first three rows, the middle three rows, and the last three rows, thus getting stems such as 1–3, 4–6, and 7–9. We separate the leaves for each stem of a group by an asterisk (*), as shown in Figure 2.17. Thus, the leaves 3 and 5 in the first row of Figure 2.17 correspond to stem 1; the leaves 2, 5, and 6 to stem 2; and leaves 0 and 1 belong to stem 3.

1–3	3 5 * 2 5 6 * 0 1
4–6	2 3 6 * 0 * 5 6
7–9	0 3 9 * 1 5 7 * 2 6

Figure 2.17 Grouped stem-and-leaf display.

■

If a stem does not contain a leaf, this can be indicated in the grouped stem-and-leaf display by two consecutive asterisks. For example, in the following stem-and-leaf display there is no leaf for 3, that is, there is no number in the 30s. The numbers in this display are 21, 25, 43, 48, and 50.

2–5 | 1 5 * * 3 8 * 0

If a stem contains too many leaves, we can list these leaves in more than one row. All these rows will have the same stem.

EXERCISES

2.41 The following data give the time (in minutes) that 20 students took to complete a statistics test.

55	49	53	59	38	56	39	58	47	53
58	42	37	43	47	44	55	51	46	45

Construct a stem-and-leaf display for these data. Arrange the leaves for each stem in increasing order.

2.42 Following are the SAT scores (out of a maximum possible score of 1600) of 12 students who took this test recently.

785	890	610	569	881	753	995	697	773	980	749	652

Construct a stem-and-leaf display. Arrange the leaves for each stem in increasing order.

2.43 Reconsider the data on the number of dentists per 100,000 people given in Example 2-3, which is reproduced here.

41	60	43	40	61	66	77	47	43	43
65	56	62	49	62	53	55	45	50	68
76	65	71	36	56	67	73	44	53	69
48	75	43	58	57	47	75	61	53	43
55	55	44	68	64	54	70	48	71	58

Prepare a stem-and-leaf display. Arrange the leaves for each stem in increasing order.

2.44 The following data (reproduced from Exercise 2.13) give the number of hospital beds per 100,000 people for 1987 for 50 states.

593	381	394	544	389	428	476	525	507	564
351	400	538	515	631	675	521	598	514	469
639	468	610	663	603	606	696	408	454	513
448	617	463	878	522	496	378	592	517	441
803	622	482	351	457	510	363	595	531	633

Prepare a stem-and-leaf display for these data. Arrange the leaves for each stem in increasing order.

2.45 The following data give the time (in minutes) taken to commute from home to work for 20 workers.

10	50	65	33	48	5	11	23	37	26
26	32	17	7	13	19	29	43	21	22

Construct a stem-and-leaf display for these data. Arrange the leaves for each stem in increasing order. (*Hint:* To prepare a stem-and-leaf display, each number in this data set can be written as a two-digit number. For example, 5 can be written as 05 for which the stem is 0 and the leaf is 5.)

2.46 The following data give the monthly grocery expenditures (in dollars) for 35 households selected from the 1987 Diary Survey.

217	326	183	191	83	130	261
599	130	174	287	261	100	313
220	509	574	192	87	391	74
243	152	370	400	391	143	304
165	174	470	217	255	348	160

a. Prepare a stem-and-leaf display for these data using the last two digits as leaves. (*Hint:* To prepare a stem-and-leaf display, each number in this data set can be written as a three-digit number. For example, 83 can be written as 083 for which the stem is 0 and the leaf is 83.)

b. Condense the stem-and-leaf display by grouping the stems as 0–1, 2–3, and 4–5.

2.47 The following data give the earnings (rounded to thousands of dollars) of 40 households selected from the 1987 Diary Survey.

39	28	3	17	42	29	6	17
12	15	60	34	44	46	6	84
44	32	24	11	58	14	75	62
13	27	21	3	46	6	11	10
79	15	20	14	30	24	18	1

a. Prepare a stem-and-leaf display for these data.

b. Condense the stem-and-leaf display by grouping the stems as 0–2, 3–5, and 6–8.

2.48 Consider the following stem-and-leaf display.

4	3 6
5	0 1 4 5 9
6	3 4 6 7 7 7 8 9
7	2 2 3 5 6 6
8	0 7 8 9

Write the data set that is represented by this stem-and-leaf display.

2.49 Consider the following stem-and-leaf display.

2–3	18 45 56 * 29 67 83 97
4–5	04 27 33 71 * 23 37 51 63 81 92
6–8	22 36 47 55 78 89 * * 10 41

Write the data set that is represented by this stem-and-leaf display.

2.7 GRAPHING TIME-SERIES DATA

As explained in Chapter 1, time-series data give information about a variable over a period of time. Example 2-11 exhibits the graphing of time-series data. In all the frequency distributions discussed so far in this chapter, the information was given about one variable only. A time-series data set contains information about two variables, one of which is time. For instance, in Example 2-11 the two variables about which we have information are time (year) and unemployment rate.

Graphing time-series data.

EXAMPLE 2-11 The following table gives the U.S. unemployment rate (measured as percentage of the civilian labor force who were unemployed) for the period 1978 through 1989.

Year	Unemployment	Year	Unemployment
1978	6.1	1984	7.5
1979	5.8	1985	7.2
1980	7.1	1986	7.0
1981	7.6	1987	6.2
1982	9.7	1988	5.5
1983	9.6	1989	5.3

Source: Economic Report of the President, 1990.

Display these data in graphic form.

Solution The given time-series data can be displayed by drawing a *line graph* or a *bar graph*. A line graph is usually used for time-series data only. To draw a line graph, we mark the years on the horizontal axis and the unemployment rate on the vertical axis, as in Figure 2.18. Then we mark a point above each year at a height equal to the unemployment rate for that year. Finally, we join consecutive points with straight lines.

From the graph of Figure 2.18 we can observe how the unemployment rate changed or fluctuated during the period 1978 through 1989. In general, a time-series line graph reveals the trend of a variable over a period of time. Note that in Figure 2.18 the vertical axis is truncated.

Another device used to display time series data is the bar graph. Figure 2.19 shows the bar graph for the data on unemployment rates. In this bar graph, each year is represented by an interval on the horizontal axis and the unemployment rate is marked on the vertical axis. The unemployment rate for a specific year is given by the height of the bar for that year.

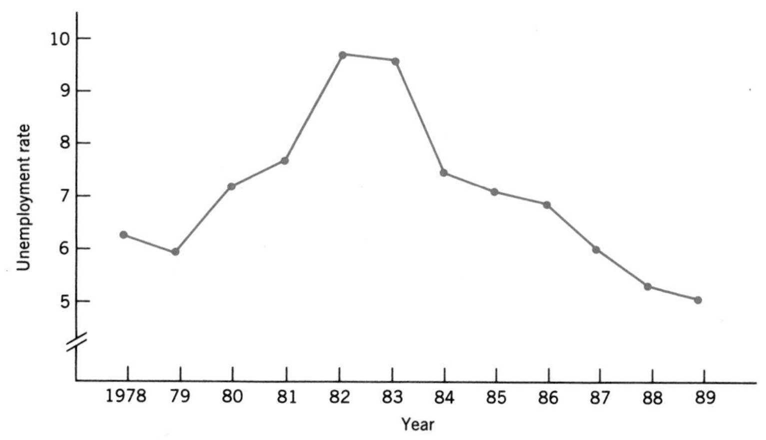

Figure 2.18 Line graph for unemployment rates.

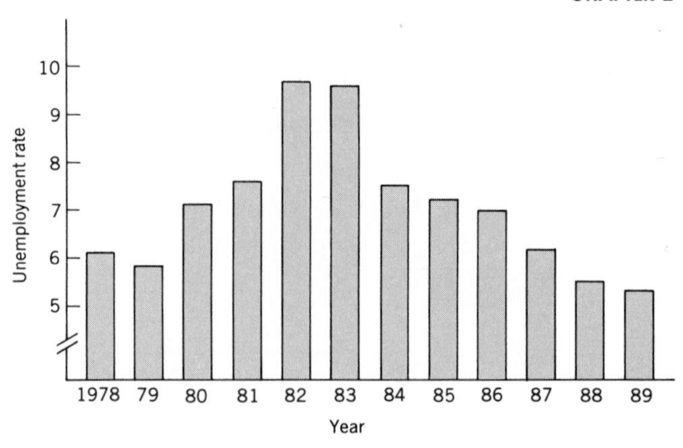

Figure 2.19 Bar graph for unemployment rates.

WARNING

Time-series data can be mistreated by manipulation in an attempt to prove different opinions. Case study 2-7 shows how time-series graphs can be used to give erroneous impressions.

CASE STUDY 2-7 WHAT IS WRONG WITH THIS PICTURE?

. . . Congress has been suspicious that this Administration, in particular, was more interested in cooking the facts to fit its ideology than in doing objective research, especially on the subject of education spending.

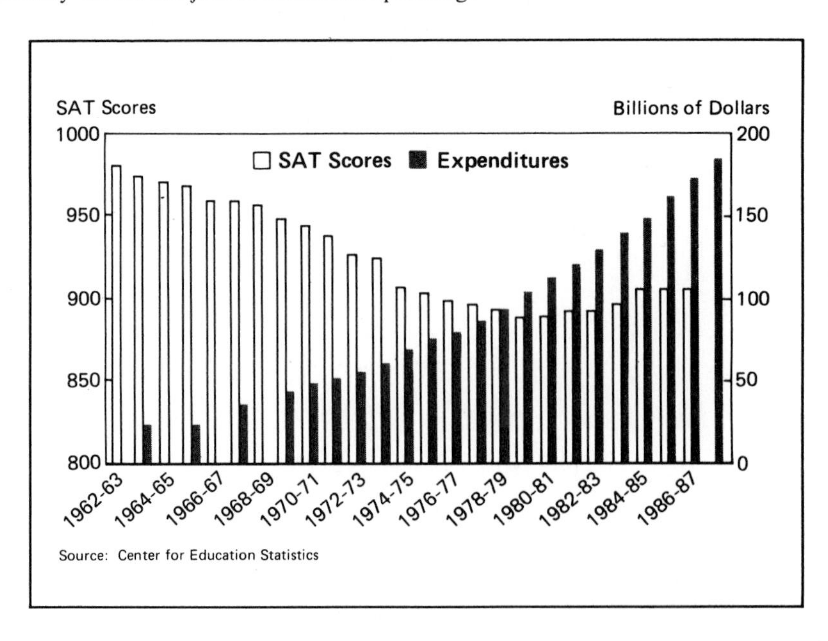

Proof that such suspicions are justified comes from a graph incorrectly attributed . . . to the U.S. Department of Education's Center for Education Statistics (CES) that was distributed at a conference by a high Department official and at a House Budget Committee hearing at which the Secretary of Education testified. The handout tries to convince us that from 1963 to 1988 there was a direct connection between the *rise* in elementary and secondary school spending and the *decline* in SAT scores. The graph makes a great visual impact. . . .

What's wrong with this picture? In the July 1988 issue of *ETS Policy Notes,* Dr. Joan Baratz-Snowden, then Director of the Policy Information Center of the Educational Testing Service, provides the answer: "The graph is as inaccurate as it is dramatic." Baratz-Snowden points out, the Administration used current dollars rather than constant dollars adjusted for inflation. A more accurate graph done by ETS shows that with inflation figured in, "expenditures have barely risen during the last 20 years." Although spending has been relatively constant, our public schools have added a host of special programs for disadvantaged, handicapped, non-English-speaking, and other students who were previously neglected and, sometimes, shut out.

Second, although SAT scores have gone down, the Education Department graph vastly exaggerated the decline. Baratz-Snowden shows us how. SAT scores go from 400–1600, but the graph uses an 800–1000 scale. . . . The Department also doesn't bother to tell us that the students who take the exam don't represent a random sample of our student population. For example, it used to be that only the brightest students were encouraged to take the SAT. Now the test-takers are more numerous and diverse, and this affects the average scores.

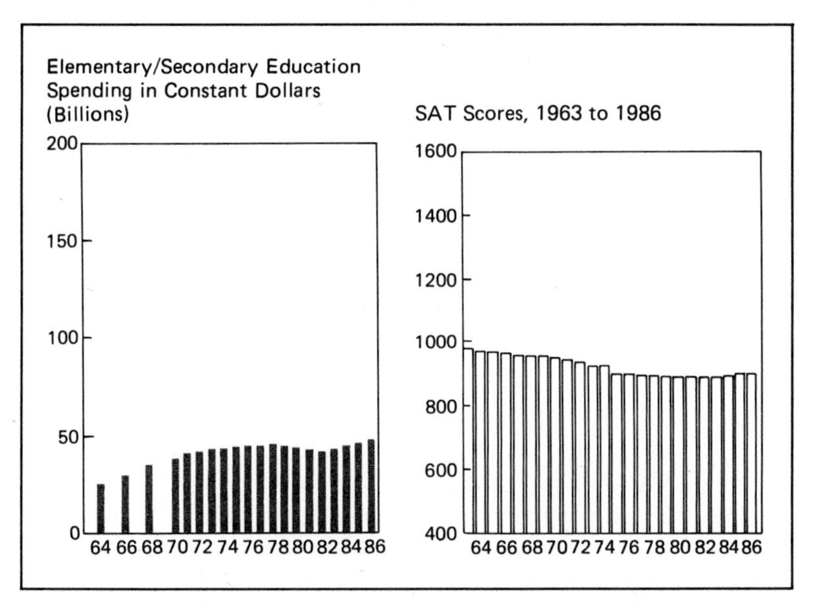

Source: Albert Shanker, "What is Wrong With This Picture? Lies, Damned Lies, and Statistics." *Chance,* 2(1), 1989, pp. 34–35. Copyright © 1989 by Springer-Verlag Heidelberg. Reprinted with permission.

EXERCISES

2.50 The following table gives the average monthly electric bill per farm for the period 1977 through 1985.

Year	Average Monthly Bill Per Farm ($)
1977	48.10
1978	53.20
1979	55.10
1980	61.90
1981	69.80
1982	83.90
1983	92.50
1984	98.00
1985	104.00

Source: National Agricultural Statistics Service.

Draw a line graph for these data.

2.51 The following table gives the percentage of youths aged 16 to 19 who were unemployed during the period 1981 through 1987.

Year	Total	White	Black
1981	19.6	17.3	41.4
1982	23.2	20.4	48.0
1983	22.4	19.3	48.5
1984	18.9	16.0	42.7
1985	18.6	15.7	40.2
1986	18.3	15.6	39.3
1987	16.9	14.4	34.7

Source: Bureau of Labor Statistics; and Susan Weber: *USA by Numbers: A Statistical Portrait of the United States*, Zero Population Growth, Washington, D.C., 1988.

Draw the line graphs for all youths, whites, and blacks on the same graph.

2.52 According to the American Bar Association, the average starting salaries for new associates in private law firms for the period 1983 through 1987 were as follows.

Year	Average Salary ($)
1983	24,000
1984	26,680
1985	26,000
1986	27,062
1987	33,000

Draw a bar graph for these data.

2.53 The following table gives the number of bachelor's degrees (in thousands) awarded to females in the United States for selected years since 1950.

Year	Bachelor's Degrees Awarded to Females (thousands)
1950	104
1955	104
1960	139
1965	219
1970	346
1975	430
1980	473
1985	522

Source: U.S. Department of Education.

Draw a bar graph for these data.

GLOSSARY

Bar graph A graph made of bars whose heights represent the frequencies of respective categories.

Class An interval that includes all the values in a (quantitative) data set that fall within two numbers, the lower and upper limits of the class.

Class boundary The midpoint of the upper limit of one class and the lower limit of the next class.

Class frequency The number of values in a data set that belong to a certain class.

Class midpoint or **mark** Obtained by dividing the sum of the lower and upper limits (or boundaries) of a class by 2.

Class width or **size** The difference between the two boundaries of a class.

Cumulative frequency The frequency of a class that includes all values in a data set that fall below the upper boundary of that class.

Cumulative frequency distribution A table that lists the total number of values that fall below the upper boundary of each class.

Cumulative relative frequency The cumulative frequency of a class divided by the total number of observations.

Cumulative percentage The cumulative relative frequency multiplied by 100.

Frequency distribution A table that lists all the categories or classes and the number of values belonging to each of these categories or classes.

Grouped data A data set presented in the form of a frequency distribution.

Histogram A graph in which classes are marked on the horizontal axis and either frequencies, relative frequencies, or percentages are marked on the vertical axis. The frequencies, relative frequencies, or percentages of various classes are represented by bars that are drawn adjacent to each other.

Ogive A curve drawn for a cumulative frequency distribution.

Percentage The percentage for a class or category is obtained by multiplying the relative frequency of that class or category by 100.

Pie chart A circle divided into portions that represent the relative frequencies or percentages of different categories or classes.

Polygon A graph formed by joining the midpoints of the tops of successive bars in a histogram by straight lines.

Raw data Data recorded in the sequence in which they are collected and before they are processed.

Relative frequency The frequency of a class or category divided by the sum of all frequencies.

Skewed to the left histogram A histogram with a longer tail on the left side.

Skewed to the right histogram A histogram with a longer tail on the right side.

Stem-and-leaf display A display of data in which each value is divided into two portions, a stem and a leaf.

Symmetric histogram A histogram that is identical on both sides of its central point.

Uniform or **rectangular histogram** A histogram with the same frequency for all classes.

KEY FORMULAS

1. **Relative frequency of a class**

$$\text{Relative frequency of a class} = \frac{\text{Frequency of that class}}{\text{Sum of all frequencies}} = \frac{f}{\Sigma f}$$

2. **Percentage of a class**

$$\text{Percentage} = (\text{Relative frequency}) \cdot 100$$

3. **Class midpoint or mark**

$$\text{Class midpoint} = \frac{\text{Upper limit} + \text{Lower limit}}{2}$$

4. **Class width or size**

$$\text{Class width} = \text{Upper boundary} - \text{Lower boundary}$$

5. **Cumulative relative frequency**

$$\text{Cumulative relative frequency} = \frac{\text{Cumulative frequency}}{\text{Total observations in the data}}$$

6. **Cumulative percentage**

$$\text{Cumulative percentage} = (\text{Cumulative relative frequency}) \cdot 100$$

SUPPLEMENTARY EXERCISES

2.54 The following data give the marital status of 30 persons selected from the 1987 Diary Survey. (M stands for married, W for widowed, D for divorced, S for separated, and N for never married.)

M	M	D	D	M	W	M	M	M	N
M	M	M	D	N	N	S	D	D	M
M	D	M	M	N	M	S	M	D	M

a. Construct a frequency distribution table for these data.
b. Prepare the relative frequency and percentage distributions.
c. What percentage of the persons in this sample are married? What percentage are either divorced or separated?

2.55 The following data give the educational attainment of 48 persons selected from the 1987 Interview Survey. In the data, E stands for elementary school education, H for high school but less than high school graduate, HG for high school graduate, S for some college but less than college graduate, C for college graduate, and M for more than 4 years of college.

H	S	HG	E	M	H	S	HG
S	HG	H	M	HG	E	HG	C
HG	C	M	H	S	S	S	H
HG	HG	M	HG	C	S	S	S
HG	M	S	C	C	E	C	S
S	HG	HG	HG	S	C	H	H

a. Prepare a frequency distribution table for these data.
b. Construct the relative frequency and percentage distributions.
c. Draw a bar graph for the relative frequency distribution and a pie chart for the percentage distribution.

2.56 The following data list the reasons why 24 (nonworking) persons, selected from the 1987 Interview Survey, were not working at the time of the survey. In the data, I stands for ill, disabled, or unable to work; T for taking care of home/family; G for going to school; C for could not find work; and R for retired.

T	R	R	I	T	I	R	C
I	R	R	R	I	G	C	R
R	G	T	R	R	R	C	I

a. Prepare a frequency distribution table for these data.
b. Calculate the relative frequencies and percentages for all categories.
c. Draw a bar graph for the frequency distribution and a pie chart for the percentage distribution.
d. What percentage of the persons in this sample are retired?

2.57 The following data give the number of earners per household for 45 households randomly selected from the 1987 Diary Survey.

2	1	2	2	2	0	1	1	2
1	3	0	2	1	2	1	2	1
2	2	1	2	1	1	0	3	1
1	1	2	2	4	2	0	1	3
1	2	2	4	1	2	0	3	2

a. Prepare a frequency distribution table for these data using single-valued classes.
b. Calculate the relative frequencies and percentages for all categories.
c. Draw a bar graph for the frequency distribution.
d. What percentage of the households in this sample have two or more earners?

2.58 The following data give the number of household members for 30 households randomly selected from the 1987 Interview Survey.

4	1	2	6	1	4	2	2	4	5
2	3	2	1	2	3	2	5	4	2
1	2	4	1	1	3	3	2	5	1

a. Prepare a frequency distribution table for these data using single-valued classes.

b. Calculate the relative frequencies and percentages for all categories.
c. What percentage of the households in this sample have 1 or 2 members?
d. Draw a bar graph for the relative frequency distribution.

2.59 The following data give the hours worked per week by 48 persons randomly selected from the 1987 Diary Survey. (These persons were selected only from those who were working at the time of the survey.)

40	30	40	48	56	60	40	20
60	50	40	40	50	45	50	40
45	70	15	45	40	65	30	34
40	40	64	60	40	40	24	45
38	40	40	46	44	40	40	50
24	50	40	40	60	40	37	40

a. Construct a frequency distribution table with the classes 14–25, 26–37, 38–49, 50–61, and 62–73.
b. Prepare the relative frequency and percentage columns for this table.
c. What percentage of the workers in this sample worked 50 or more hours?

2.60 The following data give the amount of public assistance or welfare (in hundreds of dollars) received by 45 households randomly selected from the 1987 Interview Survey. (These households were selected only from those households who received some public assistance or welfare.)

23	13	24	62	5	6	30	19	2
3	21	35	41	1	27	18	52	76
37	47	12	24	41	42	9	33	19
54	71	52	38	28	31	30	12	17
26	48	42	37	33	12	16	42	40

a. Construct a frequency table with the classes 1–16, 17–32, 33–48, 49–64, and 65–80.
b. Prepare the relative frequency and percentage columns for this table.
c. Draw a histogram and a polygon for the frequency distribution.
d. Based on the frequency histogram, can you say if the data are symmetric or skewed?

2.61 The following data give the social security contributions made during the 12-month period by 40 persons randomly selected from the 1987 Diary Survey. (These persons were selected only from those who made such contributions.)

2392	5928	780	2132	1976	1404	676	4472
3728	208	2236	4404	5132	1196	1716	1092
832	3959	3172	728	2132	1040	2028	4108
988	3936	468	2132	624	1716	4040	3174
3208	2948	1248	2704	156	208	988	624

a. Construct a frequency distribution table with the classes 1–1000, 1001–2000, 2001–3000, 3001–4000, 4001–5000, and 5001–6000.
b. Compute the relative frequencies and percentages for all classes.

c. Draw a histogram and a polygon for the relative frequency distribution.

d. What are the class boundaries and the width of the third class?

2.62 The following data give the amount of regular contributions (in hundreds of dollars) received from alimony or child support during the 12-month period by 54 persons randomly selected from the 1987 Diary Survey. (These persons were selected only from those who received such contributions.)

16	78	24	14	24	5	24	60	72
15	21	31	5	4	42	19	17	12
5	29	18	4	12	36	42	24	4
9	12	6	60	60	66	20	17	10
54	19	24	12	18	23	52	20	39
22	5	16	8	4	15	75	7	6

a. Construct a frequency table with the classes 1–16, 17–32, 33–48, 49–64, and 65–80.

b. Compute the relative frequencies and percentages for all classes.

c. What percentage of the persons in this sample received less than $3300 from alimony or child support?

d. What are the class boundaries and the width of the fifth class?

2.63 The following data give the charitable contributions (in dollars) made during 12 months before the survey by 35 households randomly selected from the 1987 Interview Survey. (These households were selected only from those that made such contributions.)

30	300	50	100	27	100	25
76	25	15	25	60	240	100
18	400	200	10	25	50	125
140	34	275	250	130	87	24
500	75	150	15	200	200	300

a. Construct a frequency distribution table with the classes 1–100, 101–200, . . . , 401–500.

b. Calculate the relative frequencies and percentages for all classes.

c. What percentage of the households in this sample made charitable contributions of more than $300?

2.64 The following data give the weekly expenditures (in dollars) on nonalcoholic beverages for 45 households randomly selected from the 1987 Diary Survey. (These households were selected only from those that incurred such expenses.)

6.5	9.0	9.2	7.2	4.6	9.0	10.5	2.4	10.9
10.4	5.4	12.7	5.4	0.9	7.1	1.4	12.3	8.2
4.7	1.3	2.5	13.5	10.1	15.9	5.6	15.1	0.7
10.1	10.3	2.2	7.1	4.6	8.0	0.9	3.3	3.1
2.2	10.6	1.3	2.7	16.5	9.8	4.9	1.6	12.7

a. Construct a frequency distribution table with the classes 0.5 to less than 3.5, 3.5 to less than 6.5, . . . , and 15.5 to less than 18.5.

b. Calculate the relative frequencies and percentages for all classes.

c. What is the width of each class?
d. Draw a histogram and a polygon for the percentage distribution.

2.65 The following data give the gross annual income (in thousands of dollars) for 60 households randomly selected from the 1987 Diary Survey.

10.6	34.0	23.3	71.7	11.0	33.0	31.5	43.5	10.1	63.8
62.0	27.9	64.8	55.3	17.2	18.3	18.0	20.0	49.1	9.0
38.3	20.1	9.0	26.2	2.6	53.1	14.9	74.3	17.8	13.9
30.0	12.5	9.0	16.7	58.1	20.3	34.7	15.7	13.0	3.6
67.7	18.2	44.5	17.9	16.5	9.9	10.3	42.0	31.3	8.3
5.1	72.0	51.6	33.7	21.1	33.6	54.1	71.2	45.1	6.1

a. Construct a frequency distribution table with the classes 0 to less than 15, 15 to less than 30, . . . , and 60 to less than 75.
b. Calculate the relative frequencies and percentages for all classes.
c. What is the width of each class?

2.66 Refer to Exercise 2.59. Prepare the cumulative frequency, cumulative relative frequency, and cumulative percentage distributions by using the frequency distribution table for the data of that exercise.

2.67 Refer to Exercise 2.60. Prepare the cumulative frequency, cumulative relative frequency, and cumulative percentage distributions by using the frequency distribution table of that exercise.

2.68 Refer to Exercise 2.61. Prepare the cumulative frequency, cumulative relative frequency, and cumulative percentage distributions by using the frequency distribution table constructed in that exercise.

2.69 Prepare the cumulative frequency, cumulative relative frequency, and cumulative percentage distributions by using the frequency distribution constructed for the data of Exercise 2.62.

2.70 Prepare the cumulative frequency, cumulative relative frequency, and cumulative percentage distributions by using the frequency distribution table constructed in Exercise 2.63.

2.71 Prepare the cumulative frequency, cumulative relative frequency, and cumulative percentage distributions by using the frequency distribution table constructed in Exercise 2.64. Draw an ogive for the cumulative percentage distribution.

2.72 Prepare the cumulative frequency, cumulative relative frequency, and cumulative percentage distributions by using the frequency distribution table constructed in Exercise 2.65. Draw an ogive for the cumulative frequency distribution.

2.73 Prepare a stem-and-leaf display for the data of Exercise 2.59.

2.74 Refer to Exercise 2.60. Prepare a stem-and-leaf display for the data given in that exercise.

2.75 Construct a stem-and-leaf display for the data of Exercise 2.62.

2.76 Refer to Exercise 2.63. Prepare a stem-and-leaf display for the data given in that exercise.

2.77 The following table gives the birth rate per 1000 persons for the United States for selected years from 1955 through 1985.

Year	Birth Rate (per 1000 persons)
1950	23.9
1955	24.9
1960	23.8
1965	19.6
1970	18.2
1975	14.6
1980	15.9
1985	15.7

Source: U.S. Bureau of the Census.

Draw a bar graph for these data.

2.78 The following table gives the mean income of families for the years 1980 through 1988.

Year	Mean Income ($)
1980	23,974
1981	25,838
1982	27,391
1983	28,820
1984	31,052
1985	32,944
1986	34,924
1987	36,884
1988	38,608

Source: U.S. Bureau of the Census.

Draw a line graph for these data.

SELF-REVIEW TEST

1. Briefly explain the difference between ungrouped and grouped data. Give one example of each type of such data.

2. The following table gives the frequency distribution of the time (in minutes) taken to run a 5-mile race for a random sample of 200 participants.

Time (minutes)	Frequency
30–39	21
40–49	36
50–59	49
60–69	61
70–79	33

Circle the correct answer for each of the following questions, which are based on this table.

a. The number of classes in the table is 5, 200, 80
b. The class width is 9, 10, 5
c. The midpoint of the third class is 53.5, 54.0, 54.5
d. The lower boundary of the second class is 40, 49, 39.5, 49.5
e. The upper limit of the second class is 40, 49, 39.5, 49.5
f. The sample size is 5, 200, 50
g. The relative frequency of the first class is .155, .172, .105

3. Briefly explain and illustrate with the help of graphs a symmetric histogram, a histogram skewed to the right, and a histogram skewed to the left.

4. Twenty elementary school children were asked if they live with both parents (B), father only (F), mother only (M), or someone else (S). The responses of the children are as follows.

M	B	B	M	F	S	B	M	F	B
B	F	B	M	M	B	B	F	B	M

a. Construct a frequency distribution table.
b. Write the relative frequencies and percentages for all categories.
c. What percentage of the children in this sample live with mother only?
d. Draw a bar graph for the frequency distribution and a pie chart for the percentages.

5. The following data set gives the teaching experience (in years) of 24 randomly selected faculty members of a university.

15	12	9	10	5	12	3	7	16	13	11	14
11	8	7	14	11	8	4	13	2	18	6	19

a. Construct a frequency distribution table with the classes 1–4, 5–8, 9–12, 13–16, and 17–20.
b. Calculate the relative frequencies and percentages for all classes.
c. What percentage of the instructors have a teaching experience of 8 or fewer years?
d. Draw the frequency histogram and polygon.

6. Refer to the frequency distribution prepared in Problem 5. Prepare the cumulative percentage distribution by using that table. Draw an ogive for this cumulative percentage distribution. Using the ogive, find the percentage of faculty members in this sample who have a teaching experience of 11 or fewer years.

7. Construct a stem-and-leaf display for the following data, which give the temperatures observed in 22 cities during December.

43	51	12	18	7	28	39	67	47	32	11
4	31	62	15	26	17	31	11	29	38	41

8. Consider the following stem-and-leaf display.

3	0 3 7
4	2 4 6 7 9
5	1 3 3 6
6	0 7 7
7	1 9

Write the data set which was used to construct this stem-and-leaf display.

9. The following table gives the expected life at birth (i.e., the number of years a person born in a particular year is expected to live) for selected years for the United States.

Year	Expected Life (years)
1920	54.1
1930	59.7
1940	62.9
1950	68.2
1960	69.7
1970	70.8
1980	73.7

Source: U.S. National Center for Health Statistics.

Draw a bar graph and a line graph for these data.

USING MINITAB

MINITAB commands HISTOGRAM and STEM-AND-LEAF are used to construct a histogram and a stem-and-leaf display, respectively, for any quantitative data. The use of these commands is explained below.

HISTOGRAM AND FREQUENCY DISTRIBUTION

To construct a histogram for a data set, first we enter the data in column C1 by using the SET command. Then we instruct MINITAB to construct a histogram for the data of column C1 by using the following command.

```
MTB > HISTOGRAM C1   ← This command instructs MINITAB to construct a histogram
                       for the data of column C1
```

The HISTOGRAM command will also list the frequencies of various classes. Illustration M2-1 describes the procedure to construct a histogram using MINITAB.

ILLUSTRATION M2-1 Refer to data on the number of dentists per 100,000 people given in Example 2-3. We construct a histogram for those data as follows.

```
MTB  > NOTE: HISTOGRAM FOR DATA ON DENTISTS
MTB  > SET C1
DATA > 41  60  43  40  61  66  77  47  43  43
DATA > 65  56  62  49  62  53  55  45  50  68
DATA > 76  65  71  36  56  67  73  44  53  69
DATA > 48  75  43  58  57  47  75  61  53  43
DATA > 55  55  44  68  64  54  70  48  71  58
DATA > END
MTB  > HISTOGRAM C1

Histogram of C1     N = 50
```

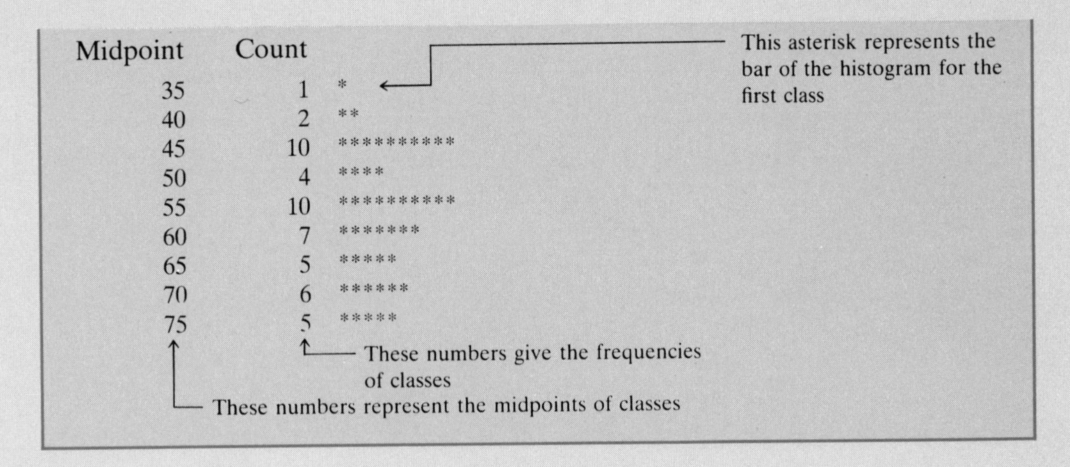

```
Midpoint   Count
      35       1  *
      40       2  **
      45      10  **********
      50       4  ****
      55      10  **********
      60       7  *******
      65       5  *****
      70       6  ******
      75       5  *****
```


In this MINITAB output, the first column gives the midpoints of various classes. Using these midpoints, we can write the classes as 33–37, 38–42, . . . , and 73–77. The second column of the MINITAB output gives the frequencies of various classes. The asterisks in the third column represent the bars of the histogram. Note that the bars are drawn horizontally and not vertically.

In Illustration M2-1, MINITAB itself decided on the number of classes and their widths. In Example 2-3 (Table 2.9), we used the classes as 31–40, 41–50, . . . , 71–80 to group these data. The midpoints of these classes are 35.5, 45.5, . . . , 75.5, and the width of each class is 10. Suppose we decide to use the same class limits and width for our MINITAB histogram. The following MINITAB commands will give us a histogram with these classes for the data entered in column C1.

```
MTB  > HISTOGRAM C1;      ← Note the semicolon at the end of this command
SUBC > INCREMENT = 10;    ← This subcommand indicates class width
SUBC > START = 35.5.      ← This subcommand instructs MINITAB that the midpoint of
                            the first class is 35.5
```

Observe these MINITAB commands for semicolons and period very carefully. Note that when we put a semicolon at the end of a MINITAB command, it instructs MINITAB that a subcommand (SUBC) is to follow with some additional information. A semicolon at the end of a subcommand indicates that another subcommand is to follow with some more information. A period at the end of a subcommand instructs MINITAB that all MINITAB commands and subcommands have been entered. In the foregoing set of MINITAB commands, we instruct MINITAB to construct a histogram for the data of column C1 with each class having a width of 10 and the first class having a midpoint of 35.5. Note that the class width in MINITAB is indicated using the INCREMENT command and the midpoint of the first class by using the START command. The following is the MINITAB output obtained by using these commands.

```
MTB  > NOTE: HISTOGRAM FOR DATA ON DENTISTS
MTB  > HISTOGRAM C1;
SUBC > INCREMENT = 10;
SUBC > START = 35.5.

Histogram of C1     N = 50

Midpoint    Count
    35.5        2  **
    45.5       15  ***************
    55.5       13  *************
    65.5       13  *************
    75.5        7  *******
```

The MINITAB DOTPLOT command plots all the data values. The following command will give a dotplot for the data entered in column C1 in Illustration M2-1.

```
MTB > DOTPLOT C1
```

Use this command to plot the data on the number of dentists.

STEM-AND-LEAF DISPLAY

To prepare a stem-and-leaf display, first we enter the data using SET command. Then we use the following MINITAB commands to construct a stem-and-leaf display for those data.

```
MTB  > STEM-AND-LEAF C1;   ← This command instructs MINITAB to make a
                             stem-and-leaf display for the data of column C1.
SUBC > INCREMENT = 10.     ← This subcommand tells MINITAB that the distance
                             between any two consecutive stems is 10 units.
```

Illustration M2-2 describes the procedure to construct a stem-and-leaf display by using MINITAB.

ILLUSTRATION M2-2 Refer to data on test scores of 30 students given in Example 2-8. A stem-and-leaf display for those data is prepared as follows.

```
MTB   > NOTE: STEM-AND-LEAF DISPLAY FOR SCORES
MTB   > SET C1
DATA > 75  52  80  96  65  79  71  87  93  95
DATA > 69  72  81  61  76  86  79  68  50  92
DATA > 83  84  77  64  71  87  72  92  57  98
DATA > END
MTB  > STEM-AND-LEAF C1;
SUBC > INCREMENT = 10.

Stem-and-leaf of C1             N = 30
Leaf unit = 1.0                     ↑
                                    └──── Number of observations
                                              in the data
    3     5  027
    8     6  14589
   (9)    7  112256799
   13     8  0134677
    6     9  223568
    ↑     ↑      ↑
    │     │      └──── These numbers give the leaves
    │     └──── These numbers represent the stems
    └──── These numbers, called depths, are explained below
```

In this MINITAB input and output, the STEM-AND-LEAF C1 command instructs MINITAB to plot a stem-and-leaf display for the data entered in column C1. The subcommand INCREMENT = 10 indicates the distance between any two consecutive stems. As a consequence of this command, the stems will be 5 (for the numbers in 50s), 6 (for the numbers in 60s), and so forth.

In this MINITAB printout, N = 30 is the number of observations in the data. LEAF UNIT = 1.0 means that the decimal point is after one leaf digit in the printout. Thus, the first number is 50, second is 52, and so on. The numbers in the first column are called *depths*, which give the cumulative frequencies from above and below. The depth, which appears in parentheses (9 in this MINITAB output), gives the number of leaves in the row that contains the median value. The depths before this row give the total number of leaves in the corresponding row and before it. Thus, the first depth (which is 3) gives the number of leaves that belong to stem 5. The second number (which is 8) gives the cumulative number of leaves that belong to the first two stems. After the stem that contains the median, the depths are cumulative from the bottom of the stem-and-leaf display. For example, the depth of 13 for the stem of 8 indicates the total number of leaves that belong to stems 8 and 9. The last depth, which is 6, indicates the number of leaves for stem 9.

COMPUTER ASSIGNMENTS

M2.1 Refer to Data Set IV of Appendix B (page 718) on time taken to run the Manchester Road Race for a sample of 500 participants. From that data set, select the 6th value and then select every 10th value after that (i.e., select the 6th, 16th, 26th, 36th, . . . values.) This subsample will give you 50 measurements. (Such a sample taken from a population is called a *systematic random sample*.) Using MINITAB, construct a histogram for these data.

M2.2 Refer to Data Set I of Appendix B on the prices of various products in different cities across the country. Using the MINITAB SAMPLE command, select a subsample of 60 from column C2 (price of half-gallon carton of whole milk). Using MINITAB, construct a histogram for these data.

M2.3 Using MINITAB, construct a histogram for the data on the number of hospital beds per 100,000 people for 50 states given in Exercise 2.13. Use the classes 300–399, 400–499, . . . , 800–899.

M2.4 Using MINITAB, construct a histogram for the data on the annual earnings (in thousands of dollars) before taxes for 36 families given in Exercise 2.23. Use the classes 0 to less than 12, 12 to less than 24, . . . , and 48 to less than 60.

M2.5 Using MINITAB, prepare a stem-and-leaf display for the data given in Exercise 2.41.

M2.6 Using MINITAB, prepare a stem-and-leaf display for the data of Exercise 2.43.

3 NUMERICAL DESCRIPTIVE MEASURES

3.1 MEASURES OF CENTRAL TENDENCY FOR UNGROUPED DATA

3.2 MEASURES OF DISPERSION FOR UNGROUPED DATA

3.3 MEAN, VARIANCE, AND STANDARD DEVIATION FOR GROUPED DATA

3.4 USE OF STANDARD DEVIATION

3.5 MEASURES OF POSITION

3.6 BOX-AND-WHISKER PLOT

APPENDIX 3.1

SELF-REVIEW TEST

USING MINITAB

In chapter 2 we discussed how to organize and display large data sets. The techniques presented in that chapter, however, are not helpful when we need to describe the main characteristics of a data set verbally. The numerical summary measures, such as the ones that give the center and spread of a distribution, provide us with the main features of a data set. For example, the techniques learned in Chapter 2 can help us to graph the data on family incomes. However, we may want to know the income of a "typical" family (given by the center of the distribution), the spread of the distribution of incomes, or the location of a family with a specific income. Figure 3.1 shows these three concepts. Such questions can be answered using the methods discussed in this chapter. Included among these are (1) measures of central tendency, (2) measures of dispersion, and (3) measures of position.

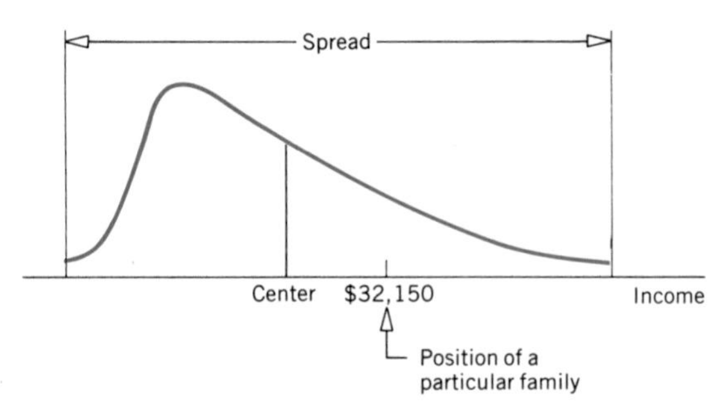

Figure 3.1

3.1 MEASURES OF CENTRAL TENDENCY FOR UNGROUPED DATA

We often represent a data set by one number, usually called the *typical value*. For example, the following excerpt from *USA Today* of May 1, 1990, gives the summary measures for the time spent watching television and the time spent doing homework by eighth-graders.

> A new study of 24,600 pupils who were eighth-graders in 1988 says the *typical pupil* at the time spent four times more hours per week watching television than doing homework.
>
> TV watching averaged 21.2 hours per week; homework, just 5.5 hours, says the National Education Longitudinal Study of 1988, released in April by the U.S. Department of Education. . . .

As we can observe, the time watching television and the time doing homework for 24,600 eighth-graders are each represented by one number, 21.2 and 5.5 hours a week, respectively. Such summary measures are called the **measures of central tendency.** A measure of central tendency gives the center of a histogram or a frequency distribution curve. This section discusses three different measures of central tendency: mean, median, and mode. We will learn how to calculate these measures for ungrouped data. We know from Chapter 2 that data that contain information on each member of the population or sample individually are called *ungrouped data,* whereas *grouped data* refer to the data presented in the form of a frequency distribution table.

3.1.1 MEAN

The **mean,** also called the *arithmetic mean*, is the most frequently used measure of central tendency. This book will use mean and *average* synonymously. For ungrouped data, the mean is obtained by dividing the sum of all values by the number of values in the data set.

$$\text{Mean} = \frac{\text{Sum of all values}}{\text{Number of values}}$$

The mean computed for sample data is denoted by $\bar{x}$ (read as "x bar"), and the mean computed for population data is denoted by μ (Greek letter mu). We know from the discussion in Chapter 2 that the number of values in a data set is denoted by n for a sample and by N for a population. In Chapter 1 we learned that a variable is denoted by x and the sum of all values of x is denoted by Σx. Using this notation, we can write the following formulas for mean.

MEAN FOR UNGROUPED DATA

Mean for population data: $\mu = \frac{\Sigma x}{N}$

Mean for sample data: $\bar{x} = \frac{\Sigma x}{n}$

Calculating sample mean for ungrouped data.

EXAMPLE 3-1 The following data give the ages of five of the employees working for Nable's Inc.

47 31 59 25 36

Find the mean.

Solution The age of employees in this example is a variable. Let us denote it by x. Then the five values of x are

$$x_1 = 47, \quad x_2 = 31, \quad x_3 = 59, \quad x_4 = 25, \quad \text{and} \quad x_5 = 36$$

where x_1 is the age of the first employee, x_2 is the age of the second employee, and so on. The sum of the ages of these five employees is

$$\Sigma x = x_1 + x_2 + x_3 + x_4 + x_5 = 47 + 31 + 59 + 25 + 36 = 198$$

Because the data set contains five values: $n = 5$.

Substituting the values of Σx and n in the sample formula, the mean age for these employees is

$$\bar{x} = \frac{\Sigma x}{n} = \frac{198}{5} = 39.6 \text{ years}$$

■

Physically, the mean is the point that balances a histogram. If we consider the mean as a fulcrum, the histogram will balance on the fulcrum. This is shown in Figure 3.2 for data from Example 3-1.

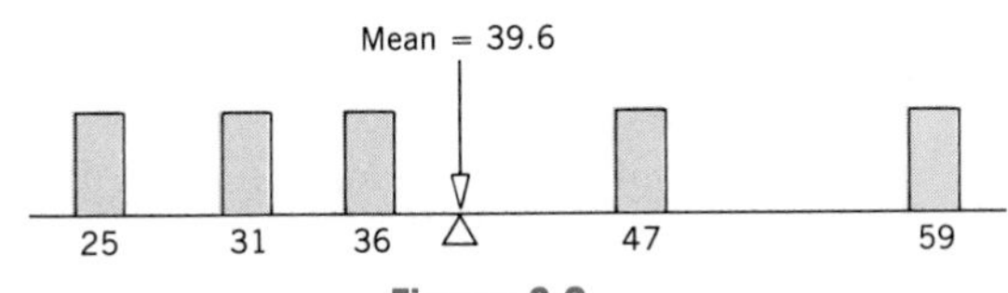

Figure 3.2

Calculating population mean for ungrouped data.

EXAMPLE 3-2 The following data give the number of marriages (in thousands) performed in the United States during each of the 12 months of 1989 (*Monthly Vital Statistics Report,* May 8, 1990).

117 126 159 185 228 291 217 245 231 210 188 208

Find the mean.

Solution This data set represents the population because it includes all 12 months of 1989. Hence, $N = 12$.

$$\begin{aligned}\Sigma x &= 117 + 126 + 159 + 185 + 228 + 291 + 217 \\ &\quad + 245 + 231 + 210 + 188 + 208 \\ &= 2405\end{aligned}$$

The mean is

$$\mu = \frac{\Sigma x}{N} = \frac{2405}{12} = 200.417 \text{ thousand}$$

Thus, the mean number of marriages performed each month during 1989 was 200,417. ■

Now if we take a sample of any three months from 1989 and calculate the mean number of marriages performed each month for those three months, this mean will be denoted by $\bar{x}$. Suppose the three values included in the sample are 126, 291, and 245. Then, the sample mean is

$$\bar{x} = \frac{(126 + 291 + 245)}{3} = 220.667 \text{ thousand} = 220{,}667 \text{ marriages}$$

If we take a second sample of three months from 1989, the value of $\bar{x}$ will (most likely) be different. Suppose the second sample includes the values 185, 231, and 188. Then, the mean for this sample is

$$\bar{x} = \frac{(185 + 231 + 188)}{3} = 201.333 \text{ thousand} = 201{,}333 \text{ marriages}$$

Consequently, we can state that the value of the population mean μ is constant. However, the value of the sample mean $\bar{x}$ varies from sample to sample. The value of $\bar{x}$ for a particular sample will depend on what values of the population are included in the sample.

Sometimes a data set may contain a few very small or a few very large values. Such values are called **outliers** or **extreme values.**

OUTLIERS OR EXTREME VALUES

Values that are very small or very large relative to the majority of the values in a data set are called outliers or extreme values.

A major shortcoming of the mean as a measure of central tendency is that it is very sensitive to outliers. Example 3-3 illustrates this point.

Illustrating the effect of an outlier on the mean.

EXAMPLE 3-3 The following table lists the 1988 population (in thousands) for the five Pacific states.

State	Population
Washington	4,648
Oregon	2,767
Alaska	524
Hawaii	1,098
California	28,314 ← An outlier

Notice that the population of California is very large compared to the populations of the other four states. Hence, it is an outlier. Show how the inclusion of this outlier affects the value of the mean.

Solution If we do not include the population of California (the outlier), the mean for the populations of the remaining four states (Washington, Oregon, Alaska, and Hawaii) is

$$\text{Mean} = \frac{(4648 + 2767 + 524 + 1098)}{4} = 2259.25 \text{ thousand}$$

Now, to see the impact of the outlier on the value of the mean, we include the population of California and find the mean population of all five Pacific states. This mean is

$$\text{Mean} = \frac{(4648 + 2767 + 524 + 1098 + 28{,}314)}{5} = 7470.20 \text{ thousand}$$

Thus, including California causes more than a threefold increase in the value of the mean as it changes from 2259.25 thousand to 7470.20 thousand. ■

This example should encourage us to be cautious. We should remember that the mean is not always the best measure of central tendency because it is heavily influenced by outliers. Sometimes other measures of central tendency give a more accurate impression of a data set.

CASE STUDY 3-1 HOOPSTERS ARE MAKING THE MOST

The chart on the next page shows how the average salaries for basketball, baseball, and football players have changed during the period 1980 to 1990. The following excerpts are quoted from *Fortune* magazine.

. . . The average hitter, not to mention the average football player, still makes less than the average hoopster. While baseball and football salaries climbed precipitously during the 1980s (see chart), the players of the National Basketball Association scored the highest jump in average pay—from $171,000 in the 1980–81 season to an estimated $900,000 for the one that begins this fall. That is 60% more than the average baseball player will make this year, and more than twice the earnings of the average gridiron working stiff. . . .

Whatever the game, the good money doesn't last long. The average baseball career lasts for four years; football, 4.4 years; basketball, 4.5.

Source: *Fortune,* May 7, 1990, p. 13. Copyright © 1990, The Time Inc. Magazine Company. Chart and excerpts reprinted with permission.

3.1.2 MEDIAN

Another important measure of central tendency is the **median.** It is defined as follows.

> **MEDIAN**
>
> The median is the value of the middle term in a data set that has been ranked in increasing order.

As is obvious from the definition of the median, it divides a ranked data set into two equal parts. The calculation of the median consists of the following two steps:

1. Rank the given data set in increasing order.
2. Find the middle term. The value of this term is the median.†

The position of the middle term in a data set with n values is given by $(n + 1)/2$. Thus, we can redefine the median as follows.

MEDIAN FOR UNGROUPED DATA

$$\text{Median} = \text{the value of the } \left(\frac{n+1}{2}\right)\text{th term in a ranked data set}$$

If the given data set represents a population, replace n by N.

If the number of observations in a data set is *odd*, then the median is given by the value of the middle term in the ranked data. If the number of observations is *even*, then the median is given by the average of the values of the two middle terms.

Calculating median for ungrouped data: odd number of data values.

EXAMPLE 3-4 The following data give the weight lost (in pounds) by a sample of five members of a health club at the end of two months of membership.

10 5 19 8 3

Find the median.

Solution First, we rank the given data in increasing order.

3 5 8 10 19

There are five observations in the data set. Consequently, $n = 5$ and

$$\frac{n+1}{2} = \frac{5+1}{2} = 3$$

Therefore, the median is the value of the third term in the ranked data.

3 5 8 10 19
↑
Median

Thus, the median weight loss for the sample of five members of this health club is eight pounds. ■

†The value of the middle term in a data set ranked in decreasing order will also give the value of the median.

Calculating median for ungrouped data: even number of data values.

EXAMPLE 3-5 The following data give the prices (in thousands of dollars) of a sample of 10 houses sold recently in a city.

215 209 187 278 216 153 191 133 356 147

Find the median.

Solution First, we rank the given data in increasing order.

133 147 153 187 191 209 215 216 278 356

There are 10 values in the data set. Hence, $n = 10$ and

$$\frac{n+1}{2} = \frac{10+1}{2} = 5.5$$

Therefore, the median is given by the mean of the fifth and sixth values in the ranked data.

133 147 153 187 191 ↑ 209 215 216 278 356

$$\text{Median} = \frac{191 + 209}{2} = \$200 \text{ thousand}$$

Thus, the median sale price of a house for this sample is $200,000. ■

The median gives the center of a histogram such that half of the data values are to the left of the median and half are to the right of the median. The advantage of using the median as a measure of central tendency is that it is not influenced by outliers. Consequently, median is preferred over mean as a measure of central tendency for data sets that contain outliers. Case Study 3-2 is an example of such a case.

CASE STUDY 3-2 MEDIAN SALE PRICES OF EXISTING SINGLE-FAMILY HOMES FOR METROPOLITAN AREAS

The Economics and Research Division of the National Association of Realtors obtains data on existing single-family home sales from more than 450 Boards of Realtors and multiple-listing systems across the country. Based on these data, the National Association of Realtors publishes descriptive statistics and graphs representing the characteristics of the existing single-family homes for the United States, for different regions of the country, for individual states, and for selected metropolitan areas.

These data appear in the Association's monthly journal *Home Sales*. The following table gives the median sale prices of existing single-family homes for the year 1988 for a few selected metropolitan areas from a recent issue of *Home Sales*.

Metropolitan Area	Median Price ($)
Baltimore, MD	88,700
Boston, MA	181,200
Chattanooga, TN	63,500
Chicago, IL	98,900
Denver, CO	81,800
Houston, TX	61,800
Los Angeles area, CA	179,400
New York/Northern New Jersey/ Long Island, NY/NJ/CT	183,800
Philadelphia, PA	102,400
San Francisco Bay area, CA	212,600
Washington, DC	132,500

The following table gives the mean and median prices of existing single-family homes for the United States and the four regions for the year 1988.

	All Regions	Northeast	Midwest	South	West
Mean price	$112,800	$155,500	$78,900	$100,800	$145,400
Median price	$89,300	$143,000	$68,400	$82,200	$124,900

As the distribution of sale prices of existing single-family homes is expected to be skewed to the right, the mean price is affected by extreme values at the upper end. Hence, the mean price is higher than the median price of single-family homes for the United States and for each of the four regions. In such a case, the median is a better measure of central tendency.

Source: *Home Sales,* 3(10), October 1989. Copyright © 1989 by National Association of Realtors. Data reproduced with permission.

3.1.3 MODE

Mode is a French word that means "fashion." In statistics, the mode represents the most common value in a data set.

MODE

The mode is the value that occurs with highest frequency in a data set.

Calculating mode for ungrouped data.

EXAMPLE 3-6 The following are the years of schooling for a sample of seven persons.

12 15 16 12 18 16 12

Find the mode.

Solution In this data set, 12 occurs three times, 16 occurs twice, and each of the remaining values occurs only once. Because 12 occurs with highest frequency, it is the mode.

$$\text{Mode} = 12 \text{ years}$$

■

A major shortcoming of the mode is that a data set may have none or more than one mode, whereas it will have only one mean and only one median. For instance, a data set with each value occurring only once has no mode. A data set with only one value occurring with highest frequency has only one mode. The data set in this case is called *unimodal*. A data set with two values occurring with the same (highest) frequency has two modes. The distribution, in this case, is said to be *bimodal*. If more than two values in a data set occur with the same (highest) frequency, then the data set is said to be *multimodal*.

Data set with no mode.

EXAMPLE 3-7 Last year's incomes of five randomly selected families were \$26,150, \$19,300, \$34,985, \$47,490, and \$13,470. Find the mode.

Solution As each value in this data set occurs only once, there is no mode. ■

Data set with two modes.

EXAMPLE 3-8 The years of marriage for a sample of eight couples are 5, 12, 7, 25, 7, 14, 12, and 6. Find the mode.

Solution In this data set, each of the values 7 and 12 occurs twice and each of the remaining values occurs only once. Therefore, this data set has two modes: 7 and 12. ■

Data set with three modes.

EXAMPLE 3-9 The ages of 10 randomly selected students from a class are 21, 19, 27, 22, 29, 19, 25, 21, 22, and 30. Find the mode.

Solution This data set has three modes: 19, 21, and 22. Each of these three values occurs with a (highest) frequency of 2. ■

One advantage of the mode is that it can be calculated for both kinds of data, quantitative and qualitative, whereas the mean and median can be calculated only for quantitative data.

Finding mode for qualitative data.

EXAMPLE 3-10 The statuses of the five students who are members of a college committee are senior, sophomore, senior, junior, senior. Find the mode.

Solution As senior occurs more frequently than the other categories, it is the mode for this data set. However, we cannot calculate the mean and median for this data set. ■

To sum up, we cannot conclude which of the three measures of central tendency is a better measure overall. Each of them may be better under different situations. Probably the mean is the most used measure of central tendency followed by the median. The mean has the advantage that its calculation includes each value of the data set. The median is a better measure when a data set includes outliers. The mode is simple to locate, but it is not of much use in practical applications.

3.1.4 RELATIONSHIP BETWEEN THE MEAN, MEDIAN, AND MODE

As discussed in Chapter 2, two of the many shapes that a histogram or a frequency distribution curve can assume are symmetric and skewed. This section describes the relationship between the mean, median, and mode for three such histograms and frequency curves. Knowing the values of the mean, median, and mode can give us some idea about the shape of a frequency curve.

1. For a symmetric histogram and frequency curve with one peak (see Fig. 3.3), the values of the mean, median, and mode are identical and they lie at the center of the distribution.

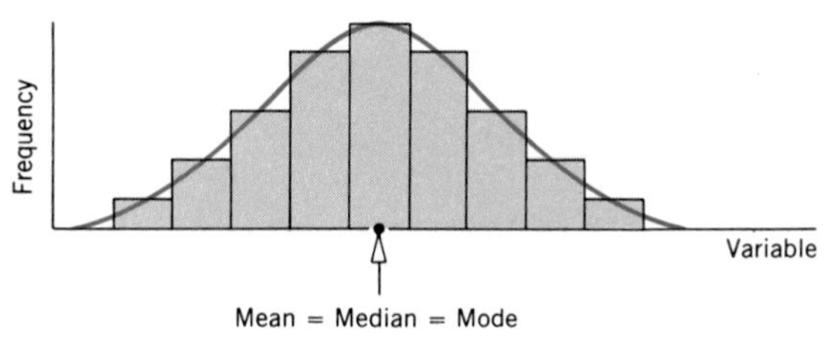

Figure 3.3 Mean, median, and mode for a symmetric histogram and frequency curve.

2. For a histogram and a frequency curve skewed to the right (see Fig. 3.4), the value of the mean is the largest, that of the mode is the smallest, and the value of the median lies between these two. (Notice that mode always occurs at the peak.) The value of the mean is the largest in this case because it is sensitive to outliers that occur in the right tail. These outliers pull the mean to the right.

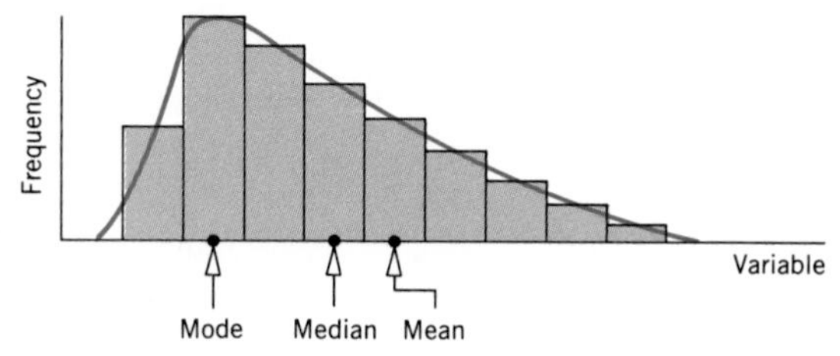

Figure 3.4 Mean, median, and mode for a histogram and a frequency curve skewed to the right.

3. If a histogram and a distribution curve are skewed to the left (see Fig. 3.5), the value of the mean is the smallest and that of the mode is the largest, with the

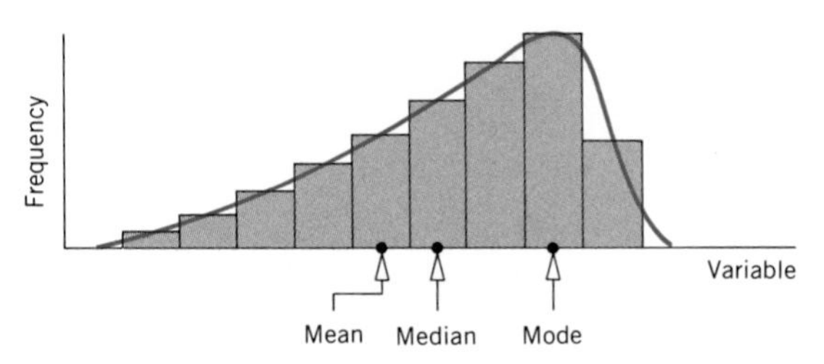

Figure 3.5 Mean, median, and mode for a histogram and a frequency curve skewed to the left.

value of the median lying between these two. In this case, the outliers in the left tail pull the mean to the left.

EXERCISES

Exercises 3.1 through 3.5 are based on the following data.

The following table, based on the American Chamber of Commerce Researchers Association Survey, gives the prices of five items in 12 urban areas across the United States. (See Data Set I of Appendix B.)

City	Apartment Rent	Phone Bill	Beauty Salon	Dry Cleaning	Price of Wine
Huntsville (AL)	422	25.81	15.29	5.20	6.62
Phoenix (AZ)	512	15.65	18.81	7.02	4.41
San Diego (CA)	899	14.26	23.48	5.98	3.83
Denver (CO)	471	17.24	20.18	5.76	4.69
Bloomington (IN)	486	23.24	15.00	5.29	5.00
New Orleans (LA)	339	22.37	22.60	5.03	4.10
Baltimore (MD)	524	23.19	16.20	6.26	5.65
Albuquerque (NM)	480	20.49	17.47	5.47	4.39
Albany (NY)	523	27.30	16.60	5.29	5.91
Charlotte (NC)	426	17.98	20.60	5.52	4.03
Salem (OR)	411	17.82	17.40	6.34	5.03
Charleston (WV)	396	29.43	16.42	5.50	5.92

Explanation of Variables

Apartment Rent Monthly rent of an unfurnished two-bedroom apartment (excluding all utilities except water), 1-½ or 2 baths, approximately 950 square feet

Phone Bill Monthly telephone charges for a private residential line (customer owns instruments)

Beauty Salon Price for woman's shampoo, trim, and blow dry

Dry Cleaning Price of dry cleaning, men's two-piece suit

Price of Wine Price of Paul Masson Chablis, 1.5-liter bottle

3.1 Calculate the mean and median for data on apartment rents.

3.2 Calculate the mean and median for data on phone bills. Do these data have a mode?

3.3 Find the mean and median for data on the price of beauty salon stylings. Do these data have a mode?

3.4 Calculate the mean and median for data on the price of dry cleaning.

3.5 Calculate the mean and median for data on the price of wine.

3.6 The following data give the number of hogs and pigs (in millions) for 1988 for 11 states with the highest number of hogs and pigs. (*Source:* National Agricultural Statistics Service. The data, entered by row, is for the states of Illinois, Indiana, Iowa, Kansas, Michigan, Minnesota, Missouri, Nebraska, North Carolina, Ohio, and South Dakota.)

5.6	4.3	13.9	1.5	1.3	4.7
2.9	4.1	2.7	2.2	1.8	

Calculate the mean and median. Do these data have a mode?

3.7 Following are the temperatures (in Fahrenheit) observed during eight wintry days in a Midwestern city.

23	14	6	−7	−2	9	16	19

Calculate the mean and median. Do these data have a mode?

3.8 The following data give the 1986 production of corn (in millions of bushels) for nine states with highest production of corn. (*Source:* National Agricultural Statistics Service. The data, entered by row, are for the states of Illinois, Indiana, Iowa, Minnesota, Missouri, Nebraska, Ohio, South Dakota, and Wisconsin.)

1404	695	1626	708	281
896	476	234	365	

Calculate the mean and median for corn production for these nine states.

3.9 The following data give the gestation period (in days) of 10 Rhesus monkeys as observed in captivity.

159	167	174	154	181
154	163	169	177	163

Find the mean, median, and mode.

3.10 The following data give the income (in dollars) from interest for 10 households selected from the 1987 Diary Survey mentioned in Chapter 2. (These households were selected only from those who had positive income from interest.)

300	125	800	30,000	15
1,200	600	60	100	1,500

a. Find the mean and median for these data.
b. Do these data contain an outlier? If yes, drop this value and recalculate the mean and median. Which of the two measures changes by a larger amount when you drop the outlier?
c. Is the mean or median a better measure for these data?

3.11 The following data give the balance (in dollars) of savings accounts at the time of the survey for nine households selected from the 1987 Interview Survey mentioned in Chapter 2. (These households were selected only from those who had savings accounts.)

47	85,000	100	2,000	1,500
6	4,000	900	3,000	

a. Calculate the mean and median for these data.
b. Do these data contain an outlier? If yes, drop this value and recalculate the mean

and median. Which of the two measures changes by a larger amount when you drop the outlier?

c. Is the mean or median a better measure for these data?

3.12 The following data give the amount (in dollars) of personal taxes paid during the 12-month period before the survey by seven households selected from the 1987 Interview Survey. (These households were selected only from those that paid taxes.)

3,536	1,125	6,136	18,480
1,506	3,354	920	

a. Find the mean and median for these data.

b. Do these data contain an outlier? If yes, drop this value and recalculate the mean and median. Which of the two measures changes by a larger amount when you drop the outlier?

3.13 The following data give the number of car thefts during the past 12 days that occurred in a city.

6 3 7 11 5 3 8 7 2 6 9 13

Find the mean, median, and mode.

3.14 The following data give the number of hours spent partying by 10 randomly selected college students during the past week.

7 14 5 0 2 7 10 4 0 8

Compute the mean, median, and mode.

***3.15** One property of the mean is that if we know the means and sample sizes of two (or more) data sets, we can compute the *combined mean* of both (or all) data sets. The combined mean for two data sets is calculated by using the formula

$$\text{Combined mean} = \bar{x} = \frac{n_1\bar{x}_1 + n_2\bar{x}_2}{n_1 + n_2}$$

where n_1 and n_2 are the sample sizes of the two data sets and $\bar{x}_1$ and $\bar{x}_2$ are the means of the two data sets respectively.

Suppose a sample of 10 statistics books gave a mean price of $41 and a sample of 8 mathematics books gave a mean price of $43. Find the combined mean. (*Hint:* For this exercise, $n_1 = 10$, $n_2 = 8$, $\bar{x}_1 = \$41$, $\bar{x}_2 = \$43$.)

***3.16** For any data, the sum of all values is equal to the product of the sample size and mean, that is, $\Sigma x = n\bar{x}$. Suppose the average money spent on shopping by 10 persons during a given week is $85.50. Find the total money spent on shopping by these 10 persons.

***3.17** The mean age of six persons is 46. The ages of five of these six persons are 57, 39, 44, 51, and 37. Find the age of the sixth person.

***3.18** Consider the following two data sets:

Data set I:	12	25	37	8	41
Data Set II:	19	32	44	15	48

Notice that each value of the second data set is obtained by adding 7 to the corresponding value of the first data set. Calculate the mean for each of these two data sets. Comment on the relationship between the two means.

***3.19** Consider the following two data sets:

Data Set I:	4	8	15	9	11
Data Set II:	8	16	30	18	22

Notice that each value of the second data set is obtained by multiplying the corresponding value of the first data set by 2. Calculate the mean for each of these two data sets. Comment on the relationship between the two means.

***3.20** *Trimmed mean* is calculated by dropping a certain percentage of values from each end of a ranked data set. Suppose the following data give the ages of 10 employees of a company.

47 53 38 26 39 49 19 67 31 23

To calculate the 10% trimmed mean, first rank these data values in increasing order and then drop 10% of the smallest values and 10% of the largest values. The mean of the remaining 80% of the values will give the 10% trimmed mean. As this data set contains 10 values, 10% of 10 is 1. Hence, drop the smallest value and the largest value from this data set. The mean of the remaining 8 values will be the 10% trimmed mean. Calculate the 10% trimmed mean for this data set.

3.2 MEASURES OF DISPERSION FOR UNGROUPED DATA

The measures of central tendency, such as the mean, median, and mode, do not reveal the whole picture of the distribution of a data set. Two data sets with the same mean may have completely different spreads. The variation among values of observations for one data set may be much larger or smaller than for the other data set. (Notice that the words dispersion, spread, and variation have the same meaning.) Consider the following two data sets on the ages of all workers working for each of the two small companies.

Company 1:	47	38	35	40	36	45	39
Company 2:	70	33	18	52	27		

The mean age of workers of each of these two companies is the same, 40 years. If we do not know the ages of individual workers for these two companies and are told only that the mean age of the workers in both companies is the same, we may deduce that the workers of these two companies have a similar age distribution. But, as we can observe, the variation in the workers' ages in each of these two companies is very different. As illustrated in the diagram, the ages of the workers of the second company have a much larger variation than the ages of the workers of the first company.

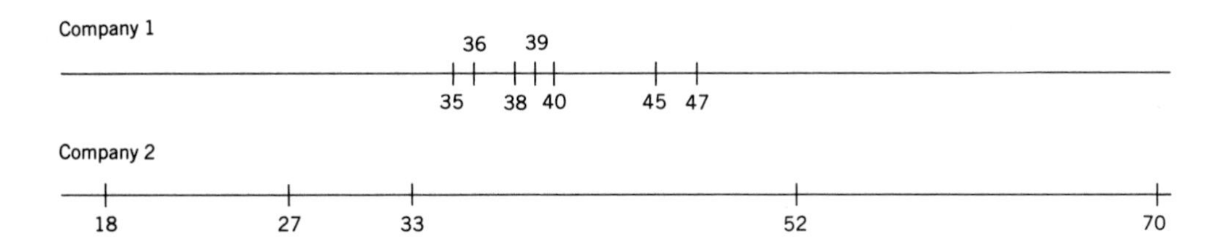

Thus the mean, or median, or mode is usually not by itself a sufficient measure to reveal the shape of the distribution of a data set. We also need a measure that can provide some information about the variation among data values. The measures that help us to know about the spread of a data set are called the **measures of dispersion.** The measures of central tendency and dispersion taken together give us a better picture of a data set than the measures of central tendency alone. This section discusses three measures of dispersion: range, variance, and standard deviation.

3.2.1 RANGE

The **range** is the simplest measure of dispersion to compute. It is calculated by finding the difference between the highest and the lowest values in a data set.

RANGE FOR UNGROUPED DATA

$$\text{Range} = \text{Highest value} - \text{Lowest value}$$

Calculating range for ungrouped data.

EXAMPLE 3-11 The following data give the farm value (in millions of dollars) of the production of wheat for 11 states for the year 1988.†

291	300	240	1147	267	234
259	430	605	305	517	

Find the range.

Solution The highest value in this data set is $1147 million and the lowest value is $234 million. Therefore,

$$\text{Range} = \text{Highest value} - \text{Lowest value} = 1147 - 234 = \$913 \text{ million} \qquad \blacksquare$$

The range, like the mean, has the disadvantage of being influenced by outliers. In Example 3-11, if the state of Kansas with a value of $1147 million is dropped, the range decreases from $913 million to $371 million. Consequently, the range is not a good measure of dispersion to use for a data set that contains outliers.

Another disadvantage of using the range as a measure of dispersion is that its calculation is based on two values only: the highest and the lowest. All other values in a data set are ignored while calculating the range. Thus, the range is not a very satisfactory measure of dispersion.

3.2.2 VARIANCE AND STANDARD DEVIATION

The **standard deviation** is the most used measure of dispersion. The value of the standard deviation tells how closely the values of observations for a data set are clustered around the mean. A lower value of the standard deviation for a data set

†The data, entered by row, are for the following states: Colorado, Idaho, Illinois, Kansas, Missouri, Montana, Nebraska, North Dakota, Oklahoma, Texas, and Washington.

indicates that the values of that data set are spread over a relatively smaller range around the mean. On the other hand, a large value of the standard deviation for a data set indicates that the values of that data set are spread over a relatively larger range around the mean.

The standard deviation is obtained by taking the positive square root of the **variance.** The variance calculated for population data is denoted by σ^2 (pronounced sigma squared),† and the variance calculated for sample data is denoted by s^2. Consequently, the standard deviation calculated for population data is denoted by σ, and the standard deviation calculated for sample data is denoted by s. Following are the *basic formulas* used to compute the variance and standard deviation.‡

$$\sigma^2 = \frac{\Sigma(x - \mu)^2}{N} \quad \text{and} \quad s^2 = \frac{\Sigma(x - \bar{x})^2}{n - 1}$$

where σ^2 is the population variance and s^2 is the sample variance.

$$\text{Standard deviation} = \sqrt{\text{Variance}}$$

The quantity $x - \mu$ or $x - \bar{x}$ is called the *deviation* of x from the mean.

The sum of the deviations of x values from the mean is always zero. That is, $\Sigma(x - \mu) = 0$ and $\Sigma(x - \bar{x}) = 0$. For example, suppose the midterm scores of four students selected from a class are 82, 95, 67, and 92. Then, the mean score for these four students is

$$\bar{x} = \frac{(82 + 95 + 67 + 92)}{4} = 84$$

The deviations of four scores from the mean are calculated in Table 3.1.

Table 3.1

x	$x - \bar{x}$
82	$82 - 84 = -2$
95	$95 - 84 = +11$
67	$67 - 84 = -17$
92	$92 - 84 = +8$
	$\Sigma(x - \bar{x}) = 0$

As we can observe from Table 3.1, the sum of the deviations of x values from the mean is zero, that is, $\Sigma(x - \bar{x}) = 0$. For this reason we square the deviations to calculate the variance and standard deviation.

From the computational point of view, it is easier and more efficient to use the *short-cut formulas* to calculate the variance and standard deviation. Use of the basic formulas for ungrouped data is illustrated in Appendix 3.1, Section A3.1.1, at the

†Σ is uppercase sigma and σ is lowercase sigma of the Greek alphabet.

‡From the basic formula for σ^2, it can be stated that the variance is the mean of the squared deviations of x values from the mean. However, this is true only for the variance calculated for the population data, and not for the variance calculated for the sample data.

end of this chapter. The short-cut formulas to calculate the variance and standard deviation are as follows.

SHORT-CUT FORMULAS FOR THE VARIANCE AND STANDARD DEVIATION FOR UNGROUPED DATA

$$\sigma^2 = \frac{\Sigma x^2 - \frac{(\Sigma x)^2}{N}}{N} \quad \text{and} \quad s^2 = \frac{\Sigma x^2 - \frac{(\Sigma x)^2}{n}}{n - 1}$$

where σ^2 is the population variance and s^2 is the sample variance.

$$\text{Standard deviation} = \sqrt{\text{Variance}}$$

Note that the denominator in the formula for the population variance is N, but in the formula for the sample variance it is $n - 1$.†

Calculating variance and standard deviation: ungrouped data.

EXAMPLE 3-12 Following are the amounts (in dollars) of electric bills for June 1990 for a sample of five families.

72.50 93.20 52.30 85.70 35.60

Find the variance and standard deviation.

Solution The computation of variance involves the following steps.

Step 1. Calculate Σx.

The sum of the entries in the first column of Table 3.2 gives the value of Σx, which is 339.30.

Table 3.2

x	x^2
72.50	5,256.25
93.20	8,686.24
52.30	2,735.29
85.70	7,344.49
35.60	1,267.36
$\Sigma x = 339.30$	$\Sigma x^2 = 25{,}289.63$

†The reason that the denominator in the sample formula is $n - 1$ and not n is the following. The sample variance underestimates the population variance when the denominator in the sample formula for variance is n. However, the sample variance does not underestimate the population variance if the denominator in the sample formula for variance is $n - 1$. In Chapter 8 we will learn that $n - 1$ is called the degrees of freedom.

Step 2. Find Σx^2.

Σx^2 is obtained by squaring each of the x values and then adding the squared values. The results of this step are shown in the second column of Table 3.2. Notice that $\Sigma x^2 = 25{,}289.63$.

Step 3. Compute the variance.

Substitute all the values in the variance formula and simplify. Because the given data belong to a sample, we use the formula for the sample variance.

$$s^2 = \frac{\Sigma x^2 - \frac{(\Sigma x)^2}{n}}{n - 1} = \frac{25{,}289.63 - \frac{(339.30)^2}{5}}{5 - 1} = 566.183$$

Step 4. Obtain the standard deviation.

The standard deviation is obtained by taking the positive square root of the variance.

$$s = \sqrt{566.183} = \$23.79$$

■

TWO OBSERVATIONS

1. **The values of the variance and standard deviation are never negative.** That is, the numerator in the formula for the variance should never give a negative value. Usually the variance and standard deviation are positive, but if the data set has no variation then the variance and standard deviation are both zero. For example, if four persons in a group are of the same age, say 35 years, then the four values in the data set are

   35 35 35 35

   If we calculate the variance and standard deviation for these data, their values will be zero. This will be so because there is no variation in the values of this data set.

2. **The measurement units of the variance are always the square of the measurement units of the original data** because the original values are squared to compute the variance. In Example 3-12, the measurement units of the original data are dollars. However, the measurement units of the variance are squared dollars, which, of course, does not make any sense. But the measurement units of the standard deviation are the same as the measurement units of the original data because the standard deviation is equal to the square root of the variance.

Calculating variance and standard deviation: ungrouped data.

EXAMPLE 3-13 The following data give the time (in minutes) taken by Jane to travel from her home to school for a sample of six days.

21 18 23 21 17 20

Calculate the variance and standard deviation.

Solution By adding all the values of x in Table 3.3, we obtain $\Sigma x = 120$. The value of Σx^2, as shown in Table 3.3, is 2424.

Table 3.3

x	x^2
21	441
18	324
23	529
21	441
17	289
20	400
$\Sigma x = 120$	$\Sigma x^2 = 2424$

Since the data are based on a sample of six days, we will use the sample formula to compute the variance. Thus, the variance is

$$s^2 = \frac{\Sigma x^2 - \frac{(\Sigma x)^2}{n}}{n - 1} = \frac{2424 - \frac{(120)^2}{6}}{6 - 1} = 4.80$$

The standard deviation is obtained by taking the (positive) square root of the variance.

$$s = \sqrt{4.80} = 2.19 \text{ minutes}$$

■

WARNING

Note that Σx^2 is not the same as $(\Sigma x)^2$. The value of Σx^2 is obtained by squaring the x values and adding them. The value of $(\Sigma x)^2$ is obtained by squaring the value of Σx.

The uses of the standard deviation are discussed in Section 3.4. Later chapters will explain how the mean and the standard deviation taken together can help in making inferences about population.

Case Study 3-3 illustrates an application of the mean and standard deviation.

CASE STUDY 3-3 DOES NECKWEAR TIGHTNESS AFFECT VISUAL PERFORMANCE?

Leonora M. Langan and Susan M. Watkins studied the tightness of neckwear for men and its effect on their visual performance. Some earlier studies stated that a shirt and tie impede the blood flow to the brain by pressing against the carotid arteries of the neck and that a tight collar may slow the pulse and lower blood pressure. Langan and Watkins defined the neckwear as "tight" if the size of the neck with buttoned shirt and knotted tie was smaller than the size of the unrestricted neck.

The authors conducted a visual discrimination test on 22 men using a tie and the shirt with the tightest collar they owned. They gave three basic tests of what is known as the critical flicker frequency (CFF) test to each man; each basic test was repeated three times. The first set of tests was given with loose tie and unbuttoned shirt collar. The second set of tests was performed with buttoned shirt collar and secured tie. The third test was given with loose tie and unbuttoned shirt collar again. Each test was comprised of looking at a pencil mark 2.54 cm in front of a lamp light on a table. The man was asked to press the response button as soon as he saw the light become solid and nonflickering. At that moment the equipment was reset in a decreasing frequency mode. The man was asked to press the response button again when the solid light seemed to flicker. The average of the two frequencies was counted as one test. The means and standard deviations of the flicker frequencies (measured in Hertz) for the three sets of tests for the 22 men are given in the table.

Test Condition	Mean (Hertz)	Standard Deviation (Hertz)
Set 1 (no neckwear pressure)	20.406	1.581
Set 2 (neckwear pressure)	19.571	1.415
Set 3 (pressure removed)	19.616	1.446

Based on this study, the authors concluded that tight neckwear can lower the visual performance of a man as measured by the CFF test. (In the table, the smaller value of the mean indicates a lower visual performance.) They also concluded that "Visual performance does not return to normal immediately after tight neckwear is removed." This research, according to the authors, holds implications and critical information for working men such as computer operators and others where precise vision is important to the work and "for bus drivers, pilots, and others whose vision is important for public safety."

Source: Leonora M. Langan and Susan M. Watkins, "Pressure of Menswear on the Neck in Relation to Visual Performance," *Human Factors* 29(1): 1987, p. 67–71.

3.2.3 POPULATION PARAMETERS AND SAMPLE STATISTICS

Numerical measures such as the mean, median, mode, range, variance, and standard deviation calculated for population data are called *population parameters* or simply **parameters.** The measures calculated for sample data are called *sample statistics* or simply **statistics.** Thus, μ and σ are population parameters and $\bar{x}$ and s are sample statistics. As an illustration, in Example 3-2 the population mean μ, which is equal to 200,417 marriages, is a population parameter. However, the sample mean $\bar{x}$, which is equal to 220,667 marriages, is a sample statistic.

EXERCISES

3.21 Range as a measure of spread has the disadvantage of being influenced by outliers. Illustrate this point with an example.

3.22 Can the value of the standard deviation be negative? When is the value of the standard deviation for a data set zero? Give one example. Calculate the standard deviation for this example and show that its value is zero.

3.23 The following data give the weekly food expenditures for a sample of five families.

$82 116 65 75 92

Find the mean for these data. Calculate the deviations of the data values from the mean. Is the sum of these deviations zero?

3.24 A sample of seven statistics books produced the following data on their prices.

$35 43 38 45 44 43 40

Calculate the range, variance, and standard deviation.

3.25 The following data give the 1987 production of strawberries (in hundred thousands of pounds) for eight states (*Source:* National Agricultural Statistics Service). The data, entered in that order, are for the following states: Arkansas, Louisiana, Michigan, New Jersey, North Carolina, Oregon, Washington, and Wisconsin.)

11 43 92 42 46 56 25 44

Calculate the range, variance, and standard deviation.

3.26 A sample of 10 professors were asked how many cups of coffee they drink each day on the average. The following are their responses.

3 8 2 7 9 5 4 2 1 5

Calculate the range, variance, and standard deviation.

3.27 The following data give the number of cars that stopped at a service station during each of the 10 hours observed.

29 35 42 31 24 18 16 27 39 34

Find the range, variance, and standard deviation.

3.28 The following data give the weight (in pounds) lost by 15 new members of a health club at the end of their first two months of membership.

5 10 8 7 25 12 5 14
11 10 21 9 8 11 18

Find the range, variance, and standard deviation.

3.29 The following data give the speeds (in miles), as measured by radar, of 13 cars traveling on interstate highway I-84.

67 72 63 66 76 69 71
76 65 79 68 67 71

Find the range, variance, and standard deviation.

3.30 The following data give the weekly expenditures (rounded to the nearest dollar) on alcoholic beverages for seven households selected from the 1987 Diary Survey. (These households were selected only from those that incurred such expenses.)

7 26 20 52 4 8 6

Calculate the range, variance, and standard deviation.

3.31 The following data give the religious contributions (in hundreds of dollars) made during the 12-month period before the survey by 10 households selected from the 1987 Interview Survey. (These households were selected only from those that made such contributions.)

8 20 7 3 2 31 10 6 3 6

Calculate the range, variance, and standard deviation.

3.32 The following data give the hourly wage for eight employees of a company.

$12 12 12 12 12 12 12 12

Find the standard deviation. Is its value zero? If yes, why?

3.33 The following data give the annual income after taxes (in thousands of dollars) for 11 households selected from the 1987 Interview Survey.

12 19 25 4 34 19 47 16 40 48 28

Determine the range, variance, and standard deviation.

***3.34** Consider the following two data sets:

Data Set I:	12	25	37	8	41
Data Set II:	19	32	44	15	48

Note that each value of the second data set is obtained by adding 7 to the corresponding value of the first data set. Calculate the standard deviation for each of these two data sets. Comment on the relationship between the two standard deviations.

***3.35** Consider the following two data sets:

Data Set I:	4	8	15	9	11
Data Set II:	8	16	30	18	22

Note that each value of the second data set is obtained by multiplying the corresponding value of the first data set by 2. Calculate the standard deviation for each of these two data sets. Comment on the relationship between the two standard deviations.

3.3 MEAN, VARIANCE, AND STANDARD DEVIATION FOR GROUPED DATA

In Sections 3.1.1 and 3.2.2, we learned how to calculate the mean, variance, and standard deviation for ungrouped data. In this section we will learn how to calculate the mean, variance, and standard deviation for grouped data. We know from Chapter 2 that grouped data are the data presented in the form of a frequency distribution table.

3.3.1 MEAN FOR GROUPED DATA

We learned in Section 3.1.1 that the mean is obtained by dividing the sum of all values in a data set by the number of values. However, if the data are given in the form of a frequency distribution table, we will no longer know the values of individual observations. Consequently, in such cases, we cannot obtain the sum of individual values. We find an approximation for this sum using the procedure explained in the next paragraph and example. The formulas used to compute the mean for grouped data are as follows.

MEAN FOR GROUPED DATA

Mean for population data: $$\mu = \frac{\Sigma mf}{N}$$

Mean for sample data: $$\bar{x} = \frac{\Sigma mf}{n}$$

where m is the midpoint and f is the frequency of a class.

To calculate the mean for grouped data, first find the midpoint of each class and then multiply the midpoints by the frequencies of the corresponding classes. The sum of these products, denoted by Σmf, gives an approximation for the sum of all values. To find the value of the mean, divide this sum by the total number of observations in the data set.

Calculating population mean for grouped data.

EXAMPLE 3-14 The following table gives the frequency distribution of time (in minutes) taken to complete a statistics test by all 50 students in a class.

Time (minutes)	f
30 to less than 34	5
34 to less than 38	10
38 to less than 42	16
42 to less than 46	15
46 to less than 50	4

Calculate the mean.

Solution Note that because the data set includes all 50 students of a class, it represents the population. In Table 3.4, m denotes the midpoints of classes.

Table 3.4

Time	f	m	mf
30 to less than 34	5	32	160
34 to less than 38	10	36	360
38 to less than 42	16	40	640
42 to less than 46	15	44	660
46 to less than 50	4	48	192
	$N = 50$		$\Sigma mf = 2012$

To calculate the mean, we first find the midpoint of each class. The class midpoints are recorded in the third column of Table 3.4. The products of the midpoints and the corresponding frequencies are listed in the fourth column of that table. The sum of the fourth column, denoted by Σmf, gives the approximate total time taken by 50 students to complete the test. We obtain the mean by dividing this sum by the total frequency. Thus,

$$\mu = \frac{\Sigma mf}{N} = \frac{2012}{50} = 40.24 \text{ minutes}$$

■

What do the numbers 160, 360, 640, 660, and 192 in the column labeled mf in Table 3.4 represent? We know from the table that five students completed this test in 30 to less than 34 minutes. Assuming that these five students are equally spread in the interval 30 to less than 34, the midpoint of this class (which is 32) gives the mean time taken to complete the test by these five students. Hence, $5(32) = 160$ is the approximate total time taken by these five students to complete the test. Similarly, 10 students who completed the test in 34 to less than 38 minutes took a total of 360 minutes to complete the test. The other numbers in this column can be interpreted in the same way. Note that these numbers give the approximate times taken to complete the test based on the assumption of equal spread within classes. The total time taken by 50 students is approximately 2012 minutes. Consequently, 40.24 is an approximation and not the exact value of the mean. We can compute the exact value of the mean only if we know the exact time taken by each of the 50 students to complete this test.

Calculating sample mean for grouped data.

EXAMPLE 3-15 The following table gives the frequency distribution of the length of time 50 employees have been with their current employers.

Length of Time (years)	Number of Employees
0 to less than 4	18
4 to less than 8	14
8 to less than 12	9
12 to less than 16	5
16 to less than 20	4

Calculate the mean.

Solution Because the data set includes only 50 employees, it represents a sample. Note that in Table 3.5, m denotes the midpoints of classes.

Table 3.5

Length of Time	f	m	mf
0 to less than 4	18	2	36
4 to less than 8	14	6	84
8 to less than 12	9	10	90
12 to less than 16	5	14	70
16 to less than 20	4	18	72
	$n = 50$		$\Sigma mf = 352$

The value of the mean is

$$\bar{x} = \frac{\Sigma mf}{n} = \frac{352}{50} = 7.04 \text{ years}$$

Thus, on average these employees have been with their current employers for 7.04 years. ■

3.3.2 VARIANCE AND STANDARD DEVIATION FOR GROUPED DATA

Following are the *basic formulas* to compute the population and sample variances for grouped data. In either case, the standard deviation is obtained by taking the positive square root of the variance.

$$\sigma^2 = \frac{\Sigma f(m - \mu)^2}{N} \quad \text{and} \quad s^2 = \frac{\Sigma f(m - \bar{x})^2}{n - 1}$$

where σ^2 is the population variance, s^2 is the sample variance, and m is the midpoint of a class.

$$\text{Standard deviation} = \sqrt{\text{Variance}}$$

Again the *short-cut formulas* are more efficient for calculating the variance and standard deviation. Use of the basic formulas is illustrated in Appendix 3.1 at the end of this chapter. Section A3.1.2 of that appendix illustrates the basic formulas for grouped data.

SHORT-CUT FORMULAS FOR THE VARIANCE AND STANDARD DEVIATION FOR GROUPED DATA

$$\sigma^2 = \frac{\Sigma m^2 f - \frac{(\Sigma mf)^2}{N}}{N} \quad \text{and} \quad s^2 = \frac{\Sigma m^2 f - \frac{(\Sigma mf)^2}{n}}{n - 1}$$

where σ^2 is the population variance, s^2 is the sample variance, and m is the midpoint of a class.

$$\text{Standard deviation} = \sqrt{\text{Variance}}$$

The following examples illustrate the use of these formulas to calculate the variance and standard deviation.

Calculating sample variance and standard deviation for grouped data.

EXAMPLE 3-16 The following table, which gives the frequency distribution for the normal annual precipitation (in inches) for 29 cities of the United States, is reproduced from Chapter 2 (Table 2.11).

Normal Annual Precipitation	f
0 to less than 12	5
12 to less than 24	6
24 to less than 36	8
36 to less than 48	6
48 to less than 60	4

Find the variance and standard deviation.

Solution The steps needed to calculate the variance and standard deviation are as follows.

Step 1. Calculate the value of Σmf.

To calculate the value of Σmf, find the midpoint m of each class (see the third column in Table 3.6), multiply the corresponding class midpoints and class frequencies (see the fourth column in Table 3.6). The value of Σmf is obtained by adding these products. Thus,

$$\Sigma mf = 846$$

Table 3.6

Normal Annual Precipitation	f	m	mf	m^2f
0 to less than 12	5	6	30	180
12 to less than 24	6	18	108	1,944
24 to less than 36	8	30	240	7,200
36 to less than 48	6	42	252	10,584
48 to less than 60	4	54	216	11,664
	$n = 29$		$\Sigma mf = 846$	$\Sigma m^2f = 31{,}572$

Step 2. Compute $\Sigma m^2 f$.

To compute $\Sigma m^2 f$, square each m value and multiply this squared value of m by the corresponding frequency (see the fifth column in Table 3.6). The sum of these products (i.e., the sum of the fifth column in Table 3.6) gives $\Sigma m^2 f = 31{,}572$.

Step 3. Compute the variance.

Because the data set is comprised of a sample of 29 cities, we will use the formula for the sample variance. Thus,

$$s^2 = \frac{\Sigma m^2 f - \dfrac{(\Sigma mf)^2}{n}}{n - 1} = \frac{31{,}572 - \dfrac{(846)^2}{29}}{29 - 1} = 246.148$$

Step 4. To obtain the standard deviation, take the square root of the variance. Thus,

$$s = \sqrt{246.148} = 15.69 \text{ inches}$$

■

Note that the values of the variance and standard deviation calculated in Example 3-16 from the grouped data are approximations. The exact values of the variance and standard deviation can be obtained only by using the 29 values for 29 cities given in Example 2–5 of Chapter 2. (See Computer Assignment M3.3 on page 156.)

Calculating population variance and standard deviation for grouped data.

EXAMPLE 3-17 The following table gives the frequency distribution of the number of visits that all 200 patients of a physician made to his office during the past year.

Visits	f
0–2	60
3–5	56
6–8	32
9–11	28
12–14	24

Calculate the variance and standard deviation.

Solution Note that the data set concerns all the patients of the given physician and, hence, constitute a population. All the information required for the calculation of the variance and standard deviation appears in Table 3.7.

Table 3.7

Visits	f	m	mf	m^2f
0–2	60	1	60	60
3–5	56	4	224	896
6–8	32	7	224	1568
9–11	28	10	280	2800
12–14	24	13	312	4056
	$N = 200$		$\Sigma mf = 1100$	$\Sigma m^2f = 9380$

From Table 3.7,

$$\Sigma mf = 1100 \quad \text{and} \quad \Sigma m^2f = 9380$$

By substituting the values in the formula for the population variance, we obtain

$$\sigma^2 = \frac{\Sigma m^2 f - \frac{(\Sigma mf)^2}{N}}{N} = \frac{9380 - \frac{(1100)^2}{200}}{200} = 16.65$$

Hence, the standard deviation is

$$\sigma = \sqrt{16.65} = 4.08 \text{ visits}$$

Thus, the standard deviation of visits that patients made to this physician's office during the past year is 4.08. ■

EXERCISES

3.36 Compute the mean, variance, and standard deviation for the following grouped data.

x	2–4	5–7	8–10	11–13	14–16
f	5	9	14	7	5

3.37 Find the mean, variance, and standard deviation for the grouped data recorded in the following table.

x	0–3	4–7	8–11	12–15	16–19
f	7	4	19	12	8

3.38 Compute the mean, variance, and standard deviation for the following grouped data.

x	f
1 to less than 5	16
5 to less than 9	27
9 to less than 13	38
13 to less than 17	14
17 to less than 21	5

3.39 Calculate the mean, variance, and standard deviation for the grouped data given in the following table.

x	f
0 to less than 4	17
4 to less than 8	23
8 to less than 12	15
12 to less than 16	11
16 to less than 20	8
20 to less than 24	6

3.40 The following table gives the grouped data on the weights of 100 newborn babies.

Weight (pounds)	Number of Babies
3 to less than 5	5
5 to less than 7	30
7 to less than 9	40
9 to less than 11	20
11 to less than 13	5

Calculate the mean, variance, and standard deviation.

3.41 The following table gives the frequency distribution of entertainment expenditures (in dollars) incurred by 50 families during the past week.

Entertainment Expenditure	Number of Families
0 to less than 10	5
10 to less than 20	10
20 to less than 30	15
30 to less than 40	12
40 to less than 50	5
50 to less than 60	3

Calculate the mean, variance, and standard deviation.

3.42 The time taken to serve each of a sample of 100 customers at a bank was observed. The following table gives the frequency distribution of service time for these 100 customers.

Service Time (minutes)	Number of Customers
0 to less than 2	18
2 to less than 4	30
4 to less than 6	24
6 to less than 8	16
8 to less than 10	8
10 to less than 12	4

Calculate the mean, variance, and standard deviation.

3.43 The following table gives the frequency distribution of total hours spent studying statistics during the semester for a sample of 40 university students enrolled in an introductory statistics course during Spring 1990.

Hours of Study	Number of Students
24 to less than 40	3
40 to less than 56	5
56 to less than 72	10
72 to less than 88	12
88 to less than 104	5
104 to less than 120	5

Find the mean, variance, and standard deviation.

3.44 The following table gives the frequency distribution of the number of personal computers sold during the past month at 40 computer stores located in New York City.

Computers Sold	Number of Stores
4–12	6
13–21	9
22–30	14
31–39	7
40–48	4

Find the mean, variance, and standard deviation.

3.45 The following table gives the frequency distribution of the number of patients that arrived at the emergency ward of a Los Angeles hospital during each of the past 60 hours.

Number of Patients	Number of Hours
2–6	11
7–11	23
12–16	13
17–21	7
22–26	6

Find the mean, variance, and standard deviation.

3.46 The following table gives the information on the amount (in dollars) of electric bills for March 1990 for a sample of 30 families.

Amount of Electric Bill (dollars)	Number of Families
0 to less than 20	4
20 to less than 40	7
40 to less than 60	9
60 to less than 80	6
80 to less than 100	4

Calculate the mean, variance, and standard deviation.

3.47 For 50 airplanes that arrived late at an airport during a week, the time by which they were late was observed. In the following table, x denotes the time (in minutes) by which an airplane was late and f denotes the number of airplanes.

x	f
0 to less than 20	14
20 to less than 40	18
40 to less than 60	9
60 to less than 80	5
80 to less than 100	4

Calculate the mean, variance, and standard deviation.

3.48 The following table gives the frequency distribution of the amount of snowfall for January 1990 for 40 cities.

Snowfall (inches)	Number of Cities
0 to less than 4	5
4 to less than 8	6
8 to less than 12	8
12 to less than 16	10
16 to less than 20	8
20 to less than 24	3

Calculate the mean, variance, and standard deviation.

3.49 The following table gives the frequency distribution of the number of cars owned by 100 households.

Number of Cars Owned	Number of Households
0	12
1	40
2	30
3	15
4	3

Find the mean, variance, and standard deviation. [*Hint:* The classes, in this example, are single-valued. The class values (the number of cars owned) will be used as values of m in the formula for the mean, variance, and standard deviation.]

3.50 The following table gives the number of television sets owned by 80 households.

Number of Television Sets Owned	Number of Households
0	4
1	33
2	28
3	10
4	5

Find the mean, variance, and standard deviation. [*Hint:* The classes, in this example, are single-valued. The class values (the number of television sets owned) will be used as values of m in the formula for the mean, variance, and standard deviation.]

3.4 USE OF STANDARD DEVIATION

By using the mean and standard deviation, we can find the proportion or percentage† of the total observations that fall within a given interval about the mean. This section will briefly discuss Chebyshev's theorem and the empirical rule, which demonstrate the use of standard deviation.

3.4.1 CHEBYSHEV'S THEOREM

Chebyshev's theorem gives the area under a curve between two points that are on opposite sides of the mean and at the same distance from the mean.

CHEBYSHEV'S THEOREM

For any number k greater than 1, at least $(1 - 1/k^2)$ of the data values lie within k standard deviations of the mean.

Figure 3.6 illustrates Chebyshev's theorem.

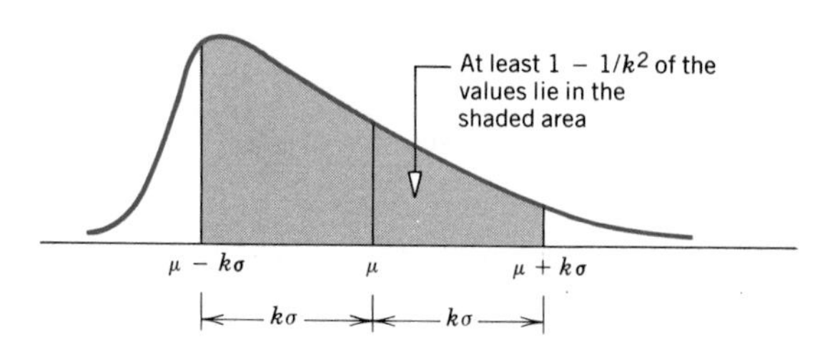

Figure 3.6 Chebyshev's theorem.

†Proportion refers to the relative frequency of a category or class and is expressed as a decimal. Percentage is the relative frequency multiplied by 100.

Thus, if $k = 2$, then

$$1 - \frac{1}{k^2} = 1 - \frac{1}{(2)^2} = 1 - \frac{1}{4} = 1 - .25 = .75 \text{ or } 75\%$$

Therefore, according to Chebyshev's theorem, at least .75 or 75% of the values for a data set lie within two standard deviations of the mean. This is shown in Figure 3.7.

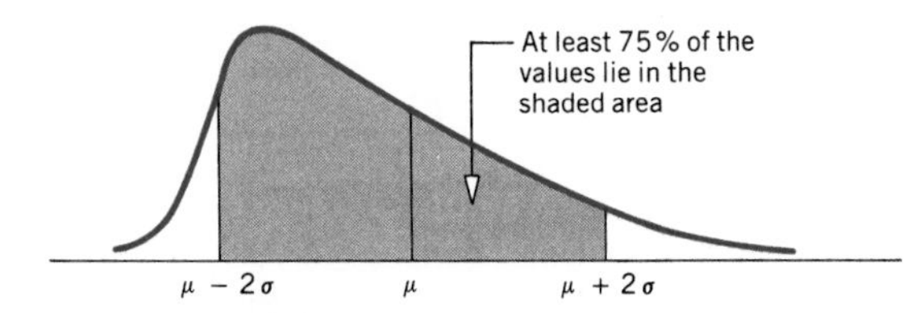

Figure 3.7 Percentage of values within two standard deviations of the mean for Chebyshev's theorem.

If $k = 3$, then,

$$1 - \frac{1}{k^2} = 1 - \frac{1}{(3)^2} = 1 - \frac{1}{9} = 1 - .11 = .89 \text{ or } 89\% \text{ approximately}$$

Therefore, according to Chebyshev's theorem, at least .89 or 89% of the values fall within three standard deviations of the mean. This is shown in Figure 3.8.

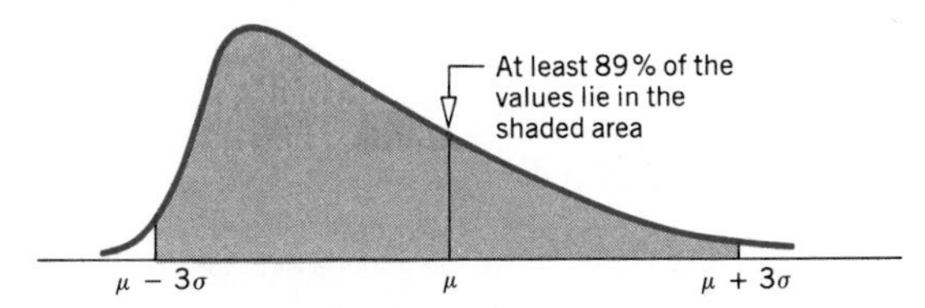

Figure 3.8 Percentage of values within three standard deviations of the mean for Chebyshev's theorem.

Although in Figures 3.6 through 3.8 we have used the population notation for the mean and standard deviation, the theorem applies to both sample and population data. Note that Chebyshev's theorem is applicable to a distribution of any shape. However, Chebyshev's theorem is used only for $k > 1$. This is so because when $k = 1$ the value of $1 - 1/k^2$ is zero, and when $k < 1$ the value of $1 - 1/k^2$ is negative.

Applying Chebyshev's theorem.

EXAMPLE 3-18 For a statistics class, the mean for the midterm scores is 75 and the standard deviation is 8. Using Chebyshev's theorem, find the percentage of students who scored between 59 and 91.

Solution We are given that $\mu = 75$ and $\sigma = 8$ for the midterm scores. To find the percentage of students who scored between 59 and 91, first we need to find k. Each of the two points, 59 and 91, is 16 units away from the mean.

$\leftarrow 59 - 75 = -16 \rightarrow \mid \leftarrow 91 - 75 = 16 \rightarrow$

59 75 91

We obtain the value of k by dividing this distance between the mean and each point by the standard deviation. Thus,

$$k = 16/8 = 2$$

$$1 - \frac{1}{k^2} = 1 - \frac{1}{(2)^2} = 1 - \frac{1}{4} = 1 - .25 = .75 \text{ or } 75\%$$

Hence, according to Chebyshev's theorem, at least 75% of the students scored between 59 and 91. This is shown in Figure 3.9.

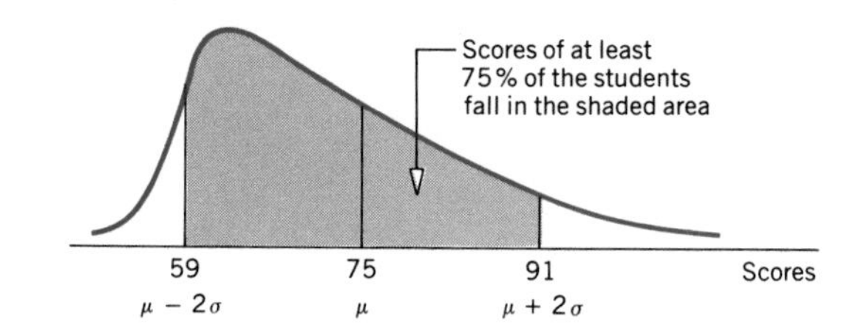

Figure 3.9 Percentage of students who scored between 59 and 91.

■

3.4.2 EMPIRICAL RULE

Whereas Chebyshev's theorem is applicable to any kind of distribution, the **empirical rule** applies only to a specific type of distribution called a *bell-shaped distribution* as shown in Figure 3.10. More will be said about such a distribution in Chapter 6, where it will be called a *normal curve*. In this section, only the following three rules for such a curve are given.

> **EMPIRICAL RULE**
>
> For a bell-shaped distribution, approximately
>
> **1.** 68% of the observations lie within one standard deviation of the mean
> **2.** 95% of the observations lie within two standard deviations of the mean
> **3.** 99.7% of the observations lie within three standard deviations of the mean

Figure 3.10 illustrates the empirical rule. Again, the empirical rule applies to population data as well as sample data.

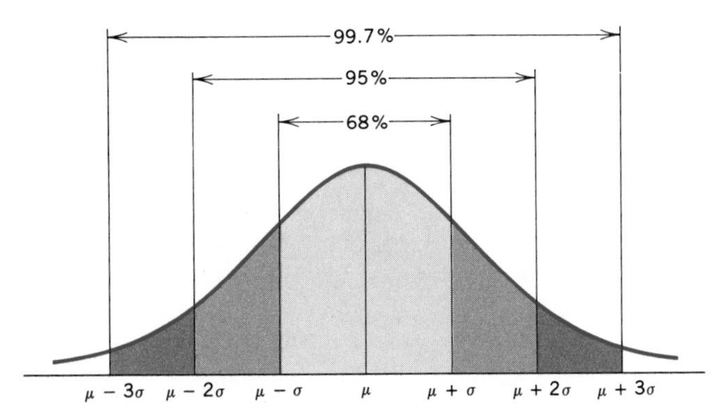

Figure 3.10 Illustration of the empirical rule.

Applying the empirical rule.

EXAMPLE 3-19 The age distribution of a sample of 5000 persons is bell-shaped with a mean of 40 years and a standard deviation of 12 years. Determine the approximate percentage of people who are 16 to 64 years old.

Solution We will use the empirical rule to find the required percentage because the distribution of ages follows a bell-shaped distribution. For this distribution

$$\bar{x} = 40 \text{ years} \quad \text{and} \quad s = 12 \text{ years}$$

Each of the two points, 16 and 64, is 24 units away from the mean. Dividing 24 by 12 we convert the distance between each of the two points and the mean in terms of standard deviation. Thus, the distance between both 16 and 40 and 40 and 64 is $2s$. Consequently, as shown in Figure 3.11, the area from 16 to 64 is the area from $\bar{x} - 2s$ to $\bar{x} + 2s$.

As the area within two standard deviations of the mean is approximately 95% for a bell-shaped curve, approximately 95% of the people in the sample are 16 to 64 years old.

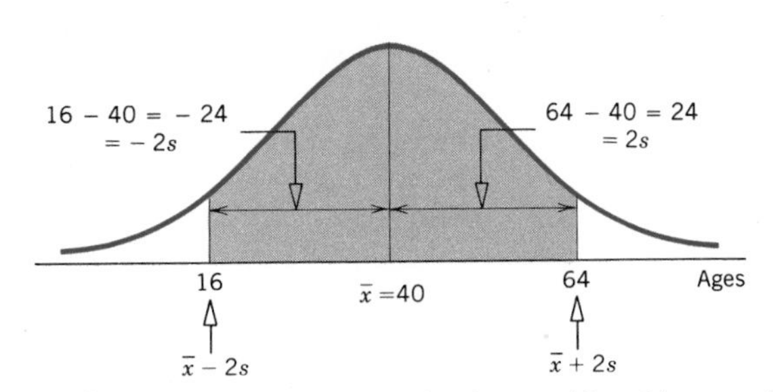

Figure 3.11 Percentage of people who are 16 to 64 years old.

■

EXERCISES

3.51 The mean time taken by all participants to run a marathon was found to be 220 minutes with a standard deviation of 20 minutes. Using Chebyshev's theorem, find the percentage of runners who ran this marathon in

a. 180 to 260 minutes **b.** 160 to 280 minutes **c.** 170 to 270 minutes

3.52 The 1989 gross sales of all firms in a large city have a mean of $1.8 million and a standard deviation of $.4 million. Using Chebyshev's theorem, find the percentage of firms with gross sales for 1989 between

a. $1 and $2.6 million **b.** $.6 and $3 million **c.** $1.2 and $2.4 million

3.53 According to a 1976–80 survey conducted by the U.S. National Center for Health Statistics, the mean height of U.S. men aged 18 through 74 was 69.1 inches. Assume that the current mean height of all such men is 69.1 inches and the standard deviation of their heights is 2.2 inches. Using Chebyshev's theorem, find the percentage of 18- through 74-year-old men who are

a. 64.7 to 73.5 inches tall **b.** 62.5 to 75.7 inches tall

3.54 According to the survey referred to in Exercise 3.53, the mean weight of U.S. men aged 18 through 74 was 172.2 pounds. Assume that the current mean weight of all such men is 172.2 pounds and the standard deviation of their weights is 22 pounds. Using Chebyshev's theorem, find the percentage of 18- through 74-year-old men who weigh between

a. 128.2 and 216.2 pounds **b.** 106.2 and 238.2 pounds

3.55 The mean life of a given brand of auto batteries is 44 months with a standard deviation of 3 months. Assume that the lives of all batteries of this brand have a bell-shaped distribution. Using empirical rule, find the percentage of batteries of this brand that will have a life of

a. 41 to 47 months **b.** 38 to 50 months **c.** 35 to 53 months

3.56 According to the American Federation of Teachers, the mean salary of public school teachers was $31,315 in 1989–90. Assume that the salaries of all public school teachers follow a bell-shaped curve with a mean of $31,315 and a standard deviation of $3,400. Using empirical rule, find the percentage of public school teachers whose 1989–90 salaries were between

a. $27,915 and $34,715 **b.** $24,515 and $38,115 **c.** $21,115 and $41,515

3.57 According to the Metropolitan Life Insurance Company's claims data for 1986, the average hospital and physician's charges for coronary bypass surgeries were $30,430 (*Statistical Bulletin*, 70(1), January–March 1989). Assume that the hospital and physician's charges for all coronary bypass surgeries for 1986 follow a bell-shaped distribution with a mean of $30,430 and a standard deviation of $4,300. Using empirical rule, find the percentage of 1986 coronary bypass surgeries for which such charges were between

a. $26,130 and $34,730 **b.** $21,830 and $39,030 **c.** $17,530 and $43,330

3.58 The ages of cars owned by all employees of a large company follow a bell-shaped distribution with a mean of 7 years and a standard deviation of 2 years. Using empirical rule, find the percentage of cars owned by these employees that are

a. 5- to 9-year-old **b.** 3- to 11-year-old **c.** 1- to 13-year-old

3.5 MEASURES OF POSITION

A **measure of position** determines the position of a single value in relation to other values in a sample or a population data set. There are many measures of position. However, only quartiles and percentiles are discussed in this section.

3.5.1 QUARTILES

Quartiles are measures that divide a ranked data set into four equal parts. Three measures will divide any data set into four equal parts. These three measures are the **first quartile** (denoted by Q_1), the **second quartile** (denoted by Q_2), and the **third**

quartile (denoted by Q_3). The data should be ranked in increasing order before the quartiles are determined. The three quartiles are defined as follows.

QUARTILES

Quartiles are three measures that divide ranked data into four equal parts. The second quartile is the same as the median of a data set. The first quartile is the value of the middle term among the observations that are less than the median, and the third quartile is the value of the middle term among the observations that are greater than the median.

Figure 3.12 describes the positions of the three quartiles.

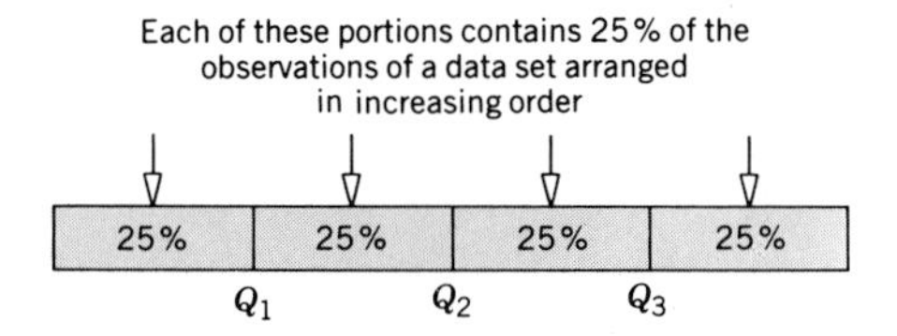

Figure 3.12 Quartiles.

Approximately 25% of the values in a ranked data set are less than Q_1 and about 75% are greater than Q_1. The second quartile Q_2 divides a ranked data set into two equal parts; hence, the second quartile and the median are the same. Approximately 75% of the data values are less than Q_3 and about 25% are greater than Q_3.

Finding quartiles for even number of data values.

EXAMPLE 3-20 The following are the scores of 12 students in a mathematics class.

75 80 68 53 99 58 76 73 85 88 91 79

Find the values of the three quartiles.

Solution First, we rank the given scores in increasing order. Then we calculate the three quartiles as follows.

Values less than the median: 53 58 68 73 75 76

Values greater than the median: 79 80 85 88 91 99

$$Q_1 = \frac{68 + 73}{2} = 70.5 \qquad Q_2 = \frac{76 + 79}{2} = 77.5 \qquad Q_3 = \frac{85 + 88}{2} = 86.5$$

($Q_2 = 77.5$ ← Also the median)

The value of Q_2, which is also the median, is given by the value of the middle term in the ranked data. In the data of this example, this value is given by the average of the sixth and seventh terms. Consequently, Q_2 is 77.5. The value of Q_1 is given by the value of the middle term of six values that fall below the median (or Q_2). Thus, it is obtained by taking the average of the third and fourth terms. So, Q_1 is 70.5. The value of Q_3 is given by the value of the middle term of the six values that fall above the median. In the data of this example, Q_3 is obtained by taking the average of the ninth and tenth terms, and it is 86.5. ■

Finding quartiles for odd number of data values.

EXAMPLE 3-21 The following are the ages of nine employees of an insurance company.

47 28 39 51 33 37 59 24 33

Find the values of the three quartiles.

Solution First we rank the given data in increasing order. Then we calculate the three quartiles as follows.

Values less than the median: 24 28 33 33 — 37 — Values greater than the median: 39 47 51 59

$$Q_1 = \frac{28 + 33}{2} = 30.5 \qquad Q_2 = 37 \qquad Q_3 = \frac{47 + 51}{2} = 49$$

($Q_2 = 37$ — Also the median)

Thus, the values of the three quartiles are

$$Q_1 = 30.5, \quad Q_2 = 37, \quad \text{and} \quad Q_3 = 49$$ ■

3.5.2 PERCENTILES

Percentiles are measures that divide a ranked data set into 100 equal parts. Each data set has 99 percentiles that divide a ranked data set into 100 equal parts. The data should be ranked in increasing order to compute percentiles. The kth percentile is denoted by P_k, where k is an integer in the range 1 to 99. For instance, the 25th percentile is denoted by P_{25}. Figure 3.13 shows the positions of the 99 percentiles.

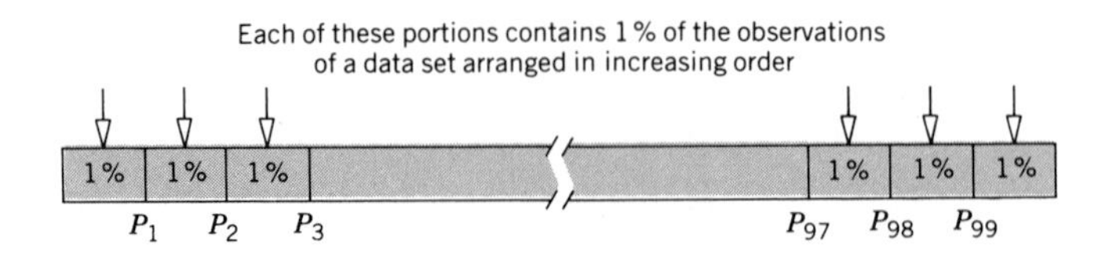

Thus, the kth percentile, P_k, can be defined as a value in a data set such that about $k\%$ of the measurements are smaller than P_k and about $(100 - k)\%$ of the measurements are greater than P_k.

The approximate value of the kth percentile is determined by the following formula.

PERCENTILES

The value of the kth percentile, denoted by P_k, is

$$P_k = \text{the value of the } \left(\frac{k\,n}{100}\right)\text{th term in a ranked data set}$$

where k denotes the number of the percentile and n represents the sample size.

Finding percentile for a data set.

EXAMPLE 3-22 The following are the scores of 12 students in a mathematics class.

75 80 68 53 99 58 76 73 85 88 91 79

Find the value of the 62nd percentile.

Solution First we arrange the scores in increasing order.

53 58 68 73 75 76 79 80 85 88 91 99

The position of the 62nd percentile is

$$\frac{kn}{100} = \frac{62(12)}{100} = 7.44\text{th term}$$

The value of 7.44th term can be approximated by the average of the 7th and 8th terms in the arranged data. Therefore, the 62nd percentile is

$$P_{62} = \frac{79 + 80}{2} = 79.5$$

Thus, approximately 62% of the scores are less than 79.5 and 38% are greater than 79.5 in the given data.

Note that if a data set contains only a few observations, then the number of values less than the 62nd percentile may not be exactly 62% and the number of values greater than the 62nd percentile may not be exactly 38%. For example, in the 12 scores of Example 3-22, 7 scores (which is approximately 58% of 12) are less than 79.5 and 5 scores (which is approximately 42% of 12) are greater than 79.5. However, these percentages tend to be more accurate in a large data set. ■

We can also calculate the **percentile rank** for a particular value x_i of a data set by using the following formula. The percentile rank of x_i gives the percentage of values in the data set that are smaller than x_i.

PERCENTILE RANK OF A VALUE

$$\text{Percentile rank of } x_i = \frac{\text{Number of values less than } x_i}{\text{Total number of values in the data set}} \cdot 100$$

Finding percentile rank of a data value.

EXAMPLE 3-23 Refer to the math scores of 12 students given in Example 3-22, which are reproduced here.

75 80 68 53 99 58 76 73 85 88 91 79

Find the percentile rank for the score 85.

Solution First we arrange the scores in increasing order.

53 58 68 73 75 76 79 80 85 88 91 99

In this data set, eight of the 12 scores are less than 85. Hence, the percentile rank of 85 is

$$\text{Percentile rank of } 85 = \frac{8}{12} \cdot 100 = 66.67$$

Rounding the answer to the nearest integral value, about 67% of the scores in the given data set are less than 85. In other words, about 67% of the students scored lower than 85. ■

EXERCISES

3.59 The following data give the time (in minutes) taken to complete a statistics test by 14 students.

93 87 96 77 73 91 82 71 98 74 95 89 79 88

a. Calculate the values of three quartiles.
b. Find the (approximate) value of the 30th percentile.
c. Compute the percentile rank of 89.

3.60 The following data give the amount (in dollars) spent on makeup during the past month by 15 women.

12	38	19	57	27	37	7	24
3	21	17	45	41	33	11	

a. Calculate the values of three quartiles.
b. Find the (approximate) value of the 53rd percentile.
c. Compute the percentile rank of 33.

3.61 The following data give the hours worked last week by 30 employees of a company.

42	45	40	38	35	47	40	27	39	43
40	53	23	51	42	48	40	36	51	40
48	34	21	40	31	34	16	39	41	36

a. Calculate the values of three quartiles.
b. Find the (approximate) value of the 79th percentile.
c. Compute the percentile rank of 39.

3.62 The following data give the scores of 19 students in a statistics class.

84	92	63	75	81	97	73	69	46	58
94	84	78	43	77	82	69	98	84	

a. Calculate the values of three quartiles.
b. Find the (approximate) value of the 93rd percentile.
c. Find the percentile rank of 82.

3.63 The following data, reproduced from Exercise 3.13, give the number of car thefts during the past 12 days that occurred in a city.

6 3 7 11 5 3 8 7 2 6 9 13

a. Calculate the values of three quartiles.
b. Find the percentile rank of 8.

3.64 The following data, reproduced from Exercise 3.14, give the number of hours spent partying by 10 randomly selected college students during the past week.

7 14 5 0 2 7 10 4 0 8

a. Calculate the values of three quartiles.
b. Find the percentile rank of 7.

3.65 The following data, reproduced from Exercise 3.28, give the weight lost by 15 members of a health club at the end of two months after joining the club.

5	10	8	7	25	12	5	14
11	10	21	9	8	11	18	

a. Calculate the values of three quartiles.
b. Calculate the (approximate) value of the 82nd percentile.
c. Find the percentile rank of 10.

3.66 The following data, reproduced from Exercise 3.29, give the speeds of 13 cars, measured by radar, traveling on interstate highway I-84.

67	72	63	66	76	69	71
76	65	79	68	67	71	

a. Find the values of three quartiles.
b. Calculate the (approximate) value of the 35th percentile.
c. Compute the percentile rank of 71.

3.6 BOX-AND-WHISKER PLOT

A **box-and-whisker plot** gives a graphic presentation of data using five measures: the median, the **lower hinge,** the **upper hinge,** the smallest value in the data set, and the largest value in the data set. A box-and-whisker plot can help us visualize the center, the spread, and the skewness of a data set. We can compare the different distributions by making box-and-whisker plots for each of them.

> **BOX-AND-WHISKER PLOT**
>
> A device that shows the center, spread, and skewness of a data set. It is constructed by drawing a box and two whiskers that use the median, the lower hinge, the upper hinge, the smallest value, and the largest value in the data set.

If the number of values in a data set is odd, then the lower hinge is the median of the values that are less than or equal to the median of the whole data set and the upper hinge is the median of the values that are greater than or equal to the median of the whole data set. If the number of values in the data set is even, then the lower hinge is the median of the values that are less than the median of the whole data set and the upper hinge is the median of the values that are greater than the median of the whole data set.

Example 3-24 explains all the steps needed to make a box-and-whisker plot.

Constructing a box-and-whisker plot for even number of data values.

EXAMPLE 3-24 The following are the incomes (in thousands of dollars) for a sample of 12 households.

23 17 21 53 16 39 26 31 48 25 29 19

Construct a box-and-whisker plot for these data.

Solution We perform the following steps to construct a box-and-whisker plot for these data.

Step 1. First, we rank the data in an increasing order and calculate the values of the median, the lower hinge, and the upper hinge.

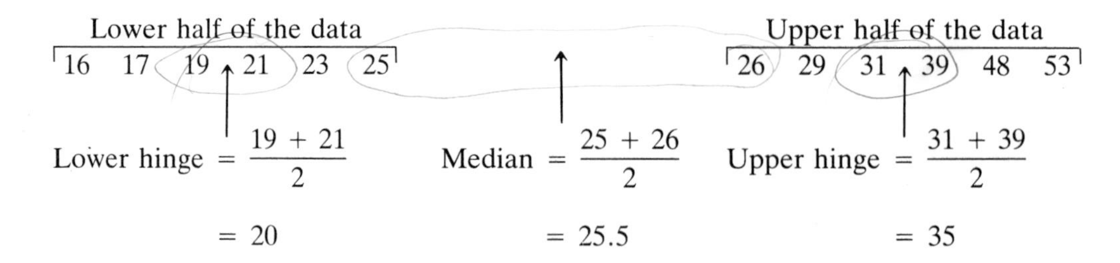

Note that when the number of observations in a data set is even (which is the same as saying that the value of the median is not one of the values in the given data set), we do not include the value of the median when calculating the hinges.

Step 2. We determine the smallest and the largest values in the given data set.

$$\text{Smallest value} = 16 \qquad \text{and} \qquad \text{Largest value} = 53$$

Step 3. We draw a horizontal line and mark the income levels on it so that all the values in the given data set are covered.

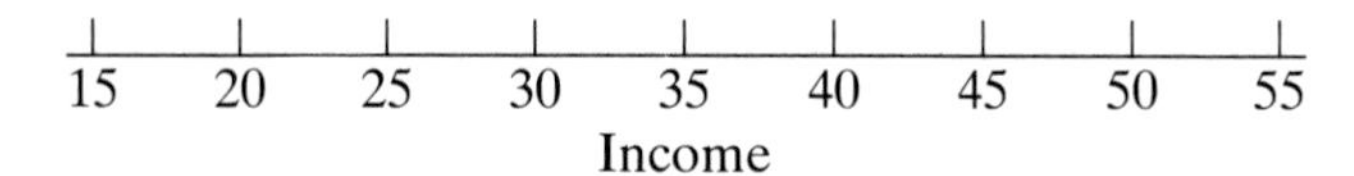

Step 4. Above the horizontal line, we draw a box with its left side at the position of the lower hinge and the right side at the position of the upper hinge. Inside the box, we draw a vertical line at the position of the median. The result of this step is shown in Figure 3.14.

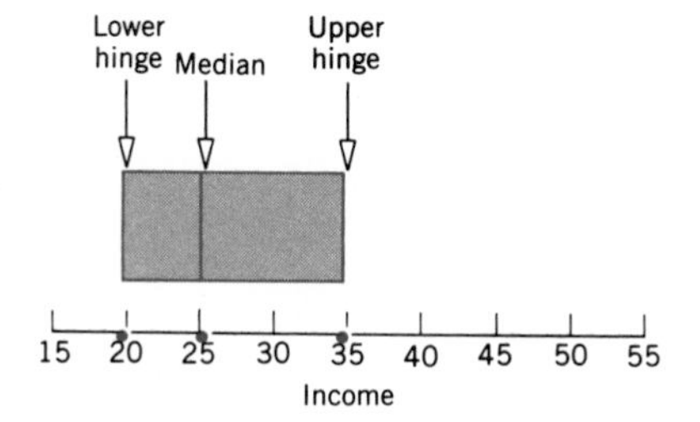

Figure 3.14

Step 5. By drawing two lines, we join the points of the smallest and the largest values to the box. These lines are called *whiskers*. This completes our box-and-whisker plot, as shown in Figure 3.15.

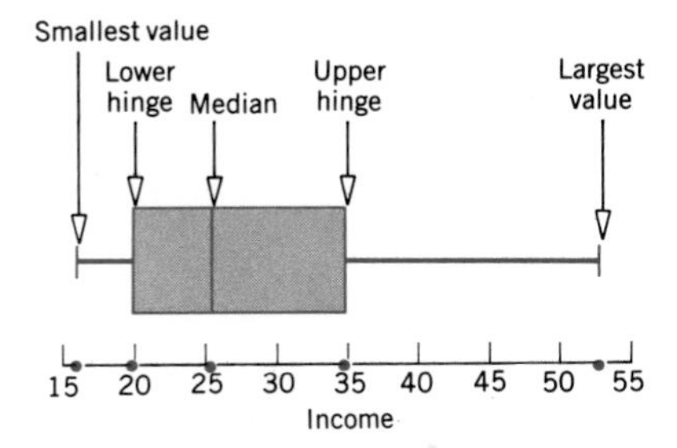

Figure 3.15 Box-and-whisker plot for incomes.

■

In Figure 3.15, about 50% of the data values fall within the box, about 25% of the values are covered by the whisker on the left side of the box, and about 25% of the values are covered by the whisker on the right side of the box. The data of this example are skewed to the right because the lower 50% of the values are spread over a smaller range than the upper 50% of the values.

For a symmetric data set, the line representing the median will be in the middle of the box and the two whiskers will be approximately of the same size. In general, when the median line is closer to the lower hinge, the data are skewed to the right; when the median line is closer to the upper hinge, the data are skewed to the left.

Making a box-and-whisker plot for odd number of data values.

EXAMPLE 3-25 The data on speeds of 13 cars traveling on interstate highway I-84 given in Exercise 3.29 are reproduced here.

67 72 63 66 76 69 71 76 65 79 68 67 71

Construct a box-and-whisker plot for these data.

Solution First, we rank the data in increasing order. Then, we calculate the values of the median, the lower hinge, and the upper hinge as follows.

63 65 66 67 67 68 69 71 71 72 76 76 79

↑

Median = 69

Values less than or equal to the median

63 65 66 67 67 68 69

↑

Lower hinge = 67

Values greater than or equal to the median

69 71 71 72 76 76 79

↑

Upper hinge = 72

Note that as the number of observations in this data set is odd (the value of the median is one of the values in the given data set), we have included the value of the median in the lower half and the upper half of the ranked data set to calculate the hinges.

The smallest and the largest values in the given data set are

Smallest value = 63 and Largest value = 79

The complete box-and-whisker plot is shown in Figure 3.16.

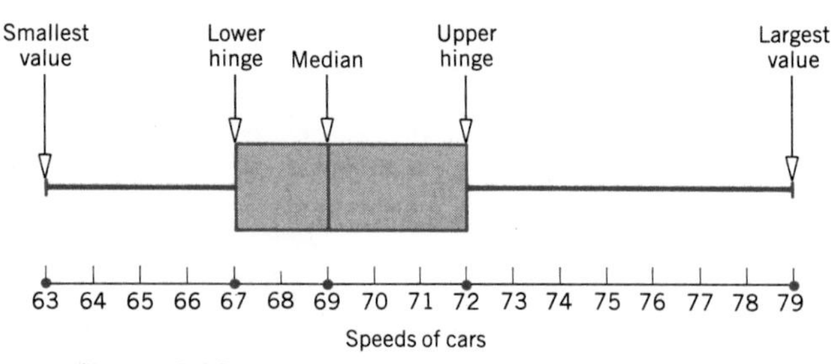

Figure 3.16 Box-and-whisker plot for speeds of cars.

■

We observe from Figure 3.16 that the whisker on the right side is much longer than the whisker on the left side of the box. The line representing the median is to the left of the midpoint of the box. Consequently, we can conclude that this data set is skewed to the right.

Sometimes, for convenience, instead of using the lower and upper hinges, quartiles are used to make a box-and-whisker plot. Actually, if the number of values in a data set is even, then the values of the first quartile and the lower hinge are the same and the values of the third quartile and the upper hinge are the same. When the number of values in a data set is odd, the values of quartiles and hinges are slightly different.

EXERCISES

3.67 The following data, reproduced from Exercise 3.59, give the time (in minutes) taken to complete a statistics test by 14 students.

93 87 96 77 73 91 82 71 98 74 95 89 79 88

Make a box-and-whisker plot. Comment on the skewness of the data.

3.68 The following data, reproduced from Exercise 3.60, give the amount (in dollars) spent on makeup during the past month by 15 women.

12	38	19	57	27	37	7	24
3	21	17	45	41	33	11	

Make a box-and-whisker plot. Comment on the skewness of the data.

3.69 The following data, reproduced from Exercise 3.61, give the hours worked last week by 30 employees of a company.

42	45	40	38	35	47	40	27	39	43
40	53	23	51	42	48	40	36	51	40
48	34	21	40	31	34	16	39	41	36

Make a box-and-whisker plot. Are the data skewed in any direction?

3.70 The following data, reproduced from Exercise 3.62, give the scores of 19 students in a statistics class.

84	92	63	75	81	97	73	69	46	58
94	84	78	43	77	82	69	98	84	

Make a box-and-whisker plot. Are the data skewed in any direction?

3.71 The following data, reproduced from Exercise 3.13, give the number of car thefts that occurred in a city during the past 12 days.

6 3 7 11 5 3 8 7 2 6 9 13

Construct a box-and-whisker plot. Are the data symmetric or skewed?

3.72 The following data, reproduced from Exercise 3.14, give the number of hours spent partying by 10 randomly selected college students during the past week.

7 14 5 0 2 7 10 4 0 8

Construct a box-and-whisker plot. Comment on the skewness of the data.

3.73 The following data, reproduced from Exercise 3.28, give the weight (in pounds) lost by 15 members of a health club at the end of two months after joining the club.

5	10	8	7	25	12	5	14
11	10	21	9	8	11	18	

Construct a box-and-whisker plot. Are the data symmetric or skewed?

3.74 The following data, reproduced from Exercise 3.33, give the annual income after taxes (in thousands of dollars) for 11 households selected from the 1987 Interview Survey.

12	19	25	4	34	19	47	16	40	48	28

Prepare a box-and-whisker plot. Comment on the skewness of the data.

GLOSSARY

Bimodal distribution A distribution that has two modes.

Box-and-whisker plot A device that shows the center, spread, and skewness of a data set by drawing a box and two whiskers using the median, the lower hinge, the upper hinge, the smallest value, and the largest value in the data set.

Chebyshev's theorem For any number k greater than 1, at least $(1 - 1/k^2)$ of the values for any distribution lie within k standard deviations of the mean.

Empirical rule For a specific bell-shaped distribution, about 68% of the observations fall in the interval $(\mu - \sigma)$ to $(\mu + \sigma)$, about 95% fall in the interval $(\mu - 2\sigma)$ to $(\mu + 2\sigma)$, and about 99.7% fall in the interval $(\mu - 3\sigma)$ to $(\mu + 3\sigma)$.

First quartile The value in a ranked data set such that about 25% of the measurements are smaller than this value and about 75% are larger. It is the median of the values that are smaller than the median of the whole data set.

Lower hinge If the number of values in a data set is odd, then the lower hinge is the median of the values that are less than or equal to the median of the whole data set. If the number of values in the data set is even, then the lower hinge is the median of the values that are less than the median of the whole data set.

Mean A measure of central tendency calculated by dividing the sum of all values by the number of values in the data set.

Measures of central tendency Measures that describe the center of a distribution. The mean, median, and mode are three of the measures of central tendency.

Measures of dispersion Measures that give the spread of a distribution. The range, variance, and standard deviation are three such measures.

Measures of position Measures that determine the position of a single value in relation to other values in a data set. Quartiles and percentiles are two examples of measures of position.

Median The value of the middle term in a ranked data set. The median divides the arranged data in two equal parts.

Mode A value (or values) that occurs with highest frequency in a data set.

Multimodal distribution A distribution that has more than two modes. Bimodal is a special case of a multimodal distribution with two modes.

Outliers or **extreme values** Values that are very small or very large relative to the majority of the values in a data set.

Parameter A summary measure calculated for population data.

Percentiles Ninety-nine values that divide a ranked data set into 100 equal parts.

Quartiles Three numbers that divide a ranked data set into four equal parts.

Range A measure of spread obtained by taking the difference between the highest and the lowest values in a data set.

Second quartile Middle or second of the three quartiles that divide a ranked data set into four equal parts. About 50% of the values in the data are smaller and about 50% are larger than the second quartile. Its value is equal to the median.

Standard deviation A measure of spread that is given by the positive square root of the variance.

Statistic A summary measure calculated for sample data.

Third quartile Third of the three quartiles that divide a ranked data set into four equal parts. About 75% of the values in a data set are smaller than the value of the third quartile and about 25% are larger. It is the median of the values greater than the median of the whole data set.

Unimodal distribution A distribution that has only one mode.

Upper hinge If the number of values in a data set is odd, then the upper hinge is the median of the values that are greater than or equal to the median of the whole data set. If the number of values in the data set is even, then the upper hinge is the median of the values that are greater than the median of the whole data set.

Variance A measure of spread.

KEY FORMULAS

1. Mean for ungrouped data

For population data:

$$\mu = \frac{\Sigma x}{N}$$

For sample data:

$$\bar{x} = \frac{\Sigma x}{n}$$

2. Mean for grouped data

For population data:

$$\mu = \frac{\Sigma mf}{N}$$

For sample data:

$$\bar{x} = \frac{\Sigma mf}{n}$$

where m is the midpoint and f is the frequency of a class.

3. **Median for ungrouped data**

$$\text{Median} = \text{Value of the } \left(\frac{n+1}{2}\right)\text{th term in a ranked data set}$$

4. **Range**

$$\text{Range} = \text{Highest value} - \text{Lowest value}$$

5. **Variance for ungrouped data**

For population data:

$$\sigma^2 = \frac{\Sigma x^2 - \frac{(\Sigma x)^2}{N}}{N}$$

For sample data:

$$s^2 = \frac{\Sigma x^2 - \frac{(\Sigma x)^2}{n}}{n - 1}$$

6. **Variance for grouped data**

For population data:

$$\sigma^2 = \frac{\Sigma m^2 f - \frac{(\Sigma mf)^2}{N}}{N}$$

For sample data:

$$s^2 = \frac{\Sigma m^2 f - \frac{(\Sigma mf)^2}{n}}{n - 1}$$

where m is the midpoint of a class.

7. **Standard deviation**

For population data:

$$\sigma = \sqrt{\text{Variance}}$$

For sample data:

$$s = \sqrt{\text{Variance}}$$

8. Percentiles

The kth percentile is given by

$$P_k = \text{Value of the } \left(\frac{k\,n}{100}\right)\text{th term in a ranked data set}$$

9. Percentile rank of a value

The percentile rank for a particular value x_i of a data set is calculated as follows.

$$\text{Percentile rank of } x_i = \frac{\text{Number of values less than } x_i}{\text{Total number of values in the data set}} \cdot 100$$

SUPPLEMENTARY EXERCISES

3.75 The following data give the charitable contributions (in dollars) made during the 12-month period by 16 households selected from the 1987 Interview Survey. (These households were selected only from those who made charitable contributions.)

30	300	50	100	27	100	25	76
25	902	240	18	400	6000	200	10

a. Calculate the mean, median, and mode for these data.
b. Do these data contain any outlier(s)? If yes, drop the outlier(s) and recalculate the mean, median, and mode. Which of these measures changes by a larger amount when you drop the outlier?

3.76 Following are the annual occupational expenses (expenses on items such as union dues, tools, uniforms, business or professional association dues, licenses, or permits) incurred by 15 households selected from the 1987 Diary Survey. (These households were selected only from those who incurred occupational expenses.)

$300	45	27	50	4000
460	300	192	784	371
200	3050	33	180	69

a. Find the mean and median for these data.
b. Do these data contain any outliers? If yes, drop the outliers and recalculate the mean and median. Which of the two measures changes by a larger amount when you drop the outliers?

3.77 The following data give the annual earnings (in thousands of dollars) before taxes for 11 households selected from the 1987 Interview Survey. (These households were selected only from those who had positive earnings.)

11.5	10.9	23.3	5.0	40.0	
20.0	47.4	16.2	48.6	55.0	22.7

a. Calculate the mean and median for these data. Do these data contain a mode?
b. Find the range, variance, and standard deviation.

3.78 The following data give the quarterly entertainment expenses (in dollars) for 14 households selected from the 1987 Interview Survey. (These households were selected only from those who incurred such expenses.)

33	37	596	86	102	322	106
450	62	55	58	91	21	949

a. Find the mean and median for these data. Do these data contain a mode?
b. Calculate the range, variance, and standard deviation.

3.79 The following data give the total number of driving citations previously received by 12 drivers.

4 8 0 3 11 7 4 14 8 13 7 9

a. Find the mean, median, and mode for these data.
b. Calculate the range, variance, and standard deviation.

3.80 The following data give the ages of nine randomly selected cars.

7 5 2 12 5 9 14 1 8

a. Calculate the mean, median, and mode for these data.
b. Find the range, variance, and standard deviation.

3.81 The following table gives the distribution of the amount of rainfall (in inches) for July 1990 for 50 cities.

Rainfall	Number of Cities
0 to less than 2	6
2 to less than 4	10
4 to less than 6	18
6 to less than 8	9
8 to less than 10	4
10 to less than 12	3

Find the mean, variance, and standard deviation.

3.82 The following table gives the distribution of hourly wages (in dollars) for all 120 employees of a company.

Hourly Wage	Number of Employees
4 to less than 6	28
6 to less than 8	36
8 to less than 10	21
10 to less than 12	14
12 to less than 14	16
14 to less than 16	5

Calculate the mean, variance, and standard deviation. Are the values of these measures the population parameters or sample statistics?

3.83 The following table gives the distribution of the amounts (in dollars) that 100 randomly selected persons said they will spend on a wedding gift for their closest friend.

Amount	Number of Persons
0 to less than 50	8
50 to less than 100	27
100 to less than 150	31
150 to less than 200	23
200 to less than 250	11

Find the mean, variance, and standard deviation. Are the values of these measures the population parameters or sample statistics?

3.84 The following table gives the distribution of the number of hours that 80 persons spent learning to drive before passing the driving test.

Number of Hours	Number of Persons
5 to 7	4
8 to 10	9
11 to 13	16
14 to 16	23
17 to 19	17
20 to 22	11

Calculate the mean, variance, and standard deviation. Are the values of these measures the population parameters or sample statistics?

3.85 The following table gives the distribution of the number of days for which all 40 employees of a company were absent during the past year.

Number of Days Absent	Number of Employees
0 to 2	13
3 to 5	14
6 to 8	6
9 to 11	4
12 to 14	3

Calculate the mean, variance, and standard deviation. Are the values of these measures the population parameters or sample statistics.

3.86 The ages of all students at a university have a mean of 23.5 years and a standard deviation of 1.6 years. Using Chebyshev's theorem, find the percentage of students at this university who are between

a. 20.3 and 26.7 years of age **b.** 19.5 and 27.5 years of age
c. 18.7 and 28.3 years of age

3.87 The weights of all babies born at a hospital have a mean of 7.4 pounds and a standard deviation of 1.4 pounds. Using Chebyshev's theorem, find the percentage of all babies born at this hospital who are between

a. 4.6 and 10.2 pounds b. 5.3 and 9.5 pounds c. 3.9 and 10.9 pounds

3.88 According to a 1976–80 survey conducted by the U.S. National Center for Health Statistics, the average height of U.S. women aged 18–74 was 63.7 inches. Assume that the current heights of all U.S. women aged 18–74 have a bell-shaped distribution with a mean of 63.7 inches and a standard deviation of 2.1 inches. Find the percentage of these women who are

a. 61.6 to 65.8 inches tall b. 59.5 to 67.9 inches tall
c. 57.4 to 70.0 inches tall

3.89 According to the survey referred to in Exercise 3.88, the average weight of the U.S. women aged 18–74 was 144.4 pounds. Assume that the current weights of all U.S. women aged 18–74 follow a bell-shaped distribution with a mean of 144.4 pounds and a standard deviation of 18 pounds. Find the percentage of these women who weigh between

a. 126.4 and 162.4 pounds b. 108.4 and 180.4 pounds
c. 90.4 and 198.4 pounds

3.90 Refer to the data of Exercise 3.79 on the total number of driving citations previously received by 12 drivers.

a. Find the values of three quartiles.
b. Calculate the (approximate) value of 24th percentile.
c. Find the percentile rank of 8.

3.91 Refer to the data on the ages of nine randomly selected cars given in Exercise 3.80.

a. Find the values of three quartiles.
b. Calculate the (approximate) value of 79th percentile.
c. Find the percentile rank of 7.

3.92 The following data give the ages of 15 employees of a company.

36	47	23	55	42	31	27	19
38	61	52	47	39	25	44	

Construct a box-and-whisker plot. Is this data set skewed in any direction? If yes, is it skewed to the right or to the left?

3.93 The following data give the prices (in thousands of dollars) of 16 recently sold houses in an area.

141	163	127	104	197	203	113	179
256	228	183	119	133	199	271	191

Make a box-and-whisker plot. Is this data set skewed in any direction? If yes, is it skewed to the right or to the left?

APPENDIX 3.1

A3.1.1 BASIC FORMULAS FOR THE VARIANCE AND STANDARD DEVIATION FOR UNGROUPED DATA

The following example demonstrates how to use the basic formulas to calculate the variance and standard deviation for ungrouped data. From Section 3.2.2, the basic formulas for variance and standard deviation are

$$\sigma^2 = \frac{\Sigma(x - \mu)^2}{N} \quad \text{and} \quad s^2 = \frac{\Sigma(x - \bar{x})^2}{n - 1}$$

where σ^2 is the population variance and s^2 is the sample variance.

$$\text{Standard deviation} = \sqrt{\text{Variance}}$$

EXAMPLE 3-26 Refer to Example 3-12. In that example, we used the short-cut formula to compute the variance and standard deviation for the data on electric bills for a sample of five families. Now we calculate the variance and standard deviation for the same data using the basic formula. The amounts of the electric bills for five families are

$72.50 93.20 52.30 85.70 35.60

All the required calculations are made in Table 3.8.

Table 3.8

x	$(x - \bar{x})$	$(x - \bar{x})^2$
72.50	4.64	21.5296
93.20	25.34	642.1156
52.30	−15.56	242.1136
85.70	17.84	318.2656
35.60	−32.26	1040.7076
$\Sigma x = 339.30$		$\Sigma(x - \bar{x})^2 = 2264.7320$

The following steps are performed to compute the variance and standard deviation.

Step 1. Find the mean as follows.

$$\bar{x} = \frac{\Sigma x}{n} = \frac{339.30}{5} = \$67.86$$

Step 2. Calculate $x - \bar{x}$, the deviation of each value of x from the mean. The results are shown in the second column of Table 3.8.

Step 3. Square each of the deviations of x from $\bar{x}$, that is, calculate each of the $(x - \bar{x})^2$ values. These are called the *squared deviations,* and they are recorded in the third column of Table 3.8.

Step 4. Add all the squared deviations to obtain $\Sigma(x - \bar{x})^2$ (i.e., sum all the values given in the third column of Table 3.8.) This gives

$$\Sigma(x - \bar{x})^2 = 2264.7320$$

Step 5. Obtain the sample variance by dividing the sum of the squared deviations by $n - 1$. Thus,

$$s^2 = \frac{\Sigma(x - \bar{x})^2}{n - 1} = \frac{2264.7320}{5 - 1} = 566.183$$

Step 6. Obtain the sample standard deviation by taking the positive square root of the variance. Hence,

$$s = \sqrt{566.183} = \$23.79$$

A3.1.2 BASIC FORMULAS FOR THE VARIANCE AND STANDARD DEVIATION FOR GROUPED DATA

The following example demonstrates how to use the basic formulas to calculate the variance and standard deviation for grouped data. The basic formulas for these calculations are

$$\sigma^2 = \frac{\Sigma f(m - \mu)^2}{N} \quad \text{and} \quad s^2 = \frac{\Sigma f(m - \bar{x})^2}{n - 1}$$

where σ^2 is the population variance, s^2 is the sample variance, m is the midpoint of a class, and f is the frequency of a class.

$$\text{Standard deviation} = \sqrt{\text{Variance}}$$

EXAMPLE 3-27 In Example 3-16, we used the short-cut formula to compute the variance and standard deviation for the data on the normal annual precipitation (in inches) for 29 cities. The frequency table containing the precipitation data is reproduced here.

Normal Annual Precipitation	f
0 to less than 12	5
12 to less than 24	6
24 to less than 36	8
36 to less than 48	6
48 to less than 60	4

All the calculations required to compute the variance and standard deviation using the basic formula are made in Table 3.9. Note that we have rounded all the calculations in this table to two decimal places to keep the calculations simple.

Table 3.9

Normal Annual Precipitation	f	m	mf	$m - \bar{x}$	$(m - \bar{x})^2$	$f(m - \bar{x})^2$
0 to less than 12	5	6	30	−23.17	536.85	2684.25
12 to less than 24	6	18	108	−11.17	124.77	748.62
24 to less than 36	8	30	240	.83	.69	5.52
36 to less than 48	6	42	252	12.83	164.61	987.66
48 to less than 60	4	54	216	24.83	616.53	2466.12
	$n = 29$		$\Sigma mf = 846$			$\Sigma f(m - \bar{x})^2 = 6892.17$

The following steps are performed to compute the variance and standard deviation using the basic formula.

Step 1. Find the midpoint of each class. Multiply the corresponding values of m and f. Find Σmf. From Table 3.9

$$\Sigma mf = 846$$

Step 2. Find the mean as follows.

$$\bar{x} = \frac{\Sigma mf}{n} = \frac{846}{29} = 29.17$$

Step 3. Calculate $m - \bar{x}$, the deviation of each value of m from the mean. These calculations are done in the fifth column of Table 3.9.

Step 4. Square each of the deviations $m - \bar{x}$, that is, calculate each of the $(m - \bar{x})^2$ values. These are called *squared deviations,* and they are recorded in the sixth column of Table 3.9.

Step 5. Multiply the squared deviations by the corresponding frequencies (see the seventh column of Table 3.9). Adding the values of the seventh column, we obtain

$$\Sigma f(m - \bar{x})^2 = 6892.17$$

Step 6. Obtain the sample variance by dividing $\Sigma f(m - \bar{x})^2$ by $n - 1$. Thus,

$$s^2 = \frac{\Sigma f(m - \bar{x})^2}{n - 1} = \frac{6892.17}{29 - 1} = 246.149$$

Step 7. Obtain the standard deviation by taking the positive square root of the variance.

$$s = \sqrt{246.149} = 15.69 \text{ inches}$$

SELF-REVIEW TEST

1. The value of the middle term in a ranked data set is called the

   a. mean **b.** median **c.** mode

2. Which of the following measures are influenced by extreme values?

   a. mean **b.** median **c.** mode **d.** range

3. Which of the following can be calculated for qualitative data?

   a. mean **b.** median **c.** mode

4. Which of the following can have more than one value?

   a. mean **b.** median **c.** mode

5. Which of the following is given by the difference between the largest and the smallest values of a data set?

   a. variance **b.** range **c.** mean

6. Which of the following is the mean of the squared deviations of x values from the mean?

   a. standard deviation **b.** population variance **c.** sample variance

7. The values of the variance and standard deviation are

   a. never negative **b.** always positive **c.** never zero

8. A summary measure calculated for the population data is called

   a. a population parameter **b.** a sample statistic **c.** an outlier

9. A summary measure calculated for the sample data is called

   a. a population parameter **b.** a sample statistic **c.** an outlier

10. Chebyshev's theorem can be applied to

    a. any distribution **b.** bell-shaped distributions only
    c. skewed distributions only

11. The empirical rule can be applied to

    a. any distribution **b.** bell-shaped distributions only
    c. skewed distributions only

12. The seventy-fifth percentile means that

    a. about 75% of the values are smaller and about 25% are larger than this value
    b. about 25% of the values are smaller and about 75% are larger than this value

13. The following data give the number of times 10 persons used their credit cards during the past three months.

    9 6 22 14 2 18 7 3 11 6

Calculate the mean, median, mode, range, variance, and standard deviation.

14. The mean, as a measure of central tendency, has a disadvantage of being influenced by extreme values. Illustrate this point with an example.

15. The range, as a measure of spread, has a disadvantage of being influenced by extreme values. Illustrate this point with an example.

16. When is the value of the standard deviation for a data set zero? Give one example of such a data set. Make the calculations and show that the standard deviation for that data set is zero.

17. The following table gives the frequency distribution of the number of items sold during a 40-day period by a mail-order company.

Items Sold	Frequency
26–30	4
31–35	13
36–40	10
41–45	6
46–50	7

a. What does the frequency column in the table represent?
b. Calculate the mean, variance, and standard deviation.

18. The cars owned by all people living in a city are on average 7.3 years old with a standard deviation of 2.2 years. Using Chebyshev's theorem, find at least what percentage of the cars in this city are

a. 2.9 to 11.7 years old **b.** .7 to 13.9 years old

19. Answer Problem 18 assuming that the ages of all cars follow a bell-shaped curve.

20. The following data give the years of schooling for a random sample of 16 adults.

12	10	15	16	12	18	20	10
8	6	12	4	18	19	18	12

Calculate the three quartiles.

21. Refer to the data of Problem 20. Calculate the (approximate) value of the 68th percentile.

22. Calculate the percentile rank of 15 for the data of Problem 20.

23. Make a box-and-whisker plot for the data of Problem 20. Comment on the skewness of the data.

*24. The mean weekly wages for a sample of 15 employees of a company are $435 and the mean weekly wages for a sample of 20 employees of another company are $490. Find the combined mean for these 35 employees.

*25. The mean GPA of five students is 3.21. The GPAs of four of these five students are 3.85, 2.67, 3.45, and 2.91. Find the GPA of the fifth student.

*26. Following are the temperatures (in Fahrenheit) observed in a city during 10 days of summer.

76	81	61	97	87	79	104	91	66	99

Calculate the 20% trimmed mean for this data set.

***27.** Consider the following two data sets.

Data set I:	8	16	20	35
Data set II:	5	13	17	32

Note that each value of the second data set is obtained by subtracting 3 from the corresponding value of the first data set.

a. Calculate the mean for each of these two data sets. Comment on the relationship between the two means.

b. Calculate the standard deviation for each of these two data sets. Comment on the relationship between the two standard deviations.

USING MINITAB

The following MINITAB commands can be used to do the statistical analysis discussed in this chapter. Assume that the data are entered in column C1.

```
MTB > MEAN C1   ← This command will compute the mean
MTB > MEDIAN C1   ← This command will compute the median
MTB > RANGE C1   ← This command will compute the range
MTB > STDEV C1   ← This command will compute the standard deviation
MTB > SUM C1   ← This command will compute the sum of all values
MTB > MAXIMUM C1   ← This command will list the maximum value
MTB > MINIMUM C1   ← This command will list the minimum value
MTB > DESCRIBE C1   ← This command will compute the various summary measures
MTB > BOXPLOT C1   ← This command will prepare the box-and-whisker plot
```

Using each of these commands displays the corresponding MINITAB solution on the screen as soon as the enter key is hit. An alternative method, with LET command, can be used to determine the mean, median, standard deviation, sum of the values, maximum value, and minimum value for a data set. (The LET command cannot be used to calculate range.) When the LET command is used, MINITAB output will not be displayed on the screen but it will be stored in the computer memory. This output can be seen on the screen at any time by entering the PRINT command. The LET command is used as follows, assuming that the data are entered in column C1.

```
MTB > LET K1 = MEAN (C1)
MTB > LET K2 = MEDIAN (C1)
MTB > LET K3 = STDEV (C1)
MTB > LET K4 = SUM (C1)
MTB > LET K5 = MAXIMUM (C1)
MTB > LET K6 = MINIMUM (C1)
```

Note that in these MINITAB commands, C1 is enclosed in parentheses. In response to the first command, MINITAB will compute the mean for the data of column C1 and store this value of the mean as K1 in the computer memory. MINITAB will do the same in response to other commands. The following PRINT command will show the values of the mean, median, standard deviation, sum, maximum, and minimum on the screen that have been stored in the computer memory as constants K1 to K6, respectively.

```
MTB > PRINT K1-K6
```

ILLUSTRATION M3-1 The speeds of 13 cars traveling on interstate highway I-84, given in Exercise 3.29, are reproduced here.

67 72 63 66 76 69 71 76 65 79 68 67 71

In the following MINITAB display, the use of aforementioned MINITAB commands is illustrated for these data.

```
MTB  > NOTE: CALCULATING DESCRIPTIVE MEASURES
MTB  > SET C1
DATA > 67 72 63 66 76 69 71 76 65 79 68 67 71
DATA > END

MTB  > MEAN C1
       MEAN = 70.000

MTB  > MEDIAN C1
       MEDIAN = 69.000

MTB  > RANGE C1
       RANGE = 16.000

MTB  > STDEV C1
       ST.DEV. = 4.7610

MTB  > SUM C1
       SUM = 910.00

MTB  > MAXIMUM C1
       MAXIMUM = 79.000

MTB  > MINIMUM C1
       MINIMUM = 63.000
```

Next, we calculate the mean, median, standard deviation, sum, maximum, and minimum for the data of column C1 using the LET command. One advantage of using the LET command is that if we save the file, the values of constants (Ks) will also be saved along with the data of column C1. Thus, when we retrieve the saved file later on, we will not have to recalculate these summary measures. Note that we have not repeated the SET command to enter the data in the following MINITAB display because we have already entered the data in column C1 in the previous MINITAB display.

```
MTB > LET K1 = MEAN (C1)
MTB > LET K2 = MEDIAN (C1)
MTB > LET K3 = STDEV (C1)
MTB > LET K4 = SUM (C1)
MTB > LET K5 = MAXIMUM (C1)
MTB > LET K6 = MINIMUM (C1)
MTB > PRINT K1-K6
K1        70.0000
K2        69.0000
K3        4.76095
K4        910.000
K5        79.0000
K6        63.0000
```

The DESCRIBE command can also be used to find some of the summary measures. The following MINITAB output is obtained in response to this command.

```
MTB > DESCRIBE C1

          N    MEAN   MEDIAN   TRMEAN   STDEV   SEMEAN
C1       13   70.00    69.00    69.82    4.76     1.32

        MIN     MAX       Q1       Q3
C1    63.00   79.00    66.50    74.00
```

Most of the entries in this MINITAB output are self-explanatory. The fourth entry, labeled TRMEAN (called the trimmed mean and briefly explained in Exercise 3.20) is the mean of the data when 5% of the values at each end of the ranked data are dropped. The sixth entry, SEMEAN, gives the standard error (also called the standard deviation) of the mean, which will be discussed in Chapter 7. The last two entries, Q1 and Q3, give the first and third quartiles, respectively. Again, note that the SET command to enter the data is not repeated.

The BOXPLOT command gives the box-and-whisker plot. In the following box-and-whisker plot for the data of column C1, the "+" sign inside the box indicates the position of the median. Note that we have used the data that have already been entered in column C1.

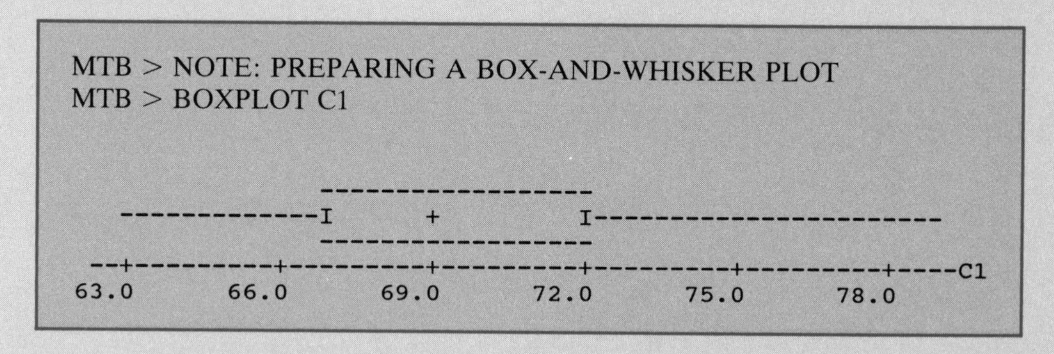

COMPUTER ASSIGNMENTS

M3.1 Refer to the subsample taken in MINITAB Computer Assignment M2.1 of Chapter 2 from the sample data on time taken to run the Manchester Road Race given in Appendix B (page 718). Using MINITAB, find the mean, median, range, and standard deviation for those data.

M3.2 Refer to Data Set I of Appendix B (page 703) on prices of various products in different cities across the country. Using MINITAB SAMPLE command, select a subsample of price of dry cleaning for 40 cities. Using MINITAB LET command, find the sum of values, maximum value, minimum value, mean, median, and standard deviation for the data of this subsample.

M3.3 Refer to data on the normal annual precipitation (in inches) for 29 cities of the United States given in Example 2-5 of Chapter 2. Those data are reproduced here.

8.12	48.61	41.84	15.36	11.71	43.81
33.34	35.40	29.46	15.31	30.97	7.82
23.47	44.76	12.08	43.56	57.55	26.36
48.49	59.74	44.12	30.34	7.11	37.39
7.49	33.91	15.31	19.71	28.61	

Using MINITAB, compute the mean and standard deviation for these data. Compare these values of the mean and standard deviation to the values of the mean and standard deviation calculated in Example 3-16.

M3.4 Refer to the data given in Exercise 3.13. Using MINITAB DESCRIBE command, find the mean, median, standard deviation, and quartiles for those data.

M3.5 Refer to Data Set II of Appendix B, which contains data on different variables for 50 states. Using MINITAB, make a box-and-whisker plot for the data on 1989 per capita income for 50 states given in column C8.

M3.6 Refer to Data Set I of Appendix B on prices of various products in different cities across the country. Using MINITAB, make a box-and-whisker plot for the data on monthly telephone charges given in column C7.

4 PROBABILITY

4.1 EXPERIMENT, OUTCOMES, AND SAMPLE SPACE

4.2 CALCULATING PROBABILITY

4.3 MARGINAL AND CONDITIONAL PROBABILITIES

4.4 MUTUALLY EXCLUSIVE EVENTS

4.5 INDEPENDENT VERSUS DEPENDENT EVENTS

4.6 COMPLEMENTARY EVENTS

4.7 INTERSECTION OF EVENTS AND THE MULTIPLICATION RULE

4.8 COUNTING RULE

4.9 UNION OF EVENTS AND THE ADDITION RULE

SELF-REVIEW TEST

People make statements about probability every day. A weather forecaster predicts that there is an 80% chance of rain tomorrow. A health-news reporter states that a smoker has a much greater chance of getting cancer than a nonsmoker. A college student may ask an instructor about the chances of passing a course or getting an A if he or she did not do well on the midterm examination.

Probability, which measures the likelihood that an event will occur, is an important part of statistics. It is the basis of inferential statistics, which will be introduced in later chapters of this book. Combining probability and probability distributions (which are discussed in Chapters 5 through 7) with descriptive statistics will help us to make decisions about populations based on the information obtained from samples. This chapter presents the basic concepts of probability and the rules for computing probability.

4.1 EXPERIMENT, OUTCOMES, AND SAMPLE SPACE

Quality control inspector Jack Cook of Tennis Products Company picks up a tennis ball from the production line to check whether it is good or defective. Jack Cook's act of inspecting a tennis ball is an example of a statistical **experiment.** The result of his inspection will be that the ball is either "good" or "defective." Each of these two observations is called an **outcome** (also called a *basic* or *final outcome*) of the experiment, and these outcomes taken together constitute the **sample space** for this experiment.

EXPERIMENT, OUTCOMES, AND SAMPLE SPACE

An experiment is a process that, when performed, results in one and only one of many observations. These observations are called the outcomes of the experiment. The collection of all outcomes for an experiment is called the sample space.

The sample space is denoted by S. The sample space for the example of inspecting a tennis ball is written as

$$S = \{\text{good, defective}\}$$

The elements of a sample space are also called the **sample points.**

Table 4.1 lists a few examples of experiments, their outcomes, and their sample spaces.

Table 4.1 An Experiment, Outcomes, and the Sample Space

Experiment	Outcomes	Sample Space
Toss a coin once	Head, Tail	S = {Head, Tail}
Roll a die once	1, 2, 3, 4, 5, 6	S = {1, 2, 3, 4, 5, 6}
Toss a coin twice	*HH, HT, TH, TT*	S = {*HH, HT, TH, TT*}
Birth of a baby	Boy, Girl	S = {Boy, Girl}
Take a test	Pass, Fail	S = {Pass, Fail}
Select a student	Male, Female	S = {Male, Female}

The sample space for an experiment can also be described by means of either a Venn diagram or a tree diagram. A **Venn diagram** is a picture (a closed geometric shape such as a rectangle, a square, or a circle) that depicts all the possible outcomes for an experiment. In a **tree diagram,** each outcome is represented by a branch of the tree. The Venn and tree diagrams help us understand probability concepts by presenting them visually. The following examples describe how to draw Venn and tree diagrams for statistical experiments.

Venn and tree diagrams: one toss of a coin.

EXAMPLE 4-1 Draw the Venn and tree diagrams for the experiment of tossing a coin once.

Solution This experiment has two possible outcomes: head and tail. Therefore, the sample space is given by

$$S = \{H, T\} \quad \text{where } H = \text{Head}, \quad T = \text{Tail}$$

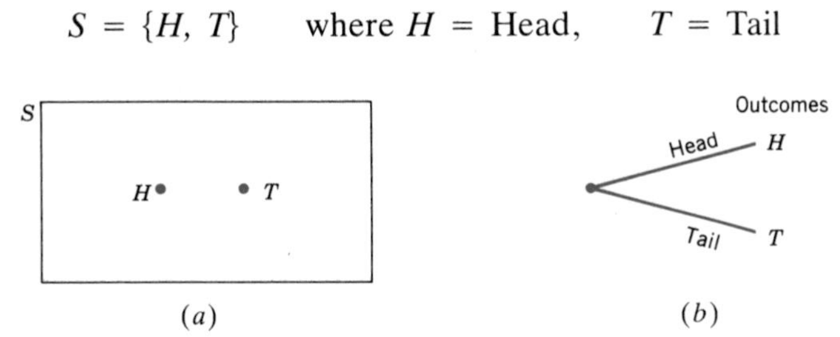

Figure 4.1 (*a*) Venn diagram and (*b*) tree diagram for one toss of a coin.

To draw a Venn diagram for this example, we draw a rectangle and mark two points inside it that represent the two outcomes, head and tail. The rectangle is labeled S because it represents the sample space (Fig. 4.1*a*). To draw a tree diagram, we draw two branches starting at the same point, one representing the head and second representing the tail. The two final outcomes are listed at the end of the branches (Fig. 4.1*b*). ■

Venn and tree diagrams: two tosses of a coin.

EXAMPLE 4-2 Draw the Venn and tree diagrams for the experiment of tossing a coin twice.

Solution This experiment can be split into two parts: the first toss and the second toss. Suppose the first time we toss the coin we obtain a head. We can still obtain a head or a tail on the second toss. This gives us two outcomes: HH (head on both tosses) and HT (head on the first toss and tail on the second toss). Now suppose we observe a tail on the first toss. Again, either a head or a tail can occur on the second toss, giving the remaining two outcomes: TH (tail on the first toss and head on the second toss) and TT (tail on both tosses). Thus, the sample space for two tosses of the coin is

$$S = \{HH, HT, TH, TT\}$$

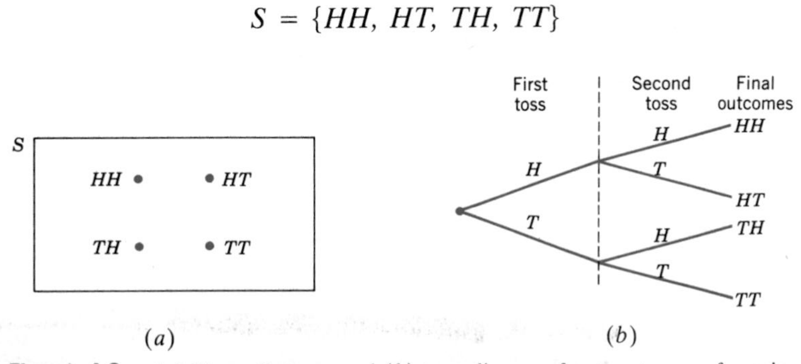

Figure 4.2 (*a*) Venn diagram and (*b*) tree diagram for two tosses of a coin.

The Venn and tree diagrams for this example are given in Figure 4.2. Both show the sample space for this experiment. ■

Venn and tree diagrams: two selections.

EXAMPLE 4-3 Suppose we randomly select two persons from the members of a club and observe whether the member selected each time is a male or a female. Write all the outcomes for this experiment. Draw the Venn and tree diagrams for this experiment.

Solution Let us denote the selection of a male by M and that of a female by F. We can compare the selection of two persons to two tosses of a coin. Since each toss of the coin can result in a head or a tail, each selection from the members of this club can result in a male or a female. As we can see from the Venn and tree diagrams of Figure 4.3, there are four final outcomes: MM, MF, FM, FF. Hence, the sample space is written as

$$S = \{MM, MF, FM, FF\}$$

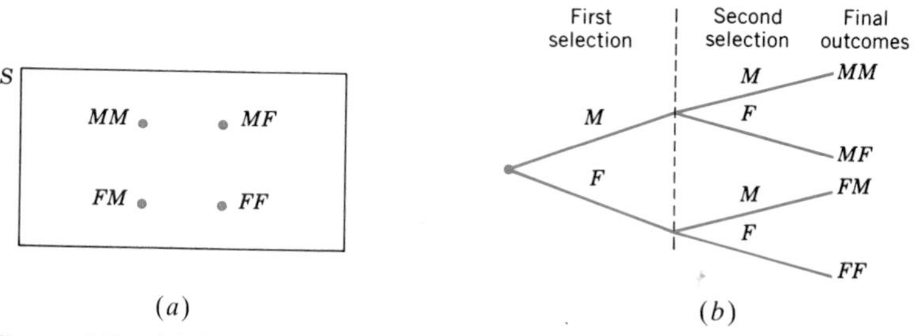

Figure 4.3 (*a*) Venn diagram and (*b*) tree diagram for selecting two persons. ■

4.1.1 SIMPLE AND COMPOUND EVENTS

An **event** consists of one or more of the outcomes of an experiment.

> **EVENT**
>
> An event is a collection of one or more of the outcomes of an experiment.

An event may be a **simple event** or a **compound event.** A simple event is also called an *elementary event,* and a compound event is also called a *composite event.*

Simple Event

Each of the final outcomes for an experiment is called a **simple event.** In other words, a simple event includes one and only one outcome. Usually, a simple event is denoted by E_1, E_2, E_3, and so forth.

> **SIMPLE EVENT**
>
> An event that includes one and only one of the (final) outcomes for an experiment is called a simple event and is usually denoted by E_i.

The following two examples describe simple events.

Illustrating simple events: two selections.

EXAMPLE 4-4 Reconsider Example 4-3 about selecting two persons from the members of a club and observing whether the person selected each time is a male or a female. Each of the final four outcomes (*MM, MF, FM,* and *FF*) for this experiment is a simple event. These four events can be denoted by E_1, E_2, E_3, and E_4, respectively. Thus,

$$E_1 = (MM) \qquad E_2 = (MF) \qquad E_3 = (FM) \qquad \text{and} \qquad E_4 = (FF)$$ ■

Illustrating simple events: one draw.

EXAMPLE 4-5 Consider the experiment of randomly drawing a ball from a box that contains seven balls of different colors. Since any of the seven balls can be drawn, this experiment has seven possible outcomes. Each of these seven outcomes is a simple event. ■

Compound Event

A **compound event** consists of more than one outcome.

> **COMPOUND EVENT**
>
> A compound event is a collection of more than one outcome.

Compound events are denoted by A, B, C, D, . . . , or by A_1, A_2, A_3, . . . , B_1, B_2, B_3, ... The following examples describe compound events.

Illustrating compound event: two tosses of a coin.

EXAMPLE 4-6 Consider again the experiment of tossing a coin twice. Define the following event for this experiment:

$$A = \text{at least one tail is observed}$$

We will observe at least one tail if either one or two tails are obtained in two tosses. Thus, event A is said to occur if either one or two tails are observed in two tosses. One tail will be observed if either HT or TH happens, and two tails will be observed if TT occurs. Hence, event A consists of a collection of three outcomes: HT, TH, and TT. We can write event A as

$$A = \{HT, TH, TT\}$$

The Venn diagram of Figure 4.4 gives a graphic presentation of compound event A.

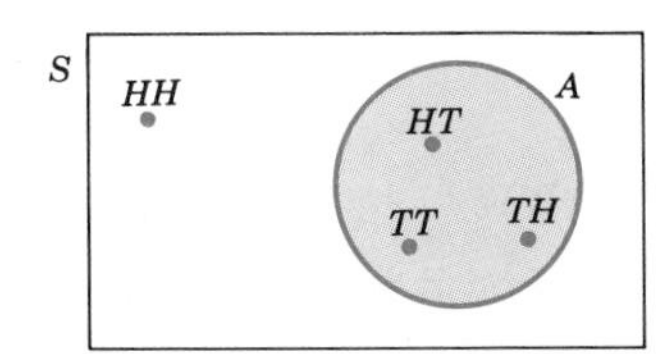

Figure 4.4 Venn diagram for event A. ■

Illustrating compound event: one roll of a die.

EXAMPLE 4-7 Consider the experiment of one roll of a die. Let B be the event that an even number is observed on the die. Then,

$$B = \{2, 4, 6\}$$

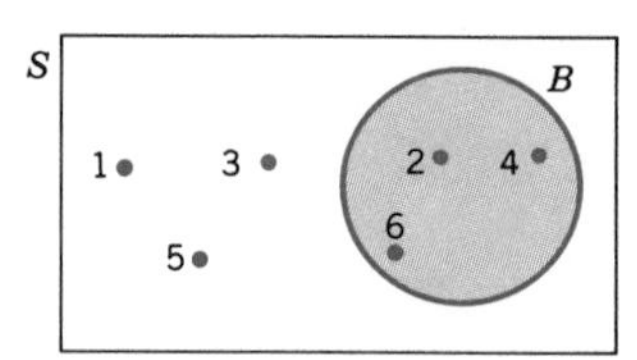

Figure 4.5 Venn diagram for event B.

The compound event B is shown in Figure 4.5. ■

Illustrating compound event: two selections.

EXAMPLE 4-8 Reconsider Example 4-3 about selecting two persons from the members of a club and observing whether the person selected at each selection is a male or a female. Let C be the event that at most one male is selected. Event C will occur if either none or one male is selected. Hence, event C is given by

$$C = \{MF, FM, FF\}$$

The Venn diagram of Figure 4.6 gives a graphic presentation of compound event C.

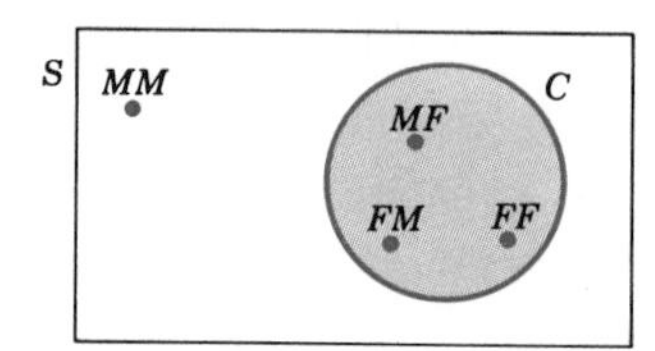

Figure 4.6 Venn diagram for event C.

■

Illustrating simple and compound events: two selections.

EXAMPLE 4-9 In a group of people, some are in favor of genetic engineering and the remaining are against it. Two persons are selected at random from this group and asked whether they are in favor of or against genetic engineering. How many distinct outcomes are possible? Draw a Venn diagram and a tree diagram for this experiment. List all the outcomes included in each of the following events and mention whether they are simple or compound events.

(a) Both persons are in favor of genetic engineering.
(b) At most one person is against genetic engineering.
(c) Exactly one person is in favor of genetic engineering.

Solution Let

$$F = \text{a person is in favor of genetic engineering}$$
$$A = \text{a person is against genetic engineering}$$

This experiment has the following four outcomes.

FF = both persons are in favor

FA = the first person is in favor and the second is against

AF = the first person is against and the second is in favor

AA = both persons are against genetic engineering

The Venn and tree diagrams in Figure 4.7 show these four outcomes.

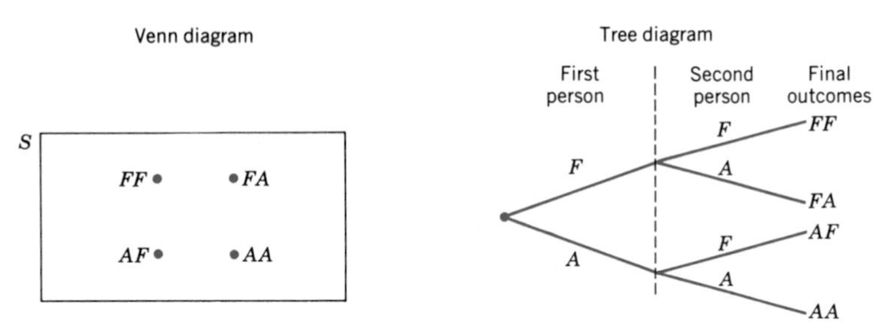

Figure 4.7 Venn and tree diagrams.

(a) The event "both persons are in favor of genetic engineering" will occur if FF is obtained. Thus,

$$\text{Both persons are in favor of genetic engineering} = \{FF\}$$

Because this event includes only one of the final four outcomes, it is a simple event.

(b) The event "at most one person is against" will occur if either none or one of the persons selected is against genetic engineering. Consequently,

$$\text{At most one person is against} = \{FF, FA, AF\}$$

Because this event includes more than one outcome, this is a compound event.

(c) The event "exactly one person is in favor" will occur if one of the two persons selected is in favor and the other is against genetic engineering. Hence, it includes the following two outcomes.

$$\text{Exactly one person is in favor} = \{FA, AF\}$$

Because this event includes more than one outcome, this is a compound event. ■

EXERCISES

4.1 Define the following terms: experiment, outcome, sample point, simple event, compound event.

4.2 Two students are randomly selected from a statistics class and it is observed whether or not they suffer from math anxiety. How many total outcomes are possible? Draw a tree diagram for this experiment. Draw a Venn diagram.

4.3 Draw a tree diagram for three tosses of a coin.

4.4 A hat contains a few red and a few green marbles. If two marbles are randomly drawn and the color of these marbles is observed, how many total outcomes are possible? Draw a tree diagram for this experiment. Show all the outcomes in a Venn diagram.

4.5 A test contains two multiple-choice questions. If a student makes a random guess to answer each question, how many outcomes are possible? Depict all these outcomes in a Venn diagram. Also draw a tree diagram for this experiment. (*Hint:* Consider two outcomes for each question: Either the answer is correct or it is wrong.)

4.6 A box contains a certain number of computer parts, a few of which are defective. Two parts are selected at random from this box and inspected to determine if they are good or defective. How many total outcomes are possible? Draw a tree diagram for this experiment.

4.7 In a group of people, some favor abortion and others do not. Three persons are selected at random from this group and their opinions in favor or against abortion are noted. How many total outcomes are possible? Draw a tree diagram for this experiment.

4.8 Refer to Exercise 4.2. List all the outcomes included in each of the following events. Indicate which are simple and which are compound events.

- **a.** Both students suffer from math anxiety.
- **b.** Exactly one student suffers from math anxiety.
- **c.** The first student does not suffer and the second suffers from math anxiety.
- **d.** None of the students suffers from math anxiety.

4.9 Refer to Exercise 4.4. List all the outcomes included in each of the following events. Indicate which are simple and which are compound events.

- **a.** Both marbles are of different colors.
- **b.** At least one marble is red.
- **c.** Not more than one marble is green.
- **d.** First marble is green and the second is red.

4.10 Refer to Exercise 4.5. List all the outcomes included in each of the following events and mention which are simple and which are compound events.

- **a.** Both answers are correct.
- **b.** At most one answer is wrong.
- **c.** The first answer is correct and the second is wrong.
- **d.** Exactly one answer is wrong.

4.11 Refer to Exercise 4.6. List all the outcomes included in each of the following events. Indicate which are simple and which are compound events.

- **a.** At least one part is good.
- **b.** Exactly one part is defective.
- **c.** The first part is good and the second is defective.
- **d.** At most one part is good.

4.12 Refer to Exercise 4.7. List all the outcomes included in each of the following events and mention which are simple and which are compound events.

- **a.** At most one person is against abortion.
- **b.** Exactly two persons are in favor of abortion.
- **c.** At least one person is against abortion.
- **d.** More than one person is against abortion.

4.2 CALCULATING PROBABILITY

Probability, which gives the likelihood of occurrence of an event, is denoted by P. The probability that a simple event E_i will occur is denoted by $P(E_i)$, and the probability that a compound event A will occur is denoted by $P(A)$.

PROBABILITY

Probability is a numerical measure of the likelihood that a specific event will occur.

TWO PROPERTIES OF PROBABILITY

The following are two important properties of probability.

1. **The probability of an event always lies in the range zero to 1.**

   Whether it is a simple or a compound event, the probability of an event is never less than zero or greater than 1. Using mathematical notation, we can write this property as follows.

$$0 \leq P(E_i) \leq 1$$
$$0 \leq P(A) \leq 1$$

An event that cannot occur has zero probability; such an event is called an **impossible event.** An event that is certain to occur has a probability equal to 1 and is called a **sure event.**

2. **The sum of the probabilities of all simple events (or final outcomes) for an experiment, denoted by $\Sigma P(E_i)$, is always 1.**

   Thus, for an experiment

$$\Sigma P(E_i) = P(E_1) + P(E_2) + P(E_3) + \cdots = 1$$

From this property, for the experiment of one toss of a coin:

$$P(H) + P(T) = 1$$

For the experiment of two tosses of a coin:

$$P(HH) + P(HT) + P(TH) + P(TT) = 1$$

For one game of football by a National Football League team:

$$P(\text{Win}) + P(\text{Loss}) + P(\text{Tie}) = 1$$

4.2.1 THREE CONCEPTUAL APPROACHES TO PROBABILITY

There are three conceptual approaches to probability: (1) classical probability, (2) the relative frequency concept of probability, and (3) the subjective probability concept. These three concepts of probability are explained next.

Classical Probability

Outcomes that have the same probability of occurrence are called **equally likely outcomes.** The classical probability rule is applied to compute the probabilities of events for an experiment all of whose outcomes are equally likely.

EQUALLY LIKELY OUTCOMES

Two or more outcomes (or events) that have the same probability of occurrence are said to be equally likely outcomes (or events).

According to the **classical probability rule,** the probability of a simple event is equal to 1 divided by the total number of outcomes for the experiment. This is obvious, as the sum of the probabilities of all final outcomes for an experiment is 1, and all the final outcomes are equally likely. On the other hand, the probability of a compound event A is equal to the number of outcomes favorable to event A divided by the total number of outcomes for the experiment.

CLASSICAL PROBABILITY RULE

$$P(E_i) = \frac{1}{\text{Total number of outcomes for the experiment}}$$

$$P(A) = \frac{\text{Number of outcomes favorable to } A}{\text{Total number of outcomes for the experiment}}$$

Examples 4-10 through 4-13 illustrate how probabilities of events are calculated using the classical probability rule.

Calculating probability of a simple event.

EXAMPLE 4-10 Find the probability of obtaining a head and the probability of obtaining a tail in one toss of a coin.

Solution The two outcomes, head and tail, are equally likely outcomes. Therefore,†

$$P(\text{head}) = \frac{1}{\text{Total number of outcomes}} = \frac{1}{2} = .50$$

†If the final answer for the probability of an event does not terminate within three decimal places, usually it will be rounded to three decimal places.

Similarly,

$$P(\text{tail}) = \frac{1}{2} = .50$$

■

Calculating probability of a compound event: one roll of a die.

EXAMPLE 4-11 Find the probability of obtaining an even number in one roll of a die.

Solution This experiment has six outcomes: 1, 2, 3, 4, 5, and 6. All these outcomes are equally likely. Let A be an event that an even number is observed on the die. Event A includes three outcomes: 2, 4, and 6. If any one of these three numbers is obtained, event A is said to occur. Hence,

$$P(A) = \frac{\text{Number of outcomes included in } A}{\text{Total number of outcomes}} = \frac{3}{6} = .50$$

■

Calculating probability of a compound event.

EXAMPLE 4-12 A club has 50 members, of whom 30 are males and 20 are females. Suppose one of these members is randomly selected for the presidency of the club. What is the probability that this member is a female?

Solution Because the selection is to be made randomly, each of the 50 members of the club has the same probability of being selected. Consequently, the experiment has a total of 50 equally likely outcomes. Twenty of these 50 outcomes are included in the event that "a female is selected." Hence,

$$P(\text{a female is selected}) = \frac{20}{50} = .40$$

■

Calculating probability of a simple and a compound event.

EXAMPLE 4-13 Consider the experiment of tossing a coin twice. Find the following probabilities.

(a) Two tails are obtained.
(b) At least one tail is obtained.

Solution The experiment of tossing a coin twice has the following four outcomes, where H denotes a head and T denotes a tail.

$$HH, \ HT, \ TH, \ TT$$

All four outcomes are equally likely.

(a) The event "two tails are obtained" includes only one outcome: TT. The probability of TT is

$$P(TT) = \frac{1}{4} = .25$$

(b) The event "at least one tail is obtained" occurs if HT, or TH, or TT happens. Thus,

$$P(\text{at least one tail is obtained}) = \frac{3}{4} = .75$$

■

Relative Frequency Concept of Probability

Suppose we want to calculate the following probabilities:

1. The probability that the next car that comes out of an auto factory is a "lemon"
2. The probability that the next baby born at a hospital is a girl
3. The probability that an 80-year-old person will live for at least one more year
4. The probability that the tossing of an unbalanced coin will result in a head
5. The probability that we will observe a 1-spot if we roll a loaded die

These probabilities cannot be computed using the classical probability rule because the various outcomes for the corresponding experiments are not equally likely. For example, the next car manufactured at an auto factory may or may not be a lemon. The two outcomes, "it is a lemon" and "it is not a lemon," are not equally likely. If they were, then (approximately) half the cars manufactured by this firm will be lemons and this may prove disastrous to the survival of the firm.

In such cases, to calculate probabilities we either use past data or generate new data by performing the experiment a large number of times. The relative frequency of an event is used as an approximation for the probability of that event. This method of assigning a probability to an event is called the **relative frequency concept of probability.** Because relative frequencies are determined by performing an experiment, the probabilities calculated using relative frequencies may change almost every time an experiment is repeated. For example, every time a new sample of 500 cars is selected from the production line of an auto factory, the number of lemons in those 500 cars is expected to be different. However, the variation in the percentage of lemons will be small if the sample size is large.

RELATIVE FREQUENCY AS AN APPROXIMATION OF PROBABILITY

If an experiment is repeated n times and an event A is observed f times, then, according to the relative frequency concept of probability:

$$P(A) = \frac{f}{n}$$

Examples 4-14 and 4-15 illustrate how the probabilities of events are approximated using the relative frequencies.

Approximating probability by relative frequency: sample data.

EXAMPLE 4-14 Ten of the 500 randomly selected cars manufactured at a certain auto factory are found to be lemons. Assuming that the lemons are manufactured randomly, what is the probability that the next car manufactured at this auto factory is a lemon?

Solution Let n denote the total number of cars and f be the number of lemons in n. Then,

$$n = 500 \quad \text{and} \quad f = 10$$

Using the relative frequency concept of probability, the probability that the next car is a lemon is

$$P(\text{next car is a lemon}) = \frac{f}{n} = \frac{10}{500} = .02$$

■

This probability is actually the relative frequency of lemons in 500 cars. Table 4.2 lists the frequency and relative frequency distributions for this example.

Table 4.2 Frequency Distribution of Cars

Car	f	Relative Frequency
Good	490	490/500 = .98
Lemon	10	10/500 = .02
	$n = 500$	Sum = 1.0

The column of relative frequencies in Table 4.2 is used as the column of approximate probabilities. Thus, from the relative frequency column:

$$P(\text{next car is a lemon}) = .02$$

and

$$P(\text{next car is a good car}) = .98$$

Note that relative frequencies are not probabilities but approximate probabilities. However, if the experiment is repeated again and again, this approximate probability of an outcome obtained from the relative frequency will approach the actual probability of that outcome. This is called the **Law of Large Numbers.**

LAW OF LARGE NUMBERS

If an experiment is repeated again and again, the probability of an event obtained from the relative frequency approaches the actual or theoretical probability.

Approximating probability by relative frequency: an experiment.

EXAMPLE 4-15 Allison has an unbalanced coin and she wants to determine the probability of obtaining a head when tossing this coin. How will she determine this probability?

Solution Because this coin is unbalanced, the events head and tail are not equally likely. Hence, the classical probability rule cannot be applied. Suppose Allison actually tosses this coin 1000 times and obtains 690 heads and 310 tails. Then,

$$n = \text{total tosses} = 1000 \qquad \text{and} \qquad f = \text{number of heads} = 690$$

Consequently

$$P(\text{head}) = \frac{f}{n} = \frac{690}{1000} = .69$$

Again, note that .69 is just an approximation of the probability of obtaining a head in one toss of this coin. Every time Allison repeats this experiment she may obtain a different probability of observing a head. However, because the number of tosses (1000) in this example is large, the variation is expected to be very small. ■

Subjective Probability

A number of experiments either do not have equally likely outcomes or cannot be repeated to generate data. In such cases, we cannot compute the probabilities of events using the classical probability rule or the relative frequency concept. For example, consider the following probabilities of events.

1. The probability that Carol, who is taking statistics this semester, will get an A in this course
2. The probability that the Dow Jones Industrial Average will be higher at the end of the next trading day
3. The probability that the Los Angeles Raiders will win the Super Bowl next season
4. The probability that Joe will lose the lawsuit that he has filed against his landlord

The classical probability rule or the relative frequency concept of probability cannot be applied to calculate these probabilities. All these examples belong to experiments that have neither equally likely outcomes nor the potential of being repeated. For example, Carol, who is taking statistics this semester, will take the test (or tests) only once and based on that she will either get an A or not. The two events "she will get an A" and "she will not get an A" are not equally likely. The probability assigned to an event in such cases is called **subjective probability.** It is based on the individual's own judgment, experience, information, and belief. Carol may assign a high probability to the event that she will get an A in statistics, whereas her instructor may assign a low probability to the same event.

SUBJECTIVE PROBABILITY

Subjective probability is the probability assigned to an event based on subjective judgment, experience, information, and belief.

Subjective probability is assigned arbitrarily. It is usually influenced by the biases, preferences, and experience of the person assigning the probability.

According to the U.S. Nuclear Regulatory Commission, the probability that a core meltdown like the one at the Chernobyl nuclear power station in the Ukraine

(USSR) will occur in the United States within the next 20 years is .50. This is an example of subjective probability because it is based on the research of the commission members and not on the performance of any experiment.

Another concept related to probability is that of odds. The odds are obtained by finding the ratio of the probability that an event will occur to the probability that an event will not occur. Case Study 4-1 is an example of the application of odds.

CASE STUDY 4-1 PROBABILITY AND ODDS

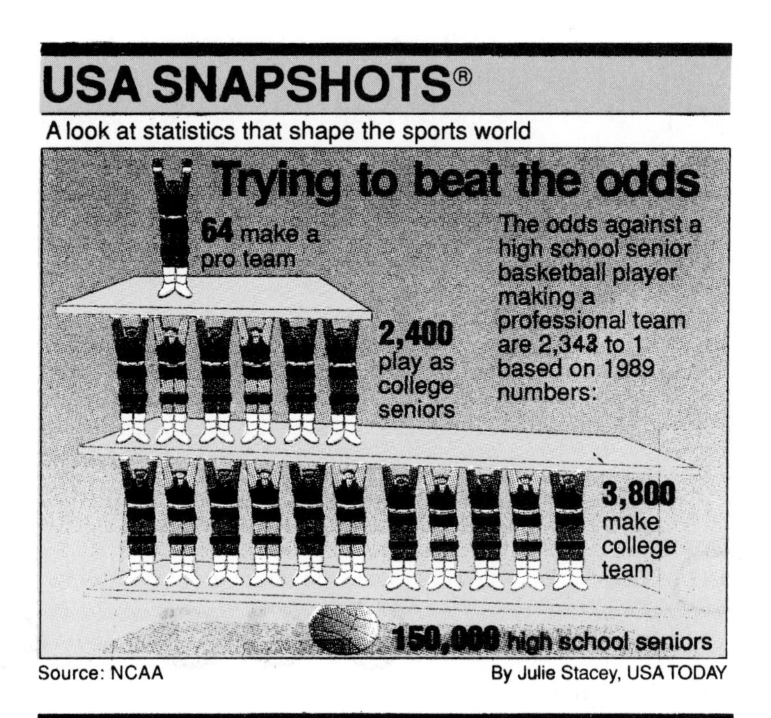

According to the information given in the graph, the odds against a high school senior basketball player making a professional team, based on 1989 data, are 2343 to 1. In other words, out of every 2344 (which is the sum of the two odds, 2343 and 1) high school senior basketball players, 2343 will not be able to make a professional team and 1 will make it. The probabilities of the two events, that a high school senior basketball player will not make a professional team and that he will make it, are calculated by dividing the respective odds by the sum of the two odds. Let

N = a high school senior basketball player will not make a pro team

M = a high school senior basketball player will make a pro team

Then

$$P(N) = \frac{2343}{2343 + 1} = .9995734$$

$$P(M) = \frac{1}{2343 + 1} = .0004266$$

If we take the ratio of $P(N)$ and $P(M)$, we obtain the odds.

How are these odds calculated from the data given in the graph? As we can observe, out of 150,000 high school senior basktball players, 64 made a pro team and 149,936 (which is 150,000 minus 64) did not. Hence, the odds against a high school senior basketball player making a pro team are 149,936 to 64. Dividing these two numbers by 64, the odds become approximately 2343 to 1.

Quiz: According to 1989 data, out of 265,000 high school senior football players, only 215 made a professional football team (*USA Today,* March 22, 1990). Calculate the odds against a high school senior football player making a professional football team.

Source: USA Today, March 21, 1990. Copyright © 1990, *USA Today*. Graph reprinted with permission.

EXERCISES

4.13 Briefly explain the two properties of probability.

4.14 Briefly describe an impossible and a sure event. What is the probability for the occurrence of each of these two events?

4.15 Which of the following values cannot be probabilities of events and why?

1/5, .97, −.35, 1.56, 5/3, 0.0, −2/7, 1.0

4.16 Which of the following values cannot be probabilities of events and why?

.46, 2/3, −.09, 1.42, .56, 9/4, −1/4, .02

4.17 In a statistics class of 45 students, 12 have a strong interest in statistics. Find the probability that a randomly selected student from this class has a strong interest in statistics.

4.18 In a group of 20 children, 10 are the only children of their parents. Find the probability that a randomly selected child from this group is the only child of his/her parents.

4.19 In a group of 50 executives, 27 have a type A personality. If one executive is selected at random from this group, what is the probability that this executive has a type A personality?

4.20 According to the U.S. Bureau of the Census, there are 13,116 thousand students aged 14 and over enrolled in colleges and universities in the United States. Of them, 5950 thousand are male and 7166 thousand are female. If one student is randomly selected from this population of students, what is the probability of selecting

a. a male **b.** a female?

4.21 According to the U.S. Bureau of the Census, there are 65,133 thousand families living in the United States. Of them, 33,213 thousand have no child under 18. If one family is selected at random from this total, what is the probability that this family has no child under 18?

4.22 A hat contains 40 marbles, 16 of which are red and 24 are green. If one marble is randomly selected out of this hat, what is the probability that this marble is

a. red **b.** green?

4.23 A die is rolled once. What is the probability that

a. a number less than 3 is obtained **b.** a number 3 to 6 is obtained?

4.24 According to the U.S. Bureau of Labor Statistics, 114,968 thousand persons aged 16 years and over are employed in the United States and 51,696 thousand of them are female. If one person is randomly selected from this group of employed people, what is the probability that this person is

a. a female **b.** a male?

4.25 According to a recent estimate by the U.S. Internal Revenue Service, there were 1650 thousand business partnerships in the United States. Of them, 890 thousand made profits and 760 thousand incurred losses. If one partnership is selected at random from these 1650 thousand partnerships, what is the probability that this partnership

a. made profits **b.** incurred losses?

4.26 According to the Administrative Office of the U.S. Courts, in 1988 a total of 594,567 bankruptcy petitions were filed in the United States. Of these, 68,501 were business petitions. If one of these 594,567 petitions is randomly selected, what is the probability that it is

a. a business bankruptcy petition **b.** a nonbusiness bankruptcy petition?

4.27 According to the American Medical Association, there are 506,474 active physicians practicing in the United States. Of them, 135,543 practice in the Northeast, 110,076 in the Midwest, 153,955 in the South, and 106,900 practice in the West. If one physician is selected at random from these 506,474 physicians, what is the probability that this physician practices in the

a. South **b.** Midwest **c.** Northeast **d.** West?

Do these four probabilities add up to 1.0? If yes, why?

4.28 A multiple-choice question in a test contains five answers. If Dianne chooses one answer based on "pure guess," what is the probability that her answer is

a. correct **b.** wrong?

Do these two probabilities add up to 1.0? If yes, why?

4.29 An NBA player has made 115 of the 130 free throws attempted so far this season. What is the (approximate) probability that he will make the next free throw he tries?

4.30 An NFL quarterback has completed 46 of the 105 passes made so far this season. What is the (approximate) probability that he will complete the next pass he makes?

4.31 According to the U.S. Bureau of the Census, there are 28,527 thousand persons aged 65 and over in the United States. Of them, 8672 thousand are living alone and the remaining are living with someone else. If one person aged 65 or over is randomly selected from these 28,527 thousand persons, what is the probability that this person is

a. living alone **b.** living with someone else?

Do these two probabilities add up to 1.0? If yes, why?

4.32 An unbalanced die is rolled 380 times and a 6-spot is obtained 95 times. What is the (approximate) probability of obtaining the 6-spot for this die?

4.33 A random sample of 1200 eighty-year-old persons was selected at the beginning of the year. Of them, 1030 were alive at the beginning of the next year. Based on the results of this survey, what is the (approximate) probability that an 80-year-old person will live for one more year?

4.34 A random sample of 800 college students showed that 240 of them are politically liberal. What is the (approximate) probability that a randomly selected college student is a liberal?

4.35 A university has a total of 320 professors and 64 of them are female. What is the probability that a randomly selected professor from this university is a female?

4.36 Out of the 3000 families living in an apartment complex in New York City, 600 paid no income tax last year. What is the probability that a randomly selected family from these 3000 families paid income tax last year?

4.3 MARGINAL AND CONDITIONAL PROBABILITIES

Suppose there are 100 doctors and nurses (let us call them medical personnel) at a hospital. Table 4.3 gives a two-way classification of these 100 medical personnel.

Table 4.3 Two-way Classification of Medical Personnel

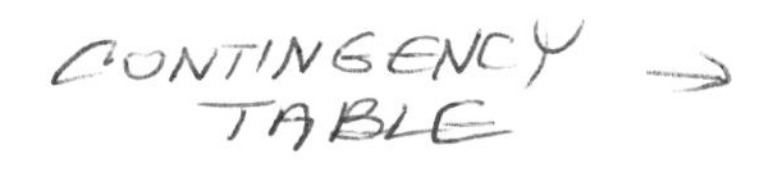

	Doctor	Nurse
Male	18	12
Female	7	63

Table 4.3 gives the distribution of 100 medical personnel based on two variables or characteristics: sex (male or female), and job (doctor or nurse). Such a table is called a *contingency table*. In Table 4.3, each box that contains a number is called a *cell*. Thus, there are four cells in this table. Each cell gives the frequency for two characteristics. For example, 18 medical personnel in this group possess two characteristics: They are male and doctors. In other words, there are 18 male doctors in this group. Similarly, there are 12 male nurses, 7 female doctors, and 63 female nurses.

By adding the row of totals and the column of totals to Table 4.3, we write Table 4.4.

Table 4.4 Two-way Classification of Medical Personnel with Totals

	Doctor	Nurse	Total
Male	18	12	30
Female	7	63	70
Total	25	75	100

Suppose one person is selected at random from this group of 100 medical personnel. This person may be classified either on the basis of sex alone or on the basis of job alone. If only one characteristic is considered at a time, the person selected can be a male, a female, a doctor, or a nurse. The probability of each of these four characteristics or events is called **marginal probability.** These probabilities are called marginal probabilities because they are calculated by dividing the corresponding row margins (totals for the rows) or column margins (totals for the columns) by the grand total.

MARGINAL PROBABILITY

Marginal probability is the probability of a single event without consideration of any other event.

For Table 4.4, the four marginal probabilities are calculated as follows.

$$P(\text{male}) = \frac{\text{Number of males}}{\text{Total number of medical personnel}} = \frac{30}{100} = .30$$

As we can observe, the probability that a male will be selected is obtained by dividing the total of the row labeled "Male"(30) by the grand total (100). Similarly,

$$P(\text{female}) = \frac{70}{100} = .70$$

$$P(\text{doctor}) = \frac{25}{100} = .25$$

and

$$P(\text{nurse}) = \frac{75}{100} = .75$$

These four marginal probabilities are shown along the right side and bottom of Table 4.5.

Table 4.5 Listing the Marginal Probabilities

	Doctor (*D*)	Nurse (*N*)	Total	
Male (*M*)	18	12	30	$P(M) = 30/100 = .30$
Female (*F*)	7	63	70	$P(F) = 70/100 = .70$
Total	25	75	100	
	$P(D) = 25/100 = .25$	$P(N) = 75/100 = .75$		

Now suppose that one person is selected at random from these 100 medical personnel. Furthermore, assume that it is known that this person (selected) is a male. In other words, the event that the person selected is a male has already occurred. What is the probability that the person selected is a doctor? This probability is written as follows.

$$P(\text{doctor}|\text{male})$$

Read as "given" (the vertical bar)

The event whose probability is to be determined (doctor)

This event has already occurred (male)

This probability, $P(\text{doctor}|\text{male})$, is called the **conditional probability** and it is read as "the probability that the person selected is a doctor given that this person is a male."

CONDITIONAL PROBABILITY

Conditional probability is the probability that an event will occur given that another event has already occurred. If A and B are two events, then the conditional probability of A is written as

$$P(A|B)$$

and read as "the probability of A given that B has already occurred."

Calculating conditional probability: two-way table.

EXAMPLE 4-16 Compute the conditional probability $P(\text{doctor}|\text{male})$ for the data on 100 medical personnel given in Table 4.4.

Solution $P(\text{doctor}|\text{male})$ is the conditional probability that a randomly selected person is a doctor given that this person is a male. It is known that the event "male" has already occurred. Based on the information that the person selected is a male, we can infer that the person selected must be one of the 30 males and, hence, must belong to the first row of Table 4.4. Therefore, we are concerned only with the first row of that table.

	Doctor	Nurse	Total
Male	18	12	30
	↑ Doctors among males		↑ Total number of males

The required conditional probability is calculated as

$$P(\text{doctor}|\text{male}) = \frac{\text{Number of doctors among males}}{\text{Total number of males}} = \frac{18}{30} = .60$$

As we can observe from this computation of conditional probability, the total number of males (the event that has already occurred) is written in the denominator and the number of doctors (the event whose probability we are to find) among males is written in the numerator. Note that we are considering the row of the event that has already occurred. The tree diagram in Figure 4.8 illustrates Example 4-16.

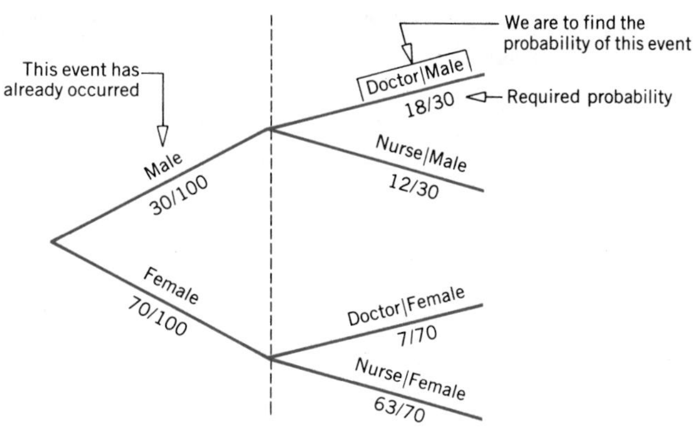

Figure 4.8 Tree diagram.

■

Calculating conditional probability: two-way table.

EXAMPLE 4-17 For the data of Table 4.4, calculate the conditional probability that a randomly selected person is a female, given that this person is a doctor.

Solution We are to compute the probability:

$$P(\text{female}|\text{doctor})$$

Since it is known that the person selected is a doctor, this person must belong to the first column (the column of doctors) and must be one of the 25 doctors.

Doctors	
18	
7	← Females among doctors
25	← Total number of doctors

Hence, the required probability is

$$P(\text{female}|\text{doctor}) = \frac{\text{Number of females among doctors}}{\text{Total number of doctors}} = \frac{7}{25} = .28$$

■

The tree diagram in Figure 4.9 illustrates this example.

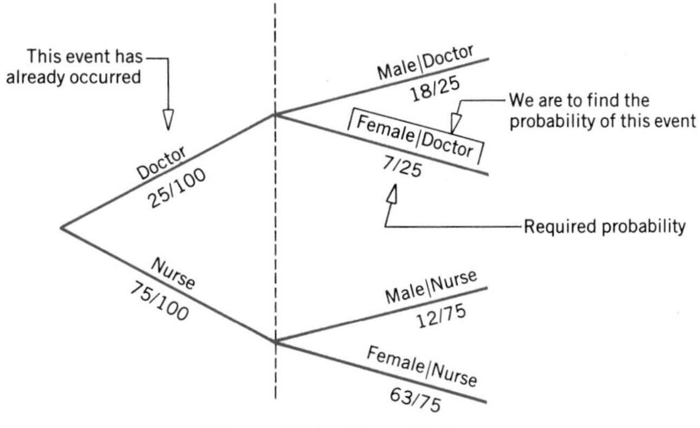

Figure 4.9 Tree diagram.

■

CASE STUDY 4-2 DOES DRINKING INCREASE THE PROBABILITY OF BREAST CANCER?

Dr. Arthur Schatzkin and his colleagues conducted a study of women to find a link between alcohol use and breast cancer. This study is based on a sample of 7188 women aged 25 to 74 years. The women were asked if they drink or not, and, if they drink, then how often they drink and how much they drink. One of the findings of this research is that the women who drink more frequently are younger, more educated, and slimmer, they smoke and eat more fat, and they have children at a later age in their lives. According to this study, the women who drink have a 50% greater chance to develop breast cancer than those who do not drink at all. Some earlier studies have determined that women overall have a 10% chance of developing breast cancer in their lives.

Based on these findings, we can calculate the following probabilities. Let us define the following events:

$$C = \text{a woman develops breast cancer}$$

$$D = \text{a woman drinks alcohol}$$

Then

$$P(C) = .10$$

Since the women who drink have a 50% greater chance to develop cancer, we obtain the following conditional probability:

$$P(C|D) = .15$$

When the women were categorized according to the amount they drink, the researchers found that those who drink more than zero and up to 1.2 grams of alcohol a day have a 40% greater chance of developing breast cancer than the nondrinkers. The chance is 50% higher for those who drink 1.3 to 4.9 grams a day and 60% higher for those who drink 5 grams or more a day. Based on this information, we can calculate the following conditional probabilities.

$$P(C|\text{drinks up to 1.2 grams a day}) = .14$$

$$P(C|\text{drinks 1.3 to 4.9 grams a day}) = .15$$

$$P(C|\text{drinks 5 grams or more a day}) = .16$$

Source: Arthur Schatzkin, Yvonne Jones, et al.: "Alcohol Consumption and Breast Cancer in the Epidemiologic Follow-up Study of the First National Health and Nutrition Examination Survey," *The New England Journal of Medicine,* 316(19):1169–1173, May 7, 1988.

CASE STUDY 4-3 DO PARENTS' PSYCHOLOGICAL PROBLEMS INCREASE THE PROBABILITIES OF CHILDREN'S PSYCHOLOGICAL PROBLEMS?

A study, conducted by Dr. Edward L. Schor over a period of six years, analyzed the effects of the psychological problems of parents on their children. Of the 372 two-child families that were studied, in 205 families at least one child had a psychological problem, and in the remaining 167 families, none of the children had a psychological problem. The following table gives the number of families with at least one child with a psychological problem, the number of families with both children without a psychological problem, and information as to whether the mother and father have or do not have psychological problems. In the table "No" means that the corresponding parent does not have a psychological problem and "Yes" means that he or she has a psychological problem.

	Mother: No		Mother: Yes		
	Father: No (N)	**Father: Yes (F)**	**Father: No (M)**	**Father: Yes (B)**	**Total**
Families with at least one child with a psychological problem (S)	47	17	65	76	205
Families with both children without a psychological problem (W)	84	19	34	30	167
Total	131	36	99	106	372

Let us define the following events for a randomly selected family:

S = at least one child has psychological problems

W = both children are without a psychological problem

N = none of the parents has psychological problems

F = father has psychological problems and mother does not

M = mother has psychological problems and father does not

B = both parents have psychological problems

Assuming that the results of the study apply, we can calculate the following (approximate) conditional probabilities:

$P(S|N) = 47/131 = .359 \qquad P(W|N) = 84/131 = .641$

$P(S|F) = 17/36 = .472 \qquad P(W|F) = 19/36 = .528$

$P(S|M) = 65/99 = .657 \qquad P(W|M) = 34/99 = .343$

$P(S|B) = 76/106 = .717 \qquad P(W|B) = 30/106 = .283$

Source: Edward L. Schor: "Families, Family Roles, and Psychological Diagnoses in Primary Care," *Journal of Developmental and Behavioral Pediatrics*, 9(6):327–332, December 1988. Copyright © by Williams & Wilkins, 1988. Data reproduced with permission.

4.4 MUTUALLY EXCLUSIVE EVENTS

Events that cannot occur together are called **mutually exclusive events.** If two or more events are mutually exclusive, then only one of them will occur every time we repeat the experiment. Two (or more) mutually exclusive events do not have any common outcome. Thus, in a case of mutually exclusive events, the occurrence of one event excludes the occurrence of the other event or events.

> **MUTUALLY EXCLUSIVE EVENTS**
>
> Events that cannot occur together are said to be mutually exclusive events.

For any experiment, the final outcomes are always mutually exclusive because one and only one of these outcomes is expected to occur in one repetition of the experiment. For example, consider tossing a coin twice. This experiment has four outcomes: *HH, HT, TH,* and *TT*. These outcomes are mutually exclusive because one and only one of these will occur when we toss this coin twice.

Illustrating mutually exclusive and mutually nonexclusive events.

EXAMPLE 4-18 Consider the following events for one roll of a die.

A = an even number is observed = $\{2, 4, 6\}$

B = an odd number is observed = $\{1, 3, 5\}$

C = a number less than 5 is observed = $\{1, 2, 3, 4\}$

Are events A and B mutually exclusive? What about events A and C?

Solution Figures 4.10 and 4.11 show the diagrams of events A and B and events A and C, respectively.

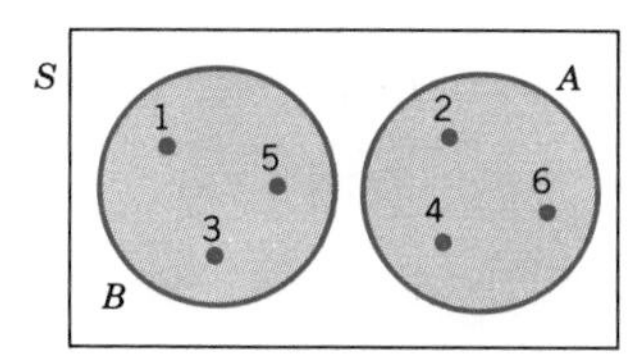

Figure 4.10 Mutually exclusive events A and B.

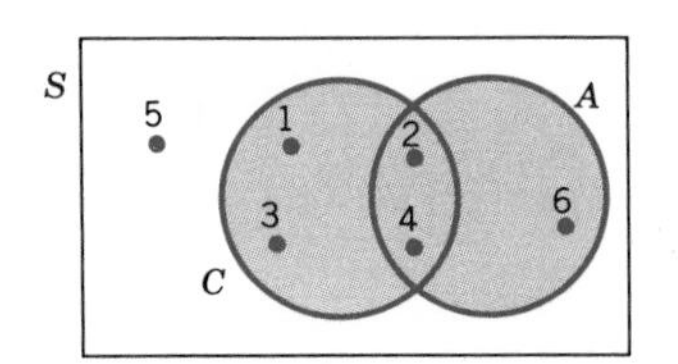

Figure 4.11 Nonmutually exclusive events A and C.

As we can observe from the definitions of events A and B and from Figure 4.10, events A and B do not have any common element. For one roll of a die, only one of the two events, A and B, can happen. Hence, these are two mutually exclusive events. On the other hand, we can observe from the definitions of events A and C and from Figure 4.11 that events A and C have two common outcomes: 2-spot and 4-spot. Thus, if we roll a die and obtain either a 2-spot or a 4-spot, then A and C both happen at the same time. Hence, events A and C are not mutually exclusive. ■

Illustrating mutually exclusive events.

EXAMPLE 4-19 Suppose an employee is selected at random from a large company. Consider the following two events.

$$N = \text{the employee selected is a nonsmoker}$$

$$S = \text{the employee selected is a smoker}$$

Are events N and S mutually exclusive?

Solution Event N consists of all those employees at this company who are nonsmokers and event S includes all those employees who are smokers. These two events are shown in Figure 4.12.

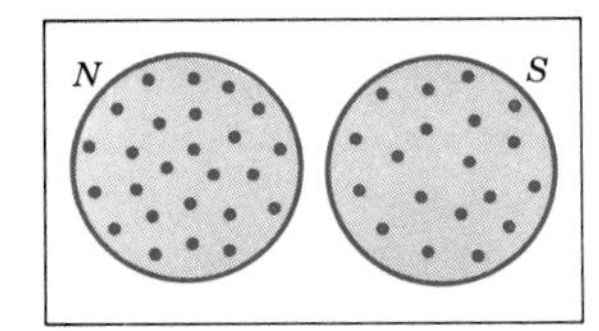

Figure 4.12 Mutually exclusive events N and S.

As we can observe from the definition of events N and S and from Figure 4.12, events N and S have no common outcome. They represent two distinct sets of employees: nonsmokers and smokers. Hence, these two events are mutually exclusive events. ■

Illustrating mutually nonexclusive events.

EXAMPLE 4-20 Refer to the information on 100 medical personnel given in Table 4.4 in Section 4.3. Are the two events "male" and "nurse" mutually exclusive?

Solution By looking at Table 4.4 we observe that 12 of the 100 persons listed in Table 4.4 are male as well as nurses. Hence, when we randomly select one person, if any of those 12 persons is selected then the events "male" and "nurse" happen together. Consequently, events "male" and "nurse"are not mutually exclusive. ■

4.5 INDEPENDENT VERSUS DEPENDENT EVENTS

In the case of two independent events, the occurrence of one event does not change the probability of the occurrence of the other event.

INDEPENDENT EVENTS

Two events are said to be independent if the occurrence of one does not affect the probability of the occurrence of the other. In other words, A and B are independent events if

$$\text{either } P(A|B) = P(A) \quad \text{or} \quad P(B|A) = P(B)$$

It can be shown that if one of these two conditions is true then the second will also be true, and if one is not true then the second will also not be true.

If the occurrence of one event affects the probability of occurrence of the other event, then the two events are said to be **dependent events.** Using probability notation, the two events will be dependent if either $P(A|B) \neq P(A)$ or $P(B|A) \neq P(B)$.

Illustrating two dependent events: two-way table.

EXAMPLE 4-21 Refer to the information on 100 medical personnel given in Table 4.4 in Section 4.3. Are the two events "female (F)" and "doctor (D)" independent?

Solution Events F and D will be independent if

$$P(F) = P(F|D)$$

Otherwise they will be dependent.

Using the information given in Table 4.4, we compute the following two probabilities.

$$P(F) = \frac{70}{100} = .70 \qquad \text{and} \qquad P(F|D) = \frac{7}{25} = .28$$

Because the two probabilities are not equal, the two events are dependent.

In this example, the dependence of D and F can also be proved by showing that the probabilities $P(D)$ and $P(D|F)$ are not equal. ■

Illustrating two dependent events.

EXAMPLE 4-22 According to the U.S. Census Bureau, there are a total of 177,677 thousand persons aged 18 and over living in the United States. Of them, 92,901 thousand are female and 111,345 thousand are married. Of all the females aged 18 and over, 56,128 thousand are married. Are events "female" and "married" independent?

Solution Based on the given information,

$$\text{Total persons} = 177{,}677 \text{ thousand}$$
$$\text{Total females} = 92{,}901 \text{ thousand}$$
$$\text{Total married persons} = 111{,}345 \text{ thousand}$$
$$\text{Female and married} = 56{,}128 \text{ thousand}$$

The two events, "female" and "married," will be independent if

$$P(\text{female}) = P(\text{female}|\text{married})$$

From this information,

$$P(\text{female}) = \frac{92{,}901}{177{,}677} = .523$$

and

$$P(\text{female}|\text{married}) = \frac{56{,}128}{111{,}345} = .504$$

Since the two probabilities are not equal, the events "female" and "married" are not independent.

Actually, using the given information, we can prepare Table 4.6 for Example 4-22. The numbers in the shaded cells are given to us. We calculate the remaining numbers by doing some arithmetic. Note that the numbers in the table are in thousands.

Table 4.6

	Married	Others	Total
Male	55,217	29,559	84,776
Female	56,128	36,773	92,901
Total	111,345	66,332	177,677

Using this table, we can find the following probabilities.

$$P(\text{female}) = \frac{92{,}901}{177{,}677} = .523$$

$$P(\text{female}|\text{married}) = \frac{56{,}128}{111{,}345} = .504$$

Hence, the two events are dependent. ■

Illustrating two independent events.

EXAMPLE 4-23 Of a total of 100 items manufactured on two machines, 15 are defective. Sixty of the total items were manufactured on the first machine and 9 of these 60 are defective. Are the events "machine type" and "defective item" independent?

Solution Let us define the following two events.

A = a randomly selected item is defective

B = a randomly selected item was manufactured on the first machine

Then, from the given information:

$$P(A) = \frac{15}{100} = .15 \quad \text{and} \quad P(A|B) = \frac{9}{60} = .15$$

Hence

$$P(A) = P(A|B)$$

Therefore, the two events, A and B, are independent.

Independence, in this case, means that the probability for any item to be defective is the same, .15, irrespective of the machine on which it was manufactured. In other words, the two machines are producing the same percentage of defective items.

We can write a two-way classification table for Example 4-23, like the one we wrote for Example 4-22, and then answer this example using the table. ■

TWO IMPORTANT OBSERVATIONS

The following are two important observations about mutually exclusive, independent, and dependent events.

1. Two events are either mutually exclusive or independent.† In other words,
   (a) mutually exclusive events are always dependent.
   (b) independent events are never mutually exclusive.
2. Dependent events may or may not be mutually exclusive.

4.6 COMPLEMENTARY EVENTS

Two mutually exclusive events that taken together include all the outcomes for an experiment are called **complementary events.**

> **COMPLEMENTARY EVENTS**
>
> The complement of event A, denoted by $\bar{A}$ and read as "A bar" or "A complement," is the event that includes all the outcomes for an experiment that are not in A.

Events A and $\bar{A}$ are complements of each other. Any two complementary events are always mutually exclusive. The Venn diagram in Figure 4.13 shows the complementary events A and $\bar{A}$.

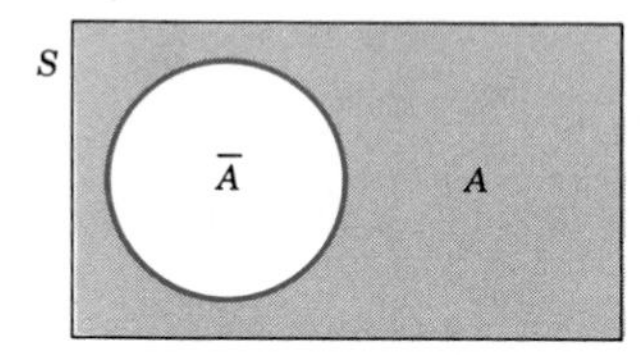

Figure 4.13 Venn diagram of two complementary events.

Because two complementary events, taken together, include all the outcomes for an experiment and because the sum of the probabilities of all outcomes is 1, it is obvious that

$$P(A) + P(\bar{A}) = 1$$

From this equation we can deduce that

$$P(A) = 1 - P(\bar{A}) \quad \text{and} \quad P(\bar{A}) = 1 - P(A)$$

Thus, if we know the probability of an event, we can find the probability of its complementary event by subtracting the given probability from 1.

†The exception to this rule occurs when at least one of the two events has a zero probability.

Calculating probabilities of complementary events.

EXAMPLE 4-24 In a lot of five washing machines, two are defective. If one machine is randomly selected, what are the two complementary events for this experiment and what are their probabilities?

Solution The two complementary events for this experiment are

$$A = \text{the machine selected is defective}$$

$$\bar{A} = \text{the machine selected is not defective}$$

Since there are two defective and three nondefective machines in this lot of five, the probabilities of events A and $\bar{A}$ are

$$P(A) = \frac{2}{5} = .40 \quad \text{and} \quad P(\bar{A}) = \frac{3}{5} = .60$$

As we can observe, the sum of the two probabilities is 1. ■

Calculating probabilities of complementary events.

EXAMPLE 4-25 There are a total of 120 professors at a college and 90 of them possess a PhD degree. If one professor is selected at random from this college, what are the two complementary events and their probabilities?

Solution The two complementary events are

$$A = \text{the randomly selected professor possesses a PhD degree}$$

$$\bar{A} = \text{the randomly selected professor does not possess a PhD degree}$$

The probabilities of the two events are

$$P(A) = \frac{90}{120} = .75 \quad \text{and} \quad P(\bar{A}) = 1 - .75 = .25$$

Figure 4.14 shows a Venn diagram for this example.

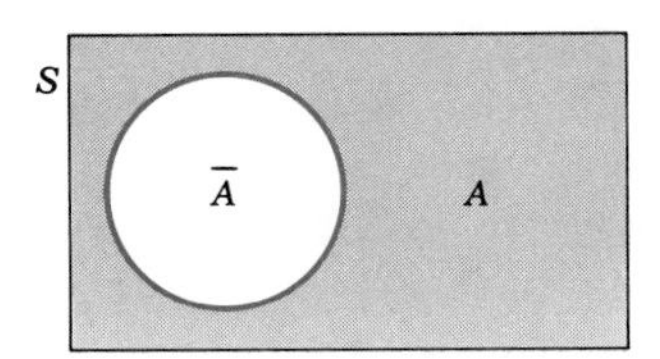

Figure 4.14 Venn diagram. ■

EXERCISES

4.37 The following table gives a two-way classification of the 1988 U.S. population (in thousands) by sex and age. (*Source:* U.S. Bureau of the Census.)

	Age in Years	
	Under 16	**16 and Over**
Male	28,914	91,289
Female	27,550	98,576

If one person is randomly selected from this population, find the probability that this person is

a. a female
b. 16 or over
c. a male given he is under 16
d. a female given she is 16 or over

4.38 All the 420 employees of a company were asked if they smoke or not and whether they are college graduates or not. Based on this information, the following two-way classification table was prepared.

	College Graduate	Not a College Graduate
Smoker	35	80
Nonsmoker	130	175

If one employee is selected at random from this company, find the probability that this employee is

a. a college graduate
b. a nonsmoker
c. a smoker given the employee is not a college graduate
d. a college graduate given the employee is a nonsmoker

4.39 The following table gives a two-way classification of all the scientists and engineers (in thousands) in the United States (*Source*: U.S. National Science Foundation).

	Male	Female
Scientist	1680	657
Engineer	2527	108

a. If one person is selected at random from these scientists and engineers, find the probability that this person is
   i. an engineer
   ii. a male
   iii. a scientist given the person is a female
   iv. a male given the person is a scientist
b. Are the events "scientist" and "engineer" mutually exclusive? What about the events "engineer" and "male"? Why or why not?
c. Are the events "female" and "engineer" independent? Why or why not?

4.40 The following table gives a two-way classification of all graduate students (in thousands) enrolled in institutions of higher education in the United States (*Source:* U.S. Department of Education).

	Type of Institute	
	Public	Private
Male	429	264
Female	516	243

a. If one graduate student is randomly selected from this population, find the probability that this student is
   i. attending a private institute
   ii. a female

iii. attending a public institute given the student is a female
iv. a male given the student is attending a private institute

b. Are the events "female" and "private" mutually exclusive? What about the events "public" and "private"? Why or why not?

c. Are the events "female" and "public" independent? Why or why not?

4.41 A group of 2000 randomly selected adults were asked if they are in favor of or against abortion. The following table gives the results of this survey.

	In Favor	Against
Male	495	405
Female	620	480

a. If one person is selected at random from these 2000 adults, find the probability that this person is
i. in favor of abortion
ii. against abortion
iii. in favor of abortion given the person is a female
iv. a male given the person is against abortion

b. Are the events "male" and "in favor" mutually exclusive? What about the events "in favor" and "against?" Why or why not?

c. Are the events "female" and "favor" independent? Why or why not?

4.42 A group of 150 randomly selected chief executive officers (CEOs) were tested for their type of personality. The following table gives the results of this survey.

	Type A	Type B
Male	78	42
Female	19	11

a. If one CEO is selected at random from this group, find the probability that this CEO
i. has a type A personality
ii. is a female
iii. is a male given he has a type A personality
iv. has a type B personality given she is a female

b. Are the events "female" and "type A personality" mutually exclusive? What about the events "type A" and "type B" personality? Why or why not?

c. Are the events "type A personality" and "male" independent? Why or why not?

4.43 Of a total of 100 items manufactured on two machines, 20 items are defective. Sixty of the total number of items were manufactured on the first machine and 10 of these 60 items are defective. Are the events "machine type" and "defective item" independent? (*Note:* Compare this exercise with Example 4-23.)

4.44 There are a total of 160 practicing physicians in a city. Of them, 55 are female and 25 are pediatricians. Of the 55 females, 8 are pediatricians. Are the events "female" and "pediatrician" independent? Are they mutually exclusive? Explain why or why not.

4.45 Define the following two events for two tosses of a coin.

$$A = \text{at least one head is obtained}$$

$$B = \text{both tails are obtained}$$

a. Are A and B mutually exclusive events? Are they independent? Explain why or why not.

b. Are A and B complementary events? If yes, first calculate the probability of B and then calculate the probability of A using the complementary event rule.

4.46 According to the U.S. Bureau of Labor Statistics, a total of 5729 thousand persons in the United States have more than one job and 3537 thousand of them are male. If one person is selected at random from these 5729 thousand persons, what are the two complementary events and their probabilities?

4.47 Let A be the event that a number less than 3 is obtained if we roll a die once. What is the probability of A? What is the complementary event of A, and what is its probability?

4.48 A hat contains 5 green, 8 red, and 7 blue marbles. Let A be the event that a red marble is drawn if we randomly select one marble out of this hat. What is the probability of A? What is the complementary event of A, and what is its probability?

4.49 Let A be the event that a driver stopped by police at night in Los Angeles is drunk. Assume that the probability of A is .15. What is the complementary event of A, and what is its probability?

4.50 Let A be the event that a randomly selected American will spend the next summer vacation in Europe. Assume this probability is .04. What is the complementary event of A, and what is its probability?

4.7 INTERSECTION OF EVENTS AND THE MULTIPLICATION RULE

This section discusses the intersection of two events and the application of the multiplication rule to compute the probability of the intersection of events.

4.7.1 INTERSECTION OF EVENTS

The intersection of two events is given by the outcomes that are common to both events.

> **INTERSECTION OF EVENTS**
>
> Let A and B be two events defined in a sample space. The intersection of A and B represents the collection of all outcomes that are common to both A and B and is denoted by:
>
> $$A \text{ and } B$$

The intersection of events A and B is also denoted by either $A \cap B$ or AB. Figure 4.15 illustrates the intersection of events A and B. The shaded area in this figure gives the intersection of events A and B, and it includes all the outcomes that are common to A and B.

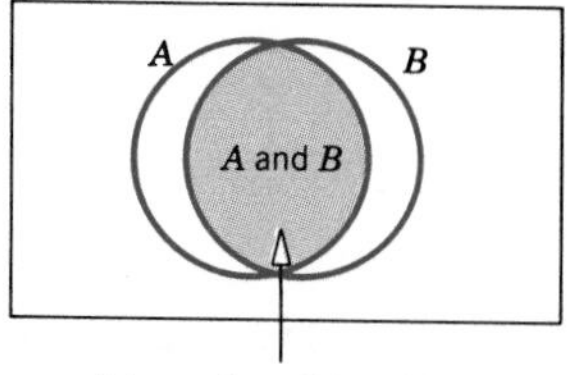

Figure 4.15 Intersection of events A and B.

Illustrating intersection of events.

EXAMPLE 4-26 According to the U.S. Department of Education, there are a total of 12,849 thousand students enrolled in institutions of higher education in the United States. Of them, 5946 thousand are male, 7371 thousand are full-time students, and 3636 thousand are male and full-time students. Show the intersection of events "male" and "full-time."

Solution Event "male and full-time" includes all students who are male and full-time. Thus, it represents the intersection of the events "male" and "full-time." This is shown in Figure 4.16.

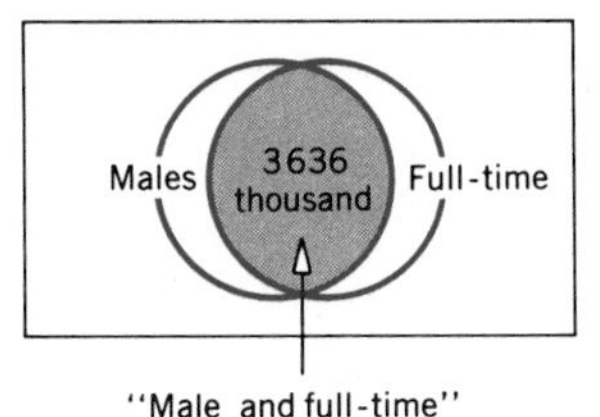

Figure 4.16 Intersection of events "male" and "full-time."

The shaded area in Figure 4.16 represents the intersection of events "male" and "full-time." As a result, it represents 3636 thousand students who are "male and full-time." ■

4.7.2 MULTIPLICATION RULE

The probability that events A and B happen together is called the **joint probability** of A and B and is written as $P(A \text{ and } B)$.

JOINT PROBABILITY

The probability of the intersection of two events is called their joint probability.

The probability of the intersection of two events is obtained by multiplying the marginal probability of one event by the conditional probability of the second event. This rule is called the *multiplication rule*.

MULTIPLICATION RULE

The probability of the intersection of two events A and B is

$$P(A \text{ and } B) = P(A)\, P(B|A)$$

The joint probability of events A and B can also be denoted by $P(A \cap B)$ or $P(AB)$.

Calculating joint probability of two events: two-way table.

EXAMPLE 4-27 The following table gives the classification of all employees of a company by sex and college degree.

CONTINGENCY TABLE

	College Graduate (C)	Not a College Graduate (D)	Total
Male (A)	7	20	27
Female (B)	4	9	13
Total	11	29	40

If one of these employees is selected at random for membership to the employee-management committee, what is the probability that this employee is a female and a college graduate?

Solution We are to calculate the probability of the intersection of events "female" (denoted by B) and "college graduate" (denoted by C). This probability will be computed using the formula:

$$P(B \text{ and } C) = P(B)\, P(C|B)$$

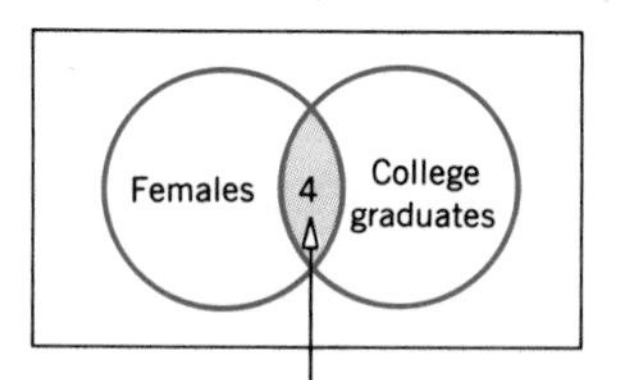

Figure 4.17

The shaded area in Figue 4.17 gives the intersection of events "female" and "college graduate." Notice that there are 13 females among 40 employees. Hence, the probability that a female is selected is

$$P(B) = \frac{13}{40}$$

To calculate the probability $P(C|B)$, we know that B has already occurred. Consequently, the employee selected is one of the 13 females. In the table, there are 4 college graduates among 13 female employees. Hence, the conditional probability of C given B is

$$P(C|B) = \frac{4}{13}$$

The joint probability of B and C is

$$P(B \text{ and } C) = P(B)\, P(C|B) = \left(\frac{13}{40}\right)\left(\frac{4}{13}\right) = .100$$

Thus, the probability is .100 that a randomly selected employee is a female and a college graduate. ■

Similarly, we can compute three other joint probabilities for the table of Example 4-27 as follows.

$$P(A \text{ and } C) = P(A)\,P(C|A) = \left(\frac{27}{40}\right)\left(\frac{7}{27}\right) = .175$$

$$P(A \text{ and } D) = P(A)\,P(D|A) = \left(\frac{27}{40}\right)\left(\frac{20}{27}\right) = .500$$

$$P(B \text{ and } D) = P(B)\,P(D|B) = \left(\frac{13}{40}\right)\left(\frac{9}{13}\right) = .225$$

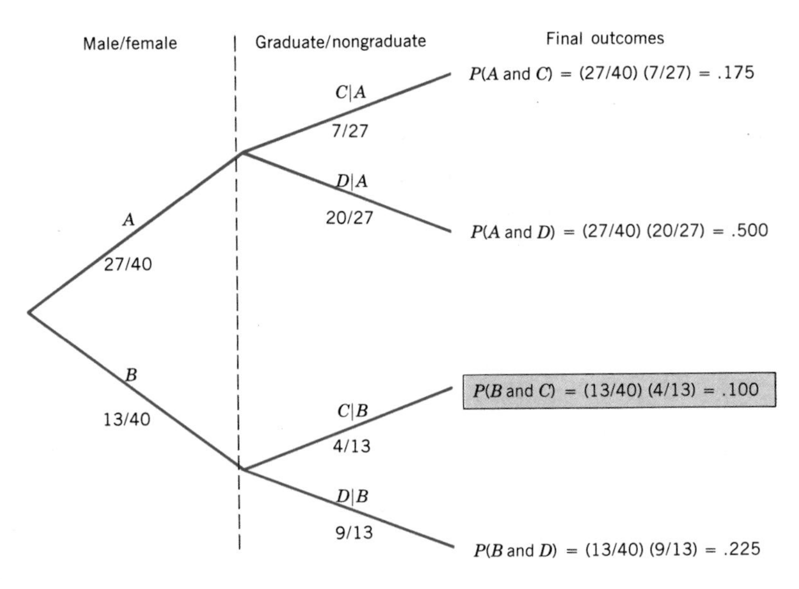

Figure 4.18 Tree diagram.

The tree diagram in Figure 4.18 shows all four joint probabilities for this example. The joint probability of B and C is highlighted in the tree diagram.

Calculating joint probability of two events.

EXAMPLE 4-28 In a lot of 20 machines, 4 are defective. If 2 machines are selected at random (without replacement) from this lot, what is the probability that both are defective?

Solution Let us define the following events for this experiment.

G_1 = event that the first machine selected is good

D_1 = event that the first machine selected is defective

G_2 = event that the second machine selected is good

D_2 = event that the second machine selected is defective

We are to calculate the joint probability of D_1 and D_2, which is given by

$$P(D_1 \text{ and } D_2) = P(D_1)\, P(D_2|D_1)$$

As we know, there are 4 defective machines in 20. Consequently, the probability of selecting a defective machine at the first selection is

$$P(D_1) = \frac{4}{20}$$

To calculate the probability $P(D_2|D_1)$, we know that the first machine selected is defective because D_1 has already occurred. Because the selections are made without replacement, there are 19 total machines and 3 of them are defective at the time of the second selection. Therefore,

$$P(D_2|D_1) = \frac{3}{19}$$

Hence the required probability is

$$P(D_1 \text{ and } D_2) = P(D_1)\, P(D_2|D_1) = \left(\frac{4}{20}\right)\left(\frac{3}{19}\right) = .032$$

The tree diagram in Figure 4.19 shows the selection procedure and the final four outcomes for this experiment together with their probabilities. The joint probability of D_1 and D_2 is highlighted in the tree diagram.

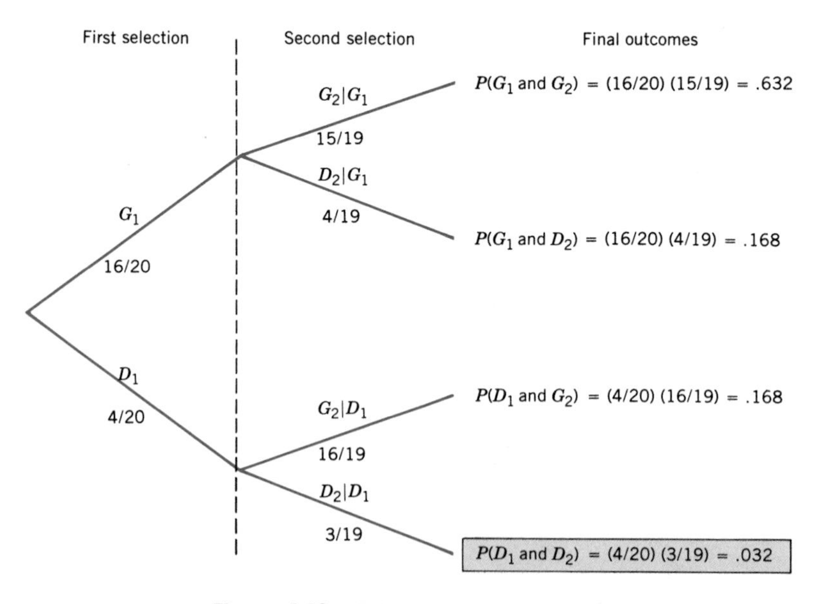

Figure 4.19 Selecting two machines.

■

Conditional probability was discussed in Section 4.3. It is obvious from the formula for joint probability that if we know the probability of an event A and the joint probability of events A and B, then we can calculate the conditional probability of B given A.

CONDITIONAL PROBABILITY

If A and B are two events, then,

$$P(B|A) = \frac{P(A \text{ and } B)}{P(A)} \quad \text{and} \quad P(A|B) = \frac{P(A \text{ and } B)}{P(B)}$$

given that $P(A) \neq 0$ and $P(B) \neq 0$.

Calculating conditional probability of an event.

EXAMPLE 4-29 The probability that a randomly selected student from a college is a senior is .20, and the joint probability that the student selected is a business major and a senior is .12. Find the conditional probability that a student selected at random is a business major given that he or she is a senior.

Solution Let us define the following two events.

$$A = \text{the student selected is a senior}$$

$$B = \text{the student selected is a business major}$$

From the given information,

$$P(A) = .20 \quad \text{and} \quad P(A \text{ and } B) = .12$$

Hence

$$P(B|A) = \frac{P(A \text{ and } B)}{P(A)} = \frac{.12}{.20} = .60$$

■

MULTIPLICATION RULE FOR INDEPENDENT EVENTS

The foregoing discussion of the multiplication rule was based on the assumption that the two events are dependent. Now suppose that events A and B are independent. Then,

$$P(A) = P(A|B) \quad \text{and} \quad P(B) = P(B|A)$$

By substituting $P(B)$ for $P(B|A)$ into the formula for the joint probability of A and B, we obtain

$$P(A \text{ and } B) = P(A)\,P(B)$$

MULTIPLICATION RULE FOR INDEPENDENT EVENTS

The probability of the intersection of two independent events A and B is

$$P(A \text{ and } B) = P(A)\,P(B)$$

Calculating joint probability of two independent events.

EXAMPLE 4-30 An office building has two fire detectors. The probability is .02 that any fire detector of this type will not go off during a fire. Find the probability that neither of the two fire detectors will go off in case of fire.

Solution In this example, the two fire detectors are independent. This is so because whether one fire detector goes off during a fire or not, it has no affect on the second fire detector. Define the following two events.

A = the first fire detector fails to go off during a fire

B = the second fire detector fails to go off during a fire

Then the joint probability of A and B is

$$P(A \text{ and } B) = P(A)\,P(B) = (.02)\,(.02) = .0004$$ ■

Calculating marginal and joint probabilities of (independent) events.

EXAMPLE 4-31 The probability that a salesperson sells more than 25 items during any week is .30, and the probability that the salesperson is a male is .65. If these two events are independent, find the probability that a randomly selected salesperson

(a) is a female
(b) sells at most 25 items during any week
(c) is a male and sells at most 25 items during a given week
(d) sells more than 25 items during any week and is a female

Solution We are given that

$$P(\text{male}) = .65 \quad \text{and} \quad P(\text{sells more than 25 items}) = .30$$

(a) The probability that the salesperson selected is a female is

$$P(\text{female}) = 1 - P(\text{male}) = 1 - .65 = .35$$

(b) The probability that the salesperson selected sells at most 25 items during any week is

$$P(\text{sells at most 25 items}) = 1 - P(\text{sells more than 25 items}) = 1 - .30 = .70$$

(c) The probability that the salesperson selected is a male and sells at most 25 items during a given week is

$$P(\text{male and sells at most 25 items}) = P(\text{male})\,P(\text{sells at most 25 items}) = (.65)\,(.70) = .455$$

Note that it is given that these two events are independent.

(d) The probability that the salesperson selected sells more than 25 items during any week and is a female is

$$P(\text{sells more than 25 items and female}) = P(\text{sells more than 25 items})\,P(\text{female})$$
$$= (.30)\,(.35) = .105$$ ■

The multiplication rule can be extended to calculate the joint probability of more than two events. The following example illustrates such a case for independent events.

Calculating joint probability of three events.

EXAMPLE 4-32 The probability that a patient is allergic to penicillin is .20. If this drug is given to three patients,

(a) find the probability that all three of them are allergic to it
(b) find the probability that at least one of them is not allergic to it

Solution

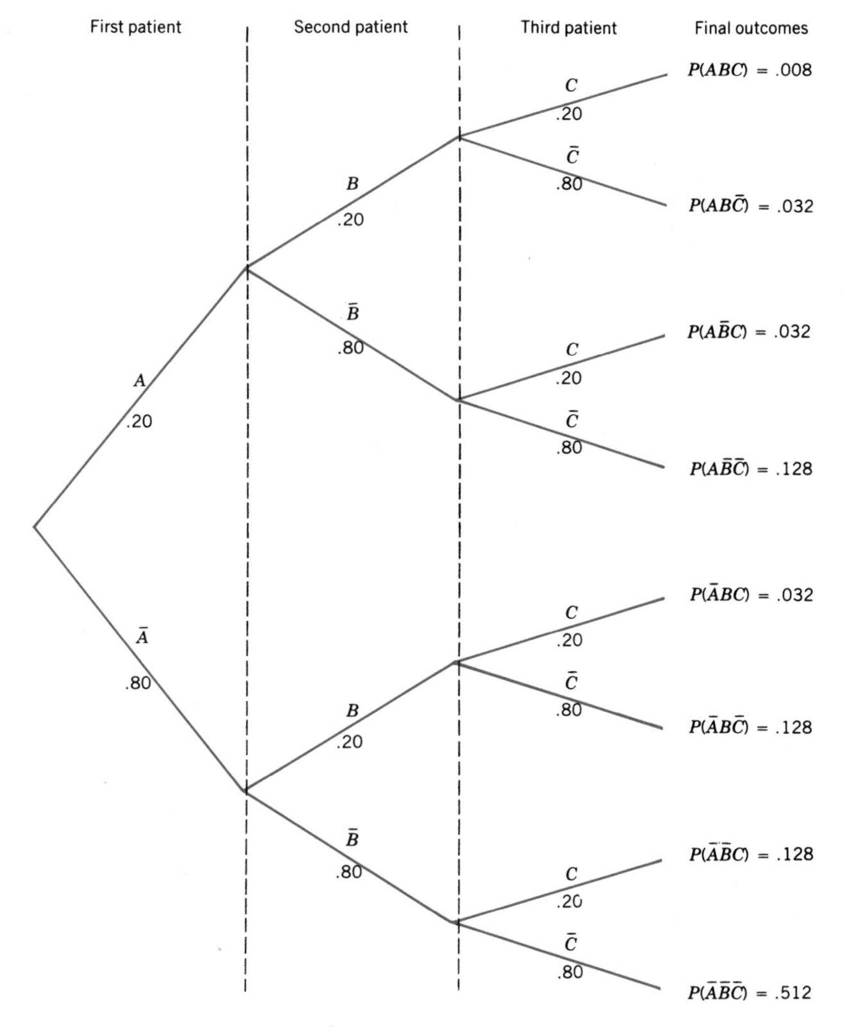

Figure 4.20

(a) Let A, B, and C denote the events that the first, second, and third patients are allergic to penicillin, respectively. We are to find the joint probability of A, B, and C. All three events are independent because whether one patient is allergic or not does not depend on whether any of the other patients is allergic. Hence,

$$P(A \text{ and } B \text{ and } C) = P(A)\,P(B)\,P(C) = (.20)\,(.20)\,(.20) = .008$$

The tree diagram in Figure 4.20 shows all the outcomes for this experiment. Events $\bar{A}$, $\bar{B}$, and $\bar{C}$ are the complementary events of A, B, and C, respectively. They represent the events that the respective patients are not allergic to penicillin. Note that the intersection of events A, B, and C is written as ABC in the tree diagram.

(b) Let us define the following events.

$$G = \text{all three patients are allergic}$$

$$H = \text{at least one patient is not allergic}$$

Events G and H are two complementary events. Event G consists of the intersection of events A, B, and C. Hence,

$$P(G) = P(A \text{ and } B \text{ and } C) = .008$$

Therefore, using the complementary event rule:

$$P(H) = 1 - P(G) = 1 - .008 = .992$$

■

CASE STUDY 4-4 BASEBALL PLAYERS HAVE "SLUMPS" AND "STREAKS"

Going "0 for July," as former infielder Bob Aspromonte once put it, is enough to make a baseball player toss out his lucky bat or start seriously searching for flaws in his hitting technique. But the culprit is usually just simple mathematics.

Statistician Harry Roberts of the University of Chicago's Graduate School of Business studied the records of major-league baseball players and found that a batter is no more likely to hit worse when he is in a slump than when he is in a hot streak. The occurrences of hits followed the same pattern as purely random events such as pulling marbles out of a hat. If there were one white marble and three black ones in the hat, for example, then a white marble would come out about one quarter of the time—a .250 average. In the same way, a player who hits .250 will in the long run get a hit every four times at bat.

But that doesn't mean the player will hit the ball exactly every fourth time he comes to the plate—just as it's unlikely that the white marble will come out exactly every fourth time.

Even a batter who goes hitless 10 times in a row might safely be able to pin the blame on statistical fluctuations. The odds of pulling a black marble out of a hat 10 times in a row is about 6 percent—not a frequent occurrence, but not impossible, either. Only in the long run do these statistical fluctuations even out. . . .

If we assume a player hits .250 in the long run, the probability that this player does not hit during any visit to the plate is .75. Hence, we can calculate the probability that he goes hitless 10 times in a row as follows.

$$P(\text{hitless 10 times in a row}) = (.75)\,(.75)\ldots(.75) \text{ ten times}$$
$$= .056$$

Note that each visit to the plate is independent and the probability that a player goes hitless 10 times in a row is given by the intersection of 10 hitless visits. This probability has been rounded off to "about 6 percent" in this illustration.

Source: *U.S. News & World Report,* July 11, 1988, p. 46.

JOINT PROBABILITY OF MUTUALLY EXCLUSIVE EVENTS

We know from an earlier discussion that two mutually exclusive events cannot happen together. Consequently, their joint probability is zero. Obviously, if two events are mutually exclusive and if one of them happens, then the second will not occur. Hence, their joint probability is zero.

JOINT PROBABILITY OF MUTUALLY EXCLUSIVE EVENTS

The joint probability of two mutually exclusive events is always zero. If A and B are two mutually exclusive events, then,

$$P(A \text{ and } B) = 0$$

Illustrating joint probability of two mutually exclusive events.

EXAMPLE 4-33 A drug is tried on a patient to determine if it cures him or not. Let

C = event that the patient is cured

N = event that the patient is not cured

What is the joint probability of C and N?

Solution The two events C and N are mutually exclusive. Either the patient will be cured or he will not be cured. Hence,

$$P(C \text{ and } N) = 0$$

■

Illustrating joint probability of two mutually exclusive events.

EXAMPLE 4-34 A coin is tossed twice. Find the joint probability of HH and TT.

Solution The events HH and TT are two mutually exclusive events. Therefore,

$$P(HH \text{ and } TT) = P(HH)\ P(TT|HH) = \left(\frac{1}{4}\right)(0) = 0$$

Note that HH and TT are two mutually exclusive events. Therefore, if HH has occurred then TT cannot occur. Consequently, the conditional probability that TT will happen given HH has already occurred is zero. ■

4.8 COUNTING RULE

The experiments dealt with so far in this chapter have had only a few outcomes, which were easy to list. However, for experiments with a large number of outcomes, it may not be easy to list all outcomes. In such cases, we may use the **counting rule** to find the total number of outcomes.

COUNTING RULE

If an experiment consists of three steps and if the first step can result in m outcomes, the second step in n outcomes, and the third step in k outcomes, then,

$$\text{Total outcomes for the experiment} = mnk$$

The counting rule can easily be extended to apply to an experiment with less or more than three steps.

Applying the counting rule: 2 steps.

EXAMPLE 4-35 Suppose we toss a coin twice. This experiment has two steps: the first toss and the second toss. Each step has two outcomes: a head and a tail. Thus,

$$\text{Total outcomes for two tosses of a coin} = 2 \cdot 2 = 4$$

We know from an earlier discussion that the four outcomes for this experiment are HH, HT, TH, and TT. ■

Applying the counting rule: 16 steps.

EXAMPLE 4-36 A National Football League team will play 16 games during a regular season. Each game can result in one of three outcomes: a win, a loss, or a tie. The total possible outcomes for 16 games are calculated as follows.

$$\begin{aligned}\text{Total outcomes} &= 3 \cdot 3 \cdot 3 \cdot 3 \cdot 3 \cdot 3 \cdot 3 \cdot 3 \cdot 3 \cdot 3 \cdot 3 \cdot 3 \cdot 3 \cdot 3 \cdot 3 \cdot 3 \\ &= 3^{16} = 43{,}046{,}721\end{aligned}$$

One of the 43,046,721 possible outcomes is 16 wins.† ■

Applying the counting rule: 3 steps.

EXAMPLE 4-37 In a collection there are 8 novels, 10 poetry books, and 6 short story books. We are to select one of each of the three kinds of books. How many total outcomes are possible?

Solution This experiment is made of three steps: selecting a novel, a poetry book, and a short story book. There are 8 outcomes for the first step because any of the 8 novels can be chosen, 10 outcomes for the second step, and 6 outcomes for the third step. Hence,

$$\text{Total outcomes} = 8 \cdot 10 \cdot 6 = 480$$ ■

Applying the counting rule: 5 steps.

EXAMPLE 4-38 Consider the experiment of tossing a coin five times. What is the probability that all five tosses will result in heads?

Solution This experiment has five steps, one for each toss. Each step has two possible outcomes. Hence,

$$\text{Total outcomes for 5 tosses of a coin} = 2 \cdot 2 \cdot 2 \cdot 2 \cdot 2 = (2)^5 = 32$$

One of these 32 outcomes gives all 5 heads: *HHHHH*. Also, all the outcomes for this experiment are equally likely. Therefore,

$$P(HHHHH) = \frac{1}{32} = .031$$ ■

EXERCISES

4.51 Find the joint probability of A and B for the following.

a. $P(A) = .40$ and $P(B|A) = .32$
b. $P(B) = .65$ and $P(A|B) = .36$
c. $P(A) = .82$ and $P(B|A) = .54$

4.52 Find the joint probability of A and B for the following.

a. $P(B) = .59$ and $P(A|B) = .77$
b. $P(A) = .28$ and $P(B|A) = .15$
c. $P(B) = .92$ and $P(A|B) = .64$

†Using a calculator to evaluate 3^{16}: If your calculator contains a y^x or an x^y key, you can use that key to simplify 3^{16} as follows: First enter 3 on the calculator, then press the y^x key, next enter 16, and finally press the = key. The screen of the calculator will display 43046721 as the answer.

4.53 Given that A and B are two independent events, find their joint probability for the following.

a. $P(A) = .61$ and $P(B) = .27$
b. $P(A) = .39$ and $P(B) = .73$
c. $P(A) = .77$ and $P(B) = .48$

4.54 Given that A and B are two independent events, find their joint probability for the following.

a. $P(A) = .20$ and $P(B) = .86$
b. $P(A) = .57$ and $P(B) = .32$
c. $P(A) = .45$ and $P(B) = .72$

4.55 Given that A, B, and C are three independent events, find their joint probability for the following.

a. $P(A) = .20$, $P(B) = .46$, and $P(C) = .15$
b. $P(A) = .44$, $P(B) = .27$, and $P(C) = .33$

4.56 Given that A, B, and C are three independent events, find their joint probability for the following.

a. $P(A) = .39$, $P(B) = .67$, and $P(C) = .75$
b. $P(A) = .71$, $P(B) = .34$, and $P(C) = .41$

4.57 Given that $P(A) = .30$ and $P(A \text{ and } B) = .24$, find $P(B|A)$.

4.58 Given that $P(B) = .65$ and $P(A \text{ and } B) = .45$, find $P(A|B)$.

4.59 The following table gives a two-way classification (by sex and marital status) of persons who hold more than one job. The numbers in the table are in thousands.

	Single	Married	Other
Male	767	2447	323
Female	681	1001	510

Source: U.S. Bureau of Labor Statistics.

If one person is randomly selected from this group, find the following probabilities.

a. P(male and other) **b.** P(female and married)

4.60 The following table gives a two-way classification of the U.S. population (in millions) based on sex and health insurance coverage.

	Covered by Health Insurance	Not Covered by Health Insurance
Male	98.4	17.5
Female	107.5	15.3

Source: U.S. Bureau of the Census.

If one person is randomly selected from this population, find the following probabilities.

a. P(male and not covered)
b. P(female and covered)

4.61 The following table, reproduced from Exercise 4.38, provides information on all 420 employees of a company who were asked if they smoke or not and whether they are college graduates or not.

	College Graduate	Not a College Graduate
Smoker	35	80
Nonsmoker	130	175

a. If one employee is randomly selected from this company, find the following probabilities.
  i. *P*(nonsmoker and nongraduate)
  ii. *P*(smoker and graduate)

b. Find *P*(graduate and nongraduate). Is this probability zero? If yes, why?

4.62 The following table gives a two-way classification of all faculty members of a university based on sex and tenure.

	Tenured	Nontenured
Male	74	28
Female	29	12

a. If one of these faculty members is selected at random, find the following probabilities.
  i. *P*(male and nontenured)
  ii. *P*(tenured and female)

b. Find *P*(tenured and nontenured). Is this probability zero? If yes, why?

4.63 In a group of 10 persons, 4 have a type A personality and 6 have a type B personality. If two persons are selected at random from this group, what is the probability that both of them have a type B personality? Draw a tree diagram for this problem.

4.64 In a statistics class of 45 students, 12 have a strong interest in statistics. If two students are selected at random from this class, what is the probability that both of them have a strong interest in statistics? Draw a tree diagram for this problem.

4.65 In a group of 15 students, 5 have liberal views. If two students are randomly selected from this group, what is the probability that the first of them has liberal views and the second does not? Draw a tree diagram for this problem.

4.66 A small company has 20 employees and 12 of them are married. If two employees are randomly selected from this company, what is the probability that the first of them is married and the second is not? Draw a tree diagram for this problem.

4.67 The probability is .76 that a family owns a house. If two families are randomly selected, what is the probability that none of them owns a house? (Note that the probability is .76 that any family owns a house. Hence, all families are independent.)

4.68 The probability is .35 that an adult has never flown on an airplane. If two adults are selected at random, what is the probability that the first of them has never flown on an airplane and the second has? Draw a tree diagram for this problem. (Note that the probability is .35 that any adult has never flown on an airplane. Hence, all adults are independent.)

4.69 The probability that any given person is allergic to a certain drug is .03. What is the probability that none of the three randomly selected persons is allergic to this drug? Assume that all three persons are independent.

4.70 The probability that a farmer is in debt is .75. What is the probability that three randomly selected farmers are all in debt? Assume independence of events.

4.71 The probability that a student is a male is .48 and that a student likes statistics is .35. If these two events are independent, what is the probability that a student selected at random is

a. a male and does not like statistics
b. a female and likes statistics

4.72 The probability that an adult man is an alcoholic is .12 and that he is self-employed is .18. If these two events are independent, what is the probability that a randomly selected man is

a. self-employed and a nonalcoholic
b. an alcoholic and not self-employed

4.73 The probability that a household owns a house is .76. The probability that a household owns a house and is a married couple is .69. Find the conditional probability that a randomly selected household is a married couple given it owns a house.

4.74 The probability that an adult has lung cancer is .01. The probability that an adult is a smoker and has lung cancer is .007. Find the conditional probability that a randomly selected adult is a smoker given that this person has lung cancer.

4.75 According to the U.S. Department of Education, there are 12,849 thousand students enrolled at all U.S. institutions of higher education. Of them, 7371 thousand are full-time students and 3735 thousand are female and full-time students. Find the probability that a student selected at random from all students enrolled at U.S. institutions of higher education is a female given that this student is a full-time student.

4.76 The probability that a randomly selected student from all students enrolled at U.S. institutions of higher education is a business major is .11 and the probability that a student is a business major and a male is .06. (*Source:* Calculated from the U.S. Department of Education data.) What is the probability that this student is a male given that he is a business major?

4.77 How many different outcomes are possible for 4 rolls of a die?

4.78 How many different outcomes are possible for 10 tosses of a coin?

4.79 A man owns 11 shirts, 9 ties, and 8 pairs of trousers. If he is to randomly select one shirt, one tie, and one pair of trousers to wear on a certain day, how many different outcomes (selections) are possible?

4.80 A student is to select three courses for next semester. If this student wants to randomly select one course from each of eight economics courses, six mathematics courses, and five computer courses, how many different outcomes are possible?

4.81 A restaurant menu has four kinds of soups, eight kinds of main courses, five kinds of desserts, and six kinds of drinks. If a customer randomly selectes one item from each of these four categories, how many different outcomes are possible?

4.9 UNION OF EVENTS AND THE ADDITION RULE

This section discusses the union of events and the addition rule that is applied to compute the probability of the union of events.

4.9.1 UNION OF EVENTS

The **union of two events** A and B includes all outcomes that are either in A or in B or in both A and B.

UNION OF EVENTS

Let A and B be two events defined in a sample space. The union of events A and B is the collection of all outcomes that belong either to A or to B or to both A and B and is denoted by

$$A \text{ or } B$$

The union of events A and B is also denoted by "$A \cup B$." The diagram in Figure 4.21 illustrates the union of events A and B.

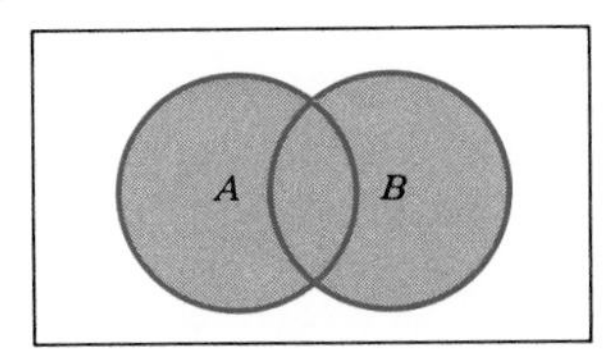

Shaded area gives the union of A and B **Figure 4.21** Union of events A and B.

Illustrating the union of two events.

EXAMPLE 4-39 According to the U.S. Department of Education, there are 11,048 thousand undergraduate students enrolled at institutions of higher education in the United States. Of them, 5979 thousand are female, 6463 thousand are full-time students, and 3299 thousand are female and full-time students. Describe the union of events "female" and "full-time."

Solution The union of events "female" and "full-time" for undergraduate students enrolled at institutions of higher education includes all students who are either females or full-time or both. The number of such students is

$$5979 + 6463 - 3299 = 9143 \text{ thousand}$$

Thus, there are a total of 9143 thousand undergraduate students enrolled at institutions of higher education who are either female or full-time students or both.

Why did we subtract 3299 thousand from the sum of 5979 thousand and 6463 thousand? The reason is that 3299 thousand students (which represent the intersection of events "female" and "full-time") are common to both events "female" and "full-time" and, hence, are counted twice. To avoid double counting, we subtracted 3299 thousand from the sum of the other two numbers. We can observe this double counting from Table 4.7, which is constructed using the given information. The sum of the

Table 4.7

	Full-time	Part-time	Total
Male	3,164	1,905	5,069
Female	3,299	2,680	5,979
Total	6,463	4,585	11,048

numbers in the three shaded cells gives the students who are either females or full-time or both. However, if we add the totals of the row labeled "Female" and the column labeled "Full-time," we count 3299 twice. Note that the numbers in the table are in thousands.

Figure 4.22 shows the diagram for the union of events "female" and "full-time."

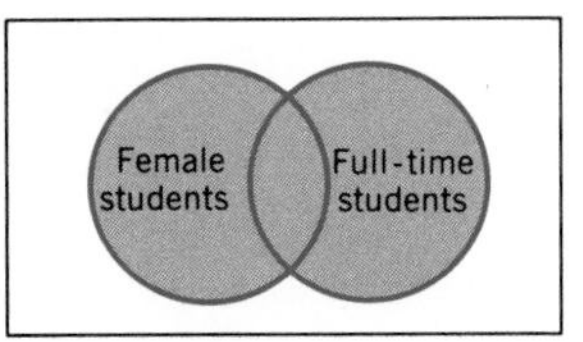

Shaded area gives the union of two events and includes 9143 thousand students **Figure 4.22** Union of events "female" and "full-time." ■

4.9.2 ADDITION RULE

The method used to calculate the probability of the union of events is called the **addition rule.** It is defined as follows.

ADDITION RULE

The probability of the union of two events A and B is

$$P(A \text{ or } B) = P(A) + P(B) - P(A \text{ and } B)$$

Thus, to calculate the probability of the union of two events A and B, we add their marginal probabilities and subtract their joint probability from this sum. We must subtract the joint probability of A and B from the sum of their marginal probabilities to avoid double counting due to common outcomes in A and B.

Calculating probability of union of two events: two-way table.

EXAMPLE 4-40 A university president has proposed that all students must do 50 hours of community service as a requirement for graduation. Three-hundred faculty members and students from this university were asked about their opinion on this issue. The following table gives a two-way classification of the responses of these faculty members and students.

	Opinion			
	Favor	**Oppose**	**Neutral**	**Total**
Faculty	45	15	10	70
Student	90	110	30	230
Total	135	125	40	300

Find the probability that one person selected at random from these 300 faculty members and students is a faculty member or favors community service.

Solution Let us define the following events.

A = the person selected is a faculty member

B = the person selected favors community service

From the information given in the table,

$$P(A) = \frac{70}{300} = .233$$

$$P(B) = \frac{135}{300} = .450$$

and

$$P(A \text{ and } B) = P(A)\,P(B|A) = \left(\frac{70}{300}\right)\left(\frac{45}{70}\right) = .150$$

Using the addition rule,

$$P(A \text{ or } B) = P(A) + P(B) - P(A \text{ and } B) = .233 + .450 - .150 = .533$$

Thus, the probability that a randomly selected person from these 300 faculty members and students is a faculty member or favors community service is .533. ■

The probability in Example 4-40 can also be calculated without using the addition rule formula. The total number of persons in the table who are either faculty members or favor community service or both is

$$45 + 15 + 10 + 90 = 160$$

Hence, the required probability is

$$P(A \text{ or } B) = \frac{160}{300} = .533$$

Calculating probability of union of two events.

EXAMPLE 4-41 There are a total of 5729 thousand persons with mutiple jobs in the United States. Of them, 3537 thousand are male, 1448 thousand are single, and 767 thousand are male and single (*Source:* U.S. Bureau of the Census). What is the probability that a randomly selected person with multiple jobs is a male or single?

Solution Let us define the following two events.

A = the randomly selected person is a male

B = the randomly selected person is single

From the given information,

$$P(A) = \frac{3537}{5729} = .617$$

$$P(B) = \frac{1448}{5729} = .253$$

and

$$P(A \text{ and } B) = \frac{767}{5729} = .134$$

Hence,

$$P(A \text{ or } B) = P(A) + P(B) - P(A \text{ and } B) = .617 + .253 - .134 = .736$$ ■

Calculating probability of union of two events.

EXAMPLE 4-42 Refer to Example 4-31. The probability that a salesperson sells more than 25 items during any week is .30 and that a salesperson is a male is .65. If these two events are independent, what is the probability that a salesperson selected at random is

(a) A female or sells at most 25 items during a given week
(b) A male or sells at most 25 items during a given week

Solution Let

M = a randomly selected salesperson is a male

F = a randomly selected salesperson is a female

A = a randomly selected salesperson sells more than 25 items

B = a randomly selected salesperson sells at most 25 items

Then,

$$P(M) = .65 \qquad P(F) = 1 - P(M) = 1 - .65 = .35$$
$$P(A) = .30 \quad \text{and} \quad P(B) = 1 - P(A) = 1 - .30 = .70$$

(a) The probability that the salesperson selected is a "female" or "sells at most 25 items during a given week" is

$$\begin{aligned} P(F \text{ or } B) &= P(F) + P(B) - P(F \text{ and } B) \\ &= P(F) + P(B) - P(F)\,P(B) \\ &= (.35) + (.70) - (.35)\,(.70) = .805 \end{aligned}$$

Note that since events F and B are independent, $P(F \text{ and } B) = P(F)P(B)$.

(b) The probability that the salesperson selected is a "male" or "sells at most 25 items during a given week" is

$$\begin{aligned} P(M \text{ or } B) &= P(M) + P(B) - P(M \text{ and } B) \\ &= P(M) + P(B) - P(M)\,P(B) \\ &= (.65) + (.70) - (.65)\,(.70) = .895 \end{aligned}$$

■

ADDITION RULE FOR MUTUALLY EXCLUSIVE EVENTS

We know from an earlier discussion that the joint probability of two mutually exclusive events is zero. When A and B are mutually exclusive events, the term $P(A \text{ and } B)$ in the addition rule becomes zero and is dropped from the formula. Thus, the probability of the union of two mutually exclusive events is given by the sum of their marginal probabilities.

ADDITION RULE FOR MUTUALLY EXCLUSIVE EVENTS

The probability of the union of two mutually exclusive events A and B is

$$P(A \text{ or } B) = P(A) + P(B)$$

Calculating probability of union of two mutually exclusive events.

EXAMPLE 4-43 According to the Census Bureau, there are a total of 84,776 thousand males 18-years-old and over in the United States. Of them, 21,517 thousand are single, 55,217 thousand are married, 2293 thousand are widowed, and 5749 thousand are divorced. What is the probability that a randomly selected male from the population of all U.S. males aged 18 or over is single or married?

Solution Let us define the following events.

A = the male selected is single

B = the male selected is married

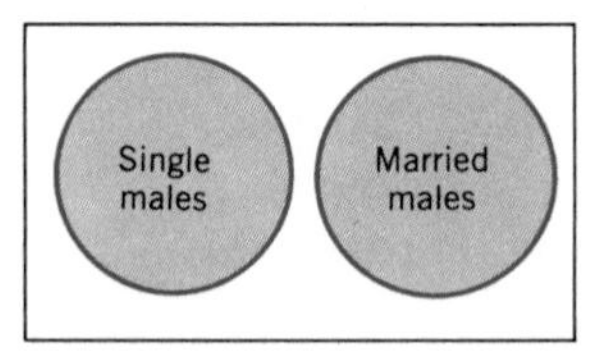

Figure 4.23

The events A and B are mutually exclusive because a male selected can be either single or married but not both. The diagram in Figure 4.23 shows the two events. From the given information,

$$P(A) = \frac{21{,}517}{84{,}776} = .254 \quad \text{and} \quad P(B) = \frac{55{,}217}{84{,}776} = .651$$

Hence

$$P(A \text{ or } B) = P(A) + P(B) = .254 + .651 = .905$$ ■

The addition rule formula can easily be extended to apply to more than two events. The following example illustrates this.

Calculating probability of union of three mutually exclusive events.

EXAMPLE 4-44 Consider the experiment of rolling a die twice. Find the probability that the sum of the numbers obtained on two rolls is either 5 or 7 or 10.

Solution The experiment of rolling a die twice has a total of 36 outcomes, which are listed in Table 4.8. Assuming that the die is balanced, these 36 outcomes are equally likely.

The events that give the sum of two numbers equal to 5 or 7 or 10 are marked in the table. As we can observe, the three events "the sum is 5," "the sum is 7," and

Table 4.8 Two Rolls of a Die

		Second Roll of the Die					
		1	**2**	**3**	**4**	**5**	**6**
First Roll of the Die	**1**	(1,1)	(1,2)	(1,3)	(1,4)	(1,5)	(1,6)
	2	(2,1)	(2,2)	(2,3)	(2,4)	(2,5)	(2,6)
	3	(3,1)	(3,2)	(3,3)	(3,4)	(3,5)	(3,6)
	4	(4,1)	(4,2)	(4,3)	(4,4)	(4,5)	(4,6)
	5	(5,1)	(5,2)	(5,3)	(5,4)	(5,5)	(5,6)
	6	(6,1)	(6,2)	(6,3)	(6,4)	(6,5)	(6,6)

"the sum is 10" are mutually exclusive. Four outcomes give a sum of 5, six give a sum of 7, and three outcomes give a sum of 10. Thus,

$$P(\text{sum is 5 or 7 or 10}) = P(\text{sum is 5}) + P(\text{sum is 7}) + P(\text{sum is 10})$$

$$= \frac{4}{36} + \frac{6}{36} + \frac{3}{36} = \frac{13}{36} = .361$$

■

Calculating probability of union of three mutually exclusive events, using a tree diagram.

EXAMPLE 4-45 The probability that a person is in favor of genetic engineering is .55 and that a person is against it is .45. Two persons are randomly selected, and it is observed whether they favor or oppose genetic engineering.

(a) Draw a tree diagram for this experiment.

(b) Find the probability that at least one of the two persons favors genetic engineering.

Solution

(a) Let

$$F = \text{a person is in favor of genetic engineering}$$

$$A = \text{a person is against genetic engineering}$$

This experiment has four outcomes: both persons are in favor (FF), the first person is in favor and the second is against (FA), the first person is against and the second is in favor (AF), and both persons are against genetic engineering (AA). The tree diagram in Figure 4.24 (on next page) shows these four outcomes and their probabilities.

(b) The probability that at least one person favors genetic engineering is given by the union of events FF, FA, and AF. These three outcomes are mutually exclusive. Hence, using the probabilities from the tree diagram,

$$P(\text{at least one person favors}) = P(FF \text{ or } FA \text{ or } AF)$$

$$= P(FF) + P(FA) + P(AF)$$

$$= .3025 + .2475 + .2475 = .7975$$

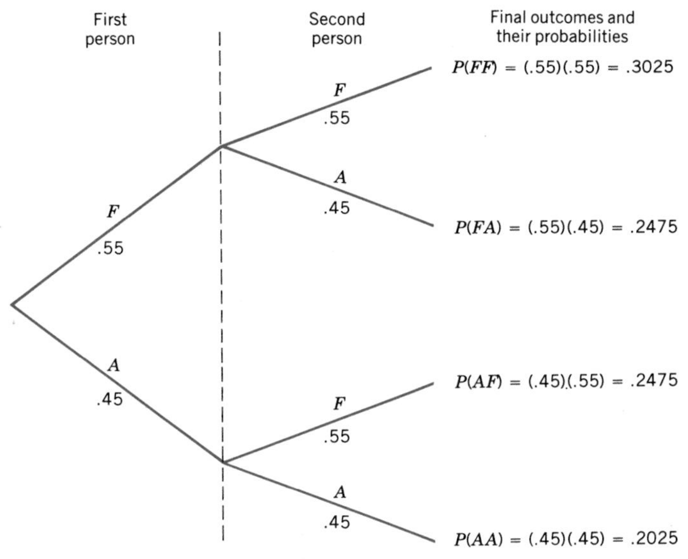

Figure 4.24 Tree diagram.

EXERCISES

4.82 Find $P(A \text{ or } B)$ for the following.

a. $P(A) = .58$, $P(B) = .66$, and $P(A \text{ and } B) = .47$
b. $P(A) = .72$, $P(B) = .42$, and $P(A \text{ and } B) = .53$

4.83 Find $P(A \text{ or } B)$ for the following.

a. $P(A) = .18$, $P(B) = .49$, and $P(A \text{ and } B) = .23$
b. $P(A) = .83$, $P(B) = .71$, and $P(A \text{ and } B) = .68$

4.84 Given that A and B are two mutually exclusive events, find $P(A \text{ or } B)$ for the following.

a. $P(A) = .57$, and $P(B) = .32$
b. $P(A) = .16$, and $P(B) = .49$

4.85 Given that A and B are two mutually exclusive events, find $P(A \text{ or } B)$ for the following.

a. $P(A) = .25$, and $P(B) = .17$
b. $P(A) = .38$, and $P(B) = .09$

4.86 The following table, reproduced from Exercise 4.62, gives a two-way classification of all faculty members of a university based on sex and tenure.

	Tenured	Nontenured
Male	74	28
Female	29	12

If one of these faculty members is selected at random, find the following probabilities.

a. $P(\text{female or nontenured})$ **b.** $P(\text{tenured or male})$.

4.87 The following table gives a two-way classification of first-time freshman enrollment (in thousands) at U.S. institutions of higher education (*Source:* U.S. Department of Education).

	Full-time	Part-time
Male	779	267
Female	848	352

If one of these students is selected at random, find the following probabilities.

a. *P*(male or full-time) **b.** *P*(part-time or female).

4.88 Refer to data from a survey of 2000 adults about their views on the issue of abortion given in Exercise 4.41. The table from that exercise is reproduced here.

	In Favor	Against
Male	495	405
Female	620	480

If one person is randomly selected from these 2000 persons, find the following probabilities.

a. *P*(female or in favor) **b.** *P*(against or male).

4.89 Refer to data on the type of personality of 150 chief executive officers given in Exercise 4.42. That table is reproduced here.

	Type A	Type B
Male	78	42
Female	19	11

If one CEO is randomly selected from these 150 CEOs, find the following probabilities.

a. *P*(type A or male) **b.** *P*(female or type B)

4.90 The probability that a family owns a washing machine is .78, that it owns a VCR is .71, and that it owns both a washing machine and a VCR is .58. What is the probability that a randomly selected family owns a washing machine or a VCR?

4.91 The probability that a football player weighs more than 230 pounds is .69, that he is at least 75 inches tall is .55, and that he weighs more than 230 pounds and is at least 75 inches tall is .43. Find the probability that a randomly selected football player weighs more than 230 pounds or is at least 75 inches tall.

4.92 The probability that a randomly selected student from a university is a senior is .18, a computer major is .14, and a senior and a computer major is .04. Find the probability that a student selected at random from this university is a senior or a computer major.

4.93 The probability that a person has a checking account is .74, a savings account is .31, and both accounts is .22. Find the probability that a randomly selected person has a checking or a savings account.

4.94 The probability that a randomly selected elementary or secondary school teacher from a city is a female is .68, holds a second job to make ends meet is .42, and a female and holds a second job is .29. Find the probability that an elementary or secondary school teacher selected at random from this city is a female or holds a second job.

4.95 According to U.S. Bureau of the Census estimates, there are a total of 177,677 thousand persons aged 18 and over. Of them, 38,881 thousand are single and 111,345 thousand are married. If one person is randomly selected from this population of people aged 18 and over, what is the probability that this person is single or married?

4.96 The probability of a student getting an A in a sociology class is .24, and that of getting a B is .28. What is the probability that a randomly selected student from this class will get an A or B in this class?

4.97 Seventy-two percent of a town's voters favor the recycling issue, 12% oppose it, and 16% are indifferent. What is the probability that a randomly selected voter from this town will either favor recycling or be indifferent?

4.98 According to U.S. Bureau of the Census estimates, there are a total of 65,133 thousand families living in the United States. Of them, 26,871 thousand have two members per family, and 13,729 thousand have four members per family. If one family is selected at random from all these families, what is the probability that it has two or four members?

4.99 The probability that a corporation makes charitable contributions is .68. Two corporations are selected at random, and it is noted whether or not they make charitable contributions.

a. Draw a tree diagram for this experiment.

b. Find the probability that at most one corporation makes charitable contributions.

4.100 The probability that an open-heart operation is successful is .72. What is the probability that in two randomly selected open-heart operations at least one will be successful? Draw a tree diagram for this experiment.

4.101 Refer to Exercise 4.71. The probability of a student being a male is .48 and that of liking statistics is .35. If these two events are independent, what is the probability that a randomly selected student is

a. a male or does not like statistics **b.** a female or likes statistics

4.102 Refer to Exercise 4.72. The probability that a male adult is alcoholic is .12 and that he is self-employed is .18. If these two events are independent, what is the probability that a randomly selected male is

a. self-employed or a nonalcoholic **b.** an alcoholic or not self-employed

GLOSSARY

Classical probability rule The method of assigning probabilities to outcomes or events of an experiment with equally likely outcomes.

Complementary events Two events that taken together include all the outcomes for an experiment but do not contain any common outcome.

Compound event An event that contains more than one outcome of an experiment. It is also called a *composite event.*

Conditional probability The probability of an event subject to the condition that another event has already occurred.

Dependent events Two events for which the occurrence of one changes the probability of the occurrence of the other.

Equally likely outcomes or events Two (or more) outcomes or events that have the same probability of occurrence.

Event A collection of one or more outcomes of an experiment.

Experiment A process with well-defined outcomes that, when performed, results in one and only one of the outcomes per repetition.

Impossible event An event that cannot occur.

Independent events Two events for which the occurrence of one does not influence the probability of the occurrence of the other.

Intersection of events Given by the outcomes that are common to two (or more) events.

Joint probability The probability that two (or more) events occur together.

Law of large numbers If an experiment is repeated again and again, the probability of an event obtained from the relative frequency approaches the actual or theoretical probability.

Marginal probability The probability of one event or characteristic without consideration of any other event.

Mutually exclusive events Two or more events that do not contain any common outcome and, hence, cannot occur together.

Outcome The result of the performance of an experiment.

Probability A numerical measure of the likelihood that a specific event will occur.

Relative frequency as an approximation of probability Probability assigned to an event based on the results of an experiment or based on historical data.

Sample point An outcome of an experiment.

Sample space The collection of all sample points or outcomes of an experiment.

Simple event An event that contains one and only one outcome of an experiment. It is also called an *elementary event.*

Subjective probability The probability assigned to an event based on the information and judgment of a person.

Sure event An event that is certain to occur.

Tree diagram A diagram in which each outcome of an experiment is represented by a branch of a tree.

Union of events All outcomes that belong either to one or to both events.

Venn diagram A picture that represents a sample space or certain events.

KEY FORMULAS

1. Classical probability rule

For a simple event E_i:

$$P(E_i) = \frac{1}{\text{Total number of outcomes for the experiment}}$$

For a compound event A:

$$P(A) = \frac{\text{Number of outcomes favorable to } A}{\text{Total number of outcomes for the experiment}}$$

2. Relative frequency as an approximation of probability

$$P(\text{an event}) = \frac{\text{Frequency of that event}}{\text{Sample size}} = \frac{f}{n}$$

3. Conditional probability of an event

$$P(B|A) = \frac{P(A \text{ and } B)}{P(A)} \quad \text{and} \quad P(A|B) = \frac{P(A \text{ and } B)}{P(B)}$$

4. Independent events
Two events A and B are independent if

$$P(A) = P(A|B) \quad \text{and/or} \quad P(B) = P(B|A)$$

5. Complementary events
For two complementary events A and $\bar{A}$:

$$P(A) + P(\bar{A}) = 1 \qquad P(A) = 1 - P(\bar{A}) \qquad \text{and} \qquad P(\bar{A}) = 1 - P(A)$$

6. Multiplication rule for joint probability of events
If A and B are dependent events, then

$$P(A \text{ and } B) = P(A)\, P(B|A)$$

If A and B are independent events, then

$$P(A \text{ and } B) = P(A)\, P(B)$$

7. Joint probability of two mutually exclusive events
For two mutually exclusive events A and B:

$$P(A \text{ and } B) = 0$$

8. Counting rule
If an experiment consists of three steps and if the first step can result in m outcomes, the second step in n outcomes, and the third step in k outcomes, then

$$\text{Total outcomes for the experiment} = mnk$$

9. Addition rule for the probability of union of events
If A and B are nonmutually exclusive events, then

$$P(A \text{ or } B) = P(A) + P(B) - P(A \text{ and } B)$$

If A and B are mutually exclusive events, then

$$P(A \text{ or } B) = P(A) + P(B)$$

SUPPLEMENTARY EXERCISES

4.103 A group is made up of Democrats and Republicans. If two persons are selected at random and the events "Democrat" and "Republican" are observed, how many total final outcomes are possible for this experiment? Draw a tree diagram for this experiment.

4.104 A box contains a certain number of apples. A few of them are bad. If two apples are selected at random from this box and it is observed whether the apples selected are good or bad, how many total final outcomes are possible for this experiment? Draw a tree diagram for this experiment.

4.105 A statistics class is given the midterm examination. If two students are selected at random from this class and it is observed whether they pass or fail the midterm examination, how many total final outcomes are possible for this experiment? Depict all the outcomes in a Venn diagram.

4.106 Refer to Exercise 4.103. List all the outcomes included in each of the following events and mention which of them are simple and which are compound events.

- **a.** Both persons selected are Democrats.
- **b.** The first person is a Republican and the second is a Democrat.
- **c.** At most one person is a Democrat.

4.107 Refer to Exercise 4.104. List all the outcomes included in each of the following events and mention which of them are simple and which are compound events.

- **a.** At most one apple selected is bad.
- **b.** Not more than one apple is good.
- **c.** Exactly one apple is good.

4.108 Refer to Exercise 4.105. List all the outcomes included in each of the following events and mention which of them are simple and which are compound events.

- **a.** Both students pass the examination.
- **b.** Exactly one student passes the examination.
- **c.** At most one student fails the examination.

4.109 In a class of 35 students, 13 are seniors, 9 are juniors, 8 are sophomores, and 5 are freshmen. If one student is randomly selected from this class, what is the probability that this student is **a.** a junior **b.** a freshman?

4.110 A lawyers' association has 80 members. Of them, 12 are corporate lawyers. If one lawyer is selected at random, what is the probability that this lawyer is

a. a corporate lawyer **b.** not a corporate lawyer?

4.111 The following table gives a two-way classification of 190 randomly selected adults on the basis of being overweight or not and having high blood pressure or normal blood pressure.

	High Blood Pressure	**Normal Blood Pressure**
Overweight	57	18
Not overweight	24	91

- **a.** If one person is selected at random from this group, find the probability that this person
  - i. is overweight
  - ii. has normal blood pressure
  - iii. has high blood pressure given the person is overweight
  - iv. is not overweight given the person has normal blood pressure
  - v. is overweight and has high blood pressure
  - vi. is not overweight or has normal blood pressure
- **b.** Are the events "overweight" and "high blood pressure" independent? Are they mutually exclusive? Explain why or why not.

4.112 A random sample of 250 adults was taken, and they were asked whether they prefer watching sports or opera on television. The following table gives the two-way classification of these 250 adults.

	Prefers Watching Sports	Prefers Watching Opera
Male	96	24
Female	45	85

a. If one adult is selected at random from this group, find the probability that this adult

i. prefers watching opera
ii. is a male
iii. prefers watching sports given the adult is a female
iv. is a male given that he prefers watching sports
v. is a female and prefers watching opera
vi. prefers watching sports or is a male

b. Are the events "female" and "prefers watching sports" independent? Are they mutually exclusive? Explain why or why not.

4.113 A random sample of 80 lawyers was taken, and they were asked if they are in favor of or against capital punishment. The following table gives the two-way classification of these 80 lawyers.

	Favors Capital Punishment	Opposes Capital Punishment
Male	32	26
Female	13	9

a. If one lawyer is randomly selected from this group, find the probability that this lawyer

i. favors capital punishment
ii. is a female
iii. opposes capital punishment given the lawyer is a female
iv. is a male given that he favors capital punishment
v. is a female and favors capital punishment
vi. opposes capital punishment or is a male

b. Are the events "female" and "opposes capital punishment" independent? Are they mutually exclusive? Explain why or why not.

4.114 The following table gives a two-way classification of 200 randomly selected purchases made at a department store.

	Paid by Cash/Check	Paid by Credit Card
Male	24	46
Female	77	53

a. If one of these 200 purchases is selected at random, find the probability that it is
   i. made by a female
   ii. paid by cash/check
   iii. paid by credit card given that this purchase is made by a male
   iv. made by a female given that it is paid by cash/check
   v. made by a female and paid by a credit card
   vi. paid by cash/check or made by a male

b. Are the events "female" and "paid by credit card" independent? Are they mutually exclusive? Explain why or why not.

4.115 A random sample of 400 college students were asked if college players should be paid or not. The following table gives a two-way classification of the responses.

	Should be Paid	Should Not be Paid
Student player	90	10
Student nonplayer	210	90

a. If one student is randomly selected from these 400 students, find the probability that this student
   i. is in favor of paying college players
   ii. favors paying college players given the student is a nonplayer
   iii. is a player and favors paying the players
   iv. is a nonplayer or is against paying the players

b. Are the events "player" and "should be paid" independent? Are they mutually exclusive? Explain why or why not.

4.116 The probability is .60 that a randomly selected person likes a specific ethnic food and .40 that he or she does not like it. If two persons are selected at random, find the probability that **a.** both of them like this ethnic food **b.** at least one likes it. Draw a tree diagram for this experiment.

4.117 The probability is .34 that a family has at least one pet. If two families are selected at random, find the probability that **a.** none of them has a pet **b.** at most one of them has pet(s). Draw a tree diagram for this experiment.

4.118 The probability that a college student has a GPA of 3.0 or higher is .37 and that a college student is overweight is .14. If these two events are independent, find the probability that a randomly selected student

a. is not overweight
b. has a GPA of less than 3.0
c. has a GPA of less than 3.0 and is overweight
d. is overweight or has a GPA of 3.0 or higher

Draw a tree diagram for this experiment.

4.119 The probability that an adult reads a newspaper everyday is .68 and that an adult is a man is .53. If these two events are independent, find the probability that a randomly selected adult

a. is a woman
b. does not read the newspaper everyday
c. is a man and reads the newspaper everyday
d. is a woman or reads the newspaper everyday

Draw a tree diagram for this experiment.

4.120 The probability that a male is a college graduate is .29 and that he is at least 69 inches tall is .66. If these two events are independent, find the probability that a randomly selected male is

a. not a college graduate
b. less than 69 inches tall
c. a college graduate and at least 69 inches tall
d. a college graduate or less than 69 inches tall

SELF-REVIEW TEST

1. The collection of all outcomes for an experiment is called

a. a sample space
b. intersection of events
c. joint probability

2. A final outcome of an experiment is called

a. a compound event
b. a simple event
c. a complementary event

3. A compound event includes

a. all final outcomes
b. exactly two outcomes
c. more than one outcome for an experiment

4. Two equally likely events

a. have the same probability of occurrence
b. cannot occur together
c. have no effect on the occurrence of each other

5. Which of the following can be applied to experiments with equally likely outcomes only?

a. Classical probability
b. Empirical probability
c. Subjective probability

6. Two mutually exclusive events

a. have the same probability
b. cannot occur together
c. have no effect on the occurrence of each other

7. Two independent events

a. have the same probability
b. cannot occur together
c. have no effect on the occurrence of each other

8. The probability of an event is always

a. greater than zero

b. in the range zero to 1.0
c. less than 1.0

9. The sum of the probabilities of all final outcomes of an experiment is always

a. 100 **b.** 1.0 **c.** zero

10. The joint probability of two mutually exclusive events is always

a. 1.0 **b.** between 0 and 1 **c.** zero

11. Two independent events are

a. always mutually exclusive
b. never mutually exclusive
c. always complementary events

12. A list of books contains 5 novels, 8 books of poetry, and 3 books of short stories. If one book is randomly selected, what is the probability that it is a book of poetry?

13. A company has 500 employees. Of them, 300 are male, and 280 are married. Of the 300 males, 190 are married.

a. Are the events "male" and "married" independent? Are they mutually exclusive? Explain why or why not.
b. If one employee of this company is selected at random, what is the probability that this employee is
i. a female
ii. a male given that he is married

14. Reconsider Problem 13. If one employee is selected at random from this company, what is the probability that this employee is a male or married?

15. Reconsider Problem 13. If two employees are selected at random from this company, what is the probability that both of them are female?

16. The probability that any adult American has ever experienced a migraine headache is .35. If two adult Americans are randomly selected, what is the probability that neither of them has ever experienced a migraine headache?

17. Five hundred married men and women were asked if they would marry their current spouses if they were given a chance to do it over again. Their responses are recorded in the following table.

	Will Marry the Current Spouse	
	Yes (Y)	**No (N)**
Male (M)	125	175
Female (F)	55	145

a. If one person is selected at random from these 500 persons, find the following probabilities.
i. Yes
ii. Yes given female
iii. Male and no
iv. Yes or female

b. Are the events "male"and "yes" independent? Are they mutually exclusive? Explain why or why not.

5 DISCRETE RANDOM VARIABLES AND THEIR PROBABILITY DISTRIBUTIONS

5.1 RANDOM VARIABLES

5.2 THE PROBABILITY DISTRIBUTION OF A DISCRETE RANDOM VARIABLE

5.3 THE MEAN OF A DISCRETE RANDOM VARIABLE

5.4 THE STANDARD DEVIATION OF A DISCRETE RANDOM VARIABLE

5.5 FACTORIALS AND COMBINATIONS

5.6 THE BINOMIAL PROBABILITY DISTRIBUTION

5.7 THE POISSON PROBABILITY DISTRIBUTION

SELF-REVIEW TEST

USING MINITAB

Chapter 4 discussed the concepts and rules of probability. This chapter extends the concept of probability to explain probability distributions. As was seen in Chapter 4, any given statistical experiment has more than one outcome. It is impossible to predict which of those outcomes will occur if that experiment is performed. Consequently, decisions are made under uncertain conditions. For example, a player who plays a lottery does not know in advance whether or not he is going to win the lottery. If he knows that he is not going to win, he will definitely not play. It is the uncertainty about winning (some positive probability of winning) that makes him play. In this chapter it is shown that if the outcomes and their probabilities for a statistical experiment are known, we can find out what will happen on average if that experiment is performed many times. For the lottery example, we can find out what a lottery player can expect to win (or lose) on average if he continues playing this lottery again and again.

First, random variables and types of random variables are explained in this chapter. Then, the concept of a probability distribution and its mean and standard deviation are discussed. Finally, two special probability distributions for a discrete random variable—the binomial probability distribution and the Poisson probability distribution—are developed.

5.1 RANDOM VARIABLES

Suppose Table 5.1 gives the frequency and relative frequency distributions of the number of children for all 2000 families living in a small town.

Table 5.1 Frequency and Relative Frequency Distributions of the Number of Children per Family

Number of Children	Frequency	Relative Frequency
0	200	200/2000 = .10
1	600	600/2000 = .30
2	800	800/2000 = .40
3	240	240/2000 = .12
4	160	160/2000 = .08
	N = 2000	Sum = 1.0

Suppose one family is randomly selected from this population. The act of randomly selecting a family is called a *random* or *chance experiment*. Let x denote the number of children in this family. Then x can assume any of the five possible values (0, 1, 2, 3, and 4) listed in the first column of Table 5.1. The value assumed by x depends on what family is selected. Thus, this value depends on the outcome of the random experiment. Consequently, x is called a **random variable** or a **chance variable.** In general, a random variable is denoted by x or y.

RANDOM VARIABLE

A random variable is a variable whose value is determined by the outcome of a random experiment.

As explained next, a random variable can be discrete or continuous.

5.1.1 DISCRETE RANDOM VARIABLE

A **discrete random variable** assumes values that can be counted. In other words, the consecutive values of a discrete random variable are separated by a certain gap.

DISCRETE RANDOM VARIABLE

A random variable whose values are countable is called a discrete random variable.

In the illustration of Table 5.1, "children per family" is an example of a discrete random variable because the values of the random variable x are countable: 0, 1, 2, 3, and 4. Some other examples of discrete random variables are

1. The number of cars sold at a dealership during a given month
2. The number of people coming to a theater on a certain day

3. The number of shoe pairs a person owns
4. The number of late flights (departures and/or arrivals) at an airport on a given day
5. The number of customers arriving at a bank during any hour
6. The number of heads obtained in three tosses of a coin

5.1.2 CONTINUOUS RANDOM VARIABLE

A random variable whose values are not countable is called a **continuous random variable.** A continuous random variable can assume any value over an interval or intervals.

> **CONTINUOUS RANDOM VARIABLE**
>
> A random variable that can assume any value contained in one or more intervals is called a continuous random variable.

Because the number of values contained in any interval is infinite, the possible number of values that a continuous random variable can assume is also infinite. Moreover, we cannot count these values. Consider the life of a battery. We can measure it as accurately as we want. For instance, the life of this battery may be 40 hours, or 40.25 hours, or 40.247 hours. Assume that the maximum life of such a battery is 200 hours. Let x denote the life of a randomly selected battery of this kind. Then x can assume any value in the interval 0 to 200. Consequently, x is a continuous random variable. As shown below, every point on the line representing the interval 0 to 200 gives a possible value of x.

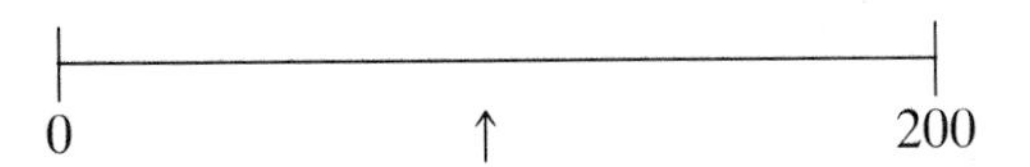

Every point on this line represents a possible value of x that denotes the life of a battery. There are an infinite number of points on this line. The values represented by all points on this line are uncountable.

The following are a few examples of continuous random variables.

1. The height of a person
2. The time taken to complete an examination
3. The amount of milk in a gallon (note that we do not expect a gallon to contain exactly 1 gallon of milk but either slightly more or slightly less than a gallon).
4. The weight of a baby
5. The ERA (earned run average) of a baseball player

This chapter is limited to the discussion of discrete random variables and their probability distributions. Continuous random variables will be discussed in Chapter 6.

EXERCISES

5.1 Explain the meaning of a random variable, a discrete random variable, and a continuous random variable. Give one example each of a discrete random variable and a continuous random variable.

5.2 Classify the following random variables as discrete or continuous.

- **a.** The number of students in a class
- **b.** The amount of soda in a 12-oz can
- **c.** The number of cattle owned by a farmer
- **d.** The age of a house
- **e.** The number of pages in a book that contain at least one error
- **f.** The time spent by a doctor examining a patient

5.3 Indicate which of the following random variables are discrete and which are continuous.

- **a.** The number of new accounts opened at a bank during a certain month
- **b.** The time taken to run a marathon
- **c.** The price of a concert ticket
- **d.** The number of rotten eggs in a box
- **e.** The points scored in a football game
- **f.** The weight of a package

5.2 THE PROBABILITY DISTRIBUTION OF A DISCRETE RANDOM VARIABLE

Let x be a discrete random variable. The **probability distribution** of x describes how the probabilities are distributed over the possible values of x.

PROBABILITY DISTRIBUTION OF A DISCRETE RANDOM VARIABLE

The probability distribution of a discrete random variable lists all the possible values that the random variable can assume and their corresponding probabilities.

Example 5-1 illustrates the concept of the probability distribution of a discrete random variable.

Probability distribution of a discrete random variable.

EXAMPLE 5-1 Recall the frequency and relative frequency distributions of the number of children for all 2000 families living in a small town given in Table 5.1. That table is reproduced as Table 5.2 on next page. Let x be the number of children for a randomly selected family. Write the probability distribution of x.

Table 5.2 Frequency and Relative Frequency Distributions of the Number of Children per Family

Number of Children	Frequency	Relative Frequency
0	200	.10
1	600	.30
2	800	.40
3	240	.12
4	160	.08
	$N = 2000$	Sum = 1.0

Solution The relative frequencies for the population give the theoretical probabilities of outcomes. Let x be the number of children in a randomly selected family. From Table 5.2, we can write Table 5.3, which lists the relative frequencies as probabilities of various values of x. Table 5.3 lists the *probability distribution* of the discrete random variable x.

Table 5.3 Probability Distribution of the Number of Children

Number of Children x	Probability $P(x)$
0	.10
1	.30
2	.40
3	.12
4	.08
	$\Sigma P(x) = 1.0$

■

As a passing note, the relative frequencies for the population data give the theoretical (or actual) probabilities of outcomes. However, if the relative frequencies belong to a sample survey, they give the approximate probabilities.

The probability distribution of a discrete random variable possesses the following two characteristics.

1. The probability assigned to each value of the random variable x lies in the range 0 to 1, that is, $0 \leq P(x) \leq 1$ for each x.
2. The sum of the probabilities assigned to all values of x is equal to 1.0, that is, $\Sigma P(x) = 1$. (Remember, if the probabilities are rounded, the sum may not be exactly 1.0.)

TWO CHARACTERISTICS OF A PROBABILITY DISTRIBUTION

The probability distribution of a discrete random variable possesses the following two characteristics.

1. $0 \leq P(x) \leq 1$ for each value of x
2. $\Sigma P(x) = 1$

These two characteristics are also called the *two conditions* that a probability distribution must satisfy. Notice that in Table 5.3, each probability listed in the column labeled $P(x)$ is between 0 and 1. Also, $\Sigma P(x) = 1.0$. Because both conditions are satisfied, Table 5.3 represents the probability distribution of x.

From Table 5.3, the probability for any value of x can be read. For example, the probability that a randomly selected family from this town has two children is .40. This probability is written as

$$P(x = 2) = .40$$

The probability that the selected family has more than two children is given by the sum of the probabilities of three and four children, respectively. This probability is .12 + .08 = .20. This probability can be written as

$$P(x > 2) = P(x = 3) + P(x = 4) = .12 + .08 = .20$$

The probability distribution of a discrete random variable can be presented in the form of a *mathematical formula, a table, or a graph*. Table 5.3 presented the probability distribution in tabular form. The histogram constructed in Figure 5.1 displays the probability distribution of Table 5.3 in graphic form. In this figure, each value of x is shown by marking an interval on the horizontal axis. The probability for each value of x is exhibited by a bar with its height equal to the corresponding probability. A histogram constructed for a probability distribution is called a *probability histogram*. This section will not discuss the presentation of a probability distribution using a mathematical formula.

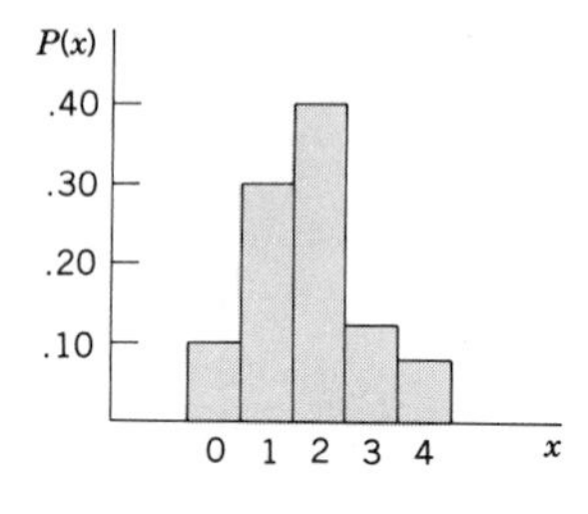

Figure 5.1 Probability histogram for Table 5.3.

Verifying conditions of a probability distribution.

EXAMPLE 5-2 Each of the following tables lists certain values of x and their probabilities. Determine whether or not each table represents a valid probability distribution.

(a)

x	$P(x)$
0	.08
1	.11
2	.39
3	.27

(b)

x	$P(x)$
2	.25
3	.34
4	.28
5	.13

(c)

x	$P(x)$
7	.70
8	.50
9	−.20

Solution

(a) Because each probability listed in this table is in the range 0 to 1, it satisfies the first condition of a probability distribution. However, the sum of all the probabilities is not equal to 1.0 because $\Sigma P(x) = .08 + .11 + .39 + .27 = .85$. Therefore, the second condition is not satisfied. Consequently, this table does not represent a valid probability distribution.

(b) Each probability listed in this table is in the range 0 to 1. Because $\Sigma P(x) = .25 + .34 + .28 + .13 = 1.0$, the sum of the probabilities for all outcomes listed in this table is also equal to 1.0. Consequently, this table represents a valid probability distribution.

(c) Although the sum of all probabilities listed in this table is equal to 1.0, one of the probabilities is negative. This violates the first condition of a probability distribution. Therefore, this table does not represent a valid probability distribution. ■

EXAMPLE 5-3 The following table lists the probability distribution of the number of breakdowns per week for a machine based on past data.

Breakdowns per week	0	1	2	3
Probability	.15	.20	.35	.30

Probability histogram.

(a) Draw a probability histogram for this distribution.

Probabilities of events for a discrete random variable.

(b) Find the probability that the number of breakdowns for this machine during a given week are

(i) exactly 2 (ii) zero to 2
(iii) more than one (iv) less than 2

Solution Let x denote the number of breakdowns for this machine during a given week. Table 5.4 lists the probability distribution of x.

Table 5.4 Probability Distribution of Breakdowns

x	$P(x)$
0	.15
1	.20
2	.35
3	.30
	$\Sigma P(x) = 1.0$

(a) Figure 5.2 shows the probability histogram for the probability distribution of Table 5.4.

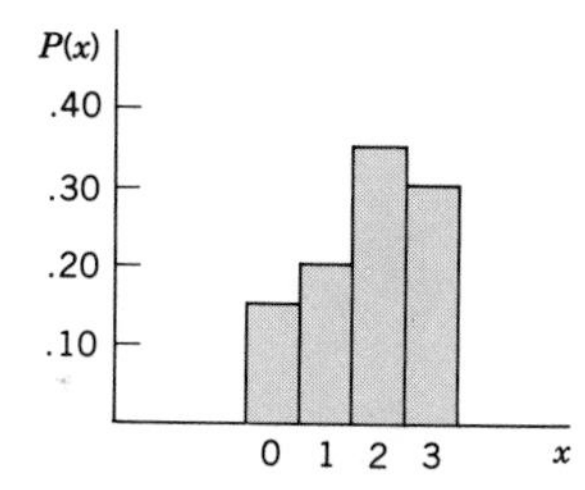

Figure 5.2 Probability histogram for Table 5.4.

(b) Using Table 5.4, we can calculate the required probabilities as follows.

(i) The probability of exactly 2 breakdowns is

$$P(\text{exactly 2 breakdowns}) = P(x = 2) = .35$$

(ii) The probability of 0 to 2 breakdowns is given by the sum of the probabilities of 0, 1, and 2 breakdowns.

$$\begin{aligned} P(\text{0 to 2 breakdowns}) &= P(0 \le x \le 2) \\ &= P(x = 0) + P(x = 1) + P(x = 2) \\ &= .15 + .20 + .35 = .70 \end{aligned}$$

(iii) The probability of more than 1 breakdown is obtained by adding the probabilities of 2 and 3 breakdowns.

$$\begin{aligned} P(\text{more than 1 breakdown}) &= P(x > 1) = P(x = 2) + P(x = 3) \\ &= .35 + .30 = .65 \end{aligned}$$

(iv) The probability of less than 2 breakdowns is given by the sum of the probabilities of 0 and 1 breakdown.

$$\begin{aligned} P(\text{less than 2 breakdowns}) &= P(x < 2) = P(x = 0) + P(x = 1) \\ &= .15 + .20 = .35 \end{aligned}$$

■

Constructing a probability distribution.

EXAMPLE 5-4 According to a survey, 60% of all students at a large university suffer from math anxiety. Two students are randomly selected from this university. Let x denote the number of students in this sample who suffer from math anxiety. Find the probability distribution of x.

Solution Let us define the following two events.

N = the student selected does not suffer from math anxiety

M = the student selected suffers from math anxiety

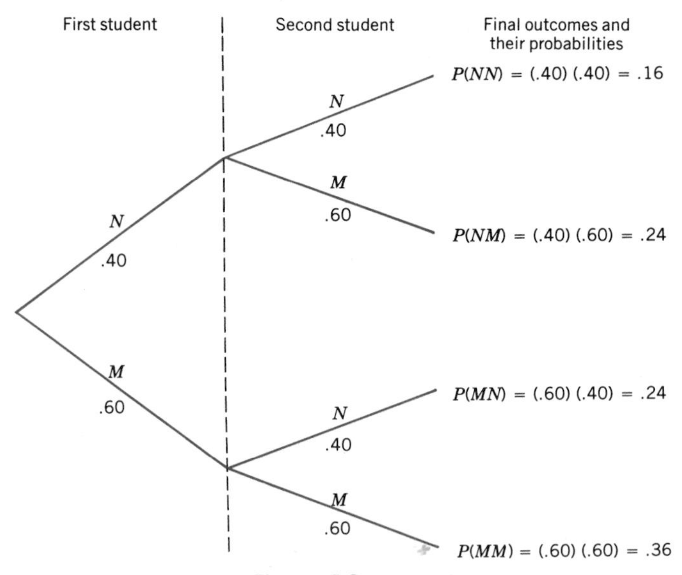

Figure 5.3 Tree diagram.

As we can observe from the tree diagram of Figure 5.3, there are four possible outcomes for this experiment: *NN* (neither of the two students suffers from math anxiety), *NM* (first student does not suffer from math anxiety and second does), *MN* (first student suffers from math anxiety and second does not), and *MM* (both students suffer from math anxiety). The probabilities of these four outcomes are listed in the tree diagram. Because 60% of students suffer from math anxiety and 40% do not, the probability is .60 that any selected student suffers from math anxiety and .40 that he or she does not.

In a sample of two students, the number who suffer from math anxiety can be 0 (*NN*), 1 (*NM* or *MN*), or 2 (*MM*). Thus, x can assume any of the three possible values: 0, 1, or 2. The probabilities of these three outcomes are calculated as follow.

$$P(x = 0) = P(NN) = .16$$

$$P(x = 1) = P(NM \text{ or } MN) = P(NM) + P(MN) = .24 + .24 = .48$$

$$P(x = 2) = P(MM) = .36$$

Using these probabilities, we write the probability distribution of x in Table 5.5.

Table 5.5 Probability Distribution of the Number of Students with Math Anxiety in a Sample of Two Students

x	$P(x)$
0	.16
1	.48
2	.36
	$\Sigma P(x) = 1.0$

■

EXERCISES

5.4 Explain the meaning of the probability distribution of a discrete random variable. Give one example of such a probability distribution. What are the three ways to present the probability distribution of a discrete random variable?

5.5 Briefly explain the two characteristics (or conditions) of the probability distribution of a discrete random variable.

5.6 Each of the following tables lists certain values of x and their probabilities. Verify if each of them represents a valid probability distribution.

a.

x	$P(x)$
0	.10
1	.05
2	.45
3	.32

b.

x	$P(x)$
2	.35
3	.28
4	.23
5	.14

c.

x	$P(x)$
7	−.25
8	.85
9	.40

5.7 Each of the following tables lists certain values of x and their probabilities. Determine if each of them satisfies the two conditions required for a valid probability distribution.

a.

x	$P(x)$
5	−.36
6	.48
7	.62
8	.26

b.

x	$P(x)$
1	.16
2	.24
3	.49

c.

x	$P(x)$
0	.15
1	.00
2	.35
3	.50

5.8 The following table, constructed using past data, lists the probability distribution of the number of exercise machines sold per day at Elmo's Sporting Goods store.

Number of exercise machines sold per day	4	5	6	7	8	9	10
Probability	.08	.11	.14	.19	.23	.16	.09

a. Draw a probability histogram for this distribution.
b. Determine the probability that the number of exercise machines sold at Elmo's on a given day is
   i. exactly 6
   ii. more than 8
   iii. between 5 and 8
   iv. less than 7

5.9 Let x denote the number of auto accidents that occur during a week in a city. The following table lists the probability distribution of x.

x	0	1	2	3	4	5	6
$P(x)$	.12	.16	.22	.18	.14	.12	.06

a. Draw a probability histogram for this distribution.
b. Determine the probability that the number of auto accidents that will occur during a given week in this city is
   i. exactly 4
   ii. more than 3
   iii. less than 3
   iv. 3 to 5

5.10 The following table lists the frequency distribution of the number of cars owned by all 2500 families living in a small town.

Number of cars owned	0	1	2	3	4
Number of families	120	970	730	410	270

a. Construct a probability distribution table for the number of cars owned by families of this town. Draw a probability histogram for this distribution.
b. Let x denote the number of cars owned by a randomly selected family from this town. Find the following probabilities.
   i. $P(x = 1)$
   ii. $P(x \geq 2)$
   iii. $P(x \leq 1)$
   iv. $P(1 \leq x \leq 3)$

5.11 Let x denote the number of shoe pairs owned by a randomly selected faculty member from a university. The following table lists the frequency distribution of x for all 750 faculty members of that university.

x	2	3	4	5	6	7	8
f	50	120	160	220	130	50	20

a. Construct a probability distribution table for the number of shoe pairs owned by faculty members at this university. Draw a probability histogram for this distribution.

b. Find the following probabilities.

i. $P(x = 5)$
ii. $P(x \geq 4)$
iii. $P(3 \leq x < 7)$
iv. $P(x < 6)$

5.12 Five percent of all cars manufactured at a large auto company are lemons. Suppose two cars are selected at random from the production line of this company. Let x denote the number of lemons in this sample. Write the probability distribution of x. Draw a tree diagram for this problem.

5.13 According to the U.S. Census Bureau, 62% of all persons aged 18 and over living in the United States are married. Suppose two persons are randomly selected from the population of all persons aged 18 and over. Let x denote the number of married persons in this sample. Construct the probability distribution table of x. Draw a tree diagram for this problem.

5.14 According to a survey, 40% of adults are against using animals for research. Assume that this result holds true for the current population of all adults. Let x be the number of adults in a random sample of two adults who are against using animals for research. Obtain the probability distribution of x. Draw a tree diagram for this problem.

5.15 According to the National Science Foundation, 16% of scientists and engineers in the United States are women (*U.S. News & World Report,* July 16, 1990). Suppose a random sample of two scientists and engineers is selected from the population of all scientists and engineers. Let x be the number of women in this sample. Construct the probability distribution table of x. Draw a tree diagram for this problem.

***5.16** A box contains 10 parts, 3 of which are known to be defective. Suppose 2 parts are randomly selected from this box. Let x denote the number of good parts in this sample. Write the probability distribution of x. Draw a tree diagram for this problem. (*Hint:* Note that the draws are made without replacement from a small population. Hence, the probabilities of outcomes do not remain constant for each draw. For a hint, refer to Example 4-28.)

***5.17** A statistics class has 20 students; 12 of them are male. Suppose two students are randomly selected from this class. Let x denote the number of males in this sample. Find the probability distribution of x. Draw a tree diagram for this problem. (*Hint:* Note that the draws are made without replacement from a small population. Hence, the probabilities of outcomes do not remain constant for each draw. For a hint, refer to Example 4-28.)

5.3 THE MEAN OF A DISCRETE RANDOM VARIABLE

The **mean of a discrete random variable,** denoted by μ, is actually the mean of its probability distribution. The mean of a discrete random variable x is also called its *expected value* and is denoted by $E(x)$. The mean (or expected value) of a discrete random variable is the value that we expect to observe per repetition on average if we perform an experiment a large number of times. For example, we may expect a car salesperson to sell on average 2.4 cars per week. This does not mean that every week this salesperson will sell exactly 2.4 cars. (Actually, he cannot sell exactly 2.4 cars.) This simply means that if we observe for many weeks, this salesperson will sell a different number of cars during different weeks. If we take the average for all these weeks, we expect this average to be 2.4 cars.

To calculate the mean of a discrete random variable x, we multiply each value of x by the corresponding probability and sum the resulting products. This sum gives the mean (or expected value) of the discrete random variable x.

MEAN OF A DISCRETE RANDOM VARIABLE

The mean of a discrete random variable x is the value that is expected to occur per repetition on average if an experiment is repeated a large number of times. It is denoted by μ and calculated as

$$\mu = \Sigma x P(x)$$

Example 5-5 illustrates the calculation of the mean of a discrete random variable.

Calculating and interpreting the mean of a discrete random variable.

EXAMPLE 5-5 Recall Example 5-3 of Section 5.2. The probability distribution Table 5.4 from that example is reproduced here. In this table, x represents the number of breakdowns for a machine during a given week and $P(x)$ is the probability of the corresponding value of x.

x	$P(x)$
0	.15
1	.20
2	.35
3	.30

Find the mean number of breakdowns per week for this machine.

Solution To find the mean number of breakdowns per week for this machine, we multiply each value of x by its probability and add these products. This sum gives the value of the mean of the probability distribution of x. The products $xP(x)$ are listed in the third column of Table 5.6. The sum of these products gives $\Sigma xP(x)$, which is the mean of x.

Table 5.6 Calculating Mean for the Probability Distribution of Breakdowns

x	$P(x)$	$xP(x)$
0	.15	$0(.15) = .00$
1	.20	$1(.20) = .20$
2	.35	$2(.35) = .70$
3	.30	$3(.30) = .90$
		$\Sigma xP(x) = 1.80$

The mean is

$$\mu = \Sigma xP(x) = 1.80$$

Thus, an average of 1.80 breakdowns are expected to occur per week over a period of time for this machine. In other words, if this machine is used for many weeks, then for certain weeks we will observe no breakdowns, for some other weeks we will observe 1 breakdown per week, and for still other weeks we will observe 2 or 3 breakdowns per week. The mean number of breakdowns is expected to be 1.80 per week. ■

Case Study 5-1 calculates the mean amount a player is expected to win by playing a lottery.

CASE STUDY 5-1 INSTANT BASEBALL LOTTERY

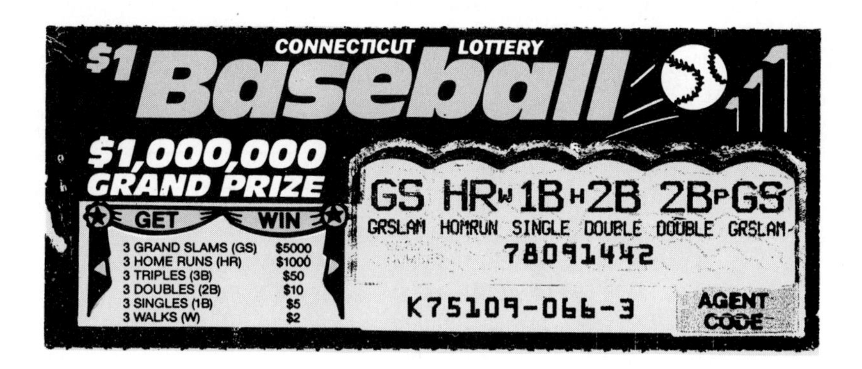

Recently the State of Connecticut had an instant lottery game called "baseball." The cost of each ticket for this lottery was $1. A player could instantly win $5000, $1000, $50, $10, $5, or $2. Each ticket had six erasable baseballs, which, when rubbed off, could reveal GS (grand slam), HR (home run), 3B (triple), 2B (double), 1B (single), or W (walk). If three of the six spots on a ticket showed the same outcome, the player won. The player won $5000 if three of the spots showed grand slams, $1000 if three showed home runs, $50 if three showed triples, $10 if three showed doubles, $5 if three showed singles, and $2 if three of them showed walks. All the winning tickets were eligible for a grand prize drawing that consisted of $1,000,000, which was to be paid in installments of $50,000 per year for twenty years. The grand prize was funded by an annuity, which cost the State approximately $500,000. The $1,000,000 prize was guaranteed and would be paid to the winner's estate in the event of the winner's death. In addition to the grand prize, there were runner-up prizes of $20,000, $15,000, $10,000 and $5,000.

Let us first consider the instant winning tickets. The following table shows the number of tickets with different prizes from a total of 26,100,000 tickets for this lottery. As is obvious from this table, out of a total of 26,100,000 tickets, 20,712,329 were nonwinning tickets (the ones with a prize of $0 in this table). Of the remaining tickets, 4,760,379 had a prize of $2 each, 522,000 had a prize of $5 each, and so on.

Prize ($)	Number of Tickets
0	20,712,329
2	4,760,379
5	522,000
10	92,655
50	12,267
1,000	261
5,000	109
	Total = 26,100,000

The net gain for each of the instant winning tickets is equal to the amount of the prize minus $1, the cost of the ticket. Thus, the net gain for each of the nonwinning tickets is −$1, the cost of the ticket. Let

$$x = \text{the net amount a player won from instant prizes}$$

The following table shows all the calculations required to compute the mean of x for the probability distribution of instant prizes. Note that for the time being we are ignoring the grand prizes. The probability of an outcome (net winnings) in case of instant prizes is calculated by dividing the number of tickets with that outcome by the total number of tickets.

x ($)	$P(x)$	$xP(x)$
−1	20,712,329/26,100,000	−.7936
1	4,760,379/26,100,000	.1824
4	522,000/26,100,000	.0800
9	92,655/26,100,000	.0320
49	12,267/26,100,000	.0230
999	261/26,100,000	.0100
4,999	109/26,100,000	.0209
		$\Sigma xP(x) = -.4453$

Hence,

$$\text{Mean of } x = \Sigma xP(x) = -\$.4453$$

Thus, if we do not consider the grand prizes, the mean of winnings from instant prizes for this lottery was −$.4453. In other words, all players taken together lost an average of $.4453 (or 44.53 cents) per ticket.

Now let us consider the grand prizes. The following table lists the amount of the grand prizes and the number of winners. Note that in this table we have used $500,000 for $1,000,000 prize because, as mentioned earlier, the cost of this prize to the State was $500,000.

Amount of Grand Prize	Number of Grand Prizes
5,000	16
10,000	1
15,000	1
20,000	1
500,000	1

All grand prizes taken together returned \$625,000 to winners. If we divide this amount by the total number of tickets (26,100,000), the result is \$.0239. In other words, \$.0239 per ticket of the players' money was returned in the form of grand prizes. Note that, for the grand prize winners, the net amount won is equal to the amount of the prize because the players did not pay any extra amount to win grand prizes. Let

$$\mu = \text{mean of the net amount a player won from all prizes}$$

Then, the mean μ is

$$\mu = \Sigma x P(x) + .0239 = -.4453 + .0239 = -.4214$$

Consequently, we can conclude that the players of this game were expected to lose an average of \$.4214 (or 42.14 cents) per ticket. This can also be interpreted as follows: only 57.86% (= 100 − 42.14) of the total money spent by players on buying these lottery tickets was returned to them in the form of prizes and 42.14% was not returned. (The money, which was not returned, covered the costs of operating the lottery, commission paid to agents, and revenue to the state of Connecticut.)

Source: Play 'Baseball', The State of Connecticut Lottery Commission. Ticket reproduced with permission.

Calculating and interpreting the mean of a discrete random variable.

EXAMPLE 5-6 Recall Example 5-4 of Section 5.2 about randomly selecting 2 students from a large university. The probability distribution table of x is reproduced here from that example. In this table, x denotes the number of students with math anxiety in a sample of two students.

x	$P(x)$
0	.16
1	.48
2	.36

Compute the mean of x and give a brief interpretation of the value of mean.

Solution Table 5.7 shows the calculations needed to compute the mean of x.

Table 5.7 Calculating the Mean for the Probability Distribution of Students with Math Anxiety

x	$P(x)$	$xP(x)$
0	.16	.00
1	.48	.48
2	.36	.72
	$\Sigma P(x) = 1.0$	$\Sigma x P(x) = 1.20$

Thus, the mean of the random variable x is

$$\mu = \Sigma x P(x) = 1.20$$

How do we interpret $\mu = 1.20$? If we select a large number of samples, each of size 2, from the population of all students at this university, then some of these samples will contain no students with math anxiety, some will contain 1, and the remaining will contain 2 students with math anxiety. Taking all samples together, we expect to observe an average of 1.20 students suffering from math anxiety in every sample of size 2. Note that in this example x cannot be fractional but the mean of x can be. ■

5.4 THE STANDARD DEVIATION OF A DISCRETE RANDOM VARIABLE

The **standard deviation of a discrete random variable,** denoted by σ, measures the spread of its probability distribution. A higher value for the standard deviation of a discrete random variable x indicates that x can assume values over a larger range about the mean. On the other hand, a smaller value for the standard deviation indicates that most of the values that x can assume are clustered closely about the mean. The basic formula to compute the standard deviation of a discrete random variable is

$$\sigma = \sqrt{\Sigma[(x - \mu)^2 \cdot P(x)]}$$

However, it is more convenient to use the following short-cut formula to compute the standard deviation of a discrete random variable.

STANDARD DEVIATION OF A DISCRETE RANDOM VARIABLE

The standard deviation of a discrete random variable x measures the spread of its probability distribution and is computed as

$$\sigma = \sqrt{\Sigma x^2 P(x) - \mu^2}$$

Note that the variance σ^2 of a discrete random variable is obtained by squaring its standard deviation.

Example 5-7 illustrates how to use the short-cut formula to compute the standard deviation of a discrete random variable.

Standard deviation of a discrete random variable.

EXAMPLE 5-7 Let x denote the number of new cars that an auto dealer will sell during a week. The following table gives the probability distribution of x.

x	$P(x)$
0	.05
1	.15
2	.25
3	.40
4	.15

Find the standard deviation of x.

Solution Table 5.8 shows all the calculations needed for the computation of the standard deviation of x.

Table 5.8 Computations to Find the Standard Deviation

x	$P(x)$	$xP(x)$	x^2	$x^2P(x)$
0	.05	.00	0	.00
1	.15	.15	1	.15
2	.25	.50	4	1.00
3	.40	1.20	9	3.60
4	.15	.60	16	2.40
		$\Sigma xP(x) = 2.45$		$\Sigma x^2P(x) = 7.15$

We perform the following steps to compute the standard deviation of x.

Step 1. Compute the mean of the discrete random variable.

The sum of the products $xP(x)$, recorded in the third column of Table 5.8, gives the mean of x.

$$\mu = \Sigma xP(x) = 2.45 \text{ cars}$$

Step 2. Compute the value of $\Sigma x^2P(x)$.

First we square each value of x and record it in the fourth column of Table 5.8. Then we multiply the values of x^2 by the corresponding values of $P(x)$. The resulting values of $x^2P(x)$ are recorded in the fifth column of Table 5.8. The sum of this column gives the value of $\Sigma x^2P(x)$.

$$\Sigma x^2P(x) = 7.15$$

Step 3. Substitute the values of μ and $\Sigma x^2P(x)$ in the formula for the standard deviation of x and simplify.

By performing this step, we obtain:

$$\sigma = \sqrt{\Sigma x^2P(x) - \mu^2} = \sqrt{7.15 - (2.45)^2} = \sqrt{1.1475} = 1.07 \text{ cars}$$

Thus, it is expected that this auto dealer will sell an average of 2.45 cars per week with a standard deviation of 1.07 cars. ■

Remember: Because the standard deviation of a discrete random variable is obtained by taking the positive square root, its value is never negative.

Standard deviation of a discrete random variable.

EXAMPLE 5-8 Let x be the number of defective bolts in a lot of 100 bolts produced on a machine. The following table gives the probability distribution of x.

x	$P(x)$
0	.02
1	.20
2	.30
3	.30
4	.10
5	.08

Compute the standard deviation of x.

Solution Table 5.9 shows all the calculations required for the computation of the standard deviation of x.

Table 5.9 Computations to Find the Standard Deviation

x	$P(x)$	$xP(x)$	x^2	$x^2P(x)$
0	.02	.00	0	.00
1	.20	.20	1	.20
2	.30	.60	4	1.20
3	.30	.90	9	2.70
4	.10	.40	16	1.60
5	.08	.40	25	2.00
		$\Sigma xP(x) = 2.50$		$\Sigma x^2P(x) = 7.70$

The mean of x is

$$\mu = \Sigma xP(x) = 2.50 \text{ defective bolts}$$

The standard deviation of x is

$$\sigma = \sqrt{\Sigma x^2P(x) - \mu^2} = \sqrt{7.70 - (2.50)^2} = \sqrt{1.45}$$
$$= 1.20 \text{ defective bolts}$$

Thus, the given machine is expected to produce an average of 2.50 defective bolts per 100 bolts with a standard deviation of 1.20. ■

EXERCISES

5.18 Find the mean and standard deviation for each of the following probability distributions.

a.

x	$P(x)$
0	.12
1	.27
2	.43
3	.18

b.

x	$P(x)$
6	.36
7	.26
8	.21
9	.17

5.19 Find the mean and standard deviation for each of the following probability distributions.

a.

x	$P(x)$
3	.09
4	.21
5	.34
6	.23
7	.13

b.

x	$P(x)$
0	.43
1	.31
2	.17
3	.09

5.20 Let x be the number of heads obtained in two tosses of a coin. The following table lists the probability distribution of x.

x	0	1	2
$P(x)$	.25	.50	.25

Calculate the mean and standard deviation of x. Give a brief interpretation of the value of the mean.

5.21 Let x be the size of a household randomly selected from a city. The following table lists the probability distribution of x.

x	1	2	3	4	5
$P(x)$	.24	.37	.19	.12	.08

Calculate the mean and standard deviation of x. Give a brief interpretation of the value of the mean.

5.22 Let x be the number of errors that a randomly selected page of a book contains. The following table lists the probability distribution of x.

x	0	1	2	3	4
$P(x)$	.73	.16	.06	.04	.01

Calculate the mean and standard deviation of x. Give a brief interpretation of the value of the mean.

5.23 Let x be the number of newborn pigs per year on a pig farm. The following table lists the probability distribution of x.

x	4	5	6	7	8	9	10
$P(x)$	.17	.22	.19	.13	.11	.10	.08

Calculate the mean and standard deviation of x.

5.24 The following table gives the probability distribution of television sets sold on a given day at an electronics store.

Televisions sold	0	1	2	3	4	5	6
Probability	.05	.12	.23	.30	.16	.10	.04

Find the mean and standard deviation for this probability distribution.

5.25 Let x be a random variable that represents the number of students who are absent on a given day from a class of 25. The following table lists the probability distribution of x.

x	0	1	2	3	4	5
$P(x)$	.08	.18	.32	.22	.14	.06

Calculate the mean and standard deviation for this probability distribution and give a brief interpretation of the value of the mean.

5.26 Refer to Exercise 5.10. Find the mean and standard deviation for the probability distribution you developed for the number of cars owned by all 2500 families living in a small town.

5.27 Refer to Exercise 5.11. Find the mean and standard deviation for the probability distribution you wrote for the number of shoe pairs owned by all 750 faculty members of that university.

5.28 Refer to the probability distribution developed in Exercise 5.12 for the number of lemons in two selected cars. Calculate the mean and standard deviation of x for that probability distribution.

5.29 Refer to the probability distribution developed in Exercise 5.13 for the number of married persons in a sample of two persons aged 18 and over. Calculate the mean and standard deviation of x for that probability distribution.

5.30 A farmer will earn a profit of $30 thousand in case of heavy rain next year, $60 thousand in case of moderate rain, and $15 thousand in case of little rain. A meteorologist forecasts that the probability is .35 for heavy rain, .40 for moderate rain, and .25 for little rain next year. Let x be the random variable that represents next year's profit in thousands of dollars for this farmer. Write the probability distribution of x. Find the mean and standard deviation of x.

5.31 An instant lottery ticket costs $2. In a total of 10,000 tickets for this lottery, 1000 tickets contain a prize of $5 each, 100 tickets have a prize of $10 each, 5 tickets a prize of $1000 each, and 1 ticket has a prize of $5000. Let x be the random variable that denotes the net amount a player wins by playing this lottery. Write the probability distribution of x. Determine the mean and standard deviation of x. How will you interpret the value of the mean of x?

***5.32** Refer to the probability distribution developed in Exercise 5.16 for the number of good parts in a random sample of two parts selected from a box. Calculate the mean and standard deviation of x for that distribution.

***5.33** Refer to the probability distribution developed in Exercise 5.17 for the number of males in a random sample of two students selected from a class. Calculate the mean and standard deviation of x for that distribution.

5.5 FACTORIALS AND COMBINATIONS

This section introduces factorials and combinations, which will be used in the binomial formula to be discussed in Section 5.6.

5.5.1 FACTORIALS

The symbol "!" (read as **"factorial"**) is used to denote factorials. The value of the factorial of a number is obtained by multiplying all integers from that number to 1. For example, "7!" is read as "seven factorial" and it is evaluated by multiplying all integers from 7 to 1.

FACTORIALS

The symbol $n!$, read as "n factorial," represents the product of all integers from n to 1. In other words:

$$n! = n(n-1)(n-2)(n-3)\ldots 3 \cdot 2 \cdot 1$$

By definition,

$$0! = 1$$

Evaluating factorial.

EXAMPLE 5-9 Evaluate 7!.

Solution To evaluate 7!, we multiply all integers from 7 to 1.

$$7! = 7 \cdot 6 \cdot 5 \cdot 4 \cdot 3 \cdot 2 \cdot 1 = 5040$$

Thus, the value of 7! is 5040.† ■

Evaluating factorial.

EXAMPLE 5-10 Evaluate 10!.

Solution The value of 10! is given by the product of all integers from 10 to 1. Thus,

$$10! = 10 \cdot 9 \cdot 8 \cdot 7 \cdot 6 \cdot 5 \cdot 4 \cdot 3 \cdot 2 \cdot 1 = 3{,}628{,}800$$ ■

Factorial of difference between two numbers.

EXAMPLE 5-11 Evaluate $(12 - 4)!$.

Solution The value of $(12 - 4)!$ is

$$(12-4)! = 8! = 8 \cdot 7 \cdot 6 \cdot 5 \cdot 4 \cdot 3 \cdot 2 \cdot 1 = 40{,}320$$ ■

Factorial of zero.

EXAMPLE 5-12 Evaluate $(5 - 5)!$.

Solution As shown here, the value of $(5 - 5)!$ is 1:

$$(5-5)! = 0! = 1$$

Note that 0! is always equal to 1. ■

We can also read the value of $n!$ for $n = 1$ to $n = 25$ from Table II of Appendix C (page 724). Example 5-13 illustrates how to read that table.

†**Using a calculator to evaluate $n!$:** Most calculators have an $n!$ function key. If your calculator has such a key, you can use it to evaluate 7! as follows: First enter 7 and then press the $n!$ key. The calculator screen will display 5040 as the answer.

Using the table of factorials.

EXAMPLE 5-13 Find the value of 15! by using Table II of Appendix C.

Solution To find the value of 15! from Table II, we locate 15 in the column labeled n. Then we read the value in the column for $n!$ entered next to 15. Thus,

$$15! = 1{,}307{,}674{,}368{,}000$$

■

5.5.2 COMBINATIONS

Quite often we face the problem of selecting a few elements from a large number of distinct elements. For example, a student may be required to attempt any 2 questions out of 4 in an examination. As another example, the faculty in a department may need to select 3 professors from 20 to form a committee. Or a lottery player may have to pick 6 numbers from 49 numbers. The question arises: In how many ways can we make the selections in each of these examples? For instance, how many possible selections exist for the student who is to choose any 2 questions out of 4? The answer is six. These six selections are

(1,2) (1,3) (1,4) (2,3) (2,4) (3,4)

The student can choose questions 1 and 2, or 1 and 3, or 1 and 4, and so on.

Each of the possible selections in this list is called a **combination.** All six combinations are distinct, that is, each combination contains a different set of questions. It is important to remember that the order in which the selections are made is not significant in case of combinations. Thus, whether we write (1,2) or (2,1), both these arrangements represent only one combination.

COMBINATIONS

Combinations give the number of ways x elements can be selected from n elements. The notation used to denote the number of combinations is

$$\binom{n}{x}$$

which is read as "combinations of n elements selected x at a time."

Suppose there are a total of n elements from which we want to select x elements. The total number of combinations is denoted by $\binom{n}{x}$.

n denotes the total number of elements

$$\binom{n}{x} = \text{the number of combinations of } n \text{ elements selected } x \text{ at a time}$$

x denotes the number of elements selected per selection

NUMBER OF COMBINATIONS

The number of combinations for selecting x elements from n distinct elements is given by the formula:

$$\binom{n}{x} = \frac{n!}{(n-x)!\,x!}$$

where $n!$, $(n-x)!$, and $x!$ are read as "n factorial," "n minus x factorial," and "x factorial," respectively.

In the combinations formula:

$$n! = n(n-1)(n-2)(n-3)\ldots 3 \cdot 2 \cdot 1$$

$$(n-x)! = (n-x)(n-x-1)(n-x-2)\ldots 3 \cdot 2 \cdot 1$$

$$x! = x(x-1)(x-2)\ldots 3 \cdot 2 \cdot 1$$

Note that in combinations, n is always greater than or equal to x. If n is smaller than x, then we cannot select x distinct elements from n distinct items.

Finding the number of combinations by using formula.

EXAMPLE 5-14 Reconsider the example of a student who is to select 2 questions from 4. Using combinations formula, find the number of ways this student can select 2 questions from 4.

Solution For this example,

$$n = \text{total number of questions} = 4$$

$$x = \text{questions to be selected} = 2$$

Therefore, the number of ways this student can select 2 questions from 4 is

$$\binom{4}{2} = \frac{4!}{(4-2)!\,2!} = \frac{4!}{2!\,2!} = \frac{4 \cdot 3 \cdot 2 \cdot 1}{2 \cdot 1 \cdot 2 \cdot 1} = 6$$

We listed these six combinations earlier in this section.† ■

Finding the number of combinations and listing them.

EXAMPLE 5-15 Three members of a jury will be randomly selected from five persons. How many different combinations are possible?

Solution There are a total of five persons and we are to select three of them. Hence,

$$n = 5 \quad \text{and} \quad x = 3$$

†**Using a calculator to evaluate** $\binom{n}{x}$: Most calculators have a Cn,r or Cn,x or $\binom{n}{r}$ or $\binom{n}{x}$ function key. (Note that all these notations are used to denote combinations.) If your calculator has such a key, read the manual that accompanies the calculator to find out how this key functions and then evaluate $\binom{4}{2}$ using that key.

Applying the combinations formula,

$$\binom{5}{3} = \frac{5!}{(5-3)!\,3!} = \frac{5!}{2!\,3!} = \frac{120}{2 \cdot 6} = 10$$

If we assume that the five persons are A, B, C, D, and E, then the 10 possible combinations for the selection of three members of the jury are

ABC ABD ABE ACD ACE ADE BCD BCE BDE CDE ■

Case Study 5-2 describes the number of ways a player can select 6 numbers from 49 in a lotto lottery game.

CASE STUDY 5-2 PLAYING LOTTO

During the past few years, many states in the United States have initiated the popular lottery game called lotto. To play lotto, a player picks any 6 numbers from 1 to 49. At the end of the lottery period, the State Lottery Commission randomly selects 6 numbers from 1 to 49. If all 6 numbers picked by a player are the same as the ones randomly selected by the lottery commission, the player wins. Let us find the probability that a player wins this game.

The total number of combinations of selecting 6 numbers from 49 numbers is

$$\binom{49}{6} = \frac{49!}{(49-6)!\,6!} = 13{,}983{,}816$$

Thus, there are a total of 13,983,816 different ways to select 6 numbers from 1 to 49. Hence, the probability that a player (who plays this lottery once) wins is

$$P(\text{player wins}) = \frac{1}{13{,}983{,}816} = .0000000715$$

5.5.3 THE TABLE OF COMBINATIONS

Table III in Appendix C lists the number of combinations of n elements selected x at a time. The following example illustrates how to read that table to find combinations.

Using the table of combinations.

EXAMPLE 5-16 Ten patients are suffering from the same disease. A physician will randomly select 3 patients on which to test a new drug. Find the total number of ways the physician can select 3 patients from the 10 patients.

Solution The total number of ways to select 3 patients from 10 is given by $\binom{10}{3}$. To find the value of $\binom{10}{3}$ from Table III, we locate 10 in the column labeled n and 3 in the row labeled x. The relevant part of that table is reproduced here as Table 5.10.

Table 5.10 Determining the Value of $\binom{10}{3}$

$x = 3$

n \ x	0	1	2	3	. . .	20
1	1	1				
2	1	2	1			
3	1	3	3	1		
.	.	.	.	.		
.	.	.	.	.		
.	.	.	.	.		
10	1	10	45	120	. . .	
.	.	.	.	.		
.	.	.	.	.	. . .	. . .

$n = 10$ →

The value of $\binom{10}{3}$

The number at the intersection of the row for $n = 10$ and the column for $x = 3$ gives the value of $\binom{10}{3}$, which is

$$\binom{10}{3} = 120$$

Thus, the physician can select 3 patients from 10 in 120 ways. ■

If the total elements and the elements to be selected are the same, then there is only one combination. In other words,

$$\binom{n}{n} = 1$$

Also, the number of combinations for selecting zero items from n is 1. That is,

$$\binom{n}{0} = 1$$

For example,

$$\binom{5}{5} = \frac{5!}{(5 - 5)!\, 5!} = \frac{5!}{0!\, 5!} = \frac{120}{(1)\,(120)} = 1$$

and

$$\binom{8}{0} = \frac{8!}{(8-0)!\,0!} = \frac{8!}{8!\,0!} = \frac{40,320}{(40,320)\,(1)} = 1$$

EXERCISES

5.34 Find the value of each of the following using the appropriate formula.

$$3! \quad (7-3)! \quad 9! \quad (14-12)! \quad \binom{5}{3} \quad \binom{7}{4} \quad \binom{9}{3} \quad \binom{6}{0} \quad \binom{3}{3}$$

Verify the calculated values by using Tables II and III of Appendix C.

5.35 Find the value of each of the following using the appropriate formula.

$$6! \quad 11! \quad (7-2)! \quad (13-5)! \quad \binom{8}{2} \quad \binom{4}{0} \quad \binom{5}{5} \quad \binom{6}{4} \quad \binom{11}{7}$$

Verify the calculated values by using Tables II and III of Appendix C.

5.36 An English department at a university has 15 faculty members. Two of the faculty members will be randomly selected to represent the department on a committee. In how many ways can the department select 2 faculty members from 15? Use the appropriate formula.

5.37 From 12 novels, you will randomly select 2. How many total selections are possible? Use the appropriate formula.

5.38 A person will randomly select 3 varieties of cookies for a party from 10 varieties. In how many ways can this person select 3 varieties from 10? Use the appropriate formula. Verify your answer by using Table III of Appendix C.

5.39 A superintendent of schools has 12 schools under her jurisdiction. She plans to visit 3 of these schools during the next week. If she randomly selects 3 schools from these 12, how many total selections are possible? Use the appropriate formula. Verify your answer by using Table III of Appendix C.

5.40 An environmental agency will randomly select 4 houses for a radon check from a block containing a total of 25 houses. How many total selections are possible? Use Table III of Appendix C.

5.41 An investor will randomly select 5 stocks from 20 for an investment. How many total combinations are possible? Use Table III of Appendix C.

5.6 THE BINOMIAL PROBABILITY DISTRIBUTION

The **binomial probability distribution** is one of the most widely used discrete probability distributions. It is used to find the probability that an outcome will occur x times in n performances of an experiment. For example, given that the probability is .05 that a VCR manufactured at a firm is defective, we may be interested in finding the probability that in a random sample of 3 VCRs manufactured at this firm, exactly 1 will be defective. As a second example, we may be interested in finding the probability that a baseball player, with a hitting percentage of 25%, will have no hits in 10 trips to the plate.

To apply the binomial probability distribution, the random variable must be a discrete dichotomous random variable. In other words, the variable must be a discrete random variable and each repetition of the experiment must result in one of two possible outcomes. The binomial distribution is applied to experiments that satisfy the four conditions (to be described in Section 5.6.1) of a *binomial experiment*. Each repetition of a binomial experiment is called a **trial** or a **Bernoulli trial** (after Jacob Bernoulli). For example, if an experiment is defined as one toss of a coin and this experiment is repeated 10 times, then each repetition (toss) is called a trial. Consequently, there are 10 total trials for this experiment.

5.6.1 THE BINOMIAL EXPERIMENT

An experiment that satisfies the following four conditions is called a **binomial experiment.**

1. There are n identical trials. In other words, the given experiment is repeated n times. All these repetitions are performed under similar conditions.
2. Each trial has two and only two outcomes. These outcomes are usually called a *success* and a *failure*.
3. The probability of success is denoted by p and that of failure by q. The sum of p and q is 1, that is, $p + q = 1$. The probabilities p and q remain constant for each trial.
4. The trials are independent. In other words, the outcome of one trial does not affect the outcome of another trial.

CONDITIONS OF A BINOMIAL EXPERIMENT

A binomial experiment must satisfy the following four conditions.

1. There are n identical trials.
2. Each trial has only two possible outcomes.
3. The probabilities of two outcomes remain constant.
4. The trials are independent.

Note that one of the two outcomes of a trial is called a *success* and the other a *failure*. When we call an outcome of a trial a success, this does not mean that the outcome is considered favorable or desirable. Similarly, a failure does not necessarily refer to an unfavorable or an undesirable outcome. Success and failure are simply the names used to denote the two possible outcomes of a trial. The outcome to which the question refers is usually called a success; the outcome to which it does not refer is called a failure.

Verifying conditions of a binomial experiment.

EXAMPLE 5-17 Consider the experiment consisting of 10 tosses of a coin. Determine if it is a binomial experiment.

Solution As described here, the experiment consisting of 10 tosses of a coin satisfies all four conditions of a binomial experiment.

1. There are a total of 10 trials (tosses), and they are all identical. All 10 tosses are performed under similar conditions.
2. Each trial (toss) has only two possible outcomes: a head and a tail. We can call a head a success and a tail a failure.
3. The probability of obtaining a head (a success) is $\frac{1}{2}$ and that of a tail (a failure) is $\frac{1}{2}$ for any toss. That is,

$$p = P(H) = \frac{1}{2} \quad \text{and} \quad q = P(T) = \frac{1}{2}$$

   The sum of these two probabilities is 1. Also, these probabilities remain the same for each toss.
4. The trials (tosses) are independent. The result of any preceding toss has no bearing on the result of any succeeding toss.

Therefore, the experiment consisting of 10 tosses is a binomial experiment. ■

Verifying conditions of a binomial experiment.

EXAMPLE 5-18 Five percent of all VCRs manufactured by a large electronics firm are defective. Three VCRs are randomly selected from the production line of this firm. The selected VCRs are inspected to determine if each of them is defective or good. Is this experiment a binomial experiment?

Solution

1. This example consists of three identical trials. A trial represents the selection and inspection of a VCR.
2. Each trial has two outcomes: A VCR is defective or a VCR is good.
3. Five percent of all VCRs are defective. So, the probability p that a VCR is defective is .05. As a result, the probability q that a VCR is good is .95. These two probabilities add up to 1.
4. Each trial (VCR) is independent. In other words, if one VCR is defective it does not affect the outcome of another VCR. This is so because the size of the population is very large as compared to the sample size.

Consequently, this is an example of a binomial experiment. ■

5.6.2 THE BINOMIAL PROBABILITY DISTRIBUTION AND BINOMIAL FORMULA

The random variable x that represents the number of successes in n trials for a binomial experiment is called a *binomial random variable*. The probability distribution of x in such experiments is called the **binomial probability distribution** or simply *binomial distribution*. Thus, the binomial probability distribution is applied to find the probability of x successes in n trials for a binomial experiment. The number of successes x in such an experiment is a discrete random variable. For example, consider 10 tosses of a coin. Suppose we call a head a success and a tail a failure. Let x be the number of successes (heads) obtained in 10 trials (tosses). Since we can obtain any number of heads from zero to 10 in 10 tosses, x can assume any of the values 0, 1, 2, . . . , 10. Hence, it is a discrete random variable.

BINOMIAL FORMULA

For a binomial experiment, the probability of exactly x successes in n trials is given by the binomial formula

$$P(x) = \binom{n}{x} p^x q^{n-x}$$

where

$$n = \text{total number of trials}$$
$$p = \text{probability of success}$$
$$q = 1 - p = \text{probability of failure}$$
$$x = \text{number of successes in } n \text{ trials}$$
$$n - x = \text{number of failures in } n \text{ trials}$$

In the binomial formula, n is the total number of trials and x is the number of successes. The difference between the total number of trials and the total number of successes, $n - x$, gives the total number of failures in n trials. The value of $\binom{n}{x}$ gives the number of ways to obtain x successes in n trials. As mentioned earlier, p and q are the probabilities of success and failure, respectively. Again, although it does not matter which of the two outcomes is called a success and which one a failure, usually the outcome to which the question refers is called a success.

To solve a binomial problem, we determine the values of n, x, $n - x$, p, and q and then substitute all these values in the binomial formula. To find the value of $\binom{n}{x}$, we can use either the combinations formula or the table of combinations (Table III of Appendix C).

To find the probability of x successes in n trials for a binomial experiment, the only values needed are those of n and p. These are called the *parameters of the binomial probability distribution* or simply the **binomial parameters.** The value of q is obtained by subtracting the value of p from 1. Thus, $q = 1 - p$.

Next, we first solve a binomial problem without using the binomial formula. Then we solve it by using the binomial formula.

Calculating probability: using a tree diagram and the binomial formula.

EXAMPLE 5-19 Five percent of all VCRs manufactured by a large electronics firm are defective. If three VCRs are randomly selected from the production line of this firm, what is the probability that exactly one of them will be defective?

Solution Let

$$D = \text{a defective VCR is selected}$$
$$G = \text{a good VCR is selected}$$

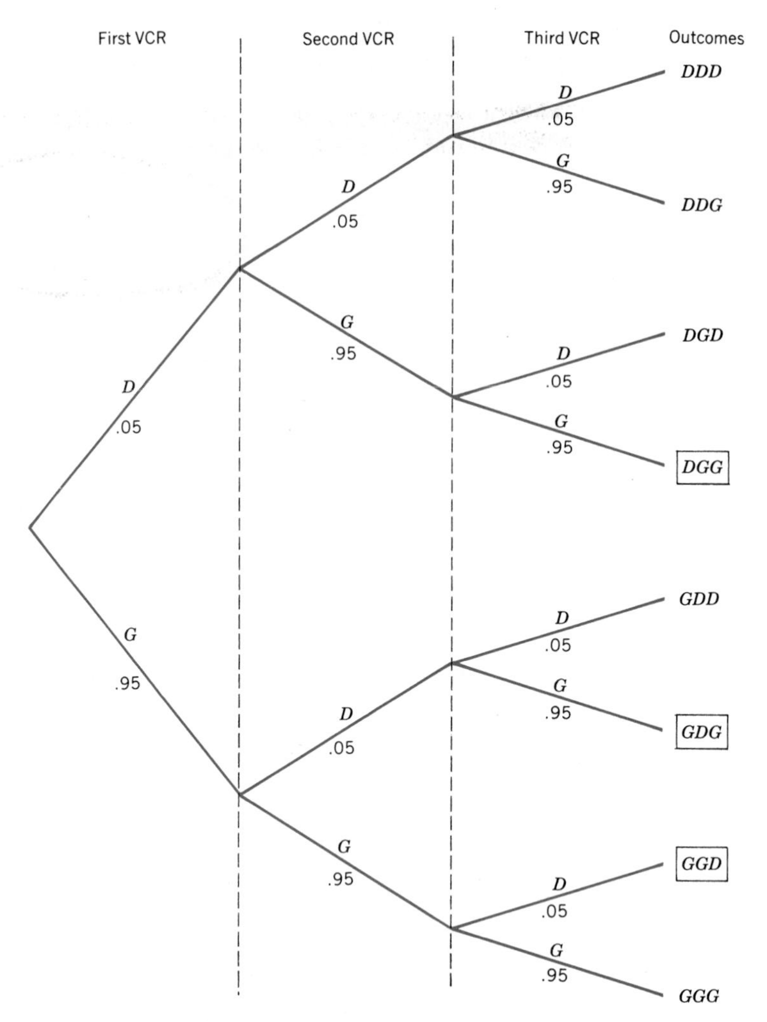

Figure 5.4 Tree diagram for selecting three VCRs.

As the tree diagram in Figure 5.4 shows, there are a total of eight outcomes and three of them contain exactly one defective VCR. These three outcomes are

$$DGG, \quad GDG, \quad \text{and} \quad GGD$$

We know that 5% of all VCRs manufactured at this firm are defective. As a result, 95% of all VCRs are good. So the probability that a randomly selected VCR is defective is .05 and the probability that it is good is .95.

$$P(D) = .05 \quad \text{and} \quad P(G) = .95$$

Because the size of the population is large (note that it is a large firm), the selections can be considered to be independent. The probability of each of the three outcomes, which give exactly one defective VCR, is calculated as follows.

$$P(DGG) = P(D) \cdot P(G) \cdot P(G) = (.05)\,(.95)\,(.95) = .0451$$
$$P(GDG) = P(G) \cdot P(D) \cdot P(G) = (.95)\,(.05)\,(.95) = .0451$$
$$P(GGD) = P(G) \cdot P(G) \cdot P(D) = (.95)\,(.95)\,(.05) = .0451$$

Note that DGG is simply the intersection of the three events D, G, and G. In other words, $P(DGG)$ is the joint probability of three events: the first VCR selected is defective, the second is good, and the third is good. To calculate this probability, we use the multiplication rule for independent events learned in Chapter 4. The same is true about the probabilities of the other two outcomes: GDG and GGD.

Exactly one defective VCR will be selected if either DGG or GDG or GGD occurs. These are three mutually exclusive outcomes. Therefore, applying the addition rule of Chapter 4, the probability of the union of these three outcomes is simply the sum of their individual probabilities.

$$\begin{aligned} P(\text{one VCR is defective in 3}) &= P(DGG \text{ or } GDG \text{ or } GGD) \\ &= P(\text{DGG}) + P(GDG) + P(GGD) \\ &= .0451 + .0451 + .0451 = .1353 \end{aligned}$$

Now let us use the binomial formula to compute this probability. Let us call the selection of a defective VCR a success and the selection of a good VCR a failure. The reason we have called a defective VCR a *success* is that the question refers to selecting exactly one defective VCR. Then,

$$\begin{aligned} n &= \text{total number of trials} = 3 \text{ VCRs} \\ x &= \text{number of successes} = \text{number of defective VCRs} = 1 \\ n - x &= \text{number of failures} = \text{number of good VCRs} = 3 - 1 = 2 \\ p &= P(\text{success}) = .05 \\ q &= P(\text{failure}) = 1 - p = .95 \end{aligned}$$

The probability of one success is denoted by $P(x = 1)$ or simply by $P(1)$. By substituting all the values in the binomial formula, we obtain:

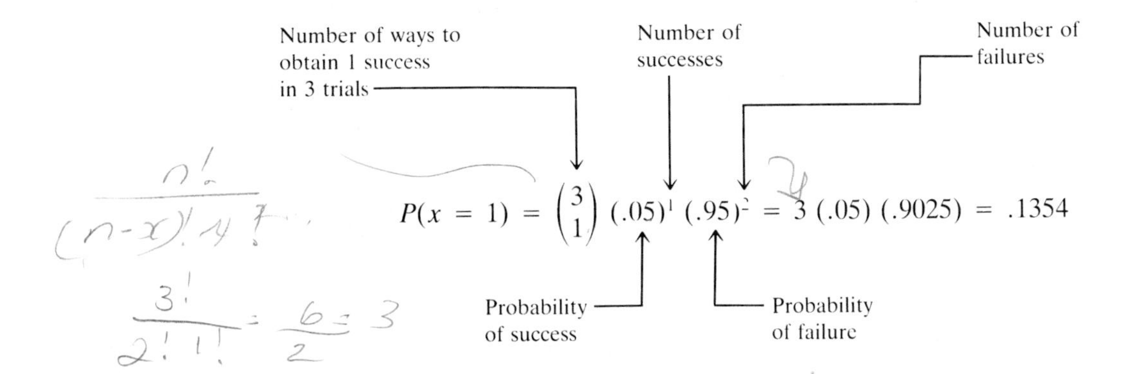

In this computation, $\binom{3}{1}$ gives the three ways to select one defective VCR in three selections. As listed earlier, these three ways to select one defective VCR are DGG, GDG, and GGD. The probability .1354 is slightly different from the earlier calculation (.1353) because of rounding. ■

Calculating probability by using the binomial formula.

EXAMPLE 5-20 According to a 1990 Roper poll, 55% of women who work outside their homes do so because of economic necessity, either to support their families or to support themselves (*The 1990 Virginia Slims Opinion Poll*). Assume that this result holds true for the current population of all working women. Find the probability that exactly 6 women in a random sample of 10 working women will say that they work outside their homes because of economic necessity.

Solution Let us call it a success if a woman says that she is working because of economic necessity and a failure if she is working for reasons other than economic necessity. Then,

$$n = \text{total number of women in the sample} = 10$$
$$x = \text{number of successes} = 6$$
$$n - x = \text{number of failures} = 10 - 6 = 4$$
$$p = P(\text{success}) = .55$$

and

$$q = P(\text{failure}) = 1 - .55 = .45$$

Substituting all the values in the binomial formula, we obtain:

$$P(x = 6) = \binom{10}{6} (.55)^6 (.45)^4 = (210)(.0277)(.0410) = .2385$$

Thus, the probability that exactly 6 women in a random sample of 10 working women will say that they work because of economic necessity is .2385.† ■

Constructing a binomial distribution and its histogram.

EXAMPLE 5-21 According to the U.S. Bureau of Labor Statistics, 56% of mothers with children under 6 years of age work outside their homes (*Newsweek*, June 4, 1990). A random sample of three mothers with children under 6 years of age is selected. Let x denote the number of mothers in this sample who work outside their homes. Assuming that 56% of all mothers with children under 6 years of age work outside their homes, write the probability distribution of x and draw a probability histogram.

Solution Let x denote the number of mothers in a sample of three who work outside their homes. Then $n - x$ is the number of mothers in the sample who do not work outside. From the given information,

$$n = \text{total number of mothers in the sample} = 3$$
$$p = P(\text{a mother works outside her home}) = .56$$
$$q = P(\text{a mother does not work outside her home}) = 1 - .56 = .44$$

The possible values that x can assume are 0, 1, 2, and 3. In other words, the number of mothers in a sample of 3 who work outside their homes can be 0, 1, 2, or

†**Using a calculator to simplify the expression for $P(x = 6)$:** As explained in Chapter 4, you can evaluate $(.55)^6$ and $(.45)^4$ by using the y^x key on your calculator. To evaluate $(.55)^6$, first enter.55, then press the y^x key, then enter 6, and finally press the = key. The calculator screen will display the answer. You can evaluate $(.45)^4$ the same way.

3. The probability of each of these four outcomes is calculated as follows.
If $x = 0$, then $n - x = 3$. Using the binomial formula, the probability of $x = 0$ is

$$P(x = 0) = \binom{3}{0} (.56)^0 (.44)^3 = (1)(1)(.0852) = .0852$$

Note that $\binom{3}{0}$ is 1 by definition and $(.56)^0$ is equal to 1 because anything raised to the power zero is always 1.
If $x = 1$, then $n - x = 2$. The probability $P(x = 1)$ is

$$P(x = 1) = \binom{3}{1} (.56)^1 (.44)^2 = (3)(.56)(.1936) = .3252$$

If $x = 2$, then $n - x = 1$. The probability $P(x = 2)$ is

$$P(x = 2) = \binom{3}{2} (.56)^2 (.44)^1 = (3)(.3136)(.44) = .4140$$

If $x = 3$, then $n - x = 0$. The probability $P(x = 3)$ is

$$P(x = 3) = \binom{3}{3} (.56)^3 (.44)^0 = (1)(.1756)(1) = .1756$$

These probabilities are written in tabular form in Table 5.11. Figure 5.5 shows the probability histogram for the probability distribution of Table 5.11.

Table 5.11 Probability Distribution of x

x	$P(x)$
0	.0852
1	.3252
2	.4140
3	.1756

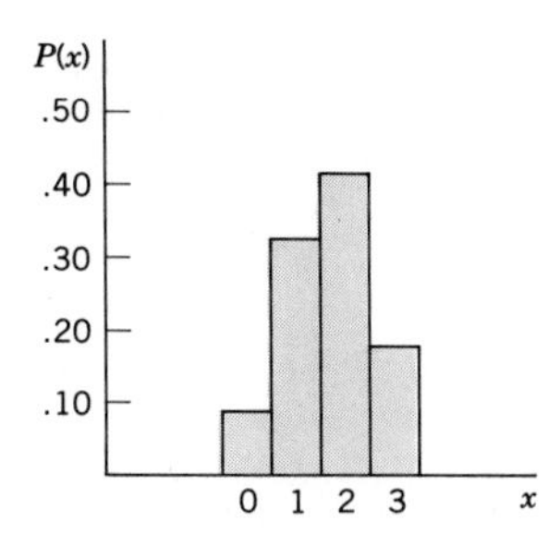

Figure 5.5 Probability histogram for Table 5.11.

■

CASE STUDY 5-3 MISSING WOMEN

[This case study is based on] the 1968 trial of the pediatrician-author Dr. Benjamin Spock and others in the U.S. District Court in Boston for conspiracy to violate the Selective Service Act by encouraging resistance to the war in Vietnam. In that trial, the defense challenged the legality of the jury-selection method. Although more than half of all eligible jurors in Boston were women, there were no women on Dr. Spock's jury. Yet he, more than any defendant, would have wanted some because so many mothers have raised their children "according to Dr. Spock"; moreover, the opinion polls showed women in general to be more opposed to the Vietnam war than men.

The question was whether this total absence of women jurors was an accident of this particular jury or whether it had resulted from systematic discrimination. Statistical reasoning was to provide the answer.

In the Boston District Court, jurors are selected in three stages. The City Directory is used for the first stage; from it, the Clerk of the Court is supposed to select 300 names at random, that is, by a lotterylike method, and put a slip with each of these names into a box. The City Directory is renewed annually by censuslike household visits of the police, and it lists all adult individuals in the Boston area. The Directory lists slightly more women than men. The second selection stage occurs when a trial is about to begin. From the 300 names in the box, the names of 30 or more potential jurors are drawn. These people are ordered to appear in court on the morning of the trial. The subgroup of 30 or more is called a venire. In the third stage, the one that most of us think of as jury selection, 12 actual jurors are selected after interrogation by both the prosecutor and the defense counsel.

The average proportion of women drawn by the six judicial colleagues of the Spock trial judge was 29%, and furthermore, the averages of these six judges bunched closely around the group average. This suggests that the proportion of women among the names in the 300-name panels in the jury box was somewhere close to that 29% mark. But . . . the Spock judge's venires had consistently lower percentages of women, with an overall average of only 14.6% women, almost exactly half of that of his colleagues.

It is possible, of course, that the selection method used by the trial judge was the same as that of his colleagues. But what is the probability that a difference as large (or larger) as that between 14.6 and 29% could arise by chance? Statistical computation revealed the probability to be 1 in 1,000,000,000,000,000,000 that the "luck of the draw" would yield the distribution of women jurors obtained by the trial judge or a more extreme one. The conclusion, therefore, was virtually inescapable: the venires for the trial judge must have been drawn from the central jury lists in a fashion that somehow systematically reduced the proportion of women jurors.

Thus the proportion of women among the potential jurors twice suffered an improper reduction—first when the court clerk reduced their share from a majority in the City Directory to 29% in the jury lists and, second, when judge managed to lower the 29% to his private average of 14.6%. In the Spock trial, only one potential woman juror came before the court, and she was easily eliminated in stage 3 by the prosecutor under his quota of peremptory challenges (for which he need not give any reasons).

For further discussion see H. Zeisel, "Dr. Spock and the Case of the Vanishing Women Jurors," *University of Chicago Law Review* 37: 1–18, 1969.

5.6.3 THE TABLE OF BINOMIAL PROBABILITIES

The probabilities for a binomial experiment can also be read from Table IV, the table of binomial probabilities, given in Appendix C. That table lists the probabilities of x for $n = 1$ to $n = 25$ and for selected values of p. Example 5-22 illustrates how to read Table IV.

Using the binomial table: x equals a specific value.

EXAMPLE 5-22 Based on data from a Peter D. Hart Research Associates' poll on consumer buying habits and attitudes, Peter Hart estimated that 5% of American shoppers are *status shoppers,* that is, the shoppers who love to buy designer labels (*The American Way of Buying,* Dow Jones & Company, Inc., 1990). Assume that this result is true for the population of all American shoppers. Using Table IV of Appendix C, find the probability that exactly three in a random sample of eight American shoppers will be status shoppers.

Solution To read the required probability from Table IV of Appendix C, first we determine the values of n, x, and p. These values are

$$n = \text{the number of shoppers in the sample} = 8$$

$$x = \text{the number of status shoppers in } 8 = 3$$

$$p = P(\text{a shopper is a status shopper}) = .05$$

Then we locate $n = 8$ in the column labeled n in Table IV. The relevant portion of Table IV with $n = 8$ is reproduced here as Table 5.12. Next, we locate 3 in the column for x in the portion of the table where $n = 8$ and locate $p = .05$ in the row for p at the top of the table. The entry at the intersection of the row for $x = 3$ and the column for $p = .05$ gives the probability of 3 successes in 8 trials when the probability of success is .05.

Table 5.12 Determining $P(x = 3)$ for $n = 8$ and $p = .05$

$p = .05$ ↓

			p			
n	x	**.05**	**.10**	**.20**	. . .	**.95**
$n = 8$ ⟶ 8	0	.6634	.4305	.1678	. . .	.0000
	1	.2793	.3826	.3355	. . .	.0000
	2	.0515	.1488	.2936	. . .	.0000
$x = 3$ ⟶	3	.0054 ←	.0331	.1468	. . .	.0000
	4	.0004	.0046	.0459	. . .	.0004
	5	.0000	.0004	.0092	. . .	.0054
	6	.0000	.0000	.0011	. . .	.0515
	7	.0000	.0000	.0001	. . .	.2793
	8	.0000	.0000	.0000	. . .	.6634

$P(x = 0) = .0054$

From Table IV or Table 5.12

$$P(x = 3) = .0054$$

■

EXAMPLE 5-23 According to a *Newsweek*/Gallup poll, 70% of adults surveyed said that "staying home with family" is their favorite way of spending an evening (*Psychology Today*, September 1988). Assume that this result holds true for the current population of all adults.

Using the binomial table: x in an interval.

(a) Using Table IV of Appendix C, find the probability that the number of adults in a sample of five adults who say staying home with family is their favorite way of spending an evening is

(i) at most 2 (ii) at least 3 (iii) 2 to 5

Writing the probability distribution for a binomial experiment.

(b) Let x be the number of adults in the sample of five who say staying home with family is their favorite way of spending an evening. Write the probability distribution of x and draw a probability histogram.

Solution Let p be the probability that an adult will say that staying home with family is his or her favorite way of spending an evening. From the given information,

$$n = \text{number of adults in the sample} = 5$$

$$p = .70$$

The portion of Table IV corresponding to $n = 5$ and $p = .70$ is reproduced here as Table 5.13.

Table 5.13 Portion of Table IV for $n = 5$ and $p = .70$

n	x	p .70
5	0	.0024
	1	.0284
	2	.1323
	3	.3087
	4	.3601
	5	.1681

(a)

(i) The event that at most two adults will say staying home with family is their favorite way of spending an evening will occur if x is equal to 0, 1, or 2. Using Table IV or Table 5.13, the required probability is

$$P(\text{at most } 2) = P(0 \text{ or } 1 \text{ or } 2) = P(x = 0) + P(x = 1) + P(x = 2)$$
$$= .0024 + .0284 + .1323 = .1631$$

If we use the binomial formula to calculate the above probability, we will have one term for each of the three probabilities as follows.

$$P(\text{at most } 2) = P(x = 0) + P(x = 1) + P(x = 2)$$
$$= \binom{5}{0}(.70)^0(.30)^5 + \binom{5}{1}(.70)^1(.30)^4 + \binom{5}{2}(.70)^2(.30)^3$$
$$= .0024 + .0284 + .1323 = .1631$$

(ii) Using Table IV or Table 5.13, the probability of at least 3 is

$$P(\text{at least } 3) = P(3 \text{ or } 4 \text{ or } 5) = P(x = 3) + P(x = 4) + P(x = 5)$$
$$= .3087 + .3601 + .1681 = .8369$$

(iii) Using Table IV or Table 5.13, the probability of 2 to 5 is

$$P(2 \text{ to } 5) = P(x = 2) + P(x = 3) + P(x = 4) + P(x = 5)$$
$$= .1323 + .3087 + .3601 + .1681 = .9692$$

(b) Using Table IV of Appendix C, or Table 5.13, we list the probability distribution of x for $n = 5$ and $p = .70$ in Table 5.14. Figure 5.6 shows the probability histogram for the probability distribution of x.

Table 5.14 Probability Distribution of x for $n = 5$ and $p = .70$

x	$P(x)$
0	.0024
1	.0284
2	.1323
3	.3087
4	.3601
5	.1681

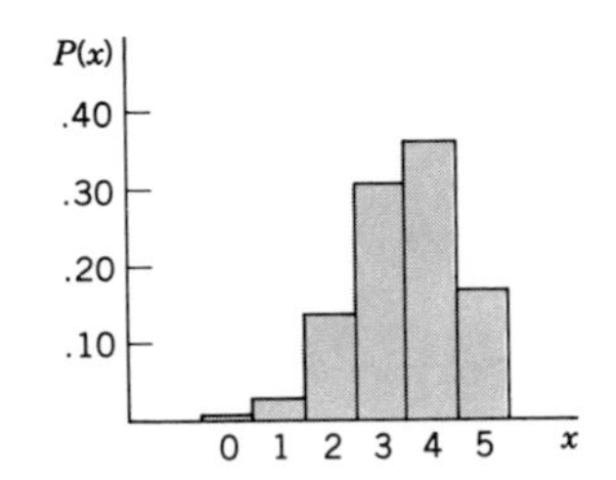

Figure 5.6 Probability histogram for Table 5.14.

■

5.6.4 THE PROBABILITY OF SUCCESS AND THE SHAPE OF THE BINOMIAL DISTRIBUTION

For any number of trials n:

1. The binomial probability distribution is symmetric if $p = .50$.
2. The binomial probability distribution is skewed to the right if p is less than .50.
3. The binomial probability distribution is skewed to the left if p is greater than .50.

These three cases are illustrated here with examples and graphs.

1. Let $n = 4$ and $p = .50$. Using Table IV of Appendix C, we have written the probability distribution of x in Table 5.15 and plotted it in Figure 5.7. As we can observe from Table 5.15 and the probability histogram of Figure 5.7, the probability distribution of x is symmetric.

Table 5.15 Probability Distribution of x for $n = 4$ and $p = .50$

x	$P(x)$
0	.0625
1	.2500
2	.3750
3	.2500
4	.0625

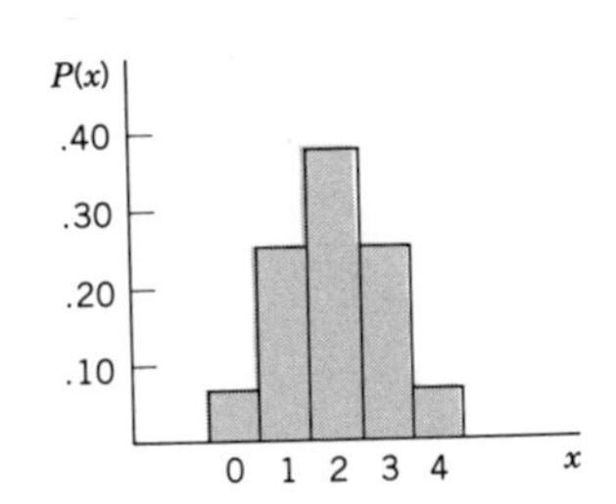

Figure 5.7 Probability histogram for Table 5.15.

2. Let $n = 4$ and $p = .30$ (which is less than .50). Table 5.16, which is written using Table IV of Appendix C, and the probability histogram of Figure 5.8 show that the probability distribution of x for $n = 4$ and $p = .30$ is skewed to the right.

Table 5.16 Probability Distribution of x for $n = 4$ and $p = .30$

x	$P(x)$
0	.2401
1	.4116
2	.2646
3	.0756
4	.0081

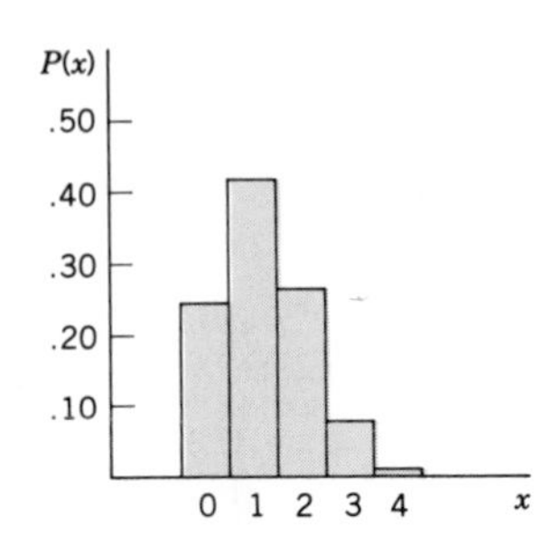

Figure 5.8 Probability histogram for Table 5.16.

3. Let $n = 4$ and $p = .80$ (which is greater than .50). Table 5.17, which is written using Table IV of Appendix C, and the probability histogram of Figure 5.9 show that the probability distribution of x for $n = 4$ and $p = .80$ is skewed to the left.

Table 5.17 Probability Distribution of x for $n = 4$ and $p = .80$

x	$P(x)$
0	.0016
1	.0256
2	.1536
3	.4096
4	.4096

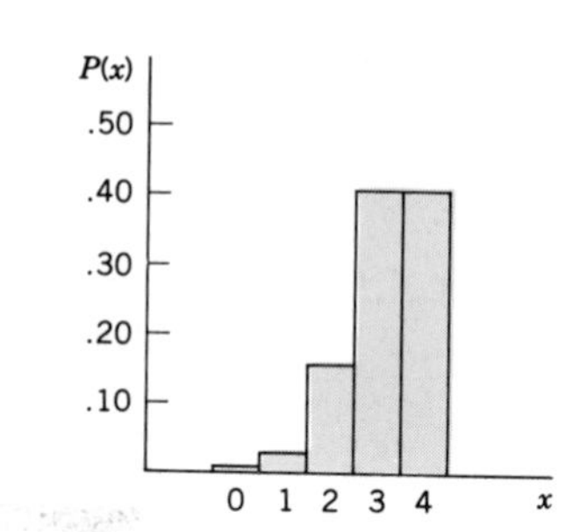

Figure 5.9 Probability histogram for Table 5.17.

5.6.5 THE MEAN AND STANDARD DEVIATION OF THE BINOMIAL DISTRIBUTION

Sections 5.3 and 5.4 explained how to compute the mean and standard deviation for the probability distribution of a discrete random variable. When a discrete random variable has a binomial distribution, the formulas learned in Sections 5.3 and 5.4 could still be used to compute its mean and standard deviation, but it would be time-consuming. It is more convenient and simpler to use the following formulas to find the mean and standard deviation in such cases.

MEAN AND STANDARD DEVIATION OF A BINOMIAL DISTRIBUTION

The mean and standard deviation for a binomial distribution are

$$\mu = np$$

and

$$\sigma = \sqrt{npq}$$

where n is the total number of trials, p is the probability of success, and q is the probability of failure.

The following example describes the computation of the mean and standard deviation for a binomial distribution.

Mean and standard deviation of a binomial random variable.

EXAMPLE 5-24 According to a 1989 Carnegie Foundation survey, 84% of college and university faculty members said that "undergraduates have become more careerist in their concerns" (*The Conditions of the Professoriate: Attitudes and Trends, 1989*, The Carnegie Foundation). Assume that this result holds true for the current population of all college and university faculty members. Let x denote the number of faculty members in a random sample of 25 college and university faculty members who hold this view. Find the mean and standard deviation of the probability distribution of x.

Solution This is a binomial experiment with 25 total trials. Each trial has two outcomes: either a faculty member holds the view that undergraduates have become more careerist in their concerns or this faculty member does not hold this view. The probabilities p and q for these two outcomes are .84 and .16, respectively. Thus,

$$n = 25, \quad p = .84, \quad \text{and} \quad q = .16$$

Using the formulas for the mean and standard deviation of the binomial distribution, we obtain:

$$\mu = np = 25(.84) = 21$$

$$\sigma = \sqrt{npq} = \sqrt{(25)(.84)(.16)} = 1.83$$

Thus, the mean of the probability distribution of x is 21 and the standard deviation is 1.83. The value of the mean is what we expect to obtain on average per repetition of the experiment. In this example, if we take many samples, each including 25 college and university faculty members, we expect that each sample will include an average of 21 faculty members who say that undergraduates have become more careerist in their concerns. ■

EXERCISES

5.42 Briefly explain the following.

a. A binomial experiment **b.** A trial

5.43 What are the parameters of the binomial probability distribution and what do they mean?

5.44 Which of the following are binomial experiments? Explain why.

a. Rolling a die many times and observing the number of spots

b. Rolling a die many times and observing whether the number obtained is even or odd

c. Selecting a few voters from a very large population of voters and observing whether or not each of them favors a certain proposition in an election when 54% of all voters are known to be in favor of this proposition

5.45 Which of the following are binomial experiments? Explain why.

a. Drawing three balls (with replacement) from a box that contains 10 balls, 6 of which are red and 4 are blue, and observing the color of the drawn balls

b. Drawing three balls (without replacement) from a box that contains 10 balls, 6 of which are red and 4 are blue, and observing the color of the drawn balls

c. Selecting a few households from New York City and observing if they own stocks when it is known that 28% of all households in New York City own stocks

5.46 Let x be a discrete random variable that possesses a binomial distribution. Using the binomial formula, find the following probabilities.

a. $P(x = 5)$ for $n = 8$ and $p = .60$
b. $P(x = 3)$ for $n = 4$ and $p = .30$
c. $P(x = 2)$ for $n = 6$ and $p = .20$

Verify your answer by using Table IV of Appendix C.

5.47 Let x be a discrete random variable that possesses a binomial distribution. Using the binomial formula, find the following probabilities.

a. $P(x = 0)$ for $n = 5$ and $p = .10$
b. $P(x = 4)$ for $n = 7$ and $p = .80$
c. $P(x = 7)$ for $n = 10$ and $p = .40$

Verify your answer by using Table IV of Appendix C.

5.48 The binomial probability distribution is symmetric when $p = .50$, skewed to the right when $p < .50$, and skewed to the left when $p > .50$. Illustrate each of these three cases by writing a probability distribution and by drawing a probability histogram. Choose any values of n and p and use the table of binomial probabilities (Table IV of Appendix C).

5.49 Forty-eight percent of adults in a *Hippocrates*/Gallup survey published in the May/June 1988 issue of the *Hippocrates* said that a woman should be held legally liable for harm done to her fetus because she smoked cigarettes or drank alcohol during her pregnancy. Assuming that this result holds true for the current population of all adults, what is the probability that exactly 7 adults in a random sample of 10 will hold this view? Use the binomial formula.

5.50 According to a survey conducted by David S. Anderson and Angelo F. Gadaleto, 71% of college and university campuses permitted drinking hard liquor on campus in 1988 (*The Chronicle of Higher Education*, November 9, 1988). Assuming this percentage is still true for all college and university campuses, find the probability that in a random sample of 15 campuses, exactly 9 permit drinking hard liquor on campus. Use the binomial formula.

5.51 According to the U.S. Bureau of Labor Statistics, 5% of people working in the United States moonlight (i.e., they hold a second job). Find the probability that the number of workers who moonlight in a random sample of 16 workers is

a. at most 5 **b.** none **c.** 3 to 6

Use the binomial table (Table IV of Appendix C) to calculate these probabilities.

5.52 According to a survey, 30% of cars in the United States were at least 10 years old in 1988 (*Business Week,* March 26, 1990). Assume that this result holds true for the current population of all cars in the United States. What is the probability that in a random sample of 20 cars, the number of cars that are at least 10 years old is

a. at least 8 **b.** 5 to 7 **c.** at most 4

Use the binomial table (Table IV of Appendix C) to calculate these probabilities.

5.53 According to the U.S. Census Bureau, 66% of households headed by single women own autos. Find the probability that in a random sample of nine such households, exactly eight own autos. Use the binomial formula.

5.54 In a poll conducted by *Parents* magazine, 65% of parents said they wished they had received more education (*Parents,* August 1988). Assume that this result is true for the current population of all parents. Find the probability that in a random sample of eight parents, exactly three will hold this view. Use the binomial formula.

5.55 According to a *Los Angeles Times* poll, 50% of women think they are overweight (*U.S. News & World Report,* February 19, 1990). Assume that this result holds true for the current population of all women. Find the probability that in a random sample of nine women, exactly three will say they are overweight. Use the binomial formula.

5.56 According to a survey, 75% of young men aged 18 to 24 are still living at home (*Time,* July 16, 1990). Assume that this result is true for the current population of all young men. What is the probability that in a random sample of seven young men, exactly six will be living at home? Use the binomial formula.

5.57 According to a survey, 40% of people in their twenties are children of divorced parents (*Time,* July 16, 1990). Assume that this result holds true for the current population of all persons in their twenties. What is the probability that in a random sample of 12 persons in their twenties, the number of persons that will be children of divorced parents is

a. at most 2 **b.** at least 7 **c.** 3 to 5

Use the binomial probabilities table to calculate these probabilities.

5.58 The Food and Drug Administration approves 20% of drugs submitted for its approval (*U.S. News & World Report,* January 23, 1989). Using the binomial table, find the probability that in a random sample of 15 drugs submitted for the FDA's approval, the number of drugs that are approved is

a. at most 3 **b.** at least 7 **c.** 2 to 5

5.59 According to a survey, 38% of young people (15- to 24-year-olds) had performed some community or neighborhood service activity within the past year (*Democracy's Next Generation,* People For the American Way, 1989). Assuming this result is true for the current population of all young people, find the probability that in a random sample of 10 such persons, exactly 5 would have performed some community or neighborhood service activity within the past year. Use the binomial formula.

5.60 According to Case Study 4-4, the probability that a baseball player will have no hits in 10 trips to the plate is .056, given that this player has a hitting percentage of 25%. Using the binomial formula, show that this probability is indeed .056.

5.61 According to the National Institute of Health, 10% of U.S. youngsters suffer from hyperactivity (*Time,* January 16, 1989). Using the binomial table, find the probability that in a random sample of 20 youngsters, the number of youngsters that will suffer from hyperactivity is

a. at most 5 **b.** 1 to 6 **c.** at least 3

5.62 In a survey conducted for *Rolling Stone* by Peter D. Hart Research Associates, 30% of adults said they are extremely satisfied with their incomes (*Rolling Stone,* April 7, 1988). Assume that this result holds true for the current population of all adults. Using the binomial table, find the probability that in a random sample of 22 adults, the number of adults that will be extremely satisfied with their incomes is

a. more than 13 **b.** 5 to 11 **c.** less than 8

5.63 According to a survey, 30% of credit card holders pay off their balances in full each month (*The Wall Street Journal,* August 20, 1990). Assume that this result holds true for the current population of credit card holders. Using the binomial table, find the probability that in a random sample of 18 credit card holders, the number of credit card holders that pay off their balances in full each month is

a. less than 3 **b.** 1 to 4 **c.** more than 6

5.64 In a survey of 35,000 women, 60% said they are very or somewhat satisfied with the way they manage their time (*Family Circle,* March 14, 1989). Assume that this result holds true for the current population of all women. Using the binomial table, find the probability that in a random sample of 18 women, the number of women that hold this view is

a. exactly 8 **b.** at most 6 **c.** 6 to 12

5.65 According to the U.S. Internal Revenue Service, 20% of the 1988 income tax returns had errors. Find the probability that in a random sample of 12 income tax returns for 1988 exactly 3 returns will have errors. Use the binomial formula.

5.66 According to a 1989 survey done by the National Highway Traffic Safety Administration, 28% of drivers in New York City use seat belts. Assume that this result holds true for the current population of all drivers in New York City. Find the probability that in a random sample of 10 drivers selected from New York City exactly 2 use seat belts. Use the binomial formula.

5.67 According to a survey conducted by the Roper Organization, 70% of households owned microwave ovens in 1989. Assume that this result holds true for the current population of all households. Let x denote the number of households who own microwave ovens in a random sample of four households. Using the binomial probabilities table, write the probability distribution of x and draw a probability histogram. Find the mean and standard deviation of x.

5.68 In a *USA Today* survey of parents with children aged 2 to 17 and living at home, 60% of the parents said that television inhibits creative thinking (*USA Today,* February 28, 1989). Assume that this result holds true for the current population of all parents. Let x denote the number of parents who hold this view in a random sample of five parents. Using the binomial probabilities table, construct the probability distribution of x and draw a probability histogram. Find the mean and standard deviation of x.

5.69 In a survey of college and university freshmen, (about) 10% of the freshmen reported that they frequently smoked cigarettes (Alexander W. Astin, et al., *The American Freshman: National Norms For Fall 1988,* American Council on Education and UCLA). Assume that this result holds true for the current population of all college and university freshmen. Let x denote the number of freshmen who frequently smoke cigarettes in a random sample of seven freshmen. Using the binomial probabilities table, write the probability distribution of x and draw a probability histogram. Find the mean and standard deviation of x.

5.70 Fifty percent of the adult population in a large city are women. A court is to randomly select a jury of five adults from the population of all adults of this city. Let x denote the number of women selected for this jury.

a. Using the binomial probabilities table, obtain the probability distribution of x and draw a probability histogram. Determine the mean and standard deviation of x.

b. What is the probability that none of the five jury members is a woman?

5.7 THE POISSON PROBABILITY DISTRIBUTION

The **Poisson probability distribution,** named after the French mathematician Simeon D. Poisson, is another important probability distribution of a discrete random variable that has a large number of applications. Suppose a washing machine in a laundromat breaks down an average of three times a month. We may want to find the probability of exactly two breakdowns during the next month. This is an example of a Poisson probability distribution problem. Each breakdown is called an *occurrence* in Poisson distribution terminology. The Poisson probability distribution is applied to experiments with random and independent occurrences. The occurrences are random in the sense that they do not follow any pattern and, hence, they are unpredictable. Independence of occurrences means that one occurrence (or nonoccurrence) of an event does not influence the successive occurrences or nonoccurrences of that event. The occurrences are always considered with respect to an interval. In the example of washing machine, the interval represented one month. The interval may be a time interval, a space interval, or a volume interval. The actual number of occurrences within an interval are random and independent. If the average number of occurrences for a given interval are known, then by using the Poisson probability distribution we can compute the probability of any number of occurrences x in that interval. Note that the number of actual occurrences in an interval are denoted by x.

CONDITIONS TO APPLY THE POISSON PROBABILITY DISTRIBUTION

The following three conditions must be satisfied to apply the Poisson probability distribution.

1. x is a discrete random variable.

2. The occurrences are random.

3. The occurrences are independent.

The following are a few examples of discrete random variables for which the occurrences are random and independent. Hence, these are examples to which the Poisson probability distribution can be applied.

1. Consider the number of patients arriving at the emergency ward of a hospital during a 1-hour interval. In this example, an occurrence is the arrival of a patient at the emergency ward, the interval is 1 hour (an interval of time), and the occurrences are random. The total number of patients who may arrive at this emergency ward during a 1-hour interval may be 0, 1, 2, 3, 4, ... The independence of occurrences in this example means that the patients arrive individually and the arrival of any two (or more) patients is not related.
2. Consider the number of defective items in the next 100 items manufactured on a machine. In this case, the interval is a volume interval (100 items). The occurrences (number of defective items) are random because there may be 0, 1, 2, 3, . . . , 100 defective items in 100 items. We can assume the occurrence of defective items to be independent of one another.
3. Consider the number of defects in a 5-foot-long iron rod. The interval, in this example, is a space interval (5 feet). The occurrences (defects) are random because there may be any number of defects in a 5-foot iron rod. We can assume these defects to be independent of one another.

The following examples also qualify for the application of the Poisson probability distribution.

1. The number of accidents that occur on a given highway during a 1-week period
2. The number of customers coming to a bank during a 1-hour interval
3. The number of television sets sold at a department store during a given week

On the other hand, arrival of patients at a physician's office will be nonrandom if the patients have to make an appointment to see the doctor. The arrival of commercial airplanes at an airport is nonrandom because all planes are scheduled to arrive at certain times and airport authorities know the exact number of arrivals for any period (although this number may change slightly because of late or early arrivals and cancellations).

In the Poisson probability distribution terminology, the *average number of occurrences* in an interval is denoted by λ (Greek letter lambda). The actual number of occurrences in that interval is denoted by x. We are to find the probability of x given λ.

POISSON PROBABILITY DISTRIBUTION FORMULA

According to the Poisson probability distribution, the probability of x occurrences in an interval is

$$P(x) = \frac{\lambda^x e^{-\lambda}}{x!}$$

where λ (pronounced lambda) is the mean number of occurrences in that interval and the value of e is approximately 2.71828.

The mean number of occurrences in an interval, denoted by λ, is called the *parameter of the Poisson probability distribution* or the **Poisson parameter.** As is obvious from the Poisson formula, we need to know only the value of λ to compute the probability of any given value of x. We can read the value of $e^{-\lambda}$ for a given λ from Table V of Appendix C. Examples 5-25 to 5-27 illustrate the use of the Poisson formula.

Using Poisson formula: x equals a specific value.

EXAMPLE 5-25 According to a U.S. National Center for Health Statistics survey, the mean number of visits to physicians by females was 6.2 in 1987. Assuming that this mean holds true for the 1987 population of all households and that the conditions of the Poisson distribution are satisfied, find the probability that a randomly selected female made exactly 5 visits to physicians in 1987.

Solution Let λ be the mean number of visits to physicians for the 1987 population of all females. Then, $\lambda = 6.2$. Let x be the number of visits to physicians by the selected female. We are to find the probability of $x = 5$. Substituting all the values in the Poisson formula, we obtain:

$$P(x = 5) = \frac{(6.2)^5 e^{-6.2}}{5!} = \frac{(9161.33)\ (.002029)}{120} = .1549$$

In these calculations, we can find the value of 5! from Table II and the value of $e^{-6.2}$ from Table V of Appendix C.†

To find the value of $e^{-6.2}$ from Table V of Appendix C, we locate 6.2 in the column for λ and read the value across from 6.2 in the column for $e^{-\lambda}$. This value is .002029. ■

Using Poisson formula.

EXAMPLE 5-26 A washing machine in a laundromat breaks down an average of three times per month. Using the Poisson formula, find the probability that during the next month this machine will have

(a) exactly 2 breakdowns
(b) at most 1 breakdown

Solution Let λ be the mean number of breakdowns per month and x be the number of breakdowns during the next month for this machine. Then,

$$\lambda = 3$$

(a) The probability of exactly 2 breakdowns is

$$P(x = 2) = \frac{(3)^2 e^{-3}}{2!} = \frac{(9)(.049787)}{2} = .2240$$

(b) The probability of at most one breakdown is given by the sum of the probabilities of zero and one breakdowns. Thus,

$$\begin{aligned} P(\text{at most 1 breakdown}) &= P(0 \text{ or } 1 \text{ breakdown}) = P(x = 0) + P(x = 1) \\ &= \frac{(3)^0 e^{-3}}{0!} + \frac{(3)^1 e^{-3}}{1!} \\ &= \frac{(1)(.049787)}{1} + \frac{(3)(.049787)}{1} \\ &= .0498 + .1494 = .1992 \end{aligned}$$

■

One important point to remember is that *the intervals for λ and x must be equal.* If they are not, the mean λ should be redefined to make them equal. Example 5-27 illustrates this point.

Poisson formula: intervals for the mean and x not equal.

EXAMPLE 5-27 On average, 3 households in 10 own answering machines. Using the Poisson formula, find the probability that in a random sample of 20 households, exactly 5 will own answering machines.

Solution Let x denote the number of households in the sample of 20 who own answering machines. We are to find $P(x = 5)$. The given mean is defined per 10

†**Using a calculator to simplify the expression for $P(x = 5)$:** In the expression for $P(x = 5)$, you can evaluate $(6.2)^5$ and 5! by using the y^x and $n!$ keys, respectively. If your calculator has the e^x key, you can evaluate $e^{-6.2}$ as follows: First enter 6.2, then press the $+/-$ key. Finally press the e^x key.

households, but x is defined for a sample of 20. As a result, we should first find the mean for 20 households. Because the mean number of households in 10 that own answering machines is 3, the mean number of households in 20 that own answering machines will be 6. Thus,

$$\lambda = 6$$

Substituting $x = 5$ and $\lambda = 6$ in the Poisson formula, we obtain:

$$P(x = 5) = \frac{(6)^5 e^{-6}}{5!} = \frac{(7776)(.002479)}{120} = .1606$$

Thus, the probability is .1606 that exactly 5 households in a random sample of 20 will own answering machines. ■

As a passing note, Example 5-27 can also be solved using the binomial probability distribution. In Exercise 5.102, Example 5-27 is presented as a binomial problem.

5.7.1 THE TABLE OF POISSON PROBABILITIES

The probabilities for a Poisson distribution can also be read from Table VI, the table of Poisson probabilities, given in Appendix C (page 735). The following example describes how to read that table.

Using the table of Poisson probabilities.

EXAMPLE 5-28 The mean number of new accounts opened at a savings bank is 2 per day. Using Table VI of Appendix C, find the probability that on a given day the number of new accounts that will be opened at this savings bank is

(a) exactly 6 (b) at most 3 (c) at least 7

Solution Let λ be the mean number of new accounts opened per day at this savings bank and let x be the number of new accounts opened at this savings bank on a given day.

(a) The values of λ and x are

$$\lambda = 2 \quad \text{and} \quad x = 6$$

In Table VI of Appendix C, we locate the column for $\lambda = 2$. In this column, we read the value that corresponds to $x = 6$. The relevant portion of that table is shown on next page as Table 5.18. The probability that exactly 6 new accounts will be opened on a given day is .0120. Therefore,

$$P(x = 6) = .0120$$

Actually, Table 5.18 gives the probability distribution of x for $\lambda = 2.0$. Note that the sum of the 10 probabilities given in Table 5.18 is .9999 and not 1.0. This is so for two reasons. First, these probabilities are rounded to four decimal places. Second, on a given day more than 9 new accounts might be opened at this bank. However, the probabilities of more than 9 new accounts are very small and they are not listed in the table.

Table 5.18 Portion of Table VI for $\lambda = 2.0$

		λ		
x	1.1	1.2	...	2.0 ← $\lambda = 2.0$
0				.1353
1				.2707
2				.2707
3				.1804
4				.0902
5				.0361
$x = 6$ → 6				.0120 ← $P(x = 6)$
7				.0034
8				.0009
9				.0002

(b) The probability of at most 3 new accounts is obtained by adding the probabilities of 0, 1, 2, or 3 new accounts being opened on a given day. Thus,

$$P(\text{at most } 3) = P(x = 0) + P(x = 1) + P(x = 2) + P(x = 3)$$
$$= .1353 + .2707 + .2707 + .1804 = .8571$$

(c) The probability of at least 7 new accounts is obtained by adding the probabilities of 7, 8, or 9 new accounts being opened at this bank on a given day. Note that 9 is the last value of x for $\lambda = 2.0$ in Table VI of Appendix C or Table 5.18. Hence, 9 is the last value of x whose probability is included in the sum. However, this does not mean that on a given day more than 9 accounts cannot be opened. It simply means that the probability of 10 or more accounts is approximately zero.

$$P(\text{at least } 7) = P(x = 7) + P(x = 8) + P(x = 9)$$
$$= .0034 + .0009 + .0002 = .0045$$

■

Constructing a Poisson probability distribution.

EXAMPLE 5-29 An auto salesperson sells an average of .9 cars per day. Let x be the number of cars sold by this salesperson on any given day. Using the Poisson probability distribution table, write the probability distribution of x.

Solution Let λ be the mean number of cars sold per day by this salesperson. Hence, $\lambda = .9$. Using the portion of Table VI corresponding to $\lambda = .9$, we write the probability distribution of x in Table 5.19.

Table 5.19 Probability Distribution of x for $\lambda = .9$

x	$P(x)$
0	.4066
1	.3659
2	.1647
3	.0494
4	.0111
5	.0020
6	.0003

Note that 6 is the highest value of x for $\lambda = .9$ listed in Table VI. However, this does not mean that this salesperson cannot sell more than 6 cars on a given day. What this means is that the probability of selling 7 cars (or any higher number) is very small. Actually, the probability of $x = 7$ for $\lambda = .9$ calculated by using the Poisson formula is .000039. When rounded to four decimal places, this probability is .0000. Hence, it is not listed in Table VI. ■

EXERCISES

5.71 What are the conditions that should be fulfilled to apply the Poisson probability distribution? What is the parameter of the Poisson distribution, and what does it mean?

5.72 Using the Poisson formula, find the following probabilities.

a. $P(x \leq 1)$ for $\lambda = 4$ **b.** $P(x = 2)$ for $\lambda = 2.2$

Verify these probabilities using Table VI of Appendix C.

5.73 Using the Poisson formula, find the following probabilities.

a. $P(x < 2)$ for $\lambda = 3$ **b.** $P(x = 8)$ for $\lambda = 5.3$

Verify these probabilities using Table VI of Appendix C.

5.74 Write the probability distribution of x for each of the following cases by using the Poisson probability distribution table.

a. $\lambda = 1.3$ **b.** $\lambda = 2.1$

5.75 Write the probability distribution of x for each of the following cases by using the Poisson probability distribution table.

a. $\lambda = .6$ **b.** $\lambda = 1.8$

5.76 An average of 8.2 crimes are reported per day to police in a city. Find the probability that exactly 3 crimes will be reported to police on a certain day in this city. Use the Poisson formula.

5.77 A mail-order company receives an average of 7.4 orders per day. Find the probability that it will receive exactly 10 orders on a certain day. Use the Poisson formula.

5.78 An average of four students fail in a statistics class of 50 students at a university. Find the probability that at most one student will fail in a given statistics class of 50 students. Use the Poisson formula.

5.79 A commuter airline receives an average of 9.7 complaints per day from its passengers. Using the Poisson formula, find the probability that on a certain day this airline will receive exactly seven complaints.

5.80 Each box of 25 apples contains an average of 4 bad apples.

a. Using the Poisson formula, find the probability that in a randomly selected box of 25 apples there will be 7 bad apples.

b. Using the Poisson probability table, find the probability that the number of bad apples in a box of 25 apples will be

i. at least 10

ii. at most 4

5.81 An average of 2.8 employees of a telephone company are absent each day.

a. Using the Poisson formula, find the probability that at most one employee will be absent on a given day at this company.

b. Using the Poisson probability table, find the probability that on a given day the number of employees that will be absent at this company is

i. 1 to 5

ii. at least 7

iii. at most 3

5.82 A certain type of cloth contains an average of .4 defects every 500 yards.

a. Using the Poisson formula, find the probability that a given piece of 500 yards of this cloth will contain exactly one defect.

b. Using the Poisson probability table, find the probability that the number of defects in a given 500-yard piece of this cloth will be

i. 2 to 4

ii. more than 3

iii. less than 2

5.83 A small post office receives an average of 2.9 phone calls per half hour.

a. Using the Poisson formula, find the probability that exactly 3 phone calls will be received at this post office during a certain hour.

b. Using the Poisson probability table, find the probability that the number of phone calls received at this post office during a certain hour will be

i. less than 4

ii. more than 12

5.84 An average of 4.5 customers come to a savings bank per half hour.

a. Using the Poisson formula, find the probability that exactly 2 customers will come to this savings bank during a given hour

b. Using the Poisson probability table, find the probability that during a given hour, the number of customers that will come to this savings bank is

i. at most 2

ii. at least 17

5.85 An average of 1.5 patients arrive per 10 minutes at the emergency ward of a small hospital.

a. Using the Poisson formula, find the probability that no patient will arrive at this ward during the next 10 minutes.

b. Let x denote the number of patients who will come to this emergency ward during a given 10-minute period. Using the Poisson probability table, write the probability distribution of x.

5.86 A newspaper contains an average of 1.1 typographical errors per page.

a. Using the Poisson formula, find the probability that a randomly selected page of this newspaper will contain no typographical error.

b. Let x denote the number of typographical errors that a randomly selected page of this newspaper contains. Using the Poisson probability table, write the probability distribution of x.

5.87 An insurance salesperson sells an average of 1.2 insurance policies per day.

a. Using the Poisson formula, find the probability that this salesperson will sell exactly one insurance policy on a certain day.

b. Let x denote the number of insurance policies that this salesperson will sell on a given day. Using the Poisson probability table, write the probability distribution of x.

5.88 An average of .6 accidents occur per day in a city.

a. Using the Poisson formula, find the probability that 4 accidents will occur in this city on a given day.

b. Let x denote the number of accidents that will occur in this city on a given day. Using the Poisson probability table, write the probability distribution of x.

GLOSSARY

Bernoulli Trial One repetition of an experiment. Also called a *trial.*

Binomial experiment An experiment that contains n identical trials such that each of these n trials has only two possible outcomes, that the probabilities of these two outcomes remain constant for each trial, and that the trials are independent.

Binomial parameters The total number of trials n and the probability of success p for the binomial probability distribution.

Binomial probability distribution The probability distribution that gives the probability of x successes in n trials.

Combinations The number of ways x elements can be selected from n elements.

Continuous random variable A random variable that can assume any value in one or more intervals.

Discrete random variable A random variable whose values are countable.

Factorial Denoted by the symbol "!." The product of all integers from a given number to 1. For example, "$n!$" (read as "n factorial") represents the product of all integers from n to 1.

Mean of a discrete random variable The mean of a discrete random variable x is the value that is expected to occur per repetition on average if an experiment is performed a large number of times. The mean of a discrete random variable is also called its *expected value*.

Poisson parameter The average number of occurrences, denoted by λ, during an interval for a Poisson probability distribution.

Poisson probability distribution The probability distribution that gives the probability of x occurrences in an interval.

Probability distribution of a discrete random variable A list of all the possible values that a discrete random variable can assume and their corresponding probabilities.

Random variable A variable, denoted by x, whose value is determined by the outcome of a random experiment. Also called a *chance variable*.

Standard deviation of a discrete random variable A measure of spread for the probability distribution of a discrete random variable.

KEY FORMULAS

1. **Mean of a discrete random variable x**

$$\mu = \Sigma x P(x)$$

The mean of a discrete random variable x is also called its expected value and is denoted by $E(x)$.

2. **Standard deviation of a discrete random variable x**

$$\sigma = \sqrt{\Sigma x^2 P(x) - \mu^2}$$

3. **Factorials**

$$n! = n\,(n - 1)\,(n - 2) \ldots 3 \cdot 2 \cdot 1$$

4. **Number of combinations of n items selected x at a time**

$$\binom{n}{x} = \frac{n!}{(n-x)!\,x!}$$

5. **Binomial probability formula**

$$P(x) = \binom{n}{x} p^x q^{n-x}$$

6. **Mean of the binomial probability distribution**

$$\mu = np$$

7. **Standard deviation of the binomial probability distribution**

$$\sigma = \sqrt{npq}$$

8. **Poisson probability formula**

$$P(x) = \frac{\lambda^x e^{-\lambda}}{x!}$$

SUPPLEMENTARY EXERCISES

5.89 The following table lists the probability distribution of the number of breakdowns per day for a certain machine.

Number of breakdowns	0	1	2	3
Probability	.55	.33	.08	.04

Let x denote the number of breakdowns per day for this machine. Determine the following probabilities.

a. $P(x = 2)$
b. $P(x \leq 1)$
c. $P(x \geq 1)$
d. $P(x < 3)$

5.90 The following table lists the probability distribution of the number of phone calls received per half hour at an office of Shopaholic Anonymous.

Number of phone calls	0	1	2	3	4
Probability	.12	.26	.34	.18	.10

Let x denote the number of phone calls received during a certain half-hour period at this office of Shopaholic Anonymous. Find the following probabilities.

a. $P(x = 1)$
b. $P(x < 2)$

Let x denote the number of phone calls received during a certain half-hour period at

c. $P(x > 2)$
d. $P(1 \leq x \leq 3)$

5.91 The probability that a family owns a house is .66. Let x be the number of families in a random sample of two who own a house. Write the probability distribution of x. (Solve without using the binomial distribution.) Draw a tree diagram for this problem.

5.92 According to the Electronic Industries Association, 69% of U.S. households own VCRs (*The Wall Street Journal,* September 6, 1990). Let x be the number of households in a random sample of two who own VCRs. Write the probability distribution of x. (Solve without using the binomial distribution.) Draw a tree diagram for this problem.

5.93 Let x be the number of cars that a randomly selected auto mechanic repairs on a given day. The following table lists the probability distribution of x.

x	2	3	4	5	6
$P(x)$	.05	.22	.35	.28	.10

Calculate the mean and standard deviation of x. Give a brief interpretation of the value of mean.

5.94 Let x be the number of times it rains during a given week in a city. The following table lists the probability distribution of x.

x	0	1	2	3	4	5
$P(x)$	.31	.24	.19	.14	.10	.02

Find the mean and standard deviation of x. Give a brief interpretation of the value of mean.

5.95 Let x be the number of accidents that occur in a factory during a given week. The following table lists the probability distribution of x.

x	0	1	2	3	4
$P(x)$	.44	.26	.19	.07	.04

Find the mean and standard deviation of x. Give a brief interpretation of the value of mean.

5.96 Let x be the number of times the mainframe computer at a university goes down during a given week. The following table lists the probability distribution of x.

x	0	1	2	3	4
$P(x)$	.65	.24	.06	.03	.02

Determine the mean and standard deviation of x. Give a brief interpretation of the value of mean.

5.97 Determine the value of each of the following using the appropriate formula.

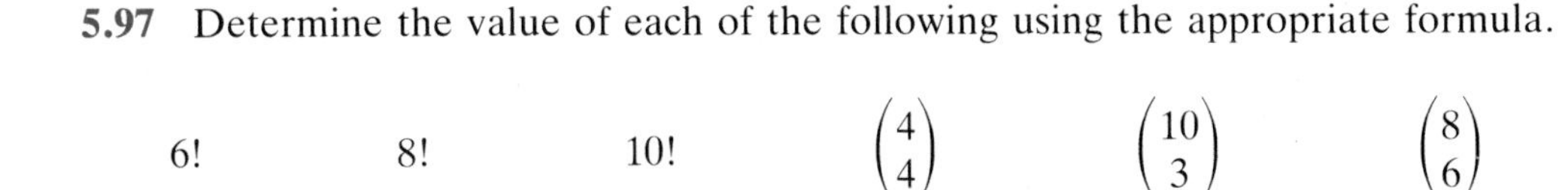

Verify the calculated values by using Tables II and III of Appendix C.

5.98 Determine the value of each of the following using the appropriate formula.

$$9! \qquad 7! \qquad 11! \qquad \binom{6}{3} \qquad \binom{11}{7} \qquad \binom{9}{0}$$

Verify the calculated values by using Tables II and III of Appendix C.

5.99 An agency is to select 4 wells from a suburban subunit to test for water contamination. This subunit contains a total of 16 wells. How many total combinations are possible?

5.100 Amanda will choose four courses from a list of eight courses for next semester. If she decides to select these courses randomly, how many combinations are possible?

5.101 According to a 1989 survey by the Roper Organization, 55% of married couples have a happy marriage. Assuming this percentage is true for the current population of all married couples, find the probability that in a random sample of six married couples exactly four will have a happy marriage.

5.102 In a city, 30% of households own answering machines. Find the probability that in 20 households selected at random from this city, exactly 5 will own answering machines. Use the binomial formula.

5.103 According to the National Education Association, 17% of public school teachers were moonlighting (that is, holding a second job) in 1988–89. Assuming this percentage is true for the current population of all public school teachers, find the probability that in a random sample of five teachers exactly two will be moonlighting. Use the binomial formula.

5.104 According to a 1988 Gallup poll, 37% of adults abstain from drinking. Assume that this percentage is true for the current population of all adults. What is the probability that in a random sample of seven adults, exactly seven abstain from drinking? Use the binomial formula.

5.105 According to the U.S. Bureau of the Census, 24% of children under 18 years of age were living with one parent in 1989. Assuming this percentage is true for the current population of all children under 18 years of age, find the probability that in a random sample of four such children none will be living with one parent. Use the binomial formula.

5.106 According to the U.S. Bureau of the Census, 30% of people aged 65 and over live alone. Using the binomial probabilities table, find the probability that in a random sample of 16 persons aged 65 and over, the number of persons living alone will be

a. none
b. at most 3
c. at least 8

5.107 According to a survey, 60% of seventh graders reported that they were never asked to write up a science experiment independently (*Crossroads in American Education,* Educational Testing Service, February 1989). Assume that this percentage is true for the current population of all seventh graders. Using the binomial probabilities table, find the probability that in a random sample of 20 seventh graders, the number of seventh graders who were never asked to write up a science experiment independently will be

- **a.** at most 4
- **b.** at least 14
- **c.** 9 to 15

5.108 According to a survey, only 40% of 13-year-old American school children can use fractions, decimals, and percents (*A World of Differences,* Educational Testing Service, January 1989). Assume that this percentage is true for the current population of all 13-year-old American school children. Using the binomial probabilities table, find the probability that in a random sample of 25 such children, the number of children who can use fractions, decimals, and percents will be

- **a.** less than 5
- **b.** more than 16

5.109 According to an estimate, 20% of children in America come from homes in which English is not spoken (*Everybody Counts,* National Research Council, 1989). Assume that this percentage is true for the current population of all children. Using the binomial probabilities table, find the probability that in a random sample of 18 children, the number of children who come from homes in which English is not spoken is

- **a.** less than 4
- **b.** more than 7

5.110 An average of 5.3 robberies occur per day in a large city.

- **a.** Using the Poisson formula, find the probability that on a given day exactly 3 robberies will occur in this city.
- **b.** Using the Poisson probability table, find the probability that on a given day the number of robberies that will occur in this city is
  - i. at least 12
  - ii. at most 3

5.111 An average of eight videos are rented per day at a video-rental store.

- **a.** Using the Poisson formula, find the probability that on a given day exactly five videos will be rented at this store.
- **b.** Using the Poisson probability table, find the probability that on a given day the number of videos that will be rented at this store is
  - i. at least 17
  - ii. at most 4

5.112 An average of 1.2 private airplanes arrive per hour at an airport.

- **a.** Using the Poisson formula, find the probability that during a given hour no private airplane will arrive at this airport.
- **b.** Let x denote the number of private airplanes that will arrive at this airport during a given hour. Using the Poisson probability table, write the probability distribution of x.

5.113 A machine produces an average of .8 items per minute.

a. Using the Poisson formula, find the probability that during a certain minute this machine will produce exactly 3 items.

b. Let x denote the number of items that this machine will produce during a given minute. Using the Poisson probability table, write the probability distribution of x.

SELF-REVIEW TEST

1. Briefly explain the meaning of a random variable, a discrete random variable, and a continuous random variable. Give one example each of a discrete and a continuous random variable.

2. What name is given to a table that lists all the values that a discrete random variable x can assume and their corresponding probabilities?

3. For the probability distribution of a discrete random variable, the probability of any single value of x is always

a. in the range zero to 1 **b.** 1.0 **c.** less than 1.0

4. For the probability distribution of a discrete random variable, the sum of the probabilities of all values of x is always

a. greater than zero **b.** 1.0 **c.** less than 1.0

5. The number of combinations of 10 items taken 7 at a time is

a. 120 **b.** 200 **c.** 80

6. State the four conditions of a binomial experiment. Give one example of such an experiment.

7. The parameters of the binomial probability distribution are

a. n, p, and q **b.** n and p **c.** n, p, and x

8. The mean and standard deviation of the binomial probability distribution with $n = 25$ and $p = .20$ are

a. 5 and 2 **b.** 8 and 4 **c.** 4 and 3

9. **a.** The binomial probability distribution is symmetric when

i. $p < .5$ **ii.** $p = .5$ **iii.** $p > .5$

b. The binomial probability distribution is skewed to the right when

i. $p < .5$ **ii.** $p = .5$ **iii.** $p > .5$

c. The binomial probability distribution is skewed to the left when

i. $p < .5$ **ii.** $p = .5$ **iii.** $p > .5$

10. The parameter/parameters of the Poisson probability distribution is/are

a. λ **b.** λ and x **c.** λ and e

11. Describe the three conditions that must be satisfied to apply the Poisson probability distribution.

12. According to an estimate, 25% of households own a foreign car. Let x denote the number of households in a random sample of two households that own a foreign car. Write the probability distribution of x. Solve without using the binomial distribution. Draw a tree diagram for this problem.

13. Refer to Problem 12. Calculate the mean and standard deviation of x.

14. Refer to Problem 12. Write the probability distribution of x using the binomial formula.

15. According to a survey, 70% of adults believe that every college student should be required to take at least one course in ethics. Assume that this percentage is true for the current population of all adults. Find the probability that the number of adults in a random sample of 12 who hold this view is

- **a.** exactly 10 (use the appropriate formula)
- **b.** at least 7 (use the appropriate table from Appendix C)
- **c.** less than 4 (use the appropriate table from Appendix C)

16. A department store sells an average of two electric appliances per day.

- **a.** What is the probability that on a given day this store will sell
  - **i.** exactly 5 electric appliances (use the appropriate formula)
  - **ii.** at most 4 electric appliances (use the appropriate table from Appendix C)
  - **iii.** 5 to 9 electric appliances (use the appropriate table from Appendix C)
- **b.** Let x be the number of electric appliances sold at this store on a given day. Write the probability distribution of x. Use the appropriate table from Appendix C.

17. The binomial probability distribution is symmetric when $p = .50$, it is skewed to the right when $p < .50$, and it is skewed to the left when $p > .50$. Illustrate these three cases by writing three probability distributions and graphing them. Choose any values of n and p and use the table of binomial probabilities (Table IV of Appendix C).

USING MINITAB

Using MINITAB, we can find the probability of a single outcome for the binomial and Poisson probability distributions or we can list the probabilities for all outcomes of a binomial or a Poisson experiment. The following explains how the MINITAB commands for binomial and Poisson experiments can be used to find probabilities.

THE BINOMIAL PROBABILITY DISTRIBUTION

The following MINITAB command and subcommand will give the probability of a specific value of x for a binomial experiment.

MTB > PDF $x=k$; ← Note the semicolon at the end of this command. This semicolon instructs MINITAB that a subcommand with the values of n and p is to follow. k is the value of x whose probability is to be determined.

SUBC > BINOMIAL $n=a$ $p=b$. ← The period at the end of this subcommand indicates the end of commands. $n=a$ is the total number of trials and $p=b$ is the probability of success.

In the first MINITAB command, **PDF** stands for the *probability density function*, which is another name for the probability. The following illustration shows the use of these commands.

ILLUSTRATION M5-1 Reconsider Example 5-19. Five percent of all VCRs manufactured by an electronics firm are defective. Let us find the probability that if we randomly select three VCRs from the production line of that firm, exactly one of them will be defective. From the given information,

$$n = \text{total number of VCRs} = 3$$

$$x = \text{defective VCRs in the sample} = 1$$

$$p = P(\text{a VCR is defective}) = .05$$

MINITAB commands to find the probability of selecting exactly one defective in three VCRs are as follow.

```
MTB  > NOTE: PROBABILITY OF x = 1 FOR n = 3 AND p = .05
MTB  > PDF  x=1;
SUBC > BINOMIAL  n=3  p=.05.

         k                     P(x = k)  ←  MINITAB output gives:
      1.00                       0.1354     P(x = 1) = 0.1354
```

Thus, the probability of selecting one defective VCR in a sample of three VCRs is .1354.

If we put a semicolon after PDF in the first MINITAB command and do not indicate the value of x, MINITAB will list the probabilities of all possible values of x for the binomial experiment with n and p values entered in the subcommand. Illustration M5-2 shows how we can obtain the probability distribution of x for a binomial experiment with $n = 3$ and $p = .05$ using MINITAB.

ILLUSTRATION M5-2 Reconsider Illustration M5-1. Let x denote the number of defective VCRs in a random sample of three. Suppose we want to list the probability distribution of x. As we know,

$$n = 3 \quad \text{and} \quad p = .05$$

The following are the MINITAB commands and the MINITAB output for this binomial problem.

```
MTB  > NOTE: BINOMIAL DISTRIBUTION FOR n = 3 AND p = .05
MTB  > PDF;  ← Note that we have not entered the value of x in this command.
SUBC > BINOMIAL  n = 3  p = .05.

     BINOMIAL WITH  N = 3  P = 0.050000
         k                 P(x = k)
         0                  0.8574
         1                  0.1354       ←  MINITAB output
         2                  0.0071
         3                  0.0001
```

In this MINITAB output, the column labeled k lists the values of x and the column labeled $P(x = k)$ lists their probabilities. Using this table, we can write the probability of any value of x for this example. For instance,

$$P(\text{at most 1 defective VCR}) = P(0) + P(1) = .8574 + .1354 = .9928$$

$$P(\text{at least 2 defective VCRs}) = P(2) + P(3) = .0071 + .0001 = .0072$$

THE POISSON PROBABILITY DISTRIBUTION

The following MINITAB command and subcommand will give the probability of a specific value of x for a Poisson problem.

```
MTB  > PDF  x=k;          ← This command instructs MINITAB that we need to determine
                            the probability of x = k where k is a specific value of x.
SUBC > POISSON  MEAN=a.   ← This command instructs MINITAB to use the Pois-
                            son distribution with λ = a.
```

Again, note the semicolon at the end of the first command. That semicolon instructs MINITAB that a subcommand with a value of λ is to follow. The following illustration shows how to use these commands to find the probability of a specific value of x for a Poisson distribution.

ILLUSTRATION M5-3 On average, a worker is absent from work for 1.2 days per month. Suppose we want to find the probability that this worker will be absent for 3 days during the next month. From the given information,

$$\lambda = \text{mean absences per month} = 1.2$$

MINITAB commands to find the probability $P(x = 3)$ are as follows.

```
MTB  > NOTE: PROBABILITY OF x = 3 FOR λ = 1.2
MTB  > PDF  x=3;
SUBC > POISSON  MEAN=1.2.

        k              P(x=k) }   ←  MINITAB output gives:
      3.00             0.0867 }      P(x = 3) = 0.0867
```

Again, if we put a semicolon after PDF in the first MINITAB command and do not provide the value of x, MINITAB will list the probabilities of all possible values of x for the Poisson problem with a value of λ given in the subcommand. The following illustration shows how we can obtain the probability distribution of x for a Poisson problem.

ILLUSTRATION M5-4 Reconsider Illustration M5-3. Let x be the number of days for which this worker will be absent from work next month. Suppose we want to list the probability distribution of x. The following MINITAB commands will do so.

```
MTB  > NOTE: POISSON DISTRIBUTION OF x FOR MEAN = 1.2
MTB  > PDF;
SUBC > POISSON   MEAN=1.2.

        POISSON WITH MEAN = 1.200
        k              P(x = k)
        0               0.3012
        1               0.3614
        2               0.2169
        3               0.0867      <—— MINITAB output
        4               0.0260
        5               0.0062
        6               0.0012
        7               0.0002
        8               0.0000
```

In the MINITAB output, the possible values of x are listed in the column labeled k and the probabilities of those values of x are listed in the column labeled $P(x = k)$. Using this output, we can determine the probability of any value of x for this Poisson problem. For example,

$$P(x \leq 3) = P(0) + P(1) + P(2) + P(3)$$
$$= .3012 + .3614 + .2169 + .0867 = .9662$$
$$P(3 \leq x \leq 5) = P(3) + P(4) + P(5) = .0867 + .0260 + .0062 = .1189$$

COMPUTER ASSIGNMENTS

M5.1 Forty-five percent of the adult population in a large city are women. A court is to randomly select a jury of 5 adults from the population of all adults of this city.

a. Using MINITAB, find the probability that none of the 5 jury members is a woman.
b. Let x denote the number of women selected for this jury. Using MINITAB, obtain the probability distribution of x.

M5.2 According to a 1989 *Newsweek* poll, 77% of adults said they support the suspension of driver's license for one to three years for casual users of drugs (*Newsweek*, September 18, 1989). Assume that this result holds true for the current population of all adults. Let x be the number in 13 randomly selected adults who hold this view.

a. Using MINITAB, find $P(x = 10)$.
b. Using MINITAB, obtain the probability distribution of x.

M5.3 A mail-order company receives an average of 7.4 orders per day.

a. Using MINITAB, find the probability that it will receive exactly 10 orders on a certain day.
b. Let x denote the number of orders received by this company on a given day. Using MINITAB, obtain the probability distribution of x.

M5.4 A commuter airline receives an average of 3.7 complaints per day from its passengers. Let x denote the number of complaints received by this airline on a given day.

a. Using MINITAB, find $P(x = 0)$.
b. Using MINITAB, construct the probability distribution of x.

6 CONTINUOUS RANDOM VARIABLES AND THE NORMAL DISTRIBUTION

6.1 A CONTINUOUS PROBABILITY DISTRIBUTION

6.2 THE NORMAL DISTRIBUTION

6.3 THE STANDARD NORMAL DISTRIBUTION

6.4 STANDARDIZING A NORMAL DISTRIBUTION

6.5 APPLICATIONS OF THE NORMAL DISTRIBUTION

6.6 DETERMINING THE z VALUE WHEN AN AREA UNDER THE STANDARD NORMAL CURVE IS KNOWN

6.7 THE NORMAL APPROXIMATION TO THE BINOMIAL DISTRIBUTION

SELF-REVIEW TEST

USING MINITAB

Discrete random variables and their probability distributions were presented in Chapter 5. Section 5.1 of that chapter defined a continuous random variable as a variable that can assume any value in one or more intervals.

The possible values that a continuous random variable can assume are infinite and uncountable. For example, the time taken to play a football game is a continuous random variable. Suppose 2.5 hours is the minimum time and 4.5 hours is the maximum time taken to play a football game. Let x be a continuous random variable that denotes the time taken to play a football game. Then x can assume any value in the interval 2.5 to 4.5 hours. This interval contains an infinite number of values that are uncountable.

A continuous random variable can have one of many probability distributions. The *normal probability distribution* (or simply, the normal distribution) is one such distribution, which will be discussed in this chapter.

6.1 A CONTINUOUS PROBABILITY DISTRIBUTION

Let x be a continuous random variable representing the height of a female student at a university. Suppose there are 5000 female students at this university. Table 6.1 lists the frequency and relative frequency distributions of x.

Table 6.1 Frequency and Relative Frequency Distributions of Heights of Female Students

Height of a Female Student (inches) x	f	Relative Frequency
60 to less than 61	90	.018
61 to less than 62	170	.034
62 to less than 63	460	.092
63 to less than 64	750	.150
64 to less than 65	970	.194
65 to less than 66	760	.152
66 to less than 67	640	.128
67 to less than 68	440	.088
68 to less than 69	320	.064
69 to less than 70	220	.044
70 to less than 71	180	.036
	$N = 5000$	Sum = 1.0

The relative frequencies listed in Table 6.1 can be used as approximate probabilities of respective classes.

Figure 6.1 displays the histogram and polygon for the relative frequency distribution of Table 6.1. Figure 6.2 shows the smoothed polygon for the data of Table 6.1. The smoothed polygon is an approximation of the *probability distribution curve* of the continuous random variable x. Note that each class in Table 6.1 has a width equal to 1 unit. If the width of classes is more than 1 unit, we first obtain the *relative frequency densities* and then graph these *relative frequency densities* to obtain the probability distribution curve. To obtain the relative frequency density of a class, we divide the relative frequency of that class by the class width. The relative frequency densities are calculated to make the sum of the areas of all rectangles in the histogram equal to 1.0. Case Study 6-1, which appears later in this section, illustrates this procedure. The probability distribution curve of a continuous random variable is also called its *probability density function*.

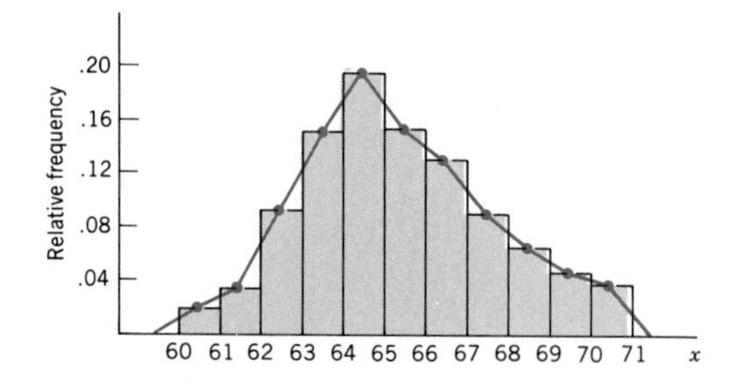

Figure 6.1 Histogram and polygon for heights.

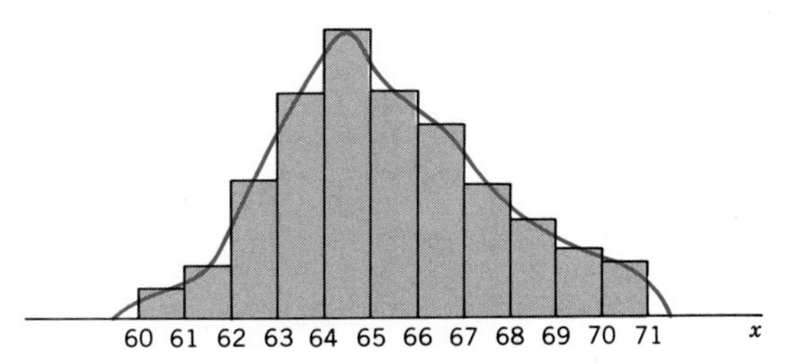

Figure 6.2 Probability distribution curve for heights.

The probability distribution of a continuous random variable possesses two characteristics:

1. The probability that x assumes a value in any interval is in the range 0 to 1.
2. The total probability of all the (mutually exclusive) intervals within which x can assume a value is 1.0.

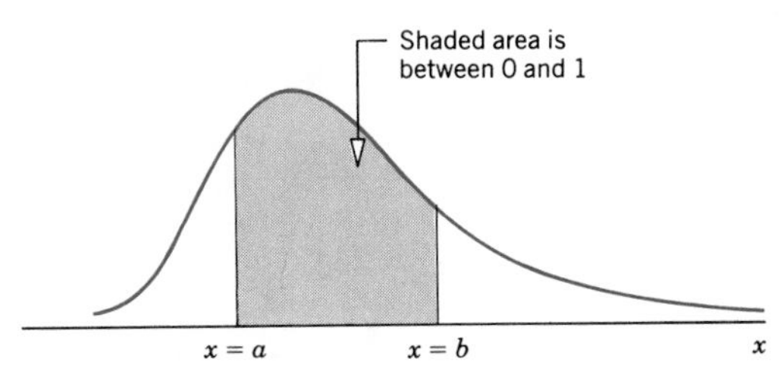

Figure 6.3 Area between two points.

The first characteristic states that the area under the probability distribution curve of a continuous random variable between any two points is between 0 and 1, as shown in Figure 6.3. The second characteristic indicates that the total area under the probability distribution curve of a continuous random variable is always 1.0 or 100%, as shown in Figure 6.4.

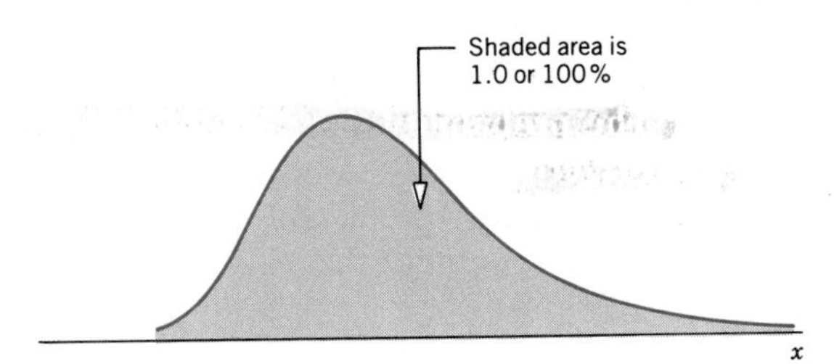

Figure 6.4 Total area under a probability distribution curve.

The probability that a continuous random variable x assumes a value within a certain interval is given by the area under the curve between two limits of the interval, as shown in Figure 6.5 on the next page. The shaded area under the curve between points a and b in this figure gives the probability that x falls in the interval a to b. Thus,

$$P(a \leq x \leq b) = \text{Area under the curve from } a \text{ to } b$$

Note that the interval $a \leq x \leq b$ states that x is greater than or equal to a but less than or equal to b.

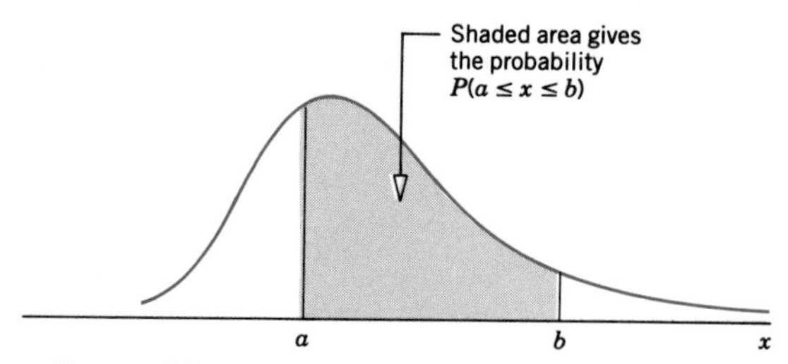

Figure 6.5 Area under the curve as probability.

Reconsider the example on the heights of all female students at a university. The probability that the height of a randomly selected female student from this university lies in the interval 65 to 68 inches is given by the area under the distribution curve of heights of all female students from $x = 65$ to $x = 68$, as shown in Figure 6.6. This probability is written as

$$P(65 \leq x \leq 68)$$

which states that x is greater than or equal to 65 but less than or equal to 68.

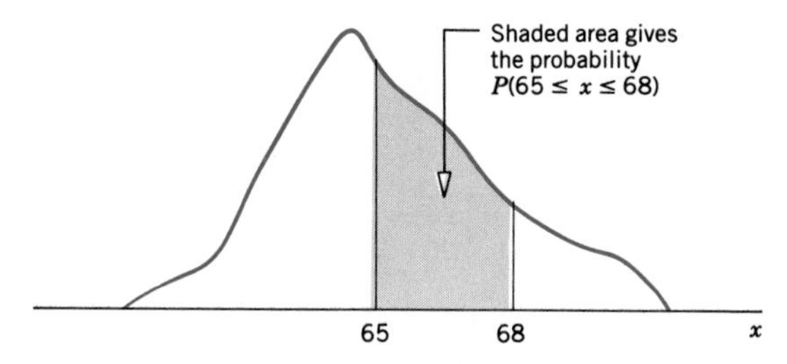

Figure 6.6 Probability that x lies in the interval 65 to 68.

For a continuous probability distribution, the probability is always calculated for an interval. For example, in Figure 6.6, the interval representing the shaded area is from 65 to 68. Consequently, the shaded area in that figure gives the probability for the interval $65 \leq x \leq 68$.

The probability that a continuous random variable *x* assumes a single value is always zero. This is so because the area of a line, which represents a single point, is zero. Thus, if x is the height of a randomly selected female student from that university, then the probability that this student is exactly 67 inches tall is zero. That is,

$$P(x = 67) = 0$$

This probability is shown in Figure 6.7. Similarly, the probability for x to assume any other single value is zero.

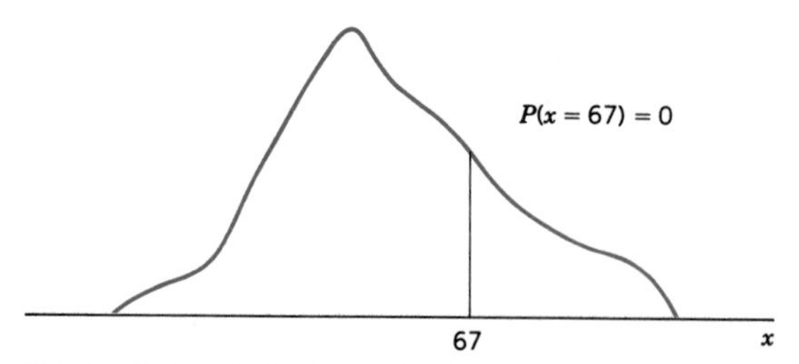

Figure 6.7 Probability of a single value of x is zero.

In general, if a and b are two of the values that x can assume, then,

$$P(a) = 0 \qquad \text{and} \qquad P(b) = 0$$

From this we can deduce that

$$P(a \le x \le b) = P(a < x < b)$$

Thus, the probability that x assumes a value in the interval a to b is the same whether or not the values a and b are included in the interval. For the example on the heights of female students, the probability that a randomly selected female student is between 65 and 68 inches tall is the same as the probability that this female is 65 to 68 inches tall. This probability is shown in Figure 6.8.

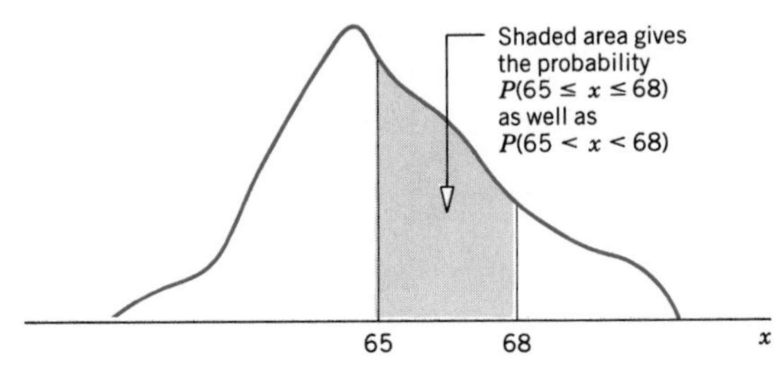

Figure 6.8 Probability "from 65 to 68" and "between 65 and 68."

Note that the interval "between 65 and 68" represents "$65 < x < 68$" and it does not include 65 and 68. On the other hand, the interval "from 65 to 68" represents "$65 \le x \le 68$" and it does include 65 and 68. However, as mentioned previously, in the case of a continuous random variable these intervals contain the same probability or area under the curve.

Case Study 6-1 describes how we obtain the probability distribution curve of a continuous random variable.

CASE STUDY 6-1 DISTRIBUTION OF TIME TAKEN TO RUN A ROAD RACE

Table 6.2 gives the frequency and relative frequency distributions for the time (in minutes) taken to run the Fifty-third Manchester Road Race (held on November 23, 1989) for a total of 5191 participants who finished that race. This road race event is held every year on Thanksgiving day in Manchester, Connecticut. The total distance of the race course is 4.748 miles. The relative frequencies of Table 6.2 are used to construct the histogram and polygon in Figure 6.9.

To derive the probability distribution curve for these data, first we calculate the relative frequency densities that are obtained by dividing the relative frequencies by the class widths. The width of each class in Table 6.2 is 5. By dividing the relative

Table 6.2 Frequency and Relative Frequency Distributions for the Road Race Data

Class	Frequency	Relative Frequency
20 to less than 25	43	.008
25 to less than 30	254	.049
30 to less than 35	661	.127
35 to less than 40	1155	.223
40 to less than 45	1395	.269
45 to less than 50	921	.177
50 to less than 55	388	.075
55 to less than 60	192	.037
60 to less than 65	43	.008
65 to less than 70	63	.012
70 to less than 75	21	.004
75 to less than 80	23	.004
80 to less than 85	32	.006
	$\Sigma f = 5191$	

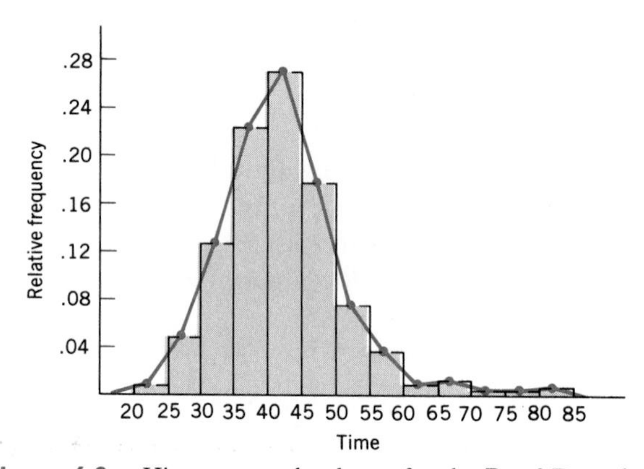

Figure 6.9 Histogram and polygon for the Road Race data.

frequencies of Table 6.2 by 5, we obtain the relative frequency densities, which are recorded in Table 6.3. Using the relative frequency densities, we draw a histogram and smoothed polygon, as shown in Figure 6.10. The curve in this figure gives the probability distribution curve for the Road Race data.

Note that the areas of rectangles in Figure 6.9 do not give probabilities (which are approximated by relative frequencies). Rather, it is the heights of these rectangles that give the probabilities. This is so because the base of each rectangle is 5 in this histogram. Consequently, the area of any rectangle is given by its height multiplied by 5. Thus, the total area of all rectangles in Figure 6.9 is 5.0 and not 1.0. However, in Figure 6.10 it is the areas and not the heights of rectangles that give the probabilities of respective classes. Thus, if we add the areas of all rectangles in Figure 6.10, we obtain the sum of all probabilities equal to 1.0. Consequently, the total area under the curve is equal to 1.0.

Table 6.3 Relative Frequency Densities

Class	Relative Frequency Density
20 to less than 25	.0016
25 to less than 30	.0098
30 to less than 35	.0254
35 to less than 40	.0446
40 to less than 45	.0538
45 to less than 50	.0354
50 to less than 55	.0150
55 to less than 60	.0074
60 to less than 65	.0016
65 to less than 70	.0024
70 to less than 75	.0008
75 to less than 80	.0008
80 to less than 85	.0012

The probability distribution of a continuous random variable has a mean and a standard deviation, which are denoted by μ and σ, respectively. The mean and standard deviation of the probability distribution curve of Figure 6.10 are 42.181 and 9.024 minutes, respectively. These values of μ and σ are calculated by using the raw data on the 5191 participants.

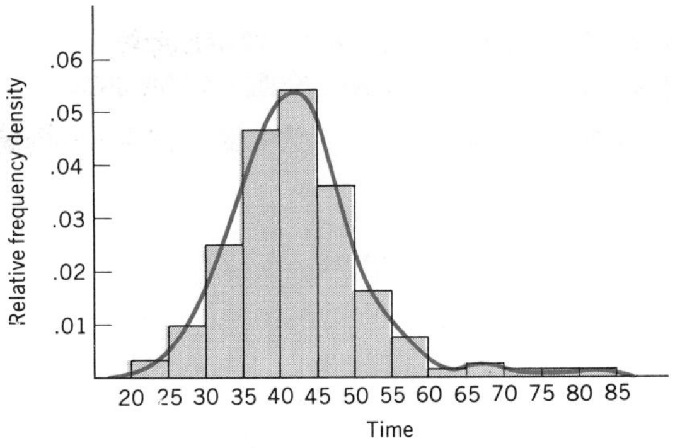

Figure 6.10 Probability distribution curve for Road Race data.

Source: This case study is based on data published in *The Hartford Courant,* November 27, 1989.

6.2 THE NORMAL DISTRIBUTION

As mentioned in Section 6.1, the normal distribution is one of the many probability distributions that a continuous random variable can have. The normal distribution is the most important and most widely used of all the probability distributions. A large number of phenomena in the real world are normally distributed either exactly or approximately. The continuous random variables representing the heights and weights of people, scores on an examination, weights of the contents of packages, amount of milk in a gallon, life of an item (such as a light bulb, a television set), and time taken to complete a certain job have all been observed to have an approximate normal distribution.

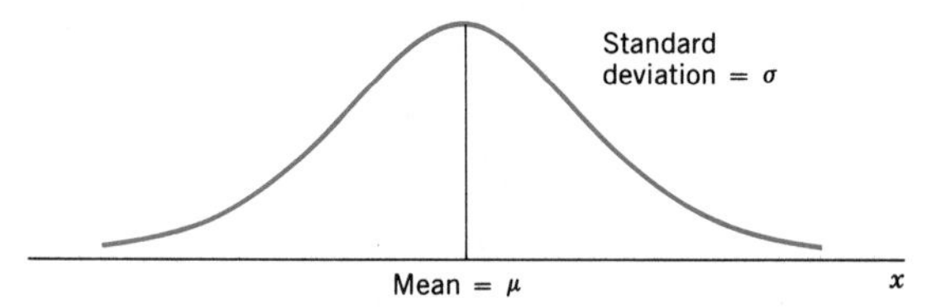

Figure 6.11 Normal distribution with mean μ and standard deviation σ.

The *normal probability distribution* or *normal curve* is given by a bell-shaped (symmetric) curve. Such a curve is shown in Figure 6.11. It has a mean of μ and a standard deviation of σ. A continuous random variable x that has a normal distribution is called a *normal random variable*. Note that not all bell-shaped curves represent a normal curve. Only a specific kind of bell-shaped curve represents a normal curve.

NORMAL PROBABILITY DISTRIBUTION

A normal probability distribution, when plotted, gives a bell-shaped curve such that

1. The total area under the curve is 1.0.
2. The curve is symmetric about the mean.
3. The two tails of the curve extend indefinitely.

A normal distribution possesses the following three characteristics.

1. The total area under a normal curve is 1.0 or 100%, as shown in Figure 6.12.

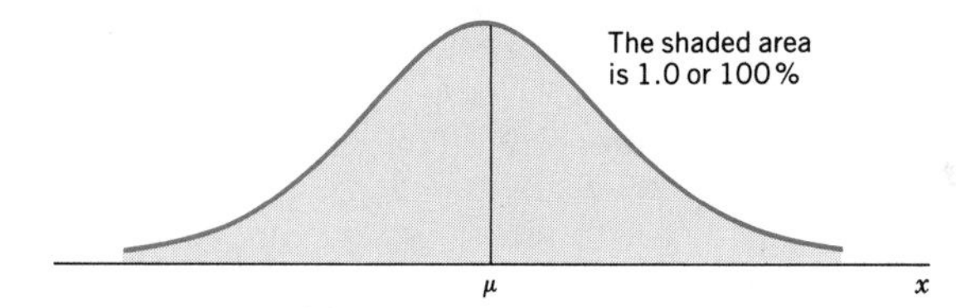

Figure 6.12 Total area under a normal curve.

2. A normal curve is symmetric about the mean, as shown in Figure 6.13 on the next page. Consequently, 1/2 of the total area under a normal curve lies on the left side of the mean and 1/2 lies on the right side of the mean.
3. The tails of a normal distribution curve extend indefinitely in both directions without touching or crossing the horizontal axis. Although a normal curve never meets the horizontal axis, beyond the points represented by $\mu - 3\sigma$ and $\mu + 3\sigma$ it becomes so close to this axis that the area under the curve beyond these points in both directions can be taken as virtually zero. These areas are shown in Figure 6.14.

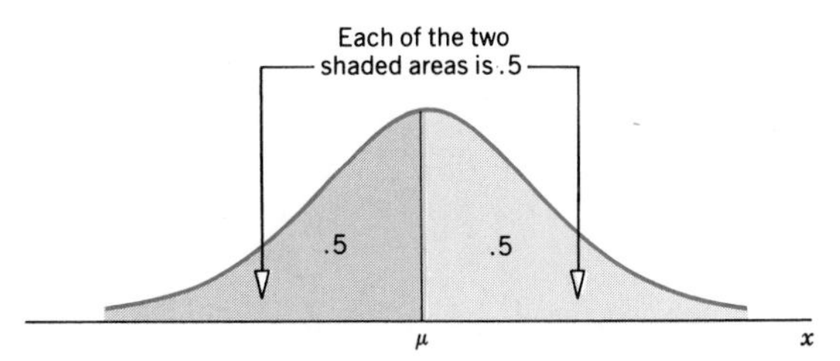

Figure 6.13 A normal curve is symmetric about the mean.

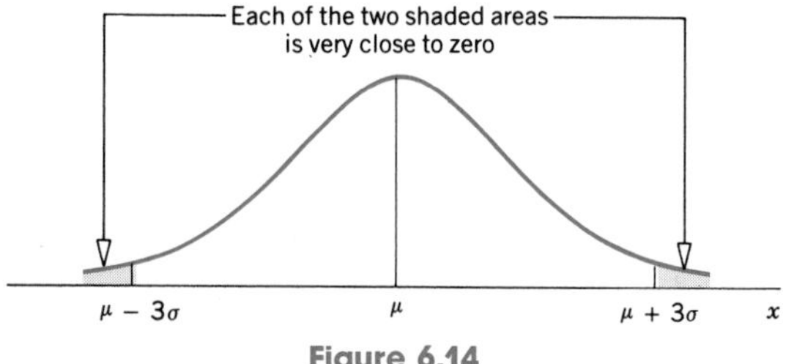

Figure 6.14

The mean μ and the standard deviation σ are the two *parameters* of the normal distribution. Given the values of these two parameters, we can find the area under a normal curve for any interval. Remember, there is not just one normal distribution curve but rather a *family* of normal distribution curves. Each different set of values of μ and σ gives a different normal distribution. The value of μ determines the center of a normal distribution curve on the horizontal axis and the value of σ gives the spread of the normal distribution curve. The three normal distribution curves drawn in Figure 6.15 have the same mean but different standard deviations. By contrast, the three normal distribution curves in Figure 6.16 (on the next page) have different means but the same standard deviation.

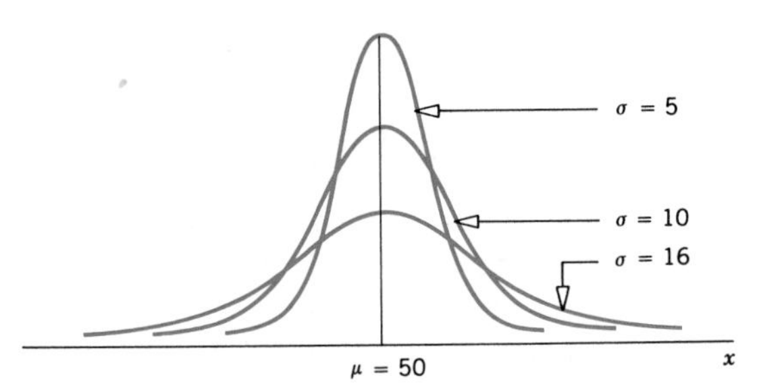

Figure 6.15 Three normal distribution curves with the same mean but different standard deviations.

Like the binomial and Poisson probability distributions discussed in Chapter 5, the normal probability distribution can also be expressed by a mathematical equation.†

†The equation of the normal distribution is

$$f(x) = \frac{1}{\sigma\sqrt{2\pi}} e^{-(1/2)\,((x-\mu)/\sigma)^2}$$

where $e = 2.71828$, $\pi = 3.14159$ approximately, and $f(x)$ gives the vertical distance between the horizontal axis and the curve at point x. For the information of those who are familiar with integral calculus, the definite integral of this equation from a to b gives the probability of x between a and b.

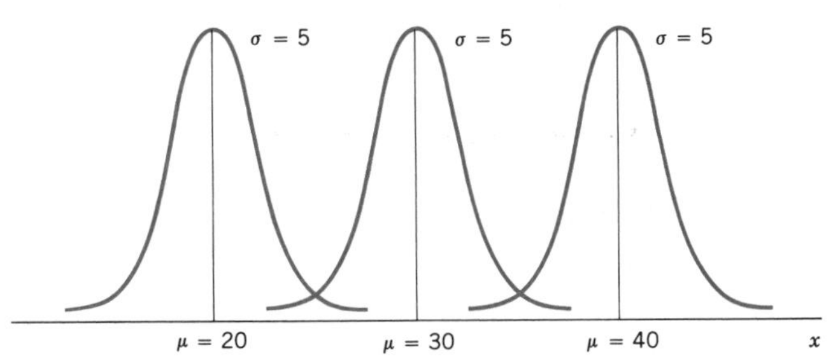

Figure 6.16 Three normal distribution curves with different means but the same standard deviation.

However, we will not use this equation to find the area under a normal curve. Instead, we will use Table VII of Appendix C (page 741).

6.3 THE STANDARD NORMAL DISTRIBUTION

The **standard normal distribution** is a special case of the normal distribution. For the standard normal distribution, the value of the mean is equal to zero and the value of the standard deviation is equal to 1.

> **STANDARD NORMAL DISTRIBUTION**
>
> The normal distribution with $\mu = 0$ and $\sigma = 1$ is called the standard normal distribution.

Figure 6.17 displays the standard normal distribution curve. The random variable that possesses the standard normal distribution is denoted by z. In other words, the units for the standard normal distribution curve are denoted by z. The units marked on the horizontal axis of the standard normal distribution are called **z values** or **z scores.** They are also called *standard units* or *standard scores*.

In Figure 6.17, the horizontal axis is labeled z. The z values on the right side of the mean (which is zero) are positive and those on the left side of the mean are negative. *The z value for a point on the horizontal axis gives the distance between the mean and that point in terms of the standard deviation*. For example, a point with a value of $z = 2$ is two standard deviations to the right of the mean. Similarly, a point with a value of $z = -2$ is two standard deviations to the left of the mean.

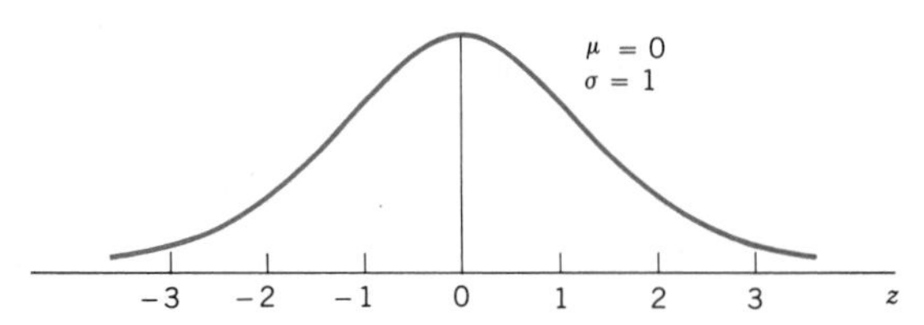

Figure 6.17 The standard normal distribution curve.

z VALUES OR z SCORES

The units marked on the horizontal axis of the standard normal curve are denoted by z and are called z values or z scores. A specific value of z gives the distance between the mean and the point represented by z in terms of the standard deviation.

The standard normal distribution table, Table VII of Appendix C, lists the areas under the standard normal curve between $z = 0$ and the values of z from 0.00 to 3.09. To read the standard normal distribution table, we always start at $z = 0$, which represents the mean of the standard normal curve. We learned earlier that the total area under a normal curve is 1.0. We also learned that, because of symmetry, the area on either side of the mean is .5. This is shown in Figure 6.18.

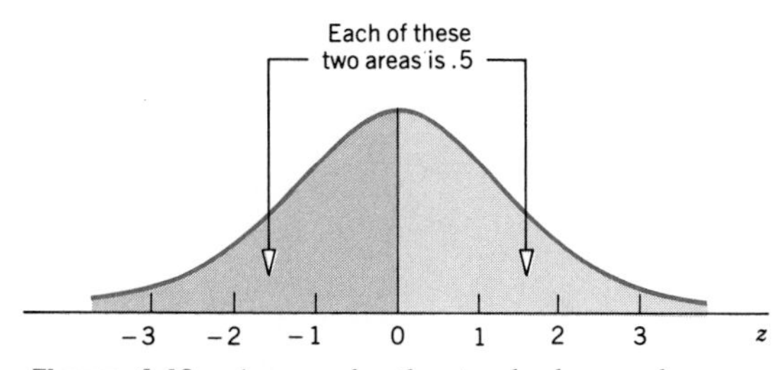

Figure 6.18 Area under the standard normal curve.

Remember Although the values of z on the left side of the mean are negative, the area is always positive.

The area under the standard normal curve between any two points can be interpreted as the probability that z assumes a value within that interval. Examples 6-1 through 6-4 describe how to read Table VII of Appendix C to find areas under the standard normal curve.

Finding area between $z = 0$ and $z > 0$.

EXAMPLE 6-1 Find the area under the standard normal curve between $z = 0$ and $z = 1.95$.

Solution We divide the number 1.95 into two portions: 1.9 (the digit before the decimal and one digit after the decimal) and .05 (the second digit after the decimal). (Note that $1.9 + .05 = 1.95$.) To find the required area, we locate 1.9 in the column for z on the left side of Table VII and .05 in the row for z at the top of Table VII. The entry where the row for 1.9 and the column for .05 intersect gives the area under the standard normal curve between $z = 0$ and $z = 1.95$. The relevant portion of Table VII is reproduced on the next page as Table 6.4. From Table VII or Table 6.4, the entry where the row for 1.9 and the column for .05 cross is .4744. Consequently, the area under the standard normal curve between $z = 0$ and $z = 1.95$ is .4744. This area is shown in Figure 6.19. It is always helpful to sketch the curve and mark the area we are determining.

Table 6.4 Finding Area Under the Standard Normal Curve between $z = 0$ and $z = 1.95$

z	.00	.01	...	.05	...	.09
0.0	.0000	.0040	...	.0199	...	.0359
0.1	.0398	.0438	...	.0596	...	.0753
0.2	.0793	.0832	...	.0987	...	.1141
.	.	.	...	.	...	.
.	.	.	...	.	...	.
.	.	.	...	.	...	.
1.9	.4713	.4719	...	.4744 ← Required area	...	.4767
.	.	.	...	.	...	.
.	.	.	...	.	...	.
.	.	.	...	.	...	.
3.0	.4987	.4987	...	.4989	...	.4990

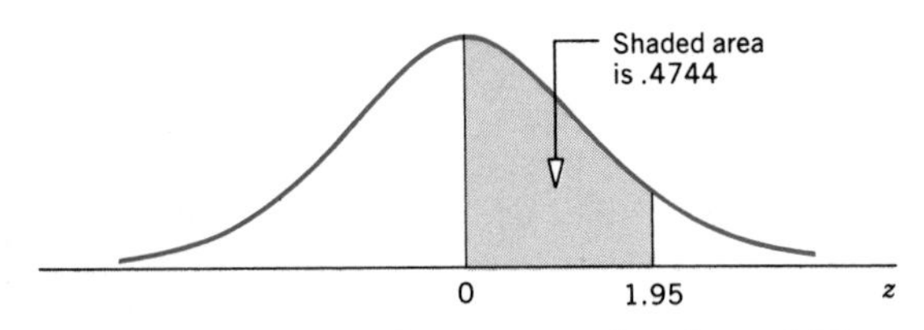

Figure 6.19 Area between $z = 0$ and $z = 1.95$.

The area between $z = 0$ and $z = 1.95$ can be interpreted as the probability that z assumes a value between 0 and 1.95. Thus,

$$\text{Area between 0 and 1.95} = P(0 < z < 1.95) = .4744$$

As mentioned in Section 6.1, the probability that a continuous random variable assumes a single value is zero. Therefore,

$$P(z = 0) = 0 \quad \text{and} \quad P(z = 1.95) = 0$$

Hence

$$P(0 < z < 1.95) = P(0 \le z \le 1.95) = .4744$$

■

Finding area between $z < 0$ and $z = 0$.

EXAMPLE 6-2 Find the area under the standard normal curve from $z = -2.17$ to $z = 0$.

Solution Because the normal distribution is symmetric about the mean, the area from $z = -2.17$ to $z = 0$ is the same as the area from $z = 0$ to $z = 2.17$, as shown in Figure 6.20.

To find the area from $z = -2.17$ to $z = 0$, we look for the area from $z = 0$ to $z = 2.17$ in the standard normal distribution table. To do so, first we locate 2.1 in the column for z and .07 in the row for z in the normal distribution table. Then, we read the number at the intersection of the row for 2.1 and the column for .07. The

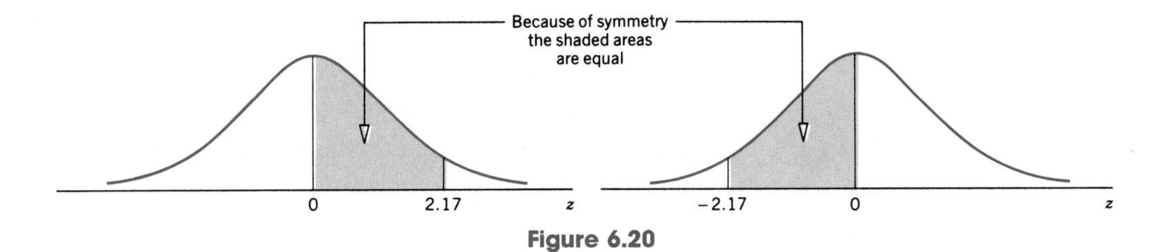

Figure 6.20

relevant portion of Table VII is reproduced here as Table 6.5. As shown in Table 6.5 and Figure 6.21, this number is .4850.

Table 6.5 Finding Area Under the Standard Normal Curve from $z = 0$ to $z = 2.17$

z	.00	.01	...	.07	...	.09
0.0	.0000	.0040	...	.0279	...	.0359
0.1	.0398	.0438	...	.0675	...	.0753
0.2	.0793	.0832	...	.1064	...	.1141
.	.	.	...	.	...	.
.	.	.	...	.	...	.
.	.	.	...	.	...	.
2.1	.4821	.4826	...	.4850 ←	...	.4857
.	.	.	...	.	...	.
.	.	.	...	.	...	.
.	.	.	...	.	...	.
3.0	.4987	.4987	...	.4989	...	.4990

Required area

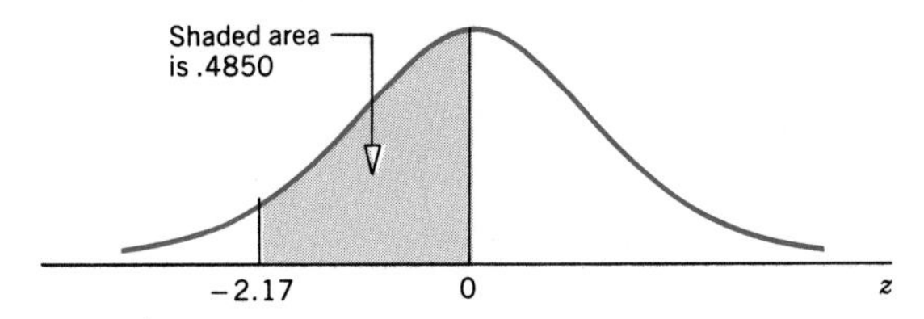

Figure 6.21 Area from $z = -2.17$ to $z = 0$.

The area from $z = -2.17$ to $z = 0$ gives the probability that z lies in the interval -2.17 to 0. This is written as $P(-2.17 \leq z \leq 0)$. Thus,

$$\text{Area from } -2.17 \text{ to } 0 = P(-2.17 \leq z \leq 0) = .4850$$ ■

EXAMPLE 6-3 Find the following areas under the standard normal curve.

(a) Area to the right of $z = 2.32$

(b) Area to the left of $z = -1.54$.

Solution

Finding area in the right tail.

(a) As mentioned earlier, to read the normal table we must start with $z = 0$. To find the area to the right of $z = 2.32$, first we find the area between $z = 0$ and $z = 2.32$. Then we subtract this area from .5, which is the total area to the right of $z = 0$. This is shown in Figure 6.22.

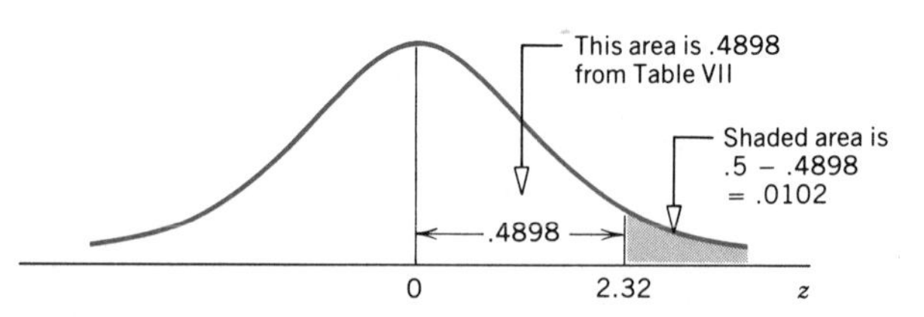

Figure 6.22 Area to the right of $z = 2.32$.

The area to the right of $z = 2.32$ gives the probability that $z > 2.32$. Thus,

$$\text{Area to the right of } 2.32 = P(z > 2.32) = .5 - .4898 = .0102$$

Area in the left tail.

(b) To find the area under the standard normal curve to the left of $z = -1.54$, first we find the area between $z = -1.54$ and $z = 0$ and then we subtract this area from .5, which is the total area to the left of $z = 0$. This area is shown in Figure 6.23.

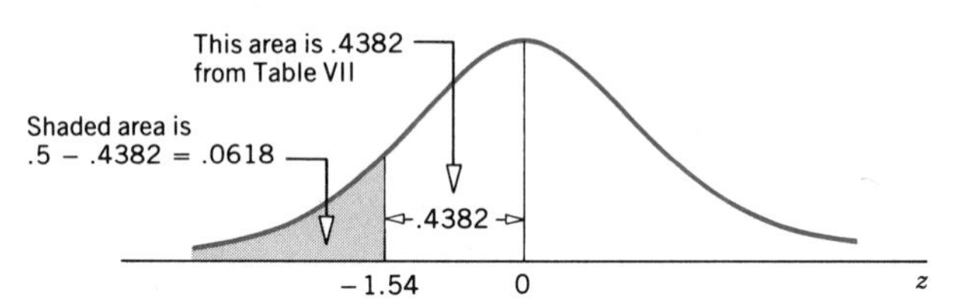

Figure 6.23 Area to the left of $z = -1.54$.

The area to the left of $z = -1.54$ gives the probability that $z < -1.54$. Thus,

$$\text{Area to the left of } -1.54 = P(z < -1.54) = .5 - .4382 = .0618$$ ■

EXAMPLE 6-4 Find the following probabilities for the standard normal curve.

(a) $P(1.19 < z < 2.12)$
(b) $P(-1.56 < z < 2.31)$
(c) $P(z > -.75)$

Solution

Finding area between two positive values of z.

(a) The probability $P(1.19 < z < 2.12)$ is given by the area under the standard normal curve between $z = 1.19$ and $z = 2.12$, which is the shaded area in Figure 6.24.

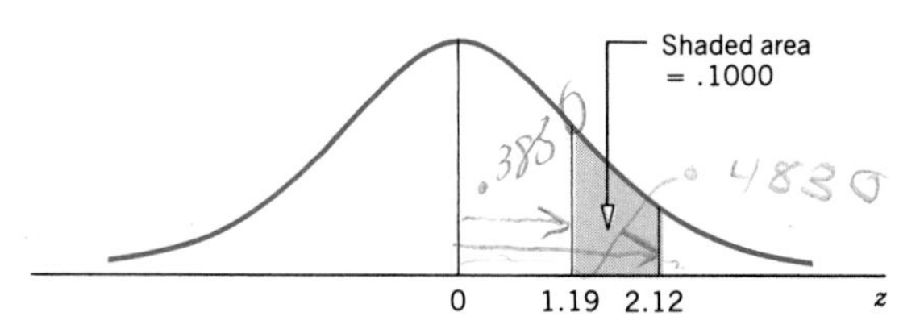

Figure 6.24 Area between $z = 1.19$ and $z = 2.12$.

Both the points, $z = 1.19$ and $z = 2.12$, are on the same (right) side of $z = 0$. To find the area between $z = 1.19$ and $z = 2.12$, first we find the areas between $z = 0$ and $z = 1.19$ and between $z = 0$ and $z = 2.12$. Then, we subtract the smaller area (the area between $z = 0$ and $z = 1.19$) from the larger area (the area between $z = 0$ and $z = 2.12$).

From Table VII,

$$\text{Area between 0 and 1.19} = P(0 < z < 1.19) = .3830$$

$$\text{Area between 0 and 2.12} = P(0 < z < 2.12) = .4830$$

Hence

$$\begin{aligned}\text{Area between 1.19 and 2.12} &= P(1.19 < z < 2.12) \\ &= P(0 < z < 2.12) - P(0 < z < 1.19) \\ &= .4830 - .3830 = .1000\end{aligned}$$

As a general rule, when the two points are on the same side of the mean, first find the areas between the mean and each of the two points. Then subtract the smaller area from the larger area.

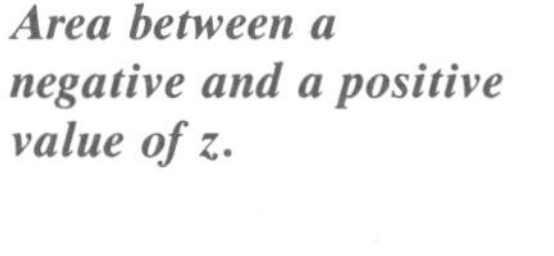
Area between a negative and a positive value of z.

(b) The probability $P(-1.56 < z < 2.31)$ is given by the area under the standard normal curve between $z = -1.56$ and $z = 2.31$, which is the shaded area in Figure 6.25.

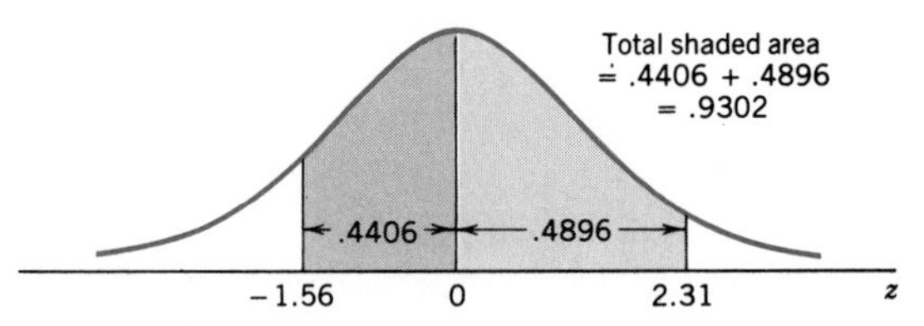

Figure 6.25 Area between $z = -1.56$ and $z = 2.31$.

The two points, $z = -1.56$ and $z = 2.31$, are on different sides of $z = 0$. Hence, the area between $z = -1.56$ and $z = 2.31$ is given by the sum of the areas between $z = -1.56$ and $z = 0$ and between $z = 0$ and $z = 2.31$, respectively.

From Table VII,

$$\text{Area between } -1.56 \text{ and 0} = P(-1.56 < z < 0) = .4406$$

$$\text{Area between 0 and 2.31} = P(0 < z < 2.31) = .4896$$

Hence

$$\text{Area between } -1.56 \text{ and } 2.31 = P(-1.56 < z < 2.31)$$
$$= P(-1.56 < z < 0) + P(0 < z < 2.31)$$
$$= .4406 + .4896 = .9302$$

As a general rule, when the two points are on different sides of the mean, first find the areas between the mean and each of the two points. Then add the two areas.

Area to the right of a negative value of z.

(c) The probability $P(z > -.75)$ is given by the area under the standard normal curve to the right of $z = -.75$, which is the shaded area in Figure 6.26.

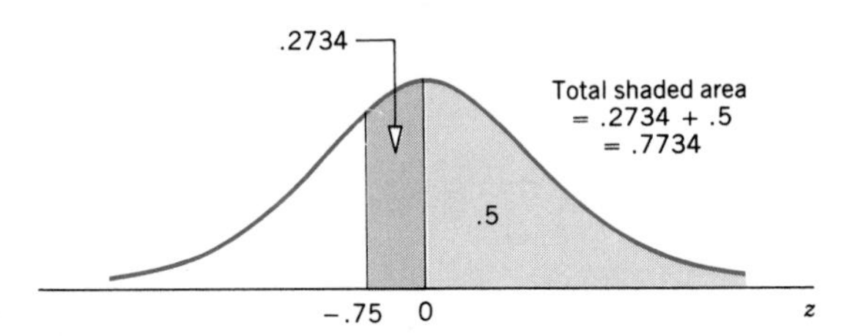

Figure 6.26 Area to the right of $z = -.75$.

The area to the right of $z = -.75$ is obtained by adding the area between $z = -.75$ and $z = 0$ and the area to the right of $z = 0$.

$$\text{Area to the right of } 0 = P(z > 0) = .5$$

From Table VII,

$$\text{Area between } -.75 \text{ and } 0 = P(-.75 < z < 0) = .2734$$

The required area is

$$\text{Area to the right of } -.75 = P(z > -.75)$$
$$= P(-.75 < z < 0) + P(z > 0)$$
$$= .2734 + .5 = .7734$$

■

In Section 3.4 of Chapter 3, while discussing the use of standard deviation, we discussed the empirical rule for a bell-shaped curve. The empirical rule is based on the standard normal distribution table. By using the normal distribution table, we can now verify the empirical rule as follows.

1. The total area within one standard unit of the mean is 68.26%.

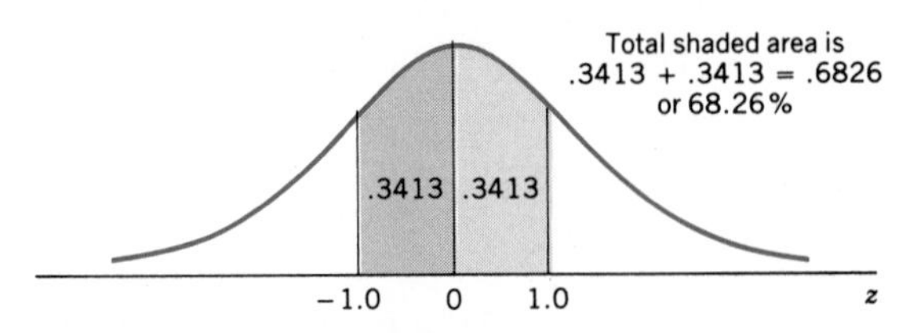

Figure 6.27 Area within one standard unit of the mean.

This area is given by the sum of the areas between $z = -1.0$ and $z = 0$ and between $z = 0$ and $z = 1.0$. As shown in Figure 6.27, each of these two areas is .3413 or 34.13%. Consequently, the total area between $z = -1.0$ and $z = 1.0$ is 68.26%.

2. The total area within two standard units of the mean is 95.44%.

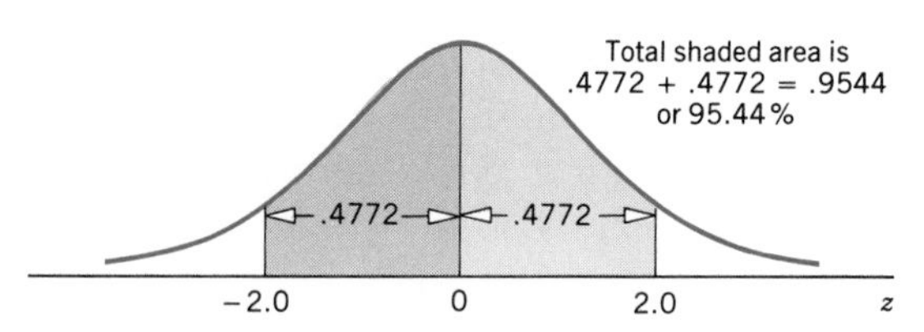

Figure 6.28 Area within two standard units of the mean.

This area is given by the sum of the areas between $z = -2.0$ and $z = 0$ and between $z = 0$ and $z = 2.0$. As shown in Figure 6.28, each of these two areas is .4772 or 47.72%. Hence, the total area between $z = -2.0$ and $z = 2.0$ is 95.44%.

3. The total area within three standard units of the mean is 99.74%.

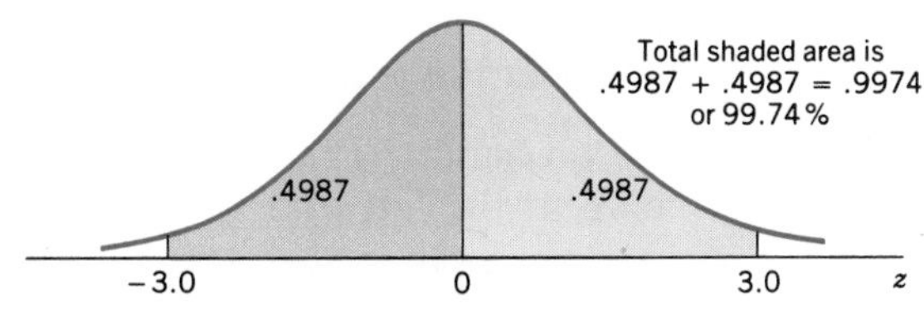

Figure 6.29 Area within three standard units of the mean.

This area is given by the sum of the areas between $z = -3.0$ and $z = 0$ and between $z = 0$ and $z = 3.0$. As shown in Figure 6.29, each of these two areas is .4987 or 49.87%. Therefore, the total area between $z = -3.0$ and $z = 3.0$ is 99.74%.

Again, note that only a specific bell-shaped curve represents the normal distribution. Now we can state that a bell-shaped curve that contains (about) 68.26% of the total area within one standard deviation of the mean, (about) 95.44% of the total area within two standard deviations of the mean, and (about) 99.74% of the total area within three standard deviations of the mean represents a normal distribution curve.

The standard normal distribution table, Table VII of Appendix C, goes only up to $z = 3.09$. In other words, that table can be read only for $z = 0$ to $z = 3.09$ (or $z = -3.09$). Consequently, if we need to find the area between $z = 0$ and $z = 4.56$ (or between $z = -4.56$ and $z = 0$) under the standard normal curve, we cannot obtain it from the normal table because the table does not contain $z = 4.56$. In such cases, the area under the normal curve between $z = 0$ and any z value greater than 3.09 (or less than -3.09) is approximated by .5. From the normal distribution table, the area between $z = 0$ and $z = 3.09$ is .4990. Hence, the area between $z = 0$ and any value of z greater than 3.09 is larger than .4990 and can be approximated by .5. Example 6-5 illustrates this procedure.

EXAMPLE 6-5 Find the following probabilities for the standard normal curve.

(a) $P(0 < z < 5.67)$
(b) $P(z < -5.35)$

Solution

Finding area between $z = 0$ and a value of $z > 3.09$.

(a) The probability $P(0 < z < 5.67)$ is given by the area under the standard normal curve between $z = 0$ and $z = 5.67$. Because $z = 5.67$ is greater than 3.09 and is not in Table VII, the area under the standard normal curve between $z = 0$ and $z = 5.67$ is approximated by .5. This area is shown in Figure 6.30.

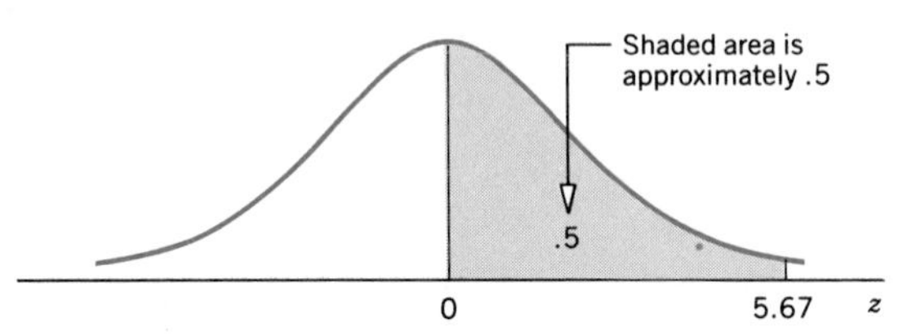

Figure 6.30 Area between $z = 0$ and $z = 5.67$.

Hence

$$\text{Area between 0 and } 5.67 = P(0 < z < 5.67) = .5 \text{ approximately}$$

Note that the area between $z = 0$ and $z = 5.67$ is not .5 but very close to .5.

Finding area to the left of $z < -3.09$.

(b) The probability $P(z < -5.35)$ represents the area under the standard normal curve to the left of $z = -5.35$. The area between $z = -5.35$ and $z = 0$ is approximately .5. Consequently, the area under the standard normal curve to the left of $z = -5.35$ is approximately zero, as shown in Figure 6.31.

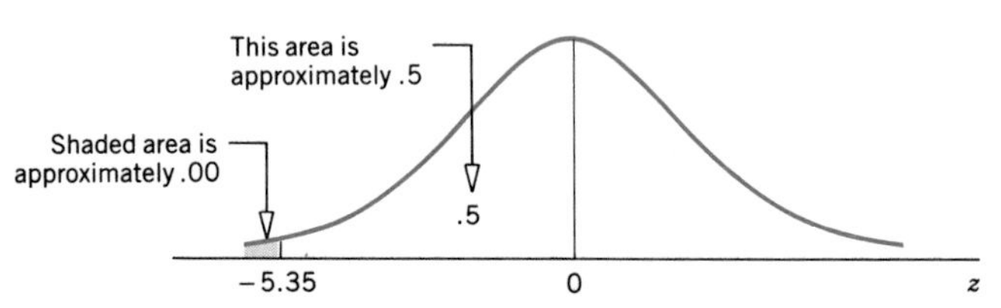

Figure 6.31 Area to the left of $z = -5.35$.

$$\begin{aligned}\text{Area to the left of } -5.35 &= P(z < -5.35) \\ &= P(z < 0) - P(-5.35 < z < 0) \\ &= .5 - .5 = .00 \text{ approximately}\end{aligned}$$

Again, note that the area to the left of $z = -5.35$ is not exactly .00 but very close to .00. ■

EXERCISES

6.1 Briefly explain the main characteristics of a normal distribution. Illustrate with the help of graphs.

6.2 Briefly describe the standard normal distribution curve.

6.3 Find the area under the standard normal curve

a. between 0 and 1.90 b. between 0 and -1.75
c. between 1.25 and 2.37 d. from -1.53 to -2.78
e. between -1.67 and 2.34

6.4 Find the area under the standard normal curve

a. from 0 to 2.34 b. between 0 and -2.78
c. from .84 to 1.95 d. between $-.57$ and -2.39
e. between -2.15 and 1.67

6.5 Find the area under the standard normal curve

a. to the right of 1.56 b. to the left of -1.97
c. to the right of -2.05 d. to the left of 1.86

6.6 Obtain the area under the standard normal curve

a. to the right of 1.83 b. to the left of -1.65
c. to the right of $-.55$ d. to the left of .79

6.7 Obtain the area under the standard normal curve

a. between 0 and 4.28 b. from 0 to -3.75
c. to the right of 7.43 d. to the left of -4.49

6.8 Find the area under the standard normal curve

a. from 0 to 3.94 b. between 0 and -5.16
c. to the right of 5.42 d. to the left of -3.68

6.9 Determine the following probabilities for the standard normal distribution.

a. $P(-1.83 \leq z \leq 2.67)$ b. $P(0 \leq z \leq 2.12)$
c. $P(-1.89 \leq z \leq 0)$ d. $P(z > 1.38)$

6.10 Determine the following probabilities for the standard normal distribution.

a. $P(-2.46 \leq z \leq 1.68)$ b. $P(0 \leq z \leq 1.86)$
c. $P(-2.58 \leq z \leq 0)$ d. $P(z \geq .83)$

6.11 Find the following probabilities for the standard normal distribution.

a. $P(z < -2.04)$ b. $P(.67 \leq z \leq 2.39)$
c. $P(-2.07 \leq z \leq -.83)$ d. $P(z < 1.71)$

6.12 Find the following probabilities for the standard normal distribution.

a. $P(z < -1.21)$ b. $P(1.03 \leq z \leq 2.79)$
c. $P(-2.34 \leq z \leq -1.09)$ d. $P(z \leq 2.02)$

6.13 Determine the following probabilities for the standard normal distribution.

a. $P(z > -.78)$ b. $P(-2.47 \leq z \leq 1.09)$
c. $P(0 \leq z \leq 4.25)$ d. $P(-5.36 \leq z \leq 0)$
e. $P(z \geq 6.07)$ f. $P(z \leq -5.27)$

6.14 Determine the following probabilities for the standard normal distribution.

a. $P(z > -1.26)$ b. $P(-.68 \leq z \leq 1.74)$
c. $P(0 \leq z \leq 3.85)$ d. $P(-4.34 \leq z \leq 0)$
e. $P(z \geq 4.82)$ f. $P(z \leq -6.12)$

6.4 STANDARDIZING A NORMAL DISTRIBUTION

As was shown in the previous section, Table VII of Appendix C can be used to find areas under the standard normal curve. However, in real-world applications, the given (continuous) random variable may have a normal distribution with values of the mean and standard deviation different from 0 and 1, respectively. The first step, in such a case, is to convert the given normal distribution to the standard normal distribution. This procedure is called *standardizing a normal distribution*. The units of a normal distribution (which is not the standard normal distribution) are denoted by x. We know from an earlier discussion that units of the standard normal distribution are denoted by z.

CONVERTING AN x VALUE TO A z VALUE

If x is a normal random variable, then a value of x can be converted to a z value by using the formula

$$z = \frac{x - \mu}{\sigma}$$

where μ and σ are the mean and standard deviation of the normal distribution of x.

Thus, to find the z value for an x value, we calculate the difference between the x value and the mean μ and divide this difference by the standard deviation σ. If the value of x is equal to μ, then its z value is equal to zero. Note that we will always round z values to two decimal places.

Remember The z value for the mean of a normal distribution is always zero.

Examples 6-6 to 6-10 describe how to convert x values to the corresponding z values and how to find areas under a normal curve.

Converting x values to the corresponding z values.

EXAMPLE 6-6 Let x be a continuous random variable that has a normal distribution with a mean of 50 and a standard deviation of 10. Convert the following x values to z values.

(a) $x = 55$
(b) $x = 35$

Solution For the given normal distribution: $\mu = 50$ and $\sigma = 10$.

(a) The z value for $x = 55$ is computed as follows.

$$z = \frac{x - \mu}{\sigma} = \frac{55 - 50}{10} = .50$$

Thus, the z value for $x = 55$ is .50. The z values for $\mu = 50$ and $x = 55$ are shown in Figure 6.32. Note that the z value for $\mu = 50$ is zero. The value $z = .50$ for $x = 55$ indicates that the distance between the mean $\mu = 50$ of the given normal distribution and the point given by $x = 55$ is 1/2 of the standard deviation $\sigma = 10$. Consequently, we can state that the z value represents the distance between μ and x in terms of the standard deviation. Because $x = 55$ is greater than $\mu = 50$, its z value is positive.

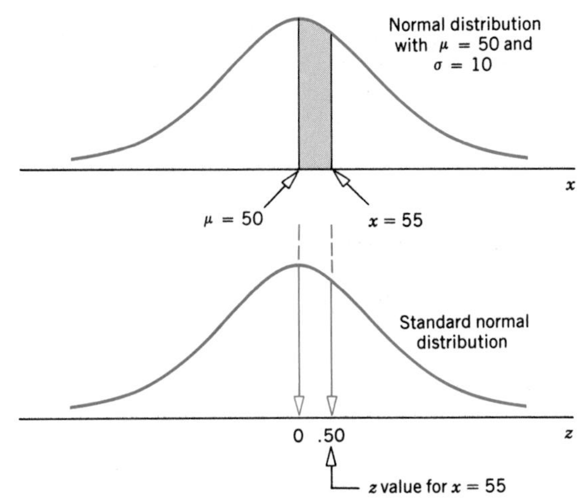

Figure 6.32 z value for $x = 55$.

Henceforth, only the z axis below the x axis will be shown, and not the standard normal curve itself.

(b) The z value for $x = 35$ is computed as follows and is shown in Figure 6.33.

$$z = \frac{x - \mu}{\sigma} = \frac{35 - 50}{10} = -1.50$$

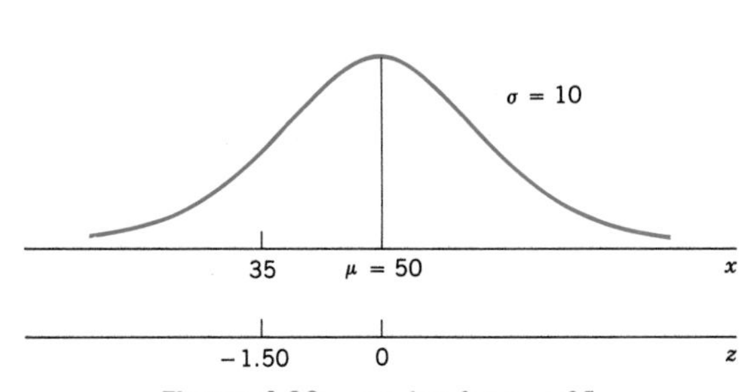

Figure 6.33 z value for $x = 35$.

Because $x = 35$ is on the left side of the mean (i.e., 35 is less than $\mu = 50$), its z value is negative. As a general rule, whenever an x value is less than the value of μ, its z value is negative. ■

Remember The z value for an x value that is greater than μ is positive; the z value for an x value that is equal to μ is zero; and the z value for an x value that is less than μ is negative.

To find the area between two values of x for a normal distribution, we first convert both values of x to their respective z values. Then we find the area under the standard normal curve between those two z values. The area between the two z values gives the area between the corresponding x values.

EXAMPLE 6-7 Let x be a continuous random variable that is normally distributed with a mean of 25 and a standard deviation of 4. Find the area

(a) between $x = 25$ and $x = 32$
(b) between $x = 18$ and $x = 34$

Solution For the given normal distribution: $\mu = 25$ and $\sigma = 4$.

Area between the mean and a point to its right.

(a) The first step to find the required area is to standardize the given normal distribution by converting $x = 25$ and $x = 32$ to respective z values using the formula

$$z = \frac{x - \mu}{\sigma}$$

The z value for $x = 25$ is zero because it is the mean of the normal distribution. The z value for $x = 32$ is

$$z = \frac{32 - 25}{4} = 1.75$$

As shown in Figure 6.34, the area between $x = 25$ and $x = 32$ under the given normal curve is equivalent to the area between $z = 0$ and $z = 1.75$ under the standard normal curve. This area from Table VII is .4599. The area between $x = 25$ and $x = 32$ under the normal curve gives the probability that x assumes a value between 25 and 32. This probability can be written as

$$P(25 < x < 32) = P(0 < z < 1.75) = .4599$$

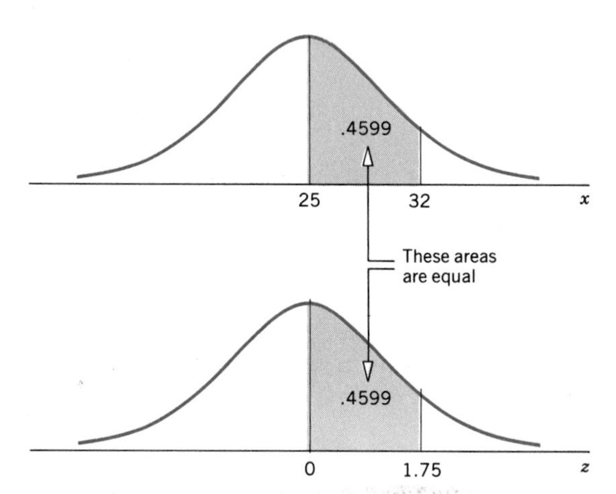

Figure 6.34 Area between $x = 25$ and $x = 32$.

Area between two points on different sides of the mean.

(b) First we calculate the z values for $x = 18$ and $x = 34$ as follows.
For $x = 18$,

$$z = \frac{18 - 25}{4} = -1.75$$

For $x = 34$,

$$z = \frac{34 - 25}{4} = 2.25$$

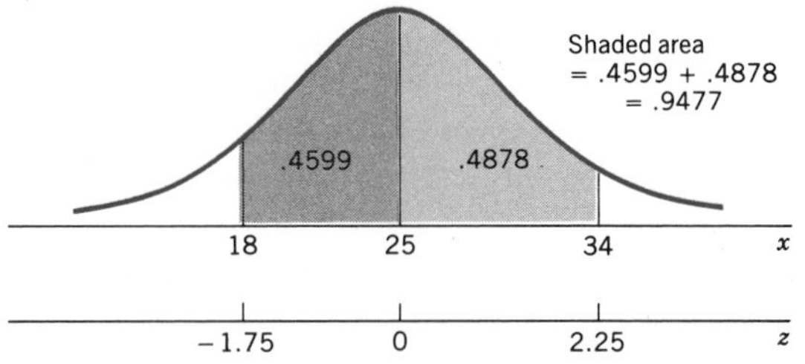

Figure 6.35 Area between $x = 18$ and $x = 34$.

The area under the given normal curve between $x = 18$ and $x = 34$ is given by the area under the standard normal curve between $z = -1.75$ and $z = 2.25$. This area is shown in Figure 6.35. The two values of z are on different sides of $z = 0$. Consequently, the total area is obtained by adding the areas between $z = -1.75$ and $z = 0$ and between $z = 0$ and $z = 2.25$. Hence,

$$\begin{aligned} P(18 < x < 34) &= P(-1.75 < z < 2.25) \\ &= P(-1.75 < z < 0) + P(0 < z < 2.25) \\ &= .4599 + .4878 = .9477 \end{aligned}$$

■

EXAMPLE 6-8 Let x be a normal random variable with its mean equal to 40 and standard deviation equal to 5. Find the following probabilities for this normal distribution.

(a) $P(x > 55)$
(b) $P(x < 49)$

Solution

Finding area in the right tail.

(a) The probability that x assumes a value greater than 55 is given by the area under the normal curve to the right of $x = 55$, as shown in Figure 6.36. This area is calculated by subtracting the area between the mean and $x = 55$ from .5, which is the total area to the right of the mean. For $x = 55$,

$$z = \frac{55 - 40}{5} = 3.00$$

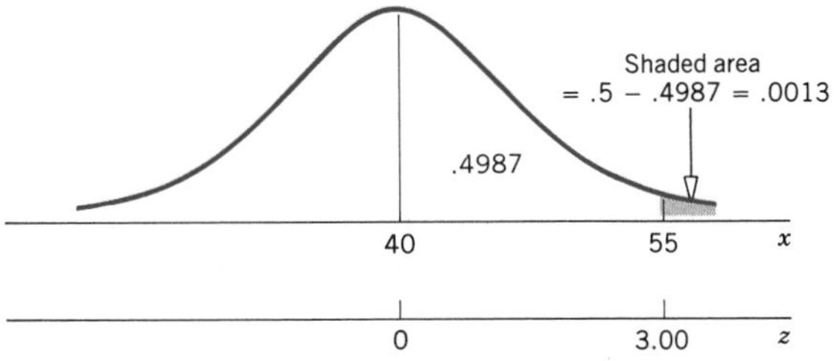

Figure 6.36 Area to the right of $x = 55$.

The required probability is

$$P(x > 55) = P(z > 3.00) = P(z > 0) - P(0 < z < 3.00)$$
$$= .5 - .4987 = .0013$$

Area to the left of an x value that is to the right of the mean.

(b) The probability that x will assume a value less than 49 is given by the area under the normal curve to the left of 49, which is the shaded area in Figure 6.37. This area is given by the sum of the area to the left of the mean and the area between the mean and $x = 49$. For $x = 49$,

$$z = \frac{49 - 40}{5} = 1.80$$

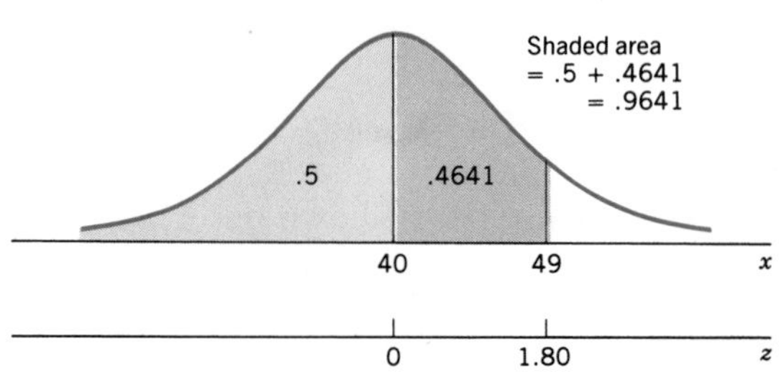

Figure 6.37 Area to the left of $x = 49$.

The required probability is

$$P(x < 49) = P(z < 1.80) = P(z < 0) + P(0 < z < 1.80)$$
$$= .5 + .4641 = .9641$$

■

Area between two x values that are less than the mean.

EXAMPLE 6-9 Let x be a continuous random variable that has a normal distribution with $\mu = 50$ and $\sigma = 8$. Find the probability $P(30 \leq x \leq 39)$.

Solution For the given normal distribution: $\mu = 50$ and $\sigma = 8$. The probability $P(30 \leq x \leq 39)$ is given by the area from $x = 30$ to $x = 39$ under the normal curve. As shown in Figure 6.38 on the next page, this area is given by the difference between the area from $x = 30$ to $x = 50$ and the area from $x = 39$ to $x = 50$. For $x = 30$,

$$z = \frac{30 - 50}{8} = -2.50$$

For $x = 39$,

$$z = \frac{39 - 50}{8} = -1.38$$

To find the required area, we first find the areas from $z = -2.50$ to $z = 0$ and from $z = -1.38$ to $z = 0$ and then take the difference between these two areas. Thus, the required probability is

$$P(30 \leq x \leq 39) = P(-2.50 \leq z \leq -1.38)$$
$$= P(-2.50 \leq z \leq 0) - P(-1.38 \leq z \leq 0)$$
$$= .4938 - .4162 = .0776$$

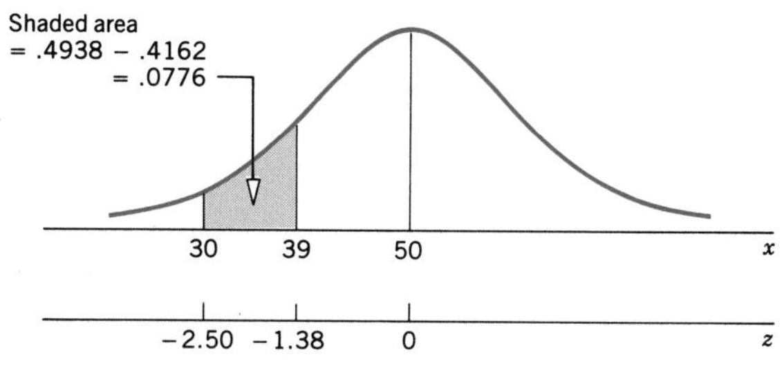

Figure 6.38 Area from $x = 30$ to $x = 39$.

■

EXAMPLE 6-10 Let x be a continuous random variable that has a normal distribution with a mean of 80 and a standard deviation of 12. Find the area under this normal curve

(a) from $x = 70$ to $x = 135$
(b) to the left of 27

Solution For the given normal distribution: $\mu = 80$ and $\sigma = 12$.

Area between two values that are on different sides of the mean.

(a) The area from $x = 70$ to $x = 135$ is obtained by adding the areas from $x = 70$ to $x = 80$ and from $x = 80$ to $x = 135$. This total area is given by the sum of the two shaded areas in Figure 6.39.
For $x = 70$,

$$z = \frac{70 - 80}{12} = -.83$$

For $x = 135$,

$$z = \frac{135 - 80}{12} = 4.58$$

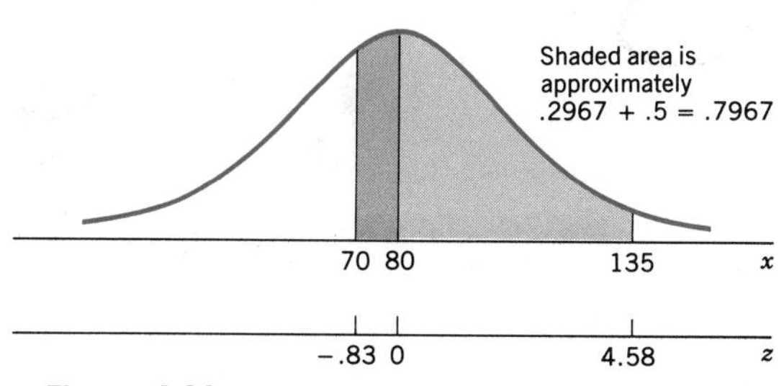

Figure 6.39 Area between $x = 70$ and $x = 135$.

Thus, the required area is obtained by adding the areas from $z = -.83$ to $z = 0$ and from $z = 0$ to $z = 4.58$ under the standard normal curve. Hence,

$$\begin{aligned} P(70 \leq x \leq 135) &= P(-.83 \leq z \leq 4.58) \\ &= P(-.83 \leq z \leq 0) + P(0 \leq z \leq 4.58) \\ &= .2967 + .5 = .7967 \text{ approximately} \end{aligned}$$

Finding area in the left tail.

(b) The area to the left of $x = 27$ is obtained by subtracting the area from $x = 27$ to $x = 80$ from .5, which is the total area to the left of the mean. This area is calculated as follows. For $x = 27$,

$$z = \frac{27 - 80}{12} = -4.42$$

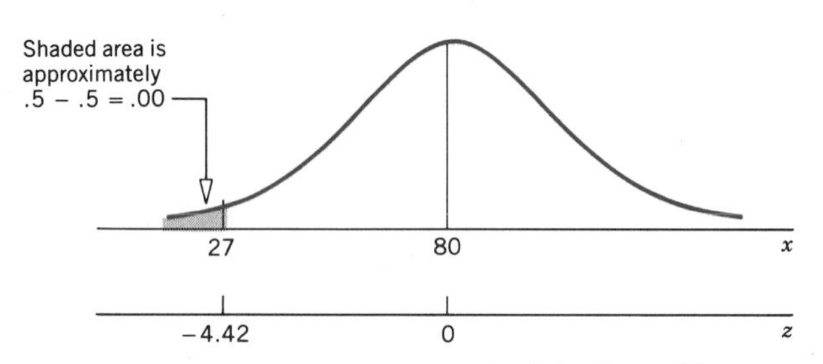

Figure 6.40 Area to the left of $x = 27$.

As shown in Figure 6.40, the required area is given by the area under the normal curve to the left of $z = -4.42$. This area is

$$P(x < 27) = P(z < -4.42) = .5 - P(-4.42 < z < 0)$$
$$= .5 - .5 = .00 \text{ approximately}$$

■

EXERCISES

6.15 Find the z value for each of the following x values for a normal curve with $\mu = 30$ and $\sigma = 5$.

a. $x = 37$ **b.** $x = 19$ **c.** $x = 23$ **d.** $x = 44$

6.16 Determine the z value for each of the following x values for a normal curve with $\mu = 16$ and $\sigma = 3$.

a. $x = 11$ **b.** $x = 22$ **c.** $x = 18$ **d.** $x = 14$

6.17 Find the area under a normal curve with $\mu = 20$ and $\sigma = 4$

a. between 20 and 27 **b.** from 23 to 25 **c.** between 9.5 and 17

6.18 Find the area under a normal curve with $\mu = 12$ and $\sigma = 2$

a. between 7.76 and 12 **b.** between 14.48 and 16.34 **c.** from 8.22 to 11.06

6.19 Determine the area under a normal curve with $\mu = 55$ and $\sigma = 7$

a. to the right of 58 **b.** to the right of 43
c. to the left of 67 **d.** to the left of 24

6.20 Find the area under a normal curve with $\mu = 37$ and $\sigma = 3$

a. to the left of 29 **b.** to the right of 53
c. to the left of 42 **d.** to the right of 35

6.21 Let x be a continuous random variable that is normally distributed with a mean of 25 and a standard deviation of 6. Find the probability that x assumes a value

a. between 29 and 36 **b.** between 22 and 33

6.22 Let x be a continuous random variable that has a normal distribution with a mean of 40 and a standard deviation of 4. Find the probability that x assumes a value

a. between 29 and 35 **b.** from 34 to 51

6.23 Let x be a continuous random variable that is normally distributed with a mean of 80 and a standard deviation of 12. Find the probability that x assumes a value

a. greater than 70 **b.** less than 75
c. greater than 100 **d.** less than 89

6.24 Let x be a continuous random variable that is normally distributed with a mean of 65 and a standard deviation of 15. Find the probability that x assumes a value

a. less than 43 **b.** greater than 74
c. greater than 56 **d.** less than 71

6.5 APPLICATIONS OF THE NORMAL DISTRIBUTION

Sections 6.2 through 6.4 discussed the normal distribution, how to convert a normal distribution to the standard normal distribution, and how to find the area under a normal distribution curve. This section presents examples that illustrate the applications of the normal distribution.

Application of the normal distribution.

EXAMPLE 6-11 According to the U.S. Census Bureau, the mean earnings of persons with 5 or more years of college were $46,853 in 1987. Assuming that the 1987 earnings of all (working) people with 5 or more years of college have a normal distribution with a mean of $46,853 and a standard deviation of $6100, find the probability that the 1987 earnings of a randomly selected person with 5 or more years of college were

(a) between $37,400 and $50,460
(b) less than $33,000

Solution Let x denote the 1987 earnings of a randomly selected person with 5 or more years of college. Then x is normally distributed with

$$\mu = \$46{,}853 \quad \text{and} \quad \sigma = \$6100$$

Area between two points that are on different sides of the mean.

(a) The probability that the 1987 earnings of a randomly selected person with 5 or more years of college were between $37,400 and $50,460 is given by the area under the normal curve between $x = \$37{,}400$ and $x = \$50{,}460$. Because these two points are on different sides of the mean, the required probability is obtained by adding the two shaded areas shown in Figure 6.41. For $x = 37{,}400$,

$$z = \frac{37{,}400 - 46{,}853}{6100} = -1.55$$

For $x = 50{,}460$,

$$z = \frac{50{,}460 - 46{,}853}{6100} = .59$$

The required probability is

$$\begin{aligned} P(37{,}400 < x < 50{,}460) &= P(-1.55 < z < .59) \\ &= P(-1.55 < z < 0) + P(0 < z < .59) \\ &= .4394 + .2224 = .6618 \end{aligned}$$

Converting this probability to a percentage, we can also state that about 66.18% of all persons with 5 or more years of college earned between \$37,400 and \$50,460 in 1987.

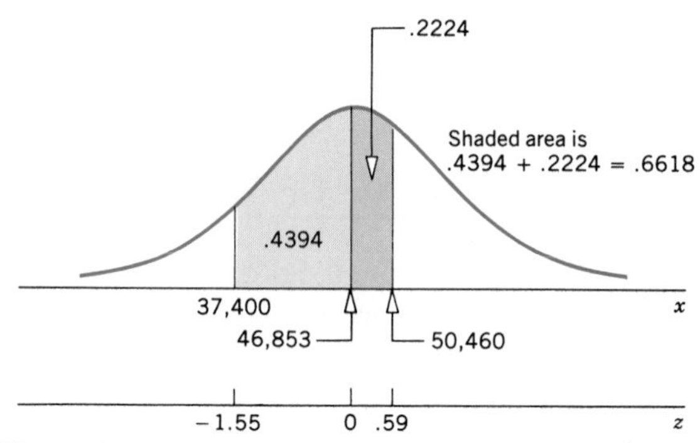

Figure 6.41 Area between $x = 37{,}400$ and $x = 50{,}460$.

Area in the left tail.

(b) The probability that the 1987 earnings of a randomly selected person with 5 or more years of college were less than \$33,000 is given by the area under the normal curve to the left of $x =$ \$33,000. This probability is .0116, as calculated below and shown in Figure 6.42. For $x = 33{,}000$,

$$z = \frac{33{,}000 - 46{,}853}{6100} = -2.27$$

The required probability is

$$P(x < 33{,}000) = P(z < -2.27) = P(z < 0) - P(-2.27 < z < 0)$$
$$= .5 - .4884 = .0116$$

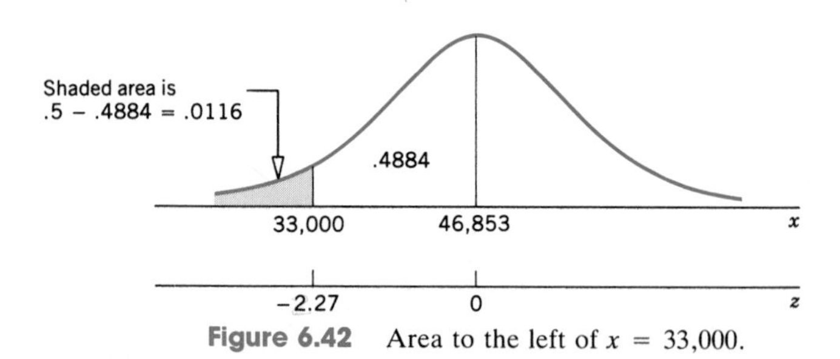

Figure 6.42 Area to the left of $x = 33{,}000$.

Converting this probability to a percentage, we can also state that approximately 1.16% of all persons with 5 or more years of college earned less than \$33,000 in 1987. ■

Application of the normal distribution.

EXAMPLE 6-12 According to a National Health and Nutrition Examination Survey, the mean weight of American women aged 18 to 74 is 144.2 pounds. Assume that the weights of the current population of American women aged 18 to 74 have a normal distribution with a mean of 144.2 pounds and a standard deviation of 21 pounds. Find the probability that the weight of a randomly selected 18- to 74-year-old American woman is

(a) less than 156 pounds
(b) between 110 and 230 pounds

Solution Let x denote the weight of a randomly selected woman aged 18 to 74. Then x is normally distributed with

$$\mu = 144.2 \text{ pounds} \quad \text{and} \quad \sigma = 21 \text{ pounds}$$

Finding area to the left of a value that is greater than the mean.

(a) The probability that the weight of a randomly selected woman aged 18 to 74 is less than 156 pounds is given by the area under the normal curve to the left of $x = 156$. This area is obtained by adding the two shaded areas shown in Figure 6.43. For $x = 156$,

$$z = \frac{156 - 144.2}{21} = .56$$

$$P(x < 156) = P(z < .56) = P(z < 0) + P(0 < z < .56)$$
$$= .5 + .2123 = .7123$$

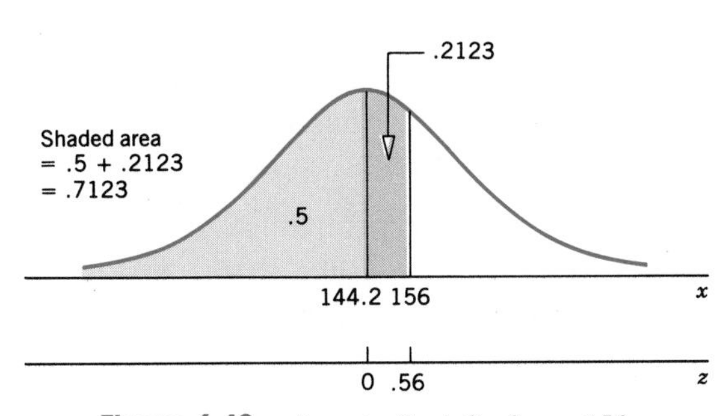

Figure 6.43 Area to the left of $x = 156$.

Converting this probability to a percentage, we can also state that approximately 71.23% of 18- to 74-year-old American women weigh less than 156 pounds.

Finding area between two points that are on different sides of the mean.

(b) The probability that the weight of a randomly selected woman aged 18 to 74 is between 110 and 230 pounds is given by the area under the normal curve between $x = 110$ and $x = 230$. This area is obtained by adding the two shaded areas shown in Figure 6.44. For $x = 110$,

$$z = \frac{110 - 144.2}{21} = -1.63$$

For $x = 230$,

$$z = \frac{230 - 144.2}{21} = 4.09$$

The required probability is

$$\begin{aligned} P(110 < x < 230) &= P(-1.63 < z < 4.09) \\ &= P(-1.63 < z < 0) + P(0 < z < 4.09) \\ &= .4484 + .5 = .9484 \text{ approximately} \end{aligned}$$

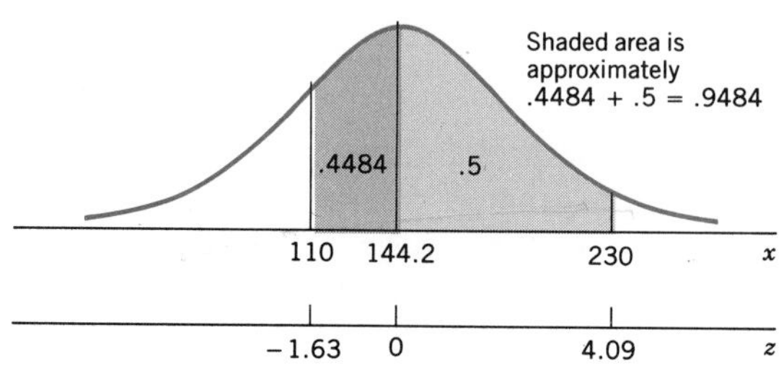

Figure 6.44 Area between $x = 110$ and $x = 230$.

Converting this probability to a percentage, we can also state that approximately 94.84% of 18- to 74-year-old American women weigh between 110 and 230 pounds. ■

Application of the normal distribution.

EXAMPLE 6-13 The net amount of soda in a can of orange cola has a normal distribution with a mean of 12 ounces and a standard deviation of .015 ounces.

(a) What is the probability that a randomly selected can of orange cola contains 11.97 to 11.99 ounces of soda?

(b) What percentage of the orange cola cans contain more than 12.02 ounces of soda?

Solution Let x be the net amount of soda in a can of orange cola. Then x has a normal distribution with $\mu = 12$ ounces and $\sigma = .015$ ounces.

Area between two points that are to the left of the mean.

(a) The probability that a randomly selected can contains 11.97 to 11.99 ounces of soda is given by the area under the normal curve from $x = 11.97$ to $x = 11.99$. This area is shown in Figure 6.45. For $x = 11.97$,

$$z = \frac{11.97 - 12}{.015} = -2.00$$

For $x = 11.99$,

$$z = \frac{11.99 - 12}{.015} = -.67$$

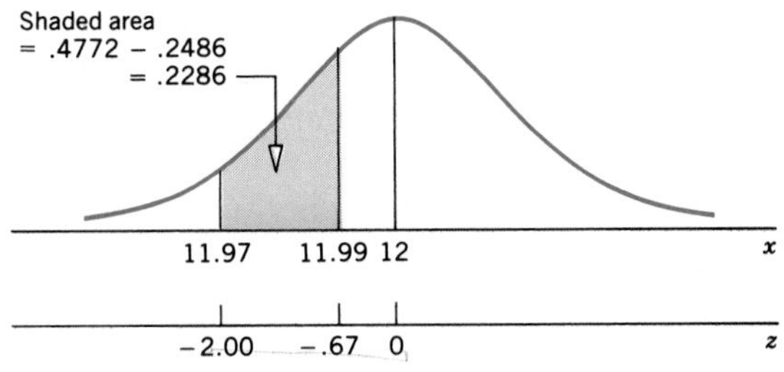

Figure 6.45 Area between $x = 11.97$ and 11.99.

The required probability is

$$\begin{aligned} P(11.97 \leq x \leq 11.99) &= P(-2.00 \leq z \leq -.67) \\ &= P(-2.00 \leq z \leq 0) - P(-.67 \leq z \leq 0) \\ &= .4772 - .2486 = .2286 \end{aligned}$$

Thus, the probability is .2286 that any randomly selected can of orange cola will contain 11.97 to 11.99 ounces of soda. We can also state that about 22.86% of the orange cola cans contain 11.97 to 11.99 ounces of soda.

Area to the right of a point that is to the right of the mean.

(b) To find the percentage of orange cola cans that contain more than 12.02 ounces of soda, we first find the probability that a randomly selected can contains more than 12.02 ounces of soda. This probability is given by the area under the normal curve to the right of 12.02, as shown in Figure 6.46. For $x = 12.02$,

$$z = \frac{12.02 - 12}{.015} = 1.33$$

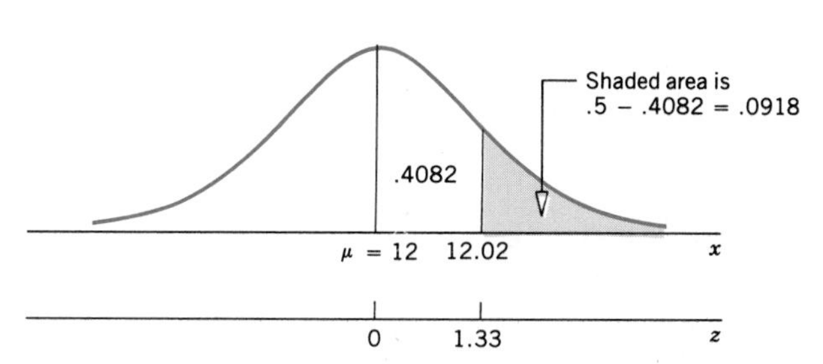

Figure 6.46 Area to the right of $x = 12.02$.

The required probability is

$$\begin{aligned} P(x > 12.02) &= P(z > 1.33) = P(z > 0) - P(0 < z < 1.33) \\ &= .5 - .4082 = .0918 \end{aligned}$$

Converting this probability to a percentage, we can state that approximately 9.18% of all the orange cola cans are expected to contain more than 12.02 ounces of soda. ■

Area to the left of a point that is to the left of the mean.

EXAMPLE 6-14 The life span of a calculator manufactured by Calcu Company has a normal distribution with a mean of 54 months and a standard deviation of 8 months. The company guarantees that any calculator that starts malfunctioning within 36 months of the purchase will be replaced by a new calculator. About what percentage of such calculators made by this company are expected to be replaced?

Solution Let x be the life span of such a calculator. Then x has a normal distribution with $\mu = 54$ and $\sigma = 8$ months. The probability that a randomly selected calculator will start malfunctioning within 36 months is given by the area under the normal curve to the left of $x = 36$, as shown in Figure 6.47. This area is approximately .0122 as calculated below. For $x = 36$,

$$z = \frac{36 - 54}{8} = -2.25$$

$$P(x < 36) = P(z < -2.25) = P(z < 0) - P(-2.25 < z < 0)$$
$$= .5 - .4878 = .0122$$

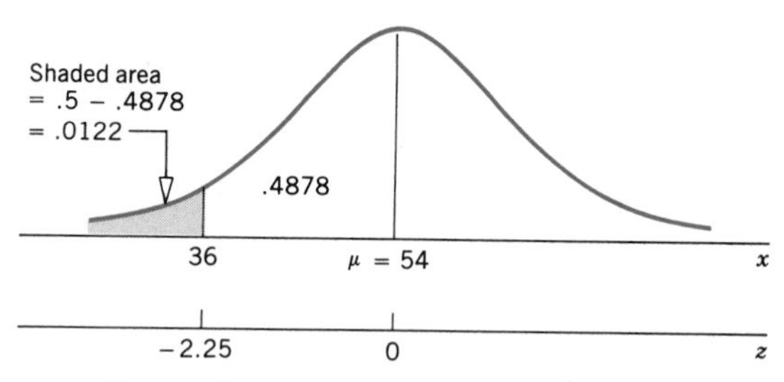

Figure 6.47 Area to the left of $x = 36$.

The probability that any randomly selected calculator manufactured by Calcu Company will start malfunctioning within 36 months is .0122. Converting this probability to a percentage, we can state that approximately 1.22% of all such calculators manufactured by this company are expected to start malfunctioning within 36 months. Hence, 1.22% of the calculators are expected to be replaced. ■

EXERCISES

6.25 According to the Health Insurance Association of America, the mean daily charge for a hospital room in 1989 was \$262. Assume that the 1989 daily charges for all hospital rooms have a normal distribution with a mean of \$262 and a standard deviation of \$35. Find the probability that the 1989 daily charge for a randomly selected hospital room was

a. greater than \$293 b. between \$242 and \$279

6.26 According to the U.S. Department of Agriculture, the mean consumption of dairy products (milk equivalent) in 1987 was 598 pounds per person in the United States. Assume that the 1987 consumption of dairy products for all U.S. people has an approximate normal distribution with a mean of 598 pounds and a standard deviation of 110 pounds. Find the probability that 1987 consumption of dairy products for a randomly selected person was

a. between 440 and 655 pounds b. more than 565 pounds

6.27 The mean SAT (Scholastic Aptitude Test) score in mathematics for all students who took this test in 1990 was 476. Suppose that the 1990 SAT scores in mathematics for all examinees have a normal distribution with a mean of 476 and a standard deviation of 80. Find the probability that the score of a randomly selected examinee who took this test in 1990 was

a. less than 525 **b.** 375 to 425

6.28 Let x denote the time taken to run a road race. Suppose x is approximately normally distributed with a mean of 195 minutes and a standard deviation of 21 minutes. If one runner is selected at random, what is the probability that this runner will complete this road race

a. in less than 150 minutes **b.** in 205 to 245 minutes

6.29 According to the American Hospital Association, the mean cost per stay in a hospital was $3850 in 1987. Suppose that the 1987 costs for all stays in all hospitals are approximately normally distributed with a mean of $3850 and a standard deviation of $950. Find the probability that the cost of a randomly selected 1987 stay in a hospital was

a. less than $2350 **b.** more than $2400

6.30 The U.S. Bureau of Labor Statistics estimated the 1989 mean annual expenditure on entertainment per person to be $1401. Assume that the 1989 entertainment expenditures for all people are approximately normally distributed with a mean of $1401 and a standard deviation of $345. Find the percentage of people that had the 1989 entertainment expenditures

a. less than $1520 **b.** more than $1690

6.31 According to the U.S. Department of Agriculture, the mean yield of corn in 1988 was 84.6 bushels (one bushel is equal to 56 pounds) per acre. Assume that the 1988 yield of corn for all acres has a normal distribution with a mean of 84.6 bushels and a standard deviation of 5.5 bushels. Find the percentage of acres with 1988 yield of corn

a. between 92 and 100 bushels **b.** between 79 and 88 bushels

6.32 According to the U.S. Energy Information Administration, the mean expenditure on residential energy consumption in 1987 was $1080 per household. Assume that the 1987 expenditures on residential energy consumption for all households are normally distributed with a mean of $1080 and a standard deviation of $235. Find the probability that the 1987 expenditure on residential energy consumption for a randomly selected household was

a. between $580 and $970 **b.** between $1015 and $1360

6.33 According to the U.S. Department of Agriculture, the mean milk production per cow was 14,200 pounds in 1988. Suppose that the 1988 milk production for all cows is normally distributed with a mean of 14,200 pounds and a standard deviation of 1300 pounds. Find the perentage of 1988 cows with a milk production of

a. less than 12,750 pounds **b.** between 13,530 and 16,360 pounds

6.34 According to the National Education Association, the mean salary of public school teachers was $29,648 in 1989. Suppose that the 1989 salaries of all teachers are approximately normally distributed with a mean of $29,648 and a standard deviation of $2500. Find the probability that the 1989 salary of a randomly selected teacher was

a. more than $31,700 **b.** between $27,625 and $32,830

6.35 The management of a supermarket wants to adopt a new promotional policy of giving a free gift to every customer who spends more than a certain amount per visit at this supermarket. The expectation of the management is that after this promotional policy is advertised, the expenditures for all customers at this supermarket will be normally distributed with a mean of $95 and a standard deviation of $21. If the management decides to give free gifts to all those customers who spend more than $130 at this supermarket during a visit, what percentage of the customers are expected to get free gifts?

6.36 The stress scores (on a scale of 1 to 10) of students before a mathematics test are found to be approximately normally distributed with a mean of 7.08 and a standard deviation of .63. Find the probability that the stress score before a mathematics test for a randomly selected student will be

a. more than 6.25 **b.** between 7.40 and 8.90

6.37 The job satisfaction scores (on a scale of 1 to 20) of workers at an insurance company are normally distributed with a mean of 13.10 and a standard deviation of 1.95. Find the probability that the job satisfaction score for a randomly selected worker from this company is

a. less than 11.25 **b.** between 8.50 and 11.70

6.38 The mean proficiency score in mathematics for 17-year-olds was 302 in 1986 (*Crossroads in American Education,* Educational Testing Service, 1989). Assume that the 1986 proficiency scores in mathematics for all 17-year-olds have a normal distribution with a mean of 302 and a standard deviation of 45. Find what percentage of the 17-year-olds had mathematics proficiency scores

a. between 260 and 380 **b.** between 330 and 425

6.39 According to a National Health and Nutrition Examination Survey, the mean height of American men aged 18 to 74 was 69.1 inches. Assume that the heights of the current population of all American men aged 18 to 74 are approximately normally distributed with a mean of 69.1 inches and a standard deviation of 2.1 inches. Find what percentage of the 18- to 74-year-old men are

a. less than 71 inches tall **b.** more than 67 inches tall

6.40 The speeds of cars traveling on interstate highway I-15 are normally distributed with a mean of 69 miles per hour and a standard deviation of 3.5 miles per hour. Find what percentage of the cars traveling on this highway have a speed of

a. 61 to 66 miles per hour **b.** 65 to 74 miles per hour

***6.41** The lengths of 3-inch nails manufactured on a machine are normally distributed with a mean of 3.0 inches and a standard deviation of .009 inches. The nails that are either less than 2.98 inches long or more than 3.02 inches long are unusable. What percentage of all the nails produced by this machine are unusable? (*Hint:* The required percentage is given by the sum of the areas under two tails of the normal curve, one to the left of $x = 2.98$ and the second to the right of $x = 3.02$.)

6.6 DETERMINING THE z VALUE WHEN AN AREA UNDER THE STANDARD NORMAL CURVE IS KNOWN

So far this chapter has discussed how to find the area under a normal curve for an interval of z or x. Now we will reverse this procedure and learn how to find the corresponding value of z when an area under the standard normal curve is known. Examples 6-15 through 6-17 describe this procedure.

Finding z when area between mean and z is known.

EXAMPLE 6-15 Find a point z such that the area under the standard normal curve between 0 and z is .4251 and the value of z is positive.

Solution As shown in Figure 6.48 on the next page, we are to find the z value such that the area between 0 and z is .4251. To find the required value of z, we look for .4251 in the body of the normal table, Table VII of Appendix C. The relevant part of that table is reproduced as Table 6.6 on the next page. Next we read the numbers, in the column and row for z, which correspond to .4251. These numbers are 1.4 and .04 respectively. Combining these two numbers, we obtain the required value of $z = 1.44$.

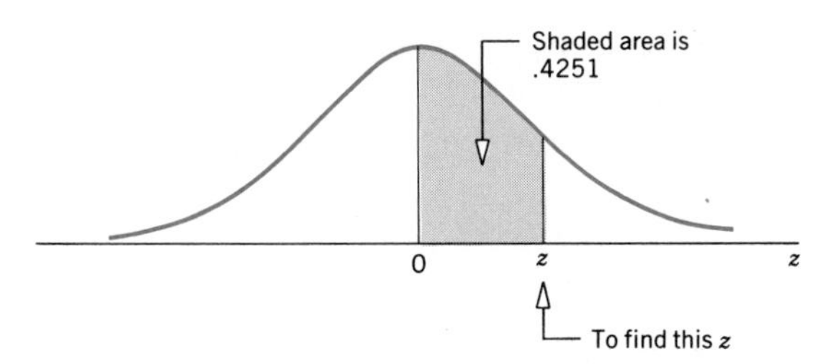

Figure 6.48 Finding the z value.

Table 6.6 Finding the z Value When Area is Given

z	.00	.01	...	.04	...	.09
0.0	.0000	.0040	...	↑	...	.0359
0.1	.0398	.0438	...		...	.0753
0.2	.0793	.0832	...		...	.1141
.	.	.	...		...	.
.	.	.	...		...	.
.	.	.	...		...	.
1.4	←			.4251 ←	...	...
.	.	.	...	.	...	.
.	.	.	...	.	...	.
.	.	.	...	.	...	.
3.0	.4987	.4987	...	.4988	...	.4990

We locate this value in Table VII

Finding z when area in the right tail is known.

EXAMPLE 6-16 Find the value of z such that the area under the standard normal curve in the right tail is .0050.

Solution to find the required value of z, first we find the area between 0 and z. The total area to the right of $z = 0$ is .5. Hence,

$$\text{Area between 0 and } z = .5 - .0050 = .4950$$

This area is shown in Figure 6.49.

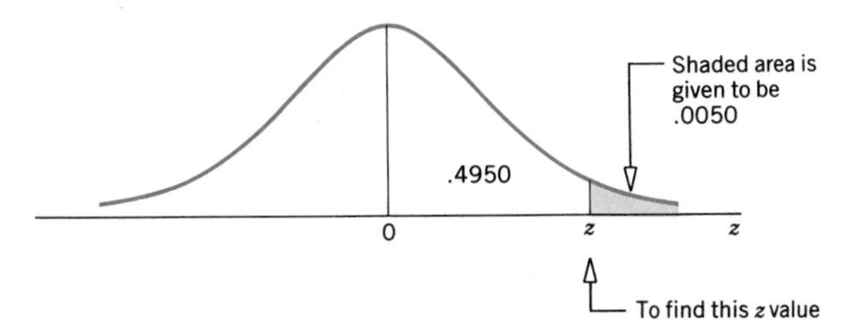

Figure 6.49 Finding the z value.

Now we look for .4950 in the body of the normal table. Table VII does not contain .4950. So we find the value closest to .4950, which is either .4949 or .4951. We can use either of these two values. If we choose .4951, the corresponding z value is 2.58. Hence, the required value of z is 2.58 and the area to the right of $z = 2.58$ is approximately .0050. Note that there is no apparent reason to choose .4951 and not to choose .4949. We can use either of these two values. If we choose .4949, the corresponding z value will be 2.57. ■

Finding z when area in the left tail is known.

EXAMPLE 6-17 Find the value of z such that the area under the standard normal curve in the left tail is .05.

Solution Because .05 is smaller than .5 and it is the area in the left tail, z is negative. To find the required value of z, first we find the area between 0 and z. The total area to the left of $z = 0$ is .5. Hence,

$$\text{Area between 0 and } z = .5 - .05 = .4500$$

This area is shown in Figure 6.50.

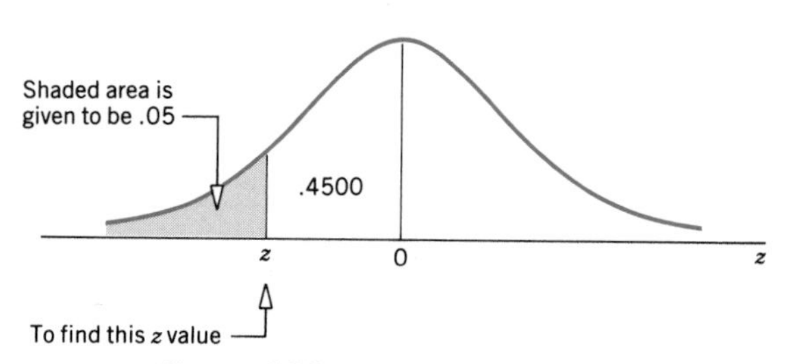

Figure 6.50 Finding the z value.

Next, we look for .4500 in the body of the normal table. The value closest to .4500 in the normal table is either .4495 or .4505. Suppose we use the value .4505. The corresponding z value is 1.65. Because z lies to the left of 0 (see Fig. 6.50), the required value of z is -1.65 and the area to the left of $z = -1.65$ is approximately .05. ■

EXERCISES

6.42 Find the value of z so that the area under the standard normal curve

a. from 0 to z is .4772 and z is positive
b. between 0 and z is (approximately) .4785 and z is negative
c. in the left tail is (approximately) .3565
d. in the right tail is (approximately) .1530

6.43 Find the value of z so that the area under the standard normal curve

a. from 0 to z is (approximately) .1965 and z is positive
b. between 0 and z is (approximately) .2740 and z is negative
c. in the left tail is (approximately) .2050
d. in the right tail is (approximately) .1053

6.44 Determine the value of z so that the area under the standard normal curve

a. in the right tail is .0500
b. in the left tail is .0250
c. in the left tail is .0100
d. in the right tail is .0050

6.45 Determine the value of z so that the area under the standard normal curve

a. in the right tail is .0250
b. in the left tail is .0500
c. in the left tail is .0010
d. in the right tail is .0100

6.7 THE NORMAL APPROXIMATION TO THE BINOMIAL DISTRIBUTION

Recall from Chapter 5 that

1. The binomial distribution is applied to a discrete random variable.
2. Each repetition, called a trial, of a binomial experiment results in one of the two possible outcomes, which are called a success and a failure.
3. The probabilities of the two (possible) outcomes remain the same for each repetition of the experiment.
4. The trials are independent.

The binomial formula, which gives the probability of x successes in n trials, is

$$P(x) = \binom{n}{x} p^x q^{n-x}$$

However, the use of the binomial formula becomes very tedious for large n. In such cases, the normal distribution can be used to approximate the binomial probability. Note that, for a binomial problem, the exact probability is obtained by using the binomial formula. If we apply the normal distribution to solve a binomial problem, the probability that we obtain is an approximation to the exact probability. The approximation obtained by using the normal distribution is very close to the exact probability when n is large and p is very close to .50. However, this does not mean that we should not use the normal approximation when p is not close to .50. The reason for the approximation being closer to the exact probability when p is close to .50 is that the binomial distribution is symmetric when $p = .50$. The normal distribution is always symmetric. Hence, the two distributions are very close to each other when n is large and p is close to .50. However (because not every symmetric bell-shaped curve is a normal curve), this does not mean that whenever $p = .50$ the binomial distribution is the same as the normal distribution.

NORMAL DISTRIBUTION AS AN APPROXIMATION TO BINOMIAL DISTRIBUTION

Usually, the normal distribution is used as an approximation to the binomial distribution when np and nq are both greater than 5, that is, when

$$np > 5 \quad \text{and} \quad nq > 5$$

Table 6.7 gives the binomial probability distribution of x for $n = 12$ and $p = .50$. This table is constructed using Table IV of Appendix C. Figure 6.51 shows the histogram and the smoothed polygon for the probability distribution of Table 6.7. As we can observe, the histogram in Figure 6.51 is symmetric and the curve obtained by joining the upper midpoints of the rectangles is approximately bell-shaped.

Table 6.7 The Binomial Probability Distribution for $n = 12$ and $p = .50$

x	$P(x)$
0	.0002
1	.0029
2	.0161
3	.0537
4	.1208
5	.1934
6	.2256
7	.1934
8	.1208
9	.0537
10	.0161
11	.0029
12	.0002

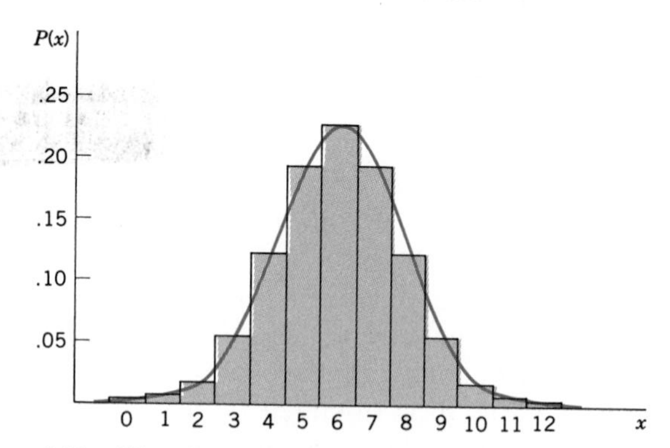

Figure 6.51 Histogram for the probability distribution of Table 6.7.

Examples 6-18 through 6-20 illustrate the application of the normal distribution as an approximation to the binomial distribution.

Normal approximation: x equals a specific value and it is to the right of the mean.

EXAMPLE 6-18 According to an estimate, 50% of the people in America have at least one credit card. If a random sample of 30 persons is taken, what is the probability that 19 of them will have at least one credit card?

Solution Let n be the total number of persons in the sample, x be the number of persons in the sample who have at least one credit card, and p be the probability that a person has at least one credit card. Then, this is a binomial problem with

$$n = 30, \quad p = .50, \quad q = 1 - p = .50,$$
$$x = 19, \quad \text{and} \quad n - x = 30 - 19 = 11$$

Using the binomial formula, the exact probability that 19 persons in a sample of 30 have at least one credit card is

$$P(x = 19) = \binom{30}{19} (.50)^{19} (.50)^{11} = .0509$$

Now let us solve this problem using the normal distribution as an approximation to the binomial distribution. For this example,

$$np = 30\,(.50) = 15 \qquad \text{and} \qquad nq = 30\,(.50) = 15$$

Thus, because np and nq are both greater than 5, we can use the normal approximation to solve this binomial problem.

To use the normal distribution, we need to know the mean and standard deviation of the distribution. Hence, *the first step in using the normal approximation to the binomial distribution is to compute the mean and standard deviation of the binomial distribution.* As we know from Chapter 5, the mean and standard deviation of a binomial distribution are given by np and $\sqrt{npq}$, respectively. Using these formulas, we obtain:

$$\mu = np = 30\,(.50) = 15$$

$$\sigma = \sqrt{npq} = \sqrt{30\,(.50)\,(.50)} = 2.739$$

The normal distribution applies to a continuous random variable, whereas the binomial distribution applies to a discrete random variable. *The second step in applying the normal approximation to the binomial distribution is to make the correction for continuity.*

As shown in Figure 6.52, the probability of 19 successes in 30 trials is given by the area of the rectangle for $x = 19$. To make the correction for continuity, we use the interval 18.5 to 19.5 for 19 persons. This interval is actually given by the two boundaries of the rectangle for $x = 19$, which is obtained by subtracting .5 from 19 and adding .5 to 19. Thus, $P(x = 19)$ for the binomial problem will be approximately equal to $P(18.5 \le x \le 19.5)$ for the normal distribution.

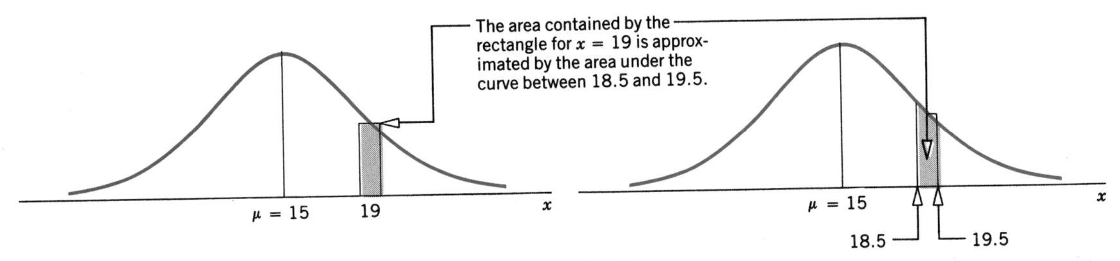

Figure 6.52

As shown in Figure 6.53, the area under the normal curve between $x = 18.5$ and $x = 19.5$ will give us the approximate probability that 19 persons possess at least one credit card. We calculate this probability as follows. For $x = 18.5$,

$$z = \frac{18.5 - 15}{2.739} = 1.28$$

For $x = 19.5$,

$$z = \frac{19.5 - 15}{2.739} = 1.64$$

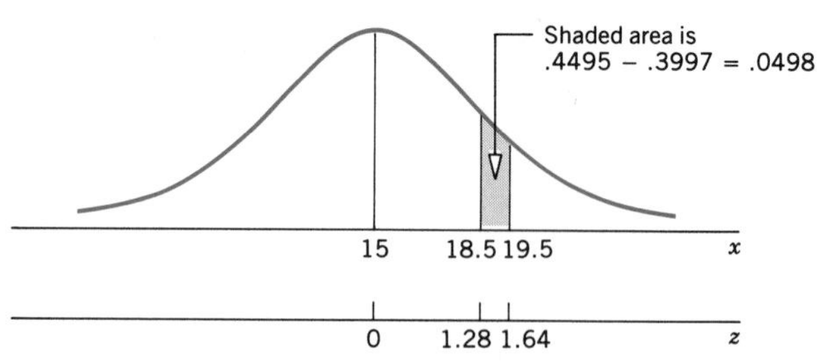

Figure 6.53 Area between $x = 18.5$ and $x = 19.5$.

The required probability is

$$\begin{aligned} P(18.5 \leq x \leq 19.5) &= P(1.28 \leq z \leq 1.64) \\ &= P(0 \leq z \leq 1.64) - P(0 \leq z \leq 1.28) \\ &= .4495 - .3997 = .0498 \end{aligned}$$

Therefore, based on the normal approximation, the probability that 19 persons in a sample of 30 will possess at least one credit card is approximately .0498. Earlier, using the binomial formula, we obtained the exact probability .0509. Hence, the error due to using the normal approximation is $.0509 - .0498 = .0011$. Thus, the exact probability is underestimated by .0011 if the normal approximation is used. ■

Normal approximation: x equals a specific value that is to the left of the mean.

EXAMPLE 6-19 Fifty-eight percent of adults polled by the Gallup Organization for the 20th Annual Gallup Poll of the Public's Attitudes Towards the Public Schools, published in the September 1988 issue of *Phi Delta Kappan,* said that they would like their children to take up teaching in the public schools as a career. Assuming that this result holds true for the current population of all adults, what is the probability that in a random sample of 50 adults exactly 22 will hold this view?

Solution Let n be the total number of adults in this sample, x be the number of adults in the sample who hold this view, and p be the probability that an adult holds this view. Then, this is a binomial problem with

$$n = 50, \quad p = .58, \quad q = .42, \quad x = 22, \quad \text{and} \quad n - x = 28$$

We are to find the probability of $x = 22$. Because n is large, it is easier to apply the normal approximation than to use the binomial formula. We can check that np and nq are both greater than 5. The mean and standard deviation of the binomial distribution are

$$\mu = np = 50\,(.58) = 29$$

$$\sigma = \sqrt{npq} = \sqrt{50\,(.58)\,(.42)} = 3.490$$

For the continuity correction, we subtract .5 from 22 and add .5 to 22 to obtain the interval 21.5 to 22.5. Thus, the probability that 22 adults will hold the given view will be approximated by the area under the normal curve from 21.5 to 22.5. This area is shown in Figure 6.54. For $x = 21.5$,

$$z = \frac{21.5 - 29}{3.490} = -2.15$$

For $x = 22.5$,

$$z = \frac{22.5 - 29}{3.490} = -1.86$$

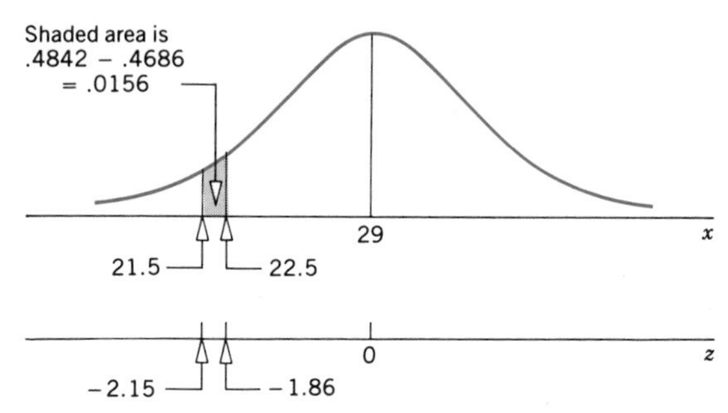

Figure 6.54 Area between $x = 21.5$ and $x = 22.5$.

The required probability is

$$\begin{aligned} P(21.5 \leq x \leq 22.5) &= P(-2.15 \leq z \leq -1.86) \\ &= P(-2.15 \leq z \leq 0) - P(-1.86 \leq z \leq 0) \\ &= .4842 - .4686 = .0156 \end{aligned}$$

Thus, the probability that exactly 22 adults in a sample of 50 would like their children to take up teaching in the public schools as a career is approximately .0156.

■

Normal approximation: $x \geq$ a value that is to the right of the mean.

EXAMPLE 6-20 According to a *Newsweek* poll conducted by the Gallup Organization and published in *Newsweek* on December 5, 1988, 63% of the women interviewed said that they do most of their clothes shopping at department stores. Assume that this percentage is true for the population of all women. What is the probability that in a random sample of 100 women, 70 or more will say that they do most of their clothes shopping at department stores?

Solution Let n be the total number of women in the sample, x be the number of women in the sample who do most of their clothes shopping at department stores, and p be the probability that a woman does most of her clothes shopping at department stores. Then, this is a binomial problem with

$$n = 100, \quad p = .63, \quad \text{and} \quad q = .37$$

We are to find the probability of 70 or more successes in 100 trials. The mean and standard deviation of the binomial distribution are

$$\mu = np = 100\,(.63) = 63$$

$$\sigma = \sqrt{npq} = \sqrt{100\,(.63)\,(.37)} = 4.828$$

For continuity correction, we subtract .5 from 70, which gives 69.5. Thus, the probability that 70 or more women out of a random sample of 100 will say that they do most of their clothes shopping at department stores is approximated by the area under the normal curve to the right of $x = 69.5$, as shown in Figure 6.55.

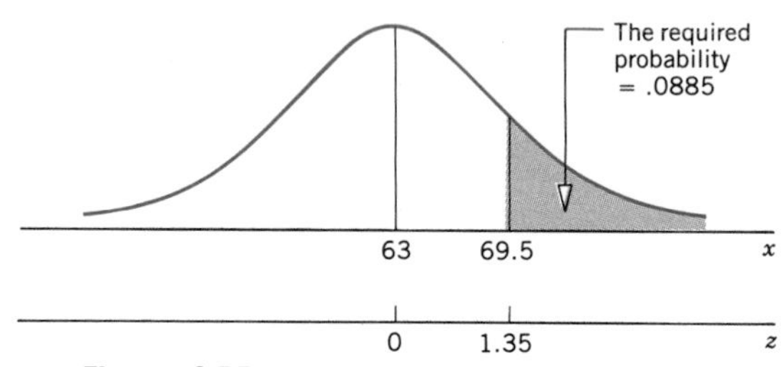

Figure 6.55 Area to the right of $x = 69.5$.

For $x = 69.5$,

$$z = \frac{69.5 - 63}{4.828} = 1.35$$

To find the required probability, we find the area between $z = 0$ and $z = 1.35$ and subtract that area from .5, which is the total area to the right of $z = 0$. Thus,

$$P(x \geq 69.5) = P(z \geq 1.35) = .5 - P(0 \leq z \leq 1.35) = .5 - .4115 = .0885$$

Thus, the probability that 70 or more women in a random sample of 100 will say that they do most of their clothes shopping at department stores is approximately .0885. ■

EXERCISES

6.46 Find the following binomial probabilities using the normal approximation.

a. $n = 140$, $p = .45$, $P(x = 67)$
b. $n = 100$, $p = .55$, $P(52 \leq x \leq 60)$
c. $n = 90$, $p = .42$, $P(x \geq 40)$
d. $n = 104$, $p = .75$, $P(x \leq 72)$

6.47 Find the following binomial probabilities using the normal approximation.

a. $n = 70$, $p = .30$, $P(x = 18)$
b. $n = 200$, $p = .70$, $P(133 \leq x \leq 145)$
c. $n = 85$, $p = .40$, $P(x \geq 30)$
d. $n = 150$, $p = .38$, $P(x \leq 62)$

6.48 Based on data from a Peter D. Hart Research Associates' poll on consumer buying habits and attitudes, Peter Hart estimated that 21% of American shoppers are *practical shoppers,* that is, the "smart shoppers who research their purchases and look for the best deal" (*The American Way of Buying,* Dow Jones & Company, Inc., 1990). Assume that this result is true for the current population of all American shoppers. Find the probability that in a random sample of 200 American shoppers, 37 to 46 are practical shoppers.

6.49 In a *Time*/CNN poll of New York City residents, 59% said "somewhere else" in response to the question, "If you could choose where you live, would you stay in New York City or move somewhere else?" (*Time,* September 17, 1990.) Assuming that this result is true for the current population of all New York City residents, find the probability that in a random sample of 300 New York City residents, 170 to 189 will say somewhere else in response to this question.

6.50 According to the Electronic Industry Association, 35% of U.S. households own telephone answering machines (*The Wall Street Journal,* September 6, 1990). Assuming that this percentage is true for the current population of all households, find the probability that in a random sample of 150 households, exactly 54 will own telephone answering machines.

6.51 In a survey of more than 8000 career women, 65% said they most want to be seen as skilled by top management (*Working Woman,* October 1989). Assuming that this percentage is true for the current population of all career women, find the probability that in a random sample of 100 career women, exactly 59 will most want to be seen as skilled by top management.

6.52 In 1989, 18% of physicians in the United States were women (*The 1990 Virginia Slims Opinion Poll*). Assume that this result holds true for the current population of all physicians. Find the probability that in a random sample of 100 physicians, 16 or more will be women.

6.53 In a *Family Circle* magazine survey, 75% of married women said that finding enough time alone with their husbands is "often" or "sometimes" a major stress in their relationship (*Family Circle,* March 14, 1989). Assuming that this percentage is true for the current population of all married women, find the probability that in a random sample of 150 married women, 105 or less will hold this view.

6.54 In a survey conducted by the Peter D. Hart Research Associates for *Rolling Stone* magazine, 47% of the 18- to 44-year-old people said that they are satisfied with their jobs (*Rolling Stone,* May 5, 1988). Assuming that this percentage is true for the current population of all 18- to 44-year-old people, find the probability that in a random sample of 120 such persons, exactly 58 will hold this view.

6.55 According to the U.S. Bureau of the Census, 24% of all persons aged 25 and over did not have a high school degree in 1988. Assuming that this percentage is true for the current population of all persons aged 25 and over, find the probability that in a random sample of 100 such persons, less than 22 will not have a high school degree.

6.56 In a 1989 survey conducted by *Fortune* magazine and Allstate Insurance, 46% of executives said that their company's productivity has been reduced because of poorly educated workers. Assuming that this percentage is true for the current population of all executives, find the probability that in a random sample of 80 executives, exactly 39 will hold this view.

6.57 According to a 1989 Roper survey, 14% of households owned computers. Assuming that this percentage is true for the current population of all households, find the probability that in a random sample of 150 households, 18 to 20 will own computers.

GLOSSARY

Continuous random variable A random variable that can assume any value in one or more intervals.

Continuity correction factor The addition of .5 and/or the subtraction of .5 from x when the normal distribution is used as an approximation to the binomial distribution, where x is the number of successes in n trials.

Normal probability distribution The probability distribution of a continuous random variable that, when plotted, gives a specific bell-shaped curve. The parameters of the normal distribution are the mean μ and the standard deviation σ.

Standard normal distribution The normal distribution with a mean of zero and a standard deviation of 1. The units of the standard normal distribution are denoted by z.

z value or z score The units of the standard normal distribution that are denoted by z.

KEY FORMULAS

1. The z Value for an x Value to Standardize a Normal Distribution

$$z = \frac{x - \mu}{\sigma}$$

2. Normal Approximation to the Binomial Distribution

The normal distribution can be used as an approximation to the binomial distribution when

$$np > 5 \quad \text{and} \quad nq > 5$$

3. Mean and Standard Deviation of the Binomial Distribution

$$\mu = np \quad \text{and} \quad \sigma = \sqrt{npq}$$

SUPPLEMENTARY EXERCISES

6.58 Let x be a continuous random variable that has a normal distribution with a mean of 40 and a standard deviation of 3. Find the probability that x assumes a value

- **a.** between 32 and 37
- **b.** between 38 and 46

6.59 Let x be a continuous random variable that has a normal distribution with a mean of 70 and a standard deviation of 12. Find the probability that x assumes a value

- **a.** greater than 87
- **b.** less than 62

6.60 According to a 1989 Decision Center Inc. survey, college students spend an average of 20 hours per week listening to the radio and watching television. Assume that the hours spent per week by all college students listening to the radio and watching television are normally distributed with a mean of 20 and a standard deviation of 4.8.

Find the probability that the number of hours spent per week listening to the radio and watching television by a randomly selected college student is

a. less than 9
b. between 13 and 17

6.61 According to the American Medical Association's Center for Health Policy Research survey for 1988, the average waiting time for patients at doctors' offices was 19.9 minutes. Assume that the waiting times for all patients at doctors' offices are normally distributed with a mean of 19.9 minutes and a standard deviation of 4.6 minutes. Find the probability that a randomly selected patient will have to wait at a doctor's office for

a. 26 to 30 minutes
b. more than 28 minutes

6.62 According to a Harris poll conducted for the Philip Morris Family Survey II of 1989, the mean monthly cost of day-care per child for single mothers was $211. Assume that the current monthly day-care costs per child for all single mothers have a normal distribution with a mean of $211 and a standard deviation of $22. Find the probability that the monthly day-care cost for a child of a randomly selected single mother will be

a. between $194 and $248
b. $228 or less

6.63 According to a survey conducted by Runzheimer International, the mean daily fixed cost (e.g., insurance, license and registration, depreciation, and interest) of owning a car was $8.39 in 1988. Assume that the current daily fixed costs for all cars have an approximate normal distribution with a mean of $8.39 and a standard deviation of $1.30. Find the probability that the daily fixed cost for a randomly selected car will be

a. between $5.50 and $7.50
b. $10.20 or more

6.64 According to an A. C. Nielsen 1989 survey, the mean number of hours spent watching television by 6- to 11-year-olds are 19.63 per week. Assume that the current weekly hours spent watching television by 6- to 11-year-olds follow a normal distribution with a mean of 19.63 and a standard deviation of 4.25. Find the probability that the weekly hours spent watching television by a randomly selected 6- to 11-year-old are

a. more than 26.70
b. between 9.5 and 14

6.65 According to the National Association of Secondary School Principals, the average salary of high school principals was $52,987 in 1989. Assume that the current salaries of all high school principals follow an approximate normal distribution with a mean of $52,987 and a standard deviation of $4900. Find the probability that the salary of a randomly selected high school principal is

a. less than $43,000
b. between $55,900 and $65,000

6.66 According to a 1989 survey conducted by R. H. Bruskin, the mean number of hours spent by adults on chores during a weekend are 14. Assume that the hours spent by all adults on chores during a weekend are approximately normally distributed with a mean of 14 and a standard deviation of 2.80. Find the probability that the hours spent by a randomly selected adult on chores during a weekend are

a. more than 11.5
b. between 10 and 19

6.67 The mean SAT (Scholastic Aptitude Test) score for 1990 for all students who took this test is 900. Assume that the 1990 SAT scores of all examinees have a normal distribution with a mean of 900 and a standard deviation of 175. Find the probability that the SAT score of a randomly selected student from the 1990 examinees is

a. between 620 and 1040
b. less than 550

6.68 The mean expenditure on clothing and shoes for all people in the United States was $799 in 1989 (*Family Economic Review,* May 1990). Assume that such 1989 expenditures for all U.S. people are approximately normally distributed with a mean of $799 and a standard deviation of $200. Find the percentage of the 1989 U.S. population who had such expenditures

a. between $940 and $1210
b. more than $970

6.69 The life of a certain type of battery is normally distributed with a mean of 90 hours and a standard deviation of 7 hours. Find the percentage of this type of batteries that will have a life of

a. less than 79 hours
b. 103 to 111 hours

6.70 The time taken by an employee of a company to do a specific job is normally distributed with a mean of 47 hours and a standard deviation of 3.5 hours. Find what percentage of the employees of this company will do this job in

a. 40 to 44 hours
b. less than 53 hours

6.71 According to an NCAA study done by Robert A. Rossi and others, women basketball players spend on average 29 hours a week on classroom-related activities (e.g., studying and going to classes) (*The Chronicle of Higher Education,* August 16, 1989). Assume that the time spent per week on classroom-related activities by all women basketball players is normally distributed with a mean of 29 hours and a standard deviation of 4 hours. Find the percentage of women basketball players for whom the number of hours spent per week on classroom-related activities are

a. 24 to 32 hours
b. more than 36 hours

6.72 According to Metropolitan Life Insurance Company's claims data, the mean charges (hospital and physician's) for coronary bypass surgeries done in 1986 were $30,430 (*Statistical Bulletin,* 70(1), January–March 1989). Assume that the charges for all such surgeries done in 1986 have a normal distribution with a mean of $30,430 and a standard deviation of $4180. Find what percentage of all such surgeries done in 1986 had charges of

a. $23,750 to $33,120
b. less than $25,600

6.73 According to Metropolitan Life Insurance Company's claims data, the mean charges (hospital and physician's) for a cesarean birth in 1986 were \$3240 (*Statistical Bulletin,* 69(3), July–September 1988). Assume that the charges for all such births for 1986 are normally distributed with a mean of \$3240 and a standard deviation of \$470. Find what percentage of all such births in 1986 had charges of

a. \$2490 to \$2900
b. more than \$2850

6.74 In a survey conducted by the Roper Organization for the Television Information Office, 72% of adults interviewed said that commercials are "a fair price to pay for free TV" (*America's Watching,* 1989). Assuming that this percentage is true for the current population of all adults, find the probability that in a random sample of 100 adults, exacty 76 will hold this view.

6.75 According to a 1990 report of the U.S. Department of Education, 71% of high school students graduate. Assuming that this percentage is true for the current population of all high school students, find the probability that in a random sample of 200 high school students, exactly 147 will graduate.

6.76 According to the U.S. Department of Labor Statistics, 65% of married women with children held jobs in 1988. Assuming that this percentage is true for the current population of all married women with children, find the probability that in a random sample of 140 such women, more than 84 will be holding jobs.

6.77 In a 1989 Louis Harris poll, 25% of adults said that children receive quality care while their parents are at work. Assuming that this percentage is true for the current population of all adults, find the probability that in a random sample of 150 adults, 26 to 39 will hold this view.

6.78 In a Roper Organization poll, when asked what would make life better, 60% of women responded "more money" (*The 1990 Virginia Slims Opinion Poll*). Assume that this result holds true for the current population of all women. Find the probability that in a random sample of 250 women, less than 154 hold this view.

6.79 In a Peter D. Hart Research Associates' survey, 47% of adults said that they "find solace in food when they're depressed" (*The American Way of Buying,* Dow Jones & Company, Inc., 1990). Assume that this result is true for the current population of all American adults. Find the probability that in a random sample of 400 American adults, exactly 176 will hold this view.

6.80 In a survey of more than 18,000 women, 46% said that religion plays "a central part" in their lives (*McCall's,* May 1989). Assuming that this percentage is true for the current population of all women, find the probability that in a random sample of 300 women, exactly 150 will say that religion plays "a central part" in their lives.

6.81 According to the U.S. Department of Labor, 25% of all workers in the United States are college graduates. Find the probability that in a random sample of 200 workers, 45 to 59 will be college graduates.

6.82 In a *Time*/CNN poll, 64% of respondents said that having an extramarital affair should not disqualify someone from holding a high position in government. Assuming that this percentage is true for the current population of all adults, find the probability that in a random sample of 80 adults, 42 to 56 will hold this view.

SELF-REVIEW TEST

1. The normal probability distribution is applied to

a. a continuous random variable
b. a discrete random variable
c. any random variable

2. For a continuous random variable, the probability of a single value of x is always

a. zero **b.** 1.0 **c.** between 0 and 1

3. Which of the following is not a characteristic of the normal distribution?

a. The total area under the curve is 1.0.
b. The curve is symmetric about the mean.
c. The two tails of the curve extend indefinitely.
d. The value of the mean is always greater than the value of the standard deviation.

4. The parameters of a normal distribution are

a. μ, z, and σ **b.** μ and σ **c.** μ, x, and σ

5. For the standard normal distribution,

a. $\mu = 0$ and $\sigma = 1$ **b.** $\mu = 1$ and $\sigma = 0$ **c.** $\mu = 100$ and $\sigma = 10$

6. The z value for μ for a normal curve is always

a. positive **b.** negative **c.** zero

7. For a normal curve, the z value for an x value that is less than μ is always

a. positive **b.** negative **c.** zero

8. Usually the normal distribution is used as an approximation to the binomial distribution when

a. $n \geq 30$ **b.** $np > 5$ and $nq > 5$ **c.** $n > 20$ and $p = .50$

9. Find the following probabilities for the standard normal distribution.

a. $P(.87 \leq z \leq 2.33)$ **b.** $P(-2.97 \leq z \leq 1.46)$
c. $P(z \leq -1.19)$ **d.** $P(z > -.71)$

10. Find the value of z for the standard normal curve such that the area

a. in the left tail is .1000
b. between 0 and z is .2291 and z is positive
c. in the right tail is .0500
d. between 0 and z is .3571 and z is negative

11. According to *presstime,* the mean base salary of newspaper editors was \$62,998 in 1989 (*Working Women,* January 1990). Assume that the current base salaries of all newspaper editors have an approximate normal distribution with a mean of \$62,998 and a standard deviation of \$10,145. Find the probability that the base salary of a randomly selected newspaper editor is

a. between \$51,800 and \$77,650 **b.** more than \$81,575
c. less than \$65,920 **d.** between \$37,000 and \$49,225

12. In a *Time*/CNN poll of 520 employed adults conducted by Yankelovich Clancy Shulman, 57% said that companies today are less loyal to their employees than 10 years ago (*Time*, September 11, 1989). Assuming that this percentage is true for the current population of all employed adults, find the probability that in a random sample of 200 such adults, the number who will hold this view is

a. 105 to 111 **b.** more than 128 **c.** exactly 110

USING MINITAB

To find a certain area under a normal curve by using MINITAB, we use the CDF command. CDF is an abbreviation of cumulative probability density. This command gives the total area under the normal curve to the left of an x or z value.

The following MINITAB commands will give the area under the standard normal curve to the left of a z value.

```
MTB  > CDF   z = k;   ← This command provides information about the value of z
SUBC > NORMAL   MEAN=0   SD=1.   ← This subcommand instructs MINITAB
                                   to use the normal distribution with
                                   μ = 0 and σ = 1
```

In the first command, k is a specific value of z. The semicolon at the end of the first command instructs MINITAB that a subcommand with more information is to follow. Illustrations M6-1 and M6-2 describe how to find area under the standard normal curve to the left of a point by using MINITAB.

ILLUSTRATION M6-1 Suppose we want to find the area under the standard normal curve to the left of $z = -2.25$. The MINITAB commands to find this area and the resulting MINITAB output are given in the following MINITAB display.

```
MTB  > NOTE: FINDING AREA TO THE LEFT OF z = -2.25
MTB  > CDF   z=-2.25;
SUBC > NORMAL   MEAN=0   SD=1.

         -2.2500      0.0122   ←—— This is the required area
```

Thus, area in the left tail of the standard normal curve to the left of $z = -2.25$ is .0122.

ILLUSTRATION M6-2 The following MINITAB commands will give us the area under the standard normal curve to the left of $z = 1.89$. The MINITAB output shows that this area is .9706.

```
MTB  > NOTE: FINDING AREA TO THE LEFT OF z = 1.89
MTB  > CDF   z=1.89;
SUBC > NORMAL   MEAN=0   SD=1.

         1.8900          0.9706   <—— This is the required area
```

Note that MINITAB gives the whole area to the left of a z value whereas Table VII of Appendix C gives the area between the mean and a z value. The area obtained in Illustration M6-2 is shown in Figure M6-1.

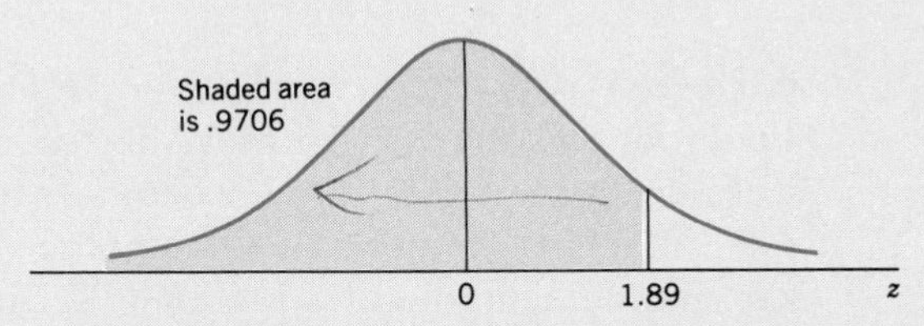

Figure M6-1 Area to the left of $z = 1.89$.

Now if we want to find the area under the standard normal curve between $z = 0$ and $z = 1.89$, we obtain this area by subtracting .5 from .9706. If we need to find the area to the right of $z = 1.89$, we subtract .9706 from 1.0, which is the total area under the normal curve.

The following MINITAB command and subcommand will help us to find the area under any normal curve.

```
MTB  > CDF   x = k;   <— This command provides information about the value of x
SUBC > NORMAL   MEAN=b   SD=c.   <— This subcommand instructs MINITAB
                                     to use the normal distribution with
                                     μ = b and σ = c
```

The first command lists the value of x. The semicolon at the end of the first command instructs MINITAB that a subcommand is to follow. This subcommand provides information about the mean and standard deviation of the normal distribution. In these MINITAB commands, k is a particular value of x, b is the value of the mean of the normal distribution, and c is the value of the standard deviation of the normal distribution. Again, note that MINITAB will give the area under the normal curve to the left of $x = k$. Illustrations M6-3 and M6-4 describe how we can find area under a normal curve to the left of a point by using MINITAB.

ILLUSTRATION M6-3 Let x denote the time taken to run a road race. Suppose x is approximately normally distributed with a mean of 195 minutes and a standard deviation of 21 minutes. Suppose we want to find the probability that a randomly selected runner will complete this road race in less than 150 minutes.

The probability of completing the road race in less than 150 minutes will be given by the area under the normal curve to the left of $x = 150$. The following MINITAB commands will give this area under a normal curve with $\mu = 195$ and $\sigma = 21$.

```
MTB  > NOTE: FINDING AREA TO THE LEFT OF x = 150
MTB  > CDF x=150;
SUBC > NORMAL  MEAN=195  SD=21.

   150.0000              0.0161  <--- This is the required area
      ↑
      └── This is the value of x
```

Thus, the probability that a randomly selected runner will complete this race in less than 150 minutes is .0161.

ILLUSTRATION M6-4 The mean SAT (Scholastic Aptitude Test) score in mathematics for all students who took this test in 1990 was 476. Assume that the 1990 SAT scores in mathematics for all examinees are normally distributed with a mean of 476 and a standard deviation of 80. Suppose we want to find the probability that the SAT score in mathematics of a randomly selected examinee who took this test in 1990 is less than 525.

The required probability is given by the area to the left of 525 under the given normal curve. The following MINITAB commands give the area to the left of $x = 525$ under the normal curve with $\mu = 476$ and $\sigma = 80$.

```
MTB  > NOTE: FINDING AREA TO THE LEFT OF x = 525
MTB  > CDF  x=525;
SUBC > NORMAL  MEAN=476  SD=80.

   525.0000              0.7299  <--- This is the required area
      ↑
      └── The value of x
```

Thus, the probability that a randomly selected examinee's 1990 SAT score in mathematics is less than 525 is .7299, as shown in Figure M6-2.

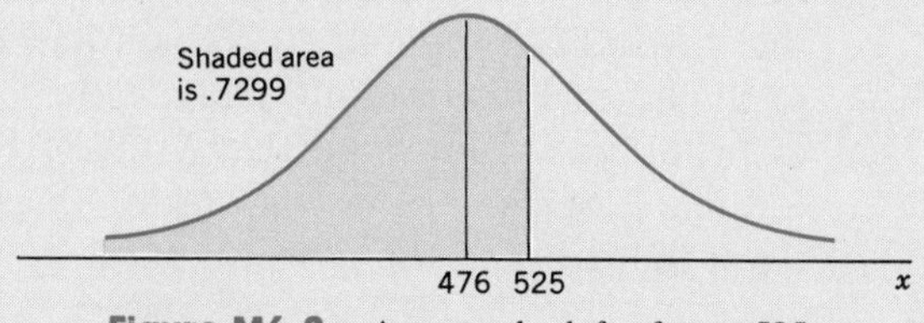

Figure M6-2 Area to the left of $x = 525$.

COMPUTER ASSIGNMENTS

M6.1 Find the area under the standard normal curve

a. to the left of -1.94 **b.** to the left of .83

M6.2 Find the following areas under a normal curve with $\mu = 86$ and $\sigma = 14$.

a. Area to the left of $x = 71$ **b.** Area to the left of $x = 96$

M6.3 The mean proficiency score in mathematics for 17-year-olds was 302 in 1986 (*Crossroads in American Education,* Educational Testing Service, 1989). Assume that the 1986 proficiency scores in mathematics for all 17-year-olds are normally distributed with a mean of 302 and a standard deviation of 45. Using MINITAB, find the percentage of 17-year-olds with mathematics proficiency scores

a. less than 270 **b.** less than 395

M6.4 According to a National Health and Nutrition Examination Survey, the mean height of American men aged 18 to 74 is 69.1 inches. Assume that the current heights of all American men aged 18 to 74 have a normal distribution with a mean of 69.1 inches and a standard deviation of 2.1 inches. Using MINITAB, find the percentage of 18- to 74-year-old men who are

a. less than 64.4 inches tall **b.** less than 71 inches tall

7 SAMPLING DISTRIBUTIONS

7.1 THE POPULATION AND SAMPLING DISTRIBUTIONS
7.2 THE SAMPLING ERROR
7.3 THE MEAN AND STANDARD DEVIATION OF $\bar{x}$
7.4 THE SHAPE OF THE SAMPLING DISTRIBUTION OF $\bar{x}$
7.5 CALCULATING THE PROBABILITY OF $\bar{x}$
7.6 POPULATION AND SAMPLE PROPORTIONS
7.7 THE MEAN, STANDARD DEVIATION, AND SHAPE OF THE SAMPLING DISTRIBUTION OF $\hat{p}$
7.8 CALCULATING THE PROBABILITY OF $\hat{p}$
SELF-REVIEW TEST

Chapters 5 and 6 discussed probability distributions of discrete and continuous random variables. This chapter extends the concept of probability distribution to that of a sample statistic. As was discussed in Chapter 3, a sample statistic is a numerical summary measure calculated for sample data. The mean, median, mode, and standard deviation calculated using sample data are called *sample statistics*. On the other hand, the same numerical summary measures calculated for population data are called *population parameters*. A population parameter is always a constant, whereas a sample statistic is always a random variable. The probability distribution of a sample statistic is more commonly called its *sampling distribution*. This chapter discusses the sampling distributions of the sample mean and the sample proportion. The concepts covered in this chapter are the foundation of the inferential statistics discussed in succeeding chapters.

7.1 THE POPULATION AND SAMPLING DISTRIBUTIONS

This section introduces the concepts of the population distribution and sampling distribution. Subsection 7.1.1 explains the population distribution, and subsection 7.1.2 describes the sampling distribution of $\bar{x}$.

7.1.1 THE POPULATION DISTRIBUTION

The **population distribution** is the probability distribution derived from the information on all elements of a population.

POPULATION DISTRIBUTION

The population distribution is the probability distribution of the population data.

Suppose there are only five students in an advanced statistics class and the midterm scores of these five students are

70 78 80 80 95

Let x denote the score of a student. Using single-valued classes (as there are only five data values, there is no need to group them), we can write the frequency distribution of scores as in Table 7.1. Dividing the frequencies of classes by the population

Table 7.1 The Population Frequency Distribution

x	f
70	1
78	1
80	2
95	1
	$N = 5$

size, we obtain the probabilities of those classes. Table 7.2, which lists the probabilities of various x values, presents the probability distribution of the population.

Table 7.2 The Population Probability Distribution

x	$P(x)$
70	$1/5 = .20$
78	$1/5 = .20$
80	$2/5 = .40$
95	$1/5 = .20$
	Sum $= 1.0$

The values of the mean and standard deviation calculated for the probability distribution of Table 7.2 gives the values of the population parameters μ and σ. These values are $\mu = 80.60$ and $\sigma = 8.09$. The values of μ and σ for the probability distribution of Table 7.2 can be calculated using the formulas given in Sections 5.3 and 5.4 of Chapter 5 (see Exercise 7.3).

7.1.2 THE SAMPLING DISTRIBUTION

As mentioned in the beginning of this chapter, the value of a population parameter is always constant. For example, for any population data set, there is only one value of the population mean μ. However, we cannot say the same about the sample mean $\bar{x}$. We would expect different samples of the same size drawn from the same population to yield different values of the sample mean $\bar{x}$. The value of the sample mean for any one sample will depend on the elements included in that sample. Consequently, *the sample mean $\bar{x}$ is a random variable*. Therefore, like other random variables, the sample mean $\bar{x}$ possesses a probability distribution, which is more commonly called the **sampling distribution of $\bar{x}$**. Other sample statistics such as the median, mode, and standard deviation also possess sampling distributions.

SAMPLING DISTRIBUTION OF $\bar{x}$

The probability distribution of $\bar{x}$ is called the sampling distribution of $\bar{x}$. It lists the various values that $\bar{x}$ can assume and the probability for each value of $\bar{x}$.

In general, the probability distribution of a sample statistic is called its sampling distribution.

Reconsider the population of midterm scores of five students given in Table 7.1. Let us consider all the possible samples of three scores each that can be drawn without replacement, from that population. The total number of possible samples, given by the combinations formula discussed in Chapter 5, is 10, that is,

$$\text{Total number of samples} = \binom{5}{3} = 10$$

The value of $\binom{5}{3}$ is obtained from Table III of Appendix C.

Suppose we assign letters A, B, C, D, and E to the scores of five students so that:

$$A = 70, \quad B = 78, \quad C = 80, \quad D = 80, \quad E = 95$$

Then, the 10 possible samples of three scores each are

ABC, ABD, ABE, ACD, ACE, ADE, BCD, BCE, BDE, CDE

These 10 samples and their respective means are listed in Table 7.3. Note that the first two samples have the same three scores. The reason for this is that two of the students (C and D) have the same score and, hence, the samples ABC and ABD contain the same values. The mean of each sample is obtained by dividing the sum of the three scores included in that sample by 3. For instance, the mean of the first

Table 7.3 All Possible Samples and Their Means When the Sample Size is 3

Sample	Scores in the Sample	$\bar{x}$
ABC	70, 78, 80	76.00
ABD	70, 78, 80	76.00
ABE	70, 78, 95	81.00
ACD	70, 80, 80	76.67
ACE	70, 80, 95	81.67
ADE	70, 80, 95	81.67
BCD	78, 80, 80	79.33
BCE	78, 80, 95	84.33
BDE	78, 80, 95	84.33
CDE	80, 80, 95	85.00

sample is $(70 + 78 + 80)/3 = 76$. Note that the values of $\bar{x}$ in Table 7.3 are rounded to two decimal places.

Table 7.4 The Frequency Distribution of $\bar{x}$ When the Sample Size is 3

$\bar{x}$	f
76.00	2
76.67	1
79.33	1
81.00	1
81.67	2
84.33	2
85.00	1

Using the values of $\bar{x}$ given in Table 7.3, we record the frequency distribution of $\bar{x}$ in Table 7.4. By dividing the frequency of each value of $\bar{x}$ by the sum of the frequencies, we obtain the probabilities of various values of $\bar{x}$. These probabilities are listed in Table 7.5. This table gives the sampling distribution of $\bar{x}$.

Table 7.5 The Sampling Distribution of $\bar{x}$ When the Sample Size is 3

$\bar{x}$	$P(\bar{x})$
76.00	$2/10 = .20$
76.67	$1/10 = .10$
79.33	$1/10 = .10$
81.00	$1/10 = .10$
81.67	$2/10 = .20$
84.33	$2/10 = .20$
85.00	$1/10 = .10$
	$\Sigma P(\bar{x}) = 1.0$

If we draw just one sample of three scores from the population of five scores, we may draw any of the 10 possible samples. Hence, the sample mean $\bar{x}$ can assume any of the values listed in Table 7.5 with the corresponding probability. For instance, the probability that the mean of a randomly drawn sample of 3 scores is 81.67 is .20. This can be written as

$$P(\bar{x} = 81.67) = .20$$

7.2 THE SAMPLING ERROR

Usually, different samples taken from the same population will give different results because they contain different elements. This is obvious from Table 7.3, which shows that the mean of a sample of three scores depends on which three of the five scores are included in the sample. The result obtained from any one sample will generally be different from the one obtained from the population. The difference between the value of a sample statistic obtained from a sample and the value of the corresponding population parameter obtained from the population is called the **sampling error.** Note that this difference represents the sampling error only if the sample is random and no nonsampling error has been made. Otherwise only a part of this difference will be due to sampling error.

SAMPLING ERROR

Sampling error is the difference between the value of a sample statistic and the value of the corresponding population parameter. In the case of the mean,

$$\text{Sampling error} = \bar{x} - \mu$$

assuming that the sample is random and no nonsampling error has been made.

It is important to remember that a sampling error occurs because of chance. The errors that occur for other reasons, such as errors made during collection, recording, and tabulation of data, are called **nonsampling errors.**† Nonsampling errors can occur both in a sample survey and in a census, whereas the sampling error occurs in a sample survey only. Nonsampling errors can be minimized by preparing the survey questionnaire carefully and handling the data cautiously. However, it is impossible to avoid the sampling error. Example 7-1 illustrates the concept of the sampling error using the mean.

Illustrating sampling error.

EXAMPLE 7-1 Reconsider the population of five scores given in Table 7.1. The five scores are 70, 78, 80, 80, and 95. The population mean is

$$\mu = (70 + 78 + 80 + 80 + 95)/5 = 80.60$$

Now suppose we take a random sample of three scores from this population. Assume that the sample drawn is the one that includes the scores 70, 80, and 95. The mean $\bar{x}$ for this sample is

$$\bar{x} = (70 + 80 + 95)/3 = 81.67$$

Consequently

$$\text{Sampling error} = \bar{x} - \mu = 81.67 - 80.60 = 1.07$$

†The nonsampling errors are discussed in more detail in Section A.4 of Appendix A.

Now suppose we take another random sample of size 3 from the same population of five scores. Assume that this time the scores included in the sample are 78, 80, and 80. The mean for this sample is

$$\bar{x} = (78 + 80 + 80)/3 = 79.33$$

This time,

$$\text{Sampling error} = \bar{x} - \mu = 79.33 - 80.60 = -1.27$$

Note that in each of these two cases, the sampling error occurred because of chance and not because of human error. ■

Case Study 7-1 offers examples of nonsampling errors. A large portion of the discrepancies that occurred in the opinion poll results and the actual results mentioned in this case study happened because of reasons other than chance. For example, the errors (or discrepancies) arising because of factors such as voters concealing the truth, antimedia sentiments of voters, voters not making up their minds until the last moment, are nonsampling errors. Consequently, a major portion of the discrepancies discussed in this case study occurred because of nonsampling errors.

CASE STUDY 7-1 SILBER'S VICTORY RENEWS QUESTIONS ON POLL ANSWERS

Democrat John Silber's unexpected victory in the Massachusetts gubernatorial primary is renewing questions about whether voters lie to pollsters.

The same questions followed the careers of Barry Goldwater and George Wallace, whose support was understated in some polls.

Silber, on leave as president of Boston University, defeated former Attorney General Francis X. Bellotti by a comfortable 9-point margin, winning 53 percent of the vote in Tuesday's Democratic primary.

Polls published just before the election showed Bellotti ahead and "surging." One poll gave Bellotti a lead of 50 percent to 39 percent. An election-eve survey showed Bellotti leading 49 percent to 35 percent.

Gerry Chervinsky, whose polls at KRC Communications Research in Cambridge showed Bellotti in the lead close to the primary, believes some poll respondents concealed their support for Silber.

They liked Silber's combative style and confrontational rhetoric, but were not willing to admit it, he believes.

Other factors also contributed to the polling problem, such as the difficulties of predicting turnout in a primary and of determining whether independent voters would take a Democratic or Republican ballot, Chervinsky said.

But Silber, in a television interview Sunday, said his showing and election results elsewhere in the nation reflected a growing "anti-media sentiment." . . .

Exit polls indicated that as many as 20 percent of the voters made up their minds in the last days of the campaign and overwhelmingly supported Silber. But Chervinsky said voters who concealed their support for Silber could have skewed polls.

EXERCISES

7.1 Briefly explain the meaning of population distribution and sampling distribution. Give an example of each.

7.2 Briefly explain the meaning of sampling error and nonsampling errors. Give an example of sampling error.

7.3 Using the formulas of Sections 5.3 and 5.4 of Chapter 5 for the mean and standard deviation of a discrete random variable, verify that the mean and standard deviation for the probability distribution of Table 7.2 are 80.60 and 8.09, respectively.

7.4 The following data give the ages of all six members of a family.

55 53 28 25 19 15

a. Let x denote the age of a member of this family. Write the population distribution of x.

b. Take all the possible samples of size five (without replacement) from this population. Write the sampling distribution of $\bar{x}$.

c. Calculate the mean for the population data. Take a random sample of size five and calculate the sample mean $\bar{x}$. Compute the sampling error.

7.5 The following data give the years of experience for all five faculty members of a department at a university.

7 8 12 7 5

a. Let x denote the years of teaching experience for a faculty member of this department. Write the population distribution of x.

b. Take all the possible samples of size four (without replacement) from this population. Write the sampling distribution of $\bar{x}$.

c. Calculate the mean for the population data. Take a random sample of size four and calculate the sample mean $\bar{x}$. Compute the sampling error.

7.3 THE MEAN AND STANDARD DEVIATION OF $\bar{x}$

The mean and standard deviation calculated for the sampling distribution of $\bar{x}$ are called the **mean** and **standard deviation of $\bar{x}$.** Actually, the mean of $\bar{x}$ is the mean of all sample means and the standard deviation of $\bar{x}$ is the standard deviation of the means of all samples. The standard deviation of $\bar{x}$ is also called the *standard error of $\bar{x}$.*

MEAN AND STANDARD DEVIATION OF $\bar{x}$

The mean and standard deviation of the sampling distribution of $\bar{x}$ are called the mean and standard deviation of $\bar{x}$ and are denoted by $\mu_{\bar{x}}$ and $\sigma_{\bar{x}}$, respectively.

If we calculate the mean and standard deviation of the 10 values of $\bar{x}$ listed in Table 7.3, we obtain the mean $\mu_{\bar{x}}$ and the standard deviation $\sigma_{\bar{x}}$ of $\bar{x}$. Alternatively, we can calculate the mean and standard deviation of the sampling distribution of $\bar{x}$ listed in Table 7.5. These will also be the values of $\mu_{\bar{x}}$ and $\sigma_{\bar{x}}$. From these calculations, we obtain $\mu_{\bar{x}} = 80.60$ and $\sigma_{\bar{x}} = 3.30$ (see Exercise 7.13 at the end of this section).

The mean of the sampling distribution of $\bar{x}$ is always the same as the mean of the population.

MEAN OF THE SAMPLING DISTRIBUTION OF $\bar{x}$

The mean of the sampling distribution of $\bar{x}$ is equal to the mean of the population. Thus,

$$\mu_{\bar{x}} = \mu$$

Hence, if we take all possible samples (of the same size) from a population and calculate their means, the mean $\mu_{\bar{x}}$ of all these sample means will be the same as the mean μ of the population. If we calculate the mean μ for the population probability distribution of Table 7.2 and the mean $\mu_{\bar{x}}$ for the sampling distribution of Table 7.5 by using the formula learned in Section 5.3 of Chaper 5, we get the same value of 80.60 for μ and $\mu_{\bar{x}}$ (see Exercise 7.13).

However, the standard deviation $\sigma_{\bar{x}}$ of $\bar{x}$ is not the same as the standard deviation σ of the population distribution (unless $n = 1$). The standard deviation of $\bar{x}$ is equal to the standard deviation of the population divided by the square root of the sample size. That is,

$$\sigma_{\bar{x}} = \frac{\sigma}{\sqrt{n}}$$

This formula for the standard deviation of $\bar{x}$ holds true only when the sampling is done either with replacement from a finite population or with or without replacement from an infinite population. These two conditions can be replaced by the condition that the sample size is small in comparison to the population size. The sample size is considered to be small compared to the population size if the sample size is equal to or less than 5% of the population size, that is, if $n/N \leq .05$. If this condition is not satisfied, we use a different formula for $\sigma_{\bar{x}}$.† In most practical applications, the sample

†If n/N is greater than .05, then $\sigma_{\bar{x}}$ is calculated as

$$\sigma_{\bar{x}} = \frac{\sigma}{\sqrt{n}} \sqrt{\frac{N-n}{N-1}}$$

where the factor

$$\sqrt{\frac{N-n}{N-1}}$$

is called the finite population correction factor. However, in this text we will not use this formula. The interested reader should see Exercise 7.13 at the end of this section.

size is usually small compared to the population size. Hence, in this text the formula used for calculating $\sigma_{\bar{x}}$ will be $\sigma_{\bar{x}} = \sigma/\sqrt{n}$.

STANDARD DEVIATION OF THE SAMPLING DISTRIBUTION OF $\bar{x}$

The standard deviation of the sampling distribution of $\bar{x}$ is

$$\sigma_{\bar{x}} = \frac{\sigma}{\sqrt{n}}$$

where σ is the standard deviation of the population and n is the sample size. This formula is used when $n/N \leq .05$, where N is the population size.

Following are two important observations regarding the sampling distribution of $\bar{x}$.

1. *The spread of the sampling distribution of $\bar{x}$ is smaller than the spread of the corresponding population distribution.* In other words, $\sigma_{\bar{x}} < \sigma$. This is obvious from the formula for $\sigma_{\bar{x}}$. When n is greater than 1, which is usually true, the denominator in $\sigma/\sqrt{n}$ is greater than 1. Hence, $\sigma_{\bar{x}}$ is smaller than σ.

2. *The standard deviation of the sampling distribution of $\bar{x}$ decreases as the sample size increases.* This feature of the sampling distribution of $\bar{x}$ is also obvious from the formula

$$\sigma_{\bar{x}} = \frac{\sigma}{\sqrt{n}}$$

As n increases, the value of $\sqrt{n}$ also increases and, hence, $\sigma/\sqrt{n}$ decreases. Thus, the value of $\sigma_{\bar{x}}$ will decrease with an increase in the sample size. Example 7-2 illustrates this feature.

Mean and standard deviation of $\bar{x}$.

EXAMPLE 7-2 The mean wage per hour for all 5000 employees working in a large company is $13.50 with a standard deviation of $2.90. Let $\bar{x}$ be the mean wage per hour for a random sample of certain employees taken from this company. Find the mean and standard deviation of $\bar{x}$ for a sample size of

(a) 30 (b) 75 (c) 200

Solution From the given information, for the population of all employees,

$$N = 5000, \quad \mu = \$13.50, \quad \text{and} \quad \sigma = \$2.90$$

(a) The mean $\mu_{\bar{x}}$ of the sampling distribution of $\bar{x}$ is

$$\mu_{\bar{x}} = \mu = \$13.50$$

In this case, $n = 30$, $N = 5000$, and $n/N = 30/5000 = .006$. As n/N is less than .05, the standard deviation of $\bar{x}$ is obtained by using the formula $\sigma/\sqrt{n}$.

Hence,

$$\sigma_{\bar{x}} = \frac{\sigma}{\sqrt{n}} = \frac{2.90}{\sqrt{30}} = \$.53$$

Thus, we can state that if we take all possible samples of size 30 from the population of all employees of this company and prepare the sampling distribution of $\bar{x}$, the mean and standard deviation of this sampling distribution of $\bar{x}$ will be \$13.50 and \$.53, respectively.

(b) In this case, $n = 75$ and $n/N = 75/5000 = .015$. The mean and standard deviation of $\bar{x}$ are

$$\mu_{\bar{x}} = \mu = \$13.50 \quad \text{and} \quad \sigma_{\bar{x}} = \frac{\sigma}{\sqrt{n}} = \frac{2.90}{\sqrt{75}} = \$.33$$

(c) In this case, $n = 200$. Consequently, $n/N = 200/5000 = .04$, which is less than .05. Therefore, the mean and standard deviation of $\bar{x}$ are

$$\mu_{\bar{x}} = \mu = \$13.50 \quad \text{and} \quad \sigma_{\bar{x}} = \frac{\sigma}{\sqrt{n}} = \frac{2.90}{\sqrt{200}} = \$.21$$

From the above illustration we observe that the mean of the sampling distribution of $\bar{x}$ is always equal to the mean of the population whatever the size of the sample. However, the value of the standard deviation of $\bar{x}$ decreases from \$.53 to \$.33 and then to \$.21 as the sample size increases from 30 to 75 and then to 200. ■

EXERCISES

7.6 Consider a large population with $\mu = 60$ and $\sigma = 12$. Assuming $n/N \leq .05$, find the mean and standard deviation of the sample mean $\bar{x}$ for a sample size of

a. 18 **b.** 90

7.7 Consider a large population with $\mu = 90$ and $\sigma = 16$. Assuming $n/N \leq .05$, find the mean and standard deviation of the sample mean $\bar{x}$ for a sample size of

a. 10 **b.** 35

7.8 According to the U.S. Bureau of Labor Statistics, the mean number of hours worked per week by workers employed in the private (nonagricultural) industrial sector was 34.7 in 1989. Assume that for the hours worked per week by the current population of all workers employed in the private (nonagricultural) industrial sector: $\mu = 34.7$ and $\sigma = 4$. Let $\bar{x}$ be the mean number of hours worked per week by a random sample of 40 such workers. Find the mean and standard deviation of $\bar{x}$.

7.9 According to the U.S. Department of Agriculture, the mean size of a farm in 1987 was 461 acres. Assume that for the sizes of all current farms: $\mu = 461$ acres and $\sigma = 140$ acres. Let $\bar{x}$ be the mean size of a farm for a random sample of 500 farms selected from the current population of all farms. Find the mean and standard deviation of $\bar{x}$.

7.10 According to a National Health and Nutrition Examination Survey, the mean height of American women aged 18 to 74 is 63.7 inches. Assume that for the heights of the current population of all American women aged 18 to 74: $\mu = 63.7$ and $\sigma = 1.4$ inches. Let $\bar{x}$ be the mean height of a random sample of 900 women aged 18 to 74. Find the mean and standard deviation of $\bar{x}$.

7.11 According to a National Health and Nutrition Examination Survey, the mean weight of American men aged 18 to 74 is 172.2 pounds. Assume that for the weights of the current population of all American men aged 18 to 74: $\mu = 172.2$ pounds and $\sigma = 20$ pounds. Let $\bar{x}$ be the mean weight of a random sample of 70 men aged 18 to 74. Find the mean and standard deviation of $\bar{x}$.

7.12 The mean salary for all 1050 professors at a university is $47,600 per annum, and the standard deviation of their salaries is $7200. Let $\bar{x}$ be the mean salary of a random sample of 16 professors selected from this university. Find the mean and standard deviation of $\bar{x}$.

***7.13** Consider the sampling distribution of $\bar{x}$ given in Table 7.5 on page 342.

a. Calculate the value of $\mu_{\bar{x}}$ using the formula: $\mu_{\bar{x}} = \Sigma \bar{x} P(\bar{x})$. Is the value of μ calculated in Exercise 7.3 the same as the value of $\mu_{\bar{x}}$ calculated in this exercise?

b. Calculate the value of $\sigma_{\bar{x}}$ using the formula

$$\sigma_{\bar{x}} = \sqrt{\Sigma \bar{x}^2 P(\bar{x}) - (\mu_{\bar{x}})^2}$$

c. From Exercise 7.3, $\sigma = 8.09$. Also, our sample size is 3 so that $n = 3$. Therefore, $\sigma/\sqrt{n} = 8.09/\sqrt{3} = 4.67$. From part (b), you should get $\sigma_{\bar{x}} = 3.30$. Why does $\sigma/\sqrt{n}$ not equal $\sigma_{\bar{x}}$ in this case?

d. In our example (given in the beginning of Section 7.1.1) on scores, $N = 5$, and $n = 3$. Hence, $n/N = 3/5 = .60$. Because n/N is greater than .05, the formula for $\sigma_{\bar{x}}$ will be

$$\sigma_{\bar{x}} = \frac{\sigma}{\sqrt{n}} \sqrt{\frac{N - n}{N - 1}}$$

Show that the value of $\sigma_{\bar{x}}$ calculated by using this formula will give the same value as the one calculated in part (b).

7.4 THE SHAPE OF THE SAMPLING DISTRIBUTION OF $\bar{x}$

The shape of the sampling distribution of $\bar{x}$ can be discussed with regard to the following two cases.

1. The population from which the samples are drawn is normally distributed.
2. The population from which the samples are drawn is nonnormally distributed.

7.4.1 SAMPLING FROM A NORMALLY DISTRIBUTED POPULATION

When the population from which samples are drawn is normally distributed with its mean equal to μ and standard deviation equal to σ, then

1. The mean of $\bar{x}$ is $\mu_{\bar{x}}$, which is equal to μ.
2. The standard deviation of $\bar{x}$ is $\sigma_{\bar{x}}$, which is equal to $\sigma/\sqrt{n}$. (We assume that $n/N \leq .05$.)
3. The shape of the sampling distribution of $\bar{x}$ is normal, whatever the value of n.

SAMPLING DISTRIBUTION OF $\bar{x}$ WHEN THE POPULATION IS NORMALLY DISTRIBUTED

If the population from which the samples are drawn is normally distributed with mean μ and standard deviation σ, then the sample mean $\bar{x}$ will also be normally distributed with the following mean and standard deviation, irrespective of the sample size.

$$\mu_{\bar{x}} = \mu \quad \text{and} \quad \sigma_{\bar{x}} = \frac{\sigma}{\sqrt{n}}$$

Remember, for $\sigma_{\bar{x}} = \sigma/\sqrt{n}$ to be true, n/N must be less than or equal to .05.

Figure 7.1*a* shows the probability distribution curve for a population. The curves in Figure 7.1*b* through Figure 7.1*e* show the sampling distributions of $\bar{x}$ for different sample sizes taken from the population of Figure 7.1*a*. As we can observe, the population is normally distributed. Because of this, the sampling distribution of $\bar{x}$ is normal for each of the four cases illustrated in parts *b* through *e*. Also notice from Figure

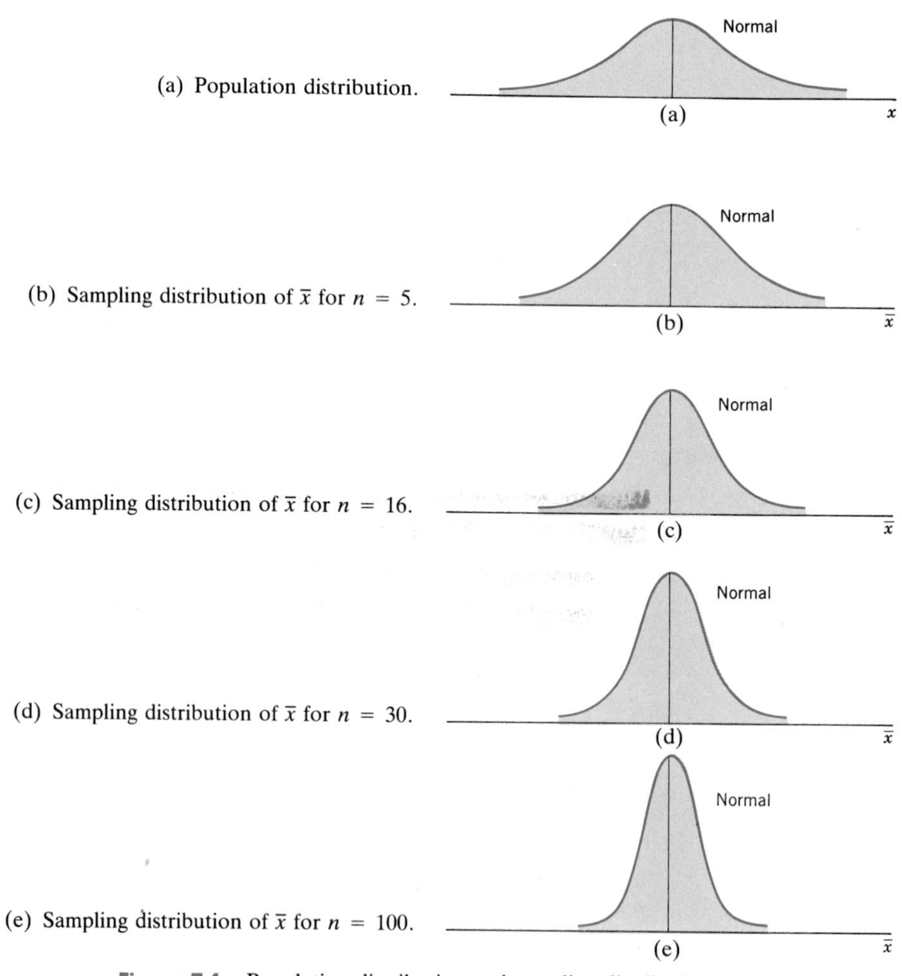

Figure 7.1 Population distribution and sampling distributions of $\bar{x}$.

7.1*b* through Figure 7.1*e* that the spread of the sampling distribution of $\bar{x}$ decreases as the sample size increases.

Example 7-3 illustrates the calculation of the mean and standard deviation of $\bar{x}$ and the description of the shape of its sampling distribution.

Sampling distribution, mean, and standard deviation of $\bar{x}$: normal population.

EXAMPLE 7-3 According to the Hertz Corporation, Americans traveled an average of 15,257 miles in 1986, counting commutation, business, vacation, and personal trips. Suppose that the distribution of miles traveled by all Americans in 1986 is normally distributed with a mean of 15,257 miles and a standard deviation of 4250 miles. Let $\bar{x}$ be the mean number of miles traveled by a random sample of certain Americans. Calculate the mean and standard deviation of $\bar{x}$ and describe the shape of its sampling distribution when the sample size is

(a) 10 (b) 50 (c) 1000

Solution

(a) The mean and standard deviation of $\bar{x}$ are

$$\mu_{\bar{x}} = \mu = 15{,}257 \text{ miles}$$

$$\sigma_{\bar{x}} = \frac{\sigma}{\sqrt{n}} = \frac{4250}{\sqrt{10}} = 1343.97 \text{ miles}$$

As the mileage traveled by all Americans is assumed to be normally distributed, the sampling distribution of $\bar{x}$ for samples of 10 people will also be normal. Figure 7.2 shows the population distribution and the sampling distribution of $\bar{x}$. Note that because the value of σ is greater than that of $\sigma_{\bar{x}}$, the population distribution has a wider spread and a smaller height than the sampling distribution of $\bar{x}$ in Figure 7.2.

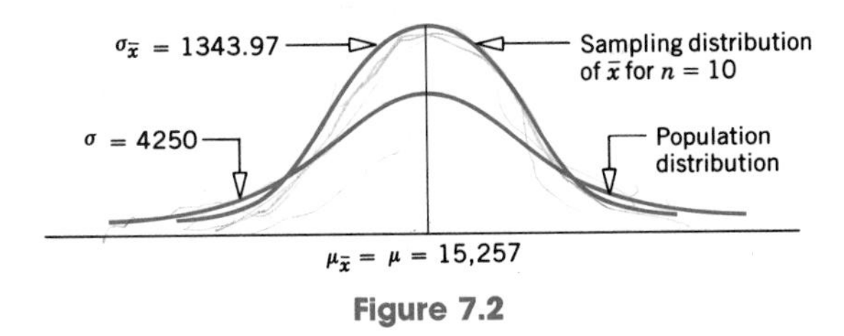

Figure 7.2

(b) The mean and standard deviation of $\bar{x}$ are

$$\mu_{\bar{x}} = \mu = 15{,}257 \text{ miles}$$

$$\sigma_{\bar{x}} = \frac{\sigma}{\sqrt{n}} = \frac{4250}{\sqrt{50}} = 601.04 \text{ miles}$$

Again, because the mileage traveled by all Americans is assumed to be normally distributed, the sampling distribution of $\bar{x}$ for samples of 50 people will also be normal. The population distribution and the sampling distribution of $\bar{x}$ are shown in Figure 7.3.

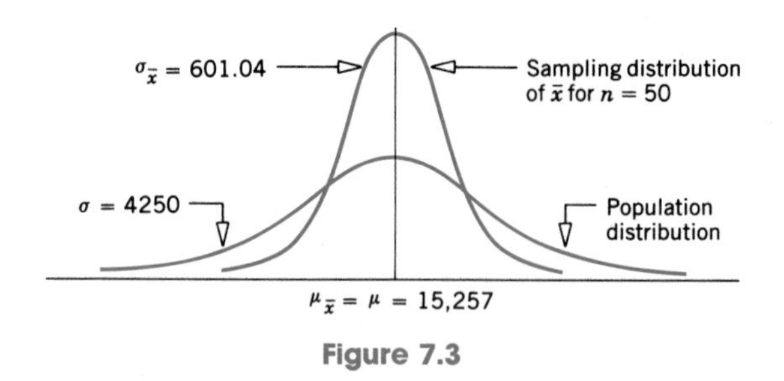

Figure 7.3

(c) The mean and standard deviation of $\bar{x}$ are

$$\mu_{\bar{x}} = \mu = 15{,}257 \text{ miles}$$

$$\sigma_{\bar{x}} = \frac{\sigma}{\sqrt{n}} = \frac{4250}{\sqrt{1000}} = 134.40 \text{ miles}$$

Again, because the mileage traveled by all Americans is assumed to be normally distributed, the sampling distribution of $\bar{x}$ for samples of 1000 people will also be normal. The two distributions are shown in Figure 7.4.

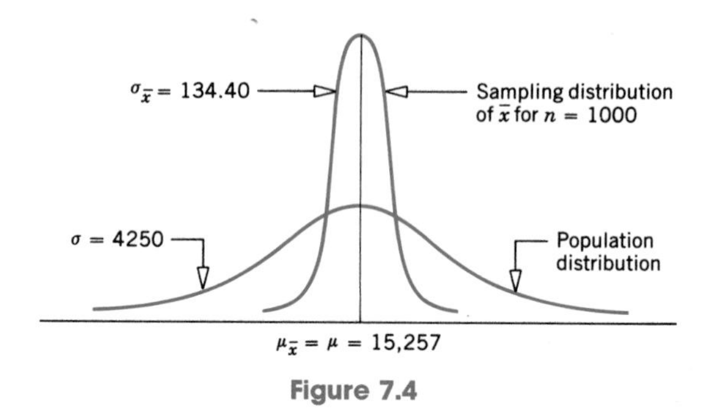

Figure 7.4

Thus, whatever the sample size, the sampling distribution of $\bar{x}$ will have a normal distribution if the population from which the samples are drawn is normally distributed. ■

7.4.2 SAMPLING FROM A NONNORMALLY DISTRIBUTED POPULATION

Most of the time the population from which the samples are taken is not normally distributed. In such cases, the shape of the sampling distribution of $\bar{x}$ is inferred from a very important theorem called the **central limit theorem.**

CENTRAL LIMIT THEOREM

According to the central limit theorem, for a large sample size, the sampling distribution of the sample mean $\bar{x}$ is approximately normal, irrespective of the shape of the population distribution. The mean and standard deviation of the sampling distribution of $\bar{x}$ are

$$\mu_{\bar{x}} = \mu \qquad \text{and} \qquad \sigma_{\bar{x}} = \frac{\sigma}{\sqrt{n}}$$

The sample size is usually considered to be large if $n \geq 30$.

Note that when the population is not normally distributed, the shape of the sampling distribution is not exactly normal but approximately normal. The approximation becomes more accurate as the sample size increases. Another point to remember is that the central limit theorem applies to *large* samples only. Usually, if the sample size is 30 or more, it is considered sufficiently large to apply the central limit theorem. Thus, according to the central limit theorem:

1. The shape of the sampling distribution of $\bar{x}$ is approximately normal, irrespective of the shape of the population distribution, when $n \geq 30$.

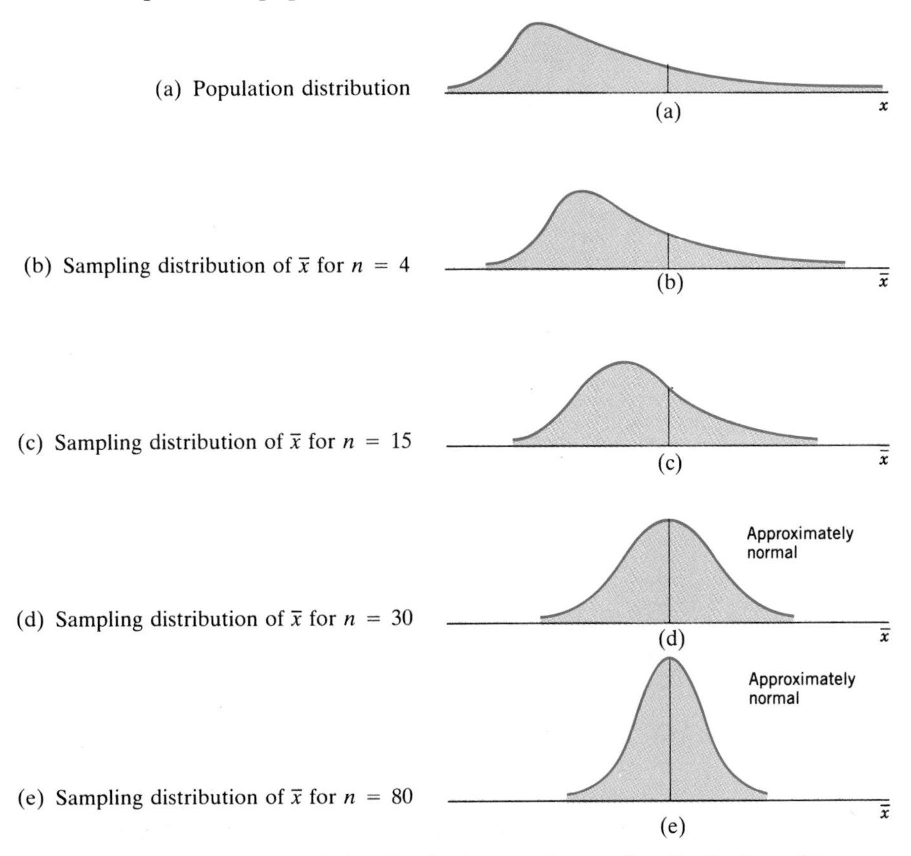

Figure 7.5 The population distribution and the sampling distributions of $\bar{x}$.

2. The mean of $\bar{x}$ is $\mu_{\bar{x}}$, which is equal to μ.
3. The standard deviation of $\bar{x}$ is $\sigma_{\bar{x}}$, which is equal to $\sigma/\sqrt{n}$.

Again, remember that for $\sigma_{\bar{x}} = \sigma/\sqrt{n}$ to be true, n/N must be less than or equal to .05.

Figure 7.5*a* shows the probability distribution curve for a population. The curves in Figure 7.5*b* through Figure 7.5*e* show the sampling distributions of $\bar{x}$ for different sample sizes taken from the population of Figure 7.5*a*. As we can observe, the population is not normally distributed. The sampling distributions of $\bar{x}$ shown in parts *b* and *c*, when $n < 30$, are not normal. However the sampling distributions of $\bar{x}$ shown in parts *d* and *e*, when $n \geq 30$, are (approximately) normal. Also notice that the spread of the sampling distribution of $\bar{x}$ decreases as the sample size increases.

Example 7-4 illustrates the calculation of the mean and standard deviation of $\bar{x}$ and describes the shape of the sampling distribution of $\bar{x}$ when the sample size is large.

Sampling distribution, mean, and standard deviation of $\bar{x}$: nonnormal population.

EXAMPLE 7-4 The mean rent paid by all tenants in a large city is \$950 with a standard deviation of \$225. However, the population distribution of rents for all tenants in this city is skewed to the right. Calculate the mean and standard deviation of $\bar{x}$ and describe the shape of its sampling distribution when the sample size is

(a) 30 (b) 100

Solution Although the population distribution of rents paid by all tenants is nonnormal, in each case the sample size is large ($n \geq 30$). Hence, the central limit theorem can be applied to infer about the shape of the sampling distribution of $\bar{x}$.

(a) Let $\bar{x}$ be the mean rent paid by a sample of 30 tenants. Then, the sampling distribution of $\bar{x}$ is approximately normal with the values of the mean and standard deviation as

$$\mu_{\bar{x}} = \mu = \$950 \quad \text{and} \quad \sigma_{\bar{x}} = \frac{\sigma}{\sqrt{n}} = \frac{225}{\sqrt{30}} = \$41.08$$

Figure 7.6 shows the population distribution and the sampling distribution of $\bar{x}$.

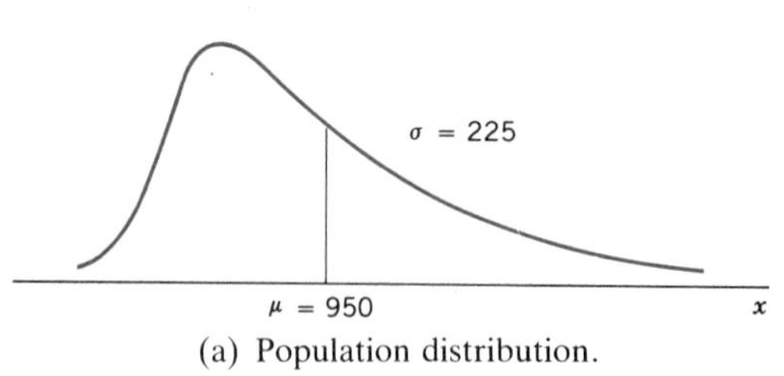

(a) Population distribution.

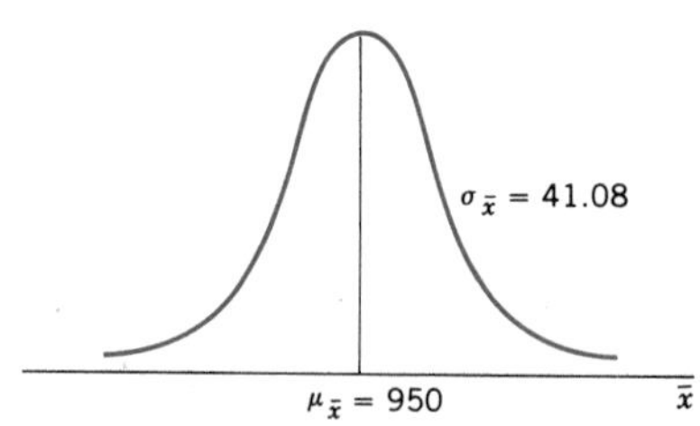

(b) Sampling distribution of $\bar{x}$ for $n = 30$.

Figure 7.6

(b) Let $\bar{x}$ be the mean rent paid by a sample of 100 tenants. Then, the sampling distribution of $\bar{x}$ is approximately normal with the values of the mean and standard deviation as

$$\mu_{\bar{x}} = \mu = \$950 \quad \text{and} \quad \sigma_{\bar{x}} = \frac{\sigma}{\sqrt{n}} = \frac{225}{\sqrt{100}} = \$22.50$$

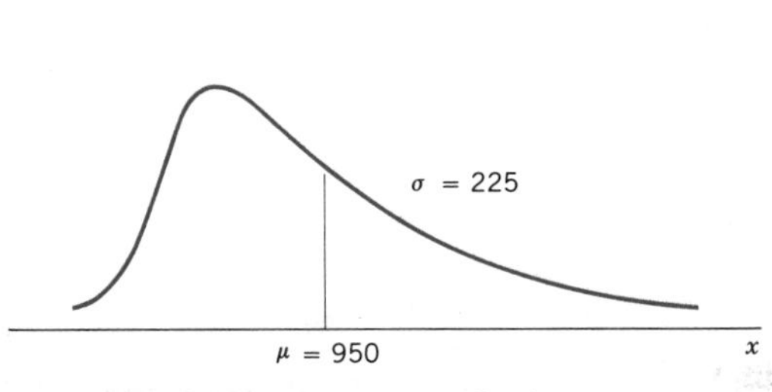

(a) Population distribution.

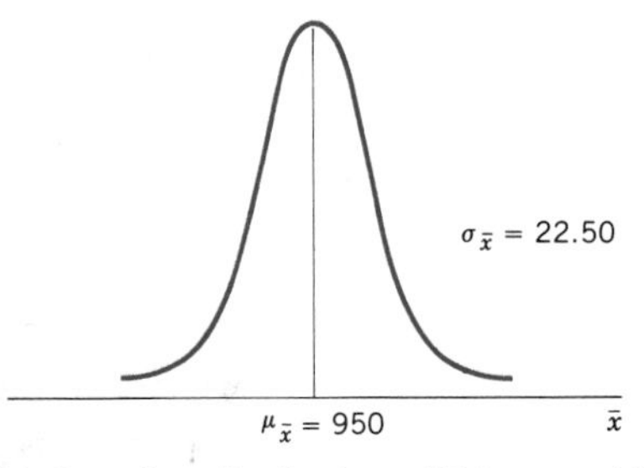

(b) Sampling distribution of $\bar{x}$ for $n = 100$.

Figure 7.7

Figure 7.7 shows the population distribution and the sampling distribution of $\bar{x}$. ■

EXERCISES

7.14 The time taken to complete a history test by all students is normally distributed with a mean of 120 minutes and a standard deviation of 10 minutes. Let $\bar{x}$ be the mean time taken to complete this history test for a random sample of 16 students. Calculate the mean and standard deviation of $\bar{x}$ and describe the shape of its sampling distribution.

7.15 The speeds of all cars traveling on a stretch of Interstate Highway I-95 are normally distributed with a mean of 68 miles per hour and a standard deviation of 3 miles. Let $\bar{x}$ be the mean speed of a random sample of 20 cars traveling on this highway. Calculate the mean and standard deviation of $\bar{x}$ and describe the shape of its sampling distribution.

7.16 The amounts of electric bills for all households in a city have an approximate normal distribution with a mean of \$42 and a standard deviation of \$7. Let $\bar{x}$ be the mean amount of electric bills for a random sample of 25 households selected from this city. Find the mean and standard deviation of $\bar{x}$ and comment on the shape of its sampling distribution.

7.17 The GPAs of all 5540 students at a university have an approximate normal distribution with a mean of 3.02 and a standard deviation of .29. Let $\bar{x}$ be the mean GPA of a random sample of 48 students selected from this university. Find the mean and standard deviation of $\bar{x}$ and comment on the shape of its sampling distribution.

7.18 The weights of all people living in a town have a distribution that is skewed to the right with a mean of 133 pounds and a standard deviation of 24 pounds. Let $\bar{x}$ be the mean weight of a random sample of 45 persons selected from this town. Find the mean and standard deviation of $\bar{x}$ and comment on the shape of its sampling distribution.

7.19 The amounts of telephone bills for all households in a large city have a distribution that is skewed to the right with a mean of \$70 and a standard deviation of \$25. Let $\bar{x}$ be the mean amount of telephone bills for a random sample of 90 households selected from this city. Calculate the mean and standard deviation of $\bar{x}$ and describe the shape of its sampling distribution.

7.20 The balances of checking accounts at a local bank have a distribution that is skewed to the right with its mean equal to \$350 and standard deviation equal to \$140. Let $\bar{x}$ be the mean balance of a random sample of 60 checking accounts selected from this bank. Calculate the mean and standard deviation of $\bar{x}$ and describe the shape of its sampling distribution.

7.21 According to the U.S. Department of Education, the mean 1987–88 salary of full-time instructional faculty at institutions of higher education was \$36,011. Suppose the distribution of 1987–88 salaries of these faculty is skewed to the right with its mean equal to \$36,011 and standard deviation equal to \$5600. Let $\bar{x}$ be the mean 1987–88 salary of a random sample of

80 faculty members selected from the institutions of higher education. Calculate the mean and standard deviation of $\bar{x}$ and describe the shape of its sampling distribution.

7.5 CALCULATING THE PROBABILITY OF $\bar{x}$

From the central limit theorem, for large samples, the sampling distribution of $\bar{x}$ is approximately normal. Based on this result, we can make the following statements about $\bar{x}$ for large samples. (The areas are found from the normal distribution table.)

1. *If we take all possible samples of the same (large) size from a population and calculate the mean for each of these samples, then about 68.26% of the sample means will be within one standard deviation of the population mean.* Or we can state that if we take one sample (of $n \geq 30$) from a population and calculate the mean for this sample, the probability that this sample mean will be within one standard deviation of the population mean is .6826. That is,

$$P(\mu - 1\sigma_{\bar{x}} \leq \bar{x} \leq \mu + 1\sigma_{\bar{x}}) = .6826$$

This probability is shown in Figure 7.8.

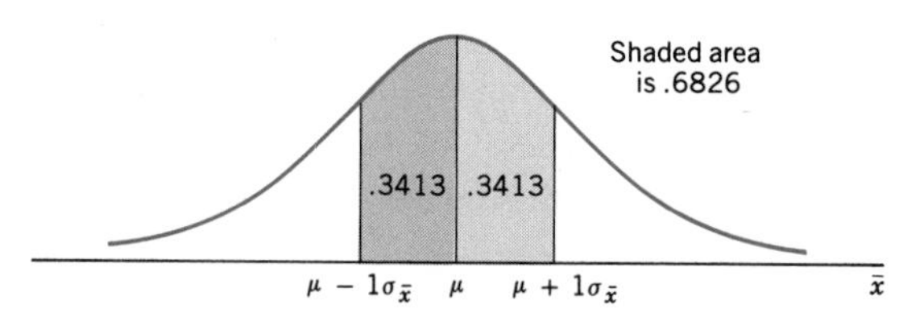

Figure 7.8 $P(\mu - 1\sigma_{\bar{x}} \leq \bar{x} \leq \mu + 1\sigma_{\bar{x}})$.

2. *If we take all possible samples of the same (large) size from a population and calculate the mean for each of these samples, then about 95.44% of the sample means will be within two standard deviations of the population mean.* Or we can state that if we take one sample (of $n \geq 30$) from a population and calculate the mean for this sample, the probability that this sample mean will be within two standard deviations of the population mean is .9544. That is,

$$P(\mu - 2\sigma_{\bar{x}} \leq \bar{x} \leq \mu + 2\sigma_{\bar{x}}) = .9544$$

This probability is shown in Figure 7.9.

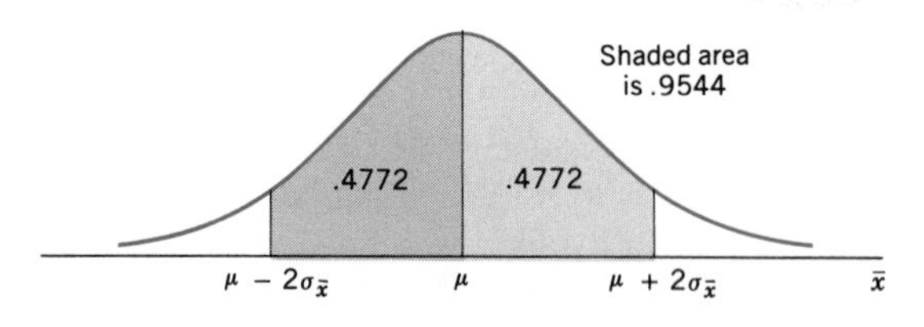

Figure 7.9 $P(\mu - 2\sigma_{\bar{x}} \leq \bar{x} \leq \mu + 2\sigma_{\bar{x}})$.

3. *If we take all possible samples of the same (large) size from a population and calculate the mean for each of these samples, then about 99.74% of the sample means* *will be within three standard deviations of the population mean.* Or we can state that if we take one sample (of $n \geq 30$) from a population and calculate the mean for this sample, the probability that this sample mean will be within three standard deviations of the population mean is .9974. That is,

$$P(\mu - 3\sigma_{\bar{x}} \leq \bar{x} \leq \mu + 3\sigma_{\bar{x}}) = .9974$$

This probability is shown in Figure 7.10.

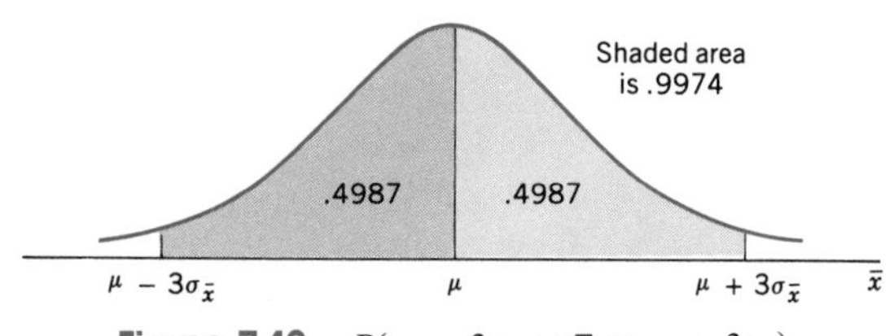

Figure 7.10 $P(\mu - 3\sigma_{\bar{x}} \leq \bar{x} \leq \mu + 3\sigma_{\bar{x}})$

When conducting a survey, we usually take one sample and compute the value of $\bar{x}$ based on that sample. We never take all possible samples of the same size and then prepare the sampling distribution of $\bar{x}$. Rather, we are more interested in finding the probability that the value of $\bar{x}$ computed from one sample falls within a given interval. Examples 7-5 and 7-6 illustrate this procedure.

Calculating probability of $\bar{x}$ in an interval: normal population.

EXAMPLE 7-5 Assume that the weights of all packages of a certain brand of cookies are normally distributed with a mean of 32 ounces and a standard deviation of .3 ounces. Find the probability that the mean weight $\bar{x}$ of a random sample of 20 packages of this brand of cookies will be between 31.8 and 31.9 ounces.

Solution Although the sample size is small, $n < 30$, the shape of the sampling distribution of $\bar{x}$ is normal because the population is normally distributed. The mean and standard deviation of $\bar{x}$ are

$$\mu_{\bar{x}} = \mu = 32 \text{ ounces} \quad \text{and} \quad \sigma_{\bar{x}} = \frac{\sigma}{\sqrt{n}} = \frac{.3}{\sqrt{20}} = .067$$

We are to compute the probability that the value of $\bar{x}$ calculated for one randomly drawn sample of 20 packages is between 31.8 and 31.9 ounces, that is

$$P(31.8 < \bar{x} < 31.9)$$

This probability is given by the area under the normal curve for $\bar{x}$ between the points $\bar{x} = 31.8$ and $\bar{x} = 31.9$. The first step in finding this area is to convert the two $\bar{x}$ values to respective z values.

z VALUE FOR A VALUE OF $\bar{x}$

The z value for a value of $\bar{x}$ is calculated as

$$z = \frac{\bar{x} - \mu}{\sigma_{\bar{x}}}$$

The z values for $\bar{x} = 31.8$ and $\bar{x} = 31.9$ are computed below and they are shown on the z scale below the normal curve for $\bar{x}$ in Figure 7.11.

For $\bar{x} = 31.8$,

$$z = \frac{31.8 - 32}{.067} = -2.99$$

For $\bar{x} = 31.9$,

$$z = \frac{31.9 - 32}{.067} = -1.49$$

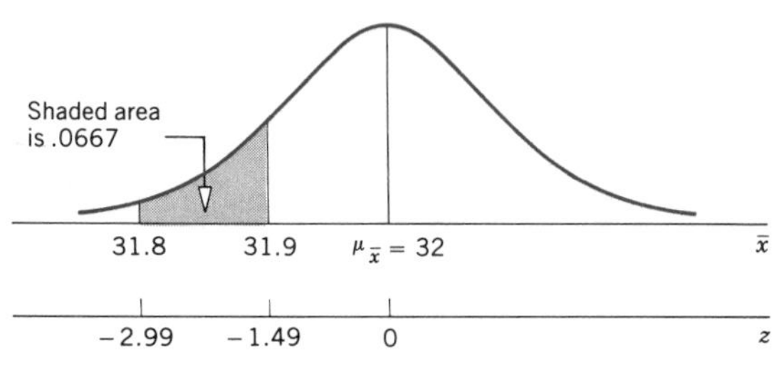

Figure 7.11 $P(31.8 < \bar{x} < 31.9)$.

The probability that $\bar{x}$ is between 31.8 and 31.9 is given by the area under the standard normal curve between $z = -2.99$ and $z = -1.49$. Thus, the required probability is

$$\begin{aligned} P(31.8) < \bar{x} < 31.9) &= P(-2.99 < z < -1.49) \\ &= P(-2.99 < z < 0) - P(-1.49 < z < 0) \\ &= .4986 - .4319 = .0667 \end{aligned}$$

Therefore, the probability is .0667 that the mean weight of a sample of 20 packages will be between 31.8 and 31.9 ounces. ■

Calculating probability of $\bar{x}$ in an interval: $n > 30$.

EXAMPLE 7-6 The mean speed of all cars traveling on interstate highway I-15 is 69 miles per hour with a standard deviation of 3.2 miles. What is the probability that the mean speed of a random sample of 64 cars traveling on this highway will be between 68 and 69.5 miles per hour?

Solution Although the shape of the probability distribution of the population is unknown, the sampling distribution of $\bar{x}$ is approximately normal because the sample size is large ($n > 30$). Remember that when the sample size is larger, the central limit theorem applies. The mean and standard deviation of the sampling distribution of $\bar{x}$ are

$$\mu_{\bar{x}} = \mu = 69 \quad \text{and} \quad \sigma_{\bar{x}} = \frac{\sigma}{\sqrt{n}} = \frac{3.2}{\sqrt{64}} = .40$$

We need to find the probability

$$P(68 < \bar{x} < 69.5)$$

This probability is given by the area under the normal curve for $\bar{x}$ between $\bar{x} = 68$ and $\bar{x} = 69.5$, as shown in Figure 7.12. We find this area as follows. For $\bar{x} = 68$,

$$z = \frac{68 - 69}{.40} = -2.50$$

For $\bar{x} = 69.5$,

$$z = \frac{69.5 - 69}{.40} = 1.25$$

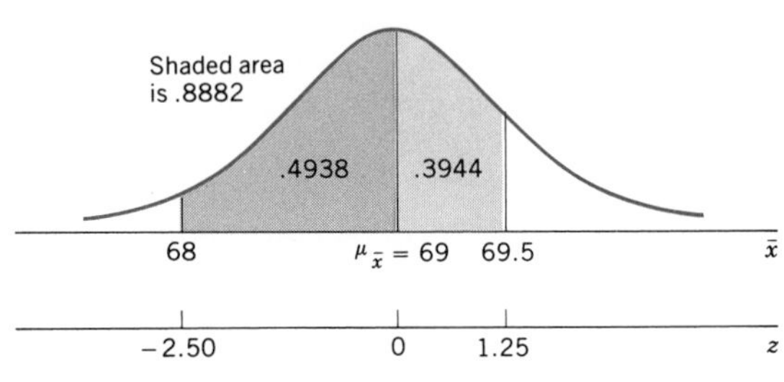

Figure 7.12 $P(68 < \bar{x} < 69.5)$.

Hence, the required probability is

$$\begin{aligned} P(68 < \bar{x} < 69.5) &= P(-2.50 < z < 1.25) \\ &= P(-2.50 < z < 0) + P(0 < z < 1.25) \\ &= .4938 + .3944 = .8882 \end{aligned}$$

Therefore, the probability that the mean speed will be between 68 and 69.5 miles per hour for a random sample of 64 cars traveling on this highway is .8882. ■

EXERCISES

7.22 A population has a mean of 80 and a standard deviation of 14. Assuming $n/N \le .05$, find the following probabilities for a sample size of 110.

a. $P(81.4 \le \bar{x} \le 83.6)$ **b.** $P(\bar{x} \ge 82.3)$

7.23 A population has a mean of 58 and a standard deviation of 12. Assuming $n/N \le .05$, find the following probabilities for a sample size of 50.

a. $P(53.7 \le \bar{x} \le 56.3)$ **b.** $P(\bar{x} \le 59.2)$

7.24 Let x be a continuous random variable that has a normal distribution with $\mu = 75$ and $\sigma = 15$. Assuming $n/N \le .05$, find the probability that the sample mean $\bar{x}$ for a random sample of 20 taken from this population will be

a. between 68.5 and 77.3 **b.** less than 72.4

7.25 Let x be a continuous random variable that has a normal distribution with $\mu = 48$ and $\sigma = 8$. Assuming $n/N \le .05$, find the probability that the sample mean $\bar{x}$ for a random sample of 16 taken from this population will be

a. between 49.6 and 52.2 **b.** more than 45.7

7.26 According to the U.S. Bureau of Labor Statistics, the mean hourly wage for construction workers was \$13.37 in 1989. Assuming that the hourly wages for all construction workers for 1989 are normally distributed with a mean of \$13.37 and a standard deviation of \$1.65, what is the probability that the mean hourly wage for a random sample of 25 construction workers taken from this population will be

a. between \$13 and \$13.75 **b.** less than \$13

7.27 The heights of all adults in the United States are normally distributed with a mean of 68 inches and a standard deviation of 3 inches. What is the probability that the mean height for a random sample of 20 adults will be

a. more than 70 inches **b.** between 65.75 and 67 inches

7.28 Let x be a continuous random variable that has a distribution skewed to the right with $\mu = 60$ and $\sigma = 10$. Assuming $n/N \leq .05$, find the probability that the sample mean $\bar{x}$ for a random sample of 40 taken from this population will be

a. less than 62.20 **b.** between 61.4 and 64.2

7.29 Let x be a continuous random variable that follows a distribution skewed to the right with $\mu = 90$ and $\sigma = 16$. Assuming $n/N \leq .05$, find the probability that the sample mean $\bar{x}$ for a random sample of 64 taken from this population will be

a. less than 82.3 **b.** more than 86.7

7.30 According to the U.S. Bureau of the Census, the mean annual income for households in 1988 was \$34,017. Assume that the 1988 incomes of all households have a distribution that is skewed to the right with a mean of \$34,017 and a standard deviation of \$12,000. Find the probability that the 1988 mean income for a random sample of 100 households will be

a. less than \$34,800 **b.** between \$35,000 and \$36,000

7.31 According to the U.S. National Center for Health Statistics, the mean hospital stay for patients in 1985 was 6.9 days. Assume that the hospital stays for all patients for this year have a distribution that is skewed to the right with a mean of 6.9 days and a standard deviation of 3 days. Find the probability that the mean hospital stay for a random sample of 50 patients will be

a. between 6 and 8 days **b.** more than 8 days

7.32 The time that college students spend studying per week has a distribution that is skewed to the right with a mean of 8.4 hours and a standard deviation of 2.7 hours. Find the probability that the mean time spent studying per week for a random sample of 45 students will be

a. between 8 and 9 hours **b.** less than 8 hours

7.33 The ages of all college students follow a distribution that is skewed to the right with a mean of 26 years and a standard deviation of 4 years. Find the probability that the mean age for a random sample of 36 students will be

a. between 25 and 27 years **b.** less than 25 years

7.6 POPULATION AND SAMPLE PROPORTIONS

The **population proportion,** denoted by p, is obtained by taking the ratio of the number of elements in a population with a specific characteristic to the total number of elements in the population. The **sample proportion,** denoted by $\hat{p}$ (read "p hat"), gives a similar ratio for a sample.

POPULATION AND SAMPLE PROPORTIONS

The population and sample proportions, denoted by p and $\hat{p}$, respectively, are calculated as

$$p = \frac{\text{Number of elements in the population with a specific characteristic}}{\text{Total number of elements in the population}}$$

$$\hat{p} = \frac{\text{Number of elements in the sample with a specific characteristic}}{\text{Total number of elements in the sample}}$$

The concept of proportion is the same as the concept of relative frequency discussed in Chapter 2 and the concept of probability of success in a binomial problem. The relative frequency of a category or class gives the proportion of the sample or population that belongs to that category or class. Similarly, the probability of success in a binomial problem represents the proportion of the sample or population that possesses a given characteristic. Example 7-7 illustrates the calculation of the population and sample proportions.

Calculating population and sample proportions.

EXAMPLE 7-7 Suppose there are 789,654 families living in a city, of which 563,282 own a house. Then, the proportion of all families in this city that own a house is

$$p = \frac{563,282}{789,654} = .71$$

Now, suppose a sample of 240 families is taken from this city and 158 of them are homeowners. Then, the sample proportion is given by

$$\hat{p} = \frac{158}{240} = .66$$

■

7.7 THE MEAN, STANDARD DEVIATION, AND SHAPE OF THE SAMPLING DISTRIBUTION OF $\hat{p}$

This section discusses the sampling distribution of the sample proportion and the mean, standard deviation, and shape of this sampling distribution.

7.7.1 THE SAMPLING DISTRIBUTION OF $\hat{p}$

Just like the sample mean $\bar{x}$, the sample proportion $\hat{p}$ is also a random variable. Hence, it possesses a probability distribution, which is called its **sampling distribution.**

SAMPLING DISTRIBUTION OF THE SAMPLE PROPORTION $\hat{p}$

The probability distribution of the sample proportion $\hat{p}$ is called its sampling distribution. It lists the various values that $\hat{p}$ can assume and their probabilities.

The value of $\hat{p}$ calculated for a particular sample depends on what elements of the population are included in that sample. Example 7-8 illustrates the concept of the sampling distribution of $\hat{p}$.

Illustrating sampling distribution of $\hat{p}$.

EXAMPLE 7-8 Boe Consultant Associates has five employees. Table 7.6 gives the names of these five employees and the information concerning their knowledge of statistics.

Table 7.6 Information on the Five Employees of Boe Consultant Associates

Name	Knows Statistics
Ally	Yes
John	No
Susan	No
Lee	Yes
Tom	Yes

If we define the population proportion p as the proportion of employees who know statistics, then,

$$p = \frac{3}{5} = .60$$

Now, suppose we take all possible samples of three employees each and compute the proportion of employees, for each sample, who know statistics. The total number of samples of size three that can be drawn from the population of five employees is $\binom{5}{3} = 10$. Table 7.7 lists these 10 possible samples and the proportion of employees who know statistics for each of those samples. Note that we have rounded the values of $\hat{p}$ to two decimal places.

Table 7.7 All Possible Samples of Size 3 and the Value of $\hat{p}$ for Each Sample

Sample	Proportion Who Know Statistics, $\hat{p}$
Ally, John, Susan	1/3 = .33
Ally, John, Lee	2/3 = .67
Ally, John, Tom	2/3 = .67
Ally, Susan, Lee	2/3 = .67
Ally, Susan, Tom	2/3 = .67
Ally, Lee, Tom	3/3 = 1.00
John, Susan, Lee	1/3 = .33
John, Susan, Tom	1/3 = .33
John, Lee, Tom	2/3 = .67
Susan, Lee, Tom	2/3 = .67

Using Table 7.7, we prepare the frequency distribution of $\hat{p}$ as recorded in Table 7.8. Dividing the frequencies of classes by the sum of the frequencies, we obtain the probabilities of classes. These probabilities are listed in Table 7.9, which gives the sampling distribution of $\hat{p}$.

Table 7.8 The Frequency Distribution of $\hat{p}$

$\hat{p}$	f
.33	3
.67	6
1.00	1
	$\Sigma f = 10$

Table 7.9 The Sampling Distribution of $\hat{p}$

$\hat{p}$	$P(\hat{p})$
.33	$3/10 = .30$
.67	$6/10 = .60$
1.00	$1/10 = .10$
	$\Sigma P(\hat{p}) = 1.0$

■

7.7.2 THE MEAN AND STANDARD DEVIATION OF $\hat{p}$

The mean of the sampling distribution of $\hat{p}$ is always equal to the population proportion p just as the mean of the sampling distribution of $\bar{x}$ is always equal to the population mean μ.

MEAN OF THE SAMPLE PROPORTION

The mean of the sample proportion $\hat{p}$ is denoted by $\mu_{\hat{p}}$ and is equal to the population proportion p. Thus,

$$\mu_{\hat{p}} = p$$

The standard deviation of $\hat{p}$, denoted by $\sigma_{\hat{p}}$, is given by the formula given on the next page. This formula is true only when the sample size is small as compared to the population size. As we know from Section 7.3, the sample size is said to be small compared to the population size if $n/N \leq .05$.†

†If n/N is greater than .05, then $\sigma_{\hat{p}}$ is calculated as

$$\sigma_{\hat{p}} = \sqrt{\frac{p\,q}{n}}\sqrt{\frac{N-n}{N-1}}$$

where the factor

$$\sqrt{\frac{N-n}{N-1}}$$

is called the finite population correction factor. However, this formula will not be used in the text.

STANDARD DEVIATION OF THE SAMPLE PROPORTION

The standard deviation of the sample proportion $\hat{p}$ is denoted by $\sigma_{\hat{p}}$ and is given by the formula

$$\sigma_{\hat{p}} = \sqrt{\frac{p\,q}{n}}$$

where p is the population proportion, $q = 1 - p$, and n is the sample size. This formula is used when $n/N \leq .05$ where N is the population size.

7.7.3 THE SHAPE OF THE SAMPLING DISTRIBUTION OF $\hat{p}$

The shape of the sampling distribution of $\hat{p}$ is inferred from the central limit theorem.

CENTRAL LIMIT THEOREM FOR SAMPLE PROPORTION

According to the central limit theorem, the sampling distribution of $\hat{p}$ is approximately normal for a sufficiently large sample size. In the case of proportion, the sample size n is considered to be sufficiently large if np and nq are both greater than 5, that is, if

$$np > 5 \quad \text{and} \quad nq > 5$$

Note that the sampling distribution of $\hat{p}$ will be normal if $np > 5$ and $nq > 5$. This is the same condition that was required for the application of the normal approximation to the binomial probability distribution in Chapter 6.

Example 7-9 shows the calculation of the mean and standard deviation of $\hat{p}$ and describes the shape of its sampling distribution.

Sampling distribution, mean, and standard deviation of $\hat{p}$.

EXAMPLE 7-9 Forty percent of all students at a university hold a (full-time or part-time) job. Let $\hat{p}$ be the proportion of students who hold a job in a random sample of 40 students selected from this university. Find the mean and standard deviation of $\hat{p}$ and describe the shape of its sampling distribution.

Solution Let p be the proportion of all students at this university who hold a job. Then,

$$p = .40 \quad \text{and} \quad q = 1 - p = 1 - .40 = .60$$

The mean of the sampling distribution of $\hat{p}$ is

$$\mu_{\hat{p}} = p = .40$$

The standard deviation of $\hat{p}$ is

$$\sigma_{\hat{p}} = \sqrt{\frac{p\,q}{n}} = \sqrt{\frac{(.40)\,(.60)}{40}} = \sqrt{.006} = .077$$

The values of np and nq are

$$np = 40\,(.40) = 16 \quad \text{and} \quad nq = 40\,(.60) = 24$$

As np and nq are both greater than 5, we can apply the central limit theorem to make an inference about the shape of the sampling distribution of $\hat{p}$. Therefore, the sampling distribution of $\hat{p}$ is approximately normal with a mean of .40 and a standard deviation of .077, as shown in Figure 7.13.

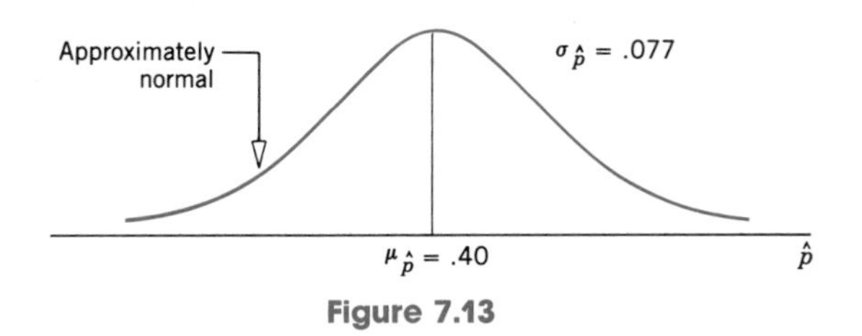

Figure 7.13

■

EXERCISES

7.34 A class has six students. The following data give information about whether or not each of these students is a senior.

Yes No No Yes Yes No

a. What proportion of the students in this class are seniors?
b. How many total samples of size 5 can be taken from this class?
c. List all possible samples of size 5 that can be taken from this class and calculate the sample proportion $\hat{p}$ of students who are seniors for each sample. Prepare the sampling distribution of $\hat{p}$.

7.35 The following data give the information on all five employees of a company.

Male Female Female Male Female

a. What proportion of employees of this company are female?
b. How many total samples of size 3 can be taken from this population?
c. List all possible samples of size 3 that can be taken from this population and calculate the sample proportion $\hat{p}$ of the employees who are female for each sample. Prepare the sampling distribution of $\hat{p}$.

7.36 According to the U.S. National Center for Health Statistics, 6.8% of all births occur with low birth weights (less than 2500 grams). Let $\hat{p}$ be the proportion of births with low birth weight in a random sample of 100 births. Find the mean and standard deviation of $\hat{p}$ and describe the shape of its sampling distribution.

7.37 According to the U.S. Department of Education, (about) 69% of teachers in all public schools are female. Let $\hat{p}$ be the proportion in a random sample of 200 public school teachers who are female. Calculate the mean and standard deviation of $\hat{p}$ and comment on the shape of its sampling distribution.

7.38 In a survey of 14- to 21-year-olds conducted by *Seventeen* magazine, 46% of the respondents said "their generation is too materialistic" (*Seventeen,* October 1989). Assume that this percentage is true for the current population of 14- to 21-year-olds. Let $\hat{p}$ be the proportion of 14- to 21-year-olds in a random sample of 500 who hold this view. Calculate the mean and standard deviation of $\hat{p}$ and comment on the shape of its sampling distribution.

7.39 According to a 1989 survey by the National Highway Traffic Safety Administration, 59.1% of the drivers in San Diego use seat belts. Assume that this percentage is true for the current population of all San Diego drivers. Let $\hat{p}$ be the proportion of drivers in a random sample of 80 drivers selected from San Diego who use seat belts. Calculate the mean and standard deviation of $\hat{p}$ and describe the shape of its sampling distribution.

7.8 CALCULATING THE PROBABILITY OF $\hat{p}$

As mentioned in Section 7.5, when we conduct a study we usually take only one sample and make all decisions or inferences on the basis of the results of that one sample. We use the concepts of the mean, standard deviation, and shape of the sampling distribution of $\hat{p}$ to compute the probability that the value of $\hat{p}$ computed from one sample falls within a given interval. Examples 7-10 and 7-11 illustrate this application.

Probability that $\hat{p}$ assumes a value in an interval.

EXAMPLE 7-10 According to the U.S. Census Bureau, approximately 66% of the households headed by single women own autos. Assume that this percentage is true for the population of all households headed by single women. Let $\hat{p}$ be the proportion of households who own autos in a random sample of 120 households headed by single women. Find the probability that the value of $\hat{p}$ is between .68 and .72.

Solution From the given information,

$$p = .66 \quad \text{and} \quad q = 1 - p = 1 - .66 = .34$$

where p is the proportion of all households headed by single women who own autos.

The mean of the sample proportion $\hat{p}$ is

$$\mu_{\hat{p}} = p = .66$$

The standard deviation of $\hat{p}$ is

$$\sigma_{\hat{p}} = \sqrt{\frac{p\,q}{n}} = \sqrt{\frac{(.66)\,(.34)}{120}} = .043$$

The values of np and nq are

$$np = 120\,(.66) = 79.2 \quad \text{and} \quad nq = 120\,(.34) = 40.8$$

As np and nq are both greater than 5, we can infer from the central limit theorem that the sampling distribution of $\hat{p}$ is approximately normal. The probability that $\hat{p}$ is between .68 and .72 is given by the area under the normal curve for $\hat{p}$ between $\hat{p} = .68$ and $\hat{p} = .72$, as shown in Figure 7.14.

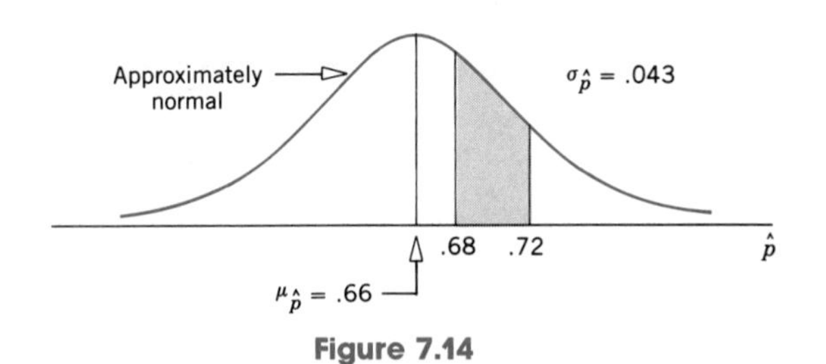

Figure 7.14

The first step in finding the area under the normal curve between $\hat{p} = .68$ and $\hat{p} = .72$ is to convert these two values to respective z values. The z value for $\hat{p}$ is computed using the following formula.

z VALUE FOR A VALUE OF $\hat{p}$

The z value for a value of $\hat{p}$ is calculated as

$$z = \frac{\hat{p} - p}{\sigma_{\hat{p}}}$$

Next, the two values of $\hat{p}$ are converted to their respective z values and then the area under the normal curve between these two points is found using the normal distribution table. For $\hat{p} = .68$,

$$z = \frac{.68 - .66}{.043} = .47$$

For $\hat{p} = .72$,

$$z = \frac{.72 - .66}{.043} = 1.40$$

The probability that $\hat{p}$ is between .68 and .72 is given by the area under the standard normal distribution between $z = .47$ and $z = 1.40$. This area is shown in Figure 7.15.

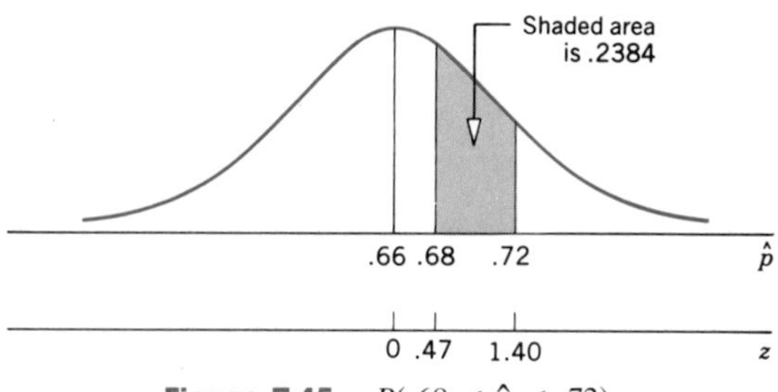

Figure 7.15 $P(.68 < \hat{p} < .72)$.

The required probability is

$$
\begin{aligned}
P(.68 < \hat{p} < .72) &= P(.47 < z < 1.40) \\
&= P(0 < z < 1.40) - P(0 < z < .47) \\
&= .4192 - .1808 = .2384
\end{aligned}
$$

Thus, the probability is .2384 that between 68% and 72% of households in a sample of 120 households headed by single women own autos. ■

Probability of $\hat{p}$ being less than a certain value.

EXAMPLE 7-11 John Smith, who is contesting for the mayor's position in a large city, claims that he is favored by 53% of all the eligible voters of that city. Assume that this claim is true. What is the probability that less than 49% in a random sample of 400 registered voters taken from this city will favor John Smith?

Solution Let p be the proportion of all eligible voters who favor John Smith. Then,

$$p = .53 \quad \text{and} \quad q = 1 - p = 1 - .53 = .47$$

The mean of the sample proportion $\hat{p}$ is

$$\mu_{\hat{p}} = p = .53$$

The population of all voters is large and the sample size is small as compared to the population. Consequently, we can assume that $n/N \leq .05$. Hence, the standard deviation of $\hat{p}$ is calculated as

$$\sigma_{\hat{p}} = \sqrt{\frac{p\,q}{n}} = \sqrt{\frac{(.53)\,(.47)}{400}} = .025$$

From the central limit theorem, the shape of the sampling distribution of $\hat{p}$ is approximately normal. The probability that $\hat{p}$ is less than.49 is given by the area under the normal curve for $\hat{p}$ to the left of $\hat{p} = .49$, as shown in Figure 7.16.

The z value for $\hat{p} = .49$ is

$$z = \frac{\hat{p} - p}{\sigma_{\hat{p}}} = \frac{.49 - .53}{.025} = -1.60$$

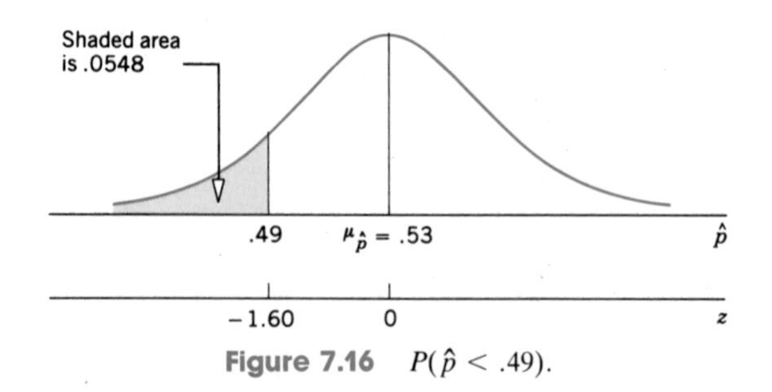

Figure 7.16 $P(\hat{p} < .49)$.

Thus, the required probability is

$$P(\hat{p} < .49) = P(z < -1.60) = .5 - P(-1.60 < z < 0)$$
$$= .5 - .4452 = .0548$$

Hence, the probability that less than 49% of the voters in a random sample of 400 voters will favor John Smith is .0548. ■

EXERCISES

7.40 Let the population proportion p be .70. Assuming $n/N \leq .05$, find the following probabilities for a sample size of 100.

a. $P(.61 \leq \hat{p} \leq .66)$ **b.** $P(\hat{p} \geq .73)$

7.41 Let the population proportion p be .40. Assuming $n/N \leq .05$, find the following probabilities for a sample size of 80.

a. $P(.44 \leq \hat{p} \leq .50)$ **b.** $P(\hat{p} \leq .43)$

7.42 Twenty-nine percent of the U.S. adult population were smokers in 1988 (*U.S. News & World Report*, January 23, 1989). Assume that this result holds true for the current population of all adults. Let $\hat{p}$ be the proportion of people who smoke in a random sample of 100 adults. Find the probability that the value of $\hat{p}$ will be

a. between .20 and .32 **b.** less than .26

7.43 In a Roper Organization survey, 65% of adults said that they are generally optimistic about their personal future (*The American Dream*, A National Survey conducted by The Roper Organization in 1987 for the *Wall Street Journal*). Assuming that this percentage is true for the current population of all adults, find the probability that the proportion of adults in a random sample of 200 adults who hold this view will be

a. between .61 and .69 **b.** more than .63

7.44 Thirty-nine percent of married women in a survey said that they married their husbands because they shared their vision of the future (*The Ladies' Home Journal*, June 1988). Assuming that this percentage is true for the current population of all married women, find the probability that the proportion of married women in a random sample of 150 who hold this view will be

a. between .32 and .37 **b.** more than .44

7.45 According to a 1989 Roper Organization Survey, 64% of households own VCRs. Assume that this percentage is true for the current population of all households and let $\hat{p}$ be the proportion of households who own VCRs in a random sample of 400 households. Find the probability that the value of $\hat{p}$ will be

a. between .61 and .68 **b.** less than .67

7.46 In a *Time*/CNN poll of adult Americans conducted by Yankelovich Clancy Shulman, 44% of respondents said that they would be willing to pay an extra $500 in taxes to clean up the environment (*Time*, December 24, 1990). Assume that this percentage is true for the current population of all adult Americans, and let $\hat{p}$ be the proportion of adult Americans in a random sample of 300 adults who will hold this view. Find the probability that the value of $\hat{p}$ will be

a. between .41 and .48 **b.** less than .39

GLOSSARY

Central limit theorem The theorem from which it is inferred that, for a large sample size ($n \geq 30$), the shape of the sampling distribution of $\bar{x}$ is approximately normal. Also, by the same theorem, the shape of the sampling distribution of $\hat{p}$ is approximately normal for a sample for which $np > 5$ and $nq > 5$.

Mean of $\bar{x}$ The mean of the sampling distribution of $\bar{x}$, denoted by $\mu_{\bar{x}}$, is equal to the population mean μ.

Mean of $\hat{p}$ The mean of the sampling distribution of $\hat{p}$, denoted by $\mu_{\hat{p}}$, is equal to the population proportion p.

Nonsampling errors The errors that occur during the collection, recording, and tabulation of data.

Population distribution The probability distribution of the population data.

Population proportion The ratio of the number of elements in a population with a specific characteristic to the total number of elements in the population.

Sample proportion The ratio of the number of elements in a sample with a specific characteristic to the total number of elements in that sample.

Sampling distribution of $\bar{x}$ The probability distribution of all values of $\bar{x}$ calculated from all possible samples of the same size taken from a population.

Sampling distribution of $\hat{p}$ The probability distribution of all values of $\hat{p}$ calculated from all possible samples of the same size taken from a population.

Sampling error The difference between the value of a sample statistic calculated from a random sample and the value of the corresponding population parameter, which occurs due to chance.

Standard deviation of $\bar{x}$ The standard deviation of the sampling distribution of $\bar{x}$, denoted by $\sigma_{\bar{x}}$, is equal to $\sigma/\sqrt{n}$ when $n/N \leq .05$.

Standard deviation of $\hat{p}$ The standard deviation of the sampling distribution of $\hat{p}$, denoted by $\sigma_{\hat{p}}$, is equal to $\sqrt{pq/n}$ when $n/N \leq .05$.

KEY FORMULAS

1. **Mean of $\bar{x}$**

$$\mu_{\bar{x}} = \mu$$

2. **Standard deviation of $\bar{x}$**

$$\sigma_{\bar{x}} = \frac{\sigma}{\sqrt{n}}$$

3. **The z value for $\bar{x}$**

$$z = \frac{\bar{x} - \mu}{\sigma_{\bar{x}}}$$

4. **Population proportion**

$$p = \frac{\text{Number of elements in a population with a specific characteristic}}{\text{Total number of elements in the population}}$$

5. **Sample proportion**

$$\hat{p} = \frac{\text{Number of elements in a sample with a specific characteristic}}{\text{Total number of elements in the sample}}$$

6. **Mean of $\hat{p}$**

$$\mu_{\hat{p}} = p$$

7. **Standard deviation of $\hat{p}$**

$$\sigma_{\hat{p}} = \sqrt{\frac{p\,q}{n}}$$

8. **The z value for $\hat{p}$**

$$z = \frac{\hat{p} - p}{\sigma_{\hat{p}}}$$

SUPPLEMENTARY EXERCISES

7.47 According to the American Medical Association's Center for Health Policy Research, in 1988 patients had to wait an average of 7.6 days to get an appointment with their doctors. Assume that currently the mean waiting time for all patients to get an appointment with their doctors is 7.6 days and standard deviation is 1.7 days. Let $\bar{x}$ be the mean waiting time to get an appointment with their doctors for a random sample of 300 patients. Find the mean and standard deviation of $\bar{x}$.

7.48 According to a 1989 Harris poll done for the Philip Morris Family Survey II, the average monthly cost of day care for executives was $244 per child. Assume that the mean for monthly costs of day care for all executives is $244 per child and the standard deviation of these costs is $20. Let $\bar{x}$ be the mean monthly cost of day care per child for a random sample of 20 executives. Find the mean and standard deviation of $\bar{x}$.

7.49 The time taken to learn a specific job follows a normal distribution with a mean of 80 minutes and a standard deviation of 8 minutes. Let $\bar{x}$ be the mean time taken to learn this job for a random sample of 23 persons. Find the mean and standard deviation of $\bar{x}$ and describe the shape of its sampling distribution.

7.50 The IQ scores for all students at a large university have a normal distribution with a mean of 121 and a standard deviation of 11. Let $\bar{x}$ be the mean IQ score for a random sample of 28 students selected from this university. Find the mean and standard deviation of $\bar{x}$ and describe the shape of its sampling distribution.

7.51 The weights of all 1560 students at an elementary school have a distribution that is skewed to the right with its mean equal to 70 pounds and standard deviation equal to 6 pounds. Let $\bar{x}$ be the mean weight of a random sample of 40 students selected from this elementary school. Calculate the mean and standard deviation of $\bar{x}$ and comment on the shape of its sampling distribution.

7.52 The weekly earnings of all 2480 employees of a company have a distribution that is skewed to the right with its mean equal to $438 and standard deviation equal to $40. Let $\bar{x}$ be the mean weekly earnings of a random sample of 95 employees selected from this company. Calculate the mean and standard deviation of $\bar{x}$ and comment on the shape of its sampling distribution.

7.53 The weights of apples grown at a Washington farm have a normal distribution with a mean of 4.1 ounces and a standard deviation of .45 ounces. Find the probability that the mean weight for a random sample of 24 apples taken from this farm will be

a. between 4.20 and 4.30 ounces **b.** more than 4.25 ounces

7.54 The amount of ice cream in a one-pound carton has a normal distribution with a mean of 16 ounces and a standard deviation of .18 ounces. Find the probability that the mean amount of ice cream in a random sample of 15 such cartons will be

a. between 15.90 and 15.95 ounces **b.** less than 15.95 ounces

7.55 The life of a specific brand of auto battery is normally distributed with a mean of 60 months and a standard deviation of 5 months. Find the probability that the mean life of a random sample of 28 such batteries will be

a. less than 59 months **b.** between 61 and 62 months

7.56 The money spent by customers at a grocery store has a distribution that is skewed to the right with a mean of $72 and a standard deviation of $26. Find the probability that the mean money spent at this grocery store by 100 randomly selected customers will be

a. less than $75 **b.** more than $70

7.57 The waiting times for patients at the emergency ward of a hospital have a distribution that is skewed to the right with a mean of 18 minutes and a standard deviation of 8 minutes. Find the probability that the mean waiting time for a random sample of 75 patients, who will come to this emergency ward, will be

a. less than 17 minutes **b.** between 19 and 20.5 minutes

7.58 The time taken by a bank teller to serve a customer has a nonnormal distribution with a mean of 4 minutes and a standard deviation of 3 minutes. Find the probability that the mean time taken by this teller to serve 50 randomly selected customers will be

a. between 4.5 and 5 minutes **b.** more than 4.5 minutes

7.59 The number of hours spent partying per week by all students at a large university has a nonnormal distribution with a mean of 7.5 hours and a standard deviation of 4 hours. Find the probability that the mean number of hours spent partying per week for a random sample of 40 students selected from this university will be

a. between 6 and 8 hours **b.** more than 7 hours

7.60 Ten percent of all items produced on a machine are defective. Let $\hat{p}$ be the proportion of defective items in a random sample of 80 items selected from the production line. Calculate the mean and standard deviation of $\hat{p}$ and describe the shape of the sampling distribution of $\hat{p}$.

7.61 According to a 1989 Roper Organization survey, 82% of households own cars. Assume that this percentage is true for the current population of all households. Let $\hat{p}$ be the proportion of households in a random sample of 400 who own cars. Calculate the mean and standard deviation of $\hat{p}$ and describe the shape of the sampling distribution of $\hat{p}$.

7.62 In a Roper Organization survey, 66% of working women said that they would continue working even if they were financially secure (*The 1985 Virginia Slims American Women's Opinion Poll,* 1985). Assume that this percentage is true for the current population of all working women. Let $\hat{p}$ be the proportion of women in a random sample of 100 working women who hold this view. Find the probability that the value of $\hat{p}$ will be

a. between .58 and .64 **b.** more than .68

7.63 In a Roper Organization survey, 48% of adults favored banning all advertisements for beer and wine from television (*The American Way of Buying,* The Dow Jones & Co., 1990). Assume that this percentage is true for the current population of all adults. Let $\hat{p}$ be the proportion of adults who hold this view in a random sample of 100 adults. Find the probability that the value of $\hat{p}$ will be

a. between .51 and .58 **b.** less than .44

7.64 In a 1990 Gallup poll, 43% of adults said that boys are easier to rear than girls. Assume that this percentage is true for the current population of all adults, and let $\hat{p}$ be the proportion of adults in a random sample of 100 adults who hold this view. Find the probability that the value of $\hat{p}$ will be

a. between .40 and .48 **b.** less than .46

7.65 In a Roper Organization survey of 8- to 17-year-olds, 39% said that getting along better with siblings would make their home life better (*The American Chicle Youth Poll,* Warner-Lambert Company, A Study Conducted by the Roper Organization, 1987). Assume that this percentage is true for the current population of all 8- to 17-year-olds. Let $\hat{p}$ be the proportion of 8- to 17-year-olds in a random sample of 240 who hold this view. Find the probability that the value of $\hat{p}$ will be

a. between .33 and .44 **b.** more than .35

SELF-REVIEW TEST

1. A sampling distribution is the probability distribution of
   a. a population parameter
   b. a sample statistic
   c. any random variable
2. The nonsampling errors are
   a. the errors that occur because the sample size is too large in relation to the population size

b. the errors made while collecting, recording, and tabulating data
c. the errors that occur because an untrained person conducts the survey

3. A sampling error is

a. the difference between the value of a sample statistic and the value of the corresponding population parameter
b. the error made while collecting, recording, and tabulating data
c. the error that occurs because the sample is too small

4. The mean of the sampling distribution of $\bar{x}$ is always equal to

a. μ **b.** $\mu - 5$ **c.** $\sigma/\sqrt{n}$

5. The condition for the standard deviation of the sample mean to be $\sigma/\sqrt{n}$ is that

a. $np > 5$ **b.** $n/N \leq .05$ **c.** $n > 30$

6. The standard deviation of the sampling distribution of the sample mean decreases when

a. x increases **b.** n increases **c.** n decreases

7. When samples are taken from a normally distributed population, the sampling distribution of the sample mean has a normal distribution

a. when $n \geq 30$ **b.** when $n/N \leq .05$ **c.** always

8. When samples are taken from a nonnormally distributed population, the sampling distribution of the sample mean has a normal distribution

a. when $n \geq 30$ **b.** when $n/N \leq .05$ **c.** always

9. In a sample of 200 adults, 60 are found to be satisfied with their weight. The proportion of adults in this sample who are satisfied with their weight is

a. .60 **b.** .30 **c.** 3.33

10. The mean of the sampling distribution of $\hat{p}$ is always equal to

a. p **b.** μ **c.** $\hat{p}$

11. The condition for the standard deviation of the sampling distribution of sample proportion to be $\sqrt{pq/n}$ is

a. $np > 5$ and $nq > 5$ **b.** $n > 30$ **c.** $n/N \leq .05$

12. The sampling distribution of $\hat{p}$ is (approximately) normal if

a. $np > 5$ and $nq > 5$ **b.** $n > 30$ **c.** $n/N \leq .05$

13. Briefly state and explain the central limit theorem.

14. The weights of all students at a large university have an approximate normal distribution with a mean of 145 pounds and a standard deviation of 18 pounds. Let $\bar{x}$ be the mean weight of a random sample of certain students selected from this university. Calculate the mean and standard deviation of $\bar{x}$ and describe the shape of its sampling distribution for a sample size of

a. 25 **b.** 100

15. The time taken to run a road race for all participants has a distribution that is skewed to the right with a mean of 47 minutes and a standard deviation of 8.4 minutes. Let $\bar{x}$ be the mean time taken to run this road race for a random sample of 50 participants. Find the mean and standard deviation of $\bar{x}$ and describe the shape of the sampling distribution of $\bar{x}$.

16. According to an IRS study, it takes an average of 222 minutes for taxpayers to prepare, copy, and mail a 1040 tax form. Assume that the time taken to prepare, copy, and mail the 1040 tax form by all taxpayers has a nonnormal distribution with a mean of 222 minutes and a standard deviation of 55 minutes. Find the probability that the mean time taken to prepare, copy, and mail this tax form for a random sample of 60 taxpayers will be

a. between 205 and 218 minutes
b. more than 230 minutes
c. less than 225 minutes
d. between 216 and 228 minutes

17. In a poll of 1002 adults conducted by Yankelovich Clancy Shulman for *Time* magazine, 52% of adults said that news anchors are not worth a million dollars a year to their networks (*Time,* August 7, 1989). Assume that this percentage is true for the current population of all adults. Let $\hat{p}$ be the proportion of adults in a random sample who hold this view. Calculate the mean and standard deviation of $\hat{p}$ and describe the shape of its sampling distribution for a sample size of

a. 100 **b.** 400

18. In a *Newsweek* poll of 750 adults conducted by the Gallup Organization, 77% of adults believed in the existence of heaven (*Newsweek,* March 27, 1989). Assume that this percentage is true for the current population of all adults. Let $\hat{p}$ be the proportion of adults in a random sample of 250 adults who believe in the existence of heaven. Find the probability that $\hat{p}$ will be

a. less than .81
b. between .79 and .83
c. more than .80
d. between .73 and .82

8 ESTIMATION OF THE MEAN AND PROPORTION

8.1 ESTIMATION: AN INTRODUCTION

8.2 POINT AND INTERVAL ESTIMATES

8.3 INTERVAL ESTIMATION OF A POPULATION MEAN: LARGE SAMPLES

8.4 INTERVAL ESTIMATION OF A POPULATION MEAN: SMALL SAMPLES

8.5 INTERVAL ESTIMATION OF A POPULATION PROPORTION: LARGE SAMPLES

8.6 SAMPLE SIZE DETERMINATION FOR THE ESTIMATION OF MEAN

8.7 SAMPLE SIZE DETERMINATION FOR THE ESTIMATION OF PROPORTION

APPENDIX 8.1: RATIONALE BEHIND THE CONFIDENCE INTERVAL FORMULA FOR μ

SELF-REVIEW TEST

USING MINITAB

Now we are entering that part of statistics called *inferential statistics*. In Chapter 1 inferential statistics was defined as the part of statistics that helps us to make decisions about some characteristics of a population based on sample information. In other words, inferential statistics uses the sample results to make decisions and draw conclusions about the population from which the sample is taken. Estimation is the first topic to be considered with regard to inferential statistics. Estimation and hypothesis testing (which is discussed in Chapter 9) taken together are usually referred to as inference making. This chapter explains how to estimate the population mean and population proportion for a single population.

8.1 ESTIMATION: AN INTRODUCTION

Estimation is a procedure by which numerical value or values are assigned to a population parameter based on the information collected from a sample.

ESTIMATION

The assignment of value(s) to a population parameter based on a value of the corresponding sample statistic is called estimation.

In inferential statistics, μ is the *true population mean* and p is called the *true population proportion*. There are many other population parameters such as the median, mode, variance, and standard deviation.

Every day we face instances where we have to estimate something. For example, we may want to estimate the total cost of our next vacation trip or the approximate monthly payment on a new car before we buy one. A few more examples of estimation are (1) that a supervisor may want to estimate the average time taken by new employees to learn a job, (2) that the U.S. Census Bureau may want to find the mean family size in the United States, and (3) that the AWAH (Association of Wives of Alcoholic Husbands) may want to find the proportion (or percentage) of all husbands in the United States who are alcoholics.

The examples about estimating the average time taken to learn a job by new employees and estimating the mean family size are illustrations of estimating the *true population mean* μ. The example about estimating the proportion of alcoholic husbands is an illustration of estimating the *true population proportion* p.

If we can conduct a *census* (a survey that includes the entire population) every time we want to find the value of a population parameter, then the estimation procedures explained in this and subsequent chapters are not needed. For example, if the U.S. Census Bureau can contact every family living in the United States every time it wants to find the mean family size, the result of the survey (which will actually be a census) will give the value of μ and the procedures learned in this chapter will not be needed. However, it is too expensive, very time consuming, or virtually impossible to contact every member of a population to collect information to find the true value of a population parameter. Therefore, we usually take a sample from the population and calculate the value of the appropriate sample statistic. Then we assign a value or values to the corresponding population parameter based on the value of the sample statistic. This chapter (and subsequent chapters) explains how to assign values to population parameters based on the values of sample statistics.

For example, to estimate the mean time taken to learn a certain job by new employees, the supervisor will take a sample of new employees and record the time taken by these employees to learn the job. Using this information, she will calculate the sample mean $\bar{x}$. Then, based on the value of $\bar{x}$, she will assign certain values to μ. As another example, to estimate the mean family size μ, the U.S. Census Bureau will take a sample of families from the United States, collect the information on the size of each of these families, and compute the value of the sample mean $\bar{x}$. Based on this value of $\bar{x}$, the Bureau will then assign values to the population mean μ. Similarly, the AWAH will take a sample of husbands and determine the value of the sample proportion $\hat{p}$, which represents the proportion of husbands in the sample who are alcoholic. Then, using this value of the sample proportion $\hat{p}$, AWAH will assign values to the population proportion p.

The value(s) assigned to a population parameter based on the value of a sample statistic is called an **estimate** of the population parameter. For example, suppose the supervisor takes a sample of 40 new employees and finds that the mean time $\bar{x}$ taken to learn this job for these employees is 5.5 hours. If she assigns this value to the population mean, then 5.5 will be called an estimate of μ. The sample statistic used to estimate a population parameter is called an **estimator.** Thus, the sample mean $\bar{x}$ is an estimator of the population mean μ, and the sample proportion $\hat{p}$ is an estimator of the population proportion p.

ESTIMATE AND ESTIMATOR

The value(s) assigned to a population parameter based on the value of a sample statistic is called an estimate. The sample statistic used to estimate a population parameter is called an estimator.

The estimation procedure involves the following steps.

1. Take a sample.
2. Collect the required information from the members of the sample.
3. Calculate the value of the sample statistic.
4. Assign value(s) to the corresponding population parameter.

8.2 POINT AND INTERVAL ESTIMATES

An estimate may be a point estimate or an interval estimate. These two types of estimates are described in this section.

8.2.1 A POINT ESTIMATE

If we select a sample and compute the value of the sample statistic for this sample, then this value gives the **point estimate** of the corresponding population parameter.

POINT ESTIMATE

The value of a sample statistic used to estimate a population parameter is called a point estimate.

Thus, the value computed for the sample mean $\bar{x}$ from a sample is a point estimate of the corresponding population mean μ. In the example mentioned earlier, suppose the U.S. Census Bureau takes a sample of 60,600 families and determines that the mean family size $\bar{x}$ for this sample is 2.7. Then, using $\bar{x}$ as a point estimate of μ, the

Bureau can state that the mean family size μ for all families is about 2.7. This procedure is called *point estimation*.

However, using a point estimate is not a good way to estimate a population parameter. The reason is that each sample taken from a population is expected to yield a different value of the sample statistic. If the Census Bureau takes a second sample of 60,600 families from the United States, the value of $\bar{x}$ calculated for this sample is expected to be different from 2.7. Thus, the value assigned to a population mean μ based on a point estimate depends on which of the samples is drawn. Consequently, the point estimate will assign a value to μ that almost always differs from the true value of the population mean. To resolve this shortcoming, an interval estimate is often used instead of a point estimate.

8.2.2 AN INTERVAL ESTIMATE

In case of **interval estimation,** instead of assigning a single value to a population parameter, an interval is constructed around the point estimate and then a probabilistic statement that this interval contains the corresponding population parameter is made.

INTERVAL ESTIMATE

In interval estimation, an interval is constructed around the point estimate and it is stated that this interval is likely to contain the corresponding population parameter.

For the example about the mean family size in the United States (Section 8.2.1), instead of saying that the mean family size for the population is 2.7, we obtain an interval by subtracting some number from 2.7 and adding the same number to 2.7. Then we state that this interval contains the population mean μ. For purposes of illustration, suppose we subtract .4 from 2.7 and add .4 to 2.7. Consequently, we obtain the interval (2.7 − .4) to (2.7 + .4) or 2.3 to 3.1. Consequently, we state that the interval 2.3 to 3.1 is likely to contain the population mean μ and that the mean family size for all families in the United States is between 2.3 and 3.1. This procedure is called *interval estimation*. The value 2.3 is called the *lower limit* of the interval and 3.1 is called the *upper limit* of the interval. Figure 8.1 illustrates the concept of interval estimation.

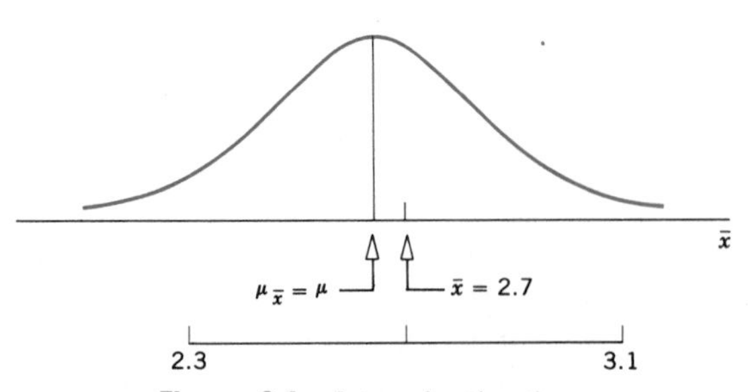

Figure 8.1 Interval estimation.

The question arises, what number should we subtract from and add to a point estimate to obtain an interval estimate? The answer to this question depends on two considerations:

1. The standard deviation $\sigma_{\bar{x}}$ of the sample mean $\bar{x}$
2. The level of confidence to be attached to the interval

First, the larger the standard deviation of $\bar{x}$, the greater the number subtracted from and added to the point estimate. Thus, it is obvious that if the range over which $\bar{x}$ can assume values is larger, the interval constructed around $\bar{x}$ must be wider to include μ.

Second, the quantity subtracted and added must be larger if we want to have a higher confidence in our interval. We always attach a probabilistic statement to the interval estimation. This probabilistic statement is given by the **confidence level.** An interval constructed based on this confidence level is called a **confidence interval.**

CONFIDENCE LEVEL AND CONFIDENCE INTERVAL

Each interval is constructed with regard to a given confidence level and is called a confidence interval. The confidence level associated with a confidence interval states how much confidence we have that this interval contains the true population parameter. The confidence level is denoted by $(1 - \alpha)100\%$.

The confidence level is denoted by $\mathbf{(1 - \alpha)100\%}$. ($\alpha$ is the Greek letter *alpha.*) When expressed as probability, it is called the *confidence coefficient* and is denoted by $1 - \alpha$. In passing, note that α is called the *significance level,* which will be explained in detail in Chapter 9.

Although any value of the confidence level can be chosen to construct a confidence interval, the more common values are 90%, 95%, and 99%. The corresponding confidence coefficients are .90, .95, and .99. The next section describes how to actually construct a confidence interval for the population mean for a large sample.

8.3 INTERVAL ESTIMATION OF A POPULATION MEAN: LARGE SAMPLES

This section explains how to construct a confidence interval for the population mean μ when the sample size is large. We may recall from the discussion in Chapter 7 that, in the case of $\bar{x}$, the sample size is considered to be large when n is 30 or larger. According to the central limit theorem, for a large sample the sampling distribution of the sample mean $\bar{x}$ is (approximately) normal irrespective of the shape of the population from which the sample is drawn. Therefore, *when the sample size is 30 or larger, we will use the normal distribution to construct a confidence interval for* μ. We also know from Chapter 7 that the standard deviation of $\bar{x}$ is $\sigma/\sqrt{n}$. However, if the population standard deviation σ is not known, then we use the sample standard deviation s for σ. Consequently, we use

$$s_{\bar{x}} = \frac{s}{\sqrt{n}}$$

for $\sigma_{\bar{x}} = \sigma/\sqrt{n}$. The value of $s_{\bar{x}}$ is a point estimate of $\sigma_{\bar{x}}$.

CONFIDENCE INTERVAL FOR μ FOR LARGE SAMPLES

The $(1 - \alpha)100\%$ confidence interval for μ is

$$\bar{x} \pm z\,\sigma_{\bar{x}} \quad \text{if } \sigma \text{ is known}$$

$$\bar{x} \pm z\,s_{\bar{x}} \quad \text{if } \sigma \text{ is not known}$$

where

$$\sigma_{\bar{x}} = \frac{\sigma}{\sqrt{n}} \quad \text{and} \quad s_{\bar{x}} = \frac{s}{\sqrt{n}}$$

The value of z is read from the standard normal distribution table for the given confidence level.

Appendix 8.1 on page 415 explains how to obtain these formulas for the confidence interval for μ. An interested reader may refer to that appendix at this point.

The quantity $z\sigma_{\bar{x}}$ (or $zs_{\bar{x}}$ when σ is not known) in the confidence interval formula is called the *maximum error of estimate* and is denoted by E.

MAXIMUM ERROR OF ESTIMATE FOR μ

The maximum error of estimate for μ, denoted by E, is

$$E = z\sigma_{\bar{x}} \quad \text{or} \quad zs_{\bar{x}}$$

The value of z in the confidence interval formula is obtained from the standard normal distribution table (Table VII of Appendix C at the end of the book) for the given confidence level. To illustrate, suppose we want to construct a 95% confidence interval for μ. A 95% confidence level means that the total area under the normal curve for $\bar{x}$ between two points (at the same distance) on both sides of μ is 95% or .95, as shown in Figure 8.2. To find the value of z, we first divide the given confidence coefficient by 2. Then we look for this number in the body of the normal table. The

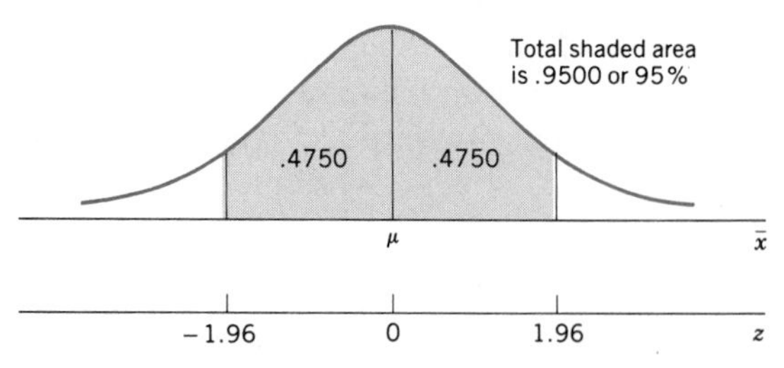

Figure 8.2 Finding z for a 95% confidence level.

corresponding value of z is the value we use in the confidence interval. Thus, to find the z value for a 95% confidence level, we perform the following two steps.

1. First we divide .95 by 2, which gives .4750.
2. Then we locate .4750 in the body of the normal table and record the corresponding value of z. This value of z is 1.96.

For a $(1 - \alpha)100\%$ confidence level, the area between $-z$ and z is $1 - \alpha$. Because the total area under the normal curve is 1.0, the total area under the curve in the two tails is α. This, as mentioned earlier, is called the significance level. In the example of Figure 8.2, $\alpha = 1 - .95 = .05$. Therefore, as shown in Figure 8.3, the area under the curve in each of the two tails is $\alpha/2$. Thus, the value of z associated with a $(1 - \alpha)100\%$ confidence level is sometimes denoted by $z_{\alpha/2}$. However, this text will denote this value simply by z.

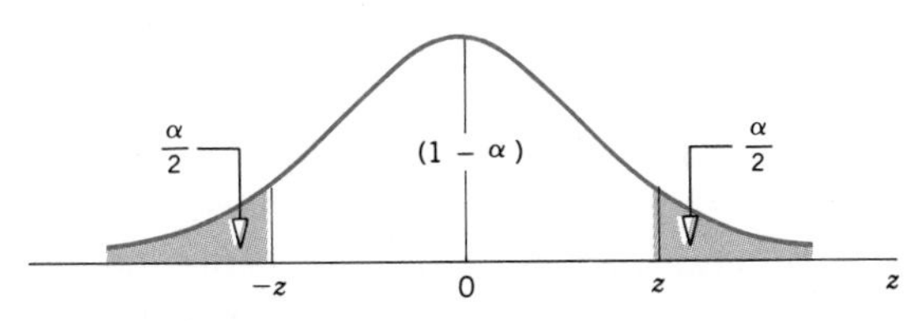

Figure 8.3 Area in the tails.

Example 8-1 describes the procedure to construct a confidence interval for μ for a large sample.

Constructing confidence interval for μ: σ known and $n \geq 30$.

EXAMPLE 8-1 A random sample of 50 college textbooks gave a mean price of \$38.40. It is known that the standard deviation of the prices of all college textbooks is \$4.75. Construct a 95% confidence interval for the mean price of all college textbooks.

Solution From the given information,

$$n = 50, \quad \bar{x} = \$38.40, \quad \sigma = \$4.75$$

and

$$\text{Confidence level} = 95\% \text{ or } .95$$

To construct a 95% confidence interval for μ, first we calculate the standard deviation of $\bar{x}$.

$$\sigma_{\bar{x}} = \frac{\sigma}{\sqrt{n}} = \frac{4.75}{\sqrt{50}} = \$.672$$

Note that we have rounded the value of $\sigma_{\bar{x}}$ to three decimal places.

Next, we find the z value for a 95% confidence level. To do so, we divide .95 by 2 to obtain .4750. From the normal table (Table VII of Appendix C), the z value for .4750 is 1.96.

Finally, we substitute all the values in the confidence interval formula for μ. The 95% confidence interval for μ is

$$\begin{aligned}\bar{x} \pm z\,\sigma_{\bar{x}} &= 38.40 \pm 1.96\,(.672)\\ &= 38.40 \pm 1.32\\ &= (38.40 - 1.32) \text{ to } (38.40 + 1.32)\\ &= \$37.08 \text{ to } \$39.72\end{aligned}$$

Thus, we are 95% confident that the mean price of all college textbooks is between \$37.08 and \$39.72. Note that we cannot say for sure whether the interval \$37.08 to \$39.72 contains the true population mean or not. Because μ is a constant, we cannot say that the probability is .95 that the interval contains μ because either it contains μ or it does not. Consequently, the probability is either 1.0 or zero that this interval contains μ. All we can say is that we are 95% confident that the mean price of all college textbooks is between \$37.08 and \$39.72. ■

How do we interpret a 95% confidence level? In terms of Example 8-1, if we take all possible samples of 50 college textbooks each and construct a 95% confidence interval for μ around each sample mean, we can expect that 95% of these intervals will include μ and 5% will not. In Figure 8.4 we show three means $\bar{x}_1$, $\bar{x}_2$, and $\bar{x}_3$ of three different samples of the same size drawn from the same population. Also shown in the figure are the 95% confidence intervals constructed around these three sample means. As we observe, the 95% confidence intervals constructed around $\bar{x}_1$ and $\bar{x}_2$ include μ, but the one constructed around $\bar{x}_3$ does not. We can state for a 95% confidence level that if we take many samples of the same size from a population and construct 95% confidence intervals around the means of these samples, then 95% of these confidence intervals will be like the ones around $\bar{x}_1$ and $\bar{x}_2$ in Figure 8.4, which include μ, and 5% will be like the one around $\bar{x}_3$, which does not include μ.

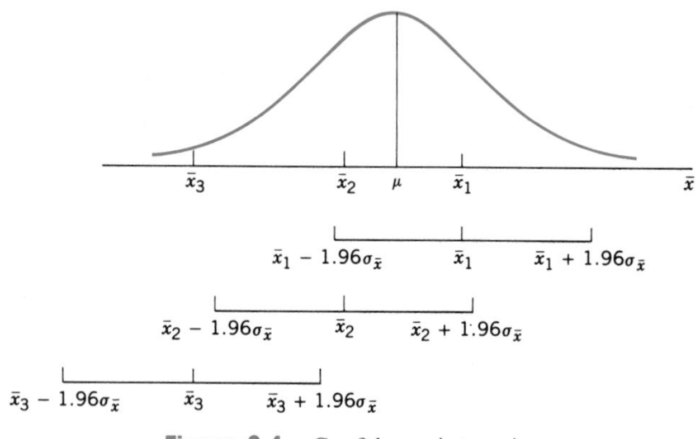

Figure 8.4 Confidence intervals.

In Example 8-1, the value of the population standard deviation σ was known. However, more often we do not know the value of σ. In such cases, we estimate the population standard deviation σ by the sample standard deviation s and estimate the standard deviation $\sigma_{\bar{x}}$ of $\bar{x}$ by $s_{\bar{x}}$. Then we use $s_{\bar{x}}$ for $\sigma_{\bar{x}}$ in the formula for the confidence interval for μ. Owing to the central limit theorem, as long as the sample size is large ($n \geq 30$), even if we do not know σ, we can use the normal distribution.

Example 8-2 illustrates the construction of a confidence interval for μ when σ is not known.

Constructing confidence interval for μ: σ not known and $n \geq 30$.

EXAMPLE 8-2 According to a survey conducted by the U.S. Bureau of Labor, the transportation and public utility workers earned on average $495.26 per week in 1989 (*Monthly Labor Review,* September 1990). Assume that this mean is based on a random sample of 1000 transportation and public utility workers and that the standard deviation of weekly earnings for this sample is $76. Find a 90% confidence interval for the 1989 mean weekly earnings of all transportation and public utility workers.

Solution From the given information,

$$n = 1000, \quad \bar{x} = \$495.26, \quad s = \$76$$

and

$$\text{Confidence level} = 90\% \text{ or } .90$$

First we find the standard deviation of $\bar{x}$. Because σ is not known, we will use $s_{\bar{x}}$ as an estimate of $\sigma_{\bar{x}}$. The value of $s_{\bar{x}}$ is

$$s_{\bar{x}} = \frac{s}{\sqrt{n}} = \frac{76}{\sqrt{1000}} = 2.403$$

Because the sample size is large ($n > 30$), we will use the normal distribution to determine the confidence interval for μ. To find z for a 90% confidence level, we divide .90 by 2 to obtain .4500. Then we locate .4500 in the body of the normal distribution table. Because .4500 is not in the normal table, we can use a number closest to .4500, which is either .4495 or .4505. If we use .4505 as an approximation for .4500, the value of z for this number is 1.65.† Substituting all the values in the formula, the 90% confidence interval for μ is

$$\bar{x} \pm z\, s_{\bar{x}} = 495.26 \pm 1.65\,(2.403) = 495.26 \pm 3.96 = \$491.30 \text{ to } \$499.22$$

Thus, we can state with 90% confidence that in 1989 the average earnings of transportation and public utility workers were between $491.30 and $499.22 per week. ■

The *width of a confidence interval* depends on the size of the maximum error $z\sigma_{\bar{x}}$, which depends on the values of z, σ, and n because $\sigma_{\bar{x}} = \sigma/\sqrt{n}$. However, the value of σ is not within the control of the investigator. Hence, the width of the confidence interval depends on

1. The value of z, which depends on the confidence level
2. The sample size n

†Note that there is no apparent reason for choosing .4505 and not choosing .4495. If we choose .4495, the z value will be 1.64. An alternative is to use the average of 1.64 and 1.65, 1.645, which we will not do in this text.

The confidence level determines the value of z, which in turn determines the size of the maximum error. The value of z increases as the confidence level increases, and it decreases as the confidence level decreases. For example, the value of z is approximately 1.65 for a 90% confidence level, 1.96 for a 95%, and approximately 2.58 for a 99% confidence level. Hence, the higher the confidence level, the larger the width of the confidence interval, other things remaining the same.

For the same value of σ, an increase in the sample size decreases the value of $\sigma_{\bar{x}}$, which in turn decreases the size of the maximum error when the confidence level remains unchanged. Therefore, an increase in the sample size decreases the width of the confidence interval.

Thus, if we want to decrease the width of a confidence interval, we have two choices:

1. Lower the confidence level
2. Increase the sample size

However, lowering the confidence level is not a good choice because a lower confidence level may give less reliable results. Therefore, we should always prefer to increase the sample size if we want to decrease the width of a confidence interval. Example 8-3 illustrates how a decrease in the confidence level decreases the width of the confidence interval.

Confidence level and width of the confidence interval.

EXAMPLE 8-3 According to claims data of the Metropolitan Life Insurance Company, 2788 cardiac catheterizations performed in the United States in 1988 revealed an average total (hospital and physician's) charge of $4820 (*Statistical Bulletin*, July–September 1990). Assume that these 2788 cases represent a random sample of all the cardiac catheterizations performed in the United States in 1988 and that the standard deviation of charges for these 2788 cases is $750. Construct a confidence interval for μ with a confidence level of

(a) 99% (b) 95%

Solution From the given information,

$$n = 2788, \quad \bar{x} = \$4820, \quad \text{and} \quad s = \$750$$

The value of $s_{\bar{x}}$ is

$$s_{\bar{x}} = \frac{s}{\sqrt{n}} = \frac{750}{\sqrt{2788}} = \$14.204$$

The sample size is large ($n > 30$), so we use the normal distribution to construct the confidence intervals.

(a) To find z for a 99% confidence level, we locate .4950 (= .99/2) in the body of the normal distribution table. Because .4950 is not in the table, we approximate it by .4951, which gives the z value of 2.58. Substituting all the values in the formula, the 99% confidence interval for μ is

$$\bar{x} \pm z\, s_{\bar{x}} = 4820 \pm 2.58\,(14.204) = 4820 \pm 36.65 = \$4783.35 \text{ to } \$4856.65$$

(b) From the normal distribution table $z = 1.96$ for a 95% confidence level. The 95% confidence interval for μ is

$$\bar{x} \pm z\, s_{\bar{x}} = 4820 \pm 1.96\,(14.204) = 4820 \pm 27.84 = \$4792.16 \text{ to } \$4847.84$$

Thus, from these two results of parts (a) and (b), we observe that the width of the confidence interval for a 99% confidence level is larger than the one for a 95% confidence level. ■

CASE STUDY 8-1 CRYING BEHAVIOR IN THE HUMAN ADULT

Dr. William H. Frey, II, of St. Paul-Ramsey Medical Center at St. Paul, Minnesota, and his associates studied the crying behavior in 286 women and 45 men aged 18 to 75 years. The subjects included in the survey were asked to keep all records for each episode of emotional and irritant crying "including date, time, duration, reason for crying (situation, thought, emotions), and components of crying episode (lump in throat, watery eyes, flowing tears, sobbing)" for a period of 30 days. The authors divided all the subjects into two groups, "those subjects who met all of the psychiatric status criteria and those who failed to meet one or more of those criteria. The criteria were as follows: no diagnosed psychiatric illness, medication for psychiatric illness, or mental health counseling in the last six months; no episode of depression lasting at least one week in the last six months; no evidence of depression . . . ; and no evidence of labile or histrionic personality disorders as indicated by answers to 11 questions. . . ." The subjects included in the first group (who met all the criteria) were called "normal subjects" in the study. Let us refer to the second group (who failed to meet one or more of the criteria) as a "special group."

Based on the records of subjects in the samples, the authors calculated the mean number of times women and men cried per month, the mean duration of crying per episode, and the corresponding standard deviations. Using the means and the standard deviations of the means calculated by the authors, we have computed the 95% confidence intervals for some of the population parameters in the following table.

Subjects	Group	Population Parameter	95% Confidence Interval
Women	Normal	Mean number of emotional crying episodes per month	5.3 ± 1.96 (.3)
Men	Normal	Mean number of emotional crying episodes per month	1.4 ± 1.96 (.4)
Women	Special	Mean number of emotional crying episodes per month	7.1 ± 1.96 (.6)
Women	Normal	Mean duration of emotional crying episodes (in minutes)	6.0 ± 1.96 (1.0)
Men	Normal	Mean duration of emotional crying episodes (in minutes)	6.0 ± 1.96 (2.0)
Women	Special	Mean duration of emotional crying episodes (in minutes)	11.0 ± 1.96 (2.0)

(*Note:* The number of subjects in different groups were: 175 in normal group for women, 111 in special group for women, and 30 in normal group for men. The special group for men is not included in the table because it included only 15 subjects, which does not make a large sample.)

The values within parentheses in the fourth column of the table are the values of $s_{\bar{x}}$.

For example, according to the authors the mean number of emotional crying episodes per month for normal women (the first group in the table) was 5.3 and the standard deviation of the mean was .3. Hence, a 95% confidence interval for the mean number of emotional crying episodes per month for women belonging to the normal group is $5.3 \pm 1.96\,(.3)$ or 4.71 to 5.89. In other words, at the 95% confidence level, we can state that all the women belonging to the normal group are expected to cry on average 4.71 to 5.89 times per month. We can interpret the other intervals for the remaining two groups the same way.

According to the study, "the stimulus for 800 crying episodes in normal women was 40% interpersonal relations (arguments, weddings, etc.), 27% media (movies, television, etc.), 6% sad thoughts, 1% physical pain, and 26% other. The stimulus distribution of episodes for normal men was 36% media, 36% interpersonal relations, 9% sad thoughts, and 19% other. . . . The primary emotion associated with female crying episodes were as follows: 49% sadness, 21% happiness, 10% anger, 7% sympathy, 5% anxiety, 3% fear, and 5% other." The 95% confidence interval for the mean duration of female emotional crying episodes where sadness was the reason was $7.0 \pm 1.96\,(.8)$ minutes and that for happiness was $2.3 \pm 1.96\,(.5)$ minutes. "Eighty-five percent of the women and 73% of the men reported they generally felt better after crying."

Source: William H. Frey II, Carrie Hoffman-Ahern, et al.: "Crying Behavior in the Human Adult," *Integrative Psychiatry*, September–October 1983, 94–100. Copyright © 1983 by Elsevier Science Publishing Co., Inc. Data and excerpts within quotes reprinted with permission of the publisher.

EXERCISES

8.1 Briefly explain the meaning of an estimator and an estimate.

8.2 Explain the meaning of a point estimate and an interval estimate.

8.3 Suppose for a data set $n = 50$, $\bar{x} = 16$, and $\sigma = 5.3$.

- **a.** Construct a 95% confidence interval for μ.
- **b.** Construct a 90% confidence interval for μ. Is the width of the 90% confidence interval smaller than the width of the 95% confidence interval calculated in part a? If yes, why is it so?
- **c.** Now suppose $n = 100$ but the values of $\bar{x}$ and σ remain the same. Find a 95% confidence interval for μ. Is the width of the 95% confidence interval for μ with $n = 100$ smaller than the width of the 95% confidence interval for μ with $n = 50$ calculated in part a? If so, why?

8.4 Suppose for a data set $n = 81$, $\bar{x} = 18.5$, and $\sigma = 3.7$.

- **a.** Construct a 90% confidence interval for μ.
- **b.** Construct a 99% confidence interval for μ. Is the width of the 99% confidence

interval larger than the width of the 90% confidence interval calculated in part a? If yes, why is it so?

c. Now suppose $n = 60$ but the values of $\bar{x}$ and σ remain the same. Find a 90% confidence interval for μ. Is the width of the 90% confidence interval for μ with $n = 60$ larger than the width of the 90% confidence interval for μ with $n = 81$ calculated in part a? If so, why?

8.5 According to the U.S. Bureau of the Census, the average (per person) annual spending on clothing and shoes in 1989 was $799 (*Family Economics Review,* May 1990). Assume that this mean is based on a random sample of 1600 persons and that the standard deviation for this sample is $176. Construct a 99% confidence interval for the average (per person) annual spending on clothing and shoes in 1989 for all persons.

8.6 According to a Gallup poll of 2727 adults conducted in May 1990, households contributed on average $734 to charitable causes in 1989. Assume that the standard deviation of the 1989 charitable contributions of these households is $185. Construct a 99% confidence interval for the mean charitable contributions made by all households in 1989.

8.7 According to a Hertz Corporation survey, the mean depreciation for compact cars was $1774 in 1987 (*Press Information,* February 15, 1988, The Hertz Corporation). Assume that this result is based on a random sample of 400 compact cars and that the standard deviation of depreciations for this sample is $290. Construct a 99% confidence interval for μ.

8.8 In a study done by Professor Harold Stevenson and others, the mean time spent on homework on weekdays by 237 American first-graders was found to be 14 minutes a day (H. W. Stevenson et al.: "Mathematics Achievement of Chinese, Japanese, and American Children," *Science,* Vol. 231, February 14, 1986). Assume that the standard deviation of time spent each day on homework during the weekdays by these 237 first-graders is 4 minutes. Make a 95% confidence interval for μ.

8.9 According to a study done by Dr. Martha S. Linet and others, the mean duration of the most recent headache was 8.2 hours for a sample of 5055 females aged 12 through 29 (Martha S. Linet et al.: "An Epidemiologic Study of Headache Among Adolescents and Young Adults," *The Journal of the American Medical Association,* 261(15): April 21, 1989). Assume that the standard deviation for this sample is 2.4 hours. Find a 95% confidence interval for the mean duration of the most recent headache for all 12- to 29-year-old females.

8.10 A random sample of 40 students taken from a university showed that their mean GPA (grade point average) is 2.94 with a standard deviation of .30. Construct a 90% confidence interval for the mean GPA of all students at this university.

8.11 According to a study done by Professor Adam Drewnowski and others on weight loss among college students, a sample of 507 female college students indicated that they desired to lose an average of 3.6 kilograms (Adam Drewnowski et al.: "The Prevalence of Bulimia Nervosa in the US College Student Population," *American Journal of Public Health,* 78(10): October 1988). Assume that the standard deviation of desired weight loss for this sample is .97 kilograms. Construct a 90% confidence interval for the mean desired weight loss for all female college students.

8.12 According to a recent Health and Nutrition Examination Survey, the mean height of women aged 18 to 74 is 63.7 inches. Assume that this result is based on a random sample of 900 women aged 18 to 74 and that the standard deviation of heights for this sample is 2.4 inches. Determine a 98% confidence interval for the mean height of all U.S. women aged 18 to 74.

8.13 According to the U.S. Bureau of the Census, the mean income for persons with a bachelor's degree is $38,973. Although the Census Bureau estimates are based on very large samples, for convenience assume that this result is based on a random sample of 5000 persons with a bachelor's degree and that the standard deviation for this sample is $6340. Find a 95% confidence interval for the mean income for all persons with a bachelor's degree.

8.14 According to a U.S. Bureau of Labor survey, workers in the private sector earned an average of \$335.20 per week in 1989 (*Monthly Labor Review,* September 1990). Assume that this mean is based on a random sample of 900 private sector workers and that the standard deviation of earnings for this sample is \$73.

a. Determine a 99% confidence interval for μ.

***b.** Suppose the confidence interval obtained in part *a* is too wide. How can the width of this interval be reduced? Discuss all alternatives. What alternative do you think is best and why?

8.15 A random sample of 100 households showed that they spent an average of \$370 on Christmas gifts last year with a standard deviation of \$105.

a. Find a 95% confidence interval for the mean expenditure on Christmas gifts for all households.

***b.** Suppose the confidence interval obtained in part *a* is too wide. How can the width of this interval be reduced? Which of these alternatives is the best and why?

***8.16** You are interested in estimating the mean commuting time from home to school for all commuter students at your school. Briefly explain the procedure you will follow to conduct this study. Collect the required data from a sample of 30 or more students and then estimate the population mean at a 99% confidence level. [*Hint:* Randomly select 30 or more students from your school (these may be your friends if they make a representative sample), collect the information on the commuting time for each of them, and then find the mean and standard deviation for these data using the formulas learned in Chapter 3. You may use MINITAB to calculate the mean and standard deviation for the sample. Finally, make a confidence interval for μ.]

8.4 INTERVAL ESTIMATION OF A POPULATION MEAN: SMALL SAMPLES

We may recall from Section 8.3 that for large samples ($n \geq 30$), whether or not σ is known, the normal distribution is used to estimate the population mean μ. We use the normal distribution in such cases because, according to the central limit theorem, the sampling distribution of $\bar{x}$ is approximately normal for large samples irrespective of the shape of the population distribution.

However, many times we can take only small samples. This may be due to the nature of the experiment or the cost involved in taking a sample. For example, to test a new drug on patients, research may have to be based on a small sample either because there are not many patients available or willing to participate or because it is too expensive to include enough patients in the research to make it a large sample.

If the sample size is small, the normal distribution can still be used to construct a confidence interval for μ if (1) the population from which the sample is taken is normally distributed, and (2) the value of σ is known. But more often we do not know σ and we have to use the sample standard deviation s as an estimator of σ. In such cases, the normal distribution cannot be used to construct confidence intervals about μ. When (1) the population from which the sample is taken is (approximately) normally distributed, (2) the sample size is small, and (3) σ is not known, the normal distribution is replaced by the *t distribution* to construct confidence intervals about μ. The t distribution is described next.

WHEN TO USE THE t DISTRIBUTION TO MAKE CONFIDENCE INTERVAL ABOUT μ

The t distribution is used to make a confidence interval about μ if

1. The population from which the sample is drawn is (approximately) normally distributed
2. The sample size is small ($n < 30$)
3. The population standard deviation σ is not known

8.4.1 THE t DISTRIBUTION

The t distribution was discovered by W. S. Gossett in 1908 and published under the pseudonym Student. Hence, the t distribution is also called *Student's t distribution*. The t distribution is similar to the normal distribution in some respects. Like the normal distribution curve, the t distribution curve is symmetric (bell-shaped) about the mean and it never meets the horizontal axis. The total area under a t distribution curve is 1.0 or 100%. However, the t distribution curve is flatter than the standard normal distribution curve. In other words, the t distribution curve has a lower height and wider spread (or, we can say, larger standard deviation) than the standard normal distribution. As the sample size increases, the t distribution approaches the standard normal distribution.

The shape of a particular t distribution curve depends on the number of **degrees of freedom (*df*)**. For the purpose of Chapters 8 and 9, the number of degrees of freedom for a t distribution is equal to the sample size minus one, that is,

$$df = n - 1$$

The number of degrees of freedom is the only parameter of the t distribution. There is a different t distribution for each number of degrees of freedom. The units of a t distribution are denoted by t. Like the standard normal distribution, the mean of the t distribution is zero. But unlike the standard normal distribution, whose standard deviation is 1, the standard deviation of a t distribution is $df/(df - 2)$, which is always greater than 1. Thus, the standard deviation of a t distribution is larger than the standard deviation of the standard normal distribution.

THE t DISTRIBUTION

The t distribution is a specific type of bell-shaped distribution with lower height and wider spread than the standard normal distribution. As the sample size becomes larger, the t distribution approaches the standard normal distribution. The t distribution has only one parameter, called the degrees of freedom (df). The mean of a t distribution is equal to zero and its standard deviation is equal to $df/(df - 2)$.

Figure 8.5 shows the standard normal distribution, and the t distribution for 9 degrees of freedom. The standard deviation of the standard normal distribution is

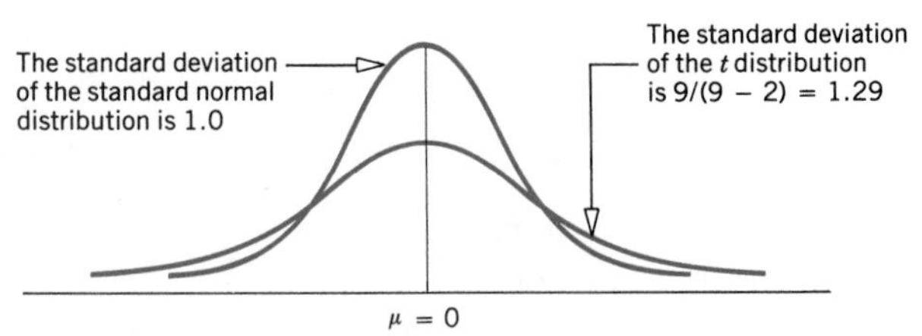

Figure 8.5 The t distribution for $df = 9$ and the standard normal distribution.

1.0, and the standard deviation of the t distribution is $df/(df - 2) = 9/(9 - 2) = 1.29$.

As stated earlier, the number of degrees of freedom for a t distribution for the purpose of this chapter is $n - 1$. **The number of degrees of freedom is defined as the number of observations that can be chosen freely.** As an example, suppose we know that the mean of four values is 20. Hence, the sum of these four values is $20(4) = 80$. Now, how many values, out of four, can we choose freely so that the sum of these four values is 80? The answer is that we can freely choose $4 - 1 = 3$ values. Suppose we choose 27, 8, and 19 as the three values. Given these three values and the information that the mean of the four values is 20, the fourth value is $80 - 27 - 8 - 19 = 26$. Thus, once we have chosen three values, the fourth value is automatically determined. Hence, the number of degrees of freedom for this example is

$$df = n - 1 = 4 - 1 = 3$$

We subtract 1 from n because we lose one degree of freedom to calculate the mean.

Table VIII of Appendix C lists the values of t for the given number of degrees of freedom and areas in the right tail of a t distribution. Because the t distribution is symmetric, these are also the values of $-t$ for the same number of degrees of freedom and the same areas in the left tail of the t distribution. Example 8-4 describes how to read Table VIII of Appendix C.

Reading the t distribution table.

EXAMPLE 8-4 Find the value of t for 16 degrees of freedom and .05 area in the right tail of a t distribution curve.

Solution In Table VIII, we locate 16 in the column of degrees of freedom (labeled df) and .05 in the row of *area in the right tail* at the top of the table. The entry at the intersection of the row of 16 and the column of .05, which is 1.746, gives the required value of t. The relevant part of Table VIII is shown as Table 8.1. The value of t read from the t distribution table is shown in Figure 8.6.

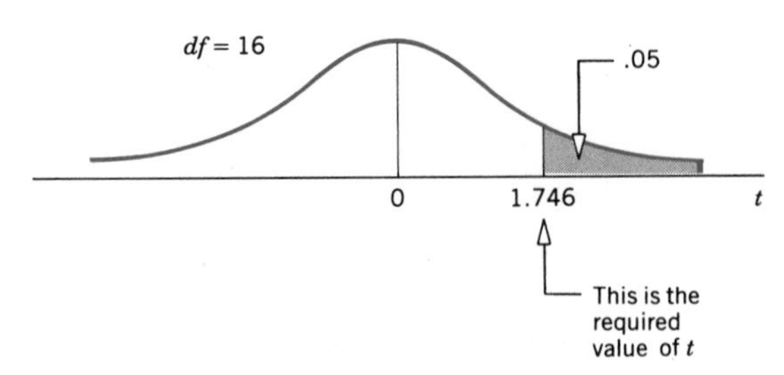

Figure 8.6 The value of t for 16 df and .05 area in the right tail.

Table 8.1 Determining t for 16 df and .05 Area in the Right Tail

	Area in the Right Tail				
df	.10	.05	.025	. . .	.001
1	3.078	6.314	12.706	. . .	318.309
2	1.886	2.920	4.303	. . .	22.327
3	1.638	2.353	3.182	. . .	10.215
.	. . .	. . .	. . .	. . .	. . .
.	. . .	. . .	. . .	. . .	. . .
.	. . .	. . .	. . .	. . .	. . .
df → 16	1.337	1.746	2.120	. . .	3.686
.	. . .	. . .	. . .	. . .	. . .
.	. . .	. . .	. . .	. . .	. . .
.	. . .	. . .	. . .	. . .	. . .
75	1.293	1.665	1.992	. . .	3.202
∞	1.282	1.645	1.960	. . .	3.090

The required value of t for 16 df and .05 area in the right tail

Because of the symmetric shape of the t distribution curve, the value of t for 16 degrees of freedom and .05 area in the left tail is -1.746. Figure 8.7 illustrates this case.

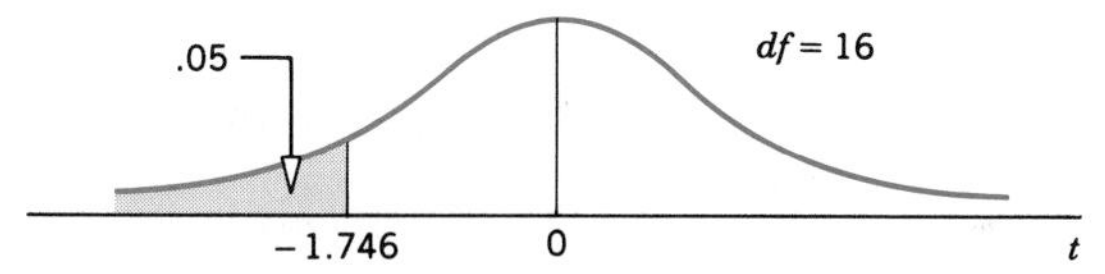

Figure 8.7 The value of t for 16 df and .05 area in the left tail.

8.4.2 THE CONFIDENCE INTERVAL FOR μ USING THE t DISTRIBUTION

To reiterate, when the following three conditions hold true, we use the t distribution to construct a confidence interval for the population mean μ.

1. The population from which the sample is taken is (approximately) normally distributed
2. The sample size is small ($n < 30$)
3. The population standard deviation σ is not known

CONFIDENCE INTERVAL FOR μ FOR SMALL SAMPLES

The $(1 - \alpha)100\%$ confidence interval for μ is

$$\bar{x} \pm t\, s_{\bar{x}}$$

where

$$s_{\bar{x}} = \frac{s}{\sqrt{n}}$$

The value of t is obtained from the t distribution table for $n - 1$ degrees of freedom and the given confidence level.

Examples 8-5 and 8-6 describe the procedure of constructing a confidence interval for μ using the t distribution.

Constructing a 95% confidence interval for μ using the t distribution.

EXAMPLE 8-5 Dr. Moore is interested in estimating the mean cholesterol level for all adult males living in Hartford. He takes a sample of 25 adult males from Hartford and finds that the mean cholesterol level for this sample is 186 with a standard deviation of 12. Assume that the cholesterol levels for all adult males in Hartford are (approximately) normally distributed. Construct a 95% confidence interval for the population mean μ.

Solution From the given information,

$$n = 25, \qquad \bar{x} = 186, \qquad s = 12$$

and

$$\text{Confidence level} = 95\% \text{ or } .95$$

The value of $s_{\bar{x}}$ is

$$s_{\bar{x}} = \frac{s}{\sqrt{n}} = \frac{12}{\sqrt{25}} = 2.40$$

To find the value of t, we need to know degrees of freedom and the area under the t distribution curve in each tail.

$$\text{Degrees of freedom} = n - 1 = 25 - 1 = 24$$

To find the area in each tail, we divide the confidence level by 2 and subtract the number obtained from .5. Hence,

$$\text{Area in each tail} = .5 - \frac{.95}{2} = .5 - .4750 = .025$$

From the t distribution table, Table VIII of Appendix C, the value of t for 24 df and .025 area in the right tail is 2.064. This value of t is shown in Figure 8.8.

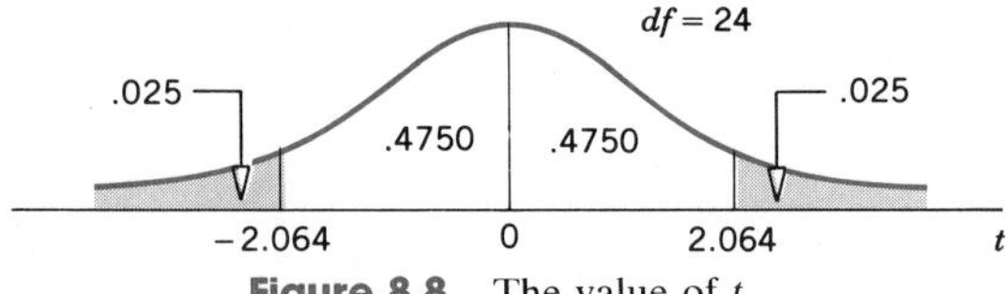

Figure 8.8 The value of t.

Substituting all the values in the formula for the confidence interval for μ, the 95% confidence interval is

$$\bar{x} \pm t\, s_{\bar{x}} = 186 \pm 2.064\ (2.40) = 186 \pm 4.95 = 181.05 \text{ to } 190.95$$

Thus, we can state with 95% confidence that the mean cholesterol level for all adult males living in Hartford lies between 181.05 and 190.95. ■

Constructing a 99% confidence interval for μ using the t distribution.

EXAMPLE 8-6 According to a survey conducted by R. L. Associates, the average cost of nursing-home care is $22,000 a year (*U.S. News & World Report,* August 29/September 5, 1988). Assume that this mean is based on a random sample of 16 nursing homes and that the standard deviation for this sample is $1850. Further assume that the nursing-home costs per year for all nursing homes have an approximate normal distribution. Find a 99% confidence interval for the population mean μ.

Solution From the given information,

$$n = 16, \quad \bar{x} = \$22{,}000, \quad s = \$1850,$$

and

$$\text{Confidence level} = 99\% \text{ or } .99$$

We calculate the standard deviation of $\bar{x}$, the number of degrees of freedom, and the area in each tail of the t distribution as follows.

$$s_{\bar{x}} = \frac{s}{\sqrt{n}} = \frac{1850}{\sqrt{16}} = 462.50$$

$$df = n - 1 = 16 - 1 = 15$$

$$\text{Area in each tail} = .5 - \frac{.99}{2} = .5 - .4950 = .005$$

From the t distribution table, $t = 2.947$ for 15 degrees of freedom and .005 area in the right tail.

The 99% confidence interval for μ is

$$\begin{aligned} \bar{x} \pm t\, s_{\bar{x}} &= 22{,}000 \pm 2.947\,(462.50) \\ &= 22{,}000 \pm 1362.99 \\ &= \$20{,}637.01 \text{ to } \$23{,}362.99 \end{aligned}$$

Thus, we can state with 99% confidence that the mean annual cost of nursing-home care is between $20,637.01 and $23,362.99. ■

Again, we can decrease the width of a confidence interval for μ either by lowering the confidence level or by increasing the sample size. Let us illustrate these two points reconsidering Example 8-6.

1. *Lowering the confidence level.* Suppose we make a 95% confidence interval for μ instead of a 99% confidence interval. The degrees of freedom are $16 - 1 = 15$, and the area in the right tail for a 95% confidence interval is .025. The value of t for 15 df and .025 area in the right tail is 2.131. Using the information given in Example 8-6, the 95% confidence interval for μ is

$$\begin{aligned} \bar{x} \pm t\, s_{\bar{x}} &= 22{,}000 \pm 2.131\,(462.50) \\ &= \$21{,}014.41 \text{ to } \$22{,}985.59 \end{aligned}$$

So we can observe that this confidence interval is narrower than the one constructed in Example 8-6.

2. *Increase the sample size*. Suppose the information given in Example 8-6 is based on a sample of 25 nursing homes. Assume that the sample mean and standard deviation are the same as mentioned in Example 8-6. Suppose the confidence level is 99%. Then, the standard deviation of the sample mean is

$$s_{\bar{x}} = \frac{s}{\sqrt{n}} = \frac{1850}{\sqrt{25}} = 370.00$$

The degrees of freedom are

$$df = n - 1 = 25 - 1 = 24$$

The t value for 24 degrees of freedom and a 99% confidence level is 2.797. Hence, the 99% confidence interval for μ is

$$\bar{x} \pm t\, s_{\bar{x}} = 22{,}000 \pm 2.797\ (370.00)$$
$$= \$20{,}965.11 \text{ to } \$23{,}034.89$$

As we can notice, this confidence interval is also narrower than the one constructed in Example 8-6.

Again, increasing the sample size is a better alternative to use to decrease the width of a confidence interval. This alternative should be preferred over lowering the confidence level.

CASE STUDY 8-2 OPENING MEDICATION CONTAINERS

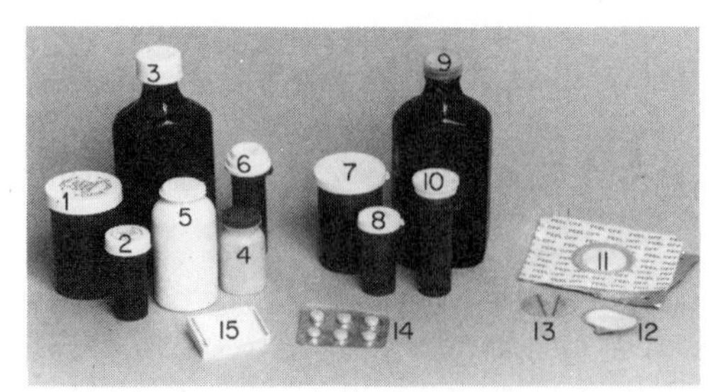

Medication containers used to evaluate the difficulty experienced by older persons in opening them. Child-resistant containers: 1—large push-and-turn; 2—small push-and-turn; 3—push-and-turn for liquids; 4—align two arrows; 5—align a tab with a notch; 6—depress tab and turn (reversible). Non-child-resistant containers: 7—large pop-off cap; 8—small pop-off cap; 9—twist off cap for liquids; 10—reversible cap in twist-off position. Nitroglycerin patches: 11—Nitro-Dur I type; 12—Transderm-Nitro type; 13—Nitro-Dur II type. Miscellaneous containers: 14—blister pack; 15—plastic pocket box.

Dr. Stevan Keram and Dr. Mark E. Williams studied how easy or difficult it is for older persons to open medication containers. They used a sample of 50 persons over 60 years of age for this study.

To measure the difficulty in opening medication containers, they timed each participant while he or she opened a series of medication containers. Each participant was asked to open 15 different containers, which included "six child-resistant and four non-child-resistant and liquid medication containers, placebo version of three transdermal nitroglycerin patches, and two miscellaneous containers" all of which are shown in the picture. To quote the authors:

> Timing began with a signal from the investigator, and ended when the container was opened (in the case of the blister pack, timing ended when the subject succeeded in removing one pill). If the subject gave up or exceeded a 70 second time limit, the trial was stopped and the outcome noted. The nitroglycerin patches were tested in a similar fashion, except that subjects were requested to remove the patch from its package and apply the exposed adhesive surface to a cork pad which had been placed on the table; the maximum time allowed for patch testing was 130 seconds. . . . Mean values and standard deviations for the opening times of the containers were calculated for all individuals who successfully opened the containers.

The first four columns of the following table list the container type, the mean opening time $\bar{x}$ (in seconds), the standard deviation s for the opening times, and the number of subjects (denoted by n) who were able to open the container, respectively. Note that the means and standard deviations are calculated only for those participants who were able to open the containers. In the fifth column are recorded the 99% confidence intervals for the mean opening times for all containers. The normal distribution has been used for cases with $n \geq 30$ and the t distribution for cases with $n < 30$. It is assumed that the assumptions required for the use of the t distribution hold true.

Container Type	$\bar{x}$	s	n	99% Confidence Interval
Non-child resistant				
1. Large pop-off cap	4.6	6.3	50	4.6 ± 2.30
2. Small pop-off cap	5.9	6.2	50	5.9 ± 2.26
3. Twist-off cap for liquids	6.0	3.1	50	6.0 ± 1.13
4. Reversible twist-off cap	8.8	6.7	50	8.8 ± 2.44
Child-resistant				
5. Large push-and-turn cap	6.1	6.0	49	6.1 ± 2.21
6. Small push-and-turn cap	9.2	12.2	44	9.2 ± 4.75
7. Push-and-turn cap for liquids	10.7	6.8	46	10.7 ± 2.59
8. Blister pack	13.3	12.6	45	13.3 ± 4.85
9. Align arrows	21.5	15.2	35	21.5 ± 6.63
10. Align tab with notch	21.7	12.6	28	21.7 ± 6.60
11. Reversible cap (depress tab and turn)	22.7	17.3	26	22.7 ± 9.46
12. Pocket box	42.7	16.7	18	42.7 ± 11.41
Nitroglycerin patches				
13. Nitro-Dur I type	36.4	18.0	23	36.4 ± 10.58
14. Nitro-Dur II type[a]	68.5	22.0	15	68.5 ± 16.91
15. Transderm type	67.7	31.7	19	67.7 ± 20.93

[a]Nitro-Dur II was included in the protocol for 25 of the 50 subjects.

What follows is an explanation of how the confidence interval for the container with the align tab with notch (number 10 in the table) is constructed. For the sample of this type of containers:

$$\bar{x} = 21.7, \quad s = 12.6, \quad \text{and} \quad n = 28$$

Because $n < 30$, we will use the t distribution.

$$\text{Confidence level} = 99\% \text{ or } .99$$

$$\text{Area in each tail} = .5 - \frac{.99}{2} = .005$$

$$df = n - 1 = 28 - 1 = 27$$

The t value for 27 df and .005 area in the right tail is 2.771. The standard deviation of $\bar{x}$ is

$$s_{\bar{x}} = \frac{s}{\sqrt{n}} = \frac{12.6}{\sqrt{28}} = 2.381$$

The 99% confidence interval for μ is

$$\bar{x} \pm t\, s_{\bar{x}} = 21.7 \pm 2.771\,(2.381) = 21.7 \pm 6.60 = 15.1 \text{ to } 28.3$$

Thus, we can state with 99% confidence that the mean time taken by older persons to open a container with the align tab with notch is between 15.1 and 28.3 seconds.

Source: Stevan Keram and Mark E. Williams, MD: "Quantifying the Ease or Difficulty Older Persons Experience in Opening Medication Containers," *Journal of the American Geriatrics Society,* 36(3): 198–201, 1988. Copyright © 1988 by The American Geriatrics Society. Data and excerpts reprinted with permission.

EXERCISES

8.17 Briefly explain the similarities and the differences between the standard normal distribution and the t distribution.

8.18 What are the parameters of a normal distribution and a t distribution?

8.19 Briefly explain the meaning of the degrees of freedom for a t distribution. Give one example to illustrate the concept of the degrees of freedom.

8.20 What assumptions must hold true to use the t distribution to make a confidence interval for μ?

8.21 Suppose for a data set $n = 16$, $\bar{x} = 65.50$, and $s = 8.6$.

- **a.** Construct a 95% confidence interval for μ.
- **b.** Construct a 90% confidence interval for μ. Is the width of the 90% confidence interval smaller than the width of the 95% confidence interval calculated in part a? If yes, why is it so?

c. Now suppose $n = 25$ but the values of $\bar{x}$ and s remain the same. Find a 95% confidence interval for μ. Is the width of the 95% confidence interval for μ with $n = 25$ smaller than the width of the 95% confidence interval for μ with $n = 16$ calculated in part a? If so, why?

8.22 Suppose for a data set $n = 27$, $\bar{x} = 21.5$, and $s = 3.9$.

a. Construct a 95% confidence interval for μ.

b. Construct a 99% confidence interval for μ. Is the width of the 99% confidence interval larger than the width of the 95% confidence interval calculated in part a? If yes, why is it so?

c. Now suppose $n = 20$ but the values of $\bar{x}$ and s remain the same. Find a 95% confidence interval for μ. Is the width of the 95% confidence interval for μ with $n = 20$ larger than the width of the 95% confidence interval for μ with $n = 27$ calculated in part a? If so, why?

8.23 According to the American Bar Association, the mean starting salary for new associates in private law firms was $33,000 in 1987. Assume that this result is based on a random sample of 20 new associates and that the standard deviation of the starting salaries of these new associates is $1800 for 1987. Further assume that the 1987 starting salaries of all new associates are normally distributed. Find a 95% confidence interval for the 1987 mean starting salary for all new associates in private law firms.

8.24 The mean weight for a random sample of 25 college students is found to be 145 pounds with a standard deviation of 15 pounds. Assume that the weights of all college students have a normal distribution. Determine a 95% confidence interval for the mean weight of all college students.

8.25 A Management Progress Study, based on assessment of the performance of general management jobholders from AT&T, found that the mean score in motivation for advancement of 15 math and science majors holding general management jobs at AT&T was 2.67 with a standard deviation of .49 (Ann Howard: "College Experiences and Managerial Performance," *Journal of Applied Psychology*, 71(3): 1986). Assume that these 15 math and science majors make a random sample of all math and science majors holding general management jobs and that the performance scores of all these managers have a normal distribution. Construct a 99% confidence interval for the population mean μ.

8.26 A random sample of 15 statistics students showed that the mean time taken to solve a computer assignment is 19 minutes with a standard deviation of 3 minutes. Construct a 99% confidence interval for the mean time taken by all statistics students to solve this computer assignment.

8.27 A random sample of 20 acres gave a mean yield of wheat equal to 38 bushels per acre with a standard deviation of 3 bushels. Assuming that the yield of wheat per acre is normally distributed, construct a 90% confidence interval for the population mean μ.

8.28 A tool manufacturing company wants to estimate the mean number of bolts produced per hour on a specific machine. The mean number of bolts produced per hour on this machine for a randomly selected 20 hours was found to be 47 with a standard deviation of 2.4. Assume that the number of bolts produced per hour on this machine has a normal distribution. Construct a 90% confidence interval for the population mean μ.

8.29 A company wants to estimate the mean net weight of boxes of its Top Taste cereals. A sample of 16 such boxes produced the net mean weight of 31.98 ounces with a standard deviation of .36 ounces. Make a 98% confidence interval for the mean net weight of all Top Taste cereal boxes. Assume that the net weights of all such cereal boxes have a normal distribution.

8.30 A random sample of 16 midsized cars, which were tested for fuel consumption, gave a mean of 26.4 miles per gallon with a standard deviation of 2.3 miles per gallon.

a. Assuming that the miles per gallon given by all midsized cars have a normal distribution, find a 99% confidence interval for the population mean μ.

*b. Suppose the confidence interval obtained in part a is too wide. How can the width of this interval be reduced? Describe all possible alternatives. Which alternative is the best and why?

8.31 The mean time taken to finish a certain job by 20 construction workers was found to be 68 minutes with a standard deviation of 6 minutes.

a. Assume that the time taken by all construction workers to finish this job is normally distributed. Construct a 95% confidence interval for the population mean μ.

*b. Suppose the confidence interval obtained in part a is too wide. How can the width of this interval be reduced? Describe all possible alternatives. Which alternative is the best and why?

***8.32** The following data give the speeds (in miles), as measured by a radar, of 10 cars traveling on Interstate Highway I-15.

66	72	62	68	76
74	65	78	67	69

Assuming that the speeds of all cars traveling on this highway have a normal distribution, construct a 90% confidence interval for the mean speed of all cars traveling on this highway. (*Hint:* First, calculate the sample mean and the sample standard deviation for these data using the formulas learned in Sections 3.1.1 and 3.2.2 of Chapter 3. Then make the confidence interval for μ.)

***8.33** A sample of eight adults was taken, and these adults were asked about the time they spend per week on leisure activities. Their responses (in hours) are as follows.

45 12 31 16 28 14 18 26

Assuming that the time spent on leisure activities by all adults is normally distributed, make a 95% confidence interval for the mean time spent per week by all adults on leisure activities. (*Hint:* First, calculate the sample mean and the sample standard deviation for these data using the formulas learned in Sections 3.1.1 and 3.2.2 of Chapter 3. Then make the confidence interval for μ.)

***8.34** You have been asked to estimate the mean time taken by a cashier to serve customers at a supermarket. Unfortunately, you have been asked to take a small sample. Briefly explain how you will conduct this study. Collect the data on time taken by a supermarket cashier to serve 10 customers. Then estimate the population mean. Choose your own confidence level. (*Hint:* Note down the time taken by a supermarket cashier to serve 10 customers. Then find the mean and standard deviation for these data using the formulas learned in Chapter 3. You may use MINITAB to calculate the mean and standard deviation for the sample. Finally, make a confidence interval for μ.)

8.5 INTERVAL ESTIMATION OF A POPULATION PROPORTION: LARGE SAMPLES

Often we want to estimate the population proportion or percentage. (*The percentage is obtained by multiplying the proportion by 100.*) For example, the production manager of a company may want to estimate the proportion of defective items produced on a machine. A candidate who is contesting an election may want to find the percentage of all voters who will vote for him. The president of a university may want to know what percentage of the students are against tuition increase.

Again, if we can conduct a census every time we want to find the value of some population proportion, there is no need to learn the procedures discussed in this section. However, we usually derive our results from sample surveys. Hence, to take into account the variability in the results obtained from different sample surveys, we need to know the procedures for estimating a population proportion discussed in this section.

As we may recall from Chapter 7, the population proportion is denoted by p and the sample proportion is denoted by $\hat{p}$. This section explains how to estimate the population proportion p using the sample proportion $\hat{p}$. The sample proportion $\hat{p}$ is a sample statistic, and it possesses a sampling distribution. From Chapter 7, we know that for large samples:

1. The sampling distribution of the sample proportion $\hat{p}$ is (approximately) normal
2. The mean $\mu_{\hat{p}}$ of the sampling distribution of $\hat{p}$ is equal to the population proportion p
3. The standard deviation $\sigma_{\hat{p}}$ of the sampling distribution of the sample proportion $\hat{p}$ is $\sqrt{pq/n}$, where $q = 1 - p$

Remember that, in the case of proportion, the sample is considered to be large if np and nq are both greater than 5.

When estimating the value of a population proportion, we do not know the values of p and q. Hence, we cannot compute $\sigma_{\hat{p}}$. Therefore, in the estimation of the population proportion, we use the value of $s_{\hat{p}}$ as an estimate of $\sigma_{\hat{p}}$. Note that the value of the sample proportion $\hat{p}$ computed from a sample is a *point estimate* of the population proportion p. The value of $s_{\hat{p}}$ is calculated using the following formula.

STANDARD DEVIATION OF $\hat{p}$

The value of $s_{\hat{p}}$, which is an estimate of $\sigma_{\hat{p}}$, is calculated as

$$s_{\hat{p}} = \sqrt{\frac{\hat{p}\,\hat{q}}{n}}$$

The $(1 - \alpha)100\%$ confidence interval for p is constructed using the following formula.

CONFIDENCE INTERVAL FOR THE POPULATION PROPORTION p

The $(1 - \alpha)100\%$ confidence interval for the population proportion p is

$$\hat{p} \pm z\, s_{\hat{p}}$$

The value of z is obtained from the normal distribution table for the given confidence level.

The rationale for using this formula for the estimation of p is similar to the one for using the formula for the estimation of the population mean μ, which is discussed in Appendix 8.1 at the end of this chapter. Examples 8-7 and 8-8 illustrate the procedure for constructing a confidence interval for p.

Constructing a 95% confidence interval for p: large sample.

EXAMPLE 8-7 In a study done by Dr. Dennis H. Novack and others about physicians' attitudes toward the use of deception to resolve ethical problems in medical practice and about how much to tell or whether or not to tell the truth in various situations, 87% of 109 physicians said that deception is acceptable on rare occasions to benefit their patients (Dennis H. Novack, et al.: "Physicians' Attitudes Toward Using Deception to Resolve Difficult Ethical Problems," *The Journal of the American Medical Association,* 261(20): May 26, 1989). Find a 95% confidence interval for the proportion of all physicians who hold this view.

Solution Let p be the proportion of all physicians who believe that deception is acceptable on rare occasions to benefit their patients, and let $\hat{p}$ be the corresponding sample proportion. From the given information,

$$n = 109, \quad \hat{p} = .87, \quad \hat{q} = 1 - \hat{p} = 1 - .87 = .13$$

and

$$\text{Confidence level} = 95\% \text{ or } .95$$

First, we calculate the value of the standard deviation $s_{\hat{p}}$ of the sample proportion as follows.

$$s_{\hat{p}} = \sqrt{\frac{\hat{p}\,\hat{q}}{n}} = \sqrt{\frac{(.87)\,(.13)}{109}} = .032$$

Second, we find the value of z for the given confidence level of 95%. To find z, we divide the confidence level by 2 and look for this number in the body of the normal distribution table and record the corresponding value of z.

$$\frac{.95}{2} = .4750$$

The z value for .4750 is 1.96 from the normal distribution table.

Finally, substituting all the values in the confidence interval formula for p, we obtain:

$$\hat{p} \pm z\, s_{\hat{p}} = .87 \pm 1.96\,(.032) = .87 \pm .063$$

$$= .807 \text{ to } .933 \quad \text{or} \quad 80.7\% \text{ to } 93.3\%$$

Thus, we can state with 95% confidence that .807 to .933 or 80.7% to 93.3% of all physicians believe that deception is acceptable on rare occasions to benefit their patients. ■

Constructing a 99% confidence interval for p: large sample.

EXAMPLE 8-8 In a survey of voters across the United States conducted by *Life* magazine and the Public Opinion Laboratory of Northern Illinois University, 52% said that they would prefer a "super morally straight person like a Sunday school teacher" to be the president of the United States ("Life Polls America: Sex and the Presidency," *Life,* August 1987). Assuming that the sample size for this study was 1500, construct a 99% confidence interval for the proportion of all voters who hold this view.

Solution Let p be the proportion of all voters who prefer a "super morally straight person like a Sunday school teacher" to be the president of the United States, and let $\hat{p}$ be the corresponding sample proportion. From the given information,

$$n = 1500, \quad \hat{p} = .52, \quad \hat{q} = 1 - \hat{p} = 1 - .52 = .48$$

and

$$\text{Confidence level} = 99\% \text{ or } .99$$

The standard deviation $s_{\hat{p}}$ of the sample proportion $\hat{p}$ is

$$s_{\hat{p}} = \sqrt{\frac{\hat{p}\,\hat{q}}{n}} = \sqrt{\frac{(.52)\,(.48)}{1500}} = .013$$

From the normal distribution table, the value of z for $.99/2 = .4950$ is 2.58. Hence, the 99% confidence interval for p is

$$\hat{p} \pm z\, s_{\hat{p}} = .52 \pm 2.58\,(.013) = .52 \pm .034 = .486 \text{ to } .554$$

Thus, we can state with 99% confidence that the proportion of all U.S. voters who would prefer a "super morally straight person like a Sunday school teacher" to be the president of the United States is between .486 and .554. The confidence interval can be converted to a percentage as 48.6% to 55.4%. ■

Again, we can decrease the width of a confidence interval for p either by lowering the confidence level or by increasing the sample size. However, lowering the confidence level is not a good choice because it simply decreases the likelihood that the confidence interval contains p. Hence, to decrease the confidence interval for p, we should always prefer to increase the sample size.

Case Study 8-3 further illustrates the application of confidence intervals for population proportion.

CASE STUDY 8-3 WHAT ENHANCES THE PROFESSIONAL IMAGE OF CAREER WOMEN?

Working Woman magazine conducted a survey of career women about their professional image. The report describing the results of the survey, based on the responses of 8033 career women, was published in the October 1988 issue of the magazine. The following excerpts are quoted from that report.

> Professional image—that complex amalgam of appearance and behavior—is often cited by career experts as one of the most important keys to the executive suite. . . . The survey reveals very definite ideas on what can enhance a woman's professional image—and what can blow it. . . . And while there's no question that career women have made great strides in the work world, the perception persists that image is more important for women than for men, say nearly two-thirds of the survey. . . . The survey provides important evidence of a new and positive feminine power style. Career women overwhelmingly are more likely to want to be seen as skilled and responsible than attractive or—horrors!—sexy at the office, anyway. . . . Flirting quashes professionalism, say 72 percent. So does little-girl behavior like giggling, say 35 percent. Ditto chewing gum, say 59 percent. . . . Crying, another traditionally female ploy, is also disdained; even more women (78%) see it as damaging than warn against acting office coquette. . . . But in general, appearing out of control is a decided drawback, and masculine forms are considered unprofessional as well. Seven out of ten women see losing your temper or using profanity as professionally damaging.
>
> How do career women embody that confident professionalism? By paying careful attention to the details of high-power polish. Nine out of ten rank clothes, hair and shoes as the most important essentials. Almost as high on the list are one's speaking voice, makeup and handshake.

The first column in the following table lists the 10 most damaging things to women's professional image according to this survey. The second column records the proportion of women in the survey who said those things were damaging to their professional image. Under the assumption that these 8033 career women comprise a random sample of all career women in the United States, the table lists the 99% confidence intervals for the corresponding population proportions. The confidence intervals are listed in the third column of the table.

Ten Most Damaging Things	Sample Proportion	99% Confidence Interval
1. Crying in the office	.78	$.78 \pm .013$
2. Miniskirt	.76	$.76 \pm .013$
3. A flirtatious manner	.72	$.72 \pm .013$
4. Losing your temper	.70	$.70 \pm .013$
5. Using off-color language	.70	$.70 \pm .013$
6. Not wearing stockings in summer	.62	$.62 \pm .014$
7. Chewing gum	.59	$.59 \pm .014$
8. Sleeveless dress or top	.45	$.45 \pm .014$
9. Giggling	.35	$.35 \pm .014$
10. Overweight	.35	$.35 \pm .014$

As an example, 78% of the 8033 career women said that crying in the office is damaging to their professional image. The 99% confidence interval for the percentage of all career women who hold this view is constructed as follows.

$$\hat{p} = .78, \qquad \hat{q} = 1 - .78 = .22, \qquad n = 8033$$

From the normal table, z for a 99% confidence level is 2.58.

$$s_{\hat{p}} = \sqrt{\frac{\hat{p}\,\hat{q}}{n}} = \sqrt{\frac{(.78)\,(.22)}{8033}} = .005$$

Therefore, the 99% confidence interval for population proportion is

$$\hat{p} \pm z\, s_{\hat{p}} = .78 \pm 2.58\,(.005) = .78 \pm .013 = .767 \text{ to } .793$$

Converting the proportion to a percentage, the 99% confidence interval for the population percentage is

76.7% to 79.3%

The remaining nine confidence intervals are constructed the same way.

Source: Leah Rosch: "Working Woman: The Professional Image Report—What Matters, What Doesn't," *Working Woman,* October 1988, pp. 108–113, 148–150. Excerpts and data reprinted with permission from *Working Woman* magazine. Copyright © 1988 by WWT Partnership.

EXERCISES

8.35 What assumption(s) must hold true to use the normal distribution to make a confidence interval for the population proportion p?

8.36 Suppose for a data set $n = 100$ and $\hat{p} = .36$.

- **a.** Construct a 95% confidence interval for p.
- **b.** Construct a 99% confidence interval for p. Is the width of the 99% confidence interval larger than the 95% confidence interval calculated in part a? If yes, why is it so?
- **c.** Now suppose $n = 200$ but the value of $\hat{p}$ remains the same. Find a 95% confidence interval for p. Is the width of the 95% confidence interval for p with $n = 200$ smaller than the 95% confidence interval for p with $n = 100$ calculated in part a? If so, why?

8.37 Suppose for a data set $n = 900$ and $\hat{p} = .76$.

- **a.** Construct a 95% confidence interval for p.
- **b.** Construct a 90% confidence interval for p. Is the width of the 90% confidence interval smaller than the 95% confidence interval calculated in part a? If yes, why is it so?
- **c.** Now suppose $n = 500$ but the value of $\hat{p}$ remains the same. Find a 95% confidence interval for p. Is the width of the 95% confidence interval for p with $n = 500$ larger than the 95% confidence interval for p with $n = 900$ calculated in part a? If so, why?

8.38 In a *Time*/CNN poll of 1009 New York City residents, 68% said that the quality of life in New York City has become worse in the past few years (*Time,* September 17, 1990). Construct a 98% confidence interval for the proportion of all New York City residents who hold this view.

8.39 In an economic literacy test given to high school students, 34% defined profits correctly as "revenues minus costs" (*Test of Economic Literacy,* Joint Council on Economic Education).

Assuming that this test was given to 2500 high school students, construct a 99% confidence interval for the proportion of all high school students who could correctly define profits.

8.40 A food company is planning to market a new ice cream. However, before marketing this ice cream, the company wants to find what percentage of the people will like it. The company's research department selected a random sample of 400 persons and asked them to taste this ice cream. Of these 400 persons, 224 said they liked it. Find, with a 99% confidence level, what percentage of all people will like this ice cream. (*Hint:* The sample proportion is $224/400 = .56$.)

8.41 According to a recent report by the U.S. Bureau of the Census, 26% of the households headed by single men own stocks, bonds, and mutual funds. Although Census Bureau estimates are based on very large samples, for convenience assume that this result is based on a random sample of 2000 households headed by single men. Find a 90% confidence interval for the proportion of all households headed by single men who own stocks, bonds, and mutual funds.

8.42 In a *New York Times*/CBS News poll of 941 registered voters, 74% of the respondents said that the Democratic party does more to help the poor (*The New York Times,* August 7, 1988). Construct a 90% confidence interval for the proportion of all registered voters who hold this view.

8.43 In a *Rolling Stone* survey of 816 adults aged 18 to 44 conducted by Peter D. Hart Research Associates, 44% of the respondents said they are worried about having enough money to live on when they retire (*Rolling Stone,* May 5, 1988). Construct a 95% confidence interval for the proportion of all adults aged 18 to 44 who are worried about having enough money to live on when they retire.

8.44 In a *Time* magazine poll of 1014 adults, 57% said it is bad for children that women work outside the home (*Time,* June 22, 1987). Construct a 95% confidence interval for the proportion of all adults who hold this view.

8.45 In a survey conducted by *Parents* magazine, 33% of parents said they play or take part in recreational activities with their children every day (*Parents,* July 1987). Assuming this survey is based on a random sample of 750 parents, construct a 90% confidence interval for the proportion of all parents who play or take part in recreational activities with their children every day.

8.46 In a survey of 1273 adults conducted by the U.S. Office of Technology Assessment, 52% of adults said it is not morally wrong to change the genetic makeup of human cells. Determine a 95% confidence interval for the proportion of all adults who hold this view.

8.47 According to a 1988 survey of 213 four-year colleges and universities conducted by David S. Anderson and Angelo F. Gadaleto, 71% of these colleges and universities permitted drinking hard liquor on campuses (*The Chronicle of Higher Education,* November 9, 1988). Find a 95% confidence interval for the proportion of all four-year colleges and universities that permitted drinking hard liquor on campuses in 1988.

8.48 In a *Hippocrates*/Gallup poll of 1000 adults, 90% said that a terminally ill person has the right to have treatment stopped so that he or she may die if the doctor agrees (*Hippocrates,* May/June 1988). Construct a 99% confidence interval for the proportion of all adults who hold this view.

8.49 In a Gallup poll of 1000 women, 33% said they experience the holidays as a time of stress (*Working Woman,* December 1988). Determine a 99% confidence interval for the proportion of all women who experience the holidays as a time of stress.

8.50 According to a *Newsweek* poll of 750 adults conducted by the Gallup Organization, 77% of adults believe in the existence of heaven (*Newsweek,* March 27, 1989). Construct a 97% confidence interval for the proportion of all adults who believe in the existence of heaven.

8.51 In a Roper Organization poll of 2002 adults, 48% favored banning all advertisements for beer and wine from TV (*The American Way of Buying,* A survey conducted for *The Wall*

Street Journal, 1990). Construct a 98% confidence interval for the percentage of all adults who favor banning all advertisements for beer and wine from TV.

*8.52 You want to estimate the proportion of students at your college who have flown on airplanes at least once in their lifetime. Briefly explain how you will make such an estimate. Collect data from 40 students at your college on whether or not they have flown on an airplane at least once. Then, calculate the proportion of students in this sample who have flown at least once. Using this information, estimate the population proportion. Select your own confidence level.

8.6 SAMPLE SIZE DETERMINATION FOR THE ESTIMATION OF MEAN

One reason why we usually conduct a sample survey and not a census is that almost always we have limited resources at our disposal. In light of this, if a smaller sample can solve our purpose, then we will be wasting our resources by taking a larger sample. For instance, suppose we want to estimate the mean life of a certain auto battery. If a sample of 40 batteries can give us the type of confidence interval that we are looking for, then we will be wasting money and time if we take a sample of a much larger size, say 500 batteries. In such cases, if we know the confidence level and the width of the confidence interval that we want, then we can find the (approximate) size of the sample that will produce the required result.

In Section 8.3 we learned that $z\sigma_{\bar{x}}$ is called the maximum error of estimate for μ and is denoted by E. As we know, the standard deviation $\sigma_{\bar{x}}$ of the sample mean $\bar{x}$ is equal to $\sigma/\sqrt{n}$. Therefore, we can write the maximum error of estimate for μ as

$$E = z \frac{\sigma}{\sqrt{n}}$$

Suppose we predetermine the size of the maximum error E and want to determine the size of the sample that will yield this maximum error. From the above expression, the following formula is obtained that determines the required sample size n.

DETERMINING THE SAMPLE SIZE FOR THE ESTIMATION OF μ

Given the confidence level and the standard deviation of the population, the sample size that will produce a predetermined maximum error E of the confidence interval estimate of μ is

$$n = \frac{z^2 \sigma^2}{E^2}$$

If we do not know σ, we can take a preliminary sample (of any arbitrarily determined size) and find the sample standard deviation s. Then we can use s for σ in the formula. However, note that using s for σ may give a sample size that eventually may produce an error much larger (or smaller) than the predetermined maximum error. This will depend on how close s and σ are.

Example 8-9 illustrates how we determine the sample size that will produce the maximum error of estimate for μ within a certain limit.

Determining sample size for the estimation of μ.

EXAMPLE 8-9 Suppose the U.S. Census Bureau wants to estimate the mean family size for all U.S. families at a 99% confidence level. It is known that the standard deviation σ for the sizes of all families in the United States is .6. How large a sample should the Bureau select if it wants the estimate to be within .01 of the population mean?

Solution The Census Bureau wants the 99% confidence interval for the mean family size to be

$$\bar{x} \pm .01$$

Hence, the maximum size of the error of estimate is to be .01, that is,

$$E = .01$$

The value of z for a 99% confidence level is 2.58. The value of σ is given to be .6. Therefore, substituting all the values in the formula and simplifying:

$$n = \frac{z^2 \sigma^2}{E^2} = \frac{(2.58)^2 (.6)^2}{(.01)^2} = \frac{(6.6564)(.36)}{(.0001)} = 23{,}963.04$$

Thus, the required sample size is 23,964. If the Census Bureau takes a sample of 23,964 families, computes the mean family size for this sample, and then constructs a 99% confidence interval around this sample mean, the maximum error of the estimate will be approximately .01. Note that we have rounded the final answer for the sample size to the next higher integer. This is always the case when determining the sample size. ■

EXERCISES

8.53 Determine the sample size for the estimate of μ for the following.

a. $E = 2.3$, $\sigma = 15.40$, confidence level = 99%
b. $E = 4.1$, $\sigma = 23.45$, confidence level = 95%
c. $E = 25.9$, $\sigma = 122.25$, confidence level = 90%

8.54 Determine the sample size for the estimate of μ for the following.

a. $E = .17$, $\sigma = .9$, confidence level = 99%
b. $E = 1.45$, $\sigma = 5.82$, confidence level = 95%
c. $E = 5.65$, $\sigma = 18.20$, confidence level = 90%

8.55 A researcher wants to determine a 95% confidence interval for the mean number of hours that high school students spend doing homework per week. She knows that the standard deviation for hours spent per week by all high school students doing homework is 7. How large a sample should the researcher select so that the estimate will be within 1.5 hours of the population mean?

8.56 A company that produces detergents wants to estimate the mean amount of detergent in 5-pound boxes at a 99% confidence level. The company knows that the standard deviation of amounts of detergent in 5-pound boxes is .8 ounces. How large a sample should the company take so that the estimate will be within .2 ounces of the population mean?

8.57 A department store manager wants to estimate, at a 90% confidence level, the mean amount spent by all customers at this store. From an earlier study, the manager knows that the standard deviation of amounts spent by all customers at this store is \$27. What sample size should he choose so that the estimate will be within \$3 of the population mean?

8.58 A U.S. government agency wants to estimate, at a 95% confidence level, the mean speed for all cars traveling on Interstate Highway I-95. From a previous study, the agency knows that the standard deviation of speeds of all cars traveling on this highway is 5 miles per hour. What sample size should the agency choose for the estimate to be within 1.5 miles of the population mean?

8.7 SAMPLE SIZE DETERMINATION FOR THE ESTIMATION OF PROPORTION

Just as we did with the mean, we can also determine the sample size for estimating the population proportion p. This sample size will yield an error of estimate that may not be larger than a predetermined maximum error. By knowing the sample size that can give us the required results, we can save our scarce resources by not taking an unnecessary large sample. From Section 8.5, the maximum error E of the interval estimation of the population proportion is

$$E = z\,\sigma_{\hat{p}} = z\sqrt{\frac{p\,q}{n}}$$

By manipulating this expression algebraically, we obtain the following formula to find the required sample size given E, p, q, and z.

DETERMINING THE SAMPLE SIZE FOR THE ESTIMATION OF *p*

Given the confidence level and the values of p and q, the sample size that will produce a predetermined maximum error E of the confidence interval estimate of p is

$$n = \frac{z^2\,p\,q}{E^2}$$

We can observe from this formula that, to find n, we need to know the values of p and q. However, the values of p and q are not known to us. In such a situation, we can choose one of the following alternatives.

1. We make the *most conservative estimate* of the sample size n by using $p = .50$ and $q = .50$. For a given E, these values of p and q will give us the largest sample size by comparison to any other pair of values of p and q because the product of $p = .50$ and $q = .50$ is greater than the product of any other pair of values for p and q.
2. We take a *preliminary sample* (of any arbitrarily determined size) and calculate $\hat{p}$ and $\hat{q}$ for this sample. Then, we use these values of $\hat{p}$ and $\hat{q}$ to find n.

Examples 8-10 and 8-11 illustrate how to determine the sample size that will produce the error of estimation for the population proportion within a predetermined maximum value. Example 8-10 gives the most conservative estimate of n and Example 8-11 uses the results from a preliminary sample to determine the required sample size.

Most conservative estimate of n for the estimation of p.

EXAMPLE 8-10 The EZ Company wants to estimate the proportion of defective items produced by a machine within .02 of the population proportion for a 95% confidence level. What is the most conservative estimate of n that will limit the maximum error to within .02 of the population proportion?

Solution The EZ Company wants the 95% confidence interval for p to be

$$\hat{p} \pm .02$$

Hence

$$E = .02$$

The value of z for a 95% confidence level is 1.96. For the most conservative estimate of n, we will use $p = .50$ and $q = .50$. Hence,

$$n = \frac{z^2\, p\, q}{E^2} = \frac{(1.96)^2\,(.50)\,(.50)}{(.02)^2} = 2401$$

Thus, if the company takes a sample of 2401 items, the estimate of p will be within .02 of the population proportion. ■

Determining n for the estimation of p using a preliminary sample.

EXAMPLE 8-11 Consider Example 8-10 again. Suppose a preliminary sample of 200 items showed that 7 percent of the items produced on this machine are defective. How large a sample should the EZ Company select so that the 95% confidence interval for p is within .02 of the population proportion?

Solution Again, the company wants the 95% confidence interval for p to be

$$\hat{p} \pm .02$$

Hence

$$E = .02$$

The value of z for a 95% confidence level is 1.96. From the preliminary sample,

$$\hat{p} = .07 \qquad \text{and} \qquad \hat{q} = 1 - .07 = .93$$

Hence, using these values of $\hat{p}$ and $\hat{q}$ as estimates of p and q, we obtain

$$n = \frac{z^2\, \hat{p}\, \hat{q}}{E^2} = \frac{(1.96)^2\,(.07)\,(.93)}{(.02)^2} = \frac{(3.8416)\,(.07)\,(.93)}{.0004} = 625.22$$

Thus, if the company takes a sample of 626 items, the estimate of p will be within .02 of the population proportion. However, we should note that this sample size will produce the maximum error within .02 points only if $\hat{p}$ is .07 or less for the new sample. But if $\hat{p}$ for the new sample happens to be much higher than .07, the maximum error will not be within .02. Hence, to avoid such a situation, we may be more conservative and take a sample of much larger size than 626 items. ■

EXERCISES

8.59 Determine the most conservative sample size for the estimation of the population proportion for the following.

a. $E = .03$, confidence level = 99%
b. $E = .04$, confidence level = 95%
c. $E = .01$, confidence level = 90%

8.60 Determine the sample size for the estimation of the population proportion for the following where $\hat{p}$ is the sample proportion based on a preliminary sample.

a. $E = .03$, $\hat{p} = .32$, confidence level = 99%
b. $E = .04$, $\hat{p} = .78$, confidence level = 95%
c. $E = .01$, $\hat{p} = .64$, confidence level = 90%

8.61 A consumer agency wants to estimate the proportion of all drivers who wear seat belts while driving. What is the most conservative estimate of the sample size that would limit the maximum error to within .02 of the population proportion for a 99% confidence interval?

8.62 Refer to Exercise 8.61. Assume that a preliminary study has shown that 65% of the drivers wear seat belts while driving. How large should the sample size be so that the 99% confidence interval for the population proportion has a maximum error of .02?

8.63 A professor of communications wants to estimate the percentage of all adults who read at least one newspaper every day. A preliminary sample taken by this professor has produced this percentage equal to 36. How large a sample should the professor take so that the maximum error for a 95% confidence interval of the population proportion is .025?

8.64 Refer to Exercise 8.63. What is the most conservative estimate of the sample size that would limit the maximum error to within 2.5% of the population percentage for a 95% confidence interval?

GLOSSARY

Confidence interval or interval estimate An interval constructed around the value of a sample statistic to estimate the corresponding population parameter.

Confidence level Confidence level, denoted by $(1 - \alpha)100\%$, states how much confidence we have that a confidence interval contains the true population parameter.

Degrees of freedom (df) The number of observations that can be chosen freely. For the estimation of μ using the t distribution, the degrees of freedom are $n - 1$.

Estimate The value of a sample statistic that is used to find the corresponding population parameter.

Estimator The sample statistic that is used to estimate a population parameter.

Maximum error of estimate The quantity that is subtracted from and added to the value of a sample statistic to obtain a confidence interval for the corresponding population parameter.

Point estimate The value of a sample statistic assigned to the corresponding population parameter.

***t* distribution** A continuous distribution with a specific type of bell-shaped curve with its mean equal to zero and standard deviation equal to $df/(df - 2)$.

KEY FORMULAS

1. **The $(1 - \alpha)100\%$ confidence interval for μ for a large sample ($n \geq 30$)**

$$\bar{x} \pm z\,\sigma_{\bar{x}} \quad \text{if } \sigma \text{ is known}$$

$$\bar{x} \pm z\,s_{\bar{x}} \quad \text{if } \sigma \text{ is not known}$$

where

$$\sigma_{\bar{x}} = \frac{\sigma}{\sqrt{n}} \quad \text{and} \quad s_{\bar{x}} = \frac{s}{\sqrt{n}}$$

2. **The $(1 - \alpha)100\%$ confidence interval for μ for a small sample ($n < 30$) when population is (approximately) normally distributed and σ is unknown**

$$\bar{x} \pm t\,s_{\bar{x}}$$

where

$$s_{\bar{x}} = \frac{s}{\sqrt{n}}$$

3. **The $(1 - \alpha)100\%$ confidence interval for p for a large sample**

$$\hat{p} \pm z\,s_{\hat{p}}$$

where

$$s_{\hat{p}} = \sqrt{\frac{\hat{p}\,\hat{q}}{n}}$$

4. **Maximum error E of the estimate for μ**

$$E = z\,\sigma_{\bar{x}}$$

where $\sigma_{\bar{x}}$ is equal to $\dfrac{\sigma}{\sqrt{n}}$

5. **Required sample size for a predetermined maximum error for estimating μ**

$$n = \frac{z^2\,\sigma^2}{E^2}$$

6. **Maximum error E of the estimate for the population proportion**

$$E = z \sqrt{\frac{p\,q}{n}}$$

7. **Required sample size for a predetermined maximum error for estimating p**

$$n = \frac{z^2\,p\,q}{E^2}$$

Use $p = .50$ and $q = .50$ for the most conservative estimate and the values of $\hat{p}$ and $\hat{q}$ if the estimate is to be based on a preliminary sample.

SUPPLEMENTARY EXERCISES

8.65 According to a recent A. C. Nielsen Television Index, the mean hours spent watching television by children aged 2 to 5 is 22.48 per week. Assume that this mean is based on a random sample of 1200 children aged 2 to 5 and that the standard deviation for this sample is 3.8 hours. Find a 99% confidence interval for the mean time spent per week watching television by all children aged 2 to 5.

8.66 According to a recent A. C. Nielsen Television Index, the mean hours spent watching television by children aged 6 to 11 is 19.63 per week. Assume that this mean is based on a random sample of 1000 children aged 6 to 11 and that the standard deviation for this sample is 3.2 hours. Construct a 99% confidence interval for the mean time spent per week watching television by all children aged 6 to 11.

8.67 A random sample of 100 movie theaters showed that the mean price of a movie ticket is $6.75 with a standard deviation of $.80. Find a 95% confidence interval for the population mean μ.

8.68 According to the study done by Professor Harold Stevenson and others mentioned in Exercise 8.8, the mean time spent on homework on weekdays by 238 American fifth-graders was found to be 46 minutes a day. Assuming that the standard deviation for this sample is 11 minutes, find a 95% confidence interval for μ.

8.69 A bank took a random sample of 100 of its delinquent credit card accounts and found that the mean amount owed on these accounts was $2130 with a standard deviation of $578. Construct a 97% confidence interval for the mean amount owed on all delinquent credit card accounts for this bank.

8.70 According to a *Newsweek* poll of 750 women conducted by the Gallup Organization, the mean weight of women is 138 pounds (*Newsweek*, December 5, 1988). Assuming that the standard deviation for this sample is 18 pounds, find a 98% confidence interval for the mean weight of all women.

8.71 A random sample of 25 wives showed that the mean time spent on housework by them is 28.9 hours a week with a standard deviation of 6.7 hours. Find a 99% confidence interval for the mean time spent on housework per week by all wives. Assume that the time spent on housework per week by all wives is (approximately) normally distributed.

8.72 According to the Metropolitan Life Insurance Company's claims data for 24 cases, the mean hospital and physician's charge for coronary bypass surgeries is $30,690 in Massachusetts (*Statistical Bulletin,* Jan–Mar 1989). Assume that these claims are representative of all such surgeries done in Massachusetts and that the standard deviation for charges for these 24 cases is $2350. Furthermore, assume that the charges for all such surgeries are normally distributed. Construct a 99% confidence interval for the mean charge for all such surgeries done in Massachusetts.

8.73 A random sample of 25 life insurance policyholders showed that the average premium they pay on their life insurance policies is $340 a year with a standard deviation of $62. Assuming that the life insurance policy premiums for all life insurance policyholders have a normal distribution, find a 95% confidence interval for the population mean μ.

8.74 A drug that provides relief from headaches was tried on 18 randomly selected patients. The experiment showed that the mean time to get relief from headaches for these patients after taking this drug was 42 minutes with a standard deviation of 8.5 minutes. Assuming that the time taken to get relief from headaches after taking this drug is (approximately) normally distributed, determine a 95% confidence interval for the mean relief time for this drug for all patients.

8.75 A survey of 20 randomly selected adult males showed that the mean time they spend per week watching sports on television is 9.50 hours with a standard deviation of 2.2 hours. Assuming that the time spent per week watching sports on television by all adult males is (approximately) normally distributed, construct a 90% confidence interval for the population mean μ.

8.76 A random sample of 20 female members of health clubs in Los Angeles showed that they spend on average 4 hours a week doing physical exercise with a standard deviation of .75 hours. Assume that the time spent doing physical exercise by all female members of health clubs in Los Angeles is (approximately) normally distributed. Find a 90% confidence interval for the population mean.

8.77 In a *Glamour* magazine survey of 926 women readers who work outside the home, 35% said that their women colleagues at work are "competitive in a sneaky, back-stabbing way" (*Glamour,* August 1988). Assuming that these 926 respondents make a random sample of all working women, find a 90% confidence interval for the population proportion.

8.78 In a Roper Organization poll, 15% of the respondents said that they believe in reincarnation (*Psychology Today,* September 1988). Assume that this poll was based on a random sample of 1500 adults. Find a 95% confidence interval for the population proportion.

8.79 In a Roper poll of 3000 working women, 56% said "they feel guilty that they don't spend more time with their families" (*The 1990 Virginia Slims Opinion Poll*). Construct a 95% confidence interval for the proportion of all working women who hold this view.

8.80 In a *Family Circle* survey of 35,000 women, 75% said that finding enough time alone with their husbands is "often" or "sometimes" a major stress in relationship (Stephanie Abarbanel and Karen Peterson: "Never Enough Time? You Can Beat the Clock," *Family Circle,* March 14, 1989). Determine a 99% confidence interval for the population proportion.

8.81 In a *Time*/CNN telephone poll of 1012 adult Americans, 11% of the respondents said that Ronald Reagan was a great president (*Time,* January 23, 1989). Construct a 99% confidence interval for the proportion of all adult Americans who think that Reagan was a great president.

8.82 In the *Time*/CNN telephone poll of 1012 adult Americans referred to in Exercise 8.81, 15% said Ronald Reagan was a poor president. Construct a 99% confidence interval for the proportion of all adult Americans who think that Reagan was a poor president.

8.83 A researcher wants to determine a 99% confidence interval for the mean number of hours that adults spend per week doing community service. How large a sample should the researcher select so that the estimate will be within 1 hour of the population mean? Assume that the standard deviation for hours spent per week by all adults doing community service is 3.

8.84 An economist wants to find a 90% confidence interval for the mean sale price of houses in a state. How large a sample should she select so that the estimate will be within \$3500 of the population mean? Assume that the standard deviation for the sale prices of all houses in this state is \$21,500.

8.85 A telephone company wants to estimate the proportion of all households that own telephone answering machines. What is the most conservative estimate of the sample size that would limit the maximum error to within .03 of the population proportion for a 95% confidence interval.

8.86 Refer to Exercise 8.85. Assume that a preliminary sample has shown that 33% of the households in the sample own telephone answering machines. How large should the sample size be so that the 95% confidence interval for the population proportion has a maximum error of .03?

APPENDIX 8.1

RATIONALE BEHIND THE CONFIDENCE INTERVAL FORMULA FOR μ

We know from Chapter 7, that for a large sample ($n \geq 30$), the sampling distribution of $\bar{x}$ is approximately normal irrespective of the shape of the population distribution. Based on the discussion of Chapter 7, we can state that 95% of all means calculated for all possible (large) samples of the same size taken from a population are expected to fall within 1.96 $\sigma_{\bar{x}}$ of μ, as shown in Figure 8.9.

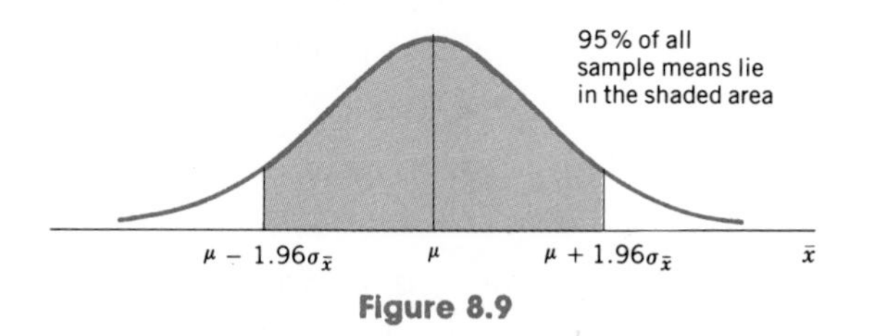

Figure 8.9

Therefore, we can state that for 95% of the (large) samples (of the same size) taken from a population, the sample means will lie in the interval $\mu - 1.96\ \sigma_{\bar{x}}$ to $\mu + 1.96\ \sigma_{\bar{x}}$. That is,

$$\mu - 1.96\ \sigma_{\bar{x}} \leq \bar{x} \leq \mu + 1.96\ \sigma_{\bar{x}} \qquad \text{for 95\% of the samples}$$

By manipulating this expression algebraically, we can state that for 95% of the samples, the population mean μ will be contained in the interval $\bar{x} - 1.96\,\sigma_{\bar{x}}$ to $\bar{x} + 1.96\,\sigma_{\bar{x}}$. That is,

$$\bar{x} - 1.96\,\sigma_{\bar{x}} \leq \mu \leq \bar{x} + 1.96\,\sigma_{\bar{x}} \qquad \text{for 95\% of the samples}$$

Generalizing this concept and replacing 95% by $(1 - \alpha)100\%$, we can state that for $(1 - \alpha)100\%$ of the samples the population mean μ will be contained in the interval $\bar{x} - z\,\sigma_{\bar{x}}$ to $\bar{x} + z\,\sigma_{\bar{x}}$. That is, for $(1 - \alpha)100\%$ of the samples, the population mean μ will be contained in the interval

$$(\bar{x} - z\,\sigma_{\bar{x}}) \text{ to } (\bar{x} + z\,\sigma_{\bar{x}}) \qquad \text{or} \qquad \bar{x} \pm z\,\sigma_{\bar{x}}$$

This gives the $(1 - \alpha)100\%$ confidence interval for the population mean μ.

SELF-REVIEW TEST

1. Complete the following sentences using the terms *population parameter* and *sample statistic*.

a. Estimation means assigning values to a ________________ based on the value of a ________________.

b. An estimator is the ________________ used to estimate a ________________.

c. The value of a ________________ is called the point estimate of the corresponding ________________.

2. A 95% confidence interval for μ can be interpreted to mean that if we take 100 samples of the same size and construct 100 such confidence intervals for μ then,

a. 95 of them will not include μ
b. 95 will include μ
c. 95 will include $\bar{x}$

3. The confidence level is denoted by

a. $(1 - \alpha)100\%$ **b.** $100\alpha\%$ **c.** α

4. The maximum error of the estimate for μ is

a. $z\sigma_{\bar{x}}$ (or $zs_{\bar{x}}$)
b. $\sigma/\sqrt{n}$ (or $s/\sqrt{n}$)
c. $\sigma_{\bar{x}}$ (or $s_{\bar{x}}$)

5. Which of the following assumptions is not required to use the t distribution to make a confidence interval for μ?

a. The population from which the sample is taken is (approximately) normally distributed.
b. $n < 30$

c. The population standard deviation σ is not known.
d. The sample size is at least 10.

6. The parameter(s) of the t distribution is(are)

a. n b. degrees of freedom c. μ and degrees of freedom

7. A sample of 50 packages mailed from a specific post office showed a mean mailing charge of $2.35 with a standard deviation of $.62. Construct a 99% confidence interval for the mean mailing charge for all packages mailed from this post office.

8. A sample of 25 malpractice lawsuits filed against doctors showed that the mean compensation awarded to the plaintiffs was $297,364 with a standard deviation of $74,820. Find a 99% confidence interval for the mean compensation awarded to plaintiffs of all such lawsuits. Assume that the compensations awarded to plaintiffs of all such lawsuits are normally distributed.

9. In a *Time*/CNN poll of 1012 adults conducted by Yankelovich Clancy Shulman, 37% said that, in general, doctors "do a good job being on time" for appointments (*Time,* July 31, 1989). Construct a 95% confidence interval for the proportion of all adults who hold this view.

10. A statistician is interested in estimating, at a 95% confidence level, the mean number of houses sold per month by all real estate agents in a large city. From an earlier study, it is known that the standard deviation of the number of houses sold per month by all real estate agents in this city is 2.1. How large a sample should be taken so that the estimate is within .65 of the population mean?

11. A company wants to estimate the proportion of all workers who hold more than one job. What is the most conservative estimate of the sample size that would limit the maximum error to within .03 of the population proportion for a 99% confidence interval?

12. Refer to Problem 11. Assume that a preliminary study has shown that 12% of adults hold more than one job. How large a sample should be taken in this case so that the maximum error is within .03 of the population proportion for a 99% confidence interval?

*13. Dr. Garcia estimated the mean stress score before a statistics test for a random sample of 25 students. She found the mean and standard deviation for this sample to be 6.8 (on a scale of 1 to 10) and 1.2, respectively. She used a 95% confidence level. However, she thinks that the confidence interval is too wide. How can she reduce the width of the confidence interval? Describe all possible alternatives. Which alternative do you think is the best and why?

USING MINITAB

This section describes how to use MINITAB commands to estimate the population mean for large and small samples. The current version of MINITAB does not have commands to construct a confidence interval for the population proportion directly. Although we can make such an interval for p by using the normal distribution as an approximation to the binomial, we will not review that procedure here.

INTERVAL ESTIMATION OF A POPULATION MEAN: LARGE SAMPLES

The following MINITAB command gives the interval estimation of the population mean for a large sample.

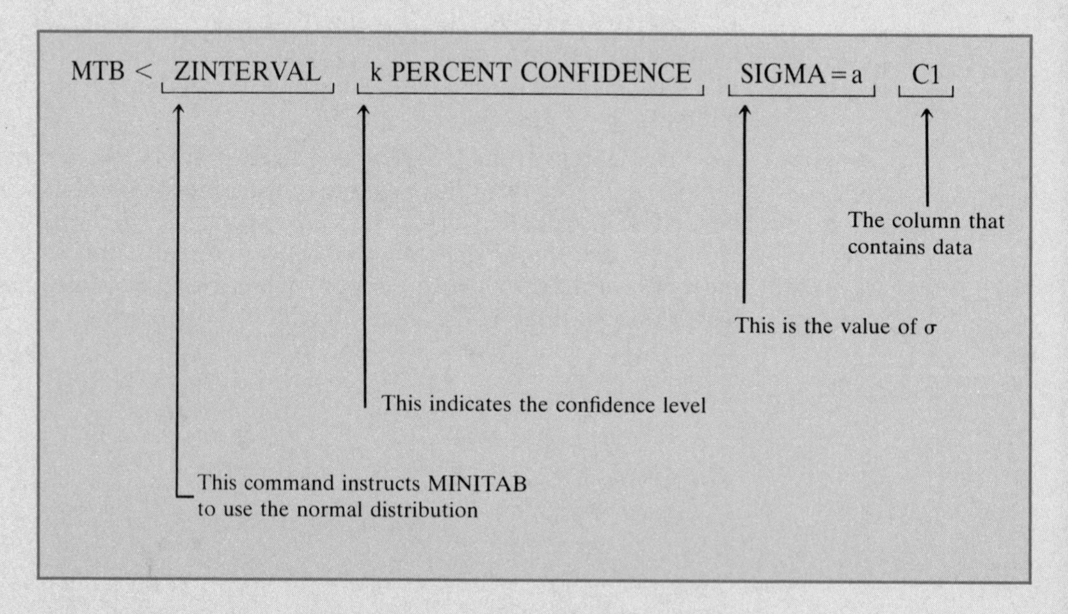

Illustration M8-1 describes the use of this procedure.

ILLUSTRATION M8-1 The following data give the heights of 36 randomly selected adults.

65 68 71 66 69 60 62 69 74 68 70 63
62 72 75 69 65 62 71 68 63 60 66 68
70 68 63 67 71 61 66 61 69 72 71 67

Using MINITAB, find a 95% confidence interval for the mean height of all adults. Assume that the standard deviation of the heights of all adults is 3.9.

Solution

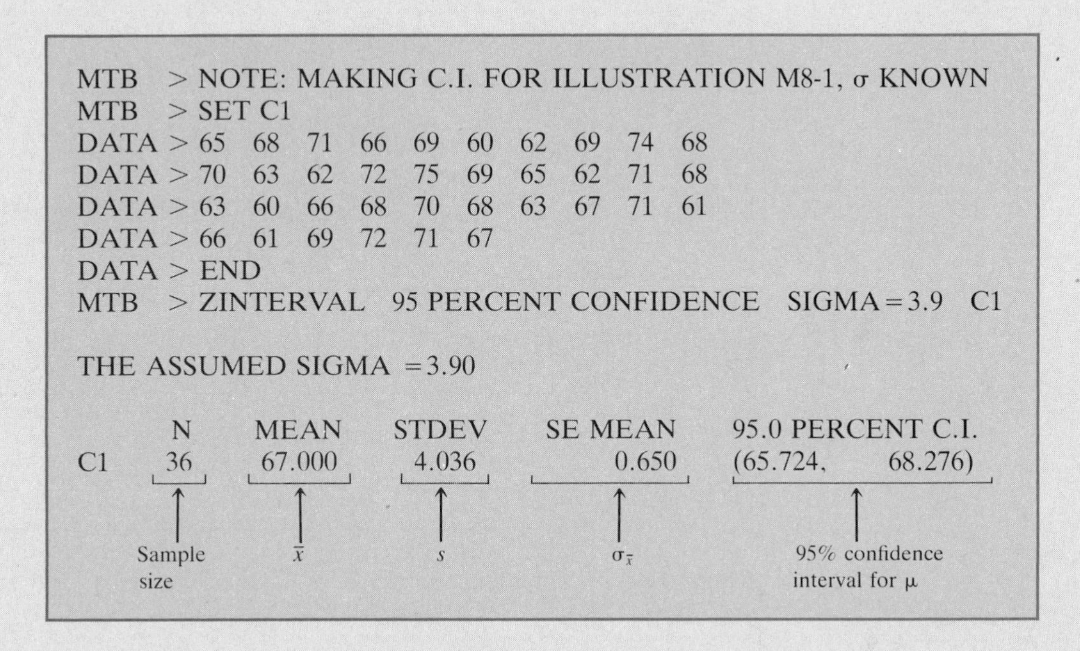

```
MTB  > NOTE: MAKING C.I. FOR ILLUSTRATION M8-1, σ KNOWN
MTB  > SET C1
DATA > 65  68  71  66  69  60  62  69  74  68
DATA > 70  63  62  72  75  69  65  62  71  68
DATA > 63  60  66  68  70  68  63  67  71  61
DATA > 66  61  69  72  71  67
DATA > END
MTB  > ZINTERVAL  95 PERCENT CONFIDENCE  SIGMA=3.9  C1

THE ASSUMED SIGMA =3.90

        N     MEAN     STDEV    SE MEAN    95.0 PERCENT C.I.
C1     36   67.000     4.036      0.650    (65.724,   68.276)
```


In the MINITAB command ZINTERVAL 95 PERCENT CONFIDENCE SIGMA = 3.9 C1 in MINITAB display, ZINTERVAL instructs MINITAB to use the normal distribution, 95 PERCENT CONFIDENCE indicates the confidence level in percentage, SIGMA = 3.9 represents the population standard deviation, and C1 instructs MINITAB to construct a confidence interval for μ using the value of $\bar{x}$ calculated for the data entered in column C1.

In the MINITAB solution, 36 gives the number of values in the data entered in column C1, 67.000 is the mean $\bar{x}$ of that data set, 4.036 is the standard deviation s of the sample data, 0.650 is the standard error (or standard deviation) of the sample mean $\bar{x}$ calculated as $\sigma/\sqrt{n} = 3.9/\sqrt{36} = .65$, 65.724 is the lower limit of the 95% confidence interval for μ, and 68.276 is the upper limit of that confidence interval.

Thus, the 95% confidence interval for the mean height of all adults based on the heights of this sample of 36 adults is

$$65.724 \quad \text{to} \quad 68.276$$

Now, suppose the population standard deviation σ is not known in Illustration M8-1. As the sample size is large, we can use the normal distribution to make a confidence interval for μ. However, when we use the ZINTERVAL command in MINITAB, we must indicate the value of the standard deviation of the population. When σ is not known, we can first calculate the standard deviation of the data in column C1 and then use this value of s as an estimate of σ in the MINITAB command

SIGMA = a. Assuming that the value of σ is not known in Illustration M8-1, we will proceed as follows to find the confidence interval for μ.

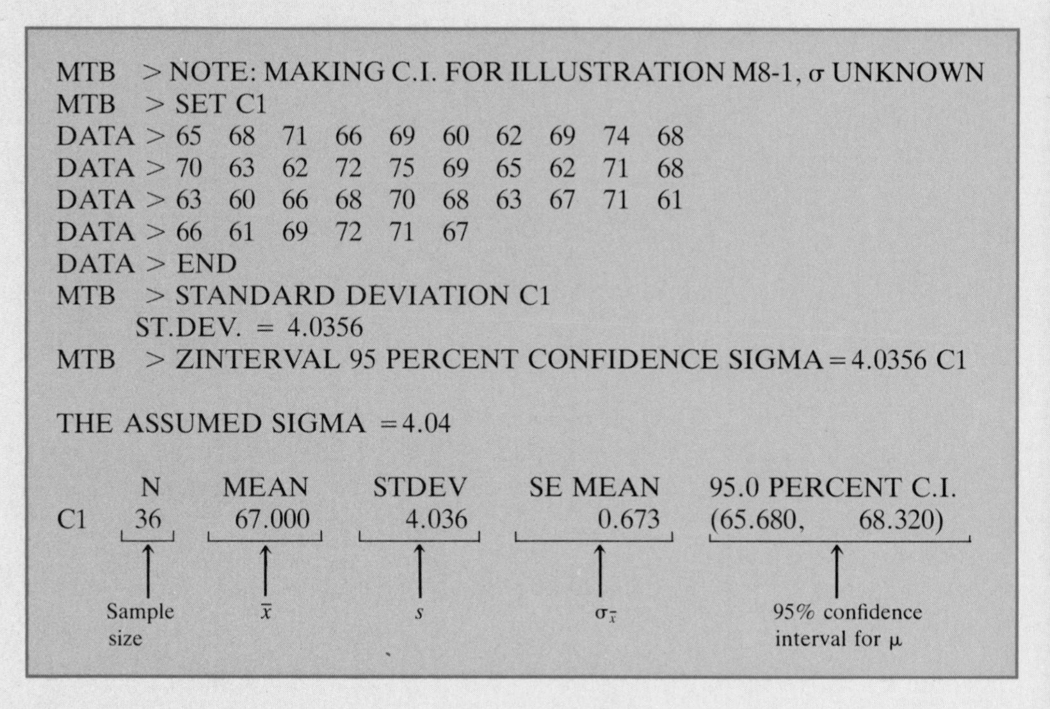

```
MTB   > NOTE: MAKING C.I. FOR ILLUSTRATION M8-1, σ UNKNOWN
MTB   > SET C1
DATA > 65  68  71  66  69  60  62  69  74  68
DATA > 70  63  62  72  75  69  65  62  71  68
DATA > 63  60  66  68  70  68  63  67  71  61
DATA > 66  61  69  72  71  67
DATA > END
MTB   > STANDARD DEVIATION C1
        ST.DEV. = 4.0356
MTB   > ZINTERVAL 95 PERCENT CONFIDENCE SIGMA = 4.0356 C1

THE ASSUMED SIGMA = 4.04

       N     MEAN     STDEV    SE MEAN    95.0 PERCENT C.I.
C1    36   67.000     4.036      0.673    (65.680,   68.320)
```


Thus, the 95% confidence interval for μ when we use s for σ is

$$65.68 \quad \text{to} \quad 68.32$$

INTERVAL ESTIMATION OF A POPULATION MEAN: SMALL SAMPLES

We know from the discussion in this chapter that when the sample size is small and the population standard deviation is not known, we use the t distribution to construct a confidence interval for μ. The MINITAB command for interval estimation of mean for a small sample when σ is unknown is as follows.

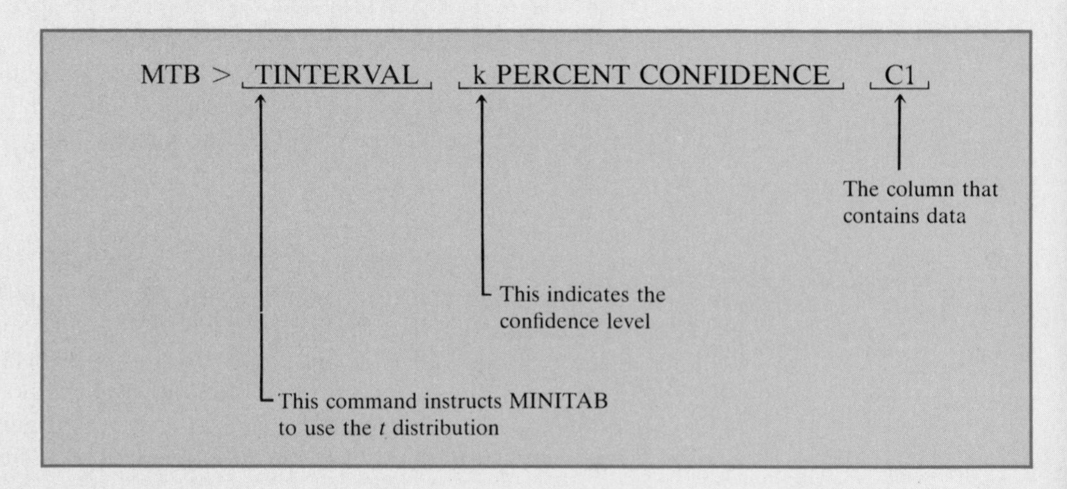

```
MTB > TINTERVAL  k PERCENT CONFIDENCE  C1
```

Illustration M8-2 explains the use of this procedure.

ILLUSTRATION M8-2 The following data give the yearly earnings (in thousands of dollars) of 12 randomly selected households.

36.75	52.43	18.82	28.45	39.50	22.65
14.30	46.75	24.48	31.70	17.25	40.27

Using MINITAB, construct a 99% confidence interval for the mean yearly earnings of all adults. Assume that the distribution of yearly earnings of all households is normal.

Solution

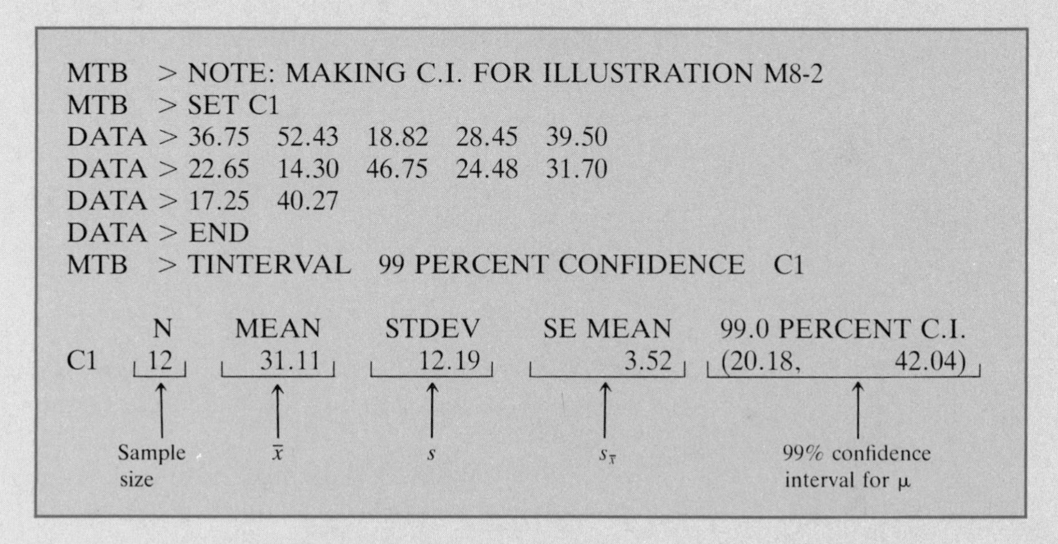

```
MTB  > NOTE: MAKING C.I. FOR ILLUSTRATION M8-2
MTB  > SET C1
DATA > 36.75  52.43  18.82  28.45  39.50
DATA > 22.65  14.30  46.75  24.48  31.70
DATA > 17.25  40.27
DATA > END
MTB  > TINTERVAL  99 PERCENT CONFIDENCE  C1

       N     MEAN     STDEV    SE MEAN    99.0 PERCENT C.I.
C1    12    31.11     12.19       3.52    (20.18,      42.04)
```


Thus, the 99% confidence interval for the yearly earnings of all households based on the earnings of 12 households is

20.18 to 42.04

Because the given data are in thousands of dollars, the confidence interval can be written as

$20,180 to $42,040

COMPUTER ASSIGNMENTS

M8.1 The data on charitable contributions (in dollars) for 35 households given in Exercise 2.63 are reproduced below.

30	300	50	100	27	100	25
76	25	15	25	60	240	100
18	400	200	10	25	50	125
140	34	275	250	130	87	24
500	75	150	15	200	200	300

Using MINITAB, construct a 99% confidence interval for μ assuming that the population standard deviation is 154.

M8.2 Refer to Data Set I of Appendix B on prices of various products in different cities across the country. Using the data on monthly telephone charges given in column C7, find a 99% confidence interval for the population mean μ.

M8.3 Refer to Manchester Road Race data for all participants that are on the floppy diskette. Using the MINITAB SAMPLE command, take a sample of 100 observation from all the participants.

a. Using the sample data and the ZINTERVAL command, make a 95% confidence interval for the mean time taken to complete this race by all participants. Use the value of the population standard deviation, which is 9.728.

b. The mean time taken to run this race by all participants is 41.97 minutes. Does the confidence interval made in part a include this population mean?

M8.4 Repeat Computer Assignment M8.3 for a sample of 25 observations. Use the TINTERVAL command and assume that σ is not known. Assume that the distribution of time taken to run this race for all participants is approximately normal.

M8.5 The data on gestation period of 10 Rhesus monkeys given in Exercise 3.9 of Chapter 3 are reproduced below.

159	167	174	154	181
154	163	169	177	163

Using MINITAB, construct a 95% confidence interval for the gestation period of all Rhesus monkeys. Assume that the distribution of gestation periods for all Rhesus monkeys is normal.

M8.6 The data on the prices (in thousands of dollars) of 16 recently sold houses given in Exercise 3.93 of Chapter 3 are reproduced here.

141	163	127	104	197	203	113	179
256	228	183	119	133	199	271	191

Using MINITAB, construct a 99% confidence interval for the mean price of all houses in this area. Assume that the distribution of prices of all houses in the given area is normal.

9 HYPOTHESIS TESTS ABOUT THE MEAN AND PROPORTION

9.1 HYPOTHESIS TESTS: AN INTRODUCTION

9.2 HYPOTHESIS TESTS ABOUT A POPULATION MEAN: LARGE SAMPLES

9.3 HYPOTHESIS TESTS USING THE *p*-VALUE APPROACH

9.4 HYPOTHESIS TESTS ABOUT A POPULATION MEAN: SMALL SAMPLES

9.5 HYPOTHESIS TESTS ABOUT A POPULATION PROPORTION: LARGE SAMPLES

SELF-REVIEW TEST

USING MINITAB

This chapter introduces the second topic in inferential statistics: tests of hypotheses. In a test of hypothesis, we test a certain given theory or belief about a population parameter. There may be some claim about a population parameter, and we may want to find out, using some sample information, if this claim is true. This chapter discusses how to make such tests of hypotheses about the population mean μ and the population proportion p.

As an example, a soft-drink company may claim that on average its cans contain 12 ounces of soda. A government agency may want to test whether the cans actually do contain on average 12 ounces of soda. As another example, according to the U.S. Bureau of Labor Statistics, 56.5% of married women in the United States were working outside their homes in 1988. An economist may want to check if this percentage is still true for this year. In the first of these two examples we are to test a hypothesis about the population mean μ, and in the second example we are to test a hypothesis about the population proportion p.

9.1 HYPOTHESIS TESTS: AN INTRODUCTION

Why do we need to test a hypothesis? Reconsider the example about soft-drink cans. Suppose we take a sample of 100 cans of the soft drink under investigation. We then find out that the mean amount of soda in these 100 cans is 11.89 ounces. Based on this result, can we state that on average the cans contain less than 12 ounces of soda and that the company is lying to the public? Not until we test the hypothesis can we make such an accusation. The reason is that the mean $\bar{x} = 11.89$ is obtained from a sample. The difference between 12 and 11.89 may have occurred only because of the sampling error. If we take another sample of 100 cans, it may give us a mean of 12.04 ounces. Therefore, we make a test of hypothesis to find out how large the difference between 12 and 11.89 is and to investigate if it is possible that this difference may have occurred as a result of chance alone. Now, if 11.89 ounces is the mean of all cans and not the mean of only 100 cans, then we do not need to make a test of hypothesis. Instead, we can immediately state that the mean amount of soda in all cans is less than 12 ounces. We make a test of hypothesis only when we are making a decision about a population parameter based on the value of a sample statistic.

9.1.1 TWO HYPOTHESES

Consider a nonstatistical example of a person who has been indicted for a crime and is being tried in a court. Based on the available evidence, the judge or jury will make one of two possible decisions.

1. The person is not guilty.
2. The person is guilty.

At the outset of the trial, the person is presumed not guilty. The prosecutor's efforts are to prove that the person has committed the crime and, hence, is guilty. In statistics, *the person is not guilty* is called the **null hypothesis** and *the person is guilty* is called the **alternative hypothesis.** The null hypothesis is denoted by H_0, and the alternative hypothesis is denoted by H_1. These two hypotheses are written as follows (notice the colon after H_0 and H_1).

Null hypothesis	H_0: The person is not guilty
Alternative hypothesis	H_1: The person is guilty

In a statistics example, the null hypothesis states that a given claim about a population parameter is true. Reconsider the example of a soft-drink company's claim that on average its cans contain 12 ounces of soda. It is possible that this claim may be true or it may be false. However, we will initially assume that the company's claim is true (i.e., the company is not guilty of lying). To test the claim of the soft-drink company, the null hypothesis will be that the company's claim is true. Let μ be the mean amount of soda in all cans. The company's claim will be true if $\mu = 12$. Hence, the null hypothesis will be written as

$$H_0: \mu = 12 \qquad \text{(the company's claim is true)}$$

In this example, the null hypothesis can also be written as $\mu \geq 12$ because if the cans contain on average more than 12 ounces of soda, the claim of the company may still be considered to be true. The company will be accused of cheating the public only if the cans contain on average less than 12 ounces of soda. However, it will not affect the test whether we use an $=$ or a $\geq$ sign in the null hypothesis as long as the alternative hypothesis has a $<$ sign. Remember that in the null hypothesis (and in the alternative hypothesis also) we use the population parameter (such as μ or p) and not the sample statistic (such as $\bar{x}$ or $\hat{p}$). In hypothesis testing, we initially assume that the company's claim (the null hypothesis) is true.

NULL HYPOTHESIS

A null hypothesis is a claim about a population parameter that is assumed to be true until it is declared false.

The alternative hypothesis in our statistics example will be that the company's claim is false and its soft-drink cans contain on average less than 12 ounces of soda, that is, $\mu < 12$. The alternative hypothesis will be written as

$$H_1: \mu < 12 \qquad \text{(the company's claim is false)}$$

ALTERNATIVE HYPOTHESIS

An alternative hypothesis is a claim about a population parameter that will be true if the null hypothesis is false.

Let us return to the example of the court trial. The trial begins with the assumption that the null hypothesis is true, that is, the person is not guilty. The prosecutor assembles all the possible evidence and presents it in the court to prove that the null hypothesis is false and the alternative hypothesis is true (i.e., the person is guilty). In the case of the statistics example, the information obtained from a sample will be used as evidence to decide whether or not the claim of the company is true. In the court case, the decision made by the judge (or jury) depends on the amount of evidence presented by the prosecutor. At the end of the trial, the judge (or jury) will consider whether the evidence presented by the prosecutor is sufficient to declare the person guilty. The amount of evidence that will be considered to be sufficient to declare the person guilty depends on the discretion of the judge (or jury).

9.1.2 REJECTION AND NONREJECTION REGIONS

In Figure 9.1, which represents the court case, the point marked "0" indicates that there is no evidence against the person being tried. The farther the point is to the right on the horizontal axis, the more convincing the evidence is that the person has committed the crime. We have arbitrarily marked a point C on the horizontal axis. Let us assume that a judge (or jury) considers any amount of evidence to the right of point C to be sufficient and any evidence to the left of C to be insufficient to declare the person guilty. Point C is called the **critical value** or **critical point** in statistics. If the amount of evidence presented by the prosecutor falls in the area to the left of

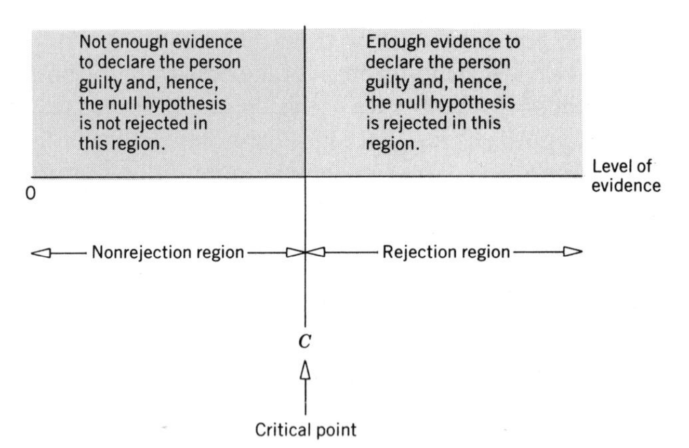

Figure 9.1 Nonrejection and rejection regions for the court case.

point C, the verdict will reflect that there is not enough evidence to declare the person guilty. Consequently, the accused person will be declared not guilty. In statistics, this decision is called *do not reject* H_0. It is equivalent to saying that there is not enough evidence to state that the null hypothesis is false. The area to the left of point C is called the *nonrejection region,* that is, this is the region where the null hypothesis is not rejected. However, if the amount of evidence falls in the area to the right of point C, the verdict will be that there is sufficient evidence to declare the person guilty. In statistics, this decision is called *reject* H_0 or *the null hypothesis is false*. Rejecting H_0 is equivalent to saying that *the alternative hypothesis is true*. The area to the right of point C is called the *rejection region,* that is, this is the region where the null hypothesis is rejected.

9.1.3 TWO TYPES OF ERRORS

We know that a court's verdict is not always correct. If a person is declared guilty at the end of a trial, there are two possibilities.

1. The person is not guilty but is declared guilty (because of what may be false evidence).
2. The person is guilty and is declared guilty.

In the first case, the court has made an error by punishing an innocent person. In statistics, this kind of error is called a **type I** or an α (*alpha*) **error.** In the second case, because the guilty person has been punished, the court has made the correct decision. The second row in the shaded portion of Table 9.1 shows these two cases.

Table 9.1

		Actual Situation	
		The Person Is Not Guilty	**The Person Is Guilty**
Court's decision	The person is not guilty	Correct decision	Type II or β error
	The person is guilty	Type I or α error	Correct decision

The two columns of Table 9.1, corresponding to *the person is not guilty* and *the person is guilty,* give the two actual situations. Which one of these is true is known only to the person being tried. The two rows in this table, corresponding to *the person is not guilty* and *the person is guilty,* show the two possible court decisions.

In our statistics example, a type I error will occur when H_0 is actually true (i.e., the cans do contain on average 12 ounces of soda), but it just happens that we draw a sample whose mean is well below 12 ounces and we wrongly reject H_0. The value of $\boldsymbol{\alpha}$, called the **significance level** of the test, represents the probability of making a type I error. In other words, α is the probability of rejecting H_0 when in fact it is true.

TYPE I ERROR

A type I error occurs when a true null hypothesis is rejected. The value of α represents the probability of committing this type of error, that is,

$$\alpha = P(H_0 \text{ is rejected} \mid H_0 \text{ is true})$$

α is called the *significance level* of the test.

The size of the rejection region in a statistics problem of hypothesis testing depends on the value assigned to α. In tests of hypotheses, we assign a value to α before making the test. Usually the maximum value assigned to α is .10 (or 10%). Although any value can be assigned to α, the commonly used values of α are .01, .05, and .10.

Now, suppose that in the court trial case the person is declared not guilty at the end of the trial. Such a verdict does not indicate that the person is indeed not guilty. It is possible that he is guilty but there is not enough evidence to prove the guilt. Hence, in this situation there are again two possibilities.

1. The person is not guilty, and he is declared not guilty.
2. The person is guilty but, *because of the lack of enough evidence,* he is declared not guilty.

In the first case, the court's decision is correct. But in the second case, the court has committed an error by setting a guilty person free. In statistics, this type of error is called a **type II** or a β (the Greek letter *beta*) **error.** These two cases are shown in the first row of the shaded portion of Table 9.1

In our statistics example, a type II error will occur when H_0 is actually false (i.e., the soda contained in all cans on average is less than 12 ounces), but it happens by chance that we draw a sample whose mean is close to or larger than 12 ounces and we wrongly conclude *do not reject* H_0. The value of β represents the probability of making a type II error. It represents the probability that H_0 is not rejected given H_0 is false.

TYPE II ERROR

A type II error occurs when a false null hypothesis is not rejected. The value of β represents the probability of committing a type II error, that is,

$$\beta = P(H_0 \text{ is not rejected} \mid H_0 \text{ is false})$$

The two types of errors that occur in hypothesis testing depend on each other. We cannot decrease α and β simultaneously for a fixed sample size. Lowering the value of α will increase the value of β, and lowering the value of β will increase the value of α. However, we can decrease both α and β at the same time by increasing the sample size. The explanation of how α and β are related and the computation of β are not within the scope of this text.

Table 9.2, which is similar to Table 9.1, is written for a statistics problem of hypothesis testing. In Table 9.2 *the person is not guilty* is replaced by H_0 *is true, the person is guilty* by H_0 *is false,* and the *court's decision* by *decision*.

Table 9.2

		Actual Situation	
		H_0 Is True	**H_0 Is False**
Decision	Do not reject H_0	Correct decision	Type II or β error
	Reject H_0	Type I or α error	Correct decision

9.1.4 TAILS OF THE TEST

The statistical hypothesis-testing procedure is similar to the trial of a person in the court but with two major differences. The first major difference is that in a statistics test of hypothesis, the partition of the total region into rejection and nonrejection regions is not arbitrary. Instead, it depends on the value assigned to α (type I error). As mentioned earlier, α is also called the significance level of the test.

The second major difference is in regard to the rejection region. In the court case the rejection region is on the right side of the critical point, as shown in Figure 9.1. However, in statistics, the rejection region for a hypothesis-testing problem can be on both sides with the nonrejection region in the middle, or it can be on the left side or on the right side of the nonrejection region. These possibilities are explained in the next three parts. A test with two rejection regions is called a **two-tailed test,** and a test with one rejection region is called a **one-tailed test.** The one-tailed test is called a **left-tailed test** if the rejection region is in the left tail of the distribution curve. It is called a **right-tailed test** if the rejection region is in the right tail of the distribution curve.

TAILS OF THE TEST

A two-tailed test has rejection regions in both tails, a left-tailed test has a rejection region in the left tail, and a right-tailed test has a rejection region in the right tail of the distribution.

A TWO-TAILED TEST

According to the U.S. Census Bureau, the mean family size in the United States was 3.19 in 1987. We want to test if this mean has changed since 1987. The key word here is *changed*. A change in the mean family size has occurred if it has either increased or decreased during this period. This is an example of a two-tailed test. Let μ be the current mean family size for all families. The two possible decisions, in this case, are

1. The mean family size has not changed, that is, $\mu = 3.19$.
2. The mean family size has changed, that is, $\mu \neq 3.19$.

We write the null and alternative hypotheses for this test as

$$H_0: \mu = 3.19 \quad \text{(the mean family size has not changed)}$$

$$H_1: \mu \neq 3.19 \quad \text{(the mean family size has changed)}$$

Whether a test is two-tailed or one-tailed is determined by the sign in the alternative hypothesis. If the alternative hypothesis has a $\neq$ sign, it is a two-tailed test. A two-tailed test has two rejection regions, one in each tail of the distribution curve, as shown in Figure 9.2. This figure shows the sampling distribution of $\bar{x}$. Assuming H_0 is true, $\bar{x}$ has a normal distribution (because of a large sample) with its mean equal to 3.19 (the value of μ in H_0). In Figure 9.2, the area of each of the two rejection regions is $\alpha/2$ and the total area of both rejection regions is α (the significance level). As shown in the graph, a two-tailed test of hypothesis has two critical values that separate the two rejection regions from the nonrejection region. We will reject H_0 if the value of $\bar{x}$ calculated for the sample falls in a rejection region. We will not reject H_0 if the value of $\bar{x}$ lies in the nonrejection region. By rejecting H_0, we are saying that the difference between the value of μ stated in H_0 and the value of $\bar{x}$ obtained from the sample is too large to have occurred because of the sampling error alone. Hence, this difference is real. By not rejecting H_0, we are saying that the difference between the value of μ stated in H_0 and the value of $\bar{x}$ obtained from the sample is small and it may have occurred because of the sampling error alone.

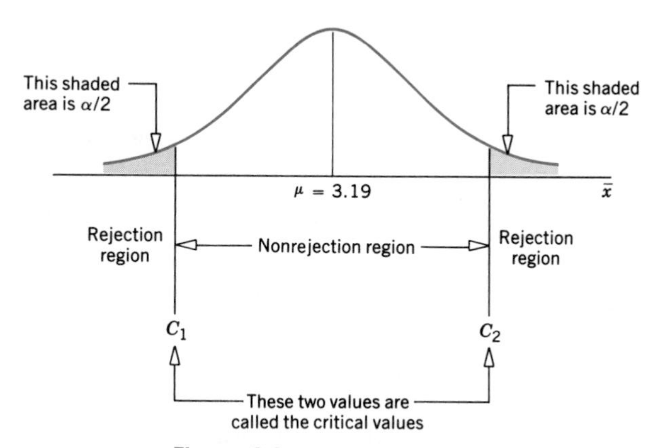

Figure 9.2 A two-tailed test.

A LEFT-TAILED TEST

Reconsider the example of mean amount of soda in all soft-drink cans produced by a company. If the company claims that these cans on average contain 12 ounces of soda but in fact they contain less than 12 ounces, then the company can be accused of cheating. Suppose a consumer agency wants to test if the mean amount of soda per can is less than 12 ounces. Note that the key phrase this time is *less than,* which indicates a left-tailed test. Let μ be the mean amount of soda in all cans. The two possible decisions are

1. The mean amount of soda in a can is not less than 12 ounces, that is, $\mu = 12$.
2. The mean amount of soda in a can is less than 12 ounces, that is, $\mu < 12$.

The null and alternative hypotheses for this test are written as

$$H_0: \mu = 12 \quad \text{(the mean is not less than 12 ounces)}$$

$$H_1: \mu < 12 \quad \text{(the mean is less than 12 ounces)}$$

In this case, we can also write the null hypothesis as H_0: $\mu \geq 12$. This will not affect the result of the test as long as the sign in H_1 is less than ($<$).

When the alternative hypothesis has a less than ($<$) sign, the test is always left-tailed. In a left-tailed test, the rejection region is in the left tail, as shown in Figure 9.3, and the area of this rejection region is equal to α (the significance level). We can observe from this figure that there is only one critical value in a left-tailed test.

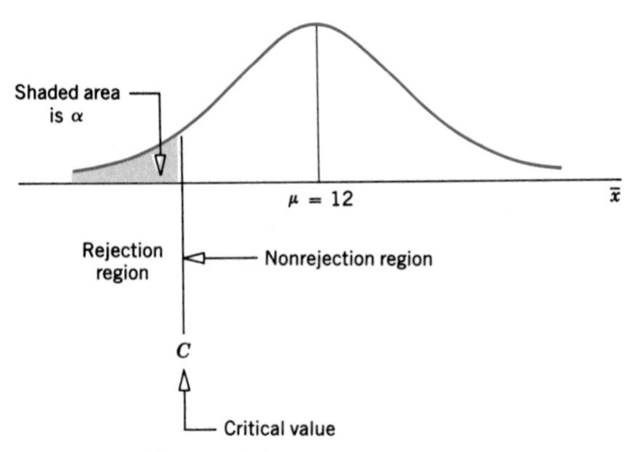

Figure 9.3 A left-tailed test.

Assuming H_0 is true, $\bar{x}$ has a normal distribution (because of a large sample) with its mean equal to 12 (the value of μ in H_0). We will reject H_0 if the value of $\bar{x}$ obtained from the sample falls in the rejection region; and we will not reject H_0 otherwise.

A RIGHT-TAILED TEST

To illustrate the third case, let us suppose that the mean income of all households in the United States was \$24,540 in 1988. Assume that we want to test if the mean income for households is higher now than in 1988. The key phrase in this case is *higher than,* which indicates a right-tailed test. Let μ be the current mean income of all households. The two possible decisions this time are

1. The current mean income of all households is not higher than \$24,540, that is, $\mu = \$24,540$.
2. The current mean income of all households is higher than \$24,540, that is, $\mu > \$24,540$.

We write the null and alternative hypotheses for this test as

H_0: μ = \$24,540 (the mean income is not higher than \$24,540)

H_1: μ > \$24,540 (the mean income is higher than \$24,540)

In this case, we can also write the null hypothesis as H_0: $\mu \leq$ \$24,540, which states that the current mean income is either the same or less than \$24,540. Again, the result of the test will not be affected whether we use an equal to (=) or a less than or equal to ($\leq$) sign in H_0 as long as the alternative hypothesis has a greater than (>) sign.

When the alternative hypothesis has a greater than (>) sign, the test is always right-tailed. In a right-tailed test, the rejection region is in the right tail, as shown in Figure 9.4. The area of this rejection region is equal to α, the significance level. Like a left-tailed test, a right-tailed test has only one critical value. This value is shown in Figure 9.4.

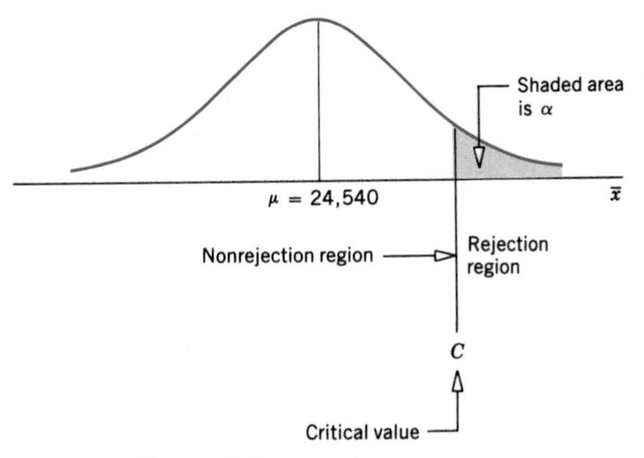

Figure 9.4 A right-tailed test.

Again, assuming H_0 is true, $\bar{x}$ has a normal distribution (because of a large sample) with its mean equal to \$24,540 (the value of μ in H_0). We will reject H_0 if the value of $\bar{x}$ obtained from the sample falls in the rejection region. Otherwise, we will not reject H_0.

Table 9.3 summarizes the foregoing discussion about the relationship between the signs in H_0 and H_1 and the tails of the test.

Note that the null hypothesis always has an *equal to* (=) or *less than or equal to* ($\leq$) or *greater than or equal to* ($\geq$) sign and the alternative hypothesis always has a *not equal to* ($\neq$) or *greater than* (>) or *less than* (<) sign.

Table 9.3

	Two-Tailed Test	Left-Tailed Test	Right-Tailed Test
Sign in the null hypothesis, H_0	=	= or $\geq$	= or $\leq$
Sign in the alternative hypothesis, H_1	$\neq$	<	>
Rejection region	On both sides	On the left side	On the right side

A test of hypothesis involves five steps, which are listed below.

STEPS OF A TEST OF HYPOTHESIS

A statistical test of hypothesis procedure contains the following five steps.

1. State the null and alternative hypotheses
2. Select the distribution to use
3. Determine the rejection and nonrejection regions
4. Calculate the value of the test statistic
5. Make a decision

With the help of examples, these steps are described in the next section.

EXERCISES

9.1 Briefly explain the meaning of each of the following terms.

a. Null hypothesis
b. Alternative hypothesis
c. Critical point(s)
d. Significance level
e. Nonrejection region
f. Rejection region
g. Tails of a test

9.2 Briefly describe type I and type II errors.

9.3 Explain how the tails of a test depend on the sign in the alternative hypothesis. Describe the signs in the null and alternative hypotheses for a two-tailed, a left-tailed, and a right-tailed test, respectively.

9.4 Write the null and alternative hypotheses for each of the following examples. Determine if each is a case of a two-tailed, a left-tailed, or a right-tailed test.

a. To test if the mean price of houses is greater than $143,000
b. To test if the mean number of hours college students spend partying per week is different from 12
c. To test if the mean life of car batteries is less than 45 months
d. To test if the mean amount of time taken to do a certain job is more than 35 minutes
e. To test if the mean age of all college students is different from 23 years

9.5 Write the null and alternative hypotheses for each of the following examples. Determine if each is a case of a two-tailed, a left-tailed, or a right-tailed test.

a. To test if the mean amount of time spent per week watching sports on television by adult males is different from 9.5 hours
b. To test if the mean GPA of all students at a university is lower than 2.9
c. To test if the mean cholesterol level of all adult males in the United States is higher than 175
d. To test if the mean amount of time spent doing homework by all fourth-graders is different from 5 hours a week

9.2 HYPOTHESIS TESTS ABOUT A POPULATION MEAN: LARGE SAMPLES

From the central limit theorem discussed in Chapter 7, the sampling distribution of $\bar{x}$ is approximately normal for large samples ($n \geq 30$). Hence, whether or not σ is known, the normal distribution is used to test hypotheses about the population mean when a sample size is large.

TEST STATISTIC

In tests of hypotheses about μ for large samples, the random variable

$$z = \frac{\bar{x} - \mu}{\sigma_{\bar{x}}} \quad \text{or} \quad \frac{\bar{x} - \mu}{s_{\bar{x}}}$$

is called the test statistic.

At the end of Section 9.1 it was mentioned that a test of hypothesis procedure involves the following five steps.

1. State the null and alternative hypotheses
2. Select the distribution to use
3. Determine the rejection and nonrejection regions
4. Calculate the value of the test statistic
5. Make a decision

Examples 9-1 through 9-3 explain these five steps for tests of hypotheses about the population mean μ. Example 9-1 is concerned with a two-tailed test and Examples 9-2 and 9-3 describe one-tailed tests.

Conducting a two-tailed test of hypothesis about μ: $n \geq 30$.

EXAMPLE 9-1 According to the U.S. Census Bureau, the mean family size was 3.19 in 1987. A sample of 900 families taken this year gave a mean family size of 3.10 with a standard deviation of .7. Test at the .05 significance level if the mean family size has changed during this period.

Solution Let μ be the current mean family size for all families and $\bar{x}$ be the mean family size for the sample. From the given information,

$$n = 900, \quad \bar{x} = 3.10, \quad \text{and} \quad s = .7$$

The mean family size for 1987 is given to be 3.19. The significance level α is .05. That is, the probability of rejecting the null hypothesis when it actually is true should not exceed .05.

Step 1. *State the null and alternative hypotheses*

Notice that we are testing for a *change* in the mean family size and not specifically for either an increase or a decrease in mean family size. Hence, we write the null and alternative hypotheses as

$$H_0: \mu = 3.19 \quad \text{(the mean family size has not changed)}$$

$$H_1: \mu \neq 3.19 \quad \text{(the mean family size has changed)}$$

Step 2. *Select the distribution to use*

Because the sample size is large ($n > 30$), the sampling distribution of $\bar{x}$ is (approximately) normal. Hence, we use the normal distribution to make the test.

Step 3. *Determine the rejection and nonrejection regions*

The significance level is .05. The $\neq$ sign in the alternative hypothesis indicates that the test is two-tailed with two rejection regions, one in each tail. Because the total area of both rejection regions is .05 (the significance level), the area of the rejection region in each tail is

$$\text{Area in each tail} = \frac{\alpha}{2} = \frac{.05}{2} = .025$$

These areas are shown in Figure 9.5. Two critical points separate the two rejection regions from the nonrejection region. To find the z values for these critical points we first find the area between the mean and one of the critical points. We obtain this area by subtracting .025 (the area in each tail) from .5, which gives .4750. Next we look for .4750 in the normal table, Table VII of Appendix C. This gives a value of $z = 1.96$. Hence, the z values of the two critical points, as shown in Figure 9.5, are -1.96 and 1.96.

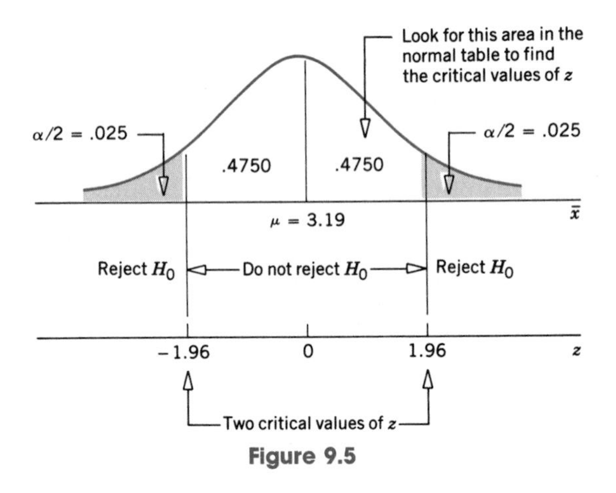

Figure 9.5

Step 4. *Calculate the value of the test statistic*

The decision to reject or not to reject the null hypothesis will depend on whether the evidence from the sample falls in the rejection or nonrejection region. If the value of the sample mean $\bar{x}$ falls in any of the two rejection regions, we reject H_0. Otherwise, we do not reject H_0. The value of $\bar{x}$ obtained from the sample is called the *observed value of* $\bar{x}$. To locate the position of $\bar{x} = 3.10$ on the distribution curve of $\bar{x}$ in Figure 9.5, we first calculate the z value for $\bar{x} = 3.10$. This is called the *value of the test statistic*. Then, we compare the value of the test statistic with the two critical values of z shown in Figure 9.5. If the value of the test statistic is between -1.96 and 1.96, we do not reject H_0. If the value of the test statistic is either greater than 1.96 or less than -1.96, we reject H_0.

CALCULATING THE VALUE OF THE TEST STATISTIC

The value of the test statistic z for $\bar{x}$ for a test of hypothesis about μ is computed as follows.

$$z = \frac{\bar{x} - \mu}{\sigma_{\bar{x}}} \quad \text{if } \sigma \text{ is known}$$

$$z = \frac{\bar{x} - \mu}{s_{\bar{x}}} \quad \text{if } \sigma \text{ is not known}$$

where

$$\sigma_{\bar{x}} = \frac{\sigma}{\sqrt{n}} \quad \text{and} \quad s_{\bar{x}} = \frac{s}{\sqrt{n}}$$

The value of $\bar{x}$ from the sample is 3.10. As σ is not known, we calculate the z value using $s_{\bar{x}}$ as follows.

$$s_{\bar{x}} = \frac{s}{\sqrt{n}} = \frac{.7}{\sqrt{900}} = .023$$

$$z = \frac{\bar{x} - \mu}{s_{\bar{x}}} = \frac{3.10 - 3.19}{.023} = -3.91$$

(3.19: From H_0)

The value of μ in the calculation of the z value is substituted from the null hypothesis. The value of $z = -3.91$ calculated for $\bar{x}$ is called the *computed value of the test statistic z*. This is the value of z that corresponds to the value of $\bar{x}$ observed from the sample. This is also called the *observed value of z*.

Step 5. *Make a decision*

In the final step we make a decision based on the location of the value of the test statistic z computed for $\bar{x}$ in Step 4. This value of $z = -3.91$ is less than the critical value of $z = -1.96$, and it falls in the rejection region in the left tail. Hence, we reject H_0 and conclude that the mean family size seems to have changed.

By rejecting the null hypothesis we are stating that the difference between the sample mean $\bar{x} = 3.10$ and the hypothesized value of the population mean $\mu = 3.19$ is too large and may not have occurred because of chance or sampling error alone. This difference seems to be real and, hence, the mean family size seems to have changed during this period. Note that the rejection of the null hypothesis does not necessarily indicate that the family size has definitely changed. It simply indicates that there is strong evidence (from the sample) that the mean family size may have changed. ■

Making a right-tailed test of hypothesis about μ: $n \geq 30$.

EXAMPLE 9-2 According to the National Association of Realtors, the mean sale price of existing single-family homes was $112,800 in 1988 (*Home Sales,* 3(10): October 1989). A random sample of 500 recently sold such homes gave a mean price of $127,400 with a standard deviation of $23,700. Test at the 1% significance level if the current mean sale price of such homes is greater than $112,800.

Solution Let μ be the current mean sale price of all existing single-family homes and $\bar{x}$ be the corresponding mean for the sample. From the given information,

$$n = 500, \quad \bar{x} = \$127{,}400, \quad \text{and} \quad s = \$23{,}700$$

The significance level is $\alpha = .01$.

Step 1. *State the null and alternative hypotheses*

We are to test if the current mean sale price of existing single-family homes is greater than \$112,800. The null and alternative hypotheses are

H_0: $\mu = \$112{,}800$ (the current mean is not greater than \$112,800)

H_1: $\mu > \$112{,}800$ (the current mean is greater than \$112,800)

Step 2. *Select the distribution to use*

Because the sample size is large ($n > 30$), we use the normal distribution to make the test.

Step 3. *Determine the rejection and nonrejection regions*

The significance level is .01. The $>$ sign in the alternative hypothesis indicates that the test is right-tailed with its rejection region in the right tail. Because there is only one rejection region, its area is $\alpha = .01$. The critical value of z, obtained from Table VII of Appendix C for .4900, is approximately 2.33, as shown in Figure 9.6.

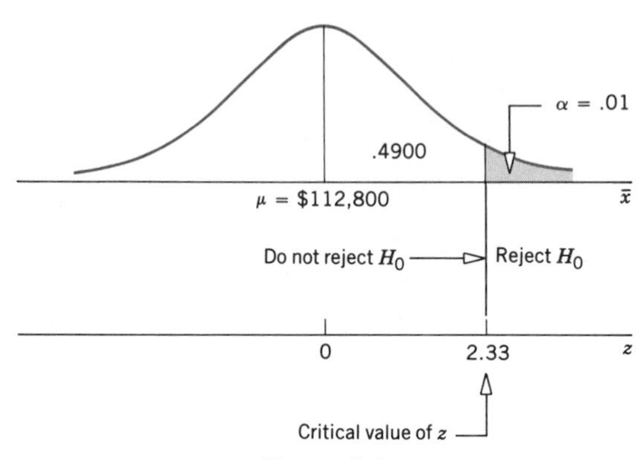

Figure 9.6

Step 4. *Calculate the value of the test statistic*

The value of the test statistic z for $\bar{x} = \$127{,}400$ is calculated as follows.

$$s_{\bar{x}} = \frac{s}{\sqrt{n}} = \frac{23{,}700}{\sqrt{500}} = 1059.896$$

$$z = \frac{\bar{x} - \mu}{s_{\bar{x}}} = \frac{127{,}400 - \overset{\text{From } H_0}{112{,}800}}{1059.896} = 13.77$$

Step 5. *Make a decision*

Because the value of the test statistic $z = 13.77$ is larger than the critical value of $z = 2.33$ and it falls in the rejection region, we reject H_0. Hence, we can state that the sample mean $\bar{x} = 127{,}400$ is too far from the hypothesized population mean $\mu = 112{,}800$. The difference between the two may not be attributed to chance or sampling error alone. Therefore, the current mean sale price of existing single-family homes is greater than $112,800. ■

Making a left-tailed test of hypothesis about μ: $n \geq 30$.

EXAMPLE 9-3 The mean length of bolts made on a machine is supposed to be 2.5 inches. A random sample of 49 such bolts gave a mean length of 2.498 inches with a standard deviation of .007 inches. Test at the 1% significance level if the mean length of bolts made on this machine is smaller than 2.5 inches.

Solution Let μ be the mean length of all bolts made on this machine and $\bar{x}$ be the corresponding mean for the sample. From the given information,

$$n = 49, \quad \bar{x} = 2.498, \quad \text{and} \quad s = .007$$

The significance level is $\alpha = .01$.

Step 1. *State the null and alternative hypotheses*

We are to test if the mean length of bolts is less than 2.5 inches. Hence, the null and alternative hypotheses are

H_0: $\mu = 2.5$ (the mean length is not less than 2.5)

H_1: $\mu < 2.5$ (the mean length is less than 2.5)

Step 2. *Select the distribution to use*

Because the sample size is large, we use the normal distribution to make the test.

Step 3. *Determine the rejection and nonrejection regions*

The significance level is .01. The $<$ sign in the alternative hypothesis indicates that the test is left-tailed with the rejection region in the left tail. The critical value of z, obtained from the normal table for .4900 is approximately -2.33. This is shown in Figure 9.7.

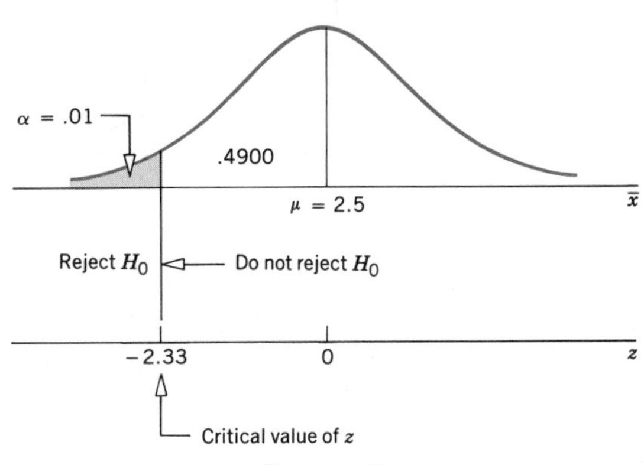

Figure 9.7

Step 4. *Calculate the value of the test statistic*

We calculate the value of the test statistic z for $\bar{x} = 2.498$ as follows.

$$s_{\bar{x}} = \frac{s}{\sqrt{n}} = \frac{.007}{\sqrt{49}} = .001$$

$$z = \frac{\bar{x} - \mu}{s_{\bar{x}}} = \frac{2.498 - 2.5}{.001} = -2.00$$

(2.5: From H_0)

Step 5. *Make a decision*

The value of the test statistic $z = -2.00$ is greater than the critical value of $z = -2.33$ and it falls in the nonrejection region. As a result, we fail to reject H_0. Hence, we can state that, based on the sample information, it does not appear that mean length of all such bolts is below the required mean. Note that we are not concluding that the mean length of all such bolts is definitely not less than 2.5 inches. By not rejecting the null hypothesis we are saying that the information obtained from the sample is strong to conclude that the mean length of all such bolts is not less than 2.5. ■

In studies published in various journals, the authors usually use the terms *significantly different* and *not significantly different* when deriving conclusions based on hypothesis tests. These terms are short versions of the terms *statistically significantly different* and *statistically not significantly different*. The statement *significantly different* means that the difference between the observed value of the sample mean $\bar{x}$ and the hypothesized value of the population mean μ is so large that it probably did not occur because of the sampling error alone. Consequently, the null hypothesis is rejected. In other words, the difference between $\bar{x}$ and μ is statistically significant. Thus, the statement *significantly different* is equivalent to saying *the null hypothesis is rejected*. In Example 9-2, we can state as a conclusion that the observed value of $\bar{x} = \$127,400$ is significantly different from the hypothesized value of $\mu = \$112,800$. That is, the current mean sale price of existing single-family homes is significantly different from \$112,800.

On the other hand, the statement *not significantly different* means that the difference between the observed value of the sample mean $\bar{x}$ and the hypothesized value of the population mean μ is so small that it may have occurred just because of chance. Hence, the null hypothesis is not rejected. Thus, the statement *not significantly different* is equivalent to saying that we *fail to reject the null hypothesis*. In Example 9-3, we can state as a conclusion that the observed value of $\bar{x} = 2.498$ is not significantly different from the hypothesized value of $\mu = 2.5$. In other words, the mean length of all bolts does not seem to be significantly different from 2.5 inches.

EXERCISES

9.6 Make the following tests of hypotheses.

a. H_0: $\mu = 25$, H_1: $\mu \neq 25$, $n = 81$, $\bar{x} = 28$, $s = 3$, $\alpha = .01$
b. H_0: $\mu = 12$, H_1: $\mu < 12$, $n = 45$, $\bar{x} = 11$, $\sigma = 4.5$, $\alpha = .05$
c. H_0: $\mu = 40$, H_1: $\mu > 40$, $n = 100$, $\bar{x} = 46$, $s = 7$, $\alpha = .10$

9.7 Make the following tests of hypotheses.

a. H_0: $\mu = 80$, H_1: $\mu \neq 80$, $n = 33$, $\bar{x} = 76$, $s = 15$, $\alpha = .10$
b. H_0: $\mu = 32$, H_1: $\mu < 32$, $n = 75$, $\bar{x} = 27$, $s = 7.4$, $\alpha = .01$
c. H_0: $\mu = 55$, H_1: $\mu > 55$, $n = 40$, $\bar{x} = 60$, $s = 4$, $\alpha = .05$

9.8 The mean earnings of young male workers (aged 20–24) were $9027 in 1986 (*The Forgotten Half: Pathways to Success for America's Youth and Young Families,* Washington, D.C.: Youth and America's Future: The William T. Grant Commission on Work, Family and Citizenship, 1988). A sample of 900 young male workers taken in 1989 showed that their real mean earnings (in 1986 dollars) are $8971 with a standard deviation of $820. Test at the 5% significance level if the real mean earnings of young male workers decreased between 1986 and 1989.

9.9 The manufacturer of a certain brand of auto batteries claims that the mean life of these batteries is 45 months. A consumer protection agency took a random sample of 36 such batteries and found that the mean life for this sample is 43.75 months with a standard deviation of 4 months. Test at the 1% significance level if the mean life of these batteries is less than 45 months.

9.10 According to the Hertz Corporation, the mean cost of owning and operating a car in 1986 was $3002. Suppose this estimate of Hertz is true for the population of all cars for 1986. A random sample of 45 cars showed that the mean cost of owning and operating these cars was $3350 for 1989 with a standard deviation of $425. Test at the 2.5% significance level if the mean cost of owning and operating a car in 1989 is greater than $3002.

9.11 According to the American Bar Association, the mean household income for lawyers was $120,000 in 1986. A random sample of 64 lawyers showed that the mean household income for these lawyers for 1989 was $135,500 with a standard deviation of $24,500. Test at the 1% significance level if the mean household income for all lawyers for 1989 is greater than $120,000.

9.12 According to the U.S. Travel Data Center, Americans spend an average of 6.1 nights away from home during summer vacations. A random sample of 200 persons taken recently showed that they spent an average of 6.3 nights away from home during summer vacations with a standard deviation of 1.8. Test at the 10% significance level if the mean number of nights spent away from home during summer vacations for all Americans is different from 6.1.

9.13 A study claims that all adults spend an average of 28 hours on chores during a weekend. A random sample of 200 adults showed that they spend an average of 27 hours on chores during a weekend with a standard deviation of 6.9 hours. Test at the 1% significance level if the mean number of hours spent by all adults on chores during a weekend is less than 28.

9.14 The mean years of schooling for all adults in a large city was 12 in 1987. A random sample of 400 adults taken recently from this city showed that the mean years of schooling for this sample is 13 with a standard deviation of 1.9. Test at the 5% significance level if the mean years of schooling for all adults in this city is now different from 12.

9.15 According to the Bureau of Labor Statistics, workers in the private sector earned an average of $9.66 an hour in 1989. A random sample of 1000 private sector workers taken recently showed that the mean hourly wage for this sample is $10.70 with a standard deviation of $1.70. Test at the 2.5% significance level if the current mean hourly wage for private sector workers is greater than $9.66.

9.16 The mean child support paid to custodial mothers by noncustodial fathers was $185 a month in 1985 (*The Forgotten Half: Pathways to Success for America's Youth and Young Families,* Washington, D.C.: Youth and America's Future: The William T. Grant Commission on Work, Family and Citizenship, 1988). A random sample of 340 custodial mothers taken recently showed that the mean child support paid to them is $216 a month with a standard deviation of $35. Test at the 1% significance level if the mean child support paid to custodial mothers is now greater than $185.

9.17 According to Louis Harris and Associates, the mean time that Americans spent "to relax, watch TV, take part in sports or hobbies, go swimming or skiing, go to the movies, theater, concerts, or other forms of entertainment, get together with friends, and so forth" was 16.6 hours a week in 1988. A recent poll of 200 Americans showed that they spend an average of 17.0 hours a week on these leisure activities with a standard deviation of 3.9 hours. Test at the 5% significance level if the mean number of hours spent per week on leisure activities is now different from 16.6.

9.18 According to the U.S. Bureau of the Census, the mean monthly income of persons with a doctorate degree was $3265 in 1984. A 1990 random sample of 650 persons with a doctorate degree gave a mean monthly income of $3725 with a standard deviation of $620. Test at the 5% significance level if the 1990 mean monthly income of doctorate degree holders is greater than $3265.

9.19 According to a 1987 study by Payment Systems Education, Americans aged 35 to 49 wrote an average of 18 personal checks a month (*ABA Banking Journal,* April 1987). A recent study of 100 randomly selected Americans aged 35 to 49 showed that they write an average of 21 checks a month with a standard deviation of 4. Test at the 1% significance level if the mean number of checks written by all Americans aged 35 to 49 is different from 18 a month.

9.20 According to John Tugman of MRCA Information Services, the mean amount spent on clothes by American women was $569 in 1987 (*Newsweek,* December 5, 1988). A random sample of 175 women taken recently showed that they spent an average of $551 on clothes last year with a standard deviation of $146. Test at the 1% significance level if the mean amount spent on clothes last year by American women was less than $569.

*9.21 A company claims that the mean net weight of its All Taste cereal boxes is 18 ounces. Suppose you want to test if the mean net weight of these cereal boxes is less than 18 ounces. Explain briefly how you would conduct this test using a large sample.

*9.22 A researcher claims that college students spend an average of 45 minutes a week on community service. You want to test if the mean time spent per week on community service by college students is different from 45 minutes. Explain briefly how you would conduct this test using a large sample.

9.3 HYPOTHESIS TESTS USING THE *p*-VALUE APPROACH

In the foregoing discussion of hypothesis testing, the value of the significance level α was determined before the test was performed. Sometimes we may prefer not to predetermine α. Instead, we may want to find a value such that a given null hypothesis will be rejected for any α greater than this value and it will not be rejected for any α smaller than this value. The *probability-value approach,* more commonly called the *p-value approach,* gives such a value. In this approach, we calculate the *p*-value for the test, which is defined as the smallest level of significance at which the null hypothesis can be rejected.

p-VALUE

The *p*-value is the smallest significance level at which the null hypothesis can be rejected.

For a one-tailed test, the *p*-value is given by the area in the tail beyond the observed value of a sample statistic such as $\bar{x}$. Figure 9.8 shows the *p*-value for a left-tailed test about μ.

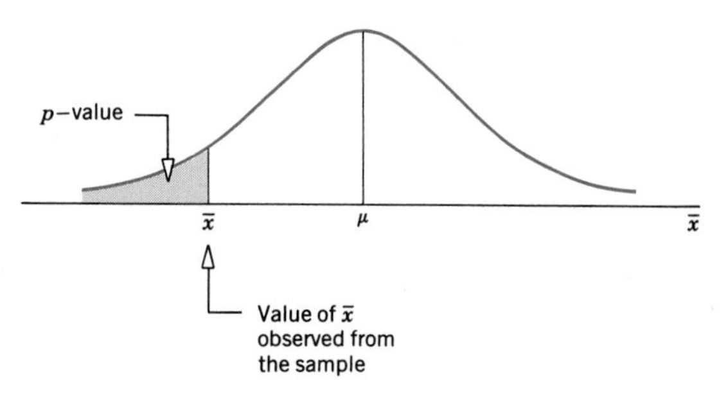

Figure 9.8 p-value for a left-tailed test.

For a two-tailed test, the p-value is twice the area in the tail beyond the observed value of a sample statistic such as $\bar{x}$. Figure 9.9 shows the p-value for a two-tailed test. Each of the areas in the two tails gives one-half the p-value.

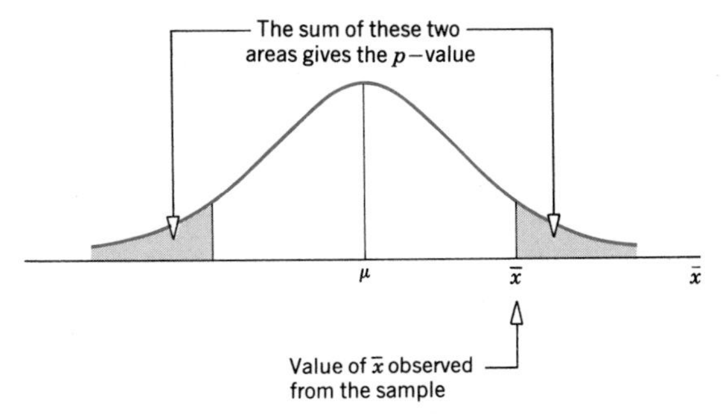

Figure 9.9 p-value for a two-tailed test.

Examples 9-4 and 9-5 illustrate the calculation and use of the p-value.

Calculating p-value for a one-tailed test of hypothesis.

EXAMPLE 9-4 A health club claims that its members lose an average of 10 pounds or more within the first month after joining the club. A consumer agency wanted to check this claim. It took a random sample of 36 members of this health club and found that they lost on average 9.2 pounds within the first month of membership with a standard deviation of 2.4 pounds. Find the p-value for this test.

Solution Let μ be the mean weight lost during the first month of membership by all members of this health club and $\bar{x}$ be the corresponding mean for the sample. Then, from the given information,

$$n = 36, \quad \bar{x} = 9.2 \text{ pounds}, \quad \text{and} \quad s = 2.4 \text{ pounds}$$

The claim of the club is that its members lose on average 10 pounds or more within the first month of membership. To calculate the p-value, we apply the following three steps.

Step 1. *State the null and alternative hypotheses*

H_0: $\mu \geq 10$ (the mean weight lost is at least 10 pounds)

H_1: $\mu < 10$ (the mean weight lost is less than 10 pounds)

Step 2. *Select the distribution to use*

Because the sample size is large, we use the normal distribution to make the test.

Step 3. *Calculate the p-value*

The $<$ sign in the alternative hypothesis indicates that the test is left-tailed. The p-value is given by the area in the left tail beyond $\bar{x} = 9.2$, as shown in Figure 9.10. To find this area, we first find the z value for $\bar{x} = 9.2$ as follows.

$$s_{\bar{x}} = \frac{s}{\sqrt{n}} = \frac{2.4}{\sqrt{36}} = .40$$

$$z = \frac{\bar{x} - \mu}{s_{\bar{x}}} = \frac{9.2 - 10}{.40} = -2.00$$

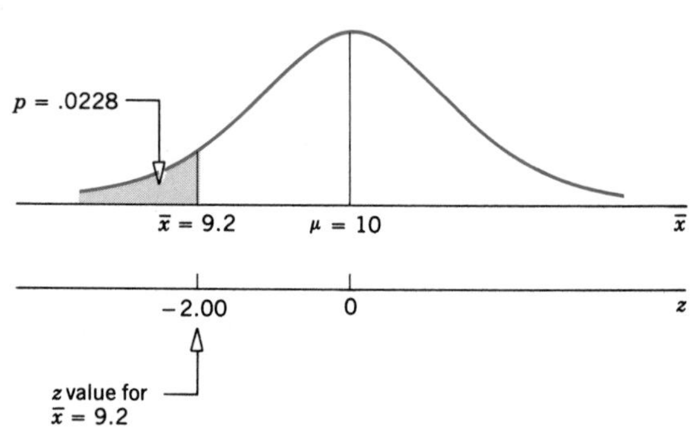

Figure 9.10

The area to the left of $\bar{x} = 9.2$ is given by the area to the left of $z = -2.00$. From the normal distribution table, the area between the mean and $z = -2.00$ is .4772. Hence, the area to the left of $z = -2.00$ is $.5 - .4772 = .0228$. Consequently,

$$p\text{-value} = .0228$$

Thus, based on the p-value of .0228 we can state that for any α (significance level) greater than .0228 we will reject the null hypothesis stated in Step 1 and for any α less than .0228 we will not reject the null hypothesis. Suppose we make the test at $\alpha = .01$. Because $\alpha = .01$ is less than the p-value of .0228, we will not reject the null hypothesis. Now, suppose we make the test at $\alpha = .05$. This time, as $\alpha = .05$ is greater than the p-value of .0228, we will reject the null hypothesis. ■

The reader should make the test of hypothesis for Example 9-4 at $\alpha = .01$ and at $\alpha = .05$ by using the five steps learned in Section 9.2. The null hypothesis will not be rejected at $\alpha = .01$ (as .01 is less than $p = .0228$) and the null hypothesis will be rejected at $\alpha = .05$ (as .05 is greater than $p = .0228$).

Calculating p-value for a two-tailed test of hypothesis.

EXAMPLE 9-5 A supervisor claims that it takes on average 50 minutes for a worker to learn a particular kind of machine-processing job. A sample of 40 workers showed that it took on average 53 minutes for them to learn this job with a standard deviation of 7 minutes. Find the p-value for the test that the mean learning time for this job is different from 50 minutes.

Solution Let μ be the mean time (in minutes) taken to learn this job by all workers and $\bar{x}$ be the corresponding sample mean. Then, from the given information,

$$n = 40, \quad \bar{x} = 53 \text{ minutes}, \quad \text{and} \quad s = 7 \text{ minutes}$$

The claim is that the mean learning time is 50 minutes. To calculate the *p*-value, we apply the following three steps.

Step 1. *State the null and alternative hypotheses*

$$H_0: \mu = 50 \qquad \text{(the claim is true)}$$

$$H_1: \mu \neq 50 \qquad \text{(the claim is not true)}$$

Step 2. *Select the distribution to use*

Because the sample size is large, we use the normal distribution to make the test.

Step 3. *Calculate the p-value*

The $\neq$ sign in the alternative hypothesis indicates that the test is two-tailed. The *p*-value is equal to twice the area in the tail of the distribution beyond $\bar{x} = 53$, as shown in Figure 9.11. To find this area, we first find the z value for $\bar{x} = 53$ as follows.

$$s_{\bar{x}} = \frac{s}{\sqrt{n}} = \frac{7}{\sqrt{40}} = 1.107$$

$$z = \frac{\bar{x} - \mu}{s_{\bar{x}}} = \frac{53 - 50}{1.107} = 2.71$$

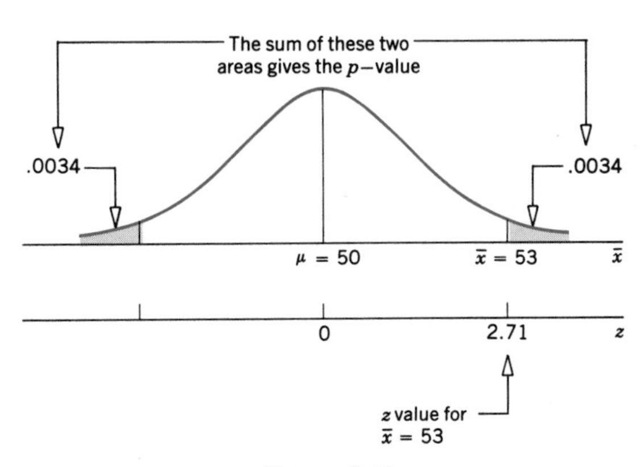

Figure 9.11

The area to the right of $\bar{x} = 53$ is given by the area to the right of $z = 2.71$. From the normal distribution table, the area between the mean and $z = 2.71$ is .4966. Hence, the area to the right of $z = 2.71$ is

$$.5 - .4966 = .0034$$

Consequently, the *p*-value is

$$p\text{-value} = 2(.0034) = .0068$$

Thus, based on the p-value of .0068, we conclude that for any α (significance level) greater than .0068 we reject the null hypothesis and for any α less than .0068 we do not reject the null hypothesis. ■

EXERCISES

9.23 Find the p-value for each of the following hypothesis tests.

a. H_0: $\mu = 23$, H_1: $\mu \neq 23$, $n = 50$, $\bar{x} = 21$, $s = 5$
b. H_0: $\mu = 15$, H_1: $\mu < 15$, $n = 80$, $\bar{x} = 13.2$, $s = 5.5$
c. H_0: $\mu = 38$, H_1: $\mu > 38$, $n = 35$, $\bar{x} = 40.6$, $s = 7.2$

9.24 Find the p-value for each of the following hypothesis tests.

a. H_0: $\mu = 46$, H_1: $\mu \neq 46$, $n = 40$, $\bar{x} = 49.43$, $s = 9.7$
b. H_0: $\mu = 26$, H_1: $\mu < 26$, $n = 33$, $\bar{x} = 24.2$, $s = 4.3$
c. H_0: $\mu = 18$, H_1: $\mu > 18$, $n = 55$, $\bar{x} = 20.4$, $s = 7.8$

9.25 According to the U.S. Department of Labor, the mean weekly salary for engineers was $720 in 1987–88. Assume that this result holds true for the population of all engineers for that year. A random sample of 45 engineers taken recently showed that their weekly salary is $751 with a standard deviation of $90. Find the p-value for the hypothesis test that the current mean weekly salary of all engineers is greater than $720.

9.26 According to the Oceanic & Atmospheric Administration, the mean consumption of seafood in the United States was 15.4 pounds per person in 1987. Assume that this result holds true for the population of all Americans for 1987. A random sample of 80 persons showed that they consumed an average of 16.1 pounds of seafood in 1989 with a standard deviation of 3.7 pounds. Find the p-value for the hypothesis test that the mean consumption of seafood per person for 1989 is different from 15.4 pounds.

9.27 According to the Hertz Corporation, in 1987 the cars were an average of 7.6 years old. A sample of 50 cars taken recently showed that they are an average of 6.9 years old with a standard deviation of 2.1 years. Find the p-value for the test that the mean age of all current cars is less than 7.6 years.

9.28 A telephone company claims that the mean duration of all long-distance phone calls is 10 minutes. A random sample of 100 long-distance calls taken from the records of this company showed that the mean duration of calls in this sample is 9.2 minutes with a standard deviation of 6.2 minutes. Find the p-value for the test that the mean duration of all long-distance calls is less than 10 minutes.

9.4 HYPOTHESIS TESTS ABOUT A POPULATION MEAN: SMALL SAMPLES

Many times the size of samples used to make the hypothesis tests about μ is small ($n < 30$). This may be the case because we have limited resources and cannot afford to take a large sample. Or it may be the nature of the experiment that we can only take a small sample. For example, to test a new model of a car for fuel efficiency (miles per gallon), the company may prefer to only take a small sample. Every car that will be used for such a purpose must be sold as a used car. In the case of a small sample, if the population from which the sample is drawn is (approximately) normally distributed and we know σ, we can still use the normal distribution to make a test of hypothesis about μ. However, if the population is (approximately) normally distributed but we do not know σ and the sample size is small ($n < 30$), then we replace the normal distribution by the t distribution to make a test of hypothesis about μ. In such a case the random variable

$$t = \frac{\bar{x} - \mu}{s_{\bar{x}}}$$

where

$$s_{\bar{x}} = \frac{s}{\sqrt{n}}$$

has a t distribution. The t is called the *test statistic* to make a hypothesis test about a population mean for small samples.

WHEN TO USE THE *t* DISTRIBUTION

The t distribution is used to test a hypothesis about μ if

1. The sample size is small, ($n < 30$)
2. The population from which the sample is taken is (approximately) normal
3. The population standard deviation σ is unknown

The procedure used to make hypothesis tests about μ in the case of small samples is similar to the one for large samples. We perform the same five steps with the only difference being the use of the t distribution in place of the normal distribution.

TEST STATISTIC

The value of the test statistic t for the sample mean $\bar{x}$ is computed as

$$t = \frac{\bar{x} - \mu}{s_{\bar{x}}}$$

where

$$s_{\bar{x}} = \frac{s}{\sqrt{n}}$$

Examples 9-6, 9-7, and 9-8 describe the procedure of testing hypotheses about the population mean by using the t distribution.

Conducting a two-tailed test of hypothesis about μ: $n < 30$.

EXAMPLE 9-6 The mean age of all CEOs (chief executive officers) for major corporations in the United States was 48 years in 1987. A random sample of 25 CEOs taken recently from major corporations showed a mean age of 46 years with a standard deviation of 5 years. Assuming that the ages of all CEOs have an approximate normal distribution, test at the 1% significance level if the current mean age of all CEOs of major corporations is different from that in 1987.

Solution Let μ be the current mean age of all CEOs of major corporations and $\bar{x}$ be the mean age of the CEOs included in the sample. Then, from the given information,

$$n = 25, \quad \bar{x} = 46 \text{ years}, \quad \text{and} \quad s = 5 \text{ years}$$

The mean age of all CEOs for 1987 is given to be 48 years. The significance level is $\alpha = .01$.

Step 1. *State the null and alternative hypotheses*

We are to test if the current mean age of CEOs is different from 48. Therefore, the null and alternative hypotheses are

H_0: $\mu = 48$ (the current mean age is not different from 48)

H_1: $\mu \neq 48$ (the current mean age is different from 48)

Step 2. *Select the distribution to use*

The sample size is small ($n < 30$) and the population has an approximate normal distribution. However, we do not know the population standard deviation σ. Hence, we use the t distribution to make the test.

Step 3. *Determine the rejection and nonrejection regions*

The $\neq$ sign in the alternative hypothesis indicates that the test is two-tailed. The significance level is .01. Hence, the total area of the two rejection regions is .01 and the area of the rejection region in each tail is

$$\text{Area in each tail} = \frac{\alpha}{2} = \frac{.01}{2} = .005$$

To find the critical values of t, we also need to know the degrees of freedom (df). The degrees of freedom for the t distribution for tests of hypotheses are $n - 1$. Therefore,

$$df = n - 1 = 25 - 1 = 24$$

From the t distribution table (Table VIII of Appendix C), the critical values of t for 24 degrees of freedom and .005 area in each tail are -2.797 and 2.797. These critical values are shown in Figure 9.12.

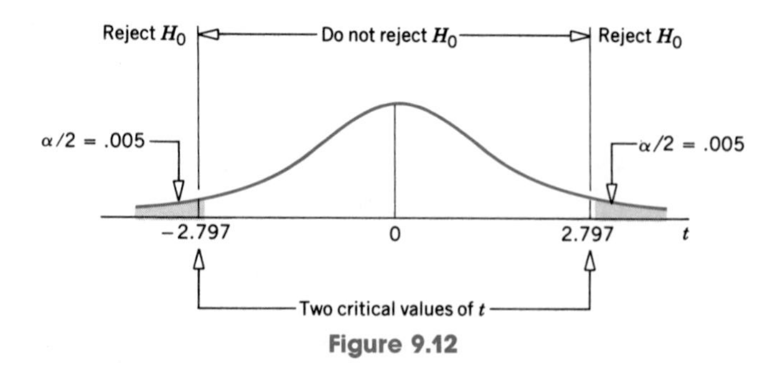

Figure 9.12

Step 4. *Calculate the value of the test statistic*

We calculate the value of the test statistic t for $\bar{x} = 46$ as follows.

$$s_{\bar{x}} = \frac{s}{\sqrt{n}} = \frac{5}{\sqrt{25}} = 1.0$$

$$t = \frac{\bar{x} - \mu}{s_{\bar{x}}} = \frac{46 - \overset{\text{From } H_0}{48}}{1.0} = -2.000$$

Step 5. *Make a decision*

The value of the test statistic $t = -2.000$ is between the two critical points, -2.797 and 2.797, which is the nonrejection region. Consequently, we fail to reject H_0. Therefore, we can state that the difference between the population mean for 1987 and the current sample mean is so small that it may have occurred because of sampling error. The mean age of the current CEOs of major corporations is not different from the mean age of CEOs of major corporations in 1987. ■

Making a left-tailed test of hypothesis about μ: $n < 30$.

EXAMPLE 9-7 A company claims that its never-die battery has a mean life of 65 months. A consumer protection agency tested 15 of these batteries to check this claim. It found the mean life of these 15 batteries to be 63 months with a standard deviation of 2 months. Test at the 5% significance level if the mean life of a never-die battery is less than 65 months. Assume that the life of such a battery has an approximate normal distribution.

Solution Let μ be the mean life of all never-die batteries and $\bar{x}$ be the corresponding mean for the sample. Then, from the given information,

$$n = 15, \quad \bar{x} = 63 \text{ months}, \quad \text{and} \quad s = 2 \text{ months}$$

The significance level is $\alpha = .05$. The company's claim is that the mean life of these batteries is 65 months.

Step 1. *State the null and alternative hypotheses*

We are to test if the mean life of never-die batteries is 65 months or less. The null and alternative hypotheses are

H_0: $\mu = 65$ (the mean life is not less than 65 months)

H_1: $\mu < 65$ (the mean life is less than 65 months)

Step 2. *Select the distribution to use*

The sample size is small, and the life of a battery is approximately normally distributed. However, we do not know the population standard deviation σ. Hence, we use the t distribution to make the test.

Step 3. *Determine the rejection and nonrejection regions*

The significance level is .05. The < sign in the alternative hypothesis indicates that the test is left-tailed with the rejection region in the left tail. To find the critical value of t, we need to know the area in the left tail and the degrees of freedom.

$$\text{Area in the left tail} = \alpha = .05$$

$$df = n - 1 = 15 - 1 = 14$$

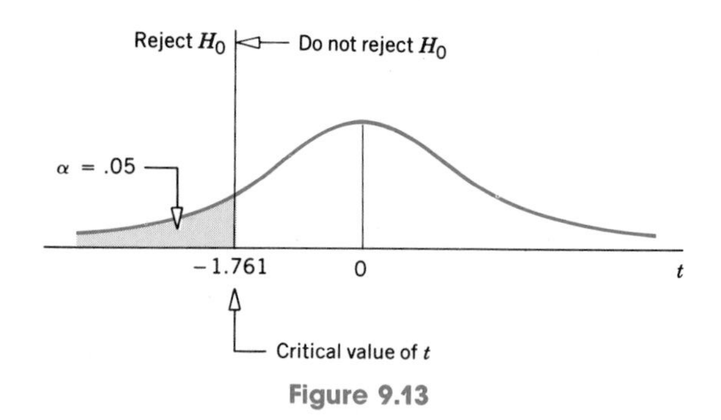

Figure 9.13

From the t distribution table, the critical value of t for 14 degrees of freedom and an area of .05 in the left tail is -1.761. This value is shown in Figure 9.13.

Step 4. *Calculate the value of the test statistic*

We calculate the value of the test statistic t for $\bar{x} = 63$ as follows.

$$s_{\bar{x}} = \frac{s}{\sqrt{n}} = \frac{2}{\sqrt{15}} = .516$$

From H_0

$$t = \frac{\bar{x} - \mu}{s_{\bar{x}}} = \frac{63 - 65}{.516} = -3.876$$

Step 5. *Make a decision*

The value of the test statistic $t = -3.876$ is less than the critical value of $t = -1.761$, and it falls in the rejection region. Therefore, we reject H_0 and conclude that the sample mean is too low compared to 65 (company's claimed value of μ) and the difference between the two may not be attributed to chance alone. The mean life of the company's never-die batteries is less than 65 months. ■

Making a right-tailed test of hypothesis about μ: $n < 30$.

EXAMPLE 9-8 A psychologist claims that the mean age at which children start walking is 12.5 months. Carol wanted to test the hypothesis that the actual mean age at which children start walking is later than 12.5 months. She took a sample of 18 children and found that the mean age at which these children started walking was 13.6 months with a standard deviation of .8 months. Test the hypothesis at the 1% significance level assuming that the ages at which all children start walking have an approximate normal distribution.

Solution Let μ be the mean age at which all children start walking and $\bar{x}$ be the corresponding mean for the sample. Then, from the given information,

$$n = 18, \quad \bar{x} = 13.6 \text{ months}, \quad s = .8 \text{ months}, \quad \text{and} \quad \alpha = .01$$

Step 1. *State the null and alternative hypotheses*

We are to test if the mean age at which all children start walking is 12.5 months or later. The null and alternative hypotheses are

H_0: $\mu = 12.5$ (the mean walking age is 12.5 months)

H_1: $\mu > 12.5$ (the mean walking age is later than 12.5 months)

Step 2. *Select the distribution to use*

The sample size is small, and the population is approximately normally distributed. However, we do not know the population standard deviation σ. Hence, we use the t distribution to make the test.

Step 3. *Determine the rejection and nonrejection regions*

The significance level is .01. The $>$ sign in the alternative hypothesis indicates that the test is right-tailed and the rejection region lies in the right tail.

$$\text{Area in the right tail} = \alpha = .01$$
$$df = n - 1 = 18 - 1 = 17$$

From the t distribution table, the critical value of t for 17 degrees of freedom and .01 area in the right tail is 2.567. This value is shown in Figure 9.14.

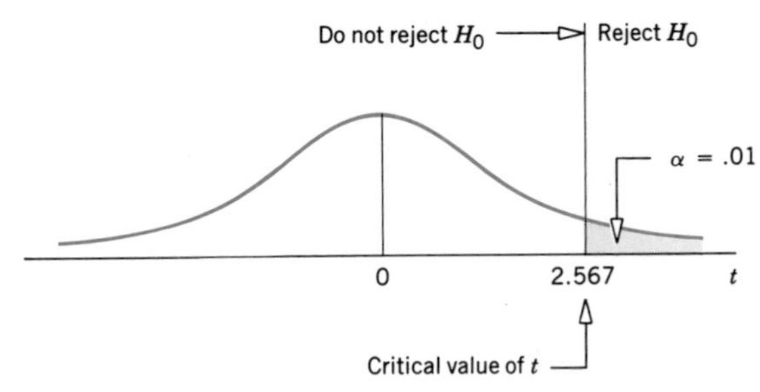

Figure 9.14

Step 4. *Calculate the value of the test statistic*

We calculate the value of the test statistic t for $\bar{x} = 13.6$ as follows.

$$s_{\bar{x}} = \frac{s}{\sqrt{n}} = \frac{.8}{\sqrt{18}} = .189$$

$$t = \frac{\bar{x} - \mu}{s_{\bar{x}}} = \frac{13.6 - 12.5}{.189} = 5.820$$

Step 5. *Make a decision*

The value of the test statistic $t = 5.820$ is larger than the critical value of $t = 2.567$, and it falls in the rejection region. Hence, we reject H_0. As a result, we conclude that the value of the sample mean is too large compared to the hypothesized value of the population mean and the difference between the two may not be attributed to chance alone. The mean age at which all children start walking is later than 12.5 months. ■

EXERCISES

9.29 Assuming that the respective populations are normally distributed, make the following hypothesis tests.

a. H_0: $\mu = 24$, H_1: $\mu \neq 24$, $n = 25$, $\bar{x} = 28$, $s = 4.9$, $\alpha = .01$
b. H_0: $\mu = 30$, H_1: $\mu < 30$, $n = 16$, $\bar{x} = 27$, $s = 6.6$, $\alpha = .025$
c. H_0: $\mu = 18$, H_1: $\mu > 18$, $n = 20$, $\bar{x} = 22$, $s = 8$, $\alpha = .10$

9.30 Assuming that the respective populations are normally distributed, make the following hypothesis tests.

a. H_0: $\mu = 60$, H_1: $\mu \neq 60$, $n = 14$, $\bar{x} = 56$, $s = 9$, $\alpha = .05$
b. H_0: $\mu = 35$, H_1: $\mu < 35$, $n = 24$, $\bar{x} = 29$, $s = 5.4$, $\alpha = .005$
c. H_0: $\mu = 47$, H_1: $\mu > 47$, $n = 18$, $\bar{x} = 51$, $s = 6$, $\alpha = .001$

9.31 The president of a university claims that the mean time spent partying by all students at this university is not more than 7 hours a week. A random sample of 20 students taken from this university showed that they spent an average of 10.7 hours partying the previous week with a standard deviation of 2.3 hours. Assuming that the time spent partying by all students at this university is approximately normally distributed, test at the 5% significance level if the president's claim is true.

9.32 According to a basketball coach, the mean height of all male college basketball players is 74 inches. A random sample of 25 such players gave a mean height of 75.2 inches with a standard deviation of 2.6 inches. Assuming that the heights of all college basketball players are normally distributed, test at the 1% significance level if their mean height is different from 74 inches.

9.33 According to the U.S. Bureau of the Census, the mean monthly income for high school diploma holders was $1045 in 1984. A random sample of 28 high school diploma holders taken recently showed that their mean monthly income is $1170 with a standard deviation of $160. Assuming that the monthly incomes of all high school diploma holders are approximately normally distributed, test at the 1% significance level if the mean monthly income of such persons has increased during this period.

9.34 A soft-drink manufacturer claims that its 12-ounce cans do not contain on average more than 30 calories. A random sample of 16 cans of this soft drink, which were checked for calories, contained a mean of 31.8 calories with a standard deviation of 3. Assuming that the number of calories in 12-ounce soda cans are normally distributed, test at the 1% significance level if the manufacturer's claim is true.

9.35 According to the National Agricultural Statistics Service, the mean yield of potatoes per acre was 292 cwt. (a cwt is equal to 100 pounds) in 1986. A random sample of 20 acres gave a mean yield of potatoes to be 301 cwt. for 1989 with a standard deviation of 22 cwt. Assuming that the yield of potatoes per acre is normally distributed, test at the 5% significance level if the mean yield of potatoes for 1989 is different from 292 cwt.

9.36 The mean balance of all checking accounts at a bank on December 31, 1986 was $850. A random sample of 25 checking accounts taken recently from this bank gave a mean balance of $775 with a standard deviation of $230. Assuming that the balances of all checking accounts at this bank are normally distributed, test at the 5% significance level if the mean balance of such accounts has decreased during this period.

9.37 The mean cost of cornea transplants was $6800 in 1986 (*Hippocrates* May/June 1988). A random sample of 25 cornea transplants performed this year gave a mean cost of $7650 with a standard deviation of $820. Test at the 1% significance level if the current mean cost of cornea transplants is greater than $6800. Assume that the costs of all cornea transplants have a normal distribution.

9.38 A paint manufacturing company claims that the mean drying time for its paints is not more than 45 minutes. A random sample of 20 gallons of paints selected from the production line of this company showed that the mean drying time for this sample is 50 minutes with a standard deviation of 3 minutes. Test at the 5% significance level if the mean drying time for this company's paints is more than 45 minutes. Assume that the drying times for these paints have a normal distribution.

9.39 According to the American Medical Association, physicians' mean income (before taxes and after expenses) was $119,500 in 1986. A random sample of 26 physicians showed that their mean income for 1989 was $136,720 with a standard deviation of $21,400. Test at the 1% significance level if physicians' mean income for 1989 is greater than $119,500. Assume that the incomes of all physicians are approximately normally distributed.

9.40 According to the Administrative Office of the U.S. Courts, the mean regular prison sentence for federal offenses was 64.6 months in 1986. A random sample of 23 federal offenses for this year showed that the mean regular prison sentence for these offenses was 60.2 months with a standard deviation of 11.7 months. Test at the 2.5% significance level if the mean regular prison sentence for federal offenses this year is less than 64.6 months. Assume that the regular prison sentences for all federal offenses are approximately normally distributed.

9.41 A business school claims that students who complete a three-month typing course can type on average at least 1200 words an hour. A random sample of 25 students who completed this course typed on average 1130 words an hour with a standard deviation of 85 words. Test at the 1% significance level if the mean typing speed for all students who complete this course is less than 1200 words an hour. Assume that the typing speeds for all students who complete this course have an approximate normal distribution.

***9.42** The manager of a service station claims that the mean amount spent on gas by its customers is $10.90. You want to test if the mean amount spent on gas at this station is different from $10.90. Briefly explain how you would conduct this test by taking a small sample.

9.5 HYPOTHESIS TESTS ABOUT A POPULATION PROPORTION: LARGE SAMPLES

Often we want to test a hypothesis about the population proportion. For example, the U.S. Census Bureau claims that 54% of young adults aged 18 to 24 lived with their parents in 1988. A sociologist may want to check if this percentage still holds. As another example, a study on drug use claims that 65% of adults aged 20 or older have never used marijuana. We may want to make a test of hypothesis to check if this claim is true.

This section presents the procedure to test hypotheses about the population proportion p for large samples. The procedure to make such tests is similar in many respects to the one for the population mean μ. The procedure includes the same five steps. Again, the test can be two-tailed or one-tailed. We know from Chapter 7 that when a sample size is large, the sample proportion $\hat{p}$ is approximately normally distributed with its mean equal to p and standard deviation equal to $\sqrt{pq/n}$. Hence, we use the normal distribution to make a hypothesis test about the population proportion p for a large sample. As was mentioned in Chapters 7 and 8, in the case of a proportion, the sample size is considered to be large when np and nq are both greater than 5.

TEST STATISTIC

The value of the test statistic z for the sample proportion $\hat{p}$ is computed as

$$z = \frac{\hat{p} - p}{\sigma_{\hat{p}}}$$

where

$$\sigma_{\hat{p}} = \sqrt{\frac{p\,q}{n}}$$

The value of p used in this formula is the one used in the null hypothesis. The value of q is equal to $1 - p$.

Examples 9-9, 9-10, and 9-11 describe the procedure of making a test of hypothesis about the population proportion p.

Conducting a two-tailed test of hypothesis about p: large sample.

EXAMPLE 9-9 An earlier study on drug use claimed that 65% of all persons aged 20 or older had never used marijuana. A researcher wanted to test if the current percentage of all persons aged 20 or older who have never used marijuana is different from 65%. He took a sample of 400 persons aged 20 or older and found that 62% of them had never used marijuana. Test the hypothesis at a .01 significance level.

Solution Let p be the proportion of all persons aged 20 or older who have never used marijuana and $\hat{p}$ be the corresponding sample proportion. Then, from the given information,

$$n = 400, \quad \hat{p} = .62, \quad \text{and} \quad \alpha = .01$$

The claim from the earlier study is that 65% of all persons aged 20 or older have never used marijuana. Hence, assuming this claim is true,

$$p = .65 \quad \text{and} \quad q = 1 - p = 1 - .65 = .35$$

Step 1. *State the null and alternative hypotheses*

The claim is true if $p = .65$, and claim is not true if $p \neq .65$. The null and alternative hypotheses are

H_0: $p = .65$ (the population proportion is not different from .65)

H^1: $p \neq .65$ (the population proportion is different from .65)

Step 2. *Select the distribution to use*

The values of np and nq are

$$np = 400\,(.65) = 260 \quad \text{and} \quad nq = 400\,(.35) = 140$$

Because both np and nq are greater than 5, the sample size is large. Consequently, we use the normal distribution to make the hypothesis test about p.

Step 3. *Determine the rejection and nonrejection regions*

The $\neq$ sign in the alternative hypothesis indicates that the test is two-tailed. The significance level is .01. Hence, the total area of the two rejection regions is .01 and the rejection region in each tail is $\alpha/2 = .01/2 = .005$. The critical values of z, obtained from the normal table for .4950, are -2.58 and 2.58, as shown in Figure 9.15.

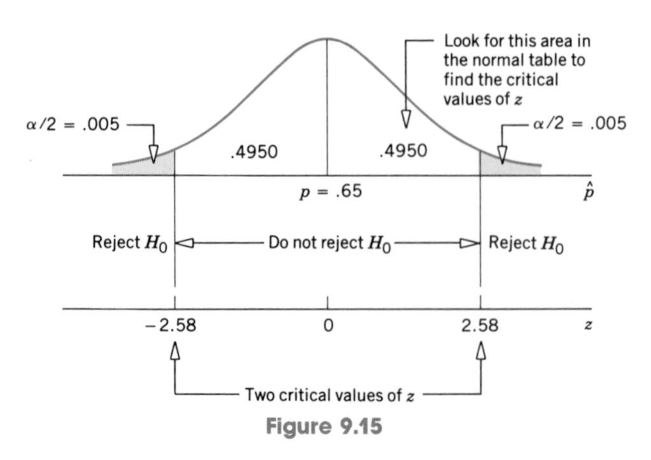

Figure 9.15

Step 4. *Calculate the value of the test statistic*

We calculate the value of the test statistic z for $\hat{p} = .62$ as follows.

$$\sigma_{\hat{p}} = \sqrt{\frac{p\,q}{n}} = \sqrt{\frac{(.65)\,(.35)}{400}} = .024$$

$$z = \frac{\hat{p} - p}{\sigma_{\hat{p}}} = \frac{.62 - \overset{\text{From } H_0}{.65}}{.024} = -1.25$$

Step 5. *Make a decision*

The value of the test statistic $z = -1.25$ for $\hat{p}$ lies between -2.58 and 2.58, and it falls in the nonrejection region. Consequently, we do not reject H_0. Therefore, we can state that the sample proportion is not too far from the hypothesized value of the population proportion and the difference between the two can be attributed to chance. Therefore, the given claim seems to be true. ■

Making a right-tailed test of hypothesis about p: large sample.

EXAMPLE 9-10 When properly working, a machine does not produce more than 4% defective items. Whenever the machine produces greater than 4% defective items, it needs adjustment. A random sample of 200 items taken from the production line by the quality control manager contained 14 defective items. Test at the 5% significance level if the machine needs adjustment.

Solution Let p be the proportion of defective items in all items produced by this machine and $\hat{p}$ be the corresponding sample proportion. Then, from the given information,

$$n = 200, \qquad \hat{p} = \frac{14}{200} = .07, \qquad \text{and} \qquad \alpha = .05$$

When the machine is working properly it produces 4% defective items. Consequently, assuming that the machine is working properly,

$$p = .04 \qquad \text{and} \qquad q = 1 - p = 1 - .04 = .96$$

Step 1. *State the null and alternative hypotheses*

The machine will not need adjustment if the percentage of defective items is 4% or less, and it will need adjustment if this percentage is greater than 4%. Hence, the null and alternative hypotheses are

H_0: $p \leq .04$ (the machine does not need adjustment)

H_1: $p > .04$ (the machine needs adjustment)

Step 2. *Select the distribution to use*

The values of np and nq are

$$np = 200\,(.04) = 8 > 5 \qquad \text{and} \qquad nq = 200\,(.96) = 192 > 5$$

Because the sample size is large, we use the normal distribution to make the hypothesis test about p.

Step 3. *Determine the rejection and nonrejection regions*

The significance level is .05. The > sign in the alternative hypothesis indicates that the test is right-tailed and the rejection region lies in the right tail with its area equal to .05. As shown in Figure 9.16, the critical value of z, obtained from the normal table for .4500, is approximately 1.65.

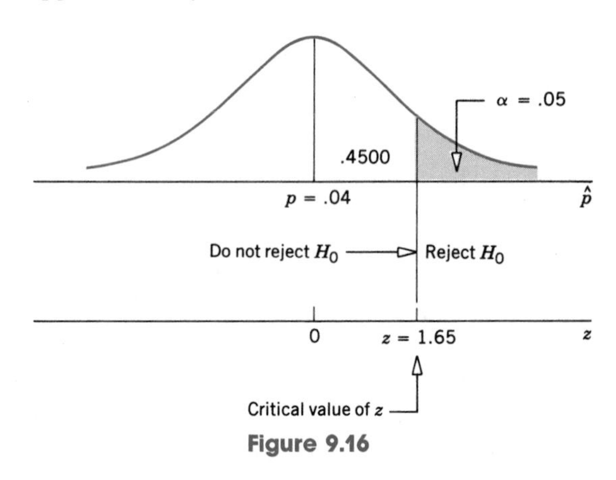

Figure 9.16

Step 4. *Calculate the value of the test statistic*

We calculate the value of the test statistic z for $\hat{p} = .07$ as follows.

$$\sigma_{\hat{p}} = \sqrt{\frac{p\,q}{n}} = \sqrt{\frac{(.04)\,(.96)}{200}} = .014$$

$$z = \frac{\hat{p} - p}{\sigma_{\hat{p}}} = \frac{.07 - \overset{\text{From } H_0}{.04}}{.014} = 2.14$$

Step 5. *Make a decision*

Because the value of the test statistic $z = 2.14$ is greater than the critical value of $z = 1.65$ and it falls in the rejection region, we reject H_0. We conclude that the sample proportion is too far from the hypothesized value of the population proportion and the difference between the two cannot be attributed to chance alone. Therefore, based on the sample information, it appears that the machine needs an adjustment. ■

Making a left-tailed test of hypothesis about p: large sample.

EXAMPLE 9-11 In 1980, when Ronald Reagan became president of the United States, 76% of American families owned homes. Assume that this percentage is true for the population of all families for 1980. In 1988, 72% of American families owned homes (*Psychology Today,* October 1988). Assume that the 1988 estimate is based on a sample of 600 families. Test at the 5% significance level if the proportion of families who own homes has decreased during this period.

Solution Let p be the proportion of all American families who owned homes in 1988 and $\hat{p}$ be the corresponding sample proportion. Then, from the given information,

$$n = 600 \quad \text{and} \quad \hat{p} = .72$$

The proportion of all families who owned homes in 1980 is given to be .76. Assuming that this proportion is also true for 1988,

$$p = .76 \quad \text{and} \quad q = 1 - p = 1 - .76 = .24$$

The significance level is $\alpha = .05$.

Step 1. *State the null and alternative hypotheses*

The null and alternative hypotheses are

H_0: $p = .76$ (the proportion has not decreased)

H_1: $p < .76$ (the proportion has decreased)

Step 2. *Select the distribution to use*

We first check whether both np and nq are greater than five.

$$np = 600\,(.76) = 456 > 5 \quad \text{and} \quad nq = 600\,(.24) = 144 > 5$$

Hence, the sample size is large. Consequently, we use the normal distribution to make the hypothesis test about p.

Step 3. *Determine the rejection and nonrejection regions*

The significance level is .05. The $<$ sign in the alternative hypothesis indicates that the test is one-tailed and the rejection region lies in the left tail with its area equal to .05. As shown in Figure 9.17, the critical value of z, obtained from the normal table for .4500, is (approximately) -1.65.

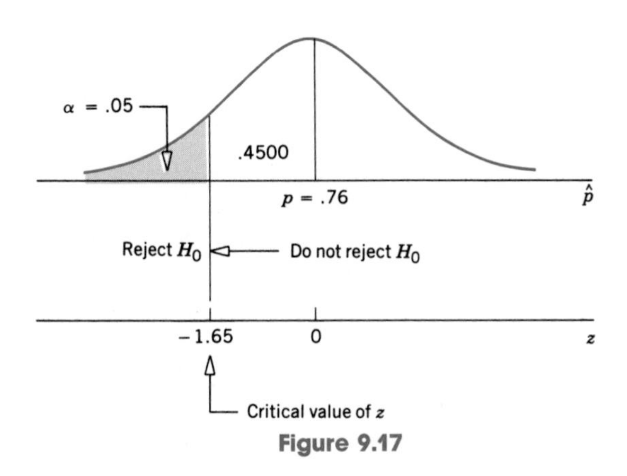

Figure 9.17

Step 4. *Calculate the value of the test statistic*

We calculate the value of the test statistic z for $\hat{p} = .72$ as follows.

$$\sigma_{\hat{p}} = \sqrt{\frac{p\,q}{n}} = \sqrt{\frac{(.76)\,(.24)}{600}} = .017$$

$$z = \frac{\hat{p} - p}{\sigma_{\hat{p}}} = \frac{.72 - .76}{.017} = -2.35$$

Step 5. *Make a decision*

The value of the test statistic $z = -2.35$ is less than the critical value of $z = -1.65$, and it falls in the rejection region. Hence, we reject H_0. We can state that the difference between the sample proportion and the hypothesized value of the population proportion is quite large, and this difference may not have occurred because of chance alone. Therefore, the proportion of homeowners among all families has decreased between 1980 and 1988. ■

EXERCISES

9.43 Make the following hypothesis tests about p.

a. H_0: $p = .45$, H_1: $p \neq .45$, $n = 100$, $\hat{p} = .48$, $\alpha = .10$
b. H_0: $p = .72$, H_1: $p < .72$, $n = 700$, $\hat{p} = .65$, $\alpha = .05$
c. H_0: $p = .30$, H_1: $p > .30$, $n = 200$, $\hat{p} = .34$, $\alpha = .01$

9.44 Make the following hypothesis tests about p.

a. H_0: $p = .57$, H_1: $p \neq .57$, $n = 800$, $\hat{p} = .51$, $\alpha = .05$
b. H_0: $p = .26$, H_1: $p < .26$, $n = 400$, $\hat{p} = .22$, $\alpha = .01$
c. H_0: $p = .84$, H_1: $p > .84$, $n = 125$, $\hat{p} = .86$, $\alpha = .025$

9.45 According to the U.S. Department of Justice, 25% of drivers arrested for driving under the influence of alcohol in 1987 were 24 years of age or younger. A random sample of 100 drivers arrested recently for driving under the influence of alcohol showed that 20% of them were 24 years of age or younger. Test at the 1% significance level if the current percentage of 24 years of age or younger among all people arrested for driving under the influence of alcohol is less than 25%.

9.46 In a 1985 study conducted by the Roper Organization, 45% of women aged 18 and older said they would "prefer to stay home and take care of a house and family" than "to have a job outside the home" (*The 1985 Virginia Slims American Women's Opinion Poll*). In a random sample of 500 women aged 18 and older taken recently, 39% held this view. Test at the 5% significance level if the current percentage of women aged 18 and older who hold this view is less than 45%.

9.47 According to the U.S. Bureau of the Census, 19% of American women had a degree beyond high school in 1984. A recent survey of 150 randomly selected women showed that 24% of them had a degree beyond high school. Test at the 2.5% significance level if the proportion of women with a degree beyond high school has increased during this period.

9.48 According to the National Education Association, 49% of school teachers in 1986 had other jobs to supplement their incomes. A random sample of 400 teachers taken this year showed that 57% of them hold other jobs. Test at the 1% significance level if the current percentage of all teachers who hold other jobs is higher than 49%.

9.49 According to a study by Professor Adam Drewnowski and others, 21.1% of female college students were on a diet at the time of the study in April 1987 (Adam Drewnowski et al.: "The Prevalence of Bulimia Nervosa in the US College Student Population," *American Journal of Public Health,* 78(10): October 1988). A recent sample of 500 female college students showed that 23% of them are on a diet. Test at the 5% significance level if the percentage of current female college students who are on a diet is different from 21.1%.

9.50 According to a study conducted during the period 1984 to 1987, 27% of mothers used marijuana during pregnancy (Barry Zuckerman: "Effects of Maternal Marijuana and Cocaine Use on Fetal Growth," *The New England Journal of Medicine,* 320(12): March 23, 1989). A recent sample of 400 mothers showed that 23% of them used marijuana during pregnancy. Test at the 1% significance level if the current percentage of mothers who have used marijuana during pregnancy is different from 27%.

9.51 According to the National Center for Health Statistics, 40% of people aged 18 and older did regular physical activities in 1985. A recent sample of 1000 persons aged 18 and older showed that 47% of them do regular physical activities. Test at the 5% significance level if the current percentage of persons aged 18 and older who do regular physical activities is higher than 40%.

9.52 In a 1987 study conducted by the Roper Organization, 66% of American adults said that factors within their personal control have determined their lot in life (*The American Dream,* A National Survey Conducted for *The Wall Street Journal* by the Roper Organization). A recent study based on a random sample of 500 American adults showed that 71% of them hold this view. Test at the 1% significance level if the percentage of American adults who hold this view is now higher than 66%.

9.53 According to a 1987 study conducted by the Roper Organization, 64% of American children lived with both biological parents (*The American Chicle Youth Poll,* The American Chicle Group, Warner-Lambert Company). A recent study based on a random sample of 700 children showed that 58% of them live with both biological parents. Test at the 2.5% significance level if the percentage of children living with both biological parents is now less than 64%.

9.54 According to a 1986 study by the National Institute of Drug Abuse, 32.4% of high school seniors had never smoked. A recently taken random sample of 800 high school seniors showed that 30% of them have never smoked. Test at the 10% significance level if the current percentage of high school seniors who have never smoked is lower than 32.4%.

9.55 According to the U.S. Bureau of the Census, 5% of couples living together are not married. A researcher took a random sample of 400 couples and found that 7.5% of them are not married. Test at the 1% significance level if the current percentage of unmarried couples is different from 5%.

***9.56** In a Gallup poll, 69% of respondents agreed that "too many children are being raised in day-care centers" (*U.S. News & World Report,* August 29/September 5, 1988). Suppose you want to test if the current percentage of adults who hold this view is different from 69%. Briefly explain how you would conduct such a test.

GLOSSARY

α The significance level of a test of hypothesis that denotes the probability of rejecting a null hypothesis when it actually is true. (The probability of committing a type I error.)

Alternative hypothesis A claim about a population parameter that will be true if the null hypothesis is false.

β The probability of not rejecting a null hypothesis when it actually is false. (The probability of committing a type II error.)

Critical value or **critical point** One or two values that divide the whole region under the distribution of a sample statistic into rejection and nonrejection regions.

Left-tailed test A test in which the rejection region lies in the left tail of the distribution.

Null hypothesis A claim about a population parameter that is assumed to be true until it is declared false.

One-tailed test A test in which there is only one rejection region.

***p*-value** The smallest significance level at which the null hypothesis can be rejected.

Right-tailed test A test in which the rejection region lies in the right tail of the distribution.

Significance level The value of α that gives the probability of committing a type I error.

Two-tailed test A test in which there are two rejection regions, one in each tail of the distribution.

Type I error An error that occurs when a true null hypothesis is rejected.

Type II error An error that occurs when a false null hypothesis is not rejected.

KEY FORMULAS

1. **Value of the test statistic z for $\bar{x}$ in a hypothesis test of μ for a large sample**

$$z = \frac{\bar{x} - \mu}{\sigma_{\bar{x}}} \quad \text{if } \sigma \text{ is known}$$

or

$$z = \frac{\bar{x} - \mu}{s_{\bar{x}}} \quad \text{if } \sigma \text{ is not known}$$

where

$$\sigma_{\bar{x}} = \frac{\sigma}{\sqrt{n}} \quad \text{and} \quad s_{\bar{x}} = \frac{s}{\sqrt{n}}$$

2. **Value of the test statistic t for $\bar{x}$ in a hypothesis test of μ for a small sample**

$$t = \frac{\bar{x} - \mu}{s_{\bar{x}}}$$

where

$$s_{\bar{x}} = \frac{s}{\sqrt{n}}$$

3. **Value of the test statistic z for $\hat{p}$ in a hypothesis test of p for a large sample**

$$z = \frac{\hat{p} - p}{\sigma_{\hat{p}}}$$

where

$$\sigma_{\hat{p}} = \sqrt{\frac{p\,q}{n}}$$

SUPPLEMENTARY EXERCISES

9.57 A manufacturer of fluorescent light bulbs claims that the mean life of these bulbs is 2500 hours. A random sample of 36 such bulbs was selected and tested. The mean life for the sample was found to be 2447 hours with a standard deviation of 180 hours. Test at the 2.5% significance level if the mean life of the fluorescent light bulbs manufactured by this company is less than 2500 hours.

9.58 According to the U.S. National Center for Health Statistics, divorced women spend an average of 11 days a year sick in bed. A random sample of 100 divorced women taken by a researcher showed that the women in this sample spent an average of 10.5 days sick in bed last year with a standard deviation of 2.4 days. Test at the 1% significance level if the mean number of days spent sick in bed by divorced women is different from 11.

9.59 A researcher claims that the mean age of men at the time of their first marriage is 27 years. A random sample of 200 men who recently were married for the first time gave a mean age of 29.4 years with a standard deviation of 3.2 years. Test at the 10% significance level if the mean age of all men at the time of their first marriage is different from 27 years.

9.60 According to the U.S. Department of Transportation, the mean number of miles driven per driver was 11,783 in 1987. The records of 200 randomly selected drivers showed that they drove an average of 12,060 miles in 1989 with a standard deviation of 3480 miles. Test at the 5% significance level if the mean number of miles driven per driver in 1989 were different from 11,783.

9.61 According to the U.S. Bureau of the Census, the mean earnings of people with 4 years of college were \$38,973 in 1987. A recently taken random sample of 400 people with 4 years of college gave the mean earnings of \$42,720 with a standard deviation of \$4645. Test at the 1% significance level if the mean earnings of people with 4 years of college are now greater than \$38,973.

9.62 According to the U.S. Bureau of Labor Statistics, the mean expenditure on food per household in 1985 was \$3394. A random sample of 500 households gave the mean food expenditure for 1989 as \$3656 with a standard deviation of \$524. Test at the 5% significance level if the mean food expenditure for all households for 1989 was greater than \$3394.

9.63 According to the U.S. Department of Labor, private sector workers earned an average of \$322.26 a week in 1988. A recently taken random sample of 400 private sector workers showed that they earn on average \$329.50 a week with a standard deviation of \$72. Find the p-value for the hypothesis test that the current mean weekly salary of private sector workers is greater than \$322.26.

9.64 According to *Newsweek* magazine, the mean weight of women is 138 pounds (*Newsweek*, December 5, 1988). A recently taken random sample of 400 women showed that their mean weight is 135 pounds with a standard deviation of 24 pounds.

Find the p-value for the hypothesis test that the current mean weight of women is different from 138 pounds.

9.65 According to the Hertz Corporation, the mean repair and maintenance cost of a car was \$1035 in 1986. A random sample of 27 cars showed that the mean repair and maintenance cost for these cars was \$1346 for 1989 with a standard deviation of \$234. Test at the 1% significance level if the 1989 mean repair and maintenance cost of a car is different from \$1035. Assume that the 1989 repair and maintenance costs for all cars are approximately normally distributed.

9.66 The mean cost of kidney transplants was \$35,000 in 1986 (*Hippocrates,* May/June, 1988). A random sample of 20 recent kidney transplants gave a mean cost of \$38,500 with a standard deviation of \$3600. Test at the 5% significance level if the current mean cost of kidney transplants is different from \$35,000. Assume that the costs of all kidney transplants have an approximate normal distribution.

9.67 The administrative office of a hospital claims that the mean waiting time for patients to get treatment in its emergency ward is at most 25 minutes. A random sample of 16 patients who received treatment in the emergency ward of this hospital gave a mean waiting time of 27.5 minutes with a standard deviation of 4.8 minutes. Test at the 1% significance level if the mean waiting time at the emergency ward of this hospital is more than 25 minutes. Assume that the waiting times for all patients at this emergency ward have a normal distribution.

9.68 According to *Dental Management* magazine, the mean gross income of orthodontists was \$243,863 in 1986 (*Working Woman,* January 1988). A random sample of 20 orthodontists showed that their mean gross income for 1989 was \$289,347 with a standard deviation of \$45,372. Test at the 5% significance level if the mean gross income of all orthodontists for 1989 was higher than \$243,863. Assume that the 1989 gross incomes of all orthodontists are approximately normally distributed.

9.69 The mean weight of all babies born in a large city is 7.6 pounds. A random sample of 23 babies selected from this city born to mothers who smoked marijuana gave a mean weight of 6.9 pounds with a standard deviation of .8 pounds. Test at the 1% significance level if the mean weight of babies born to mothers who smoke marijuana is less than 7.6 pounds. Assume that the weights of babies born to mothers who smoke marijuana have an approximate normal distribution.

9.70 A company that has introduced a new software claims that the mean time to learn how to use this software is not more than 2 hours for those somewhat familiar with computers. A random sample of 24 such persons was selected. The mean time taken by these persons to learn how to use this software was found to be 2.25 hours with a standard deviation of .57 hours. Test at the 1% significance level if the company's claim is true. Assume that the time taken by all persons who are somewhat familiar with computers to learn how to use this software is approximately normally distributed.

9.71 According to the U.S. Bureau of the Census, 41.3% of women aged 65 and over were living alone in 1986. A random sample of 400 women aged 65 and over taken recently showed that 46% of them are living alone. Test at the 1% significance level if the percentage of women aged 65 and over who are currently living alone is higher than 41.3%.

9.72 According to the U.S. Department of Labor, 57% of families had two or more wage earners in 1987. A recent poll of 900 randomly selected families showed that

64% of them have two or more wage earners. Test at the 5% significance level if the percentage of families with two or more wage earners is now higher than 57%.

9.73 According to the U.S. National Center for Health Statistics, 22% of persons aged 18 and over slept for 6 hours or less a day in 1985. A random sample of 250 persons aged 18 and over taken recently showed that 20% of them sleep for 6 hours or less a day. Test at the 5% significance level if the percentage of persons aged 18 and over who sleep for 6 hours or less a day is now less than 22%.

9.74 In a 1988 Roper poll, 88% of married women said "they'd march down the aisle again with the same man" (*Ladies' Home Journal,* June 1988). In a random sample of 1000 married women taken recently, 834 held this view. Test at the 1% significance level if the current proportion of married women who hold this view is lower than 88%. (*Hint:* First calculate the sample proportion.)

9.75 According to a Daniel Yankelovich Group and Marttila and Kiley Inc. poll, 59% of the "public sees economic competitors like Japan as a greater danger than military adversaries, like the Soviet Union, because economic competitors 'threaten our jobs and economic security' " (*Psychology Today,* September 1988). A random sample of 500 persons taken recently gave this percentage to be 63%. Test at the 5% significance level if the current percentage of people who hold this view is different from 59%.

9.76 According to the U.S. Bureau of Labor Statistics, 56% of mothers with children under the age of 6 were working outside their homes in 1988. A recently taken random sample of 900 mothers with children under the age of 6 showed that 59% of them work outside their homes. Test at the 1% significance level if the current percentage of mothers with children under the age of 6 who work outside their homes is different from 56%.

SELF-REVIEW TEST

1. A test of hypothesis is always about
   a. a population parameter
   b. a sample statistic
   c. a test statistic
2. A type I error is made when
   a. a null hypothesis is not rejected when it is actually false
   b. a null hypothesis is rejected when it is actually true
   c. an alternative hypothesis is rejected when it is actually true
3. A type II error is made when
   a. a null hypothesis is not rejected when it is actually false
   b. a null hypothesis is rejected when it is actually true
   c. an alternative hypothesis is rejected when it is actually true
4. A critical value is the value
   a. calculated from the sample data

b. determined from a table (e.g., the normal distribution table)
c. neither (a) nor (b)

5. The computed value of a test statistic is the value

a. calculated for a sample statistic
b. determined from a table (e.g., the normal distribution table)
c. neither (a) nor (b)

6. The significance level, denoted by α, is

a. the probability of committing a type I error
b. the probability of committing a type II error
c. neither (a) nor (b)

7. A two-tailed test is a test with

a. two rejection regions
b. two nonrejection regions
c. two test statistics

8. A one-tailed test

a. has one rejection region
b. has one nonrejection region
c. both (a) and (b)

9. The smallest level of significance at which a null hypothesis will be rejected is called

a. α **b.** *p*-value **c.** β

10. Which of the following is not required to apply the *t* distribution to make a test of hypothesis about μ?

a. $n < 30$
b. the population is approximately normally distributed
c. σ is unknown
d. β is known

11. The sign in the alternative hypothesis in a two-tailed test is always

a. $<$ **b.** $>$ **c.** $\neq$

12. The sign in the alternative hypothesis in a left-tailed test is always

a. $<$ **b.** $>$ **c.** $\neq$

13. The sign in the alternative hypothesis in a right-tailed test is always

a. $<$ **b.** $>$ **c.** $\neq$

14. A bank loan officer claims that the mean monthly mortgage payment made by all homeowners in a certain city is \$1365. A housing magazine wanted to test this claim. A random sample of 100 homeowners taken by this magazine gave the mean monthly mortgage as \$1489 with a standard deviation of \$278. Test at the 1% significance level if the mean monthly mortgage payment made by all homeowners in this city is different from \$1365.

15. An editor of a New York publishing company claims that the mean time taken to write a textbook is at least 36 months. A sample of 16 textbook authors showed that the mean time taken by them to write a textbook was 32 months with a standard deviation of 4.6 months. Test at the 5% significance level if the editor's claim is true. Assume that the time taken to write a textbook is normally distributed for all textbook authors.

16. In a *Time*/CNN poll of adult Americans, 73% said "there should be more government spending on educational and recreational facilities for teenagers to reduce teenage violence" (*Time,* June 12, 1989). Among a recent sample of 400 adult Americans taken by a sociologist, 77% held this view. Test at the 1% significance level if the current percentage of adult Americans who hold this view is larger than 73%.

17. According to an IRS study, it takes on average 60 minutes to prepare, copy, and mail a 1040EZ tax form. A sample of 100 taxpayers who filed the 1040EZ form last year showed that they took on average 62.6 minutes to prepare, copy, and mail this tax form. The standard deviation for the sample was 11 minutes. Find the p-value for the test that the mean time taken to prepare, copy, and mail a 1040EZ tax form is different from 60 minutes.

USING MINITAB

MINITAB, ZTEST and TTEST can be used to make hypothesis tests about the population mean using the normal and the t distributions, respectively. MINITAB does not have commands like ZTEST to perform tests of hypotheses about the population proportion. Although a program can be written to make such tests, we will not do so in this text.

HYPOTHESIS TESTS ABOUT A POPULATION MEAN: LARGE SAMPLES

When the sample size is large ($n \geq 30$), whether σ is known or not, we can use the normal distribution to make a hypothesis test about a population mean. The MINITAB commands to make such a test using the normal distribution are as follows.

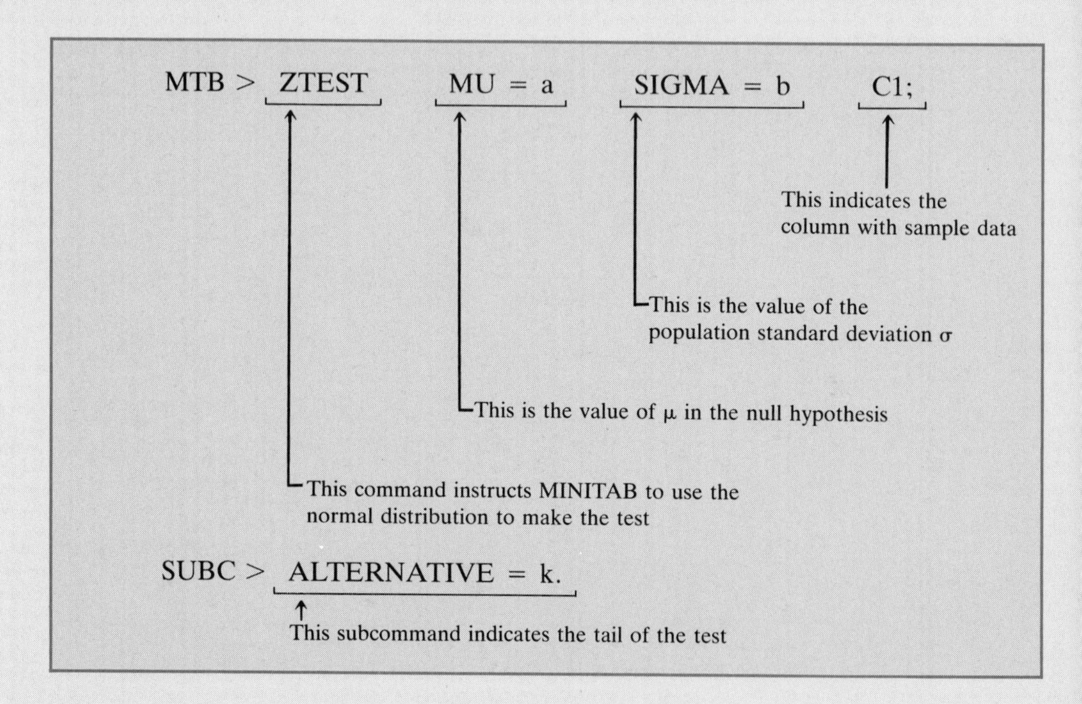

Note the semicolon at the end of the first MINITAB command. This instructs MINITAB that a subcommand is to follow. The period at the end of subcommand ALTERNATIVE = k indicates the end of subcommands.

The value of ALTERNATIVE in the subcommand is -1, 0, or 1, depending on whether the test is *left-tailed, two-tailed,* or *right-tailed,* respectively. In other words

$$\text{ALTERNATIVE} = -1 \quad \text{if the test is left-tailed}$$
$$\text{ALTERNATIVE} = 0 \quad \text{if the test is two-tailed}$$
$$\text{ALTERNATIVE} = 1 \quad \text{if the test is right-tailed}$$

Illustration M9-1 describes the use of these commands.

Illustration M9-1 An earlier study showed that the mean time spent by college students on community service is 50 minutes a week with a standard deviation of 27 minutes. The following data give the time spent on community service by each of 34 college students based on a recently taken random sample.

34	56	74	23	12	89	87	56	48	13	9	85
76	56	17	28	66	38	46	81	29	33	41	78
11	57	17	22	91	54	19	35	65	47		

Using MINITAB, test at the 5% significance level if the mean time spent doing community service by all college students is less than 50 minutes. Assume that the population standard deviation is 27.

Solution From the given information,

$$\sigma = 27 \quad \text{and} \quad \alpha = .05$$

We are to test if the mean time spent doing community service by all college students is less than 50 minutes. The null and alternative hypotheses are

$$H_0\text{: } \mu = 50$$
$$H_1\text{: } \mu < 50$$

To perform this test, we first enter the given data on 34 college students in column C1 using the SET command. Then we write the MINITAB command that indicates that the test is to be made using the normal distribution with $\mu = 50$ and $\sigma = 27$. This MINITAB command also includes the information that the test is to be made using the sample data entered in column C1. Because the test is left-tailed, the value of ALTERNATIVE in the subcommand will be -1.

In the MINITAB solution given on the next page, TEST OF MU = 50.000 VS MU L.T. 50.000 indicates that the null hypothesis is that μ is equal to 50 and the alternative hypothesis is that μ is less than (L.T.) 50. In the row of C1 in MINITAB output, 34 is the sample size, 46.853 is the mean of the sample data, 25.659 is the standard deviation of the sample data, 4.630 is the standard deviation (or standard error) of the sample mean, -0.68 is the value of the test statistic z, and 0.25 is the p-value discussed in Section 9.3 of this chapter.

To make a decision, we find the z value from the normal distribution table for a left-tailed test with $\alpha = .05$. This value (for .4500 area between the mean and z) is approximately -1.65. Because the value of the test statistic, $z = -0.68$, is greater than the critical value of $z = -1.65$ and it falls in the nonrejection region, we fail to reject the null hypothesis. (The reader is advised to draw a graph that shows the rejection and nonrejection regions.) Hence, based on this sample information, we can conclude that the mean time spent doing community service by college students does not seem to be less than 50 minutes a week.

We can also use the P VALUE, printed in the MINITAB solution, to make the decision. As we know from Section 9.3, we will reject H_0 if α is larger than the p-value and we will fail to reject H_0 if α is smaller than the p-value. In the MINITAB solution, the p-value given below P VALUE is 0.25. As $\alpha = .05$ is smaller than the p-value of 0.25, we fail to reject H_0.

```
MTB  > NOTE: TEST OF HYPOTHESIS ABOUT μ: n ≥ 30
MTB  > SET C1
DATA > 34  56  74  23  12  89  87  56  48
DATA > 13   9  85  76  56  17  28  66  38
DATA > 46  81  29  33  41  78  11  57  17
DATA > 22  91  54  19  35  65  47
DATA > END
MTB  > ZTEST  MU = 50  SIGMA = 27  C1;
SUBC > ALTERNATIVE = -1.

TEST OF MU = 50.000 VS MU L.T. 50.000
THE ASSUMED SIGMA = 27.0

      N      MEAN     STDEV    SE MEAN       Z    P VALUE
C1   34    46.853    25.659      4.630   -0.68       0.25
```

(Annotations: N = Sample size; MEAN = $\bar{x}$; STDEV = s; SE MEAN = $\sigma_{\bar{x}}$; Z = Value of the test statistic; P VALUE = p-value)

If the population standard deviation is not known but the sample size is large, we first can compute the standard deviation of C1 and then use that value as an estimate of sigma in the ZTEST command as we did in case of the ZINTERVAL command in Chapter 8. An alternative to this is to use the t distribution to make a test of hypothesis about μ irrespective of the sample size when σ is not known.

HYPOTHESIS TESTS ABOUT A POPULATION MEAN: SMALL SAMPLES

We know from the discussion in this chapter that we apply the t distribution to make a hypothesis test about the population mean when

1. The sample size is small.
2. The population is (approximately) normally distributed.
3. We do not know the population standard deviation.

MINITAB has a TTEST command to make such a hypothesis test. MINITAB commands to make such a test using the t distribution are as follows.

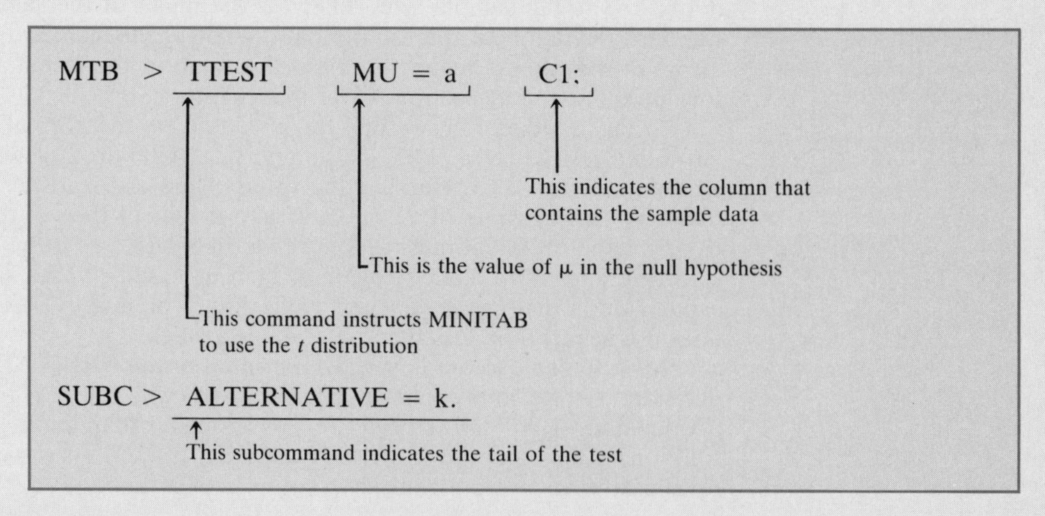

Note that, because σ is not known, its value is not mentioned in these MINITAB commands. MINITAB will use the value of the sample standard deviation calculated for the data of column C1 as an estimate of σ to make a test of hypothesis. The value of ALTERNATIVE in the subcommand will again be -1, 0, or 1, depending on whether the test is left-tailed, two-tailed, or right-tailed. Illustration M9-2 describes the use of these commands.

Illustration M9-2 A psychologist claims that the mean age at which children start walking is 12.5 months. The following data give the age at which 18 randomly selected children started walking.

15	11	13	14	15	12	15	10	16
17	14	16	13	15	15	14	11	13

Using MINITAB, test at the 1% significance level if the mean age at which children start walking is different from 12.5 months. Assume that the age at which all children start walking is normally distributed.

Solution We are to test if the mean age at which children start walking is different from 12.5 months. The null and alternative hypotheses are

$$H_0: \mu = 12.5$$

$$H_1: \mu \neq 12.5$$

The test is two-tailed. We assign a value of 0 to ALTERNATIVE in the MINITAB subcommand. To make this test, first we enter the given data values in column C1 using SET command. Then we write the MINITAB command that indicates that the test is to be made using the t distribution and that $\mu = 12.5$. This command also indicates that the test is to be made using the sample data entered in column C1.

```
MTB  > NOTE: TEST OF HYPOTHESIS ABOUT μ: n < 30
MTB  > SET C1
DATA > 15  11  13  14  15  12  15  10  16
DATA > 17  14  16  13  15  15  14  11  13
DATA > END
MTB  > TTEST  MU = 12.5  C1;
SUBC > ALTERNATIVE = 0.

TEST OF MU = 12.500 VS MU N.E. 12.500

        N     MEAN     STDEV    SE MEAN       T    P VALUE
C1     18   13.833     1.917      0.452    2.95     0.0090
        ↑        ↑         ↑          ↑       ↑          ↑
   Sample        x̄         s         s_x̄  Value of the  p-value
   size                                  test statistic
```

In the MINITAB solution, TEST OF MU = 12.500 VS MU N.E. 12.500 indicates that the null hypothesis is that the mean is equal to 12.5 and the alternative hypothesis is that the mean is not equal (N.E.) to 12.5. In the row of C1 in the MINITAB solution, 18 represents the sample size, 13.833 is the mean of the sample data, 1.917 is the standard deviation of the sample data, 0.452 is the standard deviation (or standard error) of the sample mean, 2.95 is the value of the test statistic t calculated as $(\bar{x} - \mu)/s_{\bar{x}} = (13.833 - 12.5)/0.452$, and 0.0090 is the p-value.

To make a decision, we find two (critical) values of t from the t distribution table for a two-tailed test with $\alpha/2 = .005$ and $df = 18 - 1 = 17$. These values of t are -2.898 and 2.898. Because the value of the test statistic $t = 2.95$ is greater than 2.898, it falls in the rejection region. Consequently, we reject the null hypothesis. Thus, based on this sample information, we conclude that the mean age at which all children start walking seems to be different from 12.5 months. (The reader should draw a graph showing the rejection and nonrejection regions.)

We can reach the same conclusion using the p-value. As $\alpha = .01$ is larger than the p-value of .0090, we reject the null hypothesis.

COMPUTER ASSIGNMENTS

M9.1 According to John Tugman of MRCA Information Services, the mean amount spent on clothes by American women was \$569 in 1987 (*Newsweek,* December 5, 1988). A random sample of 39 women taken recently produced the following data on the amount they spent on clothes in 1990.

671	584	328	498	827	921	425	204	382	539
1070	854	669	328	537	849	930	1234	695	738
341	189	867	923	721	125	298	473	876	932
573	931	460	1430	391	887	958	674	782	

Using MINITAB, test at the 1% significance level if the mean expenditure on clothes for American women for 1990 is different from \$569. Assume that the population standard deviation is \$132.

M9.2 The mean weight of all babies born at a hospital last year was 7.6 pounds. A random sample of 35 babies born at this hospital this year produced the following data.

8.2	9.1	6.9	5.8	6.4	10.3	12.1	9.1	5.9	7.3
11.2	8.3	6.5	7.1	8.0	9.2	5.7	9.5	8.3	6.3
4.9	7.6	10.1	9.2	8.4	7.5	7.2	8.3	7.2	9.7
6.0	8.1	6.1	8.3	6.7					

Using MINITAB, test at the 5% significance level if the mean weight of babies born at this hospital this year is higher than 7.6 pounds. Assume that the population standard deviation is 1.59 pounds.

M9.3 The president of a large university claims that the mean time spent partying by all students at this university is not more than 7 hours a week. The following data give the time spent partying during the previous week for a random sample of 16 students taken from this university.

12	9	5	15	11	13	10	6
4	11	6	9	13	6	16	8

Using MINITAB, test at the 5% significance level if the president's claim is true. Assume that the time spent partying by all students at this university has an approximate normal distribution.

M9.4 According to a basketball coach, the mean height of all male college basketball players is 74 inches. A random sample of 25 such players produced the following data on their heights.

68	76	74	83	77	76	69	67	71	74	79	85	69
78	75	78	68	72	83	79	82	76	69	70	81	

Using MINITAB, test at the 1% significance level if the mean height of male college basketball players is different from 74 inches. Assume that the heights of all male college basketball players are (approximately) normally distributed.

10 ESTIMATION AND HYPOTHESIS TESTING: TWO POPULATIONS

10.1 INFERENCES ABOUT THE DIFFERENCE BETWEEN TWO POPULATION MEANS FOR LARGE AND INDEPENDENT SAMPLES

10.2 INFERENCES ABOUT THE DIFFERENCE BETWEEN TWO POPULATION MEANS FOR SMALL AND INDEPENDENT SAMPLES: EQUAL STANDARD DEVIATIONS

10.3 INFERENCES ABOUT THE DIFFERENCE BETWEEN TWO POPULATION MEANS FOR SMALL AND INDEPENDENT SAMPLES: UNEQUAL STANDARD DEVIATIONS

10.4 INFERENCES ABOUT THE DIFFERENCE BETWEEN TWO POPULATION MEANS FOR PAIRED SAMPLES

10.5 INFERENCES ABOUT THE DIFFERENCE BETWEEN TWO POPULATION PROPORTIONS FOR LARGE AND INDEPENDENT SAMPLES

SELF-REVIEW TEST

USING MINITAB

Chapters 8 and 9 discussed the estimation and hypothesis-testing procedures for μ and p involving a single population. This chapter extends the discussion of estimation and hypothesis testing to the difference between two population parameters. Procedures to make confidence intervals and test hypotheses about the difference between two population means and the difference between two population proportions are presented in this chapter. For example, we may want to make a confidence interval for the difference between mean prices of houses in California and in New York. Or we may want to test the hypothesis that the mean price of houses in California is different from that in New York. As another example, we may want to make a confidence interval for the difference between the proportions of all male and female adults who abstain from drinking. Or we may want to test the hypothesis that the proportion of all adult males who abstain from drinking is different from the proportion of all adult females who abstain from drinking. Constructing confidence intervals and testing hypotheses about population parameters are referred to as *making inferences*.

10.1 INFERENCES ABOUT THE DIFFERENCE BETWEEN TWO POPULATION MEANS FOR LARGE AND INDEPENDENT SAMPLES

Let μ_1 be the mean of the first population and μ_2 be the mean of the second population. Then, we want to make a confidence interval and test a hypothesis about the difference between two population means $\mu_1 - \mu_2$. Let $\bar{x}_1$ be the mean of a sample taken from the first population and $\bar{x}_2$ be the mean of a sample taken from the second population. Then, $\bar{x}_1 - \bar{x}_2$ is the sample statistic used to make an interval estimate and to test a hypothesis about $\mu_1 - \mu_2$. This section discusses how to make confidence intervals and test hypotheses about $\mu_1 - \mu_2$ when the samples are large and independent. As discussed in earlier chapters, in case of μ, a sample is considered to be large if it contains 30 or more observations. The concept of independent and dependent samples is explained next.

10.1.1 INDEPENDENT VERSUS DEPENDENT SAMPLES

Two samples are independent if they are drawn from two different populations and the elements of one sample have no relationship to the elements of the second sample. If the elements of the two samples are somehow related, then the samples are said to be dependent. Thus, in two independent samples, the selection of one sample has no effect on the selection of the second sample.

INDEPENDENT VERSUS DEPENDENT SAMPLES

Two samples drawn from two populations are independent if the selection of one sample from one population does not affect the selection of the second sample from the second population. Otherwise, the samples are dependent.

Examples 10-1 and 10-2 illustrate independent and dependent samples, respectively.

Illustrating two independent samples.

EXAMPLE 10-1 Suppose we want to estimate the difference between the mean salaries of all male and all female executives. We draw two samples, one from the population of male executives and another from the population of female executives. These two samples are independent because they are drawn from two different populations and the samples have no effect on each other. ■

Illustrating two dependent samples.

EXAMPLE 10-2 Suppose we want to estimate the difference between the mean weights of all participants before and after a weight-loss program. To accomplish this, suppose we take a sample of 40 participants and measure their weights before and after the completion of this program. Note that these two samples include the same 40 participants. This is an example of two dependent samples. Such samples are also called *paired* or *matched samples*. ■

This section and Sections 10.2, 10.3, and 10.5 discuss how to make confidence intervals and test hypotheses about the difference between two population parameters when samples are independent. Section 10.4 discusses how to make confidence intervals and test hypotheses about the difference between two population means when samples are dependent.

10.1.2 THE MEAN, STANDARD DEVIATION, AND SAMPLING DISTRIBUTION OF $\bar{x}_1 - \bar{x}_2$

Suppose we draw two (independent) large samples from two different populations. Let us refer to these two populations as population 1 and population 2. Let

μ_1 = the mean of population 1

μ_2 = the mean of population 2

σ_1 = the standard deviation of population 1

σ_2 = the standard deviation of population 2

n_1 = the size of the sample drawn from population 1 ($n_1 \geq 30$)

n_2 = the size of the sample drawn from population 2 ($n_2 \geq 30$)

$\bar{x}_1$ = the mean of the sample drawn from population 1

$\bar{x}_2$ = the mean of the sample drawn from population 2

Then, from the central limit theorem, $\bar{x}_1$ is approximately normally distributed with mean μ_1 and standard deviation $\sigma_1/\sqrt{n_1}$, and $\bar{x}_2$ is approximately normally distributed with mean μ_2 and standard deviation $\sigma_2/\sqrt{n_2}$.

Using these results, we can now make the following statements about the mean, standard deviation, and shape of the sampling distribution of $\bar{x}_1 - \bar{x}_2$. Figure 10.1 shows the sampling distribution of $\bar{x}_1 - \bar{x}_2$.

1. The mean of $\bar{x}_1 - \bar{x}_2$, denoted by $\mu_{\bar{x}_1-\bar{x}_2}$, is

$$\mu_{\bar{x}_1-\bar{x}_2} = \mu_1 - \mu_2$$

2. The standard deviation of $\bar{x}_1 - \bar{x}_2$, denoted by $\sigma_{\bar{x}_1-\bar{x}_2}$, is†

$$\sigma_{\bar{x}_1-\bar{x}_2} = \sqrt{\frac{\sigma_1^2}{n_1} + \frac{\sigma_2^2}{n_2}}$$

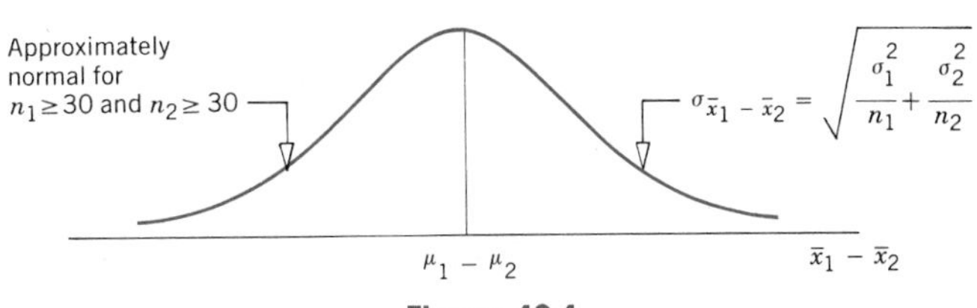

Figure 10.1

†The formula for the standard deviation of $\bar{x}_1 - \bar{x}_2$ can also be written as

$$\sigma_{\bar{x}_1-\bar{x}_2} = \sqrt{\sigma_{\bar{x}_1}^2 + \sigma_{\bar{x}_2}^2}$$

where $\sigma_{\bar{x}_1} = \sigma_1/\sqrt{n_1}$ and $\sigma_{\bar{x}_2} = \sigma_2/\sqrt{n_2}$.

3. Regardless of the shapes of two populations, the shape of the sampling distribution of $\bar{x}_1 - \bar{x}_2$ is approximately normal. This is so because the difference between two normally distributed random variables is also normally distributed. Note again that for this to hold true, both samples must be large.

THE SAMPLING DISTRIBUTION, MEAN, AND STANDARD DEVIATION OF $\bar{x}_1 - \bar{x}_2$

For two large and independent samples, taken from two different populations, the sampling distribution of $\bar{x}_1 - \bar{x}_2$ is (approximately) normal with its mean and standard deviation equal to

$$\mu_{\bar{x}_1-\bar{x}_2} = \mu_1 - \mu_2$$

and

$$\sigma_{\bar{x}_1-\bar{x}_2} = \sqrt{\frac{\sigma_1^2}{n_1} + \frac{\sigma_2^2}{n_2}}$$

However, usually we do not know the standard deviations σ_1 and σ_2 of the two populations. In such cases, we replace $\sigma_{\bar{x}_1-\bar{x}_2}$ by its point estimator $s_{\bar{x}_1-\bar{x}_2}$, which is calculated as follows.

AN ESTIMATE OF THE STANDARD DEVIATION OF $\bar{x}_1 - \bar{x}_2$

The value of $s_{\bar{x}_1-\bar{x}_2}$ is calculated as

$$s_{\bar{x}_1-\bar{x}_2} = \sqrt{\frac{s_1^2}{n_1} + \frac{s_2^2}{n_2}}$$

where s_1 and s_2 are the standard deviations of the two samples taken from the two populations.

Thus, when both sample sizes are large, the sampling distribution of $\bar{x}_1 - \bar{x}_2$ is approximately normal. Consequently, in such cases, we use the normal distribution to make a confidence interval and to test a hypothesis about $\mu_1 - \mu_2$.

10.1.3 INTERVAL ESTIMATION OF $\mu_1 - \mu_2$

By constructing a confidence interval for $\mu_1 - \mu_2$ we find the difference between the means of two populations. For example, we may want to find the difference between the mean heights of all male and female adults. The difference between the two sample means $\bar{x}_1 - \bar{x}_2$ is the point estimator of the difference between the two population means $\mu_1 - \mu_2$. Again, in this section we assume that the samples are large and independent. When these assumptions hold true, we use the normal distribution to make a confidence interval for the difference between the two population means. The following formula gives the interval estimation for $\mu_1 - \mu_2$.

CONFIDENCE INTERVAL FOR $\mu_1 - \mu_2$

The $(1 - \alpha)100\%$ confidence interval for $\mu_1 - \mu_2$ is

$$(\bar{x}_1 - \bar{x}_2) \pm z\,\sigma_{\bar{x}_1 - \bar{x}_2} \quad \text{if } \sigma_1 \text{ and } \sigma_2 \text{ are known}$$

$$(\bar{x}_1 - \bar{x}_2) \pm z\,s_{\bar{x}_1 - \bar{x}_2} \quad \text{if } \sigma_1 \text{ and } \sigma_2 \text{ are not known}$$

The value of z is obtained from the normal distribution table for the given confidence level. The values of $\sigma_{\bar{x}_1 - \bar{x}_2}$ and $s_{\bar{x}_1 - \bar{x}_2}$ are calculated as explained earlier.

Examples 10-3 and 10-4 illustrate the procedure to construct a confidence interval for $\mu_1 - \mu_2$ for large samples. In Example 10-3, the population standard deviations are known. But in Example 10-4, σ_1 and σ_2 are not known.

Constructing a confidence interval for $\mu_1 - \mu_2$: σ_1 and σ_2 known.

EXAMPLE 10-3 According to the Bureau of Labor Statistics, in 1989 construction workers earned an average of \$506 per week and manufacturing workers earned an average of \$429 per week. Assume that these mean weekly earnings have been calculated for samples of 500 and 700 workers taken from the two populations, respectively. Further assume that the standard deviations of weekly earnings of the two populations are \$60 and \$55, respectively. Construct a 95% confidence interval for the difference between the mean weekly earnings of the two populations.

Solution Refer to all construction workers as population 1 and all manufacturing workers as population 2. The respective samples, then, are samples 1 and 2. Let μ_1 and μ_2 be the means of populations 1 and 2, and let $\bar{x}_1$ and $\bar{x}_2$ be the means of the respective samples. Then, from the given information,

$$n_1 = 500 \quad \bar{x}_1 = \$506 \quad \sigma_1 = \$60$$
$$n_2 = 700 \quad \bar{x}_2 = \$429 \quad \sigma_2 = \$55$$

The confidence level is $1 - \alpha = .95$.

First, we calculate the standard deviation of $\bar{x}_1 - \bar{x}_2$ as follows.

$$\sigma_{\bar{x}_1 - \bar{x}_2} = \sqrt{\frac{\sigma_1^2}{n_1} + \frac{\sigma_2^2}{n_2}} = \sqrt{\frac{(60)^2}{500} + \frac{(55)^2}{700}} = 3.394$$

Next, we find the z value for a 95% confidence level. From the normal distribution table, this value of z is 1.96.

Finally, substituting all the values in the confidence interval formula, we obtain the 95% confidence interval for $\mu_1 - \mu_2$ as

$$(\bar{x}_1 - \bar{x}_2) \pm z\,\sigma_{\bar{x}_1 - \bar{x}_2} = (506 - 429) \pm 1.96\,(3.394) = 77 \pm 6.65$$
$$= \$70.35 \text{ to } \$83.65$$

Thus, with 95% confidence we can state that the difference between the mean weekly earnings of all construction workers and all manufacturing workers in 1989 was between \$70.35 and \$83.65. ■

Constructing a confidence interval for $\mu_1 - \mu_2$: σ_1 and σ_2 not known.

EXAMPLE 10-4 Dr. Ann Howard conducted a Management Progress Study, based on the participants from AT&T management jobholders, to assess the performance of those who possessed a college degree and those who did not (Ann Howard: "College Experiences and Managerial Performance," *Journal of Applied Psychology,* 71(3): 530–552, 1986). Let us refer to the two groups of participants as college and noncollege participants, respectively. The samples included 274 college and 148 noncollege participants. In the area of motivation for advancement, the mean scores were 2.89 for college participants and 2.70 for noncollege participants, with the two standard deviations being .57 and .64, respectively. Find a 99% confidence interval for the difference between the mean scores of the two populations in the area of motivation for advancement.

Solution Let all the management jobholders with a college degree be referred to as population 1 and the ones without a college degree be referred to as population 2. We can refer to the respective samples as samples 1 and 2. Let μ_1 and μ_2 be the means of populations 1 and 2, respectively, and let $\bar{x}_1$ and $\bar{x}_2$ be the means of the respective samples. Then, from the given information,

$$n_1 = 274 \quad \bar{x}_1 = 2.89 \quad s_1 = .57$$
$$n_2 = 148 \quad \bar{x}_2 = 2.70 \quad s_2 = .64$$

The confidence level is $1 - \alpha = .99$.

Because σ_1 and σ_2 are not known, we replace $\sigma_{\bar{x}_1-\bar{x}_2}$ by $s_{\bar{x}_1-\bar{x}_2}$ in the confidence interval formula. We calculate the value of $s_{\bar{x}_1-\bar{x}_2}$ as

$$s_{\bar{x}_1-\bar{x}_2} = \sqrt{\frac{s_1^2}{n_1} + \frac{s_2^2}{n_2}} = \sqrt{\frac{(.57)^2}{274} + \frac{(.64)^2}{148}} = .063$$

From the normal distribution table, the z value for a 99% confidence interval is (approximately) 2.58. Hence, the 99% confidence interval for $\mu_1 - \mu_2$ is

$$(\bar{x}_1 - \bar{x}_2) \pm z\, s_{\bar{x}_1-\bar{x}_2} = (2.89 - 2.70) \pm 2.58\,(.063)$$
$$= .19 \pm .16 = .03 \text{ to } .35$$

Thus, with 99% confidence we can state that the difference in the mean scores of two populations of managers in the area of motivation for advancement is between .03 and .35. ■

Case Study 10-1 further illustrates the application of the confidence interval for the difference between two population means.

CASE STUDY 10-1 INFERENCES MADE BY THE CENSUS BUREAU

The U.S. Census Bureau publishes a large number of reports every year that contain the results of various surveys done by the Bureau. Using these survey results, the Bureau estimates many population parameters such as the mean, median, and proportion. It also tests the hypotheses about a change in the values of these parameters over time. Following are two statements quoted from two of the Bureau's publications. The figures in parentheses give the 90% confidence intervals of the estimates.

1. Per capita income in 1989 was \$14,060 (± \$107), up 2.2 (± 1.0) percent from 1988, also an all-time high.
2. The average number of persons per household reached a record low of 2.62 . . . persons in 1989 which was 0.14 (± .02) fewer persons per household than in 1980.

The first statement gives two estimates. First, it gives a 90% confidence interval for the per capita income for 1989. (Per capita income is actually the mean income per person). This confidence interval is \$14,060 ± 107. This indicates that the per capita income for all people in the United States was between \$13,953 and \$14,167 in 1989. The second estimate in the first statement is that the change in the per capita income between 1988 and 1989 was (2.2 ± 1.0)%, which gives an increase of 1.2% to 3.2%. This is simply a 90% confidence interval for $\mu_2 - \mu_1$ presented as a percentage of μ_1, where μ_1 is the per capita income for 1988 and μ_2 is the per capita income for 1989. Because the percentage is positive, the Bureau called it an increase.

The second statement also gives two estimates. First, it gives the point estimate for the mean number of persons per household for 1989 as 2.62. Second, it gives a 90% confidence interval for $\mu_1 - \mu_2$, where μ_1 is the mean number of persons per household in 1980 and μ_2 is the mean number of persons per household in 1989. Thus, the 90% confidence interval for $\mu_1 - \mu_2$ is .14 ± .02 or .12 to .16. The positive values of both limits of the confidence interval indicates that μ_1 is greater than μ_2. Hence, the Bureau called it a decrease in the mean number of persons per household for this period.

Sources: 1. *Money Income and Poverty Status in the United States: 1989*. Series P-60, 168. U.S. Department of Commerce, Bureau of the Census, September 1990.
2. *Households, Families, Marital Status, and Living Arrangements: March 1989* (*Advance Report*). Series P-20, 441. U.S. Department of Commerce, Bureau of the Census, November 1989.

10.1.4 HYPOTHESIS TESTING ABOUT $\mu_1 - \mu_2$

Often we want to compare the means of two populations. For example, we may want to know if the mean price of houses in Chicago is the same as that in Los Angeles. Similarly, we may be interested in knowing if American children on average spend less hours in school than Japanese children. In both these cases we will make a test of hypothesis about $\mu_1 - \mu_2$. We may test a hypothesis that the means of two populations are different, or that the mean of the first population is greater than the mean of the second population, or that the mean of the first population is smaller than the mean of the second population. We describe these three situations as follows.

1. Testing a hypothesis that the means of two populations are different is equivalent to $\mu_1 \neq \mu_2$, which is the same as $\mu_1 - \mu_2 \neq 0$.
2. Testing a hypothesis that the mean of the first population is greater than the mean of the second population is equivalent to $\mu_1 > \mu_2$, which is the same as $\mu_1 - \mu_2 > 0$.
3. Testing a hypothesis that the mean of the first population is smaller than the mean of the second population is equivalent to $\mu_1 < \mu_2$, which is the same as $\mu_1 - \mu_2 < 0$.

The procedure followed in making a test of hypothesis about the difference between two population means is similar to the one used to test the hypotheses about one population parameters in Chapter 9. The procedure involves the same five steps we used in Chapter 9 to test hypotheses about μ and p. Because we are dealing with large (and independent) samples in this section, we will use the normal distribution to make a test of hypothesis about $\mu_1 - \mu_2$.

TEST STATISTIC z FOR $\bar{x}_1 - \bar{x}_2$

The value of the test statistic z for $\bar{x}_1 - \bar{x}_2$ is computed as

$$z = \frac{(\bar{x}_1 - \bar{x}_2) - (\mu_1 - \mu_2)}{\sigma_{\bar{x}_1 - \bar{x}_2}}$$

The value of $\mu_1 - \mu_2$ is substituted from H_0. If σ_1 and σ_2 are not known, we replace $\sigma_{\bar{x}_1 - \bar{x}_2}$ by $s_{\bar{x}_1 - \bar{x}_2}$ in the formula.

Examples 10-5 and 10-6 describe how a test of hypothesis about the difference between two population means is conducted.

Making a two-tailed test of hypothesis about $\mu_1 - \mu_2$: large samples.

EXAMPLE 10-5 Reconsider Example 10-3 on the mean weekly earnings of construction and manufacturing workers. Test at the 1% significance level if the mean weekly earnings of the two groups of workers are different.

Solution From the information given in Example 10-3,

$$n_1 = 500 \quad \bar{x}_1 = \$506 \quad \sigma_1 = \$60$$
$$n_2 = 700 \quad \bar{x}_2 = \$429 \quad \sigma_2 = \$55$$

where the subscript 1 refers to construction workers and 2 to manufacturing workers. The significance level is $\alpha = .01$. Let

μ_1 = the mean weekly earnings of all construction workers

μ_2 = the mean weekly earnings of all manufacturing workers

Step 1. *State the null and alternative hypotheses*

We are to test if the two population means are different. The two possibilities are

(a) The mean weekly earnings of all construction workers and all manufacturing workers are not different. In other words, $\mu_1 = \mu_2$, which can be written as $\mu_1 - \mu_2 = 0$.

(b) The mean weekly earnings of construction workers and manufacturing workers are different. That is, $\mu_1 \neq \mu_2$, which can be written as $\mu_1 - \mu_2 \neq 0$.

Hence, the null and alternative hypotheses are

H_0: $\mu_1 - \mu_2 = 0$ (the mean weekly earnings are not different)

H_1: $\mu_1 - \mu_2 \neq 0$ (the mean weekly earnings are different)

Step 2. *Select the distribution to use*

Because $n_1 > 30$ and $n_2 > 30$, both sample sizes are large. Hence, the distribution of $\bar{x}_1 - \bar{x}_2$ is approximately normal. Consequently, we use the normal distribution to make the hypothesis test.

Step 3. *Determine the rejection and nonrejection regions*

The significance level is given to be .01. The $\neq$ sign in the alternative hypothesis indicates that the test is two-tailed. The area under each tail is $\alpha/2 = .01/2 = .005$. Hence, the critical values of z are (approximately) 2.58 and -2.58, as shown in Figure 10.2.

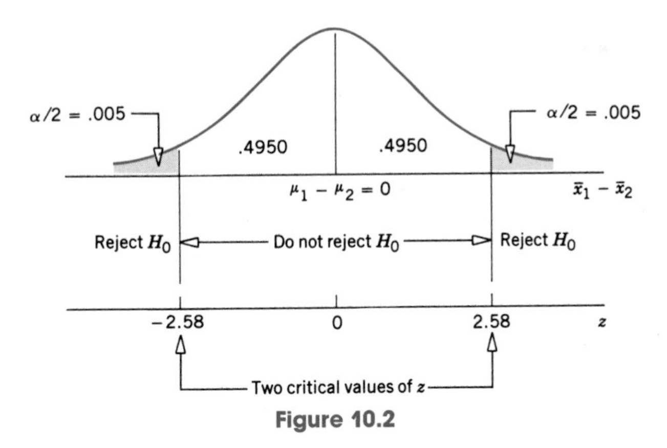

Figure 10.2

Step 4. *Calculate the value of the test statistic*

The value of the test statistic z for $\bar{x}_1 - \bar{x}_2$ is calculated as follows.

$$\sigma_{\bar{x}_1-\bar{x}_2} = \sqrt{\frac{\sigma_1^2}{n_1} + \frac{\sigma_2^2}{n_2}} = \sqrt{\frac{(60)^2}{500} + \frac{(55)^2}{700}} = 3.394$$

$$z = \frac{(\bar{x}_1 - \bar{x}_2) - (\mu_1 - \mu_2)}{\sigma_{\bar{x}_1-\bar{x}_2}} = \frac{(506 - 429) - \overset{\text{From } H_0}{0}}{3.394} = 22.69$$

Step 5. *Make a decision*

Because the value of the test statistic $z = 22.69$ falls in the rejection region, we reject the null hypothesis H_0. Therefore, we conclude that the mean weekly earnings of the two groups of workers appear to be different. Note that we cannot say for sure that the two means are different. All we can say is that the evidence from the two samples is very strong that the means are different. ■

Conducting a right-tailed test of hypothesis about $\mu_1 - \mu_2$: large samples.

EXAMPLE 10-6 Refer to Example 10-4 on the mean scores of college and noncollege participants in the management progress study done by Ann Howard. Test at the 5% significance level if the mean score in the area of motivation for advancement is higher for college degree holders than for noncollege participants.

Solution From the information given in Example 10-4,

$$n_1 = 274 \quad \bar{x}_1 = 2.89 \quad s_1 = .57$$
$$n_2 = 148 \quad \bar{x}_2 = 2.70 \quad s_2 = .64$$

where subscript 1 refers to college degree holders and 2 to noncollege participants. The significance level is $\alpha = .05$.

Step 1. *State the null and alternative hypotheses*

The two possibilities are

(a) The mean score of college degree holders is not higher than that of the noncollege participants, which can be written as $\mu_1 = \mu_2$ or $\mu_1 - \mu_2 = 0$.
(b) The mean score of college degree holders is higher than that of the noncollege participants, which can be written as $\mu_1 > \mu_2$ or $\mu_1 - \mu_2 > 0$.

Hence, the null and alternative hypotheses are

$$H_0\colon \mu_1 - \mu_2 = 0 \quad (\mu_1 \text{ is not greater than } \mu_2)$$
$$H_1\colon \mu_1 - \mu_2 > 0 \quad (\mu_1 \text{ is greater than } \mu_2)$$

Note that we can also write the null hypothesis as $\mu_1 - \mu_2 \leq 0$, which states that the mean score of college participants is less than or equal to the mean score of noncollege participants.

Step 2. *Select the distribution to use*

Because $n_1 > 30$ and $n_2 > 30$, both sample sizes are large. Hence, we use the normal distribution to make the test.

Step 3. *Determine the rejection and nonrejection regions*

The significance level is .05. The $>$ sign in the alternative hypothesis indicates that the test is right-tailed. Consequently, the critical value of z is 1.65, as shown in Figure 10.3.

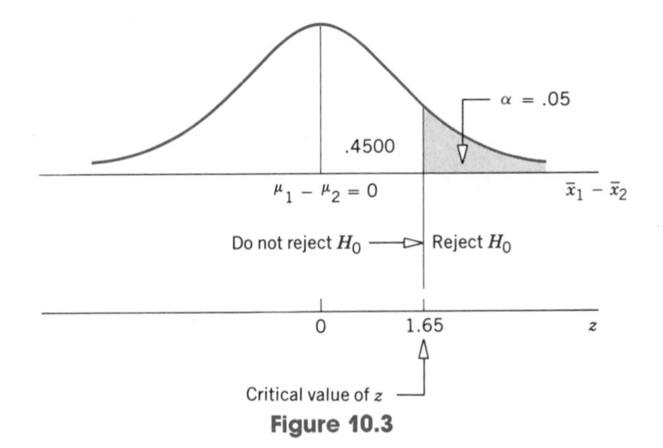

Figure 10.3

Step 4. *Calculate the value of the test statistic*

The value of the test statistic z for $\bar{x}_1 - \bar{x}_2$ is calculated as follows.

$$s_{\bar{x}_1-\bar{x}_2} = \sqrt{\frac{s_1^2}{n_1} + \frac{s_2^2}{n_2}} = \sqrt{\frac{(.57)^2}{274} + \frac{(.64)^2}{148}} = .063$$

$$z = \frac{(\bar{x}_1 - \bar{x}_2) - (\mu_1 - \mu_2)}{s_{\bar{x}_1-\bar{x}_2}} = \frac{(2.89 - 2.70) - \overset{\text{From } H_0}{0}}{.063} = 3.02$$

Step 5. *Make a decision*

Because the value of the test statistic $z = 3.02$ for $\bar{x}_1 - \bar{x}_2$ falls in the rejection region, we reject the null hypothesis H_0. Therefore, we conclude that the mean score in the area of motivation for advancement is higher for those who hold a college degree than for those who do not hold a college degree. ■

Case Study 10-2 further illustrates the application of hypothesis testing about the difference between two population means.

CASE STUDY 10-2 TIME TOGETHER AMONG DUAL-EARNER COUPLES

Professors Paul William Kingston and Steven L. Nock studied the time spent together by single- and dual-earner couples. The main hypothesis of this study was that the time spent together by couples depends on the time they (both) spend on work. Hence, in the case of couples where the husband and wife both work (i.e., dual-earner couples) the mean time spent together on various activities will be less than the mean time spent together by couples where only one of the spouses works (i.e., single-earner couples). Their study is based on samples of 177 dual-earner couples and 144 single-earner couples. In each sample, both husband and wife were asked to furnish separate records of time spent by them on various activities for many days. The results quoted in the article test the hypotheses for husbands' and wives' accounts separately. But in the following table, results are given for some of the activities based on husbands' records only. The table has been adapted from Table 2 of the article. The values of the means and standard deviations (SD) give time in minutes per day.

Activity	Single-earner Couples (sample = 144)	Dual-earner Couples (sample = 177)	Decision
Fun	Mean = 36.9	Mean = 31.6	Null not rejected
	SD = 67.9	SD = 59.3	at α = .10
Homemaking	Mean = 25.1	Mean = 16.2	Null rejected
	SD = 44.0	SD = 29.5	at α = .05
Talking	Mean = 18.5	Mean = 10.7	Null rejected
	SD = 33.9	SD = 23.9	at α = .01
Watching TV	Mean = 74.1	Mean = 57.2	Null rejected
	SD = 87.5	SD = 75.3	at α = .05
Eating meals	Mean = 47.6	Mean = 42.1	Null not rejected
	SD = 39.6	SD = 40.1	at α = .10

Let

$$\mu_1 = \text{time spent together by single-earner couples}$$

$$\mu_2 = \text{time spent together by dual-earner couples}$$

In the table, for each activity (as shown in the example that follows) the test is right-tailed with the alternative hypothesis that the single-earner couples spend more time together than dual-earner couples, that is, $\mu_1 > \mu_2$.

Let us illustrate how a test of hypothesis is performed for the category of watching TV. For this category, the mean time spent together is 74.1 minutes a day by single-earner couples and 57.2 minutes a day by dual-earner couples, with their respective standard deviations equal to 87.5 and 75.3 minutes. Then the null and alternative hypotheses for this activity are

H_0: $\mu_1 - \mu_2 = 0$ (time spent together is the same for single- and dual-earner couples)

H_1: $\mu_1 - \mu_2 > 0$ (single-earner couples spend more time together)

For single-earner couples: $n_1 = 144$, $\bar{x}_1 = 74.1$, $s_1 = 87.5$

For dual-earner couples: $n_2 = 177$, $\bar{x}_2 = 57.2$, $s_2 = 75.3$

From the normal distribution table, the critical value of z for a one-tailed test with $\alpha = .05$ is 1.65. The value of the test statistic z for $\bar{x}_1 - \bar{x}_2$ is calculated as follows.

$$s_{\bar{x}_1-\bar{x}_2} = \sqrt{\frac{s_1^2}{n_1} + \frac{s_2^2}{n_2}} = \sqrt{\frac{(87.5)^2}{144} + \frac{(75.3)^2}{177}} = 9.231$$

From H_0

$$z = \frac{(\bar{x}_1 - \bar{x}_2) - (\mu_1 - \mu_2)}{s_{\bar{x}_1-\bar{x}_2}} = \frac{(74.1 - 57.2) - 0}{9.231} = 1.83$$

Because the value of the test statistic $z = 1.83$ is greater than the critical value of $z = 1.65$ and it falls in the rejection region, we reject the null hypothesis.

Source: Paul William Kingston and Steven L. Nock: "Time Together Among Dual-earner Couples," *American Sociological Review,* 52(3): 391–400, June 1987. Copyright © 1987 by American Sociological Association. Data reproduced by permission.

Note: Some of the results reported in this case study are different from the ones reported by the authors in the article. These alterations were made after consultation with the authors and are based on a computer printout of the results supplied by the authors. The article contained some errors in reporting the significance levels due to rounding.

EXERCISES

10.1 Briefly explain the meaning of independent and dependent samples. Give one example of each.

10.2 Construct a 99% confidence interval for $\mu_1 - \mu_2$ for the following.

$$n_1 = 150 \quad \bar{x}_1 = 5.56 \quad s_1 = 1.65$$
$$n_2 = 170 \quad \bar{x}_2 = 4.80 \quad s_2 = 1.58$$

10.3 Construct a 95% confidence interval for $\mu_1 - \mu_2$ for the following.

$$n_1 = 300 \quad \bar{x}_1 = 25.0 \quad s_1 = 4.9$$
$$n_2 = 250 \quad \bar{x}_2 = 28.5 \quad s_2 = 4.5$$

10.4 Refer to Exercise 10.2. Test at the 5% significance level if the two population means are different.

10.5 Refer to Exercise 10.3. Test at the 1% significance level if the two population means are different.

10.6 Refer to Exercise 10.2. Test at the 1% significance level if μ_1 is greater than μ_2.

10.7 Refer to Exercise 10.3. Test at the 5% significance level if μ_1 is less than μ_2.

10.8 Professor Harold W. Stevenson and his colleagues studied the mathematics achievement of Chinese, Japanese, and American children (Harold W. Stevenson et al.: "Mathematics Achievement of Chinese, Japanese, and American Children," *Science,* 231, February 14, 1986).

According to that study, the mean score in a mathematics test given to 288 American kindergarten children was 37.5 with a standard deviation of 5.6. A similar test given to 280 Japanese kindergarten children produced a mean score of 42.2 with a standard deviation of 5.1.

a. Construct a 99% confidence interval for the difference between the mean scores of American and Japanese kindergarten children.

b. Test at the 5% significance level if the mean scores of American and Japanese kindergarten children are different.

10.9 According to the study done by Professor Harold W. Stevenson and his colleagues mentioned in Exercise 10.8, the mean score in a mathematics test given to 288 American kindergarten children was 37.5 with a standard deviation of 5.6. A similar test given to 286 kindergarten children in Taiwan produced a mean score of 37.8 with a standard deviation of 7.4.

a. Construct a 95% confidence interval for the difference between the mean scores of American and Taiwanese kindergarten children.

b. Test at the 1% significance level if the mean scores of American and Taiwanese kindergarten children are different.

10.10 According to the Management Progress Study done by Dr. Ann Howard mentioned in Example 10-4, the mean score in motivation for advancement for a sample of 43 management jobholders with a Master's degree was 2.92 with a standard deviation of .46. The mean score in the same area for a sample of 112 management jobholders with a Bachelor's degree was 2.81 with a standard deviation of .57.

a. Construct a 90% confidence interval for the difference between the mean scores of two populations.

b. Test at the 1% significance level if the mean score of Master's degree holders is higher than the one for Bachelor's degree holders.

10.11 According to the Bureau of Labor Statistics, the mean hourly wage in April 1990 was \$12.89 for transportation and public utility workers and \$10.74 for manufacturing workers. Assume that these two estimates are based on random samples of 1000 and 1200 workers taken respectively from the two populations. Further assume that the standard deviations of the two populations are \$1.75 and \$1.30, respectively.

a. Construct a 90% confidence interval for the difference between the mean hourly wages of two populations.

b. Test at the 5% significance level if the mean hourly wage for transportation and public utility workers in April 1990 was higher than that of manufacturing workers.

10.12 According to the Metropolitan Life Insurance Company's claims data, the mean hospital and physician's charges for coronary bypass surgery are \$24,710 for New York state and \$27,670 for Texas (*Statistical Bulletin,* January–March 1989). The two means are based on 55 claims for New York state and 186 claims for Texas. Assume that these claims are representative of all such surgeries performed in these two states. Further assume that the population standard deviations for these surgeries for these two states are \$1300 and \$1450, respectively.

a. Construct a 95% confidence interval for the difference between the mean charges for all such surgeries done in these two states.

b. Test at the 1% significance level if the mean charge for such surgeries in New York state is lower than that in Texas.

10.13 According to a study, the mean birth weight of babies born to a sample of 202 mothers who used marijuana during pregnancy was 2980 grams with a standard deviation of 662 grams. (One pound is equal to 454 grams approximately.) The mean birth weight of babies born to another sample of 895 mothers who did not use marijuana during pregnancy was 3260 grams with a standard deviation of 616 grams (Barry Zuckerman et al.: "Effects of Maternal Marijuana and Cocaine Use on Fetal Growth," *The New England Journal of Medicine,* 320(12): March 23, 1989).

a. Construct a 97% confidence interval for the difference between the mean birth weights of babies born to mothers who use marijuana and those who do not use marijuana during pregnancy.

b. Test at the 5% significance level if the mean birth weight of babies born to mothers who use marijuana during pregnancy is lower than the mean birth weight of babies born to mothers who do not use marijuana during pregnancy.

10.14 Refer to Case Study 10-2. According to the wives' records, the mean time spent together by a husband and wife watching television was 61.6 minutes a day for single-earner couples and 44.4 minutes a day for dual-earner couples. The respective sample sizes were 144 and 177 and the sample standard deviations were 79.0 and 56.8, respectively.

a. Construct a 99% confidence interval for the difference between the two population means.

b. Test at the 1% significance level if the mean time spent together watching television by single-earner couples is higher than that of dual-earner couples.

10.15 Refer to Case Study 10-2 again. According to the wives' records, the mean time spent together by a husband and wife on all activities was 236.4 minutes a day for single-earner couples and 191.5 minutes a day for dual-earner couples. The respective sample sizes were 144 and 177 and the sample standard deviations were 151.6 and 154.5, respectively.

a. Construct a 98% confidence interval for the difference between the two population means.

b. Test at the 2% significance level if the mean time spent together on all activities by single-earner couples is higher than that of dual-earner couples.

10.16 A science proficiency test was given to 859 randomly selected 13-year-old American students. The following table gives the mean scores of male and female students along with the standard deviations of the sample means (*A World of Differences: An International Assessment of Mathematics and Science,* Educational Testing Service, January 1989).

Male students	$\bar{x}_1 = 481.9$	$s_{\bar{x}_1} = 6.1$
Female students	$\bar{x}_2 = 474.9$	$s_{\bar{x}_2} = 4.7$

a. Construct a 95% confidence interval for the difference between the two population means.

b. Test at the 1% significance level if the mean scores in science proficiency test are different for male and female 13-year-old students.

(*Note:* Calculate the standard deviation of $\bar{x}_1 - \bar{x}_2$ using the formula given in the footnote on page 475, except that s is substituted for σ.)

10.17 According to the U.S. Bureau of the Census, the mean income of families was $38,608 in 1988 and $38,410 in 1987. Assume that these means are based on a sample size of 1800 families for 1988 and 1600 families for 1987. Further assume that the standard deviations for the two populations are $9569 and $9285, respectively.

a. Construct a 95% confidence interval for the difference between the two population means.

b. Test at the 5% significance level if the 1988 mean income for all families is different from that for 1987.

10.18 According to the authors of a study about employee reliability, "Although there is a major problem, it is only one factor of the larger construct of organizational delinquency. Excessive absences, tardiness, malingering, equipment damage, drug and alcohol abuse, grievances, suspensions from work, insubordination, and ordinary rule infractions are all components of the delinquency syndrome" (Joyce Hogan and Robert Hogan: "How to Measure Employee Reliability," *Journal of Applied Psychology,* 74(2): April 1989). The authors measured the employee reliability scores for male and female employees. A sample of 1637 male employees produced a mean score of 45.4 with a standard deviation of 8.0, and a sample of 590 female employees gave a mean score of 46.5 with a standard deviation of 8.3.

a. Construct a 98% confidence interval for the difference between the mean reliability scores of all male and all female employees.

b. Test at the 2% significance level if the mean reliability scores for all male and all female employees are different.

10.19 According to a study about the social interests of mothers of handicapped and normal children, "Mothers of handicapped children manifest lower self-esteem and are more reserved, emotionally less stable, humble, suspicious, apprehensive, and more tense than mothers of normal children" (Edward Mel Markowski and Lou W. Everett: "Social Interest and Mothers of Exceptional Children" *Individual Psychology,* 44(1): March 1988). A sample of 60 mothers with handicapped children produced a mean score on social interest equal to 125.62 with a standard deviation of 18.50. Another sample of 60 mothers with normal children gave a mean score on social interest of 132.02 with a standard deviation of 11.27.

a. Construct a 99% confidence interval for the difference between the mean scores on social interest for mothers with handicapped children and mothers with normal children.

b. Test at the 5% significance level if the mean score on social interest for mothers with handicapped children is lower than that of mothers with normal children.

10.2 INFERENCES ABOUT THE DIFFERENCE BETWEEN TWO POPULATION MEANS FOR SMALL AND INDEPENDENT SAMPLES: EQUAL STANDARD DEVIATIONS

Many times, either due to budget constraint or because of the nature of the populations, it may not be possible to take large samples to make inferences about the difference between two population means. This section discusses how to make a confidence interval and how to test a hypothesis about the difference between two population means when the samples are small ($n_1 < 30$ and $n_2 < 30$) and independent. Our main assumption in this case is that the two populations from which the two samples are drawn are (approximately) normally distributed. If this assumption is true, and we know the population standard deviations, we can still use the normal distribution to make inferences about $\mu_1 - \mu_2$ when samples are small and independent. However, we usually do not know the population standard deviations σ_1 and σ_2. In such cases, we replace the normal distribution by the t distribution to make inferences about $\mu_1 - \mu_2$ for small and independent samples. We will make one more assumption in this section. It is that the standard deviations of the two populations are equal. In other words, we assume that although σ_1 and σ_2 are unknown, they are equal. The case when σ_1 and σ_2 are not equal will be discussed in Section 10.3.

WHEN TO USE THE *t* DISTRIBUTION TO MAKE INFERENCES ABOUT $\mu_1 - \mu_2$

The t distribution is used to make inferences about $\mu_1 - \mu_2$ when the following assumptions hold true.

1. The two populations from which the two samples are drawn are (approximately) normally distributed.
2. The sample sizes are small ($n_1 < 30$ and $n_2 < 30$) and independent.
3. The standard deviations σ_1 and σ_2 of the two populations are unknown but they are equal, that is, $\sigma_1 = \sigma_2$.

When the standard deviations of the two populations are equal, we can use σ for both σ_1 and σ_2. Because σ is unknown, we replace it by its point estimator s_p which is called the *pooled sample standard deviation* (hence, the subscript p). The value of s_p is calculated by using the information from the two samples as follows.

THE POOLED STANDARD DEVIATION FOR TWO SAMPLES

The pooled standard deviation for two samples is calculated as

$$s_p = \sqrt{\frac{(n_1 - 1)s_1^2 + (n_2 - 1)s_2^2}{n_1 + n_2 - 2}}$$

where n_1 and n_2 are the two sample sizes and s_1^2 and s_2^2 are the variances of the two samples.

In this formula, $n_1 - 1$ are the degrees of freedom for sample 1, $n_2 - 1$ are the degrees of freedom for sample 2, and $n_1 + n_2 - 2$ are the degrees of freedom for the two samples taken together.

When s_p is used as an estimator of σ, the standard deviation of $\bar{x}_1 - \bar{x}_2$ is estimated by $s_{\bar{x}_1 - \bar{x}_2}$. The value of $s_{\bar{x}_1 - \bar{x}_2}$ is calculated by using the following formula.

ESTIMATE OF THE STANDARD DEVIATION OF $\bar{x}_1 - \bar{x}_2$

The estimate of the standard deviation of $\bar{x}_1 - \bar{x}_2$ is

$$s_{\bar{x}_1 - \bar{x}_2} = s_p \sqrt{\frac{1}{n_1} + \frac{1}{n_2}}$$

Now we are ready to learn the procedures that are used to make confidence intervals and test hypotheses about $\mu_1 - \mu_2$ for small samples taken from the two populations with unknown but equal standard deviations.

10.2.1 INTERVAL ESTIMATION OF $\mu_1 - \mu_2$

As mentioned earlier in this chapter, the difference between two sample means $\bar{x}_1 - \bar{x}_2$ is the point estimator of the difference between two population means $\mu_1 - \mu_2$. The following formula gives the confidence interval for $\mu_1 - \mu_2$ when the t distribution is used.

CONFIDENCE INTERVAL FOR $\mu_1 - \mu_2$

The $(1 - \alpha)100\%$ confidence interval for $\mu_1 - \mu_2$ is

$$(\bar{x}_1 - \bar{x}_2) \pm t\, s_{\bar{x}_1 - \bar{x}_2}$$

where the value of t is obtained from the t distribution table for a given confidence level and $n_1 + n_2 - 2$ degrees of freedom, and $s_{\bar{x}_1 - \bar{x}_2}$ is calculated as explained earlier in Section 10.2.

Example 10-7 describes the procedure to make a confidence interval for $\mu_1 - \mu_2$ using the t distribution.

Constructing a confidence interval for $\mu_1 - \mu_2$: small samples and $\sigma_1 = \sigma_2$.

EXAMPLE 10-7 A sample of 15 one-pound jars of coffee of Brand 1 showed that the mean amount of caffeine in these jars is 80 milligrams per jar with a standard deviation of 5 milligrams. Another sample of 12 one-pound coffee jars of Brand 2 gave a mean amount of caffeine of 77 milligrams per jar with a standard deviation of 6 milligrams. Construct a 95% confidence interval for the difference between the mean amounts of caffeine in one-pound coffee jars of these two brands. Assume that the two populations are normally distributed and that the standard deviations of the two populations are equal.

Solution Let μ_1 and μ_2 be the mean amounts of caffeine per jar in all one-pound jars of Brands 1 and 2, respectively, and let $\bar{x}_1$ and $\bar{x}_2$ be the means of the two respective samples. From the given information,

$$n_1 = 15 \quad \bar{x}_1 = 80 \text{ milligrams} \quad s_1 = 5 \text{ milligrams}$$
$$n_2 = 12 \quad \bar{x}_2 = 77 \text{ milligrams} \quad s_2 = 6 \text{ milligrams}$$

The confidence level is $1 - \alpha = .95$.

First we calculate the standard deviation of $\bar{x}_1 - \bar{x}_2$ as follows.

$$s_p = \sqrt{\frac{(n_1 - 1)s_1^2 + (n_2 - 1)s_2^2}{n_1 + n_2 - 2}} = \sqrt{\frac{(15 - 1)(5)^2 + (12 - 1)(6)^2}{15 + 12 - 2}} = 5.463$$

$$s_{\bar{x}_1 - \bar{x}_2} = s_p \sqrt{\frac{1}{n_1} + \frac{1}{n_2}} = (5.463)\sqrt{\frac{1}{15} + \frac{1}{12}} = 2.116$$

Next, to find the t value from the t distribution table, we need to know the area in each tail and the degrees of freedom.

$$\text{Area in each tail} = \frac{\alpha}{2} = .5 - \left(\frac{.95}{2}\right) = .025$$

$$\text{Degrees of freedom} = n_1 + n_2 - 2 = 15 + 12 - 2 = 25$$

The t value for $df = 25$ and .025 area in the right tail is 2.060. Therefore, the 95% confidence interval for $\mu_1 - \mu_2$ is

$$(\bar{x}_1 - \bar{x}_2) \pm t\, s_{\bar{x}_1-\bar{x}_2} = (80 - 77) \pm 2.060\,(2.116)$$
$$= 3 \pm 4.36 = -1.36 \text{ to } 7.36$$

Thus, with 95% confidence we can state that, based on these two sample results, the difference in the mean amounts of caffeine in one pound jars of these two brands of coffee lies between −1.36 and 7.36 milligrams. Because the lower limit of the interval is negative, it is possible that the mean amount of caffeine is greater in the second brand jars than in the first brand jars. ■

10.2.2 HYPOTHESIS TESTING ABOUT $\mu_1 - \mu_2$

When the three assumptions mentioned in Section 10.2 are satisfied, then the t distribution is applied to make a test of hypothesis about the difference between two population means. The test statistic in this case is t, which is calculated as follows.

TEST STATISTIC t FOR $\bar{x}_1 - \bar{x}_2$

The value of the test statistic t for $\bar{x}_1 - \bar{x}_2$ is computed as

$$t = \frac{(\bar{x}_1 - \bar{x}_2) - (\mu_1 - \mu_2)}{s_{\bar{x}_1-\bar{x}_2}}$$

The value of $\mu_1 - \mu_2$ in the formula is substituted from the null hypothesis and $s_{\bar{x}_1-\bar{x}_2}$ is calculated as explained in Section 10.2.

Examples 10-8 and 10-9 illustrate how a test of hypothesis about the difference between two population means is conducted using the t distribution.

Making a two-tailed test of hypothesis about $\mu_1 - \mu_2$: small samples and $\sigma_1 = \sigma_2$.

EXAMPLE 10-8 A sample of 14 cans of Brand 1 diet soda gave the mean number of calories as 23 per can with a standard deviation of 3 calories. Another sample of 16 cans of Brand 2 diet soda gave the mean number of calories as 25 per can with a standard deviation of 4 calories. At the 1% significance level, are the mean number of calories per can different for these two brands of diet soda? Assume that the calories per can of diet soda are normally distributed for each of the two brands and that the standard deviations for the two populations are equal.

Solution Let μ_1 and μ_2 be the mean number of calories per can for diet soda of Brand 1 and Brand 2, respectively, and let $\bar{x}_1$ and $\bar{x}_2$ be the means of the respective samples. From the given information,

$$n_1 = 14 \quad \bar{x}_1 = 23 \quad s_1 = 3$$
$$n_2 = 16 \quad \bar{x}_2 = 25 \quad s_2 = 4$$

The significance level is $\alpha = .01$.

Step 1. *State the null and alternative hypotheses*

We are to test for the difference in the mean number of calories per can for two brands. The null and alternative hypotheses are

H_0: $\mu_1 - \mu_2 = 0$ (the mean number of calories are not different)

H_1: $\mu_1 - \mu_2 \neq 0$ (the mean number of calories are different)

Step 2. *Select the distribution to use*

The two populations are normally distributed, the samples are small and independent, and the standard deviations of the two populations are unknown but equal. Consequently, we will use the t distribution.

Step 3. *Determine the rejection and nonrejection regions*

The $\neq$ sign in the alternative hypothesis indicates that the test is two-tailed. The significance level is .01. Hence,

$$\text{Area in each tail} = \frac{\alpha}{2} = \frac{.01}{2} = .005$$

$$\text{Degrees of freedom} = n_1 + n_2 - 2 = 14 + 16 - 2 = 28$$

The critical values of t for $df = 28$ and .005 area in each tail are -2.763 and 2.763, as shown in Figure 10.4.

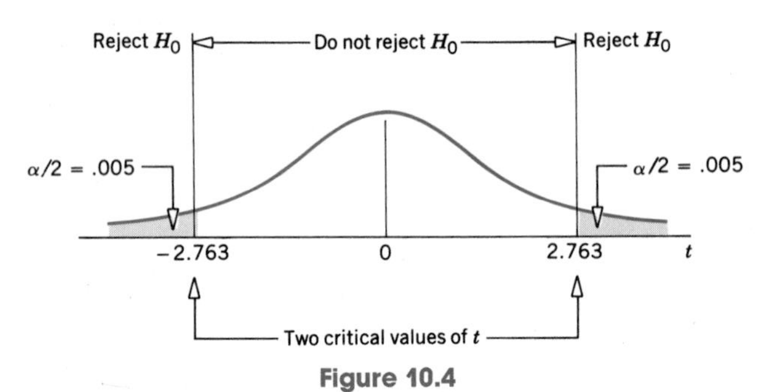

Figure 10.4

Step 4. *Calculate the value of the test statistic*

The value of the test statistic t for $\bar{x}_1 - \bar{x}_2$ is computed as follows.

$$s_p = \sqrt{\frac{(n_1 - 1)s_1^2 + (n_2 - 1)s_2^2}{n_1 + n_2 - 2}} = \sqrt{\frac{(14 - 1)(3)^2 + (16 - 1)(4)^2}{14 + 16 - 2}} = 3.571$$

$$s_{\bar{x}_1 - \bar{x}_2} = s_p \sqrt{\frac{1}{n_1} + \frac{1}{n_2}} = (3.571)\sqrt{\frac{1}{14} + \frac{1}{16}} = 1.307$$

$$t = \frac{(\bar{x}_1 - \bar{x}_2) - (\mu_1 - \mu_2)}{s_{\bar{x}_1 - \bar{x}_2}} = \frac{(23 - 25) - \overset{\text{From } H_0}{0}}{1.307} = -1.530$$

Step 5. *Make a decision*

Because the value of the test statistic $t = -1.530$ for $\bar{x}_1 - \bar{x}_2$ falls in the nonrejection region, we fail to reject the null hypothesis. Hence, there is no difference in the mean number of calories per can for the two brands of diet soda. The difference in $\bar{x}_1$ and $\bar{x}_2$ observed for two samples may have occurred because of sampling error only. ■

Making a right-tailed test of hypothesis about $\mu_1 - \mu_2$: small samples and $\sigma_1 = \sigma_2$.

EXAMPLE 10-9 A sample of 15 children from New York showed that the mean time they spend watching television is 27.50 hours a week with a standard deviation of 4 hours. Another sample of 16 children from California showed that the mean time spent by them watching television is 23.25 hours a week with a standard deviation of 5 hours. Test at the 5% significance level if the mean time spent watching television by children from New York is higher than that for children from California. Assume that the time spent watching television by children has a normal distribution for both populations and that the standard deviations for the two populations are equal.

Solution Let the children from New York be referred to as population 1 and those from California as population 2. Let μ_1 and μ_2 be the means of populations 1 and 2, respectively, and let $\bar{x}_1$ and $\bar{x}_2$ be the means of the respective samples. From the given information,

For New York:	$n_1 = 15$	$\bar{x}_1 = 27.50$	$s_1 = 4$
For California:	$n_2 = 16$	$\bar{x}_2 = 23.25$	$s_2 = 5$

The significance level is $\alpha = .05$.

Step 1. *State the null and alternative hypotheses*

The two possible decisions are

(a) The mean time spent watching television by children in New York is not higher than that for children in California. This can be written as $\mu_1 = \mu_2$ or $\mu_1 - \mu_2 = 0$.

(b) The mean time spent watching television by children in New York is higher than that for children in California. This can be written as $\mu_1 > \mu_2$ or $\mu_1 - \mu_2 > 0$.

Hence, the null and alternative hypotheses are

$$H_0: \mu_1 - \mu_2 = 0$$

$$H_1: \mu_1 - \mu_2 > 0$$

Note that the null hypothesis can also be written as $\mu_1 - \mu_2 \leq 0$.

Step 2. *Select the distribution to use*

The two populations are normally distributed, the samples are small and independent, and the standard deviations of the two populations are unknown but equal. Consequently, we will use the t distribution to make the test.

Step 3. *Determine the rejection and nonrejection regions*

The > sign in the alternative hypothesis indicates that the test is right-tailed. The significance level is .05.

$$\text{Area in the right tail} = \alpha = .05$$

$$\text{Degrees of freedom} = n_1 + n_2 - 2 = 15 + 16 - 2 = 29$$

From the t distribution table, the critical value of t for $df = 29$ and .05 area in the right tail is 1.699. This value is shown in Figure 10.5.

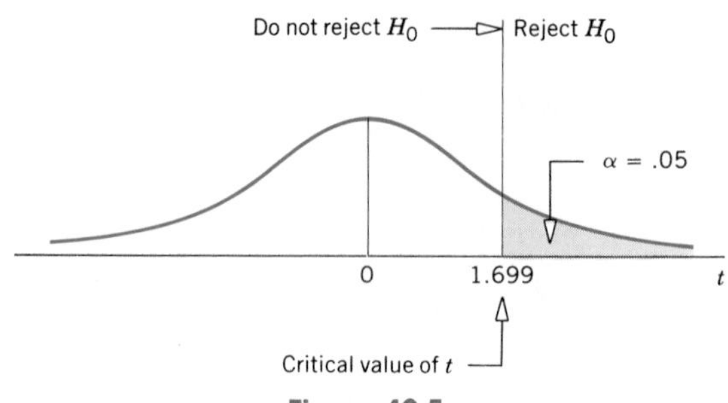

Figure 10.5

Step 4. *Calculate the value of the test statistic*

The value of the test statistic t for $\bar{x}_1 - \bar{x}_2$ is calculated as follows.

$$s_p = \sqrt{\frac{(n_1 - 1)s_1^2 + (n_2 - 1)s_2^2}{n_1 + n_2 - 2}} = \sqrt{\frac{(15 - 1)(4)^2 + (16 - 1)(5)^2}{15 + 16 - 2}} = 4.545$$

$$s_{\bar{x}_1 - \bar{x}_2} = s_p \sqrt{\frac{1}{n_1} + \frac{1}{n_2}} = (4.545)\sqrt{\frac{1}{15} + \frac{1}{16}} = 1.633$$

$$t = \frac{(\bar{x}_1 - \bar{x}_2) - (\mu_1 - \mu_2)}{s_{\bar{x}_1 - \bar{x}_2}} = \frac{(27.50 - 23.25) - 0}{1.633} = 2.603$$

(From H_0 — the 0 in the numerator)

Step 5. *Make a decision*

Because the value of the test statistic $t = 2.603$ for $\bar{x}_1 - \bar{x}_2$ falls in the rejection region, we reject the null hypothesis H_0. Hence, we conclude that children from New York state spend on average more time watching TV than children from California. ■

EXERCISES

10.20 Assuming that two populations are normally distributed with unknown but equal standard deviations, construct a 95% confidence interval for $\mu_1 - \mu_2$ for the following.

$$n_1 = 22 \quad \bar{x}_1 = 12.50 \quad s_1 = 3.75$$
$$n_2 = 18 \quad \bar{x}_2 = 14.60 \quad s_2 = 3.10$$

10.21 Assuming that two populations are normally distributed with unknown but equal population standard deviations, construct a 99% confidence interval for $\mu_1 - \mu_2$ for the following.

$$n_1 = 20 \quad \bar{x}_1 = 33.75 \quad s_1 = 5.25$$
$$n_2 = 23 \quad \bar{x}_2 = 28.50 \quad s_2 = 4.55$$

10.22 Refer to Exercise 10.20. Test at the 5% significance level if the two population means are different.

10.23 Refer to Exercise 10.21. Test at the 1% significance level if the two population means are different.

10.24 Refer to Exercise 10.20. Test at the 1% significance level if μ_1 is less than μ_2.

10.25 Refer to Exercise 10.21. Test at the 5% significance level if μ_1 is greater than μ_2.

10.26 A sample of 25 male customers who shopped at a grocery store showed that they spent an average of $58 with a standard deviation of $14.50. Another sample of 20 female customers who shopped at the same grocery store showed that they spent an average of $71 with a standard deviation of $12.40. Assume that the amounts spent at this store by all male and all female customers are both normally distributed with equal but unknown population standard deviations.

a. Construct a 99% confidence interval for the difference between the mean amounts spent by all male and all female customers at this store.
b. Test at the 5% significance level if the mean amount spent by all male customers at this store is lower than that of all female customers.

10.27 A manufacturing company is interested in buying one of two different kinds of machines. The company tested the two machines for production purposes. The first machine was run for 8 hours. It produced on average 123 items an hour with a standard deviation of 9 items. The second machine was run for 10 hours. It produced on average 114 items an hour with a standard deviation of 6 items. Assume that the production per hour for each machine is (approximately) normally distributed. Further assume that the standard deviations of the two populations are equal.

a. Make a 95% confidence interval for the difference between the two population means.
b. Test at the 5% significance level if the mean number of items produced per hour by the first machine is higher than that by the second machine.

10.28 According to a study conducted by Professors Davis and Templer, a sample of 28 children who were exposed to narcotics in utero had a mean IQ of 90.36 with a standard deviation of 11.36. Another sample of 28 children who were not exposed to narcotics in utero had a mean IQ of 96.32 with a standard deviation of 8.72 (Donald D. Davis and Donald I. Templer: "Neurobehavioral Functioning in Children Exposed to Narcotics in Utero," *Addictive Behaviors*, 13: 275–283, 1988). The children in both samples were 6 to 15 years old when tested.

a. Make a 99% confidence interval for the difference between the mean IQ scores of all children who are exposed and who are not exposed to narcotics in utero.
b. Test at the 1% significance level if the mean IQ score of children who are exposed to narcotics in utero is lower than the mean IQ score of children who are not exposed to narcotics in utero.

Assume that the IQ scores of children belonging to two populations are normally distributed with the same standard deviation.

10.29 An insurance company wants to know if the average speed at which men drive cars is higher than that for women drivers. The company took a random sample of 27 cars driven by men on a highway and found the mean speed as 68 miles per hour with a standard deviation of 2.2 miles. Another sample of 18 cars driven by women on the same highway gave a mean of 65 miles per hour with a standard deviation of 2.5 miles. Assume that the speeds at which all men and all women drive cars on this highway are both normally distributed with the same population standard deviation.

a. Construct a 95% confidence interval for the difference between the mean speeds of cars driven by all men and all women drivers on this highway.

b. Test at the 1% significance level if the mean speed of cars driven by all men drivers on this highway is higher than that of cars driven by all women drivers.

10.30 According to Metropolitan Life Insurance Company claims data, the mean hospital and physician's charges for a typical cesarean birth are $6400 for Florida and $6940 for California (*Statistical Bulletin,* July–September 1988). Assume that these two means are based on 24 claims for Florida and 28 claims for California and that the standard deviations for these two sets of claims for the two states are $980 and $1150, respectively. Further assume that such charges have a normal distribution with the same population standard deviation for each of these states.

a. Construct a 90% confidence interval for the difference between the mean charges for all cesarean births for these two states.
b. Test at the 5% significance level if the mean charges for cesarean births are different for these two states.

10.31 A company claims that its medicine, Brand A, provides faster relief from pain than another company's medicine, Brand B. A researcher tested both brands of medicine on two groups of randomly selected patients. The results of the test are given in the following table. The relief time is in minutes.

Brand	Sample Size	Relief Time	Standard Deviation
A	25	44	13
B	22	49	11

a. Construct a 99% confidence interval for the difference between the mean relief times for these two brands of medicine.
b. Test at the 1% significance level if the mean relief time for Brand A is less than that of Brand B.

Assume that the two populations are normally distributed with equal standard deviations.

10.32 A professor took two samples, one of 21 males and another of 18 females, from students at a college who were enrolled in an introductory course in statistics. The professor found that the mean score of male students in a midterm examination in statistics was 76.2 with a standard deviation of 7.3 and the mean score of female students was 78.5 with a standard deviation of 6.7.

a. Construct a 99% confidence interval for the difference between the mean scores of all male and all female students.
b. Test at the 5% significance level if the mean score of all male students is lower than that of all female students.

Assume that the two populations are normally distributed with equal but unknown standard deviations.

10.33 According to the American Association of University Professors, the mean salary of male professors at American colleges and universities was $54,340 and that of female professors was $48,080 in 1989–90. For convenience, assume that these two means are based on random samples of 28 male professors and 26 female professors. Further assume that the standard deviations for the two samples are $3100 and $2800, respectively.

a. Construct a 90% confidence interval for the difference between the two population means.
b. Test at the 1% significance level if the mean salary of all male professors for 1989–90 is higher than that of all female professors.

Assume that the salaries of all male and all female professors are both normally distributed with equal standard deviations.

10.3 INFERENCES ABOUT THE DIFFERENCE BETWEEN TWO POPULATION MEANS FOR SMALL AND INDEPENDENT SAMPLES: UNEQUAL STANDARD DEVIATIONS

Section 10.2 explained how to make inferences about the difference between two population means using the t distribution when the standard deviations of the two populations are unknown but equal and certain other assumptions hold true. Now, what if all other assumptions of Section 10.2 hold true but the population standard deviations are not only unknown but also unequal? In this case, the procedures used to make confidence intervals and to test hypotheses about $\mu_1 - \mu_2$ remain similar to the ones explained in Sections 10.2.1 and 10.2.2 except for two differences. When the population standard deviations are unknown and not equal, then the degrees of freedom are no longer given by $n_1 + n_2 - 2$ and the standard deviation of $\bar{x}_1 - \bar{x}_2$ is not calculated using the pooled standard deviation s_p.

DEGREES OF FREEDOM

When

1. The two populations from which the samples are drawn are (approximately) normally distributed
2. The two samples are small and independent
3. The two population standard deviations are unknown and unequal

then the t distribution is used to make inferences about $\mu_1 - \mu_2$ and the degrees of freedom of the t distribution are given by

$$df = \frac{\left(\frac{s_1^2}{n_1} + \frac{s_2^2}{n_2}\right)^2}{\frac{\left(\frac{s_1^2}{n_1}\right)^2}{n_1 - 1} + \frac{\left(\frac{s_2^2}{n_2}\right)^2}{n_2 - 1}}$$

The number given by this formula is always rounded down for df.

Because the standard deviations of the two populations are not known, we use $s_{\bar{x}_1-\bar{x}_2}$ as a point estimator of $\sigma_{\bar{x}_1-\bar{x}_2}$. The following formula is used to calculate the standard deviation $s_{\bar{x}_1-\bar{x}_2}$ of $\bar{x}_1 - \bar{x}_2$.

ESTIMATE OF THE STANDARD DEVIATION OF $\bar{x}_1 - \bar{x}_2$

The value of $s_{\bar{x}_1-\bar{x}_2}$ is calculated as

$$s_{\bar{x}_1-\bar{x}_2} = \sqrt{\frac{s_1^2}{n_1} + \frac{s_2^2}{n_2}}$$

10.3.1 INTERVAL ESTIMATION OF $\mu_1 - \mu_2$

Again, the difference between the two sample means $\bar{x}_1 - \bar{x}_2$ is the point estimator of the difference between the two population means $\mu_1 - \mu_2$. The following formula gives the confidence interval for $\mu_1 - \mu_2$ when the t distribution is used and the population standard deviations are unknown and presumed to be unequal.

CONFIDENCE INTERVAL FOR $\mu_1 - \mu_2$

The $(1 - \alpha)100\%$ confidence interval for $\mu_1 - \mu_2$ is

$$(\bar{x}_1 - \bar{x}_2) \pm t\, s_{\bar{x}_1 - \bar{x}_2}$$

where the value of t is obtained from the t distribution table for a given confidence level and the degrees of freedom given by the formula mentioned in Section 10.3, and $s_{\bar{x}_1 - \bar{x}_2}$ is calculated as explained earlier.

Example 10-10 describes how to construct a confidence interval for $\mu_1 - \mu_2$ when the standard deviations of the two populations are unknown and unequal.

Constructing a confidence interval for $\mu_1 - \mu_2$: small samples and $\sigma_1 \neq \sigma_2$.

EXAMPLE 10-10 According to Example 10-7 of Section 10.2.1, a sample of 15 one-pound jars of coffee of Brand 1 showed that the mean amount of caffeine in these jars is 80 milligrams per jar with a standard deviation of 5 milligrams. Another sample of 12 one-pound coffee jars of Brand 2 gave a mean amount of caffeine equal to 77 milligrams per jar with a standard deviation of 6 milligrams. Construct a 95% confidence interval for the difference between the mean amounts of caffeine in one-pound coffee jars of these two brands. Assume that the two populations are normally distributed and that the standard deviations of the two populations are not equal.

Solution Let μ_1 and μ_2 be the mean amounts of caffeine per jar in all one-pound jars of Brand 1 and Brand 2, respectively, and let $\bar{x}_1$ and $\bar{x}_2$ be the means of the two respective samples. From the given information,

$$n_1 = 15 \quad \bar{x}_1 = 80 \text{ milligrams} \quad s_1 = 5 \text{ milligrams}$$
$$n_2 = 12 \quad \bar{x}_2 = 77 \text{ milligrams} \quad s_2 = 6 \text{ milligrams}$$

The confidence level is $1 - \alpha = .95$.

First, we calculate the standard deviation of $\bar{x}_1 - \bar{x}_2$ as follows.

$$s_{\bar{x}_1 - \bar{x}_2} = \sqrt{\frac{s_1^2}{n_1} + \frac{s_2^2}{n_2}} = \sqrt{\frac{(5)^2}{15} + \frac{(6)^2}{12}} = 2.160$$

Next, to find the t value from the t distribution table, we need to know the area under the curve in each tail and the degrees of freedom.

$$\text{Area in each tail} = \frac{\alpha}{2} = .5 - \left(\frac{.95}{2}\right) = .025$$

$$df = \frac{\left(\dfrac{s_1^2}{n_1} + \dfrac{s_2^2}{n_2}\right)^2}{\dfrac{\left(\dfrac{s_1^2}{n_1}\right)^2}{n_1 - 1} + \dfrac{\left(\dfrac{s_2^2}{n_2}\right)^2}{n_2 - 1}} = \frac{\left(\dfrac{(5)^2}{15} + \dfrac{(6)^2}{12}\right)^2}{\dfrac{\left(\dfrac{(5)^2}{15}\right)^2}{(15 - 1)} + \dfrac{\left(\dfrac{(6)^2}{12}\right)^2}{(12 - 1)}} = 21.42 = 21$$

Note that the degrees of freedom are always rounded down as in this calculation.

From the t distribution table, the t value for $df = 21$ and .025 area in the right tail is 2.080. Therefore, the 95% confidence interval for $\mu_1 - \mu_2$ is

$$(\bar{x}_1 - \bar{x}_2) \pm t\, s_{\bar{x}_1 - \bar{x}_2} = (80 - 77) \pm 2.080\,(2.160)$$

$$= 3 \pm 4.49 = -1.49 \text{ to } 7.49$$

Thus, with 95% confidence we can state that, based on these two sample results, the difference in the mean amounts of caffeine in one-pound jars of these two brands of coffee is between -1.49 and 7.49 milligrams.

Comparing this confidence interval with the one obtained in Example 10-7, we observe that the two confidence intervals are very close. From this we can conclude that even if the standard deviations of the two populations are not equal and we use the procedure of Section 10.2.1 to make a confidence interval for $\mu_1 - \mu_2$, the margin of error will be small as long as the difference between the two standard deviations is not too large. ■

10.3.2 HYPOTHESIS TESTING ABOUT $\mu_1 - \mu_2$

When the standard deviations of the two populations are unknown and unequal, with other conditions of Section 10.2.2 holding true, we use the t distribution to make a test of hypothesis about $\mu_1 - \mu_2$. This procedure differs from the one in Section 10.2.2 only in the calculation of degrees of freedom for the t distribution and the standard deviation of $\bar{x}_1 - \bar{x}_2$. The df and the standard deviation of $\bar{x}_1 - \bar{x}_2$ in this case are given by the formulas presented in Section 10.3.

TEST STATISTIC t FOR $\bar{x}_1 - \bar{x}_2$

The value of the test statistic t for $\bar{x}_1 - \bar{x}_2$ is computed as

$$t = \frac{(\bar{x}_1 - \bar{x}_2) - (\mu_1 - \mu_2)}{s_{\bar{x}_1 - \bar{x}_2}}$$

The value of $\mu_1 - \mu_2$ in this formula is substituted from the null hypothesis, and $s_{\bar{x}_1 - \bar{x}_2}$ is calculated as explained in Section 10.3.

Example 10-11 illustrates the procedure to conduct a test of hypothesis about $\mu_1 - \mu_2$ when the population standard deviations are unknown and unequal.

Making a two-tailed test of hypothesis about $\mu_1 - \mu_2$: small samples and $\sigma_1 \neq \sigma_2$.

EXAMPLE 10-11 According to Example 10-8 of Section 10.2.2, a sample of 14 cans of diet soda of Brand 1 gave the mean number of calories per can as 23 with a standard deviation of 3 calories. Another sample of 16 cans of Brand 2 diet soda gave the mean number of calories as 25 per can with a standard deviation of 4 calories. Test at the 1% significance level if the mean number of calories per can are different for these two brands of diet soda. Assume that the calories per can of diet soda are normally distributed for each of these two brands and that the standard deviations for the two populations are not equal.

Solution Let μ_1 and μ_2 be the mean number of calories per can for diet soda of Brand 1 and Brand 2, respectively, and let $\bar{x}_1$ and $\bar{x}_2$ be the means of the respective samples. From the given information,

$$n_1 = 14 \qquad \bar{x}_1 = 23 \qquad s_1 = 3$$
$$n_2 = 16 \qquad \bar{x}_2 = 25 \qquad s_2 = 4$$

The significance level is $\alpha = .01$.

Step 1. *State the null and alternative hypotheses*

We are to test for the difference in the mean number of calories per can for two brands. The null and alternative hypotheses are

H_0: $\mu_1 - \mu_2 = 0$ (the mean number of calories are not different)

H_1: $\mu_1 - \mu_2 \neq 0$ (the mean number of calories are different)

Step 2. *Select the distribution to use*

The two populations are normally distributed, the samples are small and independent, and the standard deviations of the two populations are unknown and unequal. Consequently, we use the t distribution to make the test.

Step 3. *Determine the rejection and nonrejection regions*

The $\neq$ sign in the alternative hypothesis indicates that the test is two-tailed. The significance level is .01. Hence,

$$\text{Area in each tail} = \frac{\alpha}{2} = \frac{.01}{2} = .005$$

The degrees of freedom are calculated as follows.

$$df = \frac{\left(\dfrac{s_1^2}{n_1} + \dfrac{s_2^2}{n_2}\right)^2}{\dfrac{\left(\dfrac{s_1^2}{n_1}\right)^2}{n_1 - 1} + \dfrac{\left(\dfrac{s_2^2}{n_2}\right)^2}{n_2 - 1}} = \frac{\left(\dfrac{(3)^2}{14} + \dfrac{(4)^2}{16}\right)^2}{\dfrac{\left(\dfrac{(3)^2}{14}\right)^2}{(14 - 1)} + \dfrac{\left(\dfrac{(4)^2}{16}\right)^2}{(16 - 1)}} = 27.41 = 27$$

From the t distribution table, the critical values of t for $df = 27$ and .005 area in each tail are -2.771 and 2.771. These values are shown in Figure 10.6.

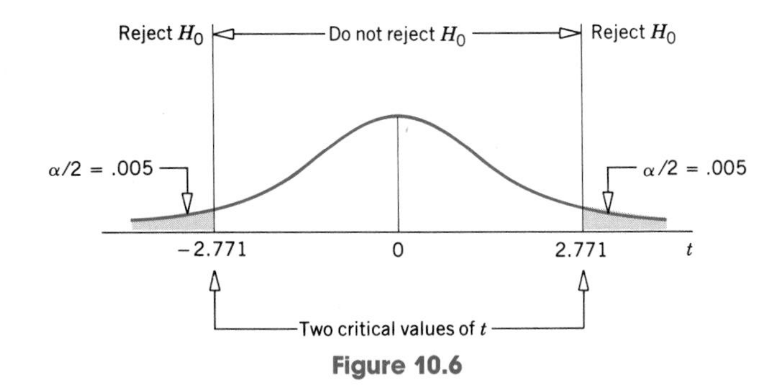

Figure 10.6

Step 4. *Calculate the value of the test statistic*

The value of the test statistic t for $\bar{x}_1 - \bar{x}_2$ is computed as follows.

$$s_{\bar{x}_1 - \bar{x}_2} = \sqrt{\frac{s_1^2}{n_1} + \frac{s_2^2}{n_2}} = \sqrt{\frac{(3)^2}{14} + \frac{(4)^2}{16}} = 1.282$$

$$t = \frac{(\bar{x}_1 - \bar{x}_2) - (\mu_1 - \mu_2)}{s_{\bar{x}_1 - \bar{x}_2}} = \frac{(23 - 25) - \overset{\text{From } H_0}{0}}{1.282} = -1.560$$

Step 5. *Make a decision*

Because the value of the test statistic $t = -1.560$ for $\bar{x}_1 - \bar{x}_2$ falls in the nonrejection region, we fail to reject the null hypothesis. Hence, there is no difference in the mean number of calories per can for these two brands of diet soda. The difference in $\bar{x}_1$ and $\bar{x}_2$ observed for two samples may have occurred because of sampling error only. ■

Remember: The degrees of freedom for the procedures to make a confidence interval and to test a hypothesis about $\mu_1 - \mu_2$ learned in Sections 10.3.1 and 10.3.2 are always rounded down.

EXERCISES

10.34 Assuming that two populations are normally distributed with unequal and unknown standard deviations, construct a 95% confidence interval for $\mu_1 - \mu_2$ for the following.

$$n_1 = 24 \quad \bar{x}_1 = 17.20 \quad s_1 = 3.90$$
$$n_2 = 16 \quad \bar{x}_2 = 19.40 \quad s_2 = 5.15$$

10.35 Assuming that two populations are normally distributed with unequal and unknown standard deviations, construct a 99% confidence interval for $\mu_1 - \mu_2$ for the following.

$$n_1 = 15 \quad \bar{x}_1 = 52.61 \quad s_1 = 3.27$$
$$n_2 = 19 \quad \bar{x}_2 = 43.75 \quad s_2 = 5.85$$

10.36 Refer to Exercise 10.34. Test at the 5% significance level if the two population means are different.

10.37 Refer to Exercise 10.35. Test at the 1% significance level if the two population means are different.

10.38 Refer to Exercise 10.34. Test at the 1% significance level if μ_1 is less than μ_2.

10.39 Refer to Exercise 10.35. Test at the 5% significance level if μ_1 is greater than μ_2.

10.40 According to the information given in Exercise 10.26, a sample of 25 male customers who shopped at a grocery store showed that they spent an average of $58 with a standard deviation of $14.50. Another sample of 20 female customers who shopped at the same grocery store showed that they spent an average of $71 with a standard deviation of $12.40. Assume that the amounts spent at this store by all male and all female customers have normal distributions with unequal and unknown population standard deviations.

- **a.** Construct a 99% confidence interval for the difference between the mean amounts spent by all male and all female customers at this store.
- **b.** Test at the 5% significance level if the mean amount spent by all male customers at this store is lower than that of all female customers.

10.41 According to Exercise 10.27, a manufacturing company is interested in buying one of two different kinds of machines. The company tested the two machines for production purposes. The first machine was run for 8 hours and produced an average of 123 items an hour with a standard deviation of 9 items. The second machine was run for 10 hours and produced an average of 114 items an hour with a standard deviation of 6 items. Assume that the production per hour for each machine is (approximately) normally distributed. Further assume that the standard deviations of the two populations are unequal.

- **a.** Make a 95% confidence interval for the difference between the two population means.
- **b.** Test at the 5% significance level if the mean number of items produced per hour by the first machine is higher than that by the second machine.

10.42 According to a study conducted by Professors Davis and Templer, a sample of 28 children who were exposed to narcotics in utero had a mean IQ of 90.36 with a standard deviation of 11.36. Another sample of 28 children who were not exposed to narcotics in utero gave a mean IQ of 96.32 with a standard deviation of 8.72 (Donald D. Davis and Donald I. Templer: "Neurobehavioral Functioning in Children Exposed to Narcotics in Utero," *Addictive Behaviors,* 13: 275–283, 1988). The children in both samples were 6 to 15 years old.

- **a.** Make a 99% confidence interval for the difference between the mean IQ scores of all children who are exposed and who are not exposed to narcotics in utero.
- **b.** Test at the 1% significance level if the mean IQ score of children who are exposed to narcotics is lower than the mean IQ score of children who are not exposed to narcotics.

Assume that the IQ scores of children belonging to two populations are normally distributed with unequal standard deviations.

10.43 According to Exercise 10.29, an insurance company wants to know if the average speed at which men drive cars is higher than that of women drivers. The company took a random sample of 27 cars driven by men on a highway and found the mean speed as 68 miles per hour with a standard deviation of 2.2 miles. Another sample of 18 cars driven by women on the same highway gave a mean of 65 miles per hour with a standard deviation of 2.5 miles. Assume that the speeds at which all men and all women drive cars on this highway both have a normal distribution with unequal population standard deviations.

- **a.** Construct a 95% confidence interval for the difference between the mean speeds of cars driven by all men and all women drivers on this highway.
- **b.** Test at the 1% significance level if the mean speed of cars driven by all men on this highway is higher than that of cars driven by all women drivers.

10.44 According to Exercise 10.31, a company claims that its Brand A medicine provides faster relief from pain than another company's medicine Brand B. A researcher tested both brands of medicine on two groups of randomly selected patients. The results of the test are given in the following table. The relief time is in minutes.

Brand	Sample Size	Relief Time	Standard Deviation
A	25	44	13
B	22	49	11

a. Construct a 99% confidence interval for the difference between the mean relief times for these two brands of medicine.
b. Test at the 1% significance level if the mean relief time for Brand A is less than that of Brand B.

Assume that the two populations are normally distributed with unequal and unknown standard deviations.

10.45 Refer to Exercise 10.32. A professor took two samples, one of 21 males and another of 18 females, from students at a college who were enrolled in an introductory course in statistics. The professor found that the mean score of these male students in a midterm examination in statistics was 76.2 with a standard deviation of 7.3 and the mean score of these female students was 78.5 with a standard deviation of 6.7.

a. Construct a 99% confidence interval for the difference between the mean scores of all male and all female students.
b. Test at the 5% significance level if the mean score of all male students is lower than that of all female students.

Assume that the two populations are normally distributed with unequal and unknown standard deviations.

10.4 INFERENCES ABOUT THE DIFFERENCE BETWEEN TWO POPULATION MEANS FOR PAIRED SAMPLES

The previous three sections of this chapter were concerned with estimation and hypothesis testing about the difference between two population means when the two samples were drawn independently from two different populations. This section describes estimation and hypothesis-testing procedures for the difference between the two population means when the samples are dependent.

In a case of two dependent samples, two data values—one in each sample—are collected from the same source (or element) and, hence, these are also called **paired** or **matched samples.** For example, we may want to make inferences about the mean weight loss for members of a health club after they have gone through an exercise program for a certain period. To do so, suppose we select a sample of 15 members of this health club and record their weights before and after the program. In this example, both sets of data are collected from the same 15 persons, once before and once after the program. Thus, although there are two samples, they both contain the same 15 persons. This is an example of paired (or dependent or matched) samples. The procedures to make confidence intervals and to test hypotheses in case of paired samples are different from the ones for independent samples discussed in earlier sections of this chapter.

PAIRED OR MATCHED SAMPLES

Two samples are said to be paired or matched samples when for each data value collected from one sample there corresponds another data value collected from the second sample and both these data values are collected from the same source.

As another example of paired samples, suppose an agronomist wants to measure the effect of a new brand of fertilizer on the yield of potatoes. To do so, he selects 10 pieces of land and divides each piece of land into two portions. Then he randomly assigns one of the two portions from each piece of land to grow potatoes without using fertilizer (or using some other brand of fertilizer). The second portion from each piece of land is used to grow potatoes using the new brand of fertilizer. Thus, he will have 10 pairs of data values. Then, using the procedure to be discussed in this section, he will make inferences about the difference in the mean yield of potatoes with the new fertilizer and without it.

The question arises, why does the agronomist not choose 10 pieces of land on which to grow potatoes without using the new brand of fertilizer and another 10 pieces of land to grow potatoes using the new brand of fertilizer? If he does so, the effect of the fertilizer might be confused with the effects due to soil differences at different locations. Thus, he will not be able to identify the effect of only the new brand of fertilizer on the yield of potatoes. Consequently, the results will not be reliable. By choosing 10 pieces of land and then dividing each of them into two portions, the researcher decreases the possibility that the difference in the productivities of different pieces of land affects the results.

In paired samples, the difference between the two data values for each element of the two samples is denoted by $\boldsymbol{d}$. This value of d is called the **paired difference.** We then treat all the values of d as one sample and make inferences applying the procedures similar to the ones used for one-sample cases in Chapters 8 and 9. Note that as each source (or element) gives a pair of values (one for each of the two data sets), each sample contains the same number of values. That is, both samples are of the same size. Hence, we denote the (common) **sample size** by $\boldsymbol{n}$, which gives the number of paired difference d values. The **degrees of freedom** for the paired samples are $\boldsymbol{n - 1}$. Let

μ_d = the mean of the paired differences for the population

σ_d = the standard deviation of the paired differences for the population

$\bar{d}$ = the mean of the paired differences for a sample

s_d = the standard deviation of the paired differences for a sample

n = the number of paired difference d values in a sample

MEAN AND STANDARD DEVIATION OF THE PAIRED DIFFERENCES FOR SAMPLES

The values of $\bar{d}$ and s_d are calculated as†

$$\bar{d} = \frac{\Sigma d}{n}$$

$$s_d = \sqrt{\frac{\Sigma d^2 - \frac{(\Sigma d)^2}{n}}{n - 1}}$$

In paired samples, instead of using $\bar{x}_1 - \bar{x}_2$ as the sample statistic to make inferences about $\mu_1 - \mu_2$, we use the sample statistic $\bar{d}$ to make inferences about μ_d. Actually the value of $\bar{d}$ is always equal to $\bar{x}_1 - \bar{x}_2$, and the value of μ_d is always equal to $\mu_1 - \mu_2$.

SAMPLING DISTRIBUTION, MEAN, AND STANDARD DEVIATION OF $\bar{d}$

If the number of paired difference values is large ($n \geq 30$), because of the central limit theorem the sampling distribution of $\bar{d}$ is approximately normal with its mean and standard deviation as

$$\mu_{\bar{d}} = \mu_d \quad \text{and} \quad \sigma_{\bar{d}} = \frac{\sigma_d}{\sqrt{n}}$$

Hence, when $n \geq 30$ the normal distribution can be used to make inferences about μ_d.

However, in cases of paired samples, the sample sizes are usually small and σ_d is unknown. In such cases, assuming that the paired differences for the population are (approximately) normally distributed, the normal distribution is replaced by the t distribution to make inferences about μ_d. When σ_d is not known, the standard deviation of $\bar{d}$ is estimated by $s_d/\sqrt{n}$.

†The basic formula to calculate s_d is

$$s_d = \sqrt{\frac{\Sigma(d - \bar{d})^2}{n - 1}}$$

However, we will not use this formula to make calculations in this chapter.

ESTIMATE OF THE STANDARD DEVIATION OF $\bar{d}$

If

1. n is less than 30
2. σ_d is not known
3. The population of paired differences is (approximately) normally distributed

then the t distribution is used to make inferences about μ_d. The standard deviation $\sigma_{\bar{d}}$ of $\bar{d}$ is estimated by $s_{\bar{d}}$, which is calculated as

$$s_{\bar{d}} = \frac{s_d}{\sqrt{n}}$$

Sections 10.4.1 and 10.4.2 describe the procedures to make a confidence interval and to test a hypothesis about μ_d when σ_d is unknown and n is small. The inferences are made using the t distribution. However, if n is large, even if σ_d is unknown, the normal distribution can be used to make inferences.

10.4.1 INTERVAL ESTIMATION OF μ_d

The mean $\bar{d}$ of paired differences for two paired samples is the point estimate for μ_d. The following formula is used to construct a confidence interval for μ_d in the case of (approximately) normally distributed populations.

CONFIDENCE INTERVAL FOR μ_d

The $(1 - \alpha)100\%$ confidence interval for μ_d is

$$\bar{d} \pm t\, s_{\bar{d}}$$

where the value of t is obtained from the t distribution table for a given confidence level and $n - 1$ degrees of freedom, and $s_{\bar{d}}$ is calculated as explained earlier in Section 10.4.

Example 10-12 illustrates the procedure to construct a confidence interval for μ_d.

Constructing a confidence interval for μ_d: paired samples.

EXAMPLE 10-12 A researcher wants to find out the effect of a special diet on systolic blood pressure. She selected a sample of seven adults and put them on the dietary program for 3 months. The following table gives the systolic blood pressure of these seven adults before and after the completion of this program.

Before	210	180	195	220	231	199	224
After	193	186	186	223	220	183	233

Let μ_d be the mean reduction in the systolic blood pressure due to this special dietary program for the population of all adults. Construct a 95% confidence interval for μ_d. Assume that the population of paired differences is (approximately) normally distributed.

Solution Because the information obtained is from paired samples, we will make the confidence interval for the paired difference mean μ_d of the population using the paired difference mean $\bar{d}$ of the sample. Let d be the difference in the systolic blood pressure of an adult before and after this special dietary program. Then, d is obtained by subtracting the systolic blood pressure after the program from the systolic blood pressure before the program. The third column of Table 10.1 lists the values of d for seven adults. The fourth column of the table records the values of d^2, which are obtained by squaring each of the d values.

Table 10.1

Before	After	Difference d	d^2
210	193	210 − 193 = 17	289
180	186	180 − 186 = −6	36
195	186	195 − 186 = 9	81
220	223	220 − 223 = −3	9
231	220	231 − 220 = 11	121
199	183	199 − 183 = 16	256
224	233	224 − 233 = −9	81
		$\Sigma d = 35$	$\Sigma d^2 = 873$

The values of $\bar{d}$ and s_d are calculated as follows.

$$\bar{d} = \frac{\Sigma d}{n} = \frac{35}{7} = 5.00$$

$$s_d = \sqrt{\frac{\Sigma d^2 - \frac{(\Sigma d)^2}{n}}{n - 1}} = \sqrt{\frac{873 - \frac{(35)^2}{7}}{7 - 1}} = 10.786$$

Hence, the standard deviation of $\bar{d}$ is

$$s_{\bar{d}} = \frac{s_d}{\sqrt{n}} = \frac{10.786}{\sqrt{7}} = 4.077$$

For a 95% confidence interval,

$$\text{Area in each tail} = \frac{\alpha}{2} = .5 - \left(\frac{.95}{2}\right) = .025$$

The degrees of freedom are

$$df = n - 1 = 7 - 1 = 6$$

From the t distribution table, the t value for $df = 6$ and .025 area in the right tail is 2.447. Therefore, the 95% confidence interval for μ_d is

$$\bar{d} \pm t\, s_{\bar{d}} = 5.00 \pm 2.447\,(4.077) = 5.00 \pm 9.98 = -4.98 \text{ to } 14.98$$

Thus, we can state with 95% confidence that the mean difference between systolic blood pressures before and after the given dietary program for all adult participants is between -4.98 and 14.98. ■

10.4.2 HYPOTHESIS TESTING ABOUT μ_d

A hypothesis about μ_d is tested by using the sample statistic $\bar{d}$. If n is 30 or larger, we can use the normal distribution to test a hypothesis about μ_d. However, if n is less than 30, we replace the normal distribution by the t distribution. To use the t distribution, we assume that the population of all paired differences is (approximately) normally distributed and that the population standard deviation σ_d of paired differences is not known. This section illustrates the case of the t distribution only. The following formula is used to calculate the value of the test statistic t when testing a hypothesis about μ_d.

TEST STATISTIC t FOR $\bar{d}$

The value of the test statistic t for $\bar{d}$ is computed as follows.

$$t = \frac{\bar{d} - \mu_d}{s_{\bar{d}}}$$

The critical value of t is found from the t distribution table for a given significance level and $n - 1$ degrees of freedom.

Examples 10-13 and 10-14 illustrate the hypothesis testing procedure for μ_d.

Conducting a right-tailed test of hypothesis about μ_d: paired samples.

EXAMPLE 10-13 The following table gives the one-week sales of 6 salespersons before and after they were given a course on "how to be a successful salesperson."

Before	12	18	25	9	14	16
After	18	24	24	14	19	20

Test at the 1% significance level if the mean weekly sales for all salespersons increase as a result of this course. Assume that the population of paired differences has a normal distribution.

Solution Because the data are for paired samples, we test a hypothesis about the paired difference mean μ_d of the population using the paired difference mean $\bar{d}$ of the sample. Let

$$d = (\text{weekly sales before the course}) - (\text{weekly sales after the course})$$

In Table 10.2, we calculate d for each of the 6 salespersons by subtracting the sales after the course from the sales before the course. The fourth column of the table lists the values of d^2.

Table 10.2

Before	After	Difference d	d^2
12	18	−6	36
18	24	−6	36
25	24	1	1
9	14	−5	25
14	19	−5	25
16	20	−4	16
		$\Sigma d = -25$	$\Sigma d^2 = 139$

The values of $\bar{d}$ and s_d are calculated as follows.

$$\bar{d} = \frac{\Sigma d}{n} = \frac{-25}{6} = -4.17$$

$$s_d = \sqrt{\frac{\Sigma d^2 - \frac{(\Sigma d)^2}{n}}{n-1}} = \sqrt{\frac{139 - \frac{(-25)^2}{6}}{6-1}} = 2.639$$

The standard deviation of $\bar{d}$ is

$$s_{\bar{d}} = \frac{s_d}{\sqrt{n}} = \frac{2.639}{\sqrt{6}} = 1.077$$

Step 1. *State the null and alternative hypotheses*

We are to test if the mean weekly sales for all salespersons increase as a result of the said course. Let μ_1 be the mean weekly sales for all salespersons before the course and μ_2 be the mean weekly sales for all salespersons after the course. Then, $\mu_d = \mu_1 - \mu_2$. The mean weekly sales for all salespersons will increase due to attending the course if μ_1 is less than μ_2, which can be written as $\mu_1 - \mu_2 < 0$ or $\mu_d < 0$. Consequently, the null and alternative hypotheses are

H_0: $\mu_d = 0$ ($\mu_1 - \mu_2 = 0$ or the mean weekly sales do not increase)

H_1: $\mu_d < 0$ ($\mu_1 - \mu_2 < 0$ or the mean weekly sales do increase)

Note that we can also write the null hypothesis as $\mu_d \geq 0$.

Step 2. *Select the distribution to use*

Because the sample size is small ($n < 30$), we use the t distribution to conduct the test.

Step 3. *Determine the rejection and nonrejection regions*

The < sign in the alternative hypothesis indicates that the test is left-tailed. The significance level is .01. Hence,

$$\text{Area in the left tail} = \alpha = .01$$

$$\text{Degrees of freedom} = n - 1 = 6 - 1 = 5$$

The critical value of t for 5 df and .01 area in the left tail is -3.365, as shown in Figure 10.7.

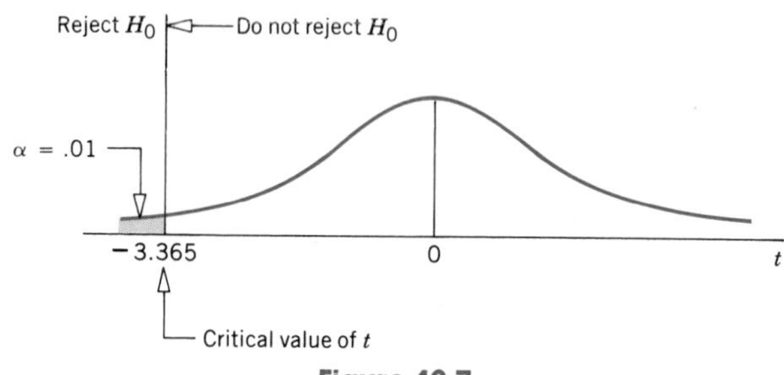

Figure 10.7

Step 4. *Calculate the value of the test statistic*

The value of the test statistic t for $\bar{d}$ is calculated as follows.

$$t = \frac{\bar{d} - \mu_d}{s_{\bar{d}}} = \frac{-4.17 - \overset{\text{From } H_0}{0}}{1.077} = -3.872$$

Step 5. *Make a decision*

Because the value of the test statistic $t = -3.872$ for $\bar{d}$ falls in the rejection region, we reject the null hypothesis. Consequently, we conclude that the mean weekly sales for all salespersons increase as a result of this course. ■

Making a two-tailed test of hypothesis about μ_d: paired samples.

EXAMPLE 10-14 Refer to Example 10-12. The table that gives the blood pressures of seven adults before and after the completion of a special dietary program is reproduced below.

Before	210	180	195	220	231	199	224
After	193	186	186	223	220	183	233

Test at the 5% significance level if the mean of paired differences μ_d is different from zero. Assume that the population of paired differences is (approximately) normally distributed.

Solution Table 10.3 gives the values of d and d^2 for each of the seven adults.

Table 10.3

Before	After	Difference d	d^2
210	193	17	289
180	186	−6	36
195	186	9	81
220	223	−3	9
231	220	11	121
199	183	16	256
224	233	−9	81
		$\Sigma d = 35$	$\Sigma d^2 = 873$

The values of $\bar{d}$ and s_d are calculated as follows.

$$\bar{d} = \frac{\Sigma d}{n} = \frac{35}{7} = 5.00$$

$$s_d = \sqrt{\frac{\Sigma d^2 - \frac{(\Sigma d)^2}{n}}{n - 1}} = \sqrt{\frac{873 - \frac{(35)^2}{7}}{7 - 1}} = 10.786$$

Hence, the standard deviation of $\bar{d}$ is

$$s_{\bar{d}} = \frac{s_d}{\sqrt{n}} = \frac{10.786}{\sqrt{7}} = 4.077$$

Step 1. *State the null and alternative hypotheses*

H_0: $\mu_d = 0$ (mean of the paired differences is not different from zero)

H_1: $\mu_d \neq 0$ (mean of the paired differences is different from zero)

Step 2. *Select the distribution to use*

Because the sample size is small, we use the t distribution to make the test.

Step 3. *Determine the rejection and nonrejection regions*

The $\neq$ sign in the alternative hypothesis indicates that the test is two-tailed. The significance level is .05.

$$\text{Area in each tail} = \frac{\alpha}{2} = \frac{.05}{2} = .025$$

$$\text{Degrees of freedom} = n - 1 = 7 - 1 = 6$$

The two critical values of t for $df = 6$ and .025 area in each tail are −2.447 and 2.447. These values are shown in Figure 10.8.

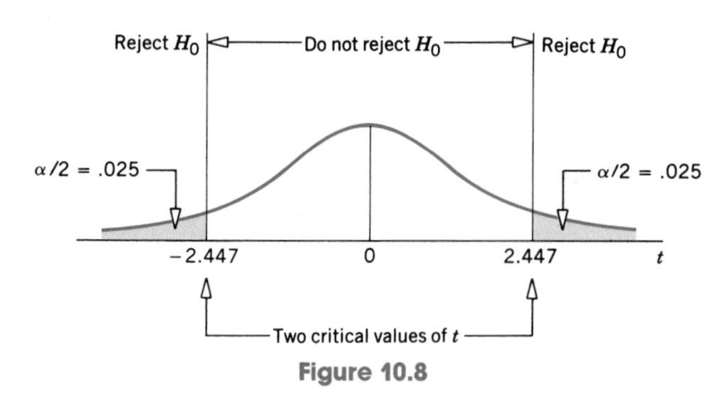

Figure 10.8

Step 4. *Calculate the value of the test statistic*

The value of the test statistic t for $\bar{d}$ is computed as follows.

$$t = \frac{\bar{d} - \mu_d}{s_{\bar{d}}} = \frac{5.00 - 0}{4.077} = 1.226$$

(0 — From H_0)

Step 5. *Make a decision*

Because the value of the test statistic $t = 1.226$ for $\bar{d}$ falls in the nonrejection region, we fail to reject the null hypothesis. Hence, we conclude that the mean of the population paired differences does not appear to be different from zero. ■

EXERCISES

10.46 Find the following confidence intervals for μ_d assuming that the populations of paired differences are normally distributed.

a. $n = 9$, $\bar{d} = 25.4$, $s_d = 13.5$, confidence level = 99%
b. $n = 26$, $\bar{d} = 13.2$, $s_d = 4.8$, confidence level = 95%
c. $n = 12$, $\bar{d} = 34.6$, $s_d = 11.7$, confidence level = 90%

10.47 Find the following confidence intervals for μ_d assuming that the populations of paired differences are normally distributed.

a. $n = 10$, $\bar{d} = 17.5$, $s_d = 6.3$, confidence level = 99%
b. $n = 24$, $\bar{d} = 55.9$, $s_d = 14.7$, confidence level = 95%
c. $n = 14$, $\bar{d} = 29.3$, $s_d = 8.3$, confidence level = 90%

10.48 Make the following tests of hypotheses assuming that the populations of paired differences are normally distributed.

a. H_0: $\mu_d = 0$, H_1: $\mu_d \neq 0$, $n = 9$, $\bar{d} = 6.7$, $s_d = 2.5$, $\alpha = .10$
b. H_0: $\mu_d = 0$, H_1: $\mu_d > 0$, $n = 22$, $\bar{d} = 14.8$, $s_d = 6.4$, $\alpha = .05$
c. H_0: $\mu_d = 0$, H_1: $\mu_d < 0$, $n = 17$, $\bar{d} = -2.3$, $s_d = 4.8$, $\alpha = .01$

10.49 Make the following hypothesis tests assuming that the populations of paired differences are normally distributed.

a. H_0: $\mu_d = 0$, H_1: $\mu_d \neq 0$, $n = 26$, $\bar{d} = 9.6$, $s_d = 3.9$, $\alpha = .05$
b. H_0: $\mu_d = 0$, H_1: $\mu_d > 0$, $n = 15$, $\bar{d} = 2.8$, $s_d = 4.7$, $\alpha = .01$
c. H_0: $\mu_d = 0$, H_1: $\mu_d < 0$, $n = 20$, $\bar{d} = -7.4$, $s_d = 2.3$, $\alpha = .10$

10.50 The following table gives the scores (on a scale of 1 to 15) of seven persons before and after they attended a course about building self-confidence.

Before	8	5	4	9	6	8	5
After	10	7	5	11	6	7	9

a. Construct a 95% confidence interval for the mean μ_d of population paired differences.

b. Test at the 1% significance level if attending this course increases the mean score.

Assume that the population of paired differences has a normal distribution.

10.51 The following table gives the IQ scores for seven pairs of identical twins who were brought up in different environments.

First	110	125	98	112	85	132	102
Second	115	119	107	102	80	114	118

a. Construct a 99% confidence interval for the mean of the differences between the scores of all pairs of twins.

b. Test at the 5% significance level if the mean of the differences between the scores of all pairs of twins is different from zero.

Assume that the population of paired differences has a normal distribution.

10.52 The following table gives the scores of seven students in a statistics course before and after they attended a course on "how to overcome math anxiety."

Before	56	69	48	74	65	71	58
After	62	73	44	85	71	70	69

a. Construct a 95% confidence interval for the mean μ_d of population paired differences.

b. Test at the 2.5% significance level if attending the course in math anxiety increases the average score in statistics.

Assume that the population of paired differences is (approximately) normally distributed.

10.53 The following table gives the scores of eight persons before and after they were given a course on how to increase their writing speed.

Before	81	75	89	91	65	70	90	69
After	97	72	93	110	78	69	115	76

a. Make a 90% confidence interval for the mean μ_d of population paired differences.

b. Test at the 5% significance level if attending the course increases writing speed.

Assume that the population of paired differences is (approximately) normally distributed.

10.54 Six persons were put on a special exercise program for 12 weeks to lose weight. The following table gives the weights (in pounds) of those six persons before and after the program.

Before	180	195	177	221	208	199
After	183	187	161	204	197	189

a. Make a 99% confidence interval for the mean μ_d of population paired differences.

b. Test at the 5% significance level if the mean weight loss for all persons due to this special exercise program is greater than zero.

Assume that the population of all paired differences is (approximately) normally distributed.

10.55 The manufacturer of a gasoline additive claims that the use of this additive increases gasoline mileage. A random sample of six cars was selected, and these cars were driven for 1 week without the gasoline additive and then for 1 week with the gasoline additive. The following table gives the miles per gallon for these cars without and with the gasoline additive.

Without	24.6	28.3	18.9	23.7	15.4	29.5
With	26.3	31.7	18.2	25.3	18.3	30.9

a. Construct a 95% confidence interval for the mean μ_d of population paired differences.

b. Test at the 1% significance level if the use of gasoline additive increases gasoline mileage.

Assume that the population of paired differences is (approximately) normally distributed.

10.5 INFERENCES ABOUT THE DIFFERENCE BETWEEN TWO POPULATION PROPORTIONS FOR LARGE AND INDEPENDENT SAMPLES

Quite often we need to construct a confidence interval and test a hypothesis about the difference between two population proportions. For instance, we may want to estimate the difference between the proportion of defective items produced on two different machines. If p_1 and p_2 are the proportions of defective items produced on the first and second machine, respectively, then we are to make a confidence interval for $p_1 - p_2$. Or we may want to test the hypothesis that the proportion of defective items produced on machine 1 is different from the proportion of defective items produced on machine 2. In this case, we are to test the null hypothesis $p_1 - p_2 = 0$ against the alternative hypothesis $p_1 - p_2 \neq 0$.

This section discusses how to make a confidence interval and test a hypothesis about $p_1 - p_2$ for two large and independent samples. The sample statistic used to make inferences about $p_1 - p_2$ is $\hat{p}_1 - \hat{p}_2$, where $\hat{p}_1$ and $\hat{p}_2$ are the proportions for two large and independent samples. As discussed in Section 7.6 of Chapter 7, we determine a sample proportion by dividing the number of elements in the sample with a given attribute by the sample size. Thus,

$$\hat{p}_1 = \frac{x_1}{n_1} \quad \text{and} \quad \hat{p}_2 = \frac{x_2}{n_2}$$

where x_1 and x_2 are the number of elements that possess a given characteristic in the two samples and n_1 and n_2 are the sizes of two samples, respectively.

10.5.1 THE MEAN, STANDARD DEVIATION, AND SAMPLING DISTRIBUTION OF $\hat{p}_1 - \hat{p}_2$

As discussed in Chapter 7, for a large sample the sample proportion $\hat{p}$ is (approximately) normally distributed with mean p and standard deviation $\sqrt{pq/n}$. Hence, for two large and independent samples of sizes n_1 and n_2, respectively, their sample proportions $\hat{p}_1$ and $\hat{p}_2$ are (approximately) normally distributed with means p_1 and p_2 and standard deviations $\sqrt{p_1q_1/n_1}$ and $\sqrt{p_2q_2/n_2}$, respectively. Using these results, we can make the following statements about the shape of the sampling distribution of $\hat{p}_1 - \hat{p}_2$ and its mean and standard deviation.

SAMPLING DISTRIBUTION, MEAN, AND STANDARD DEVIATION OF $\hat{p}_1 - \hat{p}_2$

For two large and independent samples, the sampling distribution of $\hat{p}_1 - \hat{p}_2$ is (approximately) normal with its mean and standard deviation as

$$\mu_{\hat{p}_1-\hat{p}_2} = p_1 - p_2$$

and

$$\sigma_{\hat{p}_1-\hat{p}_2} = \sqrt{\frac{p_1q_1}{n_1} + \frac{p_2q_2}{n_2}}$$

respectively, where $q_1 = 1 - p_1$ and $q_2 = 1 - p_2$.

Thus, to construct a confidence interval and to test a hypothesis about $p_1 - p_2$ for large and independent samples, we use the normal distribution. As was indicated in Chapter 7, in the case of proportion the sample is large if np and nq are both greater than 5. In the case of two samples, both sample sizes will be large if n_1p_1, n_1q_1, n_2p_2, and n_2q_2 are all greater than 5.

10.5.2 INTERVAL ESTIMATION OF $p_1 - p_2$

The difference between two sample proportions $\hat{p}_1 - \hat{p}_2$ is the point estimator for the difference between two population proportions $p_1 - p_2$. Because we do not know p_1 and p_2 when we are making a confidence interval for $p_1 - p_2$, we use $s_{\hat{p}_1-\hat{p}_2}$ as the point estimator of $\sigma_{\hat{p}_1-\hat{p}_2}$ in the interval estimation. We construct the confidence interval for $p_1 - p_2$ using the following formula.

CONFIDENCE INTERVAL FOR $p_1 - p_2$

The $(1 - \alpha)100\%$ confidence interval for $p_1 - p_2$ is

$$(\hat{p}_1 - \hat{p}_2) \pm z\, s_{\hat{p}_1-\hat{p}_2}$$

where the value of z is read from the normal distribution table for a given confidence level, and $s_{\hat{p}_1-\hat{p}_2}$ is calculated as

$$s_{\hat{p}_1-\hat{p}_2} = \sqrt{\frac{\hat{p}_1\hat{q}_1}{n_1} + \frac{\hat{p}_2\hat{q}_2}{n_2}}$$

Example 10-15 describes the procedure to make a confidence interval for the difference between two population proportions for large samples.

Constructing a confidence interval for $p_1 - p_2$: large samples.

EXAMPLE 10-15 In 1980, 76% of the families in the United States owned homes. In 1988, 72% of the families owned homes (*Psychology Today,* October 1988). Assume that the two estimates are based on sample sizes of 1500 and 1700 families, respectively. Construct a 97% confidence interval for the difference between the proportions of families who owned homes in 1980 and in 1988.

Solution Let p_1 and p_2 be the proportions of all families who owned homes in the respective years 1980 and 1988, and let $\hat{p}_1$ and $\hat{p}_2$ be the respective sample proportions. From the given information,

For 1980: $n_1 = 1500 \quad \hat{p}_1 = .76 \quad \hat{q}_1 = 1 - .76 = .24$

For 1988: $n_2 = 1700 \quad \hat{p}_2 = .72 \quad \hat{q}_2 = 1 - .72 = .28$

The confidence level is $1 - \alpha = .97$.

The standard deviation of $\hat{p}_1 - \hat{p}_2$ is

$$s_{\hat{p}_1 - \hat{p}_2} = \sqrt{\frac{\hat{p}_1\hat{q}_1}{n_1} + \frac{\hat{p}_2\hat{q}_2}{n_2}} = \sqrt{\frac{(.76)(.24)}{1500} + \frac{(.72)(.28)}{1700}} = .015$$

The z value for a 97% confidence level, obtained from the normal table for $.97/2 = .4850$, is 2.17. The 97% confidence interval for $p_1 - p_2$ is

$$(\hat{p}_1 - \hat{p}_2) \pm z\, s_{\hat{p}_1 - \hat{p}_2} = (.76 - .72) \pm 2.17\,(.015) = .04 \pm .03 = .01 \text{ to } .07$$

Thus, with 97% confidence we can state that the difference between the two proportions $p_1 - p_2$ is between .01 and .07. ■

10.5.3 HYPOTHESIS TESTING ABOUT $p_1 - p_2$

This section explains how to test a hypothesis about $p_1 - p_2$ for two large and independent samples. The procedure involves the same five steps that we have used previously. Once again, we calculate the standard deviation of $\hat{p}_1 - \hat{p}_2$ as

$$\sigma_{\hat{p}_1 - \hat{p}_2} = \sqrt{\frac{p_1 q_1}{n_1} + \frac{p_2 q_2}{n_2}}$$

When a hypothesis about $p_1 - p_2$ is tested, usually the null hypothesis is $p_1 = p_2$ and the values of p_1 and p_2 are not known. Assuming that the null hypothesis is true and $p_1 = p_2$, a common value of p_1 and p_2, denoted by $\bar{p}$, is calculated by using one of the following formulas.

$$\bar{p} = \frac{x_1 + x_2}{n_1 + n_2} \quad \text{or} \quad \bar{p} = \frac{n_1\hat{p}_1 + n_2\hat{p}_2}{n_1 + n_2}$$

Which of these formulas is used depends on whether the values of x_1 and x_2 or the values of $\hat{p}_1$ and $\hat{p}_2$ are known. Note that x_1 and x_2 are the number of elements in each of the two samples that possess a certain characteristic. This value of $\bar{p}$ is called the **pooled sample proportion.** Using the value of the pooled sample proportion, we compute an estimate of the standard deviation of $\hat{p}_1 - \hat{p}_2$ as

$$s_{\hat{p}_1 - \hat{p}_2} = \sqrt{\bar{p}\,\bar{q}\left(\frac{1}{n_1} + \frac{1}{n_2}\right)}$$

where $\bar{q} = 1 - \bar{p}$.

TEST STATISTIC z FOR $\hat{p}_1 - \hat{p}_2$

The value of the test statistic z for $\hat{p}_1 - \hat{p}_2$ is calculated as

$$z = \frac{(\hat{p}_1 - \hat{p}_2) - (p_1 - p_2)}{s_{\hat{p}_1 - \hat{p}_2}}$$

The value of $p_1 - p_2$ is substituted from H_0, which is zero.

Examples 10-16 and 10-17 illustrate the procedure to test hypotheses about the difference between two population proportions for large samples.

Making a right-tailed test of hypothesis about $p_1 - p_2$: large samples.

EXAMPLE 10-16 Reconsider Example 10-15 about the percentage of families who owned homes in 1980 and 1988. At the 1% significance level, can we conclude that the proportion of families who owned homes declined during this period?

Solution Let p_1 and p_2 be the proportions of all families who owned homes in 1980 and in 1988, respectively, and let $\hat{p}_1$ and $\hat{p}_2$ be the respective sample proportions. From the given information,

For 1980: $n_1 = 1500$ $\quad \hat{p}_1 = .76$
For 1988: $n_2 = 1700$ $\quad \hat{p}_2 = .72$

Hence

$$\hat{q}_1 = 1 - \hat{p}_1 = 1 - .76 = .24 \quad \text{and} \quad \hat{q}_2 = 1 - \hat{p}_2 = 1 - .72 = .28$$

The significance level is $\alpha = .01$.

Step 1. *State the null and alternative hypotheses*

We are to test for the decline in the proportion of home owners. The proportion of families who owned homes would have declined during this period if p_2 is less than p_1 or p_1 is greater than p_2. This can be written as $p_1 - p_2 > 0$. Thus, the two hypotheses are

$$H_0: p_1 - p_2 = 0 \quad \text{(the proportion has not declined)}$$

$$H_1: p_1 - p_2 > 0 \quad \text{(the proportion has declined)}$$

Step 2. *Select the distribution to use*

Because the samples are large, we apply the normal distribution to make the test. (Recall that, in case of proportion, a sample is large if np and nq are both greater than 5. The reader should check that in this example $n_1\hat{p}_1$, $n_1\hat{q}_1$, $n_2\hat{p}_2$, and $n_2\hat{q}_2$ are all greater than 5.)

Step 3. *Determine the rejection and nonrejection regions*

The > sign in the alternative hypothesis indicates that the test is right-tailed. From the normal distribution table, for .01 significance level, the critical value of z is 2.33. This value is shown in Figure 10.9.

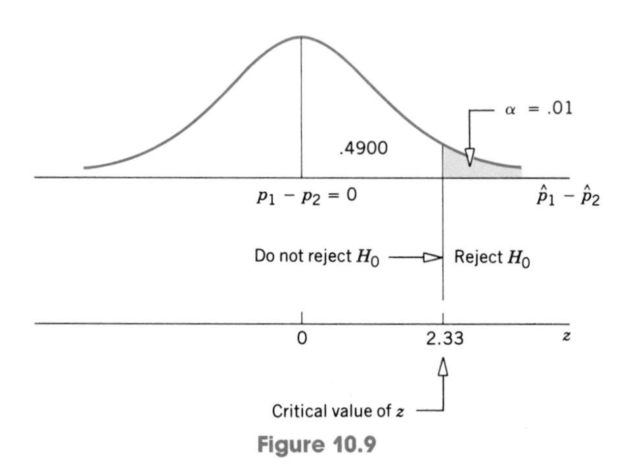

Figure 10.9

Step 4. *Calculate the value of the test statistic*

The pooled sample proportion is

$$\bar{p} = \frac{n_1\hat{p}_1 + n_2\hat{p}_2}{n_1 + n_2} = \frac{1500\ (.76) + 1700\ (.72)}{1500 + 1700} = .739$$

and

$$\bar{q} = 1 - \bar{p} = 1 - .739 = .261$$

Hence, an estimate of the standard deviation of $\hat{p}_1 - \hat{p}_2$ is

$$s_{\hat{p}_1-\hat{p}_2} = \sqrt{\bar{p}\,\bar{q}\left(\frac{1}{n_1} + \frac{1}{n_2}\right)} = \sqrt{(.739)\,(.261)\left(\frac{1}{1500} + \frac{1}{1700}\right)} = .016$$

The value of the test statistic z for $\hat{p}_1 - \hat{p}_2$ is

$$z = \frac{(\hat{p}_1 - \hat{p}_2) - (p_1 - p_2)}{s_{\hat{p}_1-\hat{p}_2}} = \frac{(.76 - .72) - \overset{\text{From } H_0}{0}}{.016} = 2.50$$

Step 5. *Make a decision*

Because the value of the test statistic $z = 2.50$ for $\hat{p}_1 - \hat{p}_2$ falls in the rejection region, we reject the null hypothesis. Therefore, we conclude that the proportion of families who owned homes in 1988 is lower than the proportion of families who owned homes in 1980. ■

Conducting a two-tailed test of hypothesis about $p_1 - p_2$: large samples.

EXAMPLE 10-17 A sample of 800 items produced on machine 1 showed that 48 of them are defective. Another sample of 900 items produced on machine 2 showed that 45 of them are defective. Test at the 1% significance level if the proportions of defective items produced on the two machines are different.

Solution Let p_1 be the proportion of defective items in all items produced on machine 1 and p_2 be the proportion of defective items in all items produced on machine 2. Let $\hat{p}_1$ and $\hat{p}_2$ be the respective sample proportions. Let x_1 and x_2 be the number of defective items in two samples, respectively. From the given information,

Machine 1:	$n_1 = 800$	$x_1 = 48$
Machine 2:	$n_2 = 900$	$x_2 = 45$

The significance level is $\alpha = .01$.

The two sample proportions are calculated as follows.

$$\hat{p}_1 = \frac{x_1}{n_1} = \frac{48}{800} = .06$$

$$\hat{p}_2 = \frac{x_2}{n_2} = \frac{45}{900} = .05$$

Step 1. *State the null and alternative hypotheses*

The null and alternative hypotheses are

$$H_0: p_1 - p_2 = 0 \quad \text{(the proportions are not different)}$$

$$H_1: p_1 - p_2 \neq 0 \quad \text{(the proportions are different)}$$

Step 2. *Select the distribution to use*

Because the samples are large and independent, we apply the normal distribution to make the test. (The reader should check that $n_1\hat{p}_1$, $n_1\hat{q}_1$, $n_2\hat{p}_2$, and $n_2\hat{q}_2$ are all greater than 5.)

Step 3. *Determine the rejection and nonrejection regions*

The $\neq$ sign in the alternative hypothesis indicates that the test is two-tailed. For a 1% significance level, the critical values of z are -2.58 and 2.58. These values are shown in Figure 10.10.

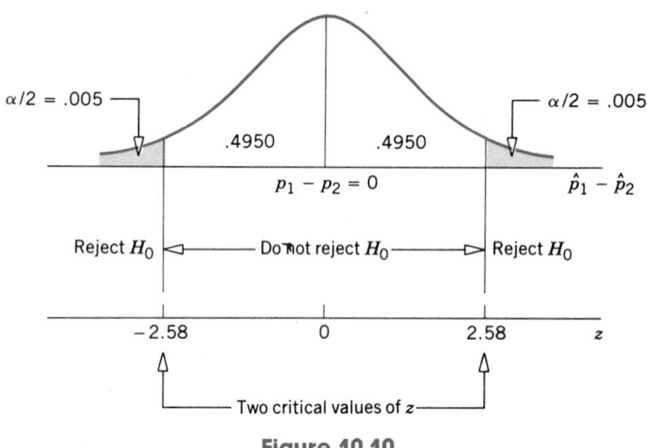

Figure 10.10

Step 4. *Calculate the value of the test statistic*

The pooled sample proportion is

$$\bar{p} = \frac{x_1 + x_2}{n_1 + n_2} = \frac{48 + 45}{800 + 900} = .055$$

and

$$\bar{q} = 1 - \bar{p} = 1 - .055 = .945$$

An estimate of the standard deviation of $\hat{p}_1 - \hat{p}_2$ is

$$s_{\hat{p}_1 - \hat{p}_2} = \sqrt{\bar{p}\,\bar{q}\left(\frac{1}{n_1} + \frac{1}{n_2}\right)} = \sqrt{(.055)\,(.945)\left(\frac{1}{800} + \frac{1}{900}\right)} = .011$$

The value of the test statistic z for $\hat{p}_1 - \hat{p}_2$ is

$$z = \frac{(\hat{p}_1 - \hat{p}_2) - (p_1 - p_2)}{s_{\hat{p}_1 - \hat{p}_2}} = \frac{(.06 - .05) - \overset{\text{From } H_0}{0}}{.011} = .91$$

Step 5. *Make a decision*

The value of the test statistic $z = .91$ for $\hat{p}_1 - \hat{p}_2$ falls in the nonrejection region. Consequently, we fail to reject the null hypothesis. As a result, the proportions of defective items produced by the two machines are not different. ■

CASE STUDY 10-3 MORE ON INFERENCES MADE BY THE CENSUS BUREAU

As was mentioned in Case Study 10-1, the U.S. Census Bureau publishes a large number of reports every year that give the results of various surveys done by the Bureau. The following are two more statements quoted from the Bureau's publications. The figures in parentheses give the 90% confidence intervals of the estimates.

1. Among 20 to 24 year olds, the proportion of males that have not married has risen 23 (±2) percentage points between 1970 and 1987, and is up 25 (±2) points for females.
2. The poverty rate was 12.8 (±0.3) percent in 1989, not significantly different from the 13.0 (±0.4) percent in 1988.

Again, the first statement gives two estimates. First, it gives a 90% confidence interval for the difference between two population proportions. Let p_1 and p_2 be the proportions of 20- to 24-year-old males who were not married in 1987 and 1970, respectively. Then the 90% confidence interval for $p_1 - p_2$ is $.23 \pm .02$ or 21% to 25%. Because both limits of this confidence interval are positive and, hence, p_1 is greater then p_2, the Bureau calls it an increase in the proportion of unmarried 20- to 24-year-old males between 1970 and 1987. Similarly, the second estimate in the first statement indicates that the corresponding 90% confidence interval for the difference between the same two proportions for 20- to 24-year-old females is $.25 \pm .02$ or 23% to 27%.

According to the second statement, at the 90% confidence level, 12.8 ± .3 or 12.5% to 13.1% of all households in the United States were living below the poverty level in 1989. The corresponding confidence interval for 1988 was 13.0 ± .4 or 12.6% to 13.4%.

The second statement also gives the result of a hypothesis test about the difference between two population proportions. Let p_1 and p_2 be the proportions of all households living below the poverty level in 1988 and 1989, respectively. Let $\hat{p}_1$ and $\hat{p}_2$ be the corresponding proportions for the two respective samples. Then, according to the two estimates,

$$\hat{p}_1 = .130 \quad \text{and} \quad \hat{p}_2 = .128$$

The Bureau tested the null hypothesis

$$H_0: p_1 - p_2 = 0$$

against the alternative hypothesis

$$H_1: p_1 - p_2 \neq 0$$

The phrase "not significantly different" in the second statement indicates that the difference between the two proportions was found to be very small (which seems to have occurred because of the sampling error). Hence, the null hypothesis was not rejected.

Source: 1. *Households, Families, Marital Status, and Living Arrangements: March 1987*. Series P-20, 417. U.S. Department of Commerce, Bureau of the Census, August 1987.
2. *Money Income and Poverty Status in the United States: 1989*. Series P-60, 168. U.S. Department of Commerce, Bureau of the Census, September 1990.

EXERCISES

10.56 Construct a 99% confidence interval for $p_1 - p_2$ for the following.

$$n_1 = 300 \quad \hat{p}_1 = .53 \quad n_2 = 200 \quad \hat{p}_2 = .59$$

10.57 Construct a 95% confidence interval for $p_1 - p_2$ for the following.

$$n_1 = 100 \quad \hat{p}_1 = .81 \quad n_2 = 150 \quad \hat{p}_2 = .76$$

10.58 Refer to Exercise 10.56. Test at the 1% significance level if the two population proportions are different.

10.59 Refer to Exercise 10.57. Test at the 5% significance level if $p_1 - p_2$ is different from zero.

10.60 Refer to Exercise 10.56. Test at the 1% significance level if p_1 is less than p_2.

10.61 Refer to Exercise 10.57. Test at the 2% significance level if p_1 is greater than p_2.

10.62 Professors Norval D. Glenn and Charles N. Weaver studied the happiness of married people. In a 1972 sample of 148 married women, 43.7% said that they were very happy with their lives. In a 1986 sample of 78 married women, 34.7% said that they were very happy with their lives (Norval D. Glenn and Charles N. Weaver, "The Changing Relationship of Marital Status to Reported Happiness," *Journal of Marriage and the Family,* 50(2): May 1988). Test at the 1% significance level if the proportions of all married women who were very happy with their lives in 1972 and 1986 are different.

10.63 According to Roper Organization surveys done in 1985 and 1990, 51% of women in 1985 and 42% of women in 1990 said that they would prefer to have a job than stay home (*The 1990 Virginia Slims Opinion Poll,* A Study Conducted by The Roper Organization, Inc.). The sample for each year included 3000 women.

- **a.** Construct a 99% confidence interval for the difference between the proportions of all women for 1985 and 1990 who would prefer to have a job.
- **b.** Test at the 1% significance level if the proportion of all women who would prefer to have a job is higher for 1985 than that for 1990.

10.64 According to the U.S. National Center for Health Statistics, 25.2% of men and 23.6% of women aged 18 and over never eat breakfast. Assume that the two estimates are based on samples of 1200 men and 1000 women aged 18 and over, respectively.

- **a.** Construct a 90% confidence interval for the difference between the proportions of all men and all women who never eat breakfast.
- **b.** At the 5% significance level, is the proportion of all men who never eat breakfast greater than that for all women?

10.65 According to a study done by Professor Adam Drewnowski and others, 21.1% of 507 female college students and 8% of 500 male college students were on a diet at the time of the study (Adam Drewnowski et al.: "The Prevalence of Bulimia Nervosa in the US College Student Population," *American Journal of Public Health,* 78(10): October 1988).

- **a.** Construct a 99% confidence interval for the difference between the proportions of all female and all male college students who were on a diet at the time of this study.
- **b.** Test at the 1% significance level if the proportion of all female college students who were on a diet at the time of this study is greater than the proportion of all male college students who were on a diet at that time.

10.66 According to the U.S. National Center for Health Statistics, 33% of men and 37% of women aged 18 and over are light drinkers. Assume that the two estimates are based on samples of 800 men and 700 women aged 18 and over, respectively.

- **a.** Construct a 97% confidence interval for the difference between the proportions of all men and all women who are light drinkers.
- **b.** Test at the 2% significance level if the proportion of all men who are light drinkers is lower than that of all women.

10.67 According to Roper Organization surveys, 35% of the people polled in 1982 and 28% of the people polled in 1988 said they were in excellent health (*Roper Reports,* January 1988). Assume that the two estimates are based on two different random samples of 1300 and 1500 people, respectively.

- **a.** Construct a 95% confidence interval for the difference between the proportions of all people who were in excellent health in 1982 and in 1988.
- **b.** At the 1% significance level, has the proportion of people who think they are in excellent health declined during the period 1982 to 1988?

10.68 Because of the emergence of cable TV, the percentage of viewers watching news on (any of) the three networks (ABC, CBS, and NBC) may be affected. According to A. C. Nielsen estimates, 72% of viewers watched network news in 1980–81. This percentage was 66% for 1985–86 (*Time,* March 16, 1987). Assume that these estimates are based on random samples of 1200 viewers for 1980–81 and 1400 viewers for 1985–86.

a. Make a 95% confidence interval for the difference between the proportions of all viewers who watched network news in 1980–81 and in 1985–86.

b. Test at the 5% significance level if the proportion of viewers who watched network news in 1985–86 is lower than that for 1980–81.

10.69 According to the U.S. Bureau of the Census, 66% of households headed by single women and 81.9% of households headed by single men own cars. Assume that these estimates are based on random samples of 1640 households headed by single women and 1800 households headed by single men.

a. Determine a 98% confidence interval for the difference between the two population proportions.

b. At the 1% significance level, can you conclude that the proportion of households headed by single women who own cars is less than the proportion of households headed by single men who own cars?

10.70 According to the U.S. Bureau of the Census, 48.8% of households headed by single women and 41.5% of households headed by single men own houses. Assume that these estimates are based on random samples of 1640 households headed by single women and 1800 households headed by single men.

a. Construct a 99% confidence interval for the difference between the two population proportions.

b. Test at the 5% significance level if the proportion of households headed by single women who own houses is higher than the proportion of households headed by single men who own houses.

GLOSSARY

d The difference between two matched values in two samples collected from the same source. It is called the paired difference.

$\bar{d}$ The mean of the paired differences for a sample.

Paired or matched samples Two samples drawn in such a way that they include the same elements and there are two data values obtained from each element, one for each sample. Also called **dependent samples.**

Independent samples Two samples drawn from two populations such that the selection of one does not affect the selection of the other.

μ_d The mean of the paired differences for the population.

s_d The standard deviation of the paired differences for a sample.

σ_d The standard deviation of the paired differences for the population.

KEY FORMULAS

1. Mean of the sampling distribution of $\bar{x}_1 - \bar{x}_2$

$$\mu_{\bar{x}_1-\bar{x}_2} = \mu_1 - \mu_2$$

2. Standard deviation of $\bar{x}_1 - \bar{x}_2$

$$\sigma_{\bar{x}_1-\bar{x}_2} = \sqrt{\frac{\sigma_1^2}{n_1} + \frac{\sigma_2^2}{n_2}}$$

3. **The $(1 - \alpha)100\%$ confidence interval for $\mu_1 - \mu_2$ for large and independent samples**

$$(\bar{x}_1 - \bar{x}_2) \pm z\, \sigma_{\bar{x}_1 - \bar{x}_2}$$

If σ_1 and σ_2 are not known, then $\sigma_{\bar{x}_1 - \bar{x}_2}$ is replaced by its point estimator $s_{\bar{x}_1 - \bar{x}_2}$, which is calculated as

$$s_{\bar{x}_1 - \bar{x}_2} = \sqrt{\frac{s_1^2}{n_1} + \frac{s_2^2}{n_2}}$$

4. **Value of the test statistic z for $\bar{x}_1 - \bar{x}_2$ for large and independent samples**

$$z = \frac{(\bar{x}_1 - \bar{x}_2) - (\mu_1 - \mu_2)}{\sigma_{\bar{x}_1 - \bar{x}_2}}$$

If σ_1 and σ_2 are not known, then $\sigma_{\bar{x}_1 - \bar{x}_2}$ is replaced by $s_{\bar{x}_1 - \bar{x}_2}$.

5. **Pooled standard deviation for two small and independent samples taken from two populations with equal standard deviations**

$$s_p = \sqrt{\frac{(n_1 - 1)s_1^2 + (n_2 - 1)s_2^2}{n_1 + n_2 - 2}}$$

6. **Estimate of the standard deviation of $\bar{x}_1 - \bar{x}_2$ for two small and independent samples taken from two populations with equal standard deviations**

$$s_{\bar{x}_1 - \bar{x}_2} = s_p \sqrt{\frac{1}{n_1} + \frac{1}{n_2}}$$

7. **The $(1 - \alpha)100\%$ confidence interval for $\mu_1 - \mu_2$ for small and independent samples taken from two populations with equal standard deviations**

$$(\bar{x}_1 - \bar{x}_2) \pm t\, s_{\bar{x}_1 - \bar{x}_2}$$

8. **Value of the test statistic t for $\bar{x}_1 - \bar{x}_2$ for small and independent samples taken from two populations with equal standard deviations**

$$t = \frac{(\bar{x}_1 - \bar{x}_2) - (\mu_1 - \mu_2)}{s_{\bar{x}_1 - \bar{x}_2}}$$

9. **Degrees of freedom to make inferences about $\mu_1 - \mu_2$ for small and independent samples taken from two populations with unequal standard deviations**

$$df = \frac{\left(\frac{s_1^2}{n_1} + \frac{s_2^2}{n_2}\right)^2}{\frac{\left(\frac{s_1^2}{n_1}\right)^2}{n_1 - 1} + \frac{\left(\frac{s_2^2}{n_2}\right)^2}{n_2 - 1}}$$

10. Estimate of the standard deviation of $\bar{x}_1 - \bar{x}_2$ for two small and independent samples taken from two populations with unequal standard deviations

$$s_{\bar{x}_1-\bar{x}_2} = \sqrt{\frac{s_1^2}{n_1} + \frac{s_2^2}{n_2}}$$

11. The $(1 - \alpha)100\%$ confidence interval for $\mu_1 - \mu_2$ for small and independent samples taken from two populations with unequal standard deviations

$$(\bar{x}_1 - \bar{x}_2) \pm t\, s_{\bar{x}_1-\bar{x}_2}$$

12. Value of the test statistic t for $\bar{x}_1 - \bar{x}_2$ for small and independent samples taken from two populations with unequal standard deviations

$$t = \frac{(\bar{x}_1 - \bar{x}_2) - (\mu_1 - \mu_2)}{s_{\bar{x}_1-\bar{x}_2}}$$

13. Sample mean for paired differences

$$\bar{d} = \frac{\Sigma d}{n}$$

14. Sample standard deviation for paired differences

$$s_d = \sqrt{\frac{\Sigma d^2 - \frac{(\Sigma d)^2}{n}}{n - 1}}$$

15. Mean and standard deviation of the sampling distribution of $\bar{d}$

$$\mu_{\bar{d}} = \mu_d$$

and

$$s_{\bar{d}} = \frac{s_d}{\sqrt{n}}$$

16. The $(1 - \alpha)100\%$ confidence interval for μ_d

$$\bar{d} \pm t\, s_{\bar{d}}$$

17. Value of the test statistic t for $\bar{d}$

$$t = \frac{\bar{d} - \mu_d}{s_{\bar{d}}}$$

18. Mean of the sampling distribution of $\hat{p}_1 - \hat{p}_2$

$$\mu_{\hat{p}_1-\hat{p}_2} = p_1 - p_2$$

19. Estimate of the standard deviation of $\hat{p}_1 - \hat{p}_2$

$$s_{\hat{p}_1-\hat{p}_2} = \sqrt{\frac{\hat{p}_1\hat{q}_1}{n_1} + \frac{\hat{p}_2\hat{q}_2}{n_2}}$$

20. The $(1 - \alpha)100\%$ confidence interval for $p_1 - p_2$

$$(\hat{p}_1 - \hat{p}_2) \pm z\, s_{\hat{p}_1-\hat{p}_2}$$

21. Pooled sample proportion for two samples

$$\bar{p} = \frac{x_1 + x_2}{n_1 + n_2} \quad \text{or} \quad \frac{n_1\hat{p}_1 + n_2\hat{p}_2}{n_1 + n_2}$$

22. Estimate of the standard deviation of $\hat{p}_1 - \hat{p}_2$ using the pooled sample proportion

$$s_{\hat{p}_1-\hat{p}_2} = \sqrt{\bar{p}\,\bar{q}\left(\frac{1}{n_1} + \frac{1}{n_2}\right)}$$

23. Value of the test statistic z for $\hat{p}_1 - \hat{p}_2$ for large and independent samples

$$z = \frac{(\hat{p}_1 - \hat{p}_2) - (p_1 - p_2)}{s_{\hat{p}_1-\hat{p}_2}}$$

SUPPLEMENTARY EXERCISES

10.71 Construct a 99% confidence interval for $\mu_1 - \mu_2$ for the following.

$$n_1 = 80 \quad \bar{x}_1 = 12.35 \quad s_1 = 2.68$$
$$n_2 = 65 \quad \bar{x}_2 = 16.40 \quad s_2 = 2.90$$

10.72 Refer to Exercise 10.71. Test at the 1% significance level if the two population means are different.

10.73 Refer to Exercise 10.71. Test at the 5% significance level if μ_1 is less than μ_2.

10.74 Assuming that the two populations are normally distributed with equal standard deviations, construct a 90% confidence interval for $\mu_1 - \mu_2$ for the following.

$$n_1 = 18 \quad \bar{x}_1 = 34.40 \quad s_1 = 6.7$$
$$n_2 = 22 \quad \bar{x}_2 = 26.50 \quad s_2 = 7.1$$

10.75 Refer to Exercise 10.74. Test at the 5% significance level if the two population means are different.

10.76 Refer to Exercise 10.74. Test at the 5% significance level if μ_1 is greater than μ_2.

10.77 Determine the following confidence intervals for μ_d assuming that the population of paired differences has a normal distribution.

a. $n = 8,\quad \bar{d} = 33.7,\quad s_d = 14.2,\quad$ confidence level = 99%
b. $n = 18,\quad \bar{d} = 9.6,\quad s_d = 2.3,\quad$ confidence level = 95%
c. $n = 26,\quad \bar{d} = 16.2,\quad s_d = 4.8,\quad$ confidence level = 90%

10.78 Perform the following tests of hypotheses assuming that the population of paired differences has a normal distribution.

a. $H_0\colon \mu_d = 0,\quad H_1\colon \mu_d \neq 0,\quad n = 7,\quad \bar{d} = 4.9,\quad s_d = 1.3,\quad \alpha = .01$
b. $H_0\colon \mu_d = 0,\quad H_1\colon \mu_d > 0,\quad n = 24,\quad \bar{d} = 11.3,\quad s_d = 5.1,\quad \alpha = .10$
c. $H_0\colon \mu_d = 0,\quad H_1\colon \mu_d < 0,\quad n = 16,\quad \bar{d} = -14.2,\quad s_d = 7.4,\quad \alpha = .05$

10.79 Construct a 99% confidence interval for $p_1 - p_2$ for the following.

$$n_1 = 250 \qquad \hat{p}_1 = .37 \qquad n_2 = 340 \qquad \hat{p}_2 = .31$$

10.80 Refer to Exercise 10.79. Test at the 1% significance level if the two population proportions are different.

10.81 Refer to Exercise 10.79. Test at the 2% significance level if p_1 is greater than p_2.

10.82 A mathematics proficiency test was given to 905 randomly selected 13-year-old American students. The following table gives the mean scores of male and female students along with the standard deviations of the sample means (*A World of Differences: An International Assessment of Mathematics and Science,* Educational Testing Service, January 1989).

Male students	$\bar{x}_1 = 474.6$	$s_{\bar{x}_1} = 6.4$
Female students	$\bar{x}_2 = 473.2$	$s_{\bar{x}_2} = 5.1$

a. Construct a 99% confidence interval for the difference between the two population means.
b. Test at the 1% significance level if the mean scores in mathematics proficiency test are different for all male and all female 13-year-old students.

(*Note:* The standard deviation of $\bar{x}_1 - \bar{x}_2$ will be calculated using the formula given in the footnote on page 475 except that we substitute s for σ.)

10.83 According to the Bureau of Labor Statistics, the mean hourly wage for mine workers was $13.14 and that for transportation and public utility workers was $12.57 in 1989. Assume that these mean hourly wages are based on samples of 1100 mine workers and 1400 transportation and public utility workers. Further assume that the standard deviations of the two samples are $1.45 and $1.68, respectively.

a. Construct a 98% confidence interval for the difference between the two population means.
b. Test at the 1% significance level if the 1989 mean hourly wage for all mine workers was higher than that for all transportation and public utility workers.

10.84 According to a study, the mean score in the verbal portion of the GRE (Graduate Record Examination) test given to a sample of 20,499 nonhandicapped students was 497 and the standard deviation was 115. Another sample of 108 physically handicapped students, who were given the same test, had a mean score of 493 with a standard deviation of 117 (Donald A. Rock et al.: "Factor Structure of the Graduate Record Examinations General Test in Handicapped and Nonhandicapped Groups," *Journal of Applied Psychology*, 73(3): August 1988).

- **a.** Construct a 99% confidence interval for the difference between the two population means.
- **b.** Test at the 1% significance level if the mean scores in this test are different for nonhandicapped and handicapped students.

10.85 According to the U.S. Department of Labor, the average weekly pay of teachers was $470 and that of registered nurses was $482 in 1987–88. Suppose these averages are calculated based on the random samples of 1300 teachers and 1500 registered nurses and that the standard deviations for these two samples are $30 and $35, respectively.

- **a.** Construct a 95% confidence interval for the difference between the two population means.
- **b.** Test at the 1% significance level if the mean weekly pay of all teachers is less than that of all registered nurses.

10.86 A researcher wants to test if the mean GPAs (grade-point averages) of all male and all female college students who actively participate in sports are different. She took a random sample of 28 male college students and 24 female college students who were actively involved in sports. She found the mean GPAs of the two groups to be 2.62 and 2.74, respectively, with the corresponding standard deviations equal to .43 and .38.

- **a.** Test at the 5% significance level if the mean GPAs of the two populations are different.
- **b.** Construct a 90% confidence interval for the difference between the two population means.

Assume that the GPAs of all male and all female college student players both have a normal distribution with the same standard deviation.

10.87 According to the National Agricultural Statistics Service, the average yield of corn per acre was 128 bushels for Ohio and 135 bushels for Iowa in 1986. Assume that these two results are based on random samples of 25 acres from Ohio and 28 acres from Iowa. Further assume that the sample standard deviations for the two states are 6 and 7 bushels, respectively.

- **a.** Construct a 95% confidence interval for the difference between the two population means.
- **b.** Test at the 5% significance level if the mean yield of potatoes per acre for Ohio is lower than that for Iowa.

Assume that the per acre yields of corn for Ohio and Iowa both have a normal distribution with the same standard deviation.

10.88 A random sample of 25 drivers insured by an insurance company called Company A showed that they paid on average a monthly insurance premium of $83 with a standard deviation of $14. Another random sample of 20 drivers insured by another insurance company called Company B showed that these drivers paid on average a monthly insurance premium of $76 with a standard deviation of $12. Assume that the insurance premiums paid by all drivers insured by companies A and B both have a normal distribution with equal standard deviations.

a. Construct a 90% confidence interval for the difference between the two population means.

b. Test at the 1% significance level if the mean monthly insurance premium paid by drivers insured by company A is higher than that of drivers insured by company B.

10.89 A random sample of 28 children taken from families with only one child gave a mean tolerance level of 2.4 (on a scale of 1 to 8) with a standard deviation of .62. Another random sample of 25 children taken from families with more than one child gave a mean tolerance level of 3.5 with a standard deviation of .47.

a. Construct a 99% confidence interval for the difference between the two population means.

b. Test at the 5% significance level if the mean tolerance level for children from families with only one child is lower than that for the children from families with more than one child.

Assume that the tolerance levels for children in both groups have a normal distribution with the same standard deviation.

10.90 Repeat Exercise 10.86, but now assume that the GPAs of all male and all female student players are both normally distributed with unequal standard deviations.

10.91 Repeat Exercise 10.87, but now assume that the per acre yields of corn for Ohio and Iowa both have a normal distribution with unequal standard deviations.

10.92 Repeat Exercise 10.88, but now assume that the insurance premiums paid by all drivers insured by companies A and B are both normally distributed with unequal standard deviations.

10.93 Repeat Exercise 10.89, but now assume that the tolerance levels for all children in both groups are normally distributed with unequal standard deviations.

10.94 A random sample of eight students was selected to test for the effectiveness of hypnosis on their academic performance. The following table gives the GPAs for the semester before and the semester after the students tried hypnosis.

Before	2.3	2.8	3.1	2.7	3.4	2.6	2.8	2.5
After	2.6	3.2	3.0	3.5	3.7	2.4	2.9	2.9

a. Construct a 99% confidence interval for the mean μ_d of population paired differences.

b. Test at the 5% significance level if there is an improvement in the academic performance of students due to hypnotism.

Assume that the population of paired differences is (approximately) normally distributed.

10.95 A random sample of nine students was selected to test for the effectiveness of a special course designed to improve memory. The following table gives the results of a memory test given to these students before and after this course.

Before	43	57	48	65	71	49	38	69	58
After	49	56	55	77	79	57	36	64	69

a. Construct a 95% confidence interval for the mean μ_d of population paired differences.
b. Test at the 1% significance level if this course makes any statistically significant improvement in the memory of all students.

Assume that the population of paired differences has a normal distribution.

10.96 In a *USA Today* poll of 1010 adults conducted by R. H. Bruskin Associates in 1989, 74.2% of men and 88.8% of women said that they are concerned about living near a nuclear power plant. Assume that there were 520 men and 490 women in this sample.

a. Construct a 99% confidence interval for the difference between the proportions of all men and all women who are concerned about living near a nuclear power plant.
b. Test at the 1% significance level if the proportion of all men who are concerned about living near a nuclear power plant is lower than that of all women.

10.97 According to a Gallup poll, 58% of adults in 1988 and 45% in 1983 said they would like their child to "take up teaching in the public schools as a career" ("The 20th Annual Gallup Poll of the Public's Attitude Towards the Public Schools," *Phi Delta Kappa,* September 1988). The sample size for the 1988 poll was 2118. Assume that the sample size for the 1983 poll was 1940.

a. Construct a 97% confidence interval for the difference between the population proportions for 1988 and 1983.
b. Test at the 2% significance level if the population proportion of adults who would like their child to take up teaching in the public schools as a career is higher for 1988 than for 1983.

10.98 According to a 1990 survey of CEOs (chief executive officers) of major corporations conducted by Korn/Ferry International and UCLA's Anderson Graduate School of Management, 48% "would choose the same career if they were starting over" again. In a similar survey conducted 10 years ago, 60% of CEOs said that they would choose the same career if they were starting over again (*The Wall Street Journal,* July 3, 1990). Assume that the 1990 survey is based on a sample of 800 CEOs and the one done 10 years ago included 600 CEOs.

a. Construct a 95% confidence interval for the difference between the two population proportions.
b. Test at the 5% significance level if the proportion of all CEOs who would choose the same career if they were starting over again has changed during the past 10 years.

SELF-REVIEW TEST

1. To test the hypothesis that university professors have mean blood pressure lower than that of company executives, which of the following will you use?

a. A left-tailed test **b.** A two-tailed test **c.** A right-tailed test.

2. Briefly explain the meaning of independent and dependent samples. Give one example of each of these cases.

3. A psychologist wanted to test if company executives have job-related stress scores higher than those of university professors. He took a sample of 40 executives and 50 professors and tested them for job-related stress. The sample of 40 executives gave a mean stress score of 7.6 with a standard deviation of .8. The sample of 50 professors produced a mean stress score of 5.4 with a standard deviation of 1.3.

- **a.** Construct a 99% confidence interval for the difference between the mean stress scores of all executives and all professors.
- **b.** Test at the 5% significance level if the mean stress score of all executives is higher than that of all professors.

4. A sample of 20 alcoholic fathers showed that they spend an average of 2.3 hours a week playing with their children with a standard deviation of .54 hours. A sample of 25 nonalcoholic fathers gave a mean of 4.6 hours a week with a standard deviation of .8 hours.

- **a.** Construct a 95% confidence interval for the difference between the mean time spent per week playing with their children by all alcoholic and all nonalcoholic fathers.
- **b.** Test at the 1% significance level if the mean time spent per week playing with their children by all alcoholic fathers is less than that of all nonalcoholic fathers.

Assume that the mean time spent per week playing with their children by all alcoholic and all nonalcoholic fathers are both normally distributed with equal but unknown standard deviations.

5. Repeat Problem 4 but now assume that the mean time spent per week playing with their children by all alcoholic and all nonalcoholic fathers are both normally distributed with unequal and unknown standard deviations.

6. The following table gives the number of items made in 1 hour by seven randomly selected workers on two different machines.

Worker	**1**	**2**	**3**	**4**	**5**	**6**	**7**
Machine I	15	18	14	20	16	18	21
Machine II	16	20	13	23	19	18	20

Let μ_d be the mean of (paired) differences between the number of items made in 1 hour on these two machines by all workers.

- **a.** Construct a 99% confidence interval for the mean μ_d of population paired differences.
- **b.** Test at the 5% significance level if the mean μ_d of population paired differences is different from zero.

Assume that the population of paired differences is (approximately) normally distributed.

7. A sample of 500 registered male voters showed that 57% of them voted in the last presidential election. Another sample of 400 registered female voters showed that 55% of them voted in the same election.

   a. Construct a 97% confidence interval for the difference between the proportion of all male and all female registered voters who voted in the last presidential election.

   b. Test at the 1% significance level if the proportion of all male voters who voted in the last presidential election is different from that of all female voters.

USING MINITAB

INFERENCES ABOUT THE DIFFERENCE BETWEEN TWO POPULATION MEANS FOR LARGE AND INDEPENDENT SAMPLES

MINITAB does not have a direct set of commands that can be used to make a confidence interval and test a hypothesis about the difference between two population means for large and independent samples using the normal distribution. The simplest way to make such inferences with MINITAB is to use the *t* distribution irrespective of the sample sizes. This procedure is explained in the following section.

INFERENCES ABOUT THE DIFFERENCE BETWEEN TWO POPULATION MEANS FOR SMALL AND INDEPENDENT SAMPLES: EQUAL POPULATION STANDARD DEVIATIONS

The first step in making a confidence interval and testing a hypothesis about the difference between two population means for small and independent samples is to enter the data for two samples in columns C1 and C2 using SET C1 and SET C2 commands. Note that if the sample sizes are the same, we can use the READ command. After the data are entered, the following MINITAB command and subcommand will give a confidence interval for $\mu_1 - \mu_2$ assuming that the standard deviations of the two populations are equal.

```
MTB  > TWOSAMPLE    99% CONFIDENCE INTERVAL    C1  C2;
SUBC > POOLED.
       └─────────┘
            ↑
       This subcommand instructs MINITAB that σ1 = σ2
```

The first command instructs MINITAB that we are to make a 99% confidence interval for the difference between the two population means using the data on two samples entered in columns C1 and C2. The subcommand POOLED instructs MINITAB that the standard deviations of the two populations are equal. MINITAB will not only give a 99% confidence interval for $\mu_1 - \mu_2$, but it will also give the results for a test of hypothesis (using the *t* distribution) for H_0: $\mu_1 - \mu_2 = 0$ against H_1: $\mu_1 - \mu_2 \neq 0$. This test of hypothesis is based on the assumption that the standard deviations of two populations are equal. Illustration M10-1 describes how to use these commands.

Illustration M10-1 A random sample of 16 men, who were driving on a highway, produced the following data on the speeds of their cars at the time of the survey.

70	67	65	72	71	54	74	69
63	57	64	76	60	55	63	69

Another random sample of 14 women, who were driving on the same highway, produced the following data on the speeds of their cars at the time of the survey.

61	55	58	66	70	54	57
60	63	72	65	63	59	67

Construct a 99% confidence interval for the difference between the mean speeds of cars driven by all men and all women drivers on this highway. Assume that the speeds at which all men and all women drive cars on this highway are both normally distributed with equal but unknown population standard deviations.

Solution Let μ_1 and μ_2 be the mean speeds of cars driven by all men and all women on this highway and let $\bar{x}_1$ and $\bar{x}_2$ be the means of the respective samples. The MINITAB commands and the MINITAB solution are as follows.

```
MTB  > NOTE: CONFIDENCE INTERVAL FOR ILLUSTRATION M10-1
MTB  > SET C1
DATA > 70 67 65 72 71 54 74 69
DATA > 63 57 64 76 60 55 63 69
DATA > END
MTB  > SET C2
DATA > 61 55 58 66 70 54 57 60 63 72 65 63 59 67
DATA > END
MTB  > TWOSAMPLE    99% CONFIDENCE INTERVAL    C1  C2;
SUBC > POOLED.

TWOSAMPLE    T    FOR    C1 VS C2
         N     MEAN     STDEV     SE MEAN
C1      16    65.56      6.64         1.7
C2      14    62.14      5.43         1.5

99 PCT   CI   FOR   MU C1 - MU C2:  (-2.8, 9.6)

TTEST   MU C1 = MU C2   (VS NE):  T = 1.53  P = 0.14  DF = 28

POOLED STDEV =      6.11
```

99% confidence interval for $\mu_1 - \mu_2$

In the MINITAB solution, the row of C1 gives

$$n_1 = 16, \quad \bar{x}_1 = 65.56, \quad s_1 = 6.64, \quad \text{and} \quad s_{\bar{x}_1} = 1.7$$

The row of C2 gives

$$n_2 = 14, \quad \bar{x}_2 = 62.14, \quad s_2 = 5.43, \quad \text{and} \quad s_{\bar{x}_2} = 1.5$$

The following portion of the MINITAB solution gives the 99% confidence interval for $\mu_1 - \mu_2$.

99 PCT CI FOR MU C1 − MU C2: (−2.8, 9.6)

From this, the 99% confidence interval for $\mu_1 - \mu_2$ is −2.8 to 9.6.

The second line from the bottom in the MINITAB solution gives the test of hypothesis (using the t distribution) for H_0: $\mu_1 = \mu_2$ against H_1: $\mu_1 \neq \mu_2$ assuming that the standard deviations of the two populations are equal. The last line gives the value of the pooled standard deviation s_p, which is 6.11.

The following MINITAB commands are used to make a test of hypothesis about $\mu_1 = \mu_2$ for small and independent samples assuming that the two samples are taken from two populations that are normally distributed with equal (but unknown) standard deviations.

```
MTB  > TWOSAMPLE     T    C1  C2;
SUBC > POOLED;
SUBC > ALTERNATIVE = k.
```

The first command, TWOSAMPLE T C1 C2, instructs MINITAB that we are to make a test of hypothesis about the difference between the two population means using the t distribution for the data from two samples entered in columns C1 and C2. The subcommand POOLED indicates that the test is to be made using the pooled standard deviation (see Section 10.2) assuming that the standard deviations of the two populations are equal. The subcommand ALTERNATIVE = k gives the alternative hypothesis. The value of k will be −1, 0, or 1, depending on whether the alternative hypothesis is $\mu_1 - \mu_2 < 0$, $\mu_1 - \mu_2 = 0$, or $\mu_1 - \mu_2 > 0$.

Illustration M10-2 shows how to make a test of hypothesis about $\mu_1 - \mu_2$ for small and independent samples when the population standard deviations are unknown but equal.

Illustration M10-2 Refer to Illustration M10-1. Test at the 1% significance level if the mean speed of cars driven by all men drivers on this highway is greater than that of cars driven by all women drivers.

Solution The null and alternative hypotheses are

$$H_0: \mu_1 = \mu_2 \quad \text{or} \quad \mu_1 - \mu_2 = 0$$

$$H_1: \mu_1 > \mu_2 \quad \text{or} \quad \mu_1 - \mu_2 > 0$$

Because we already entered the data from two samples in columns C1 and C2 in Illustration M10-1, we do not have to repeat that step. The MINITAB commands and the MINITAB solution for the test of hypothesis appear below. As the test is right-tailed, the value of k will be 1 in the subcommand ALTERNATIVE = k.

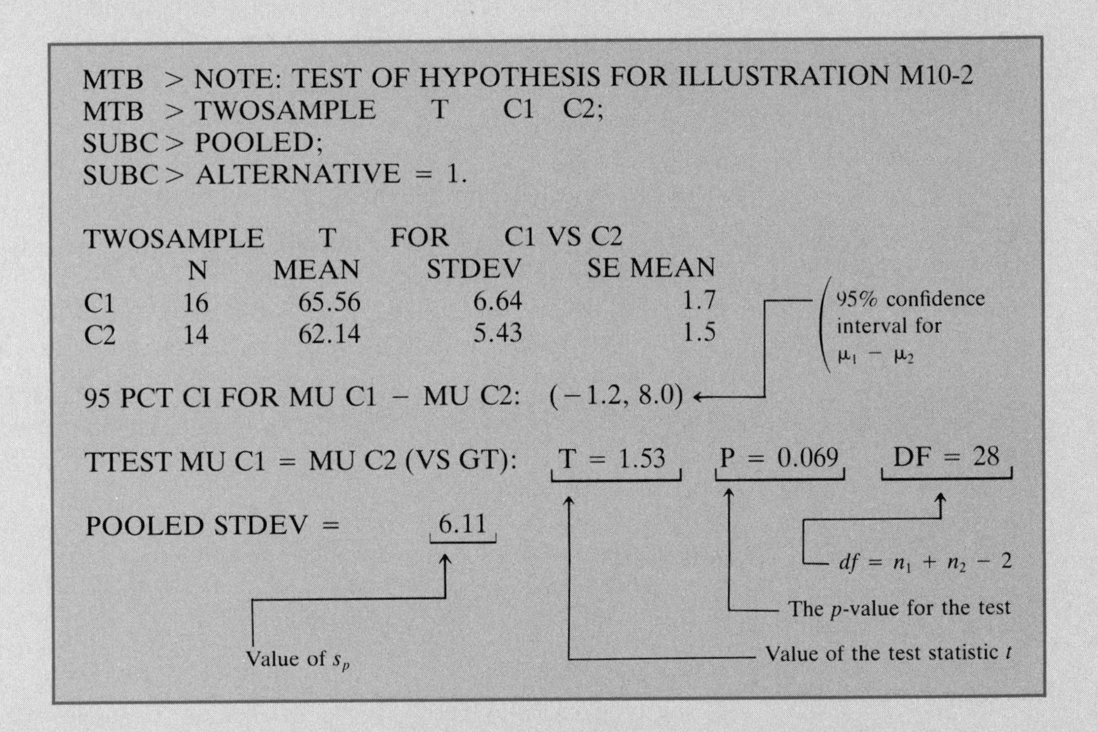

```
MTB  > NOTE: TEST OF HYPOTHESIS FOR ILLUSTRATION M10-2
MTB  > TWOSAMPLE    T    C1  C2;
SUBC > POOLED;
SUBC > ALTERNATIVE = 1.

TWOSAMPLE    T    FOR    C1 VS C2
        N      MEAN      STDEV      SE MEAN
C1     16     65.56       6.64          1.7
C2     14     62.14       5.43          1.5

95 PCT CI FOR MU C1 - MU C2:  (-1.2, 8.0)

TTEST MU C1 = MU C2 (VS GT):  T = 1.53   P = 0.069   DF = 28

POOLED STDEV =       6.11
```


From the MINITAB output, the pooled standard deviation is

$$s_p = 6.11$$

The value of the test statistic t for $\bar{x}_1 - \bar{x}_2$ is

$$t = 1.53$$

The test is right-tailed. The significance level is given to be 1%. Hence,

$$\alpha = .01 \quad \text{and} \quad df = n_1 + n_2 - 2 = 16 + 14 - 2 = 28$$

From the t distribution table, the critical value of t for $df = 28$ and .01 area in the right tail is 2.467. Because the value of the test statistic $t = 1.53$ is less than the critical value of $t = 2.467$, it falls in the nonrejection region. (The reader should draw a graph with the rejection and nonrejection regions.) Consequently, we fail to reject the null hypothesis.

We can reach the same conclusion using the p-value printed in the MINITAB solution. From the MINITAB solution, the p-value is .069. Because the value of $\alpha = .01$ is less than the p-value of .069, we do not reject the null hypothesis.

Note that this MINITAB solution also gives a 95% confidence interval for $\mu_1 - \mu_2$, which is -1.2 to 8.0.

INFERENCES ABOUT THE DIFFERENCE BETWEEN TWO POPULATION MEANS FOR SMALL AND INDEPENDENT SAMPLES: UNEQUAL POPULATION STANDARD DEVIATIONS

The procedure to make a confidence interval and test a hypothesis about $\mu_1 - \mu_2$ for small and independent samples when the standard deviations for the two populations are unknown and unequal is similar to the one just used for the case of equai population standard deviations with one exception that we do not mention POOLED in the MINITAB commands. Illustration M10-3 describes this procedure.

Illustration M10-3 Reconsider the data on the speeds of cars of 16 men and 14 women drivers driving on a highway given in Illustration M10-1.

(a) Construct a 99% confidence interval for the difference between the mean speeds of cars driven by all men and all women drivers on this highway.

(b) Test at the 1% significance level if the mean speed of cars driven by all men drivers on this highway is greater than that of cars driven by all women drivers.

Assume that the speeds at which all men and all women drivers drive cars on this highway are both normally distributed with unequal and unknown population standard deviations.

Solution

(a) Because we entered the given data in columns C1 and C2 in Illustration M10-1, we will skip the data entry commands in this illustration. The MINITAB commands and the MINITAB solution for a 99% confidence interval for $\mu_1 - \mu_2$ are presented below. Note that the only difference between these MINITAB commands and the ones in Illustration M10-1 is that we have not entered the subcommand POOLED. In the absence of this command, MINITAB assumes that the standard deviations for the two populations are not equal.

```
MTB > NOTE: SOLUTION FOR ILLUSTRATION M10-3 PART (a)
MTB > TWOSAMPLE     99% CONFIDENCE INTERVAL    C1  C2

TWOSAMPLE    T    FOR    C1 VS C2
       N    MEAN     STDEV     SE MEAN
C1    16   65.56      6.64         1.7
C2    14   62.14      5.43         1.5

99 PCT   CI   FOR   MU C1 - MU C2:  (-2.7, 9.5)

TTEST MU C1 = MU C2 (VS NE):  T = 1.55  P = 0.13  DF = 27
```

(99% confidence interval for $\mu_1 - \mu_2$)

The following portion of the MINITAB solution in MINITAB display gives the 99% confidence interval for $\mu_1 - \mu_2$.

99 PCT CI FOR MU C1 − MU C2: (−2.7, 9.5)

Thus, the 99% confidence interval for $\mu_1 - \mu_2$ is -2.7 to 9.5.

Again, the last line in this MINITAB solution gives the test of hypothesis (using the t distribution) for H_0: $\mu_1 = \mu_2$ against H_1: $\mu_1 \neq \mu_2$ assuming that the standard deviations of the two populations are not equal.

(b) The null and alternative hypotheses are

$$H_0: \mu_1 = \mu_2 \quad \text{or} \quad \mu_1 - \mu_2 = 0$$

$$H_1: \mu_1 > \mu_2 \quad \text{or} \quad \mu_1 - \mu_2 > 0$$

The following are the MINITAB commands and the MINITAB solution for a test of hypothesis about $\mu_1 - \mu_2$ using the data entered in columns C1 and C2. It is assumed that the two samples are taken from two populations that are normally distributed with unequal and unknown standard deviations.

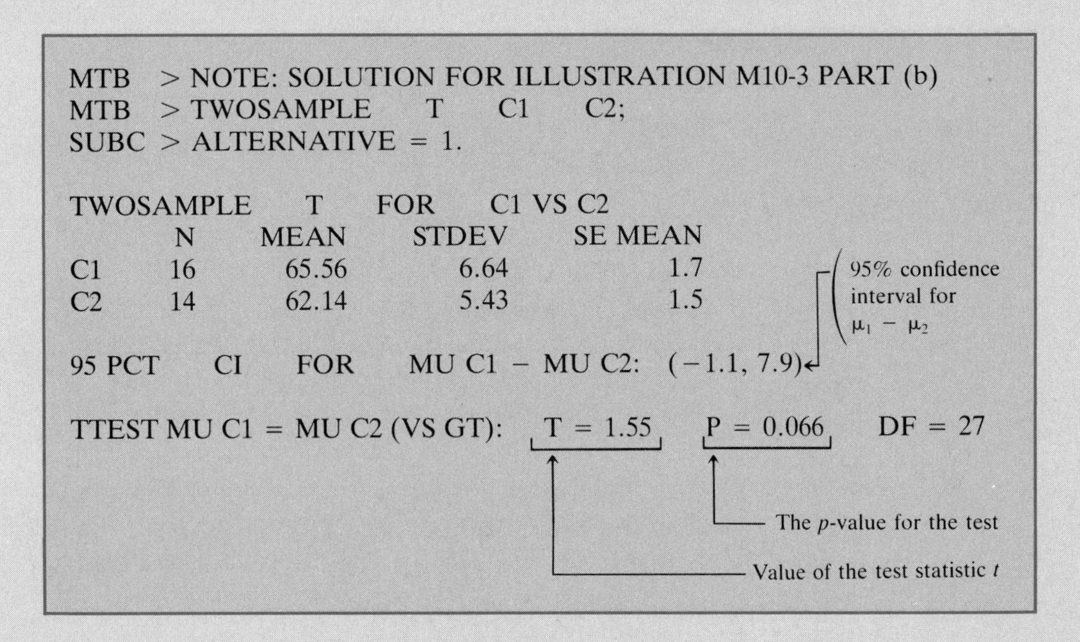
```
MTB  > NOTE: SOLUTION FOR ILLUSTRATION M10-3 PART (b)
MTB  > TWOSAMPLE    T    C1    C2;
SUBC > ALTERNATIVE = 1.

TWOSAMPLE    T    FOR    C1 VS C2
       N    MEAN    STDEV    SE MEAN
C1    16   65.56     6.64        1.7
C2    14   62.14     5.43        1.5

95 PCT   CI   FOR   MU C1 - MU C2:  (-1.1, 7.9)

TTEST MU C1 = MU C2 (VS GT):  T = 1.55   P = 0.066   DF = 27
```


The value of the test statistic t for $\bar{x}_1 - \bar{x}_2$ is

$$t = 1.55$$

The degrees of freedom for the test are 27, as shown in the MINITAB solution. The degrees of freedom are calculated using the formula given in Section 10.3. The test is right-tailed. The significance level is given to be 1%. From the t distribution table, the critical value of t for $df = 27$ and .01 area under the right tail is 2.473.

Because the value of the test statistic $t = 1.55$ is less than the critical value of $t = 2.473$, it falls in the nonrejection region. Consequently, we fail to reject the null hypothesis.

We can reach the same conclusion using the p-value printed in the MINITAB solution. From the MINITAB solution, the p-value is P = .066. Because the value of $\alpha = .01$ is smaller than the p-value of .066, we will not reject the null hypothesis.

Note that the MINITAB output in this MINITAB display also gives a 95% confidence interval for $\mu_1 - \mu_2$, which is -1.1 to 7.9.

INFERENCES ABOUT THE DIFFERENCE BETWEEN TWO POPULATION MEANS FOR PAIRED SAMPLES

To make a confidence interval and to test a hypothesis about the difference between two population means for paired samples, first we enter the given data in columns C1 and C2. Then we take the difference between the corresponding data values of columns C1 and C2 and put them in column C3 using the following command.

```
MTB > LET C3 = C1 - C2
```

The values recorded in column C3 are the paired differences denoted by d in Section 10.4.

To make a confidence interval for μ_d, we use the following MINITAB command.

```
MTB > TINTERVAL    b CONFIDENCE LEVEL    C3
```

b: Refers to the confidence level

To test a hypothesis about μ_d, we use the following MINITAB commands.

```
MTB  > TTEST DIFFERENCE = 0    C3;
SUBC > ALTERNATIVE = k.
```

← (for the TTEST line) This command instructs MINITAB that the test is to be made using the t distribution and that the null hypothesis is $\mu_d = 0$

← (for the ALTERNATIVE line) The value of k will be -1, 0, or 1 depending on whether the alternative hypothesis is less than, not equal to, or greater than zero, respectively

Illustration M10-4 describes how to use MINITAB to construct a confidence interval for μ_d using data from two paired samples.

Illustration M10-4 Recall Example 10-12. A researcher wanted to find out the effect of a special diet on systolic blood pressure. She selected a sample of seven adults and put them on this dietary program for 3 months. The following table gives the systolic blood pressures of these seven adults before and after the completion of the program.

Before	210	180	195	220	231	199	224
After	193	186	186	223	220	183	233

Construct a 95% confidence interval for μ_d where μ_d is the mean reduction in the systolic blood pressure due to this special dietary program for the population of all adults. Assume that the population of paired differences has a normal distribution.

Solution

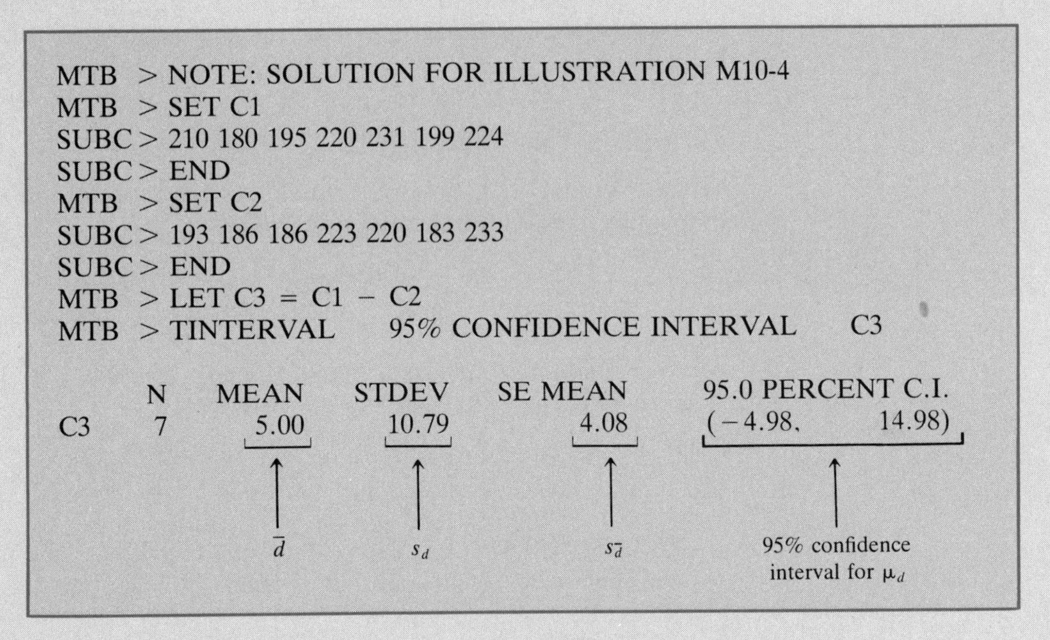

```
MTB  > NOTE: SOLUTION FOR ILLUSTRATION M10-4
MTB  > SET C1
SUBC > 210 180 195 220 231 199 224
SUBC > END
MTB  > SET C2
SUBC > 193 186 186 223 220 183 233
SUBC > END
MTB  > LET C3 = C1 - C2
MTB  > TINTERVAL    95% CONFIDENCE INTERVAL    C3

        N    MEAN    STDEV    SE MEAN    95.0 PERCENT C.I.
C3      7    5.00    10.79       4.08    (-4.98,    14.98)
```


Thus, the 95% confidence interval for μ_d is -4.98 to 14.98.

Illustration M10-5 describes the procedure to make a test of hypothesis about μ_d using MINITAB.

Illustration M10-5 Refer to Illustration M10-4. Test at the 5% significance level if the mean μ_d of the population paired differences is different from zero.

Solution We entered data in columns C1 and C2 in Illustration M10-4 and also calculated the values of d and put them in column C3. Hence, we do not repeat those steps in this illustration. The following display presents the MINITAB commands and MINITAB solution for the hypothesis test about the mean μ_d of population paired differences.

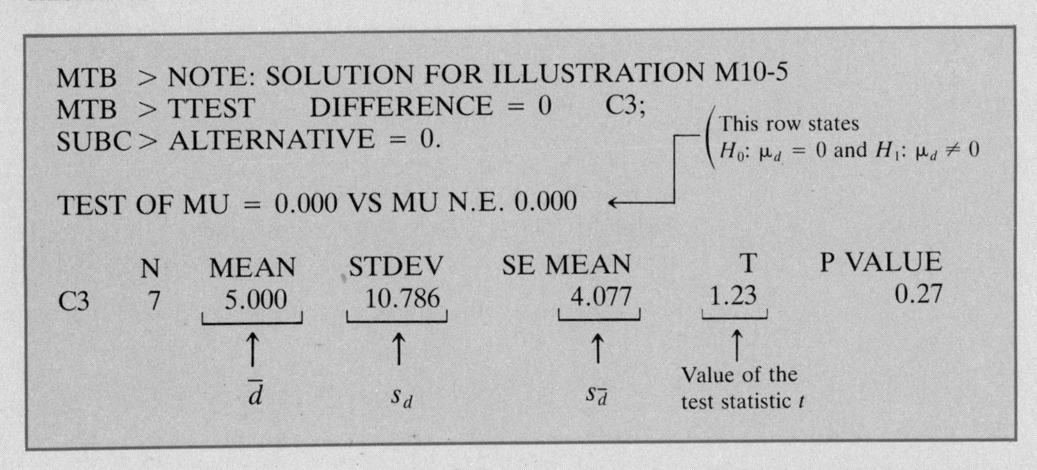

```
MTB  > NOTE: SOLUTION FOR ILLUSTRATION M10-5
MTB  > TTEST    DIFFERENCE = 0    C3;
SUBC > ALTERNATIVE = 0.

TEST OF MU = 0.000 VS MU N.E. 0.000

        N    MEAN     STDEV    SE MEAN      T    P VALUE
C3      7    5.000    10.786     4.077   1.23       0.27
```

From the MINITAB output, the value of the test statistic t for $\bar{d}$ is 1.23. The two critical values of t for $df = 7 - 1 = 6$ and $\alpha/2 = .025$ area in each tail are -2.447 and 2.447. The value of the test statistic $t = 1.23$ falls in the nonrejection region (see Figure 10.8). As a result, we fail to reject the null hypothesis.

We can reach at the same conclusion using the p-value, which is .27 in the MINITAB solution. Because $\alpha = .05$ is smaller than the p-value, we fail to reject the null hypothesis.

COMPUTER ASSIGNMENTS

M10.1 A random sample of 13 male college students who actively participate in sports gave the following data on their GPAs.

3.12	2.84	2.43	2.15	3.92	2.45	2.73
3.06	2.36	1.93	2.81	3.27	1.83	

Another random sample of 16 female college students who also actively participate in sports gave the following data on their GPAs.

2.76	3.84	2.24	2.81	1.79	3.89	2.96	3.77
2.36	2.81	3.29	2.08	3.11	1.69	2.84	3.02

a. Using MINITAB, construct a 99% confidence interval for the difference between the mean GPAs of all male and all female college students who actively participate in sports.
b. Using MINITAB, test at the 5% significance level if the mean GPAs of all male and all female college students who actively participate in sports are different.

Assume that the GPAs of all such male and female college students are both normally distributed with equal but unknown population standard deviations.

M10.2 Repeat Computer Assignment M10.1 assuming that the standard deviations of the two populations are unequal.

M10.3 Recall Exercise 10.55. The manufacturer of a gasoline additive claims that the use of this additive increases gasoline mileage. A random sample of six cars was selected. These cars were driven for 1 week without the gasoline additive and then for 1 week with the gasoline additive. The following table lists the miles per gallon given by these cars without and with the gasoline additive.

Without	24.6	28.3	18.9	23.7	15.4	29.5
With	26.3	31.7	18.2	25.3	18.3	30.9

a. Using MINITAB, construct a 99% confidence interval for the mean μ_d of population paired differences.
b. Using MINITAB, test at the 1% significance level if the use of gasoline additive increases the gasoline mileage.

Assume that the population of paired differences is (approximately) normally distributed.

11 CHI-SQUARE TESTS

11.1 THE CHI-SQUARE DISTRIBUTION
11.2 A GOODNESS-OF-FIT TEST
11.3 CONTINGENCY TABLES
11.4 A TEST OF INDEPENDENCE OR HOMOGENEITY
11.5 INFERENCES ABOUT THE POPULATION VARIANCE
SELF-REVIEW TEST
USING MINITAB

The tests of hypotheses about the mean, the difference between two means, the proportion, and the difference between two proportions were discussed in Chapters 9 and 10. The tests about proportion dealt with countable or categorical data. In the case of a proportion and the difference between two proportions, the tests concerned experiments with only two categories. Recall from Chapter 5 that such experiments are called binomial experiments.

This chapter describes three types of tests:

1. The tests of hypotheses for experiments with more than two categories, called the goodness-of-fit tests
2. The tests of hypotheses about contingency tables, called the independence and homogeneity tests
3. The tests of hypotheses about the population variance and standard deviation

All these tests are performed by using the **chi-square distribution.** The chi-square distribution is sometimes written as the χ^2 *distribution*. The symbol χ is the Greek letter *chi,* pronounced "*kī*." The values of a chi-square distribution are denoted by the symbol χ^2 just as the values of the standard normal distribution and the t distribution are denoted by z and t, respectively. Section 11.1 discusses the chi-square distribution.

11.1 THE CHI-SQUARE DISTRIBUTION

Like the t distribution, the chi-square distribution also has only one parameter called the degrees of freedom (df). The shape of a specific chi-square distribution depends on the number of degrees of freedom.† (The degrees of freedom for a chi-square distribution are calculated by using different formulas for different tests. This will be explained when we discuss those tests.) The random variable χ^2 only assumes nonnegative values. Hence, a chi-square curve starts at the origin (zero point) and lies entirely to the right of the vertical axis. Three chi-square curves are shown in Figure 11.1. They are for 2, 7, and 12 degrees of freedom.

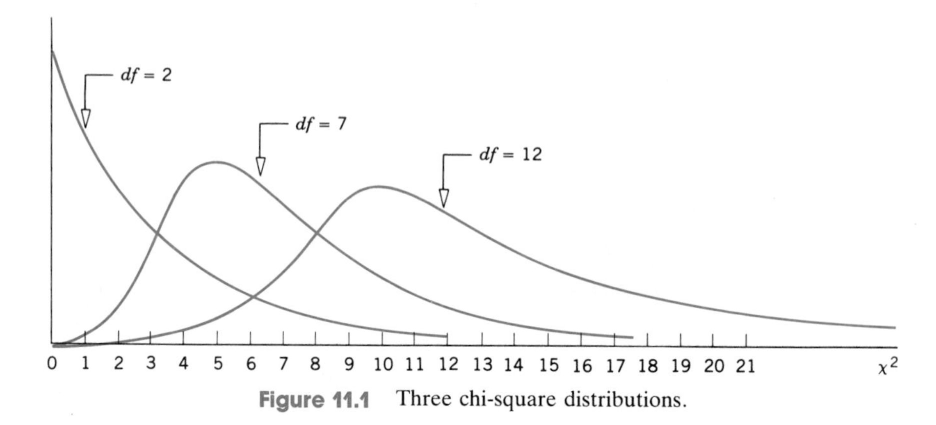

Figure 11.1 Three chi-square distributions.

As can be seen from Figure 11.1, the shape of a chi-square distribution curve is skewed for very small degrees of freedom and it changes drastically as the degrees of freedom increase. Eventually, for large degrees of freedom, the chi-square distribution curve looks like a normal curve. The peak (or mode) of a chi-square distribution curve with 1 or 2 degrees of freedom occurs at zero and for a curve with 3 or more degrees of freedom at $df - 2$. For instance, the peak of the chi-square curve with 2 df in Figure 11.1 occurs at zero. The peak for the curve with 7 df occurs at $7 - 2 = 5$. Finally, the peak for the curve with 12 df occurs at $12 - 2 = 10$. Like all other continuous distribution curves, the total area under a chi-square distribution curve is 1.0.

> **THE CHI-SQUARE DISTRIBUTION**
>
> The chi-square distribution has only one parameter called the degrees of freedom. The shape of a chi-square distribution curve is skewed to the right for small df and becomes symmetric for large df. The entire chi-square curve lies to the right of the vertical axis. The chi-square distribution assumes nonnegative values only, and these are denoted by the symbol χ^2.

If we know the degrees of freedom and the area in the right tail of a chi-square distribution, we can find the value of χ^2 from Table IX of Appendix C (page 744). Examples 11-1 and 11-2 show how to read Table IX.

†The mean of a chi-square distribution is equal to its df, and the standard deviation is equal to $\sqrt{2\,df}$.

Reading the chi-square table: area under the right tail known.

EXAMPLE 11-1 Find the χ^2 value for 7 *df* and an area of .10 in the right tail.

Solution To find the required value of χ^2, we locate 7 in the column for *df* and .10 in the top row in Table IX of Appendix C. The required χ^2 value is given by the entry at the intersection of the row for 7 and the column for .10. This value is 12.017. The relevant portion of Table IX is presented as Table 11.1 below.

Table 11.1 χ^2 for 7 *df* and .10 Area in the Right Tail

	Area in the Right Tail				
df	**.995**	. . .	**.100**	. . .	**.005**
1	0.000	. . .	2.706	. . .	7.879
2	0.010	. . .	4.605	. . .	10.597
.	. . .	. . .	. . .	. . .	. . .
.	. . .	. . .	. . .	. . .	. . .
.	. . .	. . .	. . .	. . .	. . .
7	0.989	. . .	12.017 ←	. . .	20.278
.	. . .	. . .	. . .	. . .	. . .
.	. . .	. . .	. . .	. . .	. . .
.	. . .	. . .	. . .	. . .	. . .
100	67.328	. . .	118.498	. . .	140.169

Required χ^2 value

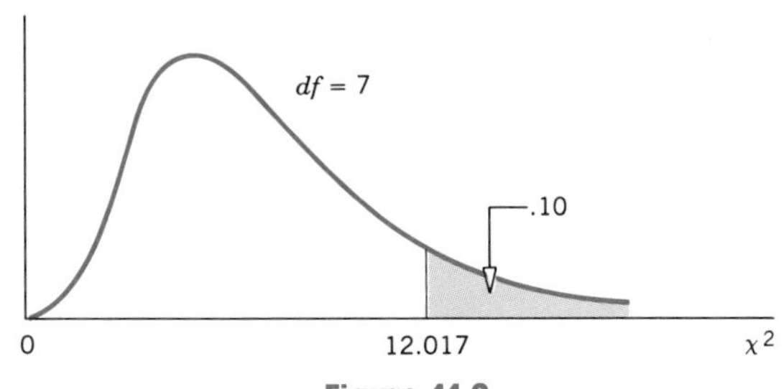

Figure 11.2

As shown in Figure 11.2, the χ^2 value for 7 *df* and an area of .10 in the right tail is 12.017. ■

Reading the chi-square table: area under the left tail known.

EXAMPLE 11-2 Find the χ^2 value for 12 *df* and an area of .05 in the left tail.

Solution We can only read Table IX for an area in the right tail. When the given area is in the left tail, as in this example, the first step is to calculate the area in the right tail as follows.

$$\text{Area in the right tail} = 1 - \text{area in the left tail}$$

Therefore, for our example,

$$\text{Area in the right tail} = 1 - .05 = .95$$

Next, we locate 12 in the column for *df* and .95 in the top row in Table IX. The required value of χ^2, given by the entry at the intersection of the row for 12 and the column for .95, is 5.226. The relevant portion of Table IX is presented as Table 11.2.

Table 11.2 χ^2 for 12 *df* and .95 Area in the Right Tail

	Area in the Right Tail				
df	**.995**	...	**.950**	...	**.005**
1	0.000	...	0.004	...	7.879
2	0.010	...	0.103	...	10.597
.	...	...	...	...	...
.	...	...	...	...	...
.	...	...	...	...	...
12	3.074	...	5.226 ← Required χ^2 value	...	28.300
.	...	...	...	...	...
.	...	...	...	...	...
.	...	...	...	...	...
100	67.328	...	77.929	...	140.169

As shown in Figure 11.3, the χ^2 value for 12 *df* and .05 area in the left tail is 5.226.

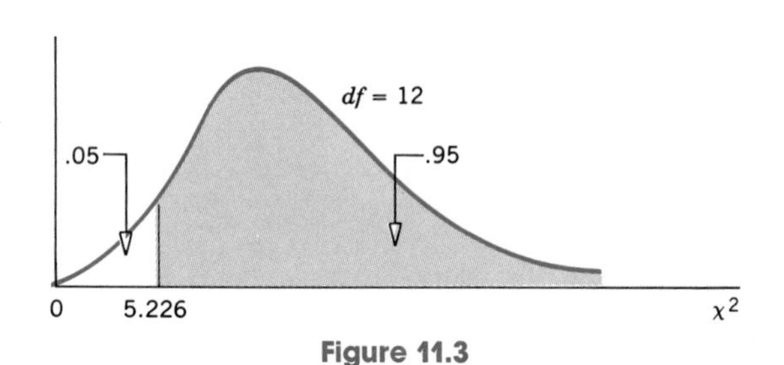

Figure 11.3

■

EXERCISES

11.1 Find the value of χ^2 for 10 degrees of freedom and an area of .025 in the right tail.

11.2 Find the value of χ^2 for 30 degrees of freedom and an area of .05 in the right tail.

11.3 Determine the value of χ^2 for 15 degrees of freedom and an area of .10 in the left tail.

11.4 Determine the value of χ^2 for 19 degrees of freedom and an area of .990 in the left tail.

11.5 Find the value of χ^2 for 4 degrees of freedom and an area of .005 in the right tail.

11.6 Determine the value of χ^2 for 13 degrees of freedom and an area of .05 in the left tail.

11.2 A GOODNESS-OF-FIT TEST

This section explains how to make tests of hypotheses about experiments with more than two possible outcomes (or categories). Such experiments, called **multinomial**

experiments, possess four characteristics. Note that a binomial experiment is a special case of a multinomial experiment.

A MULTINOMIAL EXPERIMENT

An experiment with the following characteristics is called a multinomial experiment:

1. It consists of n identical trials.
2. Each trial results in one of the k possible outcomes (or categories) where $k > 2$.
3. The trials are independent.
4. The probabilities of various outcomes remain constant for each trial.

An experiment made of many rolls of a die is an example of a multinomial experiment. It consists of many identical rolls (trials); each roll (trial) results in one of the six outcomes; each roll is independent of the other rolls; and the probabilities of six outcomes remain constant for each roll.

As a second example of a multinomial experiment, consider the percentage of people who are in favor of the death penalty, who are against it, or who have no opinion. Suppose we select many persons and ask the opinion of each about the death penalty. There will be as many trials for this experiment as the number of persons selected. Each person can belong to any of the three categories. The response of each selected person is independent of the responses of other persons. Given that the population is large, the probabilities of each person belonging to three categories remain the same. Consequently, this is an example of a multinomial experiment.

The frequencies obtained from the performance of an experiment are called the **observed frequencies.** Under a **goodness-of-fit test,** we test the null hypothesis that the observed frequencies for an experiment follow a certain pattern or theoretical distribution. The test is called a goodness-of-fit test because the hypothesis tested is how good the observed frequencies fit a given pattern.

For our first example involving the experiment of many rolls of a die, we may test the null hypothesis that the given die is fair. The die will be fair if the observed frequency for each outcome is close to one-sixth the total number of rolls.

For our second example involving opinions about the death penalty, suppose 42% of adults were in favor of the death penalty, 35% were against it, and 23% had no opinion in 1980. We want to test if these percentages are true for the current population of adults. Suppose we take a sample of 1000 adults and observe that 490 of them are in favor of the death penalty, 370 are against it, and 140 have no opinion. The frequencies 490, 370, and 140 are the observed frequencies. Now, assuming that the 1980 percentages are still true (which will be our null hypothesis), in a sample of 1000 adults we will expect 420 to be in favor of the death penalty, 350 to be against it, and 230 to have no opinion. These frequencies are obtained by multiplying the sample size (1000) by the 1980 proportions. These frequencies are called the **expected frequencies.** Then, we will make a decision to reject or not to reject the null hypothesis based on how large the difference between the observed frequencies and the expected frequencies is. To make this test, we will use the chi-square distribution. Note that, in this case, we are testing the null hypothesis that all three percentages (or proportions) are unchanged. However, if we want to make a test for only one of the three proportions, we use the procedure learned in Section 9.5 of Chapter 9. For example, if we are testing a hypothesis that the percentage of the current population of adults who are in favor of the death penalty is different from 42%, then we will test the null hypothesis H_0: $p = .42$ against the alternative hypothesis H_1: $p \neq .42$. This test will be made using the procedure of Section 9.5 of Chapter 9.

DEGREES OF FREEDOM FOR A GOODNESS-OF-FIT TEST

In a goodness-of-fit test, the degrees of freedom are

$$df = k - 1$$

where k denotes the number of possible outcomes (or categories) for the experiment.

The procedure to make a goodness-of-fit test involves the same five steps that were used in the preceding chapters. *The chi-square goodness-of-fit test is always a right-tailed test.*

As mentioned earlier, the frequencies obtained from the performance of an experiment are called the observed frequencies. They are denoted by O. To make a goodness-of-fit test, we calculate the expected frequencies for all categories of the experiment. The expected frequency for a category, denoted by E, is given by the product of n and p, where n is the total number of trials and p is the probability for that category.

OBSERVED AND EXPECTED FREQUENCIES

The frequencies obtained from the performance of an experiment are called the observed frequencies and are denoted by O. The expected frequencies, denoted by E, are the frequencies that we expect to obtain if the null hypothesis is true. The expected frequency for a category is obtained as

$$E = np$$

where n is the sample size and p is the probability that an element belongs to that category if the null hypothesis is true.

Whether the null hypothesis is rejected or not depends on how far the observed and expected frequencies are from each other. To find how large the difference between the observed frequencies and the expected frequencies is, we do not look at $\Sigma(O - E)$ because some of the $O - E$ differences will be positive and others will be negative. Consequently, the net result will always be zero. Therefore, we square each of the $O - E$ values to obtain $(O - E)^2$ and then weight them according to the reciprocals of their expected frequencies. The sum of the resulting numbers gives the computed value of the test statistic χ^2.

TEST STATISTIC FOR A GOODNESS-OF-FIT TEST

The value of the test statistic χ^2 is calculated as

$$\chi^2 = \Sigma \frac{(O - E)^2}{E}$$

where

O = observed frequency for a category

E = expected frequency for a category = np

χ^2 is called the test statistic

To make a goodness-of-fit test, the sample size should be large enough so that the expected frequency for each category is at least 5. If there is a category with expected frequency less than 5, either increase the sample size or combine two or more categories to make each expected frequency at least 5.

Examples 11-3 and 11-4 describe the procedure for performing goodness-of-fit tests using the chi-square distribution.

Goodness of fit test: equal proportions for all categories.

EXAMPLE 11-3 The following table lists the age distribution for a sample of 100 persons arrested for drunk driving.

Age	16–25	26–35	36–45	46–55	56 & older
Arrests	32	25	19	16	8

Using the 1% significance level, test the null hypothesis that the proportion of people arrested for drunk driving is the same for all age groups.

Solution To make this test, we proceed as follows.

Step 1. *State the null and alternative hypotheses*

Because there are five categories listed in the table, the proportion of drunk drivers will be the same for all age groups if each group contains one-fifth of the total drunk drivers. The null and alternative hypotheses are

H_0: The proportion of people arrested for drunk driving is the same for all age groups

H_1: The proportion of people arrested for drunk driving is not the same for all age groups

If the proportion of people arrested for drunk driving is the same for all age groups, then the probability for any randomly selected drunk driver to belong to any of the five age groups listed in the table will be $1/5 = .20$. Let p_1, p_2, p_3, p_4, and p_5 be the probabilities of any randomly selected drunk driver to belong to each of the five age groups, respectively. Then the null hypothesis can also be written as

$$H_0: p_1 = p_2 = p_3 = p_4 = p_5 = .20$$

and the alternative hypothesis can be stated as

$$H_1\text{: At least two of the five probabilities are not equal to .20}$$

Step 2. *Select the distribution to use*

Because there are five categories (i.e., five age groups listed in the table), it is a multinomial experiment. Consequently, we use the chi-square distribution to make the test.

Step 3. *Determine the rejection and nonrejection regions*

The significance level is given to be .01 and the goodness-of-fit test is always right-tailed. Therefore,

$$\text{Area in the right tail} = \alpha = .01$$

$$k = \text{number of categories} = 5$$

$$df = k - 1 = 5 - 1 = 4$$

From the chi-square table (Table IX of Appendix C), the critical value of χ^2 for 4 *df* and .01 area in the right tail is 13.277, as shown in Figure 11.4.

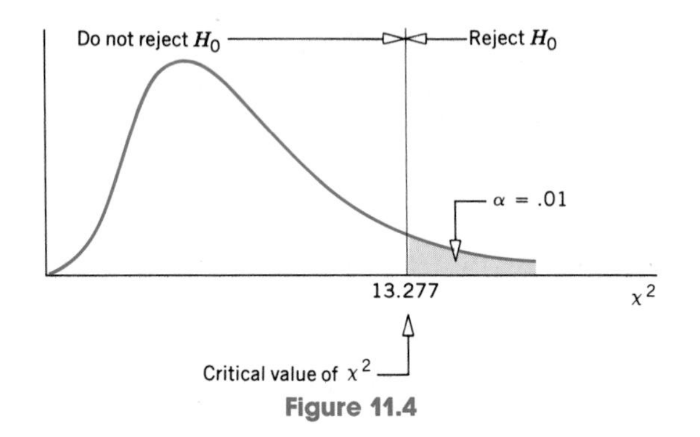

Figure 11.4

Step 4. *Calculate the value of the test statistic*

All the required calculations to find the value of the test statistic χ^2 are shown in Table 11.3.

The calculations made in Table 11.3 are explained below.

1. The first two columns in Table 11.3 list the five categories (age groups) and the observed frequencies for the sample of 100 drunk drivers, respectively. The third column contains the probabilities for the five categories assuming that the null hypothesis is true.
2. The fourth column contains the expected frequencies. These frequencies are obtained by multiplying the sample size (n = 100) by probabilities listed in the

Table 11.3

Category (age)	Observed Frequency O	p	Expected Frequency $E = np$	$(O - E)$	$(O - E)^2$	$\frac{(O - E)^2}{E}$
16–25	32	.20	100(.20) = 20	12	144	7.20
26–35	25	.20	100(.20) = 20	5	25	1.25
36–45	19	.20	100(.20) = 20	−1	1	.05
46–55	16	.20	100(.20) = 20	−4	16	.80
56 & older	8	.20	100(.20) = 20	−12	144	7.20
	$n = 100$					Sum = 16.50

third column. If the null hypothesis is true (i.e., the drunk drivers are equally distributed over all categories), then we will expect 20 out of 100 drunk drivers to belong to each category. Consequently, each category in the fourth column has the same expected frequency.

3. The fifth column lists the differences between the observed and expected frequencies, that is, $O - E$. These values are squared and recorded in the sixth column.
4. Finally, we divide the squared differences (of the sixth column) by the corresponding expected frequencies (listed in the fourth column) and write the resulting numbers in the seventh column.
5. The sum of the seventh column gives the value of the test statistic χ^2. Thus,

$$\chi^2 = \Sigma \frac{(O - E)^2}{E} = 16.50$$

Step 5. *Make a decision*

The value of the test statistic $\chi^2 = 16.50$ is greater than the critical value of $\chi^2 = 13.277$, and it falls in the rejection region. Hence, we reject the null hypothesis and state that the proportion of drunk drivers is not the same for all age groups listed in the given table. ■

Goodness of fit test: testing if results of a survey fit a given distributions.

EXAMPLE 11-4 In a 1986 Roper Poll conducted for *U.S. News & World Report,* adults were asked if they think "over the next 20 years . . . science and technology will do more good than harm for the human race, or more harm than good." Of the respondents, 64% said more good, 23% said more harm, and 13% said neither or they did not know (*Survey 1986: Issues and Reactions—Major Events of 1986 and How You Felt About Them*). A recently taken random sample of 500 adults produced the results listed in the following table in response to the same question.

Category	More Good	More Harm	Neither/Do Not Know
Number of persons	342	107	51

Test at the 1% significance level if the current percentage distribution of adults belonging to the three categories is different from that for 1986.

Solution

Step 1. *State the null and alternative hypotheses*

The null and alternative hypotheses are

H_0: The current percentage distribution of adults belonging to three categories is the same as that for 1986

H_1: The current percentage distribution of adults belonging to three categories is different from that for 1986

Step 2. *Select the distribution to use*

Because there are three categories (more good, more harm, and neither/do not know), it is a multinomial experiment. Consequently, we use the chi-square distribution to make the test.

Step 3. *Determine the rejection and nonrejection regions*

The significance level is .01. Because a goodness-of-fit test is right-tailed,

$$\text{Area in the right tail} = \alpha = .01$$
$$k = \text{number of categories} = 3$$
$$df = k - 1 = 3 - 1 = 2$$

From the chi-square table (Table IX of Appendix C), the critical value of χ^2 for 2 *df* and .01 area in the right tail is 9.210. This value is shown in Figure 11.5.

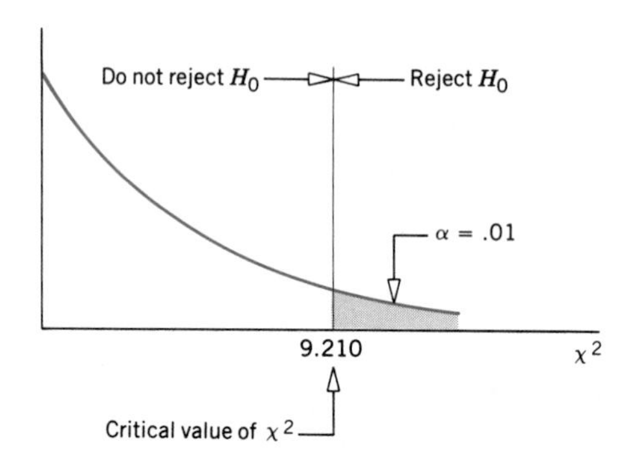

Figure 11.5

Step 4. *Calculate the value of the test statistic*

All the required calculations to find the value of the test statistic χ^2 are shown in Table 11.4.

Table 11.4

Category	Observed Frequency O	p	Expected Frequency $E = np$	$(O - E)$	$(O - E)^2$	$\frac{(O - E)^2}{E}$
More good	342	.64	500(.64) = 320	22	484	1.513
More harm	107	.23	500(.23) = 115	−8	64	.557
Neither/ Do not know	51	.13	500(.13) = 65	−14	196	3.015
	$n = 500$					Sum = 5.085

Note that the given percentages have been converted to probabilities and recorded in the third column of Table 11.4. The value of the test statistic χ^2 is given by the sum of the last column. Thus,

$$\chi^2 = \Sigma \frac{(O - E)^2}{E} = 5.085$$

Step 5. *Make a decision*

The value of the test statistic $\chi^2 = 5.085$ is smaller than the critical value of $\chi^2 = 9.210$ and it falls in the nonrejection region. Hence, we fail to reject the null hypothesis and state that the current percentage distribution of adults belonging to the three categories is not different from that for 1986. The difference between the observed frequencies and the expected frequencies seems to have occurred only because of sampling error. ■

CASE STUDY 11-1 ARE LEADERSHIP STYLES EVENLY DISTRIBUTED?

Professor Donald C. Lueder conducted a study on the distribution of leadership styles based on a sample of 95 doctoral students enrolled in a basic administration course. These students were experienced teachers and administrators. The students were administered the Leader Effectiveness and Description instrument of self-perception test (in short, called LEAD-Self) developed by Paul Hersey and Kenneth Blanchard to measure leadership style and leader adaptability.

The LEAD-Self instrument contains 12 situations. Each respondent was required to select one of the four alternative action responses for each situation. Each of the four responses corresponds to a different leadership style. The four leadership styles are "high task/low relationship (S1), or 'telling'; high task/high relationship (S2), or 'selling'; low task/high relationship (S3), or 'participating'; and low task/low relationship (S4), or 'delegating'." The author hypothesized that a significantly high number of respondents would choose S2. To test this hypothesis, he applied three statistical tests, one of which is the chi-square test.

The category having the largest number of responses of a student to the 12 situations was considered to be the leadership style of that respondent. For 10 of the 95 respondents, at least two categories had the same number of responses. Hence, the author dropped these 10 respondents from the chi-square test. Of the 85 respondents, 4 were observed to belong to the S1 leadership style, 66 to the S2, 14 to the S3, and 1 to the S4 leadership style. Table 11.5 gives all the calculations required to make the chi-square test. The expected frequencies are calculated based on the following null and alternative hypotheses.

H_0: The respondents are evenly distributed over different leadership styles

H_1: The respondents are not evenly distributed over different leadership styles

If the null hypothesis is true, then we will expect 85 respondents to be equally distributed over the four categories. Hence, the expected frequency for each category is 85/4 = 21.25.

Table 11.5

Leadership Style	Observed Frequency O	Expected Frequency E	$O - E$	$\frac{(O - E)^2}{E}$
S1	4	21.25	−17.25	14.00
S2	66	21.25	44.75	94.24
S3	14	21.25	−7.25	2.47
S4	1	21.25	−20.25	19.30
Total	85	85		130.01

The value of the test statistic χ^2 is

$$\chi^2 = \Sigma \frac{(O - E)^2}{E} = 130.01$$

The author made the test at $\alpha = .001$. The critical value of χ^2 for 3 *df* and .001 area in the right tail is 16.27. (Note that this value is not in Table IX.) Because the value of the test statistic $\chi^2 = 130.01$ is larger than the critical value of $\chi^2 = 16.27$, the null hypothesis is rejected. The decision is that the respondents are not evenly distributed over the different leadership styles.

Source: Donald C. Lueder: "Don't Be Mislead by LEAD." *The Journal of Applied Behavioral Science,* 21(2): 143–151, 1985. Copyright © 1985 by JAI Press, Inc. Data and excerpts reprinted with permission.

EXERCISES

11.7 Describe the four characteristics of a multinomial experiment.

11.8 The following table gives the percentage distribution, according to marital status, of the 1986 U.S. population of 18-year-olds and over (Source: U.S. Bureau of the Census).

Marital status	Single	Married	Widowed	Divorced
Percent	21.6	62.9	7.7	7.8

A recent random sample of 400 persons, aged 18 and over, showed that 102 of them were single, 220 were married, 35 were widowed, and 43 were divorced. Test at the 5% significance level if the percentage distribution of marital status of the U.S. population aged 18 and over has changed during this period.

11.9 The following table gives the 1986 percentage distribution of public elementary and secondary school male teachers according to age (Source: National Education Association).

Age	18–24	25–34	35–44	45–54	55–64	65 & Older
Percent	1.8	16.6	47.0	23.1	11.0	.5

A recent sample of 300 public elementary and secondary school male teachers showed that the number of teachers belonging to various age groups listed in the table are 9, 56, 133, 63, 36, and 3, respectively. Test at the 1% significance level if the age distribution of such teachers has changed during this period.

11.10 According to a 1988 survey conducted by the Roper Organization, 28% of persons aged 18 and older were in excellent health, 61% in good health, 9% in not very good health, and 2% were in poor health (*Roper Reports,* 1988). A recent sample of 400 persons aged 18 and older produced the results listed in the following table.

Health condition	Excellent	Good	Not Very Good	Poor
Number of persons	89	259	46	6

Test at the 10% significance level if the current percentage distribution of people with excellent, good, not very good, and poor health condition is different from that for 1988.

11.11 The following table lists the frequency distribution for a sample of 50 absences of college students from classes according to the day of occurrence.

Day of the week	Mon	Tue	Wed	Thu	Fri
Number of absences	14	6	3	11	16

Test at the 5% significance level if the null hypothesis that the absences are equally distributed over all days of the week is true.

11.12 The following table lists the frequency distribution for 60 rolls of a die.

Outcome	1-spot	2-spot	3-spot	4-spot	5-spot	6-spot
Frequency	7	12	8	15	11	7

Test at the 5% significance level if the null hypothesis that the given die is fair is true.

11.13 The following table gives the average number of items produced by an employee of a small company on different days of the week for a sample of 5 weeks.

Day	Mon	Tue	Wed	Thu	Fri
Items produced	8	12	14	11	7

Using the 5% significance level, test the null hypothesis that the number of items produced by this employee on different days of the week is the same.

11.14 The following table lists the frequency distribution of cars sold during the past 12 months at an auto dealership.

Month	Jan	Feb	Mar	Apr	May	Jun	Jul	Aug	Sep	Oct	Nov	Dec
Cars sold	21	17	15	12	14	12	13	19	23	26	24	28

Using the 10% significance level, test the null hypothesis that the number of cars sold at this dealership is the same for each month.

11.15 Of all students enrolled at a large undergraduate university, 19% are seniors, 23% are juniors, 27% are sophomores, and 31% are freshmen. A sample of 200 students taken from this university by the student senate to conduct a survey includes 53 seniors, 46 juniors, 52 sophomores, and 49 freshmen. Using the 10% significance level, test the null hypothesis that this sample is a random sample. (*Hint:* This sample will be a random sample if it includes approximately 19% seniors, 23% juniors, 27% sophomores, and 31% freshmen.)

11.3 CONTINGENCY TABLES

Often we may have information on more than one variable for each element. Such information can be presented using a two-way classification table. Such a table is also called a *contingency table* or *cross-tabulation*. Table 11.6 is an example of a contingency table. It gives information on the type of institution and sex of student for all students who were enrolled in institutions of higher education in 1984. Table 11.6 has two rows (one for males and one for females) and two columns (one for public and one for private institutions). Hence, it is also called a 2 × 2 (read as "two by two") contingency table.

Table 11.6 Total Enrollment in Institutions of Higher Education, by Type of Institution and Sex of Student, 1984

	Type of Institution	
	Public	**Private**
Male	4,448,502	1,375,886 ← Students who are male and enrolled in private institutions
Female	4,976,409	1,360,981

Source: U.S. Department of Education, Center for Education Statistics.

A contingency table can be of any size. For example, it can be 2 × 3, 3 × 2, 3 × 3, or 4 × 2. Note that in these notations, the first number refers to the number of rows in the table and the second number refers to the number of columns. For example, a 3 × 2 table will contain 3 rows and 2 columns. Each of the four boxes that contain numbers in Table 11.6 is called a *cell*. The number of cells for a contingency table is obtained by multiplying the number of rows by the number of columns. Thus, Table 11.6 contains 2 × 2 = 4 cells. The subjects entered in each cell of a contingency table possess two characteristics. For example, 1,375,886 students listed in the second cell of the first row of Table 11.6 are "male and enrolled in private institutions." The numbers entered in the cells are usually called the *joint frequencies*. For example, 1,375,886 students belong to the joint category of "male and enrolled in private institutions." Hence, it is the joint frequency of this category.

11.4 A TEST OF INDEPENDENCE OR HOMOGENEITY

This section is concerned with tests of independence and homogeneity, which are made using the contingency tables. Except for a few modifications, the procedure used to make such tests is almost the same as the one applied in Section 11.2 for a goodness-of-fit test.

11.4.1 A TEST OF INDEPENDENCE

In a **test of independence** for a contingency table, we test the null hypothesis that the two attributes (characteristics) of the elements of a given population are not related (that is, they are independent) against the alternative hypothesis that the two characteristics are related (that is, they are dependent). For example, we may want to test if the affiliation of people with the Democratic and Republican parties is independent of their income levels. We perform such a test by using the chi-square distribution. As another example, we may want to test if there is an association between being a male or female and a preference for watching sports or soap operas on television.

DEGREES OF FREEDOM FOR A TEST OF INDEPENDENCE

A test of independence involves a test of the null hypothesis that two attributes of a population are not related. The degrees of freedom for a test of independence are

$$df = (R - 1)(C - 1)$$

where R and C are the number of rows and the number of columns, respectively, in the given contingency table.

The value of the test statistic χ^2 in the case of a test of independence is obtained using the same formula as in a goodness-of-fit test, described in Section 11.2.

TEST STATISTIC FOR A TEST OF INDEPENDENCE

The value of the test statistic χ^2 for a test of independence is calculated as

$$\chi^2 = \Sigma \frac{(O - E)^2}{E}$$

where O and E are the observed and expected frequencies, respectively, for a cell.

The null hypothesis in a test of independence is always that the two attributes are not related. The alternative hypothesis is that the two attributes are related.

The frequencies obtained from the performance of an experiment for a contingency table are called the **observed frequencies.** The procedure to calculate the **expected frequencies** for a contingency table for a test of independence is different from the one for a goodness-of-fit test. Example 11-5 describes this procedure.

Calculating expected frequencies for a test of independence.

EXAMPLE 11-5 Suppose a sample of 300 persons is taken and their income levels and party affiliations are recorded (assume that all 300 persons are either Democrats or Republicans). Just for illustrative purposes, further assume that people with an income of more than $30,000 a year are categorized as belonging to the "high income" group and those with an income of $30,000 or less belong to the "low income" group. Table 11.7 gives the two-way classification of 300 persons. Thus, 60 of the persons are Democrats and belong to the high-income group, and 75 are Republicans and belong to the high-income group.

Table 11.7

	High Income (H)	Low Income (L)	Row Totals
Democrat (D)	60	110	170
Republican (R)	75	55	130
Column totals	135	165	300

The numbers 60, 110, 75, and 55 are called the *observed frequencies* for the four cells.

As mentioned earlier, the null hypothesis in a test of independence is that the two attributes are independent. To begin with, we assume that the null hypothesis is true and that the two attributes are independent. Assuming that the null hypothesis is true and that the political affiliation and income level are not related, we can calculate the following probabilities.

$$P(\text{a person is Democrat}) = P(D) = 170/300$$

$$P(\text{a person belongs to the high income group}) = P(H) = 135/300$$

$$P(D \text{ and } H) = P(D) \cdot P(H) = (170/300) \cdot (135/300)$$

If D and H are independent, the number of persons that are expected to be Democrats and belong to the high income group in a sample of 300 is calculated as follows.

$$\begin{aligned} E \text{ for ``Democrat and high income''} &= 300 \times P(D \text{ and } H) \\ &= 300 \times \frac{170}{300} \times \frac{135}{300} \\ &= \frac{170 \times 135}{300} \\ &= \frac{(\text{Row total})(\text{Column total})}{\text{Sample size}} \end{aligned}$$

Thus, the rule to obtain the expected frequency for a cell is to divide the product of the corresponding row and column totals by the sample size.

EXPECTED FREQUENCIES FOR A TEST OF INDEPENDENCE

The expected frequency E for a cell is calculated as

$$E = \frac{(\text{Row total})(\text{Column total})}{n}$$

Using this rule, we calculate the expected frequencies of the four cells as follows.

$$E \text{ for ``Democrat and high income'' cell} = (170)(135)/300 = 76.50$$

$$E \text{ for ``Democrat and low income'' cell} = (170)(165)/300 = 93.50$$

$$E \text{ for "Republican and high income" cell} = (130)(135)/300 = 58.50$$

$$E \text{ for "Republican and low income" cell} = (130)(165)/300 = 71.50$$

Table 11.8

	High Income	Low Income	Row Totals
Democrat	60 (76.50)	110 (93.50)	170
Republican	75 (58.50)	55 (71.50)	130
Column totals	135	165	300

The expected frequencies are usually written in parentheses below the observed frequencies within the corresponding cells, as shown in Table 11.8. ■

Like a goodness-of-fit test, a test of independence is always right-tailed. To apply a chi-square test of independence, the sample size should be large enough so that the expected frequency for each cell is at least 5. If the expected frequency for a cell is not at least 5, we either increase the sample size or combine some categories. Examples 11-6 and 11-7 describe the procedure to make tests of independence using the chi-square distribution.

A test of independence: 2 × 2 table.

EXAMPLE 11-6 Reconsider the two-way classification of 300 persons based on political affiliation and income level given in Example 11-5. Test at the 5% significance level if political affiliation and income level are dependent.

Solution The test involves the following five steps.

Step 1. *State the null and alternative hypotheses*

As mentioned earlier, the null hypothesis must be that the two attributes are independent and the alternative hypothesis that they are dependent.

H_0: Political affiliation and income level are independent

H_1: Political affiliation and income level are dependent

Step 2. *Select the distribution to use*

We use the chi-square distribution to make a test of independence for a contingency table.

Step 3. *Determine the rejection and nonrejection regions*

The significance level is 5%. Because a test of independence is always right-tailed, the area of the rejection region is .05 and it falls in the right tail. The contingency table contains two rows (of Democrats and Republicans) and two columns (of high income and low income). Note that we do not count the row and column of totals. The degrees of freedom are

$$df = (R - 1)(C - 1) = (2 - 1)(2 - 1) = 1$$

From Table IX of Appendix C, the critical value of χ^2 for 1 *df* and $\alpha = .05$ is 3.841. This value is shown in Figure 11.6.

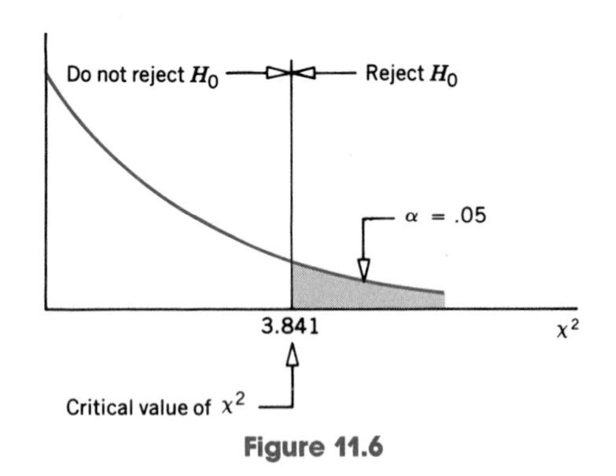

Figure 11.6

Step 4. *Calculate the value of the test statistic*

Table 11.8 with observed and expected frequencies constructed in Example 11-5 is reproduced as Table 11.9 below.

Table 11.9

	High Income	Low Income	Row Totals
Democrat	60 (76.50)	110 (93.50)	170
Republican	75 (58.50)	55 (71.50)	130
Column totals	135	165	300

To compute the value of the test statistic χ^2, we take the difference between the corresponding observed and expected frequencies listed in Table 11.9, square those differences, and then divide each of the squared differences by the respective expected frequencies. The sum of the resulting numbers gives the value of the test statistic χ^2. We make these calculations as follows.

$$\chi^2 = \Sigma \frac{(O - E)^2}{E}$$

$$= \frac{(60 - 76.50)^2}{76.50} + \frac{(110 - 93.50)^2}{93.50} + \frac{(75 - 58.50)^2}{58.50} + \frac{(55 - 71.50)^2}{71.50}$$

$$= 3.559 + 2.912 + 4.654 + 3.808 = 14.933$$

Step 5. *Make a decision*

The value of the test statistic $\chi^2 = 14.933$ is greater than the critical value of $\chi^2 = 3.841$ and it falls in the rejection region. Hence, we reject the null hypothesis and state that there is strong evidence from the sample that the two characteristics, political affiliation and income level, are dependent. ■

A test of independence: 2 × 3 table.

EXAMPLE 11-7 The information obtained from a sample of 600 persons, aged 20 and older, with known cases of diabetes is summarized in the following table.

	Age		
	20–44	**45–64**	**65 and Older**
Men	36	109	126
Women	52	117	160

Test at the 1% significance level if sex and age are related for diabetic persons aged 20 and older.

Solution

Step 1. *State the null and alternative hypotheses*

The null and alternative hypotheses are

H_0: Sex and age are independent for diabetic people aged 20 and older

H_1: Sex and age are dependent for diabetic people aged 20 and older

Step 2. *Select the distribution to use*

Because we are making a test of independence, we use the chi-square distribution to make the test.

Step 3. *Determine the rejection and nonrejection regions*

With a significance level of 1%, the area of the rejection region is .01 and it lies in the right tail. The contingency table contains two rows (of men and women) and three columns (of 20–44, 45–64, and 65 and older age groups). Hence, the degrees of freedom are

$$df = (R - 1)(C - 1) = (2 - 1)(3 - 1) = 2$$

From Table IX of Appendix C, the critical value of χ^2 for $df = 2$ and $\alpha = .01$ is 9.210, as shown in Figure 11.7.

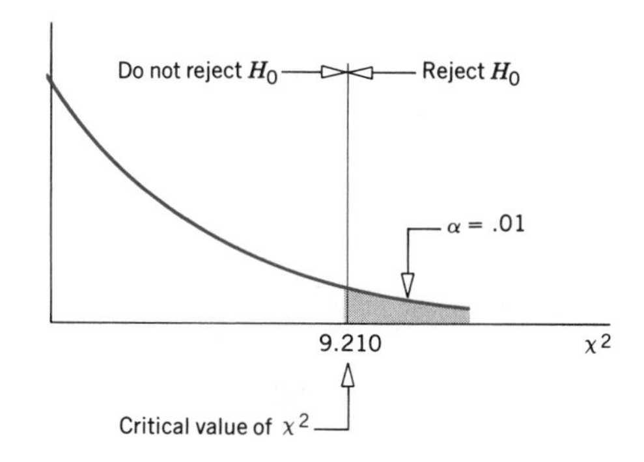

Figure 11.7

Step 4. *Calculate the value of the test statistic*

The expected frequencies for various cells are calculated as follows and they are listed within parentheses in Table 11.10. Note that we have added the row and column of totals in Table 11.10.

E for "men and 20–44" cell = (271)(88)/600 = 39.75

E for "men and 45–64" cell = (271)(226)/600 = 102.08

E for "men and 65 and older" cell = (271)(286)/600 = 129.18

E for "women and 20–44" cell = (329)(88)/600 = 48.25

E for "women and 45–64" cell = (329)(226)/600 = 123.92

E for "women and 65 and older" cell = (329)(286)/600 = 156.82

Table 11.10

	Age			
	20–44	**45–64**	**65 and older**	**Row Totals**
Men	36 (39.75)	109 (102.08)	126 (129.18)	271
Women	52 (48.25)	117 (123.92)	160 (156.82)	329
Column totals	88	226	286	600

The value of the test statistic χ^2 is computed as follows.

$$\chi^2 = \Sigma \frac{(O - E)^2}{E}$$

$$= \frac{(36 - 39.75)^2}{39.75} + \frac{(109 - 102.08)^2}{102.08} + \frac{(126 - 129.18)^2}{129.18}$$

$$+ \frac{(52 - 48.25)^2}{48.25} + \frac{(117 - 123.92)^2}{123.92} + \frac{(160 - 156.82)^2}{156.82}$$

$$= .354 + .469 + .078 + .291 + .386 + .064 = 1.642$$

Step 5. *Make a decision*

The value of the test statistic $\chi^2 = 1.642$ is smaller than the critical value of $\chi^2 = 9.210$ and it falls in the nonrejection region. Consequently, we fail to reject the null hypothesis and conclude that the two characteristics, sex and age for diabetic people aged 20 and older, do not appear to be related. ■

CASE STUDY 11-2 PREVALENCE OF SUICIDAL BEHAVIOR AMONG HIGH SCHOOL STUDENTS

Dr. Harkavy Friedman and colleagues studied the prevalence of suicidal behavior among high school students. They selected a sample of 380 students from a public high school in New York. The sample included 200 males and 180 females. It represented all four high school grades almost equally and included students from many races and religions. The authors divided the students into three categories: no suicidal behavior group (those who "had never thought about killing themselves"), suicidal ideator group (who had "thought about killing themselves but did not actually try"), and suicidal attempter group (those who "had tried to kill themselves at least once"). Then they used chi-square distribution to test for independence between suicidal behavior and various characteristics of the students. Among such characteristics were sex, grade, race, and religion. The two tables given here show the results of that study for sex of the students and suicidal behavior. The first table lists the observed frequencies for male and female students that belong to different suicidal groups.

	No Suicidal Behavior	**Suicidal Ideators**	**Suicide Attempters**
Male	90	102	8
Female	56	99	25

The second table lists the observed and expected frequencies.

	No Suicidal Behavior	**Suicidal Ideators**	**Suicide Attempters**	**Row Totals**
Male	90 (76.84)	102 (105.79)	8 (17.37)	200
Female	56 (69.16)	99 (95.21)	25 (15.63)	180
Column totals	146	201	33	380

The two hypotheses are

H_0: Suicidal behavior and sex are independent

H_1: Suicidal behavior and sex are dependent

The value of the test statistic χ^2 is found to be 15.71. (The reader is advised to perform the calculations and verify $\chi^2 = 15.71$.) The p-value for 2 degrees of freedom and the value of the test statistic $\chi^2 = 15.71$ was calculated by the authors of the article as .0001. This means that the null hypothesis is rejected for any α greater than .0001. For example, the null hypothesis will be rejected at $\alpha = .01$. Thus, the conclusion of the study is that the suicidal behavior and sex of a student appear to be related.

Source: Jill M. Harkavy Friedman et al.: "Prevalence of Specific Suicidal Behavior in a High School Sample," *The American Journal of Psychiatry*, 144(9): 1203–1206, 1987.

11.4.2 A TEST OF HOMOGENEITY

In a **test of homogeneity,** we test if two (or more) populations are homogeneous (similar) with regard to the distribution of a certain characteristic. For example, we might be interested in testing the hypothesis that the proportion of households that belong to different income groups are the same in California and Wisconsin. Or we may want to test if the preferences of people in Florida, Arizona, and Vermont are similar with regard to Coke, Pepsi, and 7-Up.

> **A TEST OF HOMOGENEITY**
>
> A test of homogeneity involves testing the null hypothesis that the proportions of elements with certain characteristics in two or more different populations are the same.

Let us consider the example of testing the null hypothesis that the proportions of households in California and Wisconsin that belong to various income groups are the same. (Note that in a test of homogeneity, the null hypothesis will always be that the proportions of elements with certain characteristics are the same in two or more populations. The alternative hypothesis will be that these proportions are not the same.) Suppose we define three income strata: high-income group (with an income of more than $50,000), medium-income group (with an income of $25,000 to $50,000), and low-income group (with an income of less than $25,000). Furthermore, assume that we take one sample of 250 households from California and another sample of 150 households from Wisconsin, collect the information on incomes of these households, and prepare the contingency Table 11.11.

Table 11.11

	California	Wisconsin	Row Totals
High income	70	34	104
Medium income	80	40	120
Low income	100	76	176
Column totals	250	150	400

Note that in this example the column totals are fixed. That is, we decide in advance to take samples of 250 households from California and 150 from Wisconsin. However, the row totals (of 104, 120, and 176) are determined randomly by the outcomes of the two samples. If we compare this example to the one about political affiliations and income levels taken in the previous section, we will note that neither of the column or row totals were fixed in that example. Instead, the researcher took just one sample of 300 persons, collected the information on their political affiliations and incomes, and prepared the contingency table. Thus, in that example, the row and column totals were all determined randomly. Hence, when the row and column totals are both determined randomly, we make a test of independence. However, when either column or row totals are fixed, we make a test of homogeneity. In the case of income groups in California and Wisconsin, we will make a test of homogeneity to test for the similarity of income groups in two states.

The procedure to make a test of homogeneity is similar to the procedure used to make a test of independence discussed earlier. Like a test of independence, a test of homogeneity is also right-tailed. Example 11-8 illustrates the procedure to make a homogeneity test.

A test of homogeneity.

EXAMPLE 11-8 Consider the data on income distribution for households in California and Wisconsin given in Table 11.11. At the 5% significance level, test the null hypothesis that the distribution of households with regard to income levels is similar (homogeneous) for the two states.

Solution

Step 1. *State the null and alternative hypotheses*

The two hypotheses are†

H_0: The proportion of households that belong to each income group is the same in both states

H_1: The proportion of households that belong to each income group is not the same in both states

Step 2. *Select the distribution to use*

We use the chi-square distribution to make a homogeneity test.

Step 3. *Determine the rejection and nonrejection regions*

The significance level is 5%. Because the homogeneity test is right-tailed, the area of the rejection region is .05 and it lies in the right tail. The contingency table for income groups in California and Wisconsin contains three rows and two columns. Hence, the degrees of freedom are

$$df = (R - 1)(C - 1) = (3 - 1)(2 - 1) = 2$$

From Table IX of Appendix C, the value of χ^2 for 2 df and .05 area in the right tail is 5.991. This value is shown in Figure 11.8.

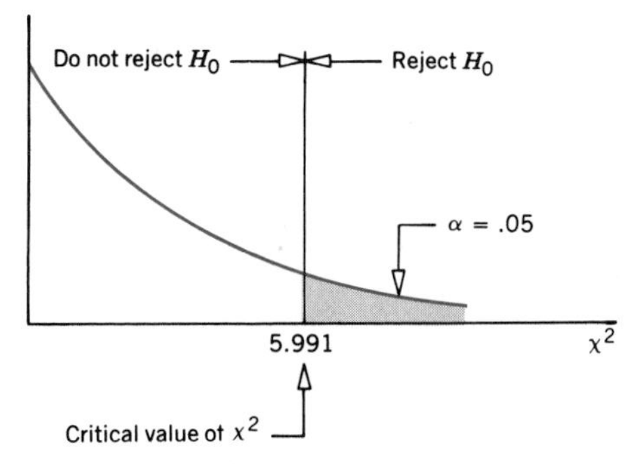

Figure 11.8

†Let p_{HC}, p_{MC}, and p_{LC} be the proportions of households in California that belong to high, middle, and low income groups, respectively. Let p_{HW}, p_{MW}, and p_{LW} be the corresponding proportions for Wisconsin. Then, we can also write the null hypothesis as

$$H_0: p_{HC} = p_{HW},\ p_{MC} = p_{MW}, \text{ and } p_{LC} = p_{LW}$$

and the alternative hypothesis as

$$H_1: \text{At least two of the equalities mentioned in } H_0 \text{ are not true}$$

Step 4. *Calculate the value of the test statistic*

To compute the value of the test statistic χ^2, we need to calculate the expected frequencies first. Table 11.12 lists the observed as well as expected frequencies.

Table 11.12

	California	Wisconsin	Row Totals
High income	70 (65)	34 (39)	104
Medium income	80 (75)	40 (45)	120
Low income	100 (110)	76 (66)	176
Column totals	250	150	400

The numbers in parentheses in Table 11.12 are expected frequencies, which are calculated using the formula

$$E = \frac{(\text{Row total})(\text{Column total})}{\text{Total of both samples}}$$

Thus, for instance,

$$E \text{ for ``high income and California'' cell} = \frac{(104)\,(250)}{(400)} = 65$$

The remaining expected frequencies are calculated in the same way. Note that the expected frequencies in a test of homogeneity are calculated in the same way as in a test of independence.

The value of the test statistic χ^2 is computed as follows.

$$\begin{aligned}\chi^2 &= \Sigma \frac{(O - E)^2}{E}\\ &= \frac{(70-65)^2}{65} + \frac{(34-39)^2}{39} + \frac{(80-75)^2}{75} + \frac{(40-45)^2}{45}\\ &\quad + \frac{(100-110)^2}{110} + \frac{(76-66)^2}{66}\\ &= .385 + .641 + .333 + .555 + .909 + 1.515 = 4.338\end{aligned}$$

Step 5. *Make a decision*

The value of the test statistic $\chi^2 = 4.338$ is less than the critical value of $\chi^2 = 5.991$, and it falls in the nonrejection region. Hence, we fail to reject the null hypothesis and state that the distribution of households with regard to income appears to be similar in California and Wisconsin. ■

EXERCISES

11.16 Two hundred adults were asked whether they prefer to watch sports or soap operas on television. The following table lists the preferences of these men and women.

	Sports	Soap Operas
Men	51	39
Women	68	42

Test at the 10% significance level if being a man or a woman and preferring to watch sports or soap operas are related.

11.17 A sample of 150 chief executive officers (CEOs) were tested for their type of personality. The following table gives the results of this survey.

	Type A	Type B
Men	78	42
Women	12	18

Test at the 5% significance level if sex and type of personality are related for all CEOs.

11.18 One hundred auto drivers, who were stopped by police for some violation, were also checked to see whether they were wearing their seat belts. The following table records the results of this survey.

	Wearing a Seat Belt	Not Wearing a Seat Belt
Male	34	21
Female	32	13

Test at the 2.5% significance level if being a male or a female and wearing or not wearing a seat belt are related.

11.19 A sample of 400 items manufactured on two machines was taken. The items were classified on the basis of the machine on which they were manufactured and whether they were good or defective. The following table gives the results of the survey.

	Machine I	Machine II
Good	196	175
Defective	20	9

Test at the 5% significance level if the machine on which an item is manufactured and an item being good or defective are associated.

11.20 The following table gives a two-way classification for a sample of 190 adults on the basis of whether or not they are overweight and whether they have high blood pressure or normal blood pressure.

	High Blood Pressure	Normal Blood Pressure
Overweight	57	18
Not overweight	24	91

Test at the 1% significance level if the two attributes listed in the table are dependent for all adults.

11.21 The following table gives the two-way classification of 400 randomly selected persons based on their status as smoker or nonsmoker and on the number of visits they made to their physicians last year.

	Visits to the Physician		
	0–1	2–4	≥5
Smoker	20	60	80
Nonsmoker	110	90	40

Test at the 5% significance level if smoking and visits to the physician are related for all persons.

11.22 The following table is based on the GPAs (grade point averages) of a sample of 300 students selected from all classes taught by one instructor during the past 4 years and how these students evaluated this instructor.

		GPA of the Student		
		Below 2.5	2.5 to 3.5	Above 3.5
Evaluation of the Instructor	Excellent	18	33	37
	Good	17	27	43
	Average	21	31	23
	Poor	25	14	11

Test at the 1% significance level if GPAs of students and instructor evaluations are dependent.

11.23 The following table gives the distribution of 200 people from the Northeast region of the United States and 300 from the Western region according to their party affiliations.

	Northeast	West
Democrats	88	121
Republicans	76	130
Others	36	49

Using the 10% significance level, test the null hypothesis that the party affiliations in the two regions are homogeneous.

11.24 Two drugs were administered to two groups of randomly assigned patients to cure the same disease. One group had 60 patients and the other 40 patients. The following table gives information about the number of patients who were cured and the ones who were not cured by each of the two drugs.

	Cured	Not Cured
Drug I	46	14
Drug II	18	22

Test at the 1% significance level if the two drugs are different in curing and not curing the patients.

11.25 The following table gives the distribution of grades for three professors for a few randomly selected classes that each of them taught during the past 2 years.

		Professor		
		Miller	Smith	Moore
Grade	A	18	36	20
	B	25	44	15
	C	85	73	82
	D & F	17	12	8

Using the 5% significance level, test the null hypothesis that the grade distributions are homogeneous for these three professors.

11.26 Two random samples, one of 95 blue-collar workers and a second of 50 white-collar workers, were taken from a large company. These workers were asked about their views on a certain company issue. The following table gives the results of the survey.

	Opinion		
	Favor	**Oppose**	**Uncertain**
Blue-collar workers	47	39	9
White-collar workers	21	26	3

Using the 2.5% significance level, test the null hypothesis that the distributions of opinion are homogeneous for the two groups of workers.

11.27 Two samples, one of 100 scientists and another of 100 engineers, produced the following two-way classification table.

	Male	**Female**
Scientist	69	31
Engineer	92	8

Using the 5% significance level, test the null hypothesis that the distributions of being a scientist or engineer are homogeneous for all males and females.

11.5 INFERENCES ABOUT THE POPULATION VARIANCE

Earlier chapters explained how to make inferences (confidence intervals and hypothesis tests) about the population mean and population proportion. However, we may often need to control the variance (or standard deviation). Hence, there may be a need to estimate and to test a hypothesis about the population variance σ^2. Section 11.5.1 describes how to make a confidence interval for the population variance and standard deviation. Section 11.5.2 explains how to test a hypothesis about the population variance and standard deviation.

As an example, suppose a machine is set up to fill packages of cookies so that the net weight of cookies per package is 32 ounces. Note that the machine will not put exactly 32 ounces of cookies in each package. Some of the packages will contain less and some will contain more than 32 ounces. However, if the variance (and, hence, the standard deviation) is too large, some of the packages will contain quite a bit below 32 ounces of cookies and some others will contain quite a bit above 32 ounces. The manufacturer will not want a large variation in the amounts of cookies put in different packages. To keep this variation within some specified acceptable limit, the machine will be adjusted from time to time. Before the manager decides to adjust the machine at any time, he must estimate the variance or test a hypothesis or do both to find out if the variance exceeds the maximum acceptable value.

Like every sample statistic, the sample variance is a random variable and it possesses a sampling distribution. If all the possible samples of a given size are taken from a population and their variances are calculated, the probability distribution of these variances is called the *sampling distribution of the sample variance*.

SAMPLING DISTRIBUTION OF $(n - 1)s^2/\sigma^2$

If the population, from which the sample is taken is (approximately) normally distributed, then

$$\frac{(n-1)\,s^2}{\sigma^2}$$

has a chi-square distribution with $n - 1$ degrees of freedom.

Thus, the chi-square distribution is used to construct a confidence interval and test a hypothesis about the population variance σ^2.

11.5.1 ESTIMATION OF THE POPULATION VARIANCE

The value of the sample variance s^2 is a point estimate of the population variance σ^2. The $(1 - \alpha)100\%$ confidence interval for σ^2 is given by the following formula.

CONFIDENCE INTERVAL FOR THE POPULATION VARIANCE σ^2

Assuming that the population from which the sample is taken is (approximately) normally distributed, the $(1 - \alpha)100\%$ confidence interval for the population variance σ^2 is

$$\frac{(n-1)\,s^2}{\chi^2_{\alpha/2}} \quad \text{to} \quad \frac{(n-1)\,s^2}{\chi^2_{1-\alpha/2}}$$

where $\chi^2_{\alpha/2}$ and $\chi^2_{1-\alpha/2}$ are obtained from the chi-square table for $\alpha/2$ and $1 - \alpha/2$ areas in the right tail, respectively, and for $n - 1$ degrees of freedom.

The procedure to make a confidence interval for σ^2 involves the following three steps.

1. Take a sample of size n and compute s^2 using the formula learned in Chapter 3. (If n and s^2 are given, then only perform steps 2 and 3.)
2. Calculate $\alpha/2$ and $1 - \alpha/2$. Find two values of χ^2 from the chi-square table (Table IX of Appendix C): one for $\alpha/2$ area in the right tail and $df = n - 1$, and the second for $1 - \alpha/2$ area in the right tail and $df = n - 1$.
3. Substitute all the values in the formula for the confidence interval for σ^2 and simplify.

The confidence interval for the population standard deviation can be obtained by simply taking the positive square root of the two limits of the confidence interval for the population variance.

Example 11-9 illustrates the estimation of the population variance and population standard deviation.

Constructing confidence intervals for σ^2 and σ.

EXAMPLE 11-9 The manufacturer of Cocoa Cookies wants to estimate the variance of the net weights of cookies in all 32-ounce packages. A random sample of 25 packages taken from the production line gave a sample variance of .029 square ounces. Construct a 95% confidence interval for σ^2. Assume that the net weights of cookies in all packages are normally distributed.

Solution

Step 1. From the given information,

$$n = 25 \quad \text{and} \quad s^2 = .029$$

Step 2. The confidence level is $1 - \alpha = .95$. Hence,

$$\alpha = 1 - .95 = .05$$

$$\alpha/2 = .05/2 = .025 \quad \text{and} \quad 1 - \alpha/2 = 1 - .025 = .975$$

$$df = n - 1 = 25 - 1 = 24$$

From Table IX,

$$\chi^2 \text{ for 24 } df \text{ and .025 area in the right tail} = 39.364$$

$$\chi^2 \text{ for 24 } df \text{ and .975 area in the right tail} = 12.401$$

These values are shown in Figure 11.9.

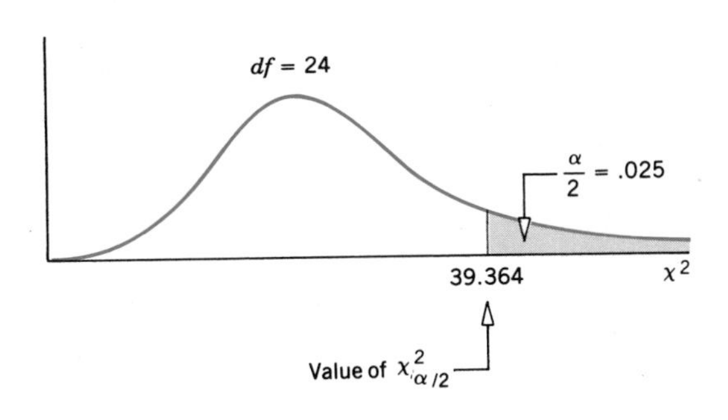

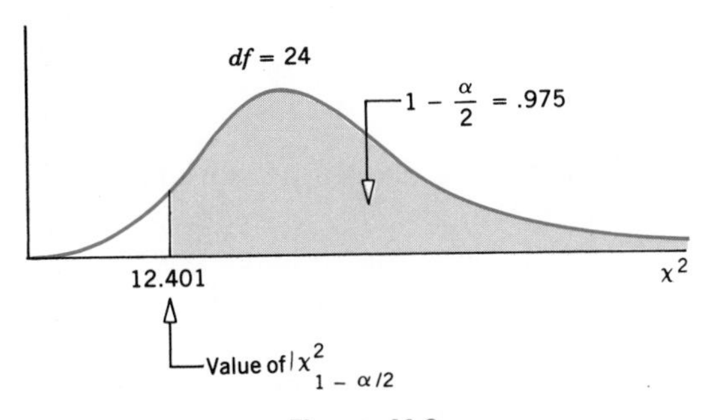

Figure 11.9

Step 3. The 95% confidence interval for σ^2 is

$$\frac{(n-1)\,s^2}{\chi^2_{\alpha/2}} \quad \text{to} \quad \frac{(n-1)\,s^2}{\chi^2_{1-\alpha/2}}$$

or

$$\frac{(25-1)\,(.029)}{39.364} \quad \text{to} \quad \frac{(25-1)\,(.029)}{12.401}$$

or

$$.0177 \quad \text{to} \quad .0561$$

Thus, with 95% confidence, we can state that the variance for all packages of Cocoa Cookies lies between .0177 and .0561 square ounces.

We can obtain the confidence interval for the population standard deviation σ by taking the positive square root of the two limits of the confidence interval for the population variance. Thus, a 95% confidence interval for the population standard deviation is

$$\sqrt{.0177} \quad \text{to} \quad \sqrt{.0561} \quad \text{or} \quad .133 \quad \text{to} \quad .237$$

Hence, the standard deviation of all packages of Cocoa Cookies is between .133 and .237 ounces at a 95% confidence level. ■

11.5.2 HYPOTHESIS TESTS ABOUT THE POPULATION VARIANCE

A test of hypothesis about the population variance can be one-tailed or two-tailed. To make a test of hypothesis about σ^2, we perform the same five steps we have used earlier in hypothesis-testing examples. The procedure to test a hypothesis about σ^2 discussed in this section is applied only when the population from which a sample is taken is (approximately) normally distributed.

TEST STATISTIC FOR A TEST OF HYPOTHESIS ABOUT σ^2

The value of the test statistic χ^2 is calculated as

$$\chi^2 = \frac{(n-1)\,s^2}{\sigma^2}$$

where s^2 is the sample variance, σ^2 is the hypothesized value of the population variance, and $n - 1$ represents the degrees of freedom. The population from which the sample is taken is assumed to be (approximately) normally distributed.

Examples 11-10 and 11-11 illustrate the procedure to make tests of hypotheses about σ^2.

Making a right-tailed test of hypothesis about σ^2.

EXAMPLE 11-10 Refer to Example 11-9 about the variance of Cocoa Cookies. Suppose the acceptable value of variance is .015 or less. The sample variance for a sample of 25 packages is .029 square ounces. At the 1% significance level, can we conclude that the population variance exceeds the acceptable limit?

Solution From the given information,

$$n = 25, \quad \alpha = .01, \quad \text{and} \quad s^2 = .029$$

The acceptable value of variance is .015 or less.

Step 1. *State the null and alternative hypotheses*

We are to test whether or not the population variance is within the acceptable limit. The population variance is within the acceptable limit if it is less than or equal to .015; otherwise, it is not. Hence, the two hypotheses are

H_0: $\sigma^2 \leq .015$ (the population variance is within the acceptable limit)

H_1: $\sigma^2 > .015$ (the population variance exceeds the acceptable limit)

Step 2. *Select the distribution to use*

We use the chi-square distribution to test a hypothesis about σ^2.

Step 3. *Determine the rejection and nonrejection regions*

The significance level is 1% and, because of the > sign in H_1, the test is right-tailed. The rejection region lies in the right tail with its area equal to .01. The degrees of freedom for a chi-square test about σ^2 are $n - 1$. Hence,

$$df = n - 1 = 25 - 1 = 24$$

From Table IX of Appendix C, the critical value of χ^2 for 24 degrees of freedom and .01 area in the right tail is 42.980. This value is shown in Figure 11.10.

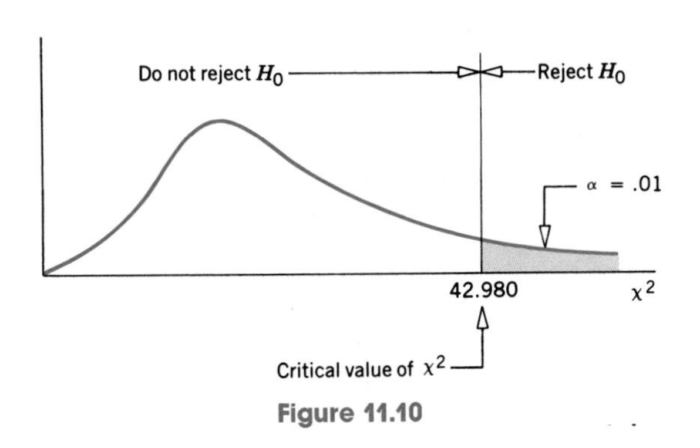

Figure 11.10

Step 4. *Calculate the value of the test statistic*

The value of the test statistic χ^2 for the sample variance is calculated as follows.

$$\chi^2 = \frac{(n-1)\,s^2}{\sigma^2} = \frac{(25-1)\,(.029)}{(.015)} = 46.400$$

↑ From H_0

Step 5. *Make a decision*

The value of the test statistic $\chi^2 = 46.400$ falls in the rejection region. Consequently, we reject H_0 and conclude that the population variance is not within the acceptable limit. ■

Conducting a two-tailed test of hypothesis about σ^2.

EXAMPLE 11-11 The variance of scores in a standardized mathematics test for high school seniors was 150 in 1988. A sample of scores for 20 high school seniors who took that test this year gave a variance of 170. Test at the 5% significance level if the variance of current scores of all high school seniors in this test is different from 150. Assume that the scores of all high school seniors in this test are (approximately) normally distributed.

Solution From the given information,

$$n = 20, \quad \alpha = .05, \quad \text{and} \quad s^2 = 170$$

The population variance was 150 in 1988.

Step 1. *State the null and alternative hypotheses*

The null and alternative hypotheses are

H_0: $\sigma^2 = 150$ (the population variance is not different from 150)

H_1: $\sigma^2 \neq 150$ (the population variance is different from 150)

Step 2. *Select the distribution to use*

We use the chi-square distribution to test a hypothesis about σ^2.

Step 3. *Determine the rejection and nonrejection regions*

The significance level is 5%. The $\neq$ sign in H_1 indicates that the test is two-tailed. The rejection region lies in both tails with its total area equal to .05. Consequently, the area in each tail is .025. Hence,

$$\frac{\alpha}{2} = \frac{.05}{2} = .025 \quad \text{and} \quad 1 - \frac{\alpha}{2} = 1 - .025 = .975$$

The degrees of freedom are $df = n - 1 = 20 - 1 = 19$.

From Table IX, the critical values of χ^2 for 19 degrees of freedom and for $\alpha/2$ and $1 - \alpha/2$ areas in the right tail are

χ^2 for 19 df and .025 area in the right tail = 32.852

χ^2 for 19 df and .025 area in the left tail

= χ^2 for 19 df and .975 area in the right tail = 8.907

These two values are shown in Figure 11.11.

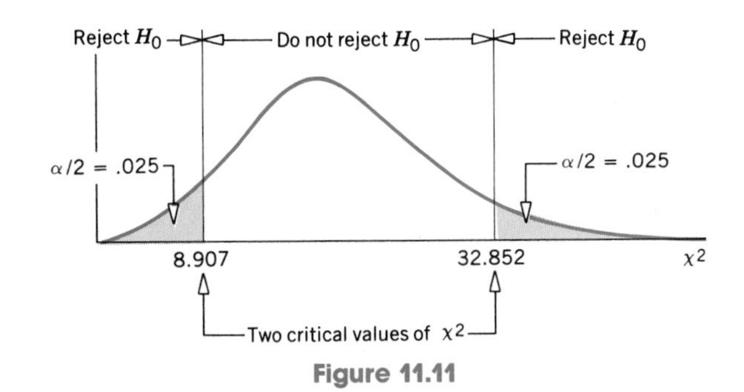

Figure 11.11

Step 4. *Calculate the value of the test statistic*

The value of the test statistic χ^2 for the sample variance is calculated as follows.

$$\chi^2 = \frac{(n-1)\,s^2}{\sigma^2} = \frac{(20-1)\,(170)}{(150)} = 21.533$$

From H_0 (points to the 150 in the denominator)

Step 5. *Make a decision*

The value of the test statistic $\chi^2 = 21.533$ falls in the nonrejection region. Hence, we fail to reject H_0 and conclude that the population variance does not appear to be different from 150. ■

EXERCISES

11.28 Construct the 95% confidence intervals for the population variance and standard deviation for the following. Assume that the respective populations are normally distributed.

a. $n = 20$, $s^2 = 14$ **b.** $n = 16$, $s^2 = .9$

11.29 Construct the 99% confidence intervals for the population variance and standard deviation for the following. Assume that the respective populations are normally distributed.

a. $n = 25$, $s^2 = 1.8$ **b.** $n = 14$, $s^2 = 3.6$

11.30 Refer to Exercise 11.28a. Test at the 1% significance level if the population variance is different from 11.0.

11.31 Refer to Exercise 11.28b. Test at the 5% significance level if the population variance is greater than .50.

11.32 Refer to Exercise 11.29a. Test at the 2.5% significance level if the population variance is greater than 1.0.

11.33 Refer to Exercise 11.29b. Test at the 5% significance level if the population variance is different from 1.7.

11.34 A random sample of 25 students taken from a university gave the variance of their GPAs equal to .19.

a. Construct the 95% confidence intervals for the population variance and standard deviation. Assume that the GPAs of all students are (approximately) normally distributed.

b. The variance of GPAs of all students at this university was .13 two years ago. Test at the 1% significance level if the variance of GPAs now is different from .13.

11.35 The management of a soft-drink company does not want the variance of the amount of soda in 12-ounce cans to be more than .01 square ounces. (Recall from Chapter 3 that the variance is always in square units.) A random sample of twenty 12-ounce cans taken from the production line of this company showed that the variance for this sample was .014 square ounces.

- **a.** Construct a 99% confidence interval for the population variance. Assume that the amount of soda in all 12-ounce cans has a normal distribution.
- **b.** Test at the 5% significance level if the variance of the amounts of soda in all 12-ounce cans for this company is greater than .01 square ounces.

11.36 An auto manufacturing company claims that model AST727 of a car manufactured by this company gives 28 miles per gallon. This claim has been proved to be true. But the manufacturer also wants to keep the variance of miles per gallon below a certain limit. A random sample of 24 cars of this model showed that the variance of miles per gallon is .58.

- **a.** Construct the 90% confidence intervals for the population variance and standard deviation. Assume that the miles per gallon for all such cars are (approximately) normally distributed.
- **b.** The manufacturer does not want the variance of miles per gallon to be larger than .30. Test at the 2.5% significance level if the sample result indicates that the population variance is larger than .30.

11.37 The 2-inch bolts manufactured by a company must have a variance of .003 square inches or less for acceptance by a buyer. A random sample of 29 bolts gave a variance of .0058 square inches.

- **a.** Test at the 1% significance level if the variance of 2-inch bolts is greater than .003 square inches. Assume that the lengths of all 2-inch bolts manufactured by this company are (approximately) normally distributed.
- **b.** Make the 98% confidence intervals for the population variance and standard deviation.

GLOSSARY

Chi-square distribution A distribution, with degrees of freedom as the only parameter, that is skewed to the right for small *df* and looks like a normal curve for large *df*.

Expected frequencies The frequencies for different categories of a multinomial experiment or for different cells of a contingency table that are expected to occur when a given null hypothesis is true.

Goodness-of-fit test A test of the null hypothesis that the observed frequencies for an experiment follow a certain pattern or theoretical distribution.

Multinomial experiment An experiment for which (1) the trials are identical, (2) there are more than two possible outcomes per trial, (3) the trials are independent, and (4) the probabilities of various outcomes remain constant for each trial.

Observed frequencies The frequencies actually obtained from the performance of an experiment.

Test of homogeneity A test of the null hypothesis that the proportions of elements that belong to different groups in two (or more) populations are similar.

Test of independence A test of the null hypothesis that the two attributes of a population are not related.

KEY FORMULAS

1. **Expected frequency for a category for a goodness-of-fit test**

$$E = np$$

where n is the sample size and p is the probability of that category.

2. **Value of the test statistic χ^2 for a goodness-of-fit test and a test of independence or homogeneity**

$$\chi^2 = \Sigma \frac{(O - E)^2}{E}$$

where O and E are the observed and expected frequencies, respectively, for a category or cell.

3. **Degrees of freedom for a goodness-of-fit test**

$$df = k - 1$$

where k denotes the number of categories for the experiment.

4. **Degrees of freedom for a test of independence or homogeneity**

$$df = (R - 1)(C - 1)$$

where R and C are, respectively, the number of rows and columns for the contingency table.

5. **Expected frequency for a cell for an independence or homogeneity test**

$$E = \frac{\text{(Row total)(Column total)}}{\text{Sample size}}$$

6. **The $(1 - \alpha)100\%$ confidence interval for the population variance σ^2**

$$\frac{(n - 1)\, s^2}{\chi^2_{\alpha/2}} \quad \text{to} \quad \frac{(n - 1)\, s^2}{\chi^2_{1-\alpha/2}}$$

where $\chi^2_{\alpha/2}$ and $\chi^2_{1-\alpha/2}$ are obtained from the chi-square table for $\alpha/2$ and $1 - \alpha/2$ areas in the right tail and $n - 1$ degrees of freedom.

7. **Value of the test statistic χ^2 in a hypothesis test about σ^2**

$$\chi^2 = \frac{(n - 1)\, s^2}{\sigma^2}$$

where s^2 is the sample variance, σ^2 is the hypothesized value of the population variance, and $n - 1$ are the degrees of freedom.

SUPPLEMENTARY EXERCISES

11.38 In a 1987 Roper poll, school children aged 8 to 17 were asked about the overall effect on children aged 12 or less of both mothers and fathers working outside the home. Twenty-five percent of the children surveyed said that the effect is good, 39% said it is bad, 29% said there is no effect, and 7% said they did not know (*The American Chicle Youth Poll,* commissioned by The American Chicle Group, Warner-Lambert Company, March 1987). A random sample of 250 school children aged 8 to 17 taken recently produced the results listed in the following table in response to the same question.

Category	Good	Bad	No Effect	Do Not Know
Number of children	56	111	61	22

Test at the 1% significance level if the current percentage distribution of school children aged 8 to 17 that belong to the four categories is different from the one for 1987.

11.39 A random sample of 100 adults is selected from each of the five different age groups and the number of persons who are alcoholic in each group is noted. The findings are listed in the following table.

Age group	21–25	26–35	36–45	46–65	66 and Older
Persons alcoholic	4	9	15	18	11

Using the 5% significance level, test the null hypothesis that the proportion of alcoholics is the same for all five age groups.

11.40 The following table lists the number of persons in a random sample of 210 according to the day of the week on which they prefer to do their grocery shopping.

Day	Mon	Tue	Wed	Thu	Fri	Sat	Sun
Number of persons	9	15	10	28	38	69	41

Using the 2.5% significance level, test the null hypothesis that the proportion of persons who prefer to do their grocery shopping on a particular day is the same for all days of the week.

11.41 In a 1986 Roper poll, conducted for *U.S. News & World Report,* people were asked how satisfied they were "with the way things are going in the United States today." Of the respondents, 14% said very satisfied, 60% said more or less satisfied, 24% said not at all satisfied, and 2% said they did not know (*Survey 1986: Issues and Reactions—Major Events of 1986 and How You Felt About Them*). A random sample of 450 persons taken recently produced the results listed in the following table in response to the same question.

Category	Very	More or Less	Not at All	Do Not Know
Number of persons	73	267	104	6

Test at the 1% significance level if the current percentage distribution of the people that belong to four categories is different from the one for 1986.

11.42 A sample of 100 persons who suffer from allergies were asked during what season they suffer the most from allergies. The results of the survey are recorded in the following table.

Season	Fall	Winter	Spring	Summer
Persons allergic	20	13	29	38

Using the 2.5% significance level, test the null hypothesis that the proportions of all allergic persons are equally distributed over the four seasons.

11.43 The following table gives the percentage distribution (according to age) of persons 20 years and older who were arrested for homicide in 1984.

Age	20–24	25–29	30–39	40 and Older
Percent	30	23	27	20

Among a random sample of 80 persons, 20 years or older, who were arrested this year for homicide, 33 belonged to the 20–24 age group, 25 to the 25–29 age group, 13 to the 30–39 age group, and 9 belonged to the 40 and older age group. Test at the 1% significance level if the age distribution of persons arrested for homicide this year is different from that for 1984.

11.44 The president of a bank selected a sample of 200 loan applications to check if the approval or rejection of an application depends on which one of the two loan officers handles that application. The information obtained from the sample is summarized in the following table.

	Approved	**Rejected**
Thurow	57	38
Webber	69	36

Test at the 2.5% significance level if the approval or rejection of a loan application depends on which loan officer handles the application.

11.45 The following table gives the two-way classification of a sample of 1000 students selected from certain colleges and universities.

	Business Major	**Nonbusiness Major**
Male	77	389
Female	52	482

Test at the 1% significance level if being a male or a female and being a business or nonbusiness major are related.

11.46 A random sample of 100 jurors was selected and they were asked whether they have ever been victims of crime. They were also asked whether they are strict, fair, or lenient on punishment for crime. The following table gives the results of the survey.

	Strict	Fair	Lenient
Have been a victim	20	8	3
Have never been a victim	22	38	9

Test at the 5% significance level if the two attributes for all jurors are dependent.

11.47 A sample of 500 persons who held more than one job produced the following two-way table.

	Single	Married	Other
Male	69	212	39
Female	33	102	45

Test at the 10% significance level if sex and marital status are related for all persons who hold more than one job.

11.48 Two samples, one of 90 adult males and another of 95 adult females, were taken. These adults were asked whether they favor or oppose the right to abortion. The following table records the results of the survey.

	Favor	Oppose
Male	41	49
Female	59	36

Using the 5% significance level, test the null hypothesis that the percentages of all adult males and all adult females who favor and oppose the right to abortion are homogeneous.

11.49 Four samples, one from each of four regions in the United States, were taken. The people polled were asked whether or not they support a certain farm subsidy program. The results of the survey are summarized in the following table.

	Favor	Oppose	Uncertain
Northeast	56	33	11
Midwest	73	23	4
South	67	28	5
West	59	35	6

Using the 1% significance level, test the null hypothesis that the percentage of people with different opinions are similar for all four regions.

11.50 Construct the 98% confidence intervals for the population variance and standard deviation for the following data assuming that the respective populations are (approximately) normally distributed.

a. $n = 21, \quad s^2 = 8.5$ **b.** $n = 17, \quad s^2 = 1.4$

11.51 Construct the 90% confidence intervals for the population variance and standard deviation for the following data assuming that the respective populations are (approximately) normally distributed.

a. $n = 12, \quad s^2 = 6.9$ **b.** $n = 19, \quad s^2 = 14.3$

11.52 Refer to Exercise 11.50a. Test at the 5% significance level if the population variance is different from 6.0.

11.53 Refer to Exercise 11.50b. Test at the 2.5% significance level if the population variance is greater than .7.

11.54 Refer to Exercise 11.51a. Test at the 1% significance level if the population variance is greater than 4.0.

11.55 Refer to Exercise 11.51b. Test at the 5% significance level if the population variance is different from 10.0.

11.56 A bank manager does not want the variance of the waiting times for her customers to be higher than 4.0 square minutes. A random sample of 25 customers taken from this bank gave the variance of the waiting times equal to 7.9 square minutes.

a. Test at the 1% significance level if the variance of the waiting times for all customers at this bank is higher than 4.0 square minutes. Assume that the waiting times for all customers are normally distributed.

b. Construct a 99% confidence interval for the population variance.

11.57 The variance of the SAT scores for all students who took that test this year is 5000. The variance of the SAT scores for a random sample of 20 students from one school is equal to 3200.

a. Test at the 2.5% significance level if the variance of the SAT scores for students from this school is lower than 5000. Assume that SAT scores for all students at this school are (approximately) normally distributed.

b. Construct the 98% confidence intervals for the variance and the standard deviation of SAT scores for all students at this school.

11.58 If the variance of the diameters of the ball bearings manufactured on a machine is larger than .025 square millimeters, the machine needs to be adjusted. A random sample of 23 ball bearings gave a variance of the diameters equal to .031 square millimeters.

a. Test at the 5% significance level if the variance of the diameters of the ball bearings manufactured on this machine is higher than .025 square millimeters. Assume that the lengths of the diameters of all ball bearings are normally distributed.

b. Construct a 95% confidence interval for the population variance.

SELF-REVIEW TEST

1. The variable χ^2 assumes only

a. positive **b.** nonnegative **c.** nonpositive values

2. The parameter(s) of the chi-square distribution is (are)

   a. degrees of freedom **b.** *df* and n **c.** χ^2

3. Which of the following is not a characteristic of a multinomial experiment?

   a. It consists of n identical trials.
   b. There are k possible outcomes for each trial and $k > 2$.
   c. The occurrences are random.
   d. The trials are independent.
   e. The probabilities of outcomes remain constant for each trial.

4. The observed frequencies for a goodness-of-fit test are

   a. the frequencies obtained from the performance of an experiment
   b. the frequencies given by the product of n and p
   c. the frequencies obtained by adding the results of a and b

5. The expected frequencies for a goodness-of-fit test are

   a. the frequencies obtained from the performance of an experiment
   b. the frequencies given by the product of n and p
   c. the frequencies obtained by adding the results of a and b

6. The degrees of freedom for a goodness-of-fit test are

   a. $n - 1$ **b.** $k - 1$ **c.** $n + k - 1$

7. A chi-square goodness-of-fit test is always

   a. two-tailed **b.** left-tailed **c.** right-tailed

8. To apply a goodness-of-fit test, the expected frequency of each category must be at least

   a. 10 **b.** 5 **c.** 8.

9. The degrees of freedom for a test of independence are

   a. $(R - 1)(C - 1)$ **b.** $n - 2$ **c.** $(n - 1)(k - 1)$

10. A study done more than a decade ago found that 63% of adults opposed living together before marriage, 27% were in favor, and 10% had no opinion. A recent sample of 300 adults showed that 118 of them are against living together before marriage, 162 are in favor, and 20 have no opinion. Test at the 1% significance level if the current percentage distribution of opinions of adults on this issue is different from the one obtained more than a decade ago.

11. The following table gives the two-way classification of 1000 persons who have been married at least once. They are classified by educational level and marital status.

	Educational Level			
	Less Than High School	**High School Degree**	**Some College**	**College Degree**
Divorced	173	158	95	53
Never divorced	162	126	116	117

Test at the 1% significance level if the educational level and being ever divorced are dependent.

12. To study if obesity among children of obese and nonobese parents is homogeneous, a psychologist took one sample of 50 obese children and another sample of

50 nonobese children. The following table gives the number of children from both samples whose parents were obese or nonobese. The parents are classified as obese if at least one of the parents is obese.

		Child	
		Obese	**Nonobese**
Parent	Obese	33	22
	Nonobese	17	28

Using the 5% significance level, test the null hypothesis that the percentages of obese and nonobese children with obese and nonobese parents are homogeneous.

13. A cough syrup drug manufacturer requires that the variance for a chemical contained in the bottles of this drug should not exceed .03 square ounces. A sample of 25 such bottles gave the variance for this chemical as .06 square ounces.

  a. Construct the 99% confidence intervals for the population variance and the population standard deviation. Assume that the amount of this chemical in all such bottles is (approximately) normally distributed.

  b. Test at the 1% significance level if the variance of this chemical in all such bottles exceeds .03 square ounces.

USING MINITAB

MINITAB does not have a direct command to make a goodness-of-fit test. However, by combining a few commands, we can make such a test. The MINITAB command CHISQUARE is used to make a test of independence or homogeneity.

A GOODNESS-OF-FIT TEST

Illustration M11-1 describes the procedure to perform a goodness-of-fit test using MINITAB.

Illustration M11-1 Refer to Example 11-4. In a 1986 Roper poll, 64% of adults said "over the next 20 years . . . science and technology will do more good than harm for the human race," 23% said it will do more harm, and 13% said neither/do not know. A recent sample of 500 adults produced the results listed in the following table in response to the same question.

Category	More Good	More Harm	Neither/Do Not Know
Number of persons	342	107	51

Test at the 1% significance level if the current percentage distribution of adults belonging to three categories is different from the one for 1986.

Solution The null and alternative hypotheses are

H_0: The current percentage distribution is the same as the one for 1986

H_1: The current percentage distribution is different from that for 1986

To make a goodness-of-fit test for this illustration, perform the following steps.

1. First, enter the data on probabilities and observed frequencies in columns C1 and C2 using the READ command.
2. Second, find the sum of column C2 (the column of observed frequencies) and put it in K1. This gives the sample size.
3. Third, create column C3 by multiplying the sum of column C2 by the probabilities of column C1. Column C3 lists the expected frequencies.

4. Fourth, create column C4 using the formula $(O - E)^2/E$.

5. Fifth, find the sum of the values listed in column C4, which gives the value of the test statistic χ^2.

6. Finally, find the value of χ^2 for the given α and df from the chi-square table and make a decision by comparing it with the value of the test statistic χ^2.

All these steps are shown in the following MINITAB input and output display.

```
MTB  > NOTE: GOODNESS-OF-FIT TEST FOR ILLUSTRATION M11-1
MTB  > READ C1 C2  ← This command enters the data on
DATA > .64  342       probabilities and observed frequencies
DATA > .23  107       in columns C1 and C2, respectively
DATA > .13  51
DATA > END
       3 ROWS READ

MTB  > SUM C2 PUT IN K1  ← This command calculates the sum of the values entered
                           in column C2 and puts it in K1
   SUM    =    500.000   ← This is the sample size

MTB  > LET C3 = K1 * C1  ← This command creates column C3 of expected frequencies
                           by multiplying K1 and column C1 values

MTB  > LET C4 = (C2 - C3)**2/C3  ← This command calculates
                                   (O - E)^2 / E for each category

MTB  > PRINT C1 C2 C3 C4

ROW    C1    C2    C3        C4
  1  0.64   342   320   1.51250   ← Compare this table
  2  0.23   107   115   0.55652     with Table 11.4 of
  3  0.13    51    65   3.01538     Example 11-4

MTB  > SUM C4 ←———— This command prints the sum of column C4,
   SUM    =    5.0844   which is the value of the test statistic χ²
```

From the MINITAB solution, the value of the test statistic χ^2 is 5.0844. For this illustration, $\alpha = .01$ and $df = 3 - 1 = 2$. The critical value of χ^2 from the chi-square table for $\alpha = .01$ and $df = 2$ is 9.210 (see Figure 11.5 of Example 11-4). The value of the test statistic $\chi^2 = 5.0844$ is smaller than the critical value of $\chi^2 = 9.210$, and it falls in the nonrejection region. Consequently, we fail to reject the null hypothesis.

Hence, we conclude that the current observed frequencies seem to fit the 1986 distribution.

A TEST OF INDEPENDENCE OR HOMOGENEITY

To make a chi-square test of independence or homogeneity, we first enter the data for all columns of a given table by using the READ command. Remember, we do not enter the data listed in the row and column of totals. MINITAB command CHISQUARE followed by the names of the data columns gives us the value of the test statistic. Finally, we compare this value of the test statistic with the value of χ^2 calculated from the chi-square table and make a decision. Illustration M11-2 describes the procedure used to make a chi-square test of independence. The procedure for a chi-square test of homogeneity is similar.

Illustration M11-2 According to Example 11-7, a sample of 600 persons, aged 20 and older, with known cases of diabetes produced the following table.

	Age		
	20–44	**45–64**	**65 and Older**
Men	36	109	126
Women	52	117	160

Test at the 1% significance level if sex and age are related for diabetic persons aged 20 and older.

Solution The null and alternative hypotheses are

H_0: Sex and age are independent for diabetic people aged 20 and older

H_1: Sex and age are dependent for diabetic people aged 20 and older

First we enter the data given in the contingency table into MINITAB. We use the MINITAB command READ C1 C2 C3 to enter the data given in three columns of the table. Note that the number of columns in the READ command is equal to the number of columns in the contingency table. There are three columns in the table, one for each age group. The MINITAB commands and the MINITAB output for a test of independence for these data are given on the next page.

From the MINITAB output, the value of the test statistic χ^2 is 1.643. The critical value of χ^2 from Table IX of Appendix C for $\alpha = .01$ and $df = (2 - 1)(3 - 1) = 2$ is 9.210. Because the value of the test statistic $\chi^2 = 1.643$ is smaller than the critical value of $\chi^2 = 9.210$ and it falls in the nonrejection region, we fail to reject the null hypothesis (see Figure 11.7 of Example 11-7). Consequently, we conclude that sex and age for diabetic persons aged 20 and older are not related.

```
MTB  > NOTE: MAKING A TEST OF INDEPENDENCE
MTB  > READ C1 C2 C3 <--  [This command enters the values given
DATA > 36  109  126       [in the contingency table
DATA > 52  117  160
DATA > END
       2 ROWS READ

MTB  > CHISQUARE C1 C2 C3 <-- [This command instructs MINITAB to make a
                              [chi-square test using the data of columns C1,
                              [C2, and C3

Expected counts are printed below observed counts

           C1       C2       C3    Total
  1        36      109      126      271
        39.75   102.08   129.18 <--
                                         --These rows give
  2        52      117      160      329   the expected
        48.25   123.92   156.82 <--        frequencies

Total      88      226      286      600

CHISQ = 0.353 + 0.470 + 0.078 +
        0.291 + 0.387 + 0.064 = 1.643 <-- [This is the value
                                          [of the test statistic
DF = 2
```

COMPUTER ASSIGNMENTS

M11.1 Forty-two percent of adults were in favor of the death penalty, 35% were against it, and 23% had no opinion in 1980. A recent sample of 1000 adults showed that 490 of them are in favor of the death penalty, 370 are against it, and 140 have no opinion. Using MINITAB, test at the 1% significance level if the percentages of the current population of all adults who are in favor of the death penalty, against it, and have no opinion are different from those for 1980.

M11.2 A sample of 4000 persons aged 18 and older produced the following two-way classification table.

	Men	Women
Single	531	357
Married	1375	1179
Widowed	55	195
Divorced	139	169

Using MINITAB, test at the 10% significance level if sex and marital status are dependent for all persons aged 18 and older.

M11.3 Two samples, one of 3000 students from urban high schools and another of 2000 students from rural high schools, were taken. These students were asked if they have ever smoked. The following table lists the summary of the results.

	Urban	**Rural**
Have never smoked	1448	1228
Have smoked	1552	772

Using the 5% significance level, test the null hypothesis that the proportions of urban and rural students who have smoked and who have never smoked are homogeneous.

12 SIMPLE LINEAR REGRESSION

12.1 THE SIMPLE LINEAR REGRESSION MODEL
12.2 THE SIMPLE LINEAR REGRESSION ANALYSIS
12.3 THE STANDARD DEVIATION OF THE RANDOM ERROR
12.4 THE COEFFICIENT OF DETERMINATION
12.5 INFERENCES ABOUT B
12.6 LINEAR CORRELATION
12.7 REGRESSION ANALYSIS: A COMPLETE EXAMPLE
12.8 USING THE REGRESSION MODEL
12.9 CAUTIONS IN USING REGRESSION

This chapter will consider the relationship between two variables in two ways: (1) by using the regression line, and (2) by computing the correlation coefficient. By using a regression line, we can evaluate the magnitude of change in one variable due to a certain change in another variable. For example, an economist may want to estimate the amount of change in food expenditure due to a certain change in the income of a household. A sociologist may want to estimate the increase in the crime rate due to a particular increase in the unemployment rate. Besides answering these questions, a regression line also helps us to predict the value of one variable for a given value of another variable. For example, by using the regression line, we can predict the (approximate) food expenditure of a household with a given income.

The correlation coefficient, on the other hand, simply tells us how strongly two variables are related. It does not provide any information about the size of change in one variable as a result of a certain change in the other variable. For example, the correlation coefficient tells us how strongly income and food expenditure or the unemployment rate and crime rate are related.

12.1 THE SIMPLE LINEAR REGRESSION MODEL

Only simple linear regression will be discussed in this chapter.† In the next two subsections the meaning of the words *simple* and *linear* as used in *simple linear regression* is explained.

12.1.1 SIMPLE REGRESSION

Let us return to the example of an economist investigating the relationship between food expenditure and income. When a household decides how much money it should spend on food every week or every month, what factors or variables does it consider? Certainly, income of the household is one factor. However, many other variables also affect food expenditure. For instance, the assets owned by the household, the size of the household, the preferences and tastes of household members, and any special dietary needs of household members are some of the variables that will influence a household's decision about food expenditure. These variables are called **independent variables** (or *explanatory variables*) because all of them vary independently and they explain the variation in food expenditure among different households. In other words, these variables explain why different households spend different amounts on food. Food expenditure is called the **dependent variable** because it depends on the independent variables. When we study the effect of two or more independent variables on a dependent variable by using regression analysis, it is called a *multiple regression*. However, if we choose only one (usually the most important) independent variable and study the effect of that single variable on a dependent variable, it is called a **simple regression.** Thus, a simple regression includes only two variables: one independent and one dependent variable. Note that whether it is a simple or a multiple regression analysis, it always includes one and only one dependent variable. It is only the number of independent variables that vary in simple and multiple regressions.

SIMPLE REGRESSION

A regression model is a mathematical equation that describes the relationship between two or more variables. A simple regression includes only two variables: one independent and one dependent. The dependent variable is the one being explained and the independent variable is the one used to explain the variation in the dependent variable.

12.1.2 LINEAR REGRESSION

The relationship between two variables in a regression analysis is expressed by a mathematical equation called a **regression equation or model.** A regression equation, when plotted, may assume one of the many possible shapes. One of those shapes is that of a straight line. A regression equation that gives a straight-line relationship between two variables is called a **linear regression model;** otherwise, it is called a *nonlinear regression model*. In this chapter, only linear regression models are studied.

†The term regression was first used by Sir Francis Galton (1822–1911), who studied the relationship between heights of children and those of their parents.

LINEAR REGRESSION

A (simple) regression model is called a linear regression model if it gives a straight-line relationship between two variables.

The two diagrams in Figure 12.1 show a linear and a nonlinear relationship between the dependent variable food expenditure and the independent variable income.

A linear relationship between income and food expenditure, which is shown in Figure 12.1*a*, indicates that as income increases the food expenditure always increases at the same rate. However, a nonlinear relationship between income and food expenditure, as depicted in Figure 12.1*b*, shows that as income increases the food expenditure increases, although, after a point, the rate of increase in food expenditure is lower for every subsequent increase in income.

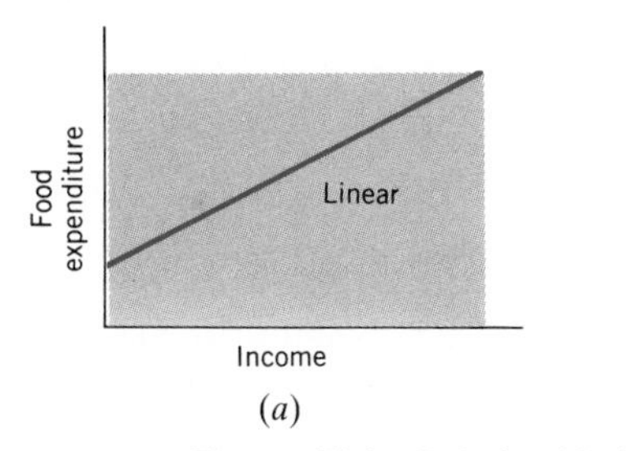

(*a*)

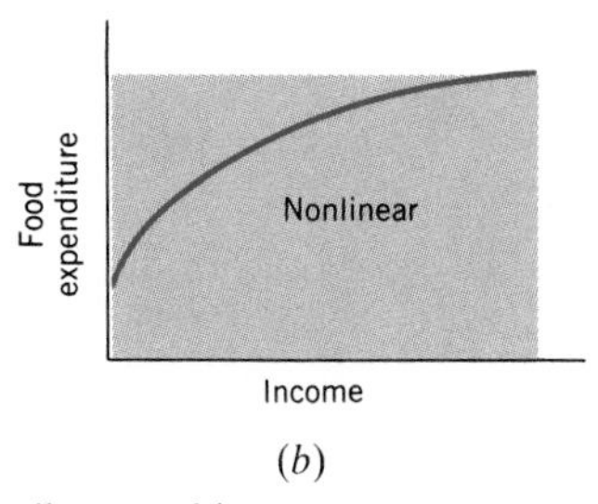

(*b*)

Figure 12.1 Relationship between food expenditure and income. (a) Linear relationship. (b) Nonlinear relationship.

The *equation of a linear relationship* between two variables x and y is written as

$$y = a + bx$$

For different straight lines, a and b assume different values. For instance, when $a = 50$ and $b = 5$, then the equation becomes

$$y = 50 + 5x$$

To plot a straight line, we need to know two points that lie on that line. We can find two points on a line by assigning any two values to x and then calculating the corresponding values of y. For the equation $y = 50 + 5x$, the two points are

1. When $x = 0$, then $y = 50 + 5(0) = 50$
2. When $x = 10$, then $y = 50 + 5(10) = 100$

These two points are plotted in Figure 12.2. By joining these two points we obtain the line representing the equation $y = 50 + 5x$.

Note that in Figure 12.2 the line intersects the y (vertical) axis at 50. Consequently, 50 is called the **y-intercept.** The y-intercept is the constant term in the equation. It is the value of y when x is zero.

In the equation $y = 50 + 5x$, 5 is called the *coefficient of* x or the **slope** of the line. It gives the amount of change in y due to a change of one unit in x. For example,

If $x = 10$, then $y = 50 + 5(10) = 100$

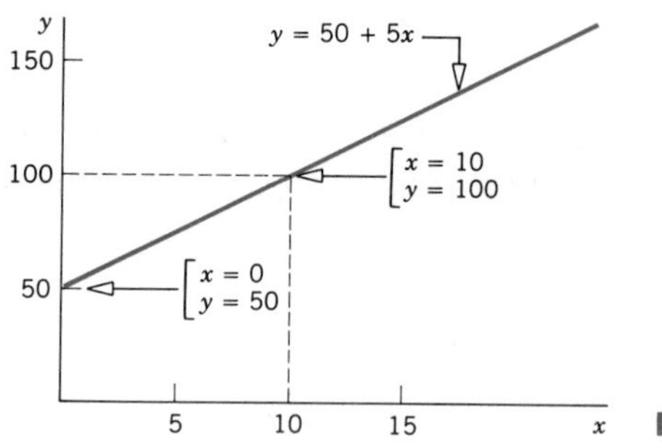

Figure 12.2

and

$$\text{If } x = 11, \text{ then} \qquad y = 50 + 5\,(11) = 105$$

Hence, as x increases by 1 unit (from 10 to 11), y increases by 5 units (from 100 to 105). These changes in x and y are shown in Figure 12.3.

In general, when an equation is written in the form

$$y = a + bx$$

a gives the y-intercept and b represents the slope of the line. In other words, a represents the point where the line intersects the y-axis and b gives the amount of change in y due to a change of one unit in x. Note that b is also called the *coefficient of x*.

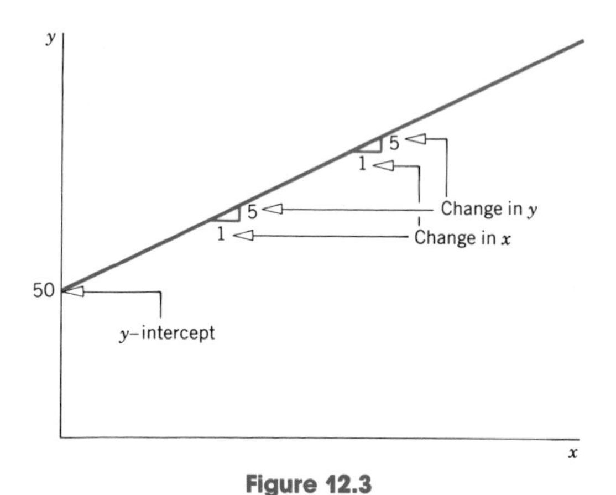

Figure 12.3

12.2 THE SIMPLE LINEAR REGRESSION ANALYSIS

In a regression model, the independent variable is usually denoted by x and the dependent variable is usually denoted by y. The x variable, with its coefficient, is written on the right side of the "=" sign whereas the y variable is written on the left side of the "=" sign. The y-intercept and the slope, which we earlier denoted by a and b, can be represented by any of the many commonly used symbols. Let us denote the y-intercept (which is also called the *constant term*) by A and the slope (or the coefficient of x variable) by B. Then, our simple linear regression model is written as

Constant term or y-intercept ↘ ↙ Slope

$$y = A + Bx \tag{1}$$

↑ Dependent variable ↑ Independent variable

In model (1), A gives the value of y for $x = 0$, and B gives the change in y due to a change of one unit in x.

Model (1) is called a **deterministic model.** It gives an *exact relationship* between x and y. This model simply states that y is determined exactly by x and for a given value of x there is one and only one (unique) value of y.

However, in many cases the relationship between variables is not exact. For instance, if y is food expenditure and x is income, then model (1) would state that food expenditure is determined by income only and that all households with the same income will spend the same amount on food. But as mentioned earlier, food expenditure is determined by many variables, only one of which is included in model (1). In reality, different households with the same income spend different amounts on food because of the differences in the size of the households, the assets they own, and their preferences and tastes. Hence, to take these variables into consideration and to make the model complete, we add another term to the right side of model (1). This term is called the **random error term.** It is denoted by ϵ (Greek letter epsilon). The complete regression model is written as

$$y = A + Bx + \epsilon \tag{2}$$

↑ Random error term

Regression model (2) is called a **probabilistic model** (or a *statistical relationship*).

EQUATION OF A REGRESSION MODEL

In the regression model $y = A + Bx + \epsilon$, A is called the y-intercept or constant term, B is the slope, and ϵ is the random error term. The dependent and independent variables are y and x, respectively.

The random error term ϵ is included in the model to represent the following two phenomena.

1. *Missing or omitted variables.* As mentioned earlier, food expenditure is affected by many variables other than income. The random error term ϵ is included to capture the effect of all those missing or omitted variables that have not been included in the model.
2. *Random variation.* Human behavior is unpredictable. For example, a household may have many parties during one month and may spend more than usual on food during that month. The same household may spend less than usual during another month because it spent quite a bit of money to buy furniture. The variation in food expenditure for such reasons may be called random variation.

In model (2), A and B are the *population parameters*. The regression line obtained for model (2) by using the population data is called the **population regression line.** The values of A and B in the population regression line are called the **true values** of the y-intercept and slope.

However, population data are difficult to obtain. As a result, we almost always use the sample data to estimate model (2). The values of the y-intercept and slope calculated from sample data on x and y are called the **estimated values** of A and B and are denoted by a and b. Using a and b, we write the estimated model as

$$\hat{y} = a + bx \qquad (3)$$

where $\hat{y}$ (pronounced "y hat") is the *estimated* or *predicted value* of y for a given value of x. Equation 3 is called the *estimated model;* it gives the regression of y on x.

ESTIMATES OF *A* AND *B*

In the model $\hat{y} = a + bx$, a and b, which are calculated by using sample data, are called the estimates of A and B.

12.2.1 SCATTER DIAGRAM

Suppose we take a sample of seven households and collect information on their incomes and food expenditures for the past month. The information obtained (in hundreds of dollars) is given in Table 12.1.

Table 12.1 Incomes and Food Expenditures of Seven Households

Income (hundreds of dollars)	Food Expenditure (hundreds of dollars)
22	7
32	8
16	5
37	10
12	4
27	6
17	6

In Table 12.1, we have a pair of observations for each of the seven households. Each pair consists of one observation on income and a second on food expenditure. By plotting all seven pairs of values, we obtain a **scatter diagram** or **scattergram.** Figure 12.4 gives the scatter diagram for the data of Table 12.1. Each dot in this diagram represents one household. A scatter diagram is helpful in detecting a relationship between two variables. For example, by looking at the scatter diagram of Figure 12.4, we can observe that there exists a strong linear relationship between food expenditure and income. If a straight line is drawn through the points, the points will be scattered closely around the line.

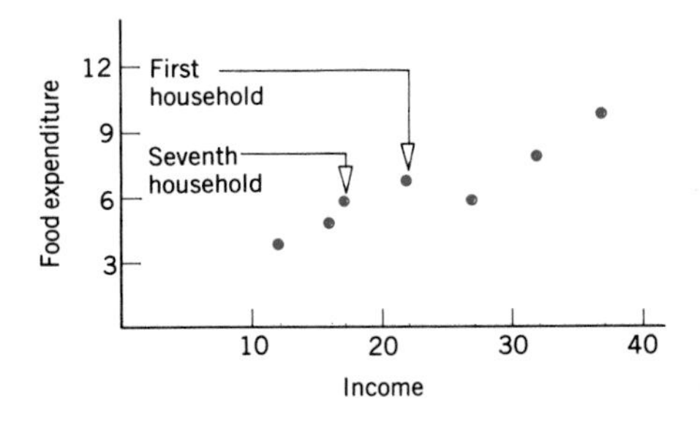

Figure 12.4 Scatter diagram.

SCATTER DIAGRAM

A plot of paired observations is called a scatter diagram.

As shown in Figure 12.5, a large number of straight lines can be drawn through the scatter diagram of Figure 12.4. Each of these lines will give different values for a and b of model (3).

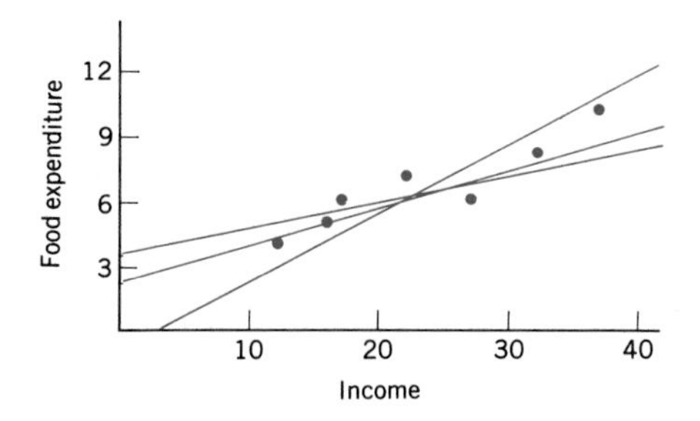

Figure 12.5

In regression analysis, we try to find a line that best fits the points in the scatter diagram. Such a line provides the best possible description of the relationship between the dependent and independent variables. The **least squares method,** to be discussed in the next section, gives such a line. The line obtained by using the least squares method is called the *least squares line*.

12.2.2 LEAST SQUARES LINE

The value of y obtained for a member from the survey is called the **observed or actual value of y.** As mentioned earlier in Section 12.2, the value of y, denoted by $\hat{y}$, obtained for a given x by using the regression line is called the **predicted value of y.** The random error ϵ denotes the difference between the actual value of y and the predicted value of y for population data. For example, for a given household, ϵ is the difference between what this household actually spent on food during the past month and what is predicted using the population regression line. The ϵ is also called the *residual,* as it measures the surplus (positive or negative) of actual food expenditure over what is predicted by using the regression line. If we estimate model (2) by using sample data, the difference between the actual y and predicted y based on this estimation cannot be denoted by ϵ. *The random error for the sample regression model is denoted by e.* Thus, e is used to estimate ϵ in the same way as a and b are used to estimate

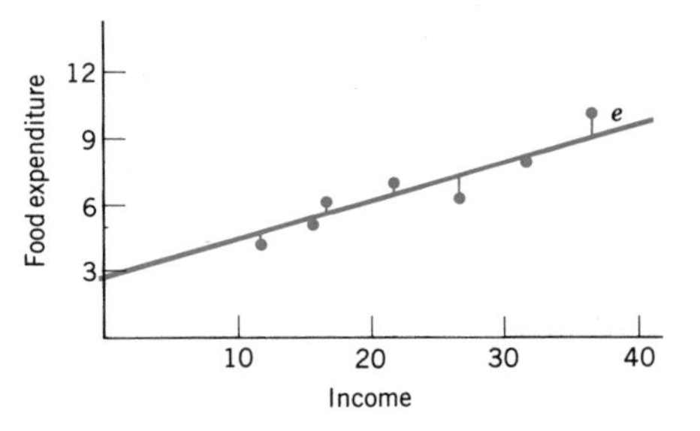

Figure 12.6

A and B, respectively. If we estimate model (2) using sample data, then the value of e is given by

$$e = \text{Actual food expenditure} - \text{Predicted food expenditure} = y - \hat{y}$$

In Figure 12.6, e is the vertical distance between the actual position of a household and the point on the regression line. Note that in such a diagram, we always measure the dependent variable on the vertical axis and independent variable on the horizontal axis.

The value of an error is positive if the point that gives the actual food expenditure is above the regression line and negative if it is below the regression line. The sum of these errors is always zero. In other words, the sum of the actual food expenditures for seven households included in the sample will be the same as the sum of the food expenditures predicted from the regression model. Thus,

$$\Sigma e = \Sigma(y - \hat{y}) = 0$$

Hence, to find the line that best fits the scatter of points, we cannot minimize the sum of errors. Instead, we define a new quantity called the **error sum of squares,** denoted by **SSE,** by adding the squares of errors.

$$\text{SSE} = \Sigma e^2 = \Sigma(y - \hat{y})^2$$

The least squares method gives the values of a and b for model (3) such that the sum of squared errors (SSE) is minimum.

ERROR SUM OF SQUARES (SSE)

The error sum of squares, denoted by SSE, is

$$\text{SSE} = \Sigma e^2 = \Sigma(y - \hat{y})^2$$

The values of a and b that give the minimum SSE are called the **least squares estimates** of A and B, and the regression line obtained with these estimates is called the least squares line.

The least squares values of a and b are computed using the following formulas.

THE LEAST SQUARES LINE

For the least squares regression line $\hat{y} = a + bx$

$$b = \frac{SS_{xy}}{SS_{xx}} \quad \text{and} \quad a = \bar{y} - b\bar{x}$$

where

$$SS_{xy} = \Sigma xy - \frac{(\Sigma x)(\Sigma y)}{n} \quad \text{and} \quad SS_{xx} = \Sigma x^2 - \frac{(\Sigma x)^2}{n}$$

and "SS" stands for "sum of squares."†

The least squares regression line $\hat{y} = a + bx$ is also called the regression of y on x.

Example 12-1 illustrates how we find the least squares regression line.

Estimating the least squares regression line.

EXAMPLE 12-1 Find the least squares regression line for the data on incomes and food expenditures of seven households given in Table 12.1 Use income as an independent variable and food expenditure as a dependent variable.

Solution We are to find the values of y-intercept a and slope b for the regression model $\hat{y} = a + bx$. Table 12.2 shows the calculations required for the computation of a and b. We denote the independent variable (income) by x and the dependent variable (food expenditure) by y.

Table 12.2

Income x	Food Expenditure y	xy	x^2
22	7	154	484
32	8	256	1024
16	5	80	256
37	10	370	1369
12	4	48	144
27	6	162	729
17	6	102	289
$\Sigma x = 163$	$\Sigma y = 46$	$\Sigma xy = 1172$	$\Sigma x^2 = 4295$

The following steps are performed to compute a and b.

†The values of SS_{xy} and SS_{xx} can also be obtained by using the following basic formulas.

$$SS_{xy} = \Sigma(x - \bar{x})(y - \bar{y}) \quad \text{and} \quad SS_{xx} = \Sigma(x - \bar{x})^2$$

However, these formulas usually take longer to make calculations.

Step 1. Compute Σx, Σy, $\bar{x}$, and $\bar{y}$.

$$\Sigma x = 163 \qquad \Sigma y = 46$$

$$\bar{x} = \frac{\Sigma x}{n} = \frac{163}{7} = 23.29$$

$$\bar{y} = \frac{\Sigma y}{n} = \frac{46}{7} = 6.57$$

Step 2. Compute Σxy and Σx^2.

To calculate Σxy, we multiply the corresponding values of x and y. Then, we sum all the products. The products of x and y are recorded in the third column of Table 12.2. To compute Σx^2, we square each of the x values and then add them. The squared values of x are listed in the fourth column of Table 12.2. Thus,

$$\Sigma xy = 1172 \qquad \text{and} \qquad \Sigma x^2 = 4295$$

Step 3. Compute SS_{xy} and SS_{xx}.

$$SS_{xy} = \Sigma xy - \frac{(\Sigma x)(\Sigma y)}{n} = 1172 - \frac{(163)(46)}{7} = 100.857$$

$$SS_{xx} = \Sigma x^2 - \frac{(\Sigma x)^2}{n} = 4295 - \frac{(163)^2}{7} = 499.429$$

Step 4. Compute a and b.

$$b = \frac{SS_{xy}}{SS_{xx}} = \frac{100.857}{499.429} = .20$$

$$a = \bar{y} - b\bar{x} = 6.57 - (.20)(23.29) = 1.91$$

Thus, our estimated regression model $\hat{y} = a + bx$ is

$$\hat{y} = 1.91 + .20x$$

This regression line is called the least squares line. It gives the *regression of food expenditure on income.* ■

Using this estimated model, we can find the predicted value of y for a specific value of x. For instance, suppose we randomly select a household whose monthly income is \$2700 so that $x = 27$ (recall that x denotes income in hundreds of dollars). The predicted value of food expenditure for this household is

$$\hat{y} = 1.91 + (.20)(27) = \$7.31 \text{ hundred}$$

In other words, based on our regression line, we predict that a household with a monthly income of \$2700 is expected to spend \$731 a month on food. This value of $\hat{y}$ can also be interpreted as the mean value of y for $x = 27$. Thus, we can state that

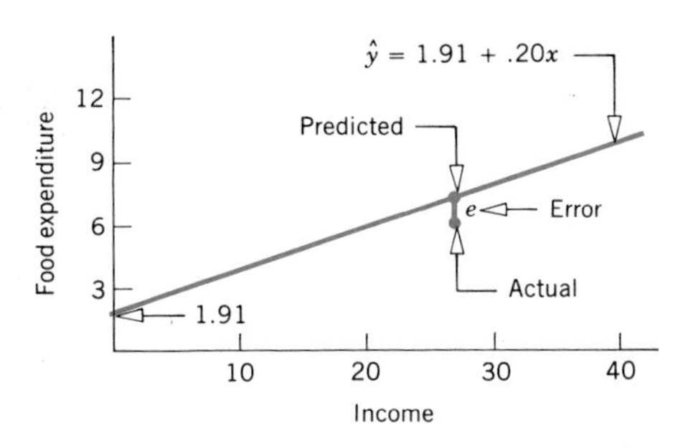

Figure 12.7 Errors of prediction.

on average all households with a monthly income of \$2700 spend \$731 per month on food.

In our data on seven households, there is one household whose income is \$2700. The actual food expenditure for that household is \$600 (see Table 12.1). The difference between actual and predicted values gives the error of prediction. Thus, the error of prediction for this household, which is shown in Figure 12.7, is

$$e = y - \hat{y} = 6.00 - 7.31 = -\$1.31 \text{ hundred}$$

Therefore, the error of prediction is −\$131. The negative error indicates that the predicted value of y is greater than the actual value of y. Thus, if we use the regression model, this household's food expenditure is overestimated by \$131.

12.2.3 INTERPRETATION OF *a* AND *b*

How do we interpret $a = 1.91$ and $b = .20$ obtained in Example 12-1 for the regression of food expenditure on income? A brief explanation of the y-intercept and slope of a regression line was given in Section 12.1.2. The next two parts explain the meaning of a and b in more detail.

Interpretation of *a*

Consider a household with zero income. Using the estimated regression line obtained in Example 12-1, the predicted value of y for $x = 0$ is

$$\hat{y} = 1.91 + .20\,(0) = \$1.91 \text{ hundred}$$

Thus, we can state that a household with zero income is expected to spend \$191 per month on food. Alternatively, we can also state that the average monthly food expenditure for all households with zero income is \$191. Thus, $a = 1.91$ gives the predicted or mean value of y for $x = 0$ based on the regression model estimated for sample data.

However, we should be careful while making this interpretation of a. In our sample of seven households, the incomes vary from a minimum of \$1200 to a maximum of \$3700. (Note that in Table 12.1 the minimum value of x is 12 and the maximum value is 37.) Hence, our regression line is only valid for the values of x between 12 and 37. If we predict y for a value of x outside this range, the prediction usually will not hold true. Thus, since $x = 0$ is outside the range of household incomes that we have in the sample data, the prediction that a household with zero income spends \$191 per month on food does not carry much credibility. The same is true if we try to predict y for an income greater than \$3700, which is the maximum value of x in Table 12.1.

Interpretation of *b*

The value of b in a regression model gives the change in y (dependent variable) due to a change of one unit in x (independent variable). For example, by using the regression line obtained in Example 12-1, when $x = 15$,

$$\hat{y} = 1.91 + .20\ (15) = 4.91$$

and when $x = 16$,

$$\hat{y} = 1.91 + .20\ (16) = 5.11$$

Hence, when x increased by one unit, from 15 to 16, $\hat{y}$ increased by $5.11 - 4.91 = .20$, which is the value of b. Because our unit of measurement is in hundreds of dollars, we can state that on average a \$100 increase in income will cause a \$20 increase in food expenditure. We can also state that on average a \$1 increase in income of a household will increase the food expenditure by \$.20. Note the phrase "on average" in these statements. The regression line is seen to be a measure of the mean value of y for a given x. If one household's income is increased by \$100, that household's food expenditure may or may not increase by \$20. But if the incomes of all households are increased by \$100 each, the average increase in their food expenditures will be very close to \$20.

Note that when b is positive, an increase in x will lead to an increase in y and a decrease in x will lead to a decrease in y. In other words, when b is positive, the movements in x and y are in the same direction. Such a relationship between x and y is called a *positive relationship*. The regression line in this case slopes upward from left to right. On the other hand, if the value of b is negative, an increase in x will cause a decrease in y and a decrease in x will cause an increase in y. The changes in x and y in this case are in opposite directions. Such a relationship between x and y is called a *negative relationship*. The regression line in this case slopes downward from left to right. The two diagrams in Figure 12.8 show these two cases.

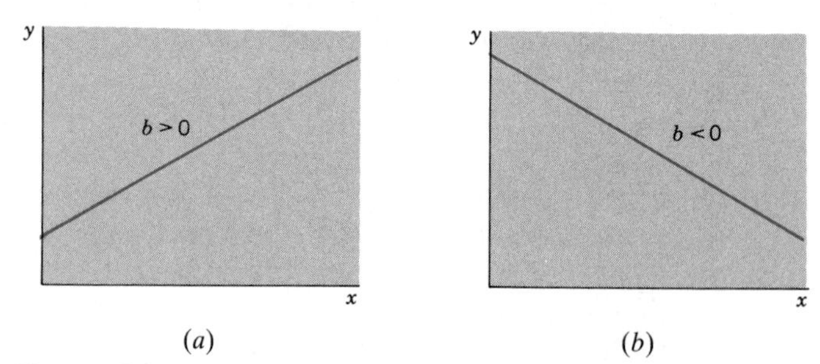

Figure 12.8 Positive (a) and negative (b) relationships between x and y.

Remember: For a regression model, b is computed as: $b = SS_{xy}/SS_{xx}$. The value of SS_{xx} is always positive and that of SS_{xy} can be positive or negative. Hence, the sign of b depends on the sign of SS_{xy}. If SS_{xy} is positive (as in the example of incomes and food expenditures of seven households), then b will be positive; and if SS_{xy} is negative, then b will be negative.

Case Study 12-1 illustrates the difference between the population regression line and a sample regression line, and between the population random error and the sample random error.

CASE STUDY 12-1 HEIGHTS AND WEIGHTS OF NBA PLAYERS

Data Set III given in Appendix B lists the heights and weights of all NBA (National Basketball Association) players who were on the rosters of all NBA teams at the beginning of 1990–91 season. These data comprise the population of NBA players for that point in time. We postulate the following simple linear regression model for these data.

$$y = A + Bx + \epsilon$$

where y is the weight and x is the height of an NBA player.

Using the population data, we obtain the regression line

$$\hat{y} = -272.40 + 6.17x$$

This equation gives the population regression line because it is obtained by using the population data. (In the population regression line, we can also write $\mu_{y|x}$ for $\hat{y}$. This is explained in Section 12.2.4.) Thus, the true values of A and B are

$$A = -272.40, \quad \text{and} \quad B = 6.17$$

The value of B indicates that for every 1 inch increase in the height of an NBA player, weight increases on average by 6.17 pounds. However, $A = -272.40$ does not make any sense. It tells that the weight of a player with zero height is -272.40 pounds. (Recall that Section 12.2.3 mentioned that we cannot apply the regression equation to predict y for values of x outside the range of data used to find the regression line.) Figure 12.9 gives the scatter diagram for the heights and weights of all NBA players.

Next, we selected a random sample of 30 players and estimated the regression model for these sample data. The estimated regression line for the sample is

$$\hat{y} = -297.80 + 6.53x$$

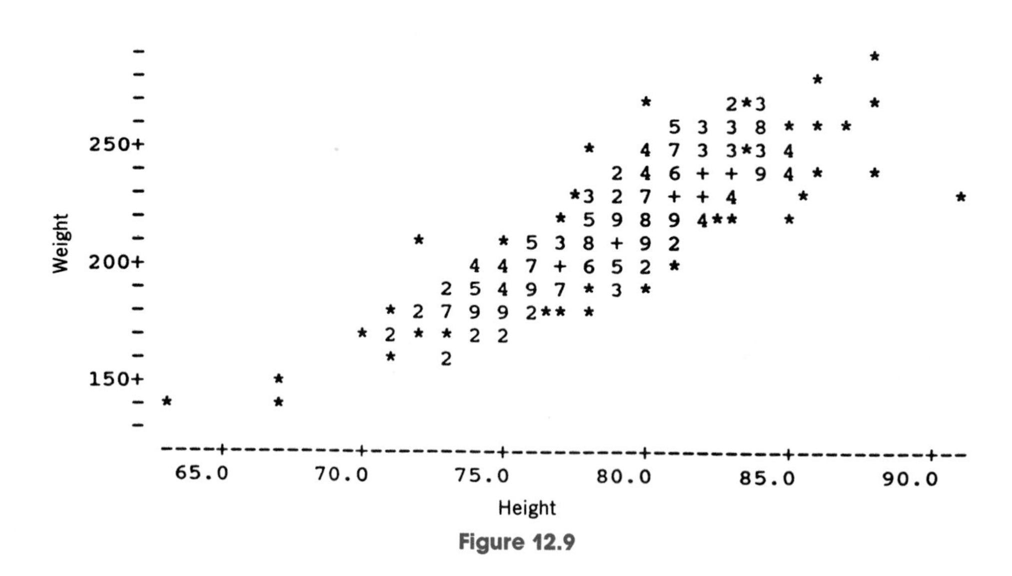

Figure 12.9

The values of a and b are: $a = -297.80$ and $b = 6.53$. These values of a and b give the estimates of A and B based on sample data. The scatter diagram for the sample observations on heights and weights is given in Figure 12.10.

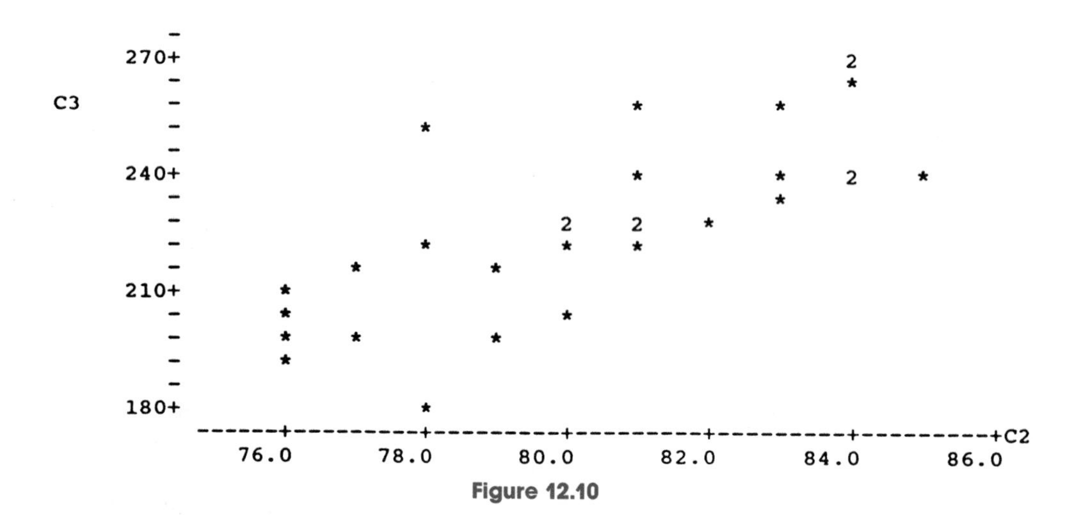

Figure 12.10

As we can observe from Figures 12.9 and 12.10, the scatter diagrams for population and sample data both show a linear relationship between heights and weights of NBA players.

Now, suppose we select one player, say, Earvin "Magic" Johnson, whose height is 81 inches and weight is 220 pounds (he is included in the sample data also). The actual value of y for him is 220. If we use the population regression line, his predicted weight is

$$\hat{y} = -272.40 + 6.17\,(81) = 227.37$$

Hence, the error ϵ is

$$\epsilon = y - \hat{y} = 220 - 227.37 = -7.37$$

Now let us calculate the error e for Johnson using the sample regression line. The predicted weight from the sample regression line is

$$\hat{y} = -297.80 + 6.53\ (81) = 231.13$$

and the error e is

$$e = y - \hat{y} = 220 - 231.13 = -11.13$$

This value of e is an estimate of ϵ.

12.2.4 ASSUMPTIONS OF THE REGRESSION MODEL

Like any other theory, the linear regression analysis is also based on certain assumptions. Consider the population regression model

$$y = A + Bx + \epsilon \tag{4}$$

There are four assumptions made about this model. These assumptions are explained with reference to the example regarding incomes and food expenditures of households. Note that these assumptions are made about the population regression model and not about the sample regression model.

Assumption 1: The random error term ϵ has a mean equal to zero for each x. In other words, among all households with the same income, some spend more than the predicted food expenditure (and, hence, have positive errors) and others spend less than the predicted food expenditure (and, hence, have negative errors). This assumption simply states that the sum of the positive errors is equal to the sum of the negative errors so that the mean of errors for all households with the same income is zero. Thus, when the mean value of ϵ is zero, the mean value of y for a given x is equal to $A + Bx$ and it is written as

$$\mu_{y|x} = A + Bx$$

$\mu_{y|x}$ is read as "**the mean value of *y* for a given *x*.**" When we estimate model (4) using the population data, the points on the regression line give the average values of y, denoted by $\mu_{y|x}$, for the corresponding values of x for the population.

Assumption 2: The errors associated with different observations are independent. According to this assumption, the errors for any two households in our example are independent. In other words, all households decide independently how much to spend on food.

Assumption 3: For any given x, the distribution of errors is normal. The corollary of this assumption is that the food expenditures for all households with the same income are normally distributed.

Assumption 4: The distribution of population errors for each x has the same (constant) standard deviation, which is denoted by σ_ϵ. This assumption indicates that the spread of points around the regression line is similar for all x values.

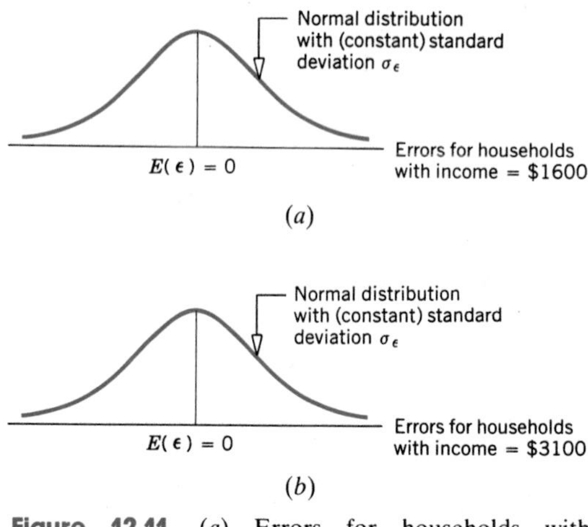

Figure 12.11 (*a*) Errors for households with income = $1600. (*b*) Errors for households with income = $3100.

Figure 12.11 illustrates the meaning of the first, third, and fourth assumptions for households with an income of $1600 and $3100 a month. The same assumptions hold true for any other income level. In the population of all households, there will be many households with a monthly income of $1600. Using the population regression line, if we calculate the errors for all these households and prepare the distribution of these errors, it will look like the distribution given in Figure 12.11*a*. Its standard deviation will be σ_ϵ. Similarly, Figure 12.11*b* gives the distribution of errors for all those households in the population whose monthly income is $3100. Its standard deviation is also σ_ϵ. These distributions are identical.

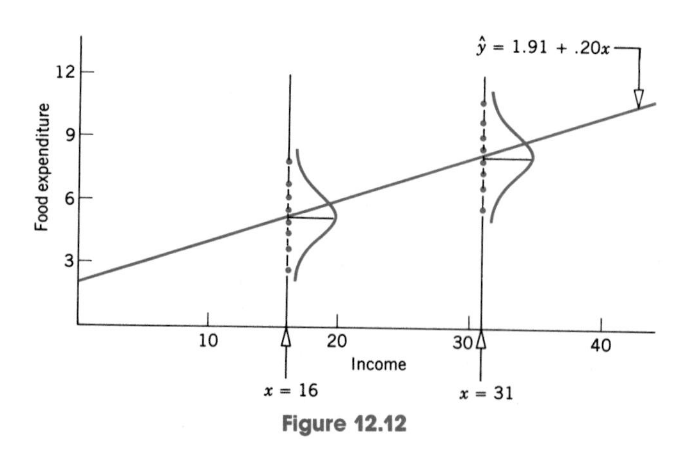

Figure 12.12

Figure 12.12 shows how these distributions look when they are imposed on the same diagram with a regression line. The points on the vertical line through $x = 16$ give the food expenditures for various households in the population, each of which has the same monthly income of \$1600. The same is true about the vertical line through $x = 31$ or any other vertical line for some other value of x.

12.2.5 A NOTE ON THE USE OF SIMPLE LINEAR REGRESSION

We should apply linear regression with caution. When we use simple linear regression, we assume that the relationship between two variables is described by a straight line. In the real world, the relationship between variables may not be linear. Hence, before we use a simple linear regression, it is better to construct a scatter diagram and look at the plot of the data points. We should estimate a linear regression model only if the scatter diagram indicates such a relationship. The scatter diagrams of Figure 12.13 give two examples where the relationship between x and y is not linear. Hence, fitting linear regression in such cases would be wrong.

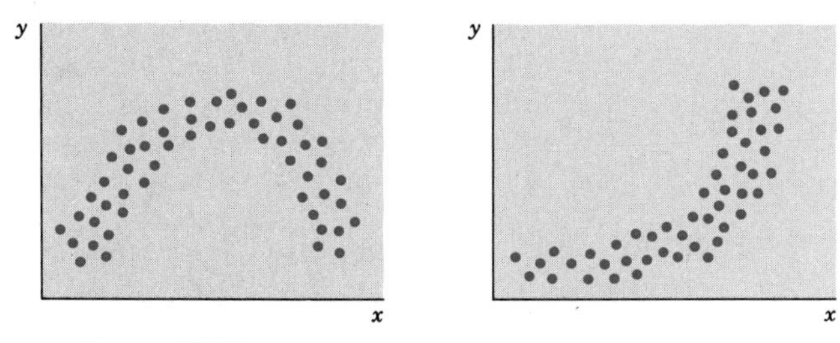

Figure 12.13 Nonlinear relationship between x and y.

EXERCISES

12.1 Explain the meaning of the words *simple* and *linear* as used in *simple linear regression*.

12.2 Plot the following straight lines. Give the values of the y-intercept and slope for each of these lines and interpret them.

a. $y = 100 - 5x$ **b.** $y = -40 + 2x$ **c.** $y = 25 + 4x$

12.3 Plot the following straight lines. Give the values of the y-intercept and slope for each of these lines and interpret them.

a. $y = -60 + 8x$ **b.** $y = 44 + 4x$ **c.** $y = 75 - 5x$

12.4 Briefly explain a deterministic and a probabilistic regression model.

12.5 Why is the random error term included in a regression model?

12.6 Explain the following.

a. Population regression line

b. True values of A and B

c. Estimated values of A and B denoted by a and b

12.7 The following table gives seven pairs of x and y values.

x	12	16	8	5	11	20	9
y	7	11	4	4	9	13	6

a. Compute SS_{xx} and SS_{xy}.
b. Find the regression line $\hat{y} = a + b\,x$.

12.8 The following table gives 10 pairs of x and y values.

x	20	11	16	18	22	12	14	9	23	25
y	12	8	10	9	12	8	9	7	13	12

a. Compute SS_{xx} and SS_{xy}.
b. Find the least squares line.

12.9 The following table gives the ages (in years) and prices (in hundreds of dollars) for eight cars of a specific model.

Age	8	3	6	9	2	5	6	3
Price	16	74	38	19	102	36	33	69

a. Construct a scatter diagram for these data. Does the scatter diagram exhibit a linear relationship between ages and prices of cars?
b. Find the regression of price on age.
c. Give a brief interpretation of the values of a and b.
d. Plot the regression line on the scatter diagram and show the errors by drawing vertical lines between scatter points and the regression line.
e. Predict the price of a 7-year-old car of this model.

12.10 Seven students were tested for stress before a mathematics test. The following table gives the stress scores (on a scale of 1 to 10) of these students and their scores in the math test.

Stress score	6.5	4.0	2.5	7.2	8.1	3.4	5.5
Test score	81	96	93	70	63	84	73

a. Construct a scatter diagram for these data. Does the scatter diagram exhibit a linear relationship between stress scores and test scores?
b. Find the regression of test scores on stress scores.
c. Give a brief interpretation of the values of a and b.
d. Plot the regression line on the scatter diagram and show the errors by drawing vertical lines between scatter points and the regression line.
e. Predict the test score of a student with a 7.5 stress score before a math test.

12.11 The following table lists the size (in hundreds of square feet) of six houses and the monthly rents (in dollars) paid by tenants for those houses.

Size of the house	21	16	19	27	34	23
Monthly rent	700	580	720	850	1050	800

a. Construct a scatter diagram for these data. Does the scatter diagram show a linear relationship between size of houses and monthly rents?
b. Find the regression line $\hat{y} = a + bx$ with size of a house as an independent variable and monthly rent as a dependent variable.
c. Give a brief interpretation of the values of a and b.
d. Plot the regression line on the scatter diagram and show the errors by drawing vertical lines between scatter points and the regression line.
e. Predict the monthly rent for a house with 2500 square feet.

12.12 The following table lists the annual incomes (in thousands of dollars) of six persons and the amounts of their life insurance policies (in thousands of dollars).

Annual income	47	54	25	37	62	18
Life insurance	250	300	100	150	500	75

a. Construct a scatter diagram for these data. Does the scatter diagram show a linear relationship between annual incomes and amounts of life insurance policies?

b. Find the regression line $\hat{y} = a + bx$ with annual income as an independent variable and the amount of life insurance policy as a dependent variable.

c. Plot the regression line on the scatter diagram and show the errors by drawing vertical lines between scatter points and the regression line.

d. Give a brief interpretation of the values of a and b.

12.13 The following table gives the total payroll (rounded to millions of dollars) as of March 1989 and the percentage of games won during the 1988 season by each of the National League baseball teams.

Team	Total Payroll	Percentage of Games Won
Atlanta Braves	9	34
Chicago Cubs	12	48
Cincinnati Reds	11	54
Houston Astros	16	51
Los Angeles Dodgers	22	58
Montreal Expos	12	50
New York Mets	20	63
Philadelphia Phillies	10	40
Pittsburgh Pirates	12	53
St. Louis Cardinals	15	47
San Diego Padres	14	52
San Francisco Giants	14	51

Source: USA Today, March 30, and April 3, 1989. Copyright © 1989, *USA Today*. Adapted with permission.

a. Find the least squares regression line with total payroll as an independent variable and percentage of games won as a dependent variable.

b. Is the regression line obtained in part (a) the population regression line? Why or why not? Do the values of y-intercept and slope in the regression line give A and B or a and b?

c. Give a brief interpretation of the values of y-intercept and slope.

d. Predict the percentage of games won for a team with a total payroll of $13.40 million.

12.14 The following table gives the percentage of games won and the average attendance (rounded to nearest thousand) per home game for the 1988 season for each of the American Leage baseball teams.

Team	Percentage of Games Won	Average Attendance per Home Game (thousands)
Baltimore Orioles	34	22
Boston Red Sox	55	31
California Angels	46	29
Chicago White Sox	44	14
Cleveland Indians	48	18
Detroit Tigers	54	26
Kansas City Royals	52	29
Milwaukee Brewers	54	24
Minnesota Twins	56	37
New York Yankees	53	34
Oakland Athletics	64	29
Seattle Mariners	42	13
Texas Rangers	44	20
Toronto Blue Jays	54	32

Source: USA Today, April 3, 1989. Copyright © 1989, *USA Today*. Adapted with permission.

a. Find the least squares regression line with percentage of games won as an independent variable and average attendance as a dependent variable.
b. Is the regression line obtained in part (a) the population regression line? Why or why not? Do the values of y-intercept and slope give A and B or a and b?
c. Give a brief interpretation of the values of y-intercept and slope.
d. Predict the average attendance per home game for a team with a 49.6 percent of games won.

12.15 Briefly explain the assumptions of the population regression model.

12.3 THE STANDARD DEVIATION OF THE RANDOM ERROR

For the example of incomes and food expenditures, all the households with the same income are expected to spend different amounts on food. Consequently, the random error ϵ will assume different values for these households. The standard deviation σ_ϵ measures the spread of these errors around the regression line. The standard deviation σ_ϵ of errors tells us how widely the errors and, hence, the values of y are spread for a given x. In Figure 12.12, which is reproduced as Figure 12.14, the points on the

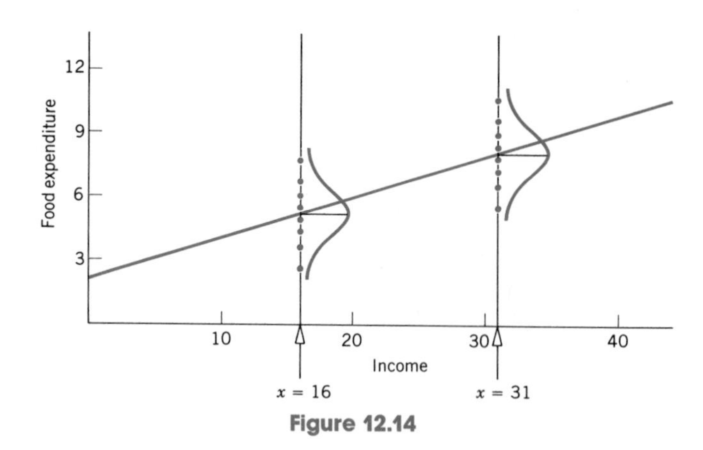

Figure 12.14

vertical line through $x = 16$ give the monthly food expenditures for all households with a monthly income of \$1600. The distance of each dot from the point on the regression line gives the value of the corresponding error. The standard deviation σ_ϵ of errors measures the spread of such points around the population regression line. The same is true for $x = 31$ or any other value of x.

Note that σ_ϵ denotes the standard deviation of errors for the population. However, usually σ_ϵ is unknown. In such cases, it is estimated by s_e, which is the standard deviation of errors for the sample data. The following is the basic formula to calculate the standard deviation s_e.

$$s_e = \sqrt{\frac{\text{SSE}}{n - 2}}$$

where

$$\text{SSE} = \Sigma(y - \hat{y})^2$$

In this formula, $n - 2$ represents the **degrees of freedom** for the regression model. The reason that $df = n - 2$ is that we lose one degree of freedom to calculate $\bar{x}$ and one for $\bar{y}$.

DEGREES OF FREEDOM FOR A SIMPLE LINEAR REGRESSION MODEL

The degrees of freedom for a simple regression model are

$$df = n - 2$$

For computational purposes, it is more convenient to use the following formula to calculate the standard deviation s_e of errors.

STANDARD DEVIATION OF ERRORS

The standard deviation s_e of errors is calculated as

$$s_e = \sqrt{\frac{\text{SS}_{yy} - b\,\text{SS}_{xy}}{n - 2}}$$

where

$$\text{SS}_{yy} = \Sigma y^2 - \frac{(\Sigma y)^2}{n}$$

and SS_{xy} is calculated as discussed earlier.†

Like the value of SS_{xx}, the value of SS_{yy} is always positive.

†The basic formula to calculate SS_{yy} is $\text{SS}_{yy} = \Sigma(y - \bar{y})^2$.

Calculating the standard deviation of errors.

EXAMPLE 12-2 Compute the standard deviation s_e of errors for the data on monthly incomes and food expenditures of seven households given in Table 12.1.

Solution To compute s_e, we need to know the values of SS_{yy}, SS_{xy}, and b. Earlier, in Example 12-1 on page 601, we computed SS_{xy} and b. These values are

$$SS_{xy} = 100.857 \quad \text{and} \quad b = .20$$

To compute SS_{yy}, we calculate Σy^2 in Table 12.3.

Table 12.3

Income x	Food Expenditure y	y^2
22	7	49
32	8	64
16	5	25
37	10	100
12	4	16
27	6	36
17	6	36
$\Sigma x = 163$	$\Sigma y = 46$	$\Sigma y^2 = 326$

The value of SS_{yy} is

$$SS_{yy} = \Sigma y^2 - \frac{(\Sigma y)^2}{n} = 326 - \frac{(46)^2}{7} = 23.714$$

Hence, the standard deviation s_e of errors is

$$s_e = \sqrt{\frac{SS_{yy} - b\,SS_{xy}}{n - 2}} = \sqrt{\frac{23.714 - .20\,(100.857)}{7 - 2}} = .842$$

■

12.4 THE COEFFICIENT OF DETERMINATION

We may ask the question, "How good is the regression model?" In other words, "How well does the independent variable explain the dependent variable in the regression model?" The **coefficient of determination** is one concept that answers this question.

For a moment, assume that we possess information only on food expenditures of households and not on their incomes. Hence, in this case, we cannot use the regression line to predict the food expenditure for any household. As we did in earlier chapters, in the absence of a regression model, we use $\bar{y}$ to estimate or predict every household's food expenditure. Consequently, the error of prediction for each household is now given by $y - \bar{y}$, which is the difference between the actual food expenditure of a household and the mean food expenditure. If we calculate such errors for all households, square and add them, the resulting sum is called the **total sum of squares** and is denoted by **SST.** Actually, SST is the same as SS_{yy} and is defined as

$$SST = SS_{yy} = \Sigma(y - \bar{y})^2$$

However, for computational purposes, SST is calculated using the following formula.

TOTAL SUM OF SQUARES (SST)

The total sum of squares, denoted by SST, is

$$SST = \Sigma y^2 - \frac{(\Sigma y)^2}{n}$$

Note that this is the same formula that was used to calculate SS_{yy}.

The value of SS_{yy}, which is 23.714, was calculated in Example 12-2. Consequently, the value of SST is

$$SST = 23.714$$

From Example 12-1, $\bar{y} = 6.57$. Figure 12.15 shows the total errors for each of the seven households of the sample.

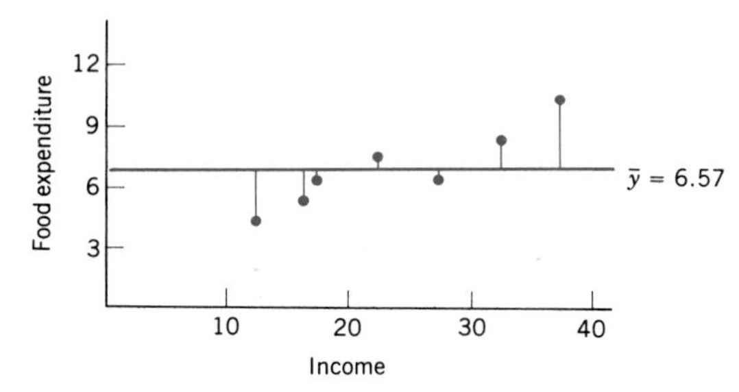

Figure 12.15 Total errors.

Now suppose we use the simple linear regression model to predict the food expenditure of each of the seven households in our sample. In this case, we predict each household's food expenditure by using the regression line we estimated earlier in Example 12-1, which is

$$\hat{y} = 1.91 + .20x$$

The predicted food expenditures, denoted by $\hat{y}$, for all households are shown in Table 12.4. Also shown are the errors and error squares.

We calculate the values of $\hat{y}$ (given in the third column of Table 12.4) by substituting the values of x in the estimated regression model. For example, the value of

Table 12.4

x	y	$\hat{y} = 1.91 + .20x$	$e = y - \hat{y}$	$e^2 = (y - \hat{y})^2$
22	7	6.31	.69	.476
32	8	8.31	−.31	.096
16	5	5.11	−.11	.012
37	10	9.31	.69	.476
12	4	4.31	−.31	.096
27	6	7.31	−1.31	1.716
17	6	5.31	.69	.476
				$\Sigma e^2 = \Sigma(y - \hat{y})^2 = 3.348$

x for the first household is 22. Substituting this value of x in the regression line, we obtain

$$\hat{y} = 1.91 + .20\,(22) = 6.31$$

Similarly, we find the other values of $\hat{y}$.

The error sum of squares SSE is given by the sum of the fifth column in Table 12.4. Thus,

$$\text{SSE} = \Sigma(y - \hat{y})^2 = 3.348$$

The errors of prediction for the regression model for seven households are shown in Figure 12.16.

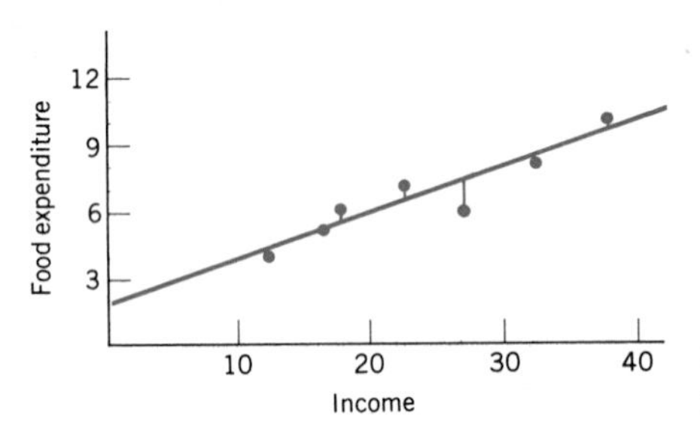

Figure 12.16 Errors of prediction when the regression model is used.

Thus, from the foregoing calculations,

$$\text{SST} = 23.714 \quad \text{and} \quad \text{SSE} = 3.348$$

These values indicate that the sum of squared errors decreased from 23.714 to 3.348 when we used $\hat{y}$ to predict food expenditures in place of $\bar{y}$. This reduction in squared errors is called the **regression sum of squares** and is denoted by **SSR.** Hence,†

$$\text{SSR} = \text{SST} - \text{SSE} = 23.714 - 3.348 = 20.366$$

REGRESSION SUM OF SQUARES (SSR)

The regression sum of squares, denoted by SSR, is

$$\text{SSR} = \text{SST} - \text{SSE}$$

†The formula for calculating SSR is $\text{SSR} = \Sigma(\hat{y} - \bar{y})^2$.

Thus, SSR is that portion of SST that is explained by the use of the regression model and SSE is that portion of SST that is *not* explained by the use of the regression model. The sum of SSR and SSE is always equal to SST. Thus,

$$\text{SST} = \text{SSR} + \text{SSE}$$

The ratio of SSR to SST is called the **coefficient of determination** and is denoted by r^2. The coefficient of determination gives the proportion of SST that is explained by the use of the regression model. The value of the coefficient of determination always lies in the range zero to 1. The coefficient of determination can be calculated by using the following formula.

$$r^2 = \frac{\text{SSR}}{\text{SST}} \quad \text{or} \quad \frac{\text{SST} - \text{SSE}}{\text{SST}}$$

However, for computational purposes, the following formula is more efficient to use to calculate r^2.

COEFFICIENT OF DETERMINATION

The coefficient of determination, denoted by r^2, represents the proportion of SST that is explained by the use of the regression model. The computational formula for r^2 is

$$r^2 = \frac{b\ \text{SS}_{xy}}{\text{SS}_{yy}}$$

and

$$0 \leq r^2 \leq 1$$

Calculating the coefficient of determination.

EXAMPLE 12-3 For the data of Table 12.1 on monthly incomes and food expenditures of seven households, calculate the coefficient of determination.

Solution From earlier calculations made in Examples 12-1 and 12-2,

$$b = .20, \quad \text{SS}_{xy} = 100.857, \quad \text{and} \quad \text{SS}_{yy} = 23.714$$

Hence

$$r^2 = \frac{b\ \text{SS}_{xy}}{\text{SS}_{yy}} = \frac{(.20)\ (100.857)}{(23.714)} = .85$$

Thus, we can state that SST is reduced by approximately 85% (from 23.714 to 3.348) when we use $\hat{y}$, instead of $\bar{y}$, to predict the food expenditures of households. Note that r^2 is usually rounded to two decimal places. ■

SST is a measure of the total variation in food expenditures, SSR is the portion of total variation explained by the regression model (or by income), and SSE is the portion of total variation not explained by the regression model. Hence, for Example

12-3 we can state that 85% of the total variation in food expenditures of households occurs because of the variation in their incomes, and the remaining 15% is due to randomness or other variables.

Usually, the higher the value of r^2, the better the regression model. This is so because if r^2 is larger, a greater portion of the total errors is explained by the included independent variable and a smaller portion of errors is attributed to other variables and to randomness.

EXERCISES

12.16 The following table gives six pairs of x and y values.

x	14	21	9	8	16	13
y	7	9	4	5	8	6

Compute the following.

a. SS_{xx}, SS_{yy}, and SS_{xy}
b. Standard deviation of errors
c. SST, SSE, and SSR
d. Coefficient of determination

12.17 The following table gives eight pairs of x and y values.

x	18	16	28	30	13	20	15	31
y	6	5	11	10	5	10	6	13

Calculate the following.

a. SS_{xx}, SS_{yy}, and SS_{xy}
b. Standard deviation of errors
c. SST, SSE, and SSR
d. Coefficient of determination

12.18 The following table gives the information on the average saturated fat (in grams) consumed per day and the cholesterol level (in milligrams per hundred milliliters) of eight males.

Fat consumption	55	65	50	34	43	58	69	36
Cholesterol level	180	210	195	165	170	204	235	150

Compute the following.

a. SS_{xx}, SS_{yy}, and SS_{xy}
b. Standard deviation of errors
c. SST, SSE, and SSR
d. Coefficient of determination

12.19 The following table gives the information on the monthly incomes (in hundreds of dollars) and the monthly telephone bills (in dollars) for a random sample of 10 households.

Income	16	45	36	32	30	13	41	15	36	40
Phone bill	35	78	102	56	75	26	130	42	59	85

Find the following.

a. SS_{xx}, SS_{yy}, and SS_{xy}
b. Standard deviation of errors
c. SST, SSE, and SSR
d. Coefficient of determination

12.20 Refer to Exercise 12.9. The following table, which gives the ages (in years) and prices (in hundreds of dollars) of eight cars of a specific model, is reproduced from that exercise.

Age	8	3	6	9	2	5	6	3
Price	16	74	38	19	102	36	33	69

a. Calculate the standard deviation of errors.
b. Compute the coefficient of determination and give a brief interpretation of it.

12.21 The following data on the stress scores before a math test and the math test scores for seven students are reproduced from Exercise 12.10.

Stress score	6.5	4.0	2.5	7.2	8.1	3.4	5.5
Test score	81	96	93	70	63	84	73

a. Calculate the standard deviation of errors.
b. Compute the coefficient of determination and give a brief interpretation of it.

12.22 The following table, reproduced from Exercise 12.11, lists the size of six houses (in hundreds of square feet) and the monthly rents (in dollars) paid by tenants for those houses.

Size of the house	21	16	19	27	34	23
Monthly rent	700	580	720	850	1050	800

a. Calculate the standard deviation of errors.
b. Compute the coefficient of determination.

12.23 The following data on annual incomes (in thousands of dollars) and amounts (in thousands of dollars) of life insurance policies for six persons is reproduced from Exercise 12.12.

Annual income	47	54	25	37	62	18
Life insurance	250	300	100	150	500	75

a. Calculate the standard deviation of errors.
b. Compute the coefficient of determination.

12.5 INFERENCES ABOUT *B*

This section is concerned with estimation and tests of hypotheses about the population regression slope B. We can also make confidence intervals and test hypotheses about the y-intercept A of the population regression line. However, making inferences about A is beyond the scope of this text.

12.5.1 SAMPLING DISTRIBUTION OF *b*

One of the main purposes for determining a regression line is to find the true value of slope B of the population regression line. However, in almost all cases, the regression line is estimated using sample data. Then, based on the sample regression line, inferences are made about the population regression line. The slope b of a sample regression line gives a point estimate of slope B of the population regression line. The different sample regression lines estimated for different samples taken from the same population will give different values of b. If only one sample is taken and the regression line for that sample is estimated, the value of b will depend on which sample is drawn. Thus, b is a random variable and it possesses a probability distribution, which is more commonly called its *sampling distribution*. The shape of the sampling distribution of b, its mean, and standard deviation are as follows.

MEAN, STANDARD DEVIATION, AND SAMPLING DISTRIBUTION OF *b*

Because of the assumption of normally distributed random errors, the sampling distribution of b is normal. The mean and standard deviation of b, denoted by μ_b and σ_b, respectively, are

$$\mu_b = B \quad \text{and} \quad \sigma_b = \frac{\sigma_\epsilon}{\sqrt{SS_{xx}}}$$

However, usually the standard deviation σ_ϵ of population errors is not known. Hence, the sample standard deviation s_e is used to estimate σ_ϵ. In such a case, when σ_ϵ is unknown, the standard deviation of b is estimated by s_b, which is calculated as

$$s_b = s_e/\sqrt{SS_{xx}}$$

If σ_ϵ is not known and the sample size is large ($n \geq 30$), the normal distribution can be used to make inferences about B. However, if σ_ϵ is not known and the sample size is small ($n < 30$), the normal distribution is replaced by the t distribution to make inferences about B.

12.5.2 ESTIMATION OF *B*

The value of b obtained from the sample regression line is a point estimate of slope B of the population regression line. As mentioned in Section 12.5.1, if σ_ϵ is not known and the sample size is small, the t distribution is used to make a confidence interval for B.

CONFIDENCE INTERVAL FOR *B*

The $(1 - \alpha)100\%$ confidence interval for B is given by

$$b \pm t\, s_b$$

where

$$s_b = \frac{s_e}{\sqrt{SS_{xx}}}$$

and the value of t is obtained from the t distribution table for $\alpha/2$ area in the right tail and $n - 2$ degrees of freedom.

The following example describes the procedure for making a confidence interval for B.

Constructing a confidence interval for B.

EXAMPLE 12-4 Construct a 95% confidence interval for B for the data on incomes and food expenditures of seven households given in Table 12.1.

Solution From the given information and earlier calculations,

$$n = 7, \quad b = .20, \quad SS_{xx} = 499.429, \quad s_e = .842, \quad \text{and} \quad 1 - \alpha = .95$$

Hence

$$s_b = \frac{s_e}{\sqrt{SS_{xx}}} = \frac{.842}{\sqrt{499.429}} = .038$$

$$df = n - 2 = 7 - 2 = 5$$

$$\alpha/2 = .5 - \left(\frac{.95}{2}\right) = .025$$

From the t distribution table, the value of t for 5 df and .025 area in the right tail is 2.571. The 95% confidence interval for B is

$$b \pm t\, s_b = .20 \pm 2.571\,(.038) = .20 \pm .10 = .10 \text{ to } .30$$

Thus, we are 95% confident that slope B of the population regression line is between .10 and .30. ■

12.5.3 HYPOTHESIS TESTING ABOUT *B*

Testing a hypothesis about B, when the null hypothesis is $B = 0$ (i.e., the slope of the regression line is zero), is equivalent to testing that x does not determine y and that the regression line is of no use in predicting y for a given x. However, we should remember that we are testing for a linear relationship between x and y. It is possible that x may determine y nonlinearly. Hence, a nonlinear relationship may exist between x and y.

To test the hypothesis that x does not determine y linearly, we will test the null hypothesis that the slope of the regression line is zero, that is, $B = 0$. The alternative hypothesis can be: (1) x determines y, that is, $B \neq 0$; (2) x determines y positively, that is, $B > 0$; or (3) x determines y negatively, that is, $B < 0$.

The procedure used to make a hypothesis test about B is similar to the one used in earlier chapters. It involves the same five steps.

TEST STATISTIC FOR *b*

The value of the test statistic t for b is calculated as

$$t = \frac{b - B}{s_b}$$

The value of B is substituted from the null hypothesis.

The following example illustrates the procedure for testing a hypothesis about B.

Conducting a test of hypothesis about B.

EXAMPLE 12-5 Test at the 1% significance level if the slope of the regression line for the example on incomes and food expenditures of seven households is positive.

Solution From the given information and earlier calculations in Examples 12-1 and 12-4,

$$n = 7, \quad b = .20, \quad \text{and} \quad s_b = .038$$

Step 1. *State the null and alternative hypotheses*

We are to test whether or not slope B of the population regression line is positive. Hence, the two hypotheses are

$$H_0: B = 0 \quad \text{(the slope is not positive)}$$
$$H_1: B > 0 \quad \text{(the slope is positive)}$$

Note that we can also write the null hypothesis as H_0: $B \leq 0$.

Step 2. *Select the distribution to use*

The sample size is small ($n < 30$) and σ_ϵ is not known. Hence, we will use the t distribution to make the test about B.

Step 3. *Determine the rejection and nonrejection regions*

The significance level is .01. The ">" sign in the alternative hypothesis indicates that the test is right-tailed. Hence,

$$\text{Area in the right tail} = \alpha = .01$$

and

$$df = n - 2 = 7 - 2 = 5$$

From the t distribution table, the critical value of t for 5 df and .01 area in the right tail is 3.365, as shown in Figure 12.17.

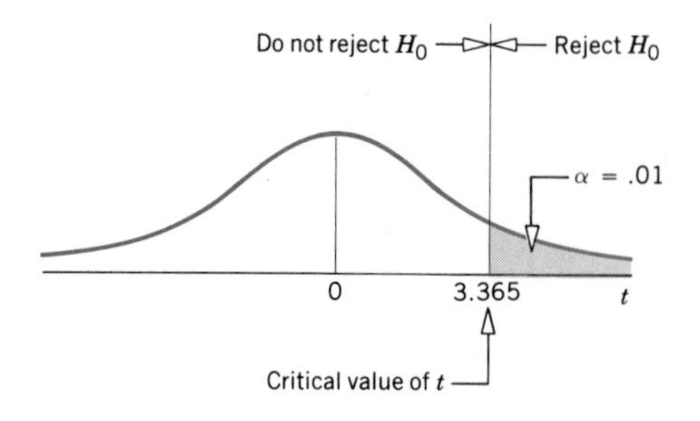

Figure 12.17

Step 4. *Calculate the value of the test statistic*

The value of the test statistic t for b is calculated as follows.

$$t = \frac{b - B}{s_b} = \frac{.20 - \overbrace{0}^{\text{From } H_0}}{.038} = 5.263$$

Step 5. *Make a decision*

The value of the test statistic $t = 5.263$ is greater than the critical value of $t = 3.365$, and it falls in the rejection region. Hence, we reject the null hypothesis and conclude that x (income) determines y (food expenditure) positively. That is, food expenditure increases with an increase in income and it decreases with a decrease in income. ■

Note that the null hypothesis does not always have to be $B = 0$. We may test the null hypothesis that B is equal to a certain value. See Exercises 12.30 and 12.31 for such cases.

EXERCISES

12.24 The following data give the experience (in years) and the monthly salaries (in hundreds of dollars) of nine randomly selected secretaries.

Experience	14	3	5	6	4	9	18	5	16
Monthly salary	22	12	15	17	15	19	24	13	27

a. Find the least squares regression line with experience as an independent variable and monthly salary as a dependent variable.
b. Construct a 95% confidence interval for B.
c. Test at the 1% significance level if B is positive.

12.25 The following data give the midterm scores in a course for a sample of 10 students and the scores of student evaluations of the instructor. (In the instructor evaluation scores, 1 is the lowest and 4 is the highest score.)

Instructor score	3	2	3	1	2	4	3	4	4	2
Midterm score	93	75	97	64	47	99	78	88	93	81

a. Find the regression of instructor scores on midterm scores.
b. Construct a 99% confidence interval for B.
c. Test at the 2.5% significance level if B is positive.

12.26 Refer to Exercise 12.9. The data on ages (in years) and prices (in hundreds of dollars) for eight cars of a specific model are reproduced below from that exercise.

Age	8	3	6	9	2	5	6	3
Price	16	74	38	19	102	36	33	69

a. Find the regression line $\hat{y} = a + bx$, where x is the age of a car and y is the price of a car.
b. Construct a 99% confidence interval for B.
c. Test at the 5% significance level if B is negative.

12.27 The following data on the stress scores before a math test and the math test scores for seven students are reproduced from Exercise 12.10.

Stress score	6.5	4.0	2.5	7.2	8.1	3.4	5.5
Test score	81	96	93	70	63	84	73

a. Find the regression line $\hat{y} = a + bx$, where x is the stress score and y is the test score.
b. Make a 95% confidence interval for B.
c. Test at the 2.5% significance level if B is negative.

12.28 The data on the size of six houses (in hundreds of square feet) and the monthly rents (in dollars) paid by tenants for those houses is reproduced below from Exercise 12.11.

Size of the house	21	16	19	27	34	23
Monthly rent	700	580	720	850	1050	800

a. Find the simple linear regression line with size of the house as an independent variable and monthly rent as a dependent variable.
b. Construct a 90% confidence interval for B.
c. Test at the 5% significance level if B is positive.

12.29 The following data on annual incomes (in thousands of dollars) and amounts (in thousands of dollars) of life insurance policies for six persons is reproduced from Exercise 12.12.

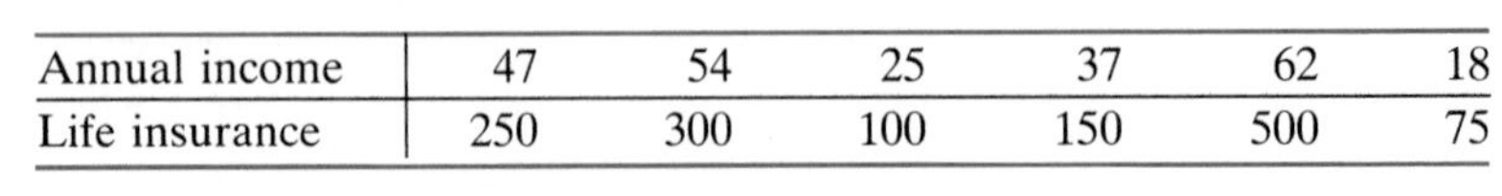

Annual income	47	54	25	37	62	18
Life insurance	250	300	100	150	500	75

a. Find the least squares regression line with annual income as an independent variable and amount of life insurance policy as a dependent variable.
b. Construct a 98% confidence interval for B.
c. Test at the 1% significance level if B is positive.

12.30 The following table, reproduced from Exercise 12.18, gives the information on the average saturated fat (in grams) consumed per day and the cholesterol level (in milligrams per hundred milliliters) of eight males.

Fat consumption	55	65	50	34	43	58	69	36
Cholesterol level	180	210	195	165	170	204	235	150

a. Find the regression line $\hat{y} = a + bx$, where x is the fat consumption and y is the cholesterol level.
b. Construct a 90% confidence interval for B.
c. An earlier study claims that B is 1.75. Test at the 5% significance level if B is different from 1.75. (*Hint:* The null hypothesis will be H_0: $B = 1.75$, and the alternative hypothesis will be H_1: $B \neq 1.75$. Notice that the value of $B = 1.75$ will be used to calculate the value of the test statistic t.)

12.31 The following table, reproduced from Exercise 12.19, gives the information on monthly incomes (in hundreds of dollars) and monthly telephone bills (in dollars) for a random sample of 10 households.

Income	16	45	36	32	30	13	41	15	36	40
Phone bill	35	78	102	56	75	26	130	42	59	85

a. Find the least squares regression line. Take income as an independent variable and phone bill as a dependent variable.
b. Make a 98% confidence interval for B.
c. An earlier study claims that B is 1.50. Test at the 1% significance level if B is greater than 1.50. (*Hint:* The null hypothesis will be H_0: $B = 1.50$, and the alternative hypothesis will be H_1: $B > 1.50$. Notice that the value of $B = 1.50$ will be used in the calculation of the value of test statistic t.)

12.6 LINEAR CORRELATION

Another measure of the relationship between two variables is the correlation coefficient. This section describes the *simple linear correlation,* for short **linear correlation,** which measures the strength of the linear association between two variables. In other words, the linear correlation coefficient measures how closely the points in a scatter diagram are spread around the regression line. The correlation coefficient calculated for the population data is denoted by ρ (pronounced "rho"), and the one calculated for sample data is denoted by r. (Notice that the square of the correlation coefficient is equal to the coefficient of determination.)

THE VALUE OF THE CORRELATION COEFFICIENT

The value of the correlation coefficient always lies in the range -1 to 1, i.e.,

$$-1 \le r \le 1$$

If $r = 1$, it is said to be the case of a *perfect positive linear correlation*. In such a case, all the points lie on a straight line that slopes upward from left to right, as shown in Figure 12.18*a*. If $r = -1$, the correlation is said to be a *perfect negative linear correlation*. In this case, all the points lie on a straight line that slopes downward from left to right, as shown in Figure 12.18*b*. If the points are scattered all over the diagram, as shown in Figure 12.18*c*, then there is *no linear correlation* between the two variables and consequently $r = 0$.

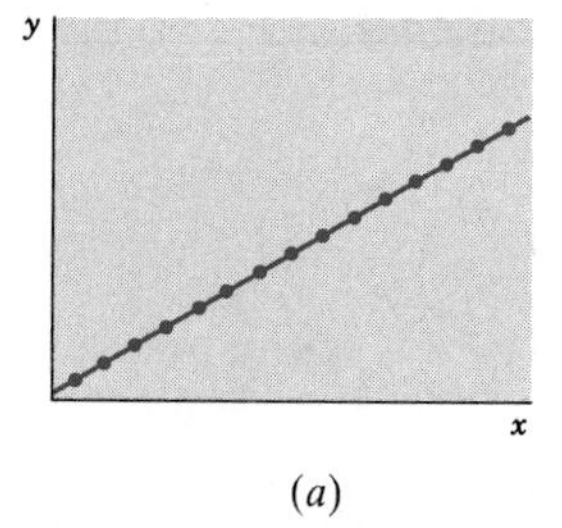

(*a*)

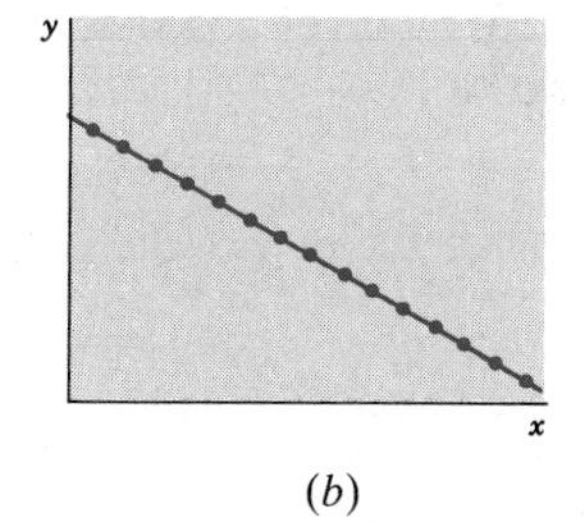

(*b*)

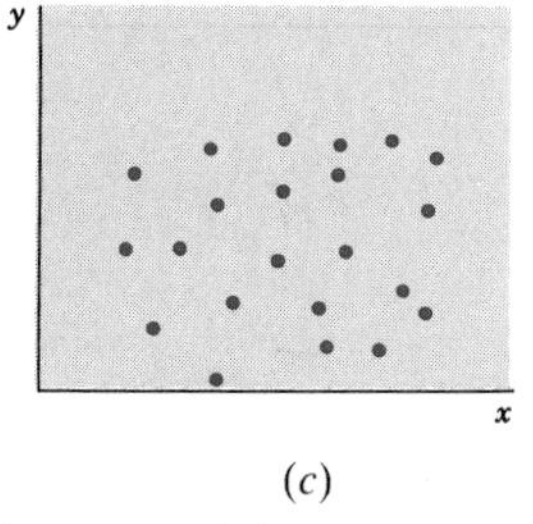

(*c*)

Figure 12.18 Linear correlation between variables. (*a*) Perfect positive linear correlation, $r = 1$. (*b*) Perfect negative linear correlation, $r = -1$. (*c*) No linear correlation, $r = 0$.

We do not usually encounter an example with perfect positive or perfect negative correlation. What we observe in real-world problems is either a positive linear correlation with $0 < r < 1$ (that is, the correlation is greater than zero but less than 1) or a negative linear correlation with $-1 < r < 0$ (that is, the correlation is greater than -1 but less than zero).

If the correlation between two variables is positive and close to 1, we say that the variables have a *strong positive linear correlation*. If the correlation between two variables is positive but close to zero, then the variables have a *weak positive linear correlation*. On the other hand, if the correlation between two variables is negative and close to -1, then the variables are said to have a *strong negative linear correlation*. Also, if the correlation between variables is negative but close to zero, there exists a *weak negative linear correlation* between the variables. Graphically, a strong correlation indicates that the points in the scatter diagram are very close to the regression line and a weak correlation indicates that the points in the scatter diagram are widely spread around the regression line. These four cases are shown in Figure 12.19.

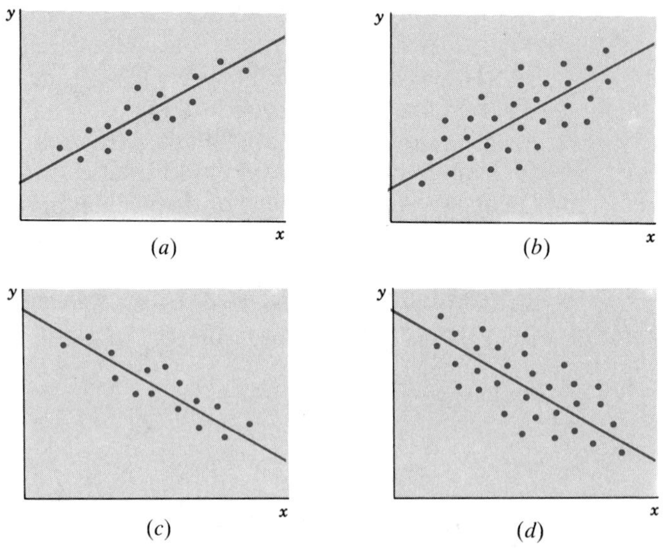

Figure 12.19 Correlation between variables. (*a*) Strong positive linear correlation (r is close to 1). (*b*) Weak positive linear correlation (r is positive but close to zero). (*c*) Strong negative linear correlation (r is close to -1). (*d*) Weak negative linear correlation (r is negative and close to zero).

The linear correlation coefficient is calculated by using the following formula. This correlation coefficient is also called the *Pearson's product moment correlation coefficient*.

LINEAR CORRELATION COEFFICIENT

The simple linear correlation, denoted by r, measures the strength of linear relationship between two variables for a sample and is calculated as

$$r = \frac{SS_{xy}}{\sqrt{SS_{xx}\,SS_{yy}}}$$

As SS_{xx} and SS_{yy} are both always positive, the sign of the correlation coefficient r depends on the sign of SS_{xy}. If SS_{xy} is positive, then r will be positive. If SS_{xy} is negative, then r will be negative. Another important observation to remember is that *r and b, calculated for the same sample, will always have the same sign. That is, r and b are both either positive or negative*. This is so because r and b both provide us with the information about the relationship between x and y. Likewise, the corresponding population parameters ρ and B will always have the same sign.

Calculating the linear correlation coefficient.

EXAMPLE 12-6 Calculate the correlation coefficient for the example on incomes and food expenditures of seven households.

Solution From earlier calculations made in Examples 12-1 and 12-2,

$$SS_{xy} = 100.857, \quad SS_{xx} = 499.429, \quad \text{and} \quad SS_{yy} = 23.714$$

Hence

$$r = \frac{SS_{xy}}{\sqrt{SS_{xx}\, SS_{yy}}} = \frac{100.857}{\sqrt{(499.429)\,(23.714)}} = .93$$

Thus, the linear correlation coefficient is .93. The correlation coefficient is usually rounded to two decimal places. ■

The linear correlation coefficient simply tells us how strongly the two variables are (linearly) related. The correlation coefficient of .93 for incomes and food expenditures of seven households indicates that income and food expenditure are very strongly and positively correlated. This correlation coefficient does not, however, provide us with any more information.

The square of the correlation coefficient gives the coefficient of determination, which was explained in Section 12.4. Thus, $(.93)^2$ is .86, which is close to the value of $r^2 = .85$ calculated in Example 12-3. (The difference is due to rounding error.)

Sometimes the calculated value of r may indicate that the two variables are very strongly linearly correlated, but in reality they are not. For example, if we calculate the correlation coefficient between the price of Coke and the size of family in the United States using data for the last 30 years, we will find a strong negative linear correlation. Over time, the price of Coke has increased and the size of family has decreased. This finding does not mean that family size and price of Coke are related. As a result, before we calculate the correlation coefficient, we must seek help from a theory or from common sense to postulate whether or not the two variables have a causal relationship.

Another point to note is that in a simple regression model one of the two variables is categorized as an independent variable and the other is classified as a dependent variable. However, no such distinction is made between the two variables when the correlation coefficient is calculated.

EXERCISES

12.32 Will you expect a positive, zero, or a negative correlation between the two variables for each of the following examples?

a. Grade of a student and hours spent studying

b. Income and entertainment expenditure of a household

c. Age of a woman and makeup expenses per month
d. Price of a computer and consumption of Coke
e. Price and consumption of wine

12.33 Will you expect a positive, zero, or a negative correlation between the two variables for each of the following examples?

a. SAT score and GPA of a student
b. Weight and blood pressure of a person
c. Amount of fertilizer used and yield of grain per acre
d. Age and price of a house
e. Height of a husband and his wife's income

12.34 Construct a scatter diagram for the following data.

x	5	8	10	12	16	20
y	55	46	40	34	22	10

a. Do all points lie on a straight line? By looking at the scatter diagram, what do you expect the value of r to be?
b. Compute the value of the correlation coefficient. Is this value of r the same as what you expected in part a?

12.35 Plot a scatter diagram for the following data.

x	50	60	30	20	10	25
y	2	100	10	90	50	5

a. By looking at the scatter diagram, do you expect the value of r to be close to zero or close to 1 or -1?
b. Compute the value of the correlation coefficient. Is this value of r close to what you expected in part a?

12.36 The following table lists the midterm and final examination scores for seven students in a statistics class.

Midterm score	79	95	81	66	87	97	59
Final exam score	85	97	78	76	94	91	67

a. Do you expect the midterm and final examination scores to be positively or negatively related?
b. Plot a scatter diagram. By looking at the scatter diagram, do you expect the correlation coefficient between these two variables to be close to zero or close to 1 or -1?
c. Find the correlation coefficient. Is the value of r consistent with what you expected in parts a and b?

12.37 The following data give the ages of husbands and wives for six couples.

Husband's age	43	57	28	19	33	39
Wife's age	37	51	32	20	31	38

a. Do you expect the ages of husbands and wives to be positively or negatively related?
b. Plot a scatter diagram. By looking at the scatter diagram, do you expect the correlation coefficient between these two variables to be close to zero or close to 1 or -1?

c. Find the correlation coefficient. Is the value of r consistent with what you expected in parts a and b?

12.38 Refer to Exercise 12.9. The following data on ages (in years) and prices (in hundreds of dollars) of eight cars of a specific model are reproduced from that exercise.

Age	8	3	6	9	2	5	6	3
Price	16	74	38	19	102	36	33	69

Find the correlation coefficient. Is the sign of the correlation coefficient the same as that of b in the regression line estimated in Exercise 12.9?

12.39 The following data on stress scores before a math test and math test scores for seven students are reproduced from Exercise 12.10.

Stress score	6.5	4.0	2.5	7.2	8.1	3.4	5.5
Test score	81	96	93	70	63	84	73

Find the correlation coefficient. Is the sign of the correlation coefficient the same as that of b in the regression line estimated in Exercise 12.10?

12.40 The following table, reproduced from Exercise 12.18, gives the information on average saturated fat (in grams) consumed per day and cholesterol level (in milligrams per hundred milliliters) of eight males.

Fat consumption	55	65	50	34	43	58	69	36
Cholesterol level	180	210	195	165	170	204	235	150

Compute the coefficient of correlation.

12.41 The following table, reproduced from Exercise 12.19, gives the information on monthly incomes (in hundreds of dollars) and monthly telephone bills (in dollars) for a random sample of 10 households.

Income	16	45	36	32	30	13	41	15	36	40
Phone bill	35	78	102	56	75	26	130	42	59	85

Calculate the coefficient of correlation.

12.42 The following table, reproduced from Exercise 12.24, gives the experience (in years) and monthly salaries (in hundreds of dollars) of nine randomly selected secretaries.

Experience	14	3	5	6	4	9	18	5	16
Monthly salary	22	12	15	17	15	19	24	13	27

a. Do you expect the experience and monthly salaries to be positively or negatively related?

b. Compute the correlation coefficient.

12.43 The following data, reproduced from Exercise 12.25, give the midterm scores in a course for 10 students and the scores of student evaluations of the instructor. (In the instructor evaluation scores, 1 is the lowest and 4 is the highest score.)

Instructor score	3	2	3	1	2	4	3	4	4	2
Midterm score	93	75	97	64	47	99	78	88	93	81

a. Do you expect the instructor scores and midterm scores to be positively or negatively related?
b. Calculate the correlation coefficient.

12.7 REGRESSION ANALYSIS: A COMPLETE EXAMPLE

This section works out an example that includes all the topics discussed so far in this chapter.

A complete example of regression analysis.

EXAMPLE 12-7 A random sample of eight auto drivers insured with a company and having similar auto insurance policies was selected. The following table lists their driving experiences (in years) and the monthly auto insurance premiums (in dollars) paid by them.

Driving Experience (years)	Monthly Auto Insurance Premium (dollars)
5	64
2	87
12	50
9	71
15	44
6	56
25	42
16	60

(a) Is it the insurance premium that depends on the driving experience or is it the driving experience that depends on the insurance premium? Do you expect a positive or a negative relationship between these two variables?
(b) Compute SS_{xx}, SS_{yy}, and SS_{xy}.
(c) Find the least squares regression line.
(d) Interpret the meaning of the values of a and b calculated in part (c).
(e) Plot the scatter diagram and the regression line.
(f) Calculate r and r^2 and explain what they mean.
(g) Predict the monthly auto insurance premium for a driver with a driving experience of 10 years.
(h) Compute the standard deviation of errors.
(i) Construct a 90% confidence interval for B.
(j) Test at the 5% significance level if B is negative.

Solution

(a) Based on theory and intuition, we expect the insurance premium to depend on the driving experience. Hence, the insurance premium would be a dependent variable and driving experience an independent variable in the regression model. A new driver is called a high risk by the insurance companies, and he or she has to pay a high premium for auto insurance. On average, the insurance premium is expected to decrease with an increase in the years of driving experience.

Therefore, we expect a negative relationship between the two variables, insurance premium and years of driving experience. In other words, the population correlation coefficient ρ and the population regression slope B are both expected to be negative.

(b) Table 12.5 shows the calculation of Σx, Σy, Σxy, Σx^2, and Σy^2.

Table 12.5

Experience x	Premium y	xy	x^2	y^2
5	64	320	25	4096
2	87	174	4	7569
12	50	600	144	2500
9	71	639	81	5041
15	44	660	225	1936
6	56	336	36	3136
25	42	1050	625	1764
16	60	960	256	3600
$\Sigma x = 90$	$\Sigma y = 474$	$\Sigma xy = 4739$	$\Sigma x^2 = 1396$	$\Sigma y^2 = 29{,}642$

The values of $\bar{x}$ and $\bar{y}$ are

$$\bar{x} = \frac{\Sigma x}{n} = \frac{90}{8} = 11.25$$

$$\bar{y} = \frac{\Sigma y}{n} = \frac{474}{8} = 59.25$$

The values of SS_{xy}, SS_{xx}, and SS_{yy} are computed as follows.

$$SS_{xy} = \Sigma xy - \frac{(\Sigma x)(\Sigma y)}{n} = 4739 - \frac{(90)(474)}{8} = -593.50$$

$$SS_{xx} = \Sigma x^2 - \frac{(\Sigma x)^2}{n} = 1396 - \frac{(90)^2}{8} = 383.50$$

$$SS_{yy} = \Sigma y^2 - \frac{(\Sigma y)^2}{n} = 29{,}642 - \frac{(474)^2}{8} = 1557.50$$

(c) To find the regression line, we calculate a and b as follows.

$$b = \frac{SS_{xy}}{SS_{xx}} = \frac{-593.50}{383.50} = -1.55$$

$$a = \bar{y} - b\bar{x} = 59.25 - (-1.55)(11.25) = 76.69$$

Thus, our estimated regression line $\hat{y} = a + bx$ is

$$\hat{y} = 76.69 - 1.55x$$

(d) The value of $a = 76.69$ gives the value of $\hat{y}$ for $x = 0$, that is, it gives the monthly auto insurance premium for a driver with no driving experience. However, as mentioned earlier in this chapter, we should not attach much importance to this statement because the sample contains only drivers with 2 or more years of experience.

The value of b gives the change in $\hat{y}$ due to a change of one unit in x. Hence, $b = -1.55$ indicates that on average for every extra year of driving experience the monthly auto insurance premium decreases by \$1.55. Note that when b is negative then y decreases as x increases.

(e) Figure 12.20 shows the scatter diagram and the regression line for the data on eight auto drivers. Note that the regression line slopes downward from left to right. This result is consistent with the negative relationship we anticipated between driving experience and insurance premium.

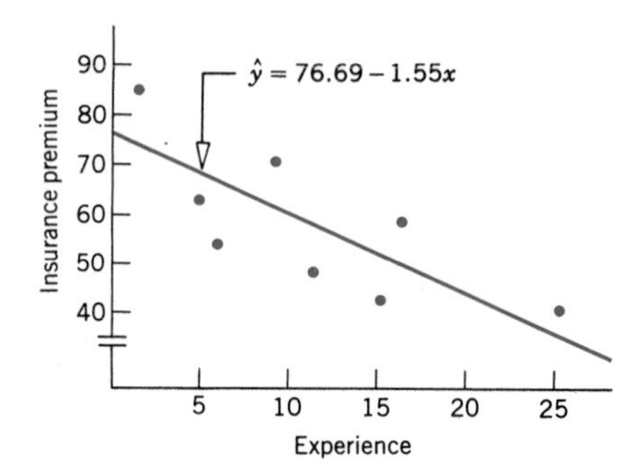

Figure 12.20

(f) The values of r and r^2 are computed as follows.

$$r = \frac{SS_{xy}}{\sqrt{SS_{xx}\ SS_{yy}}} = \frac{-593.50}{\sqrt{(383.50)\ (1557.50)}} = -.77$$

$$r^2 = \frac{b\ SS_{xy}}{SS_{yy}} = \frac{(-1.55)\ (-593.50)}{1557.50} = .59$$

The value of $r = -.77$ indicates that the driving experience and monthly auto insurance premium are negatively related. The (linear) relationship is strong but not very strong.

The value of $r^2 = .59$ states that 59% of the total variation in insurance premiums is explained by years of driving experience and 41% is not. The low value of r^2 indicates that there may be many other important variables that contribute to the determination of auto insurance premium. For example, the auto insurance premium is expected to depend on the driving record of a driver and the type and age of the car.

(g) Using the estimated regression line, the predicted value of y for $x = 10$ is

$$\hat{y} = 76.69 - 1.55x = 76.69 - 1.55\ (10) = \$61.19$$

Hence, we expect the monthly auto insurance premium of a driver with a driving experience of 10 years to be \$61.19.

(h) The standard deviation of errors is

$$s_e = \sqrt{\frac{SS_{yy} - b\ SS_{xy}}{n - 2}} = \sqrt{\frac{1557.50 - (-1.55)\ (-593.50)}{8 - 2}} = 10.308$$

(i) To construct a 90% confidence interval for B, first we calculate the standard deviation of b.

$$s_b = \frac{s_e}{\sqrt{SS_{xx}}} = \frac{10.308}{\sqrt{383.50}} = .526$$

For a 90% confidence level, the area in each tail is

$$\frac{\alpha}{2} = .5 - \left(\frac{.90}{2}\right) = .05$$

The degrees of freedom are

$$df = n - 2 = 8 - 2 = 6$$

From the t distribution table, the t value for .05 area in the right tail and 6 df is 1.943. The 90% confidence interval for B is

$$b \pm t\, s_b = -1.55 \pm 1.943\ (.526) = -1.55 \pm 1.02 = -2.57 \text{ to } -.53$$

Thus, we can state with 90% confidence that B lies in the interval -2.57 to $-.53$. That is, on average the monthly auto insurance premium of a driver decreases by an amount between \$.53 and \$2.57 for every extra year of driving experience.

(j) We perform the following five steps to test the hypothesis about B.

Step 1. *State the null and alternative hypotheses*

The null and alternative hypotheses are written as follows:

$$H_0\text{: } B = 0 \quad (B \text{ is not negative})$$

$$H_1\text{: } B < 0 \quad (B \text{ is negative})$$

Note that the null hypothesis can also be written as H_0: $B \geq 0$.

Step 2. *Select the distribution to use*

As the sample size is small and σ_ϵ is not known, we use the t distribution to make the hypothesis test.

Step 3. *Determine the rejection and nonrejection regions*

The significance level is .05. The "<" sign in the alternative hypothesis indicates that it is a left-tailed test.

$$\text{Area in the left tail} = \alpha = .05$$

$$df = n - 2 = 8 - 2 = 6$$

From the t distribution table, the critical value of t for .05 area in the left tail and 6 df is -1.943, as shown in Figure 12.21.

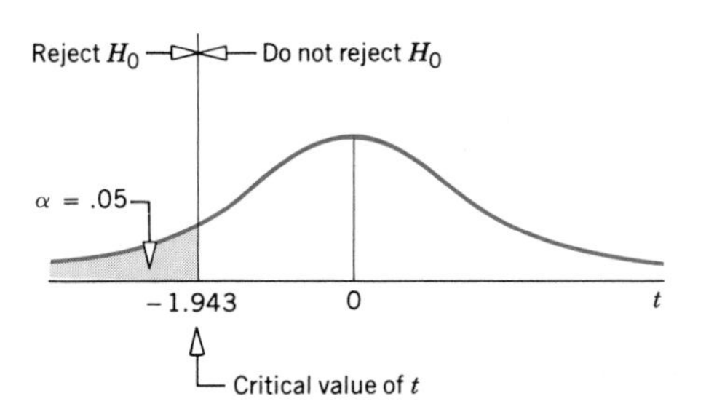

Figure 12.21

Step 4. *Calculate the value of the test statistic*

The value of the test statistic t for b is calculated as follows.

$$t = \frac{b - B}{s_b} = \frac{-1.55 - \overset{\text{From } H_0}{0}}{.526} = -2.947$$

Step 5. *Make a decision*

The value of the test statistic $t = -2.947$ falls in the rejection region. Hence, we reject the null hypothesis and conclude that B is negative. That is, the monthly auto insurance premium decreases with an increase in years of driving experience. ■

EXERCISES

12.44 Eight students, randomly selected from a large class, were asked to keep a record of hours they spent studying before the midterm examination. The following table gives the number of hours that these eight students studied before the midterm and their scores in the midterm.

Hours studied	15	7	12	8	18	6	9	11
Midterm score	97	78	87	92	89	57	74	69

a. Is it the midterm scores that depend on hours studied or hours studied that depend on midterm scores? Do you expect a positive or a negative relationship between these two variables?
b. Taking hours studied as an independent variable and midterm scores as a dependent variable, compute SS_{xx}, SS_{yy}, and SS_{xy}.
c. Find the least squares regression line.
d. Interpret the meaning of the values of a and b calculated in part (c).
e. Plot the scatter diagram and the regression line.
f. Calculate r and r^2 and briefly explain what they mean.
g. Predict the midterm score of a student who studied for 14 hours.
h. Compute the standard deviation of errors.
i. Construct a 99% confidence interval for B.
j. Test at the 1% significance level if B is positive.

12.45 Usually students who are good in mathematics do well in economics. A professor wanted to check if the scores of students in these two areas are related. He took a random sample of seven students from all students who were enrolled in one mathematics course and one economics course. The following table gives the midterm scores of these students in mathematics and economics.

Math score	75	93	82	65	73	96	67
Economics score	83	96	79	70	69	92	69

a. Taking math scores as an independent variable and economics scores as a dependent variable, compute SS_{xx}, SS_{yy}, and SS_{xy}.
b. Find the least squares regression line.
c. Interpret the meaning of the values of a and b calculated in part (b).
d. Plot the scatter diagram and the regression line.
e. Calculate r and r^2 and explain what they mean.
f. Compute the standard deviation of errors.
g. Predict the midterm score in economics of a student whose midterm score in mathematics is 90.
h. Construct a 99% confidence interval for B.
i. Test at the 5% significance level if B is positive.

12.46 The following table gives the yearly income (in thousands of dollars) and the value of the home owned (in thousands of dollars) for a random sample of nine home owners selected from a certain area.

Yearly Income ($ thousands)	**Value of the Home ($ thousands)**
36	129
64	310
49	260
21	92
28	126
47	242
58	288
19	81
32	134

a. Is it the value of the home owned that depends on the yearly income of a household or is it the yearly income that depends on the value of the home owned? Do you expect a positive or a negative relationship between these two variables?
b. With yearly income as an independent variable and value of the home owned as a dependent variable, compute SS_{xx}, SS_{yy}, and SS_{xy}.
c. Find the regression line $\hat{y} = a + bx$.
d. Briefly explain the meaning of the values of a and b.
e. Calculate r and r^2 and explain what they mean.
f. Predict the value of the home owned by a family with an income of $40,000 a year.
g. Compute the standard deviation of errors.
h. Construct a 90% confidence interval for B.
i. Test at the 10% significance level if B is different from zero.

12.47 A farmer wanted to find the relationship between the amount of fertilizer used and the yield of corn. He selected seven acres of his land on which he used different amounts of fertilizer to grow corn. The following table gives the amount of fertilizer used (in pounds) and the yield of corn (in bushels) for each of the seven acres.

Fertilizer Used (pounds)	Yield of Corn (bushels)
120	138
80	112
100	129
70	96
88	119
75	104
110	134

a. With amount of fertilizer used as an independent variable and yield of corn as a dependent variable, compute SS_{xx}, SS_{yy}, and SS_{xy}.
b. Find the least squares regression line.
c. Interpret the meaning of the values of a and b.
d. Calculate r and r^2 and explain what they mean.
e. Compute the standard deviation of errors.
f. Predict the yield of corn per acre for $x = 105$.
g. Construct a 98% confidence interval for B.
h. Test at the 5% significance level if B is different from zero.

12.48 The following table gives the information on age and cholesterol level for a random sample of 10 men.

Age	58	69	43	39	63	52	47	31	74	36
Cholesterol level	189	235	193	177	154	191	213	175	198	181

a. Taking age as an independent variable and cholesterol level as a dependent variable, compute SS_{xx}, SS_{yy}, and SS_{xy}.
b. Find the regression of cholesterol level on age.
c. Briefly explain the meaning of the values of a and b.
d. Calculate r and r^2 and explain what they mean.
e. Compute the standard deviation of errors.
f. Construct a 95% confidence interval for B.
g. Test at the 5% significance level if B is positive.

12.49 The following table gives the information on income (in thousands of dollars) and charitable contributions (in hundreds of dollars) for the past year for a random sample of 10 households.

Income (thousands of dollars)	Charitable Contributions (hundreds of dollars)
33	10
23	4
42	9
47	23
26	3
21	2
28	8
39	16
58	18
17	1

a. With income as an independent variable and charitable contributions as a dependent variable, compute SS_{xx}, SS_{yy}, and SS_{xy}.
b. Find the regression of charitable contributions on income.
c. Briefly explain the meaning of the values of a and b.
d. Calculate r and r^2 and briefly explain what they mean.
e. Compute the standard deviation of errors.
f. Construct a 99% confidence interval for B.
g. Test at the 1% significance level if B is positive.

12.50 The following table gives the information on GPAs (grade point averages) and starting salaries (rounded to the nearest thousand dollars) of seven college graduates.

GPA	2.90	3.81	3.20	2.42	3.94	2.05	2.25
Starting salary	23	28	23	21	32	19	22

a. With GPA as an independent variable and starting salary as a dependent variable, compute SS_{xx}, SS_{yy}, and SS_{xy}.
b. Find the least squares regression line.
c. Interpret the meaning of the values of a and b.
d. Calculate r and r^2 and briefly explain what they mean.
e. Compute the standard deviation of errors.
f. Construct a 98% confidence interval for B.
g. Test at the 5% significance level if B is different from zero.

12.51 The following data give the information on ages (in years) and the monthly maintenance costs (in dollars) for a sample of eight cars.

Age	6	4	11	8	13	9	6	3
Maintenance cost	64	35	116	92	125	78	72	33

a. Taking age as an independent variable and maintenance cost as a dependent variable, compute SS_{xx}, SS_{yy}, and SS_{xy}.
b. Find the regression line $\hat{y} = a + bx$.
c. Briefly explain the meaning of the values of a and b.
d. Calculate r and r^2 and briefly explain what they mean.
e. Compute the standard deviation of errors.
f. Construct a 95% confidence interval for B.
g. Test at the 1% significance level if B is different from zero.

12.52 The following data give the information on the lowest-cost ticket price (in dollars) and the average attendance (rounded to the nearest thousand) for the past year for six football teams.

Ticket price	12.50	9.50	10.00	14.50	16.00	12.00
Attendance	56	65	71	69	55	42

a. Taking ticket price as an independent variable and attendance as a dependent variable, compute SS_{xx}, SS_{yy}, and SS_{xy}.
b. Find the least squares regression line.
c. Briefly explain the meaning of the values of a and b.
d. Calculate r and r^2 and briefly explain what they mean.

e. Compute the standard deviation of errors.
f. Construct a 90% confidence interval for B.
g. Test at the 2.5% significance level if B is negative.

12.53 Seven mathematics classes were observed to find the relationship between the class size and the percentage of students who pass the course. The following data give the information on the number of students and the percentage who passed for each of these seven classes.

Number of students	57	34	25	66	43	18	51
Pass percentage	81	88	92	82	86	94	86

a. Taking number of students as an independent variable and pass percentage as a dependent variable, compute SS_{xx}, SS_{yy}, and SS_{xy}.
b. Find the least squares regression line.
c. Interpret the meaning of the values of a and b.
d. Calculate r and r^2 and briefly explain what they mean.
e. Compute the standard deviation of errors.
f. Construct a 98% confidence interval for B.
g. Test at the 1% significance level if B is negative.

12.8 USING THE REGRESSION MODEL

Let us return to the example on incomes and food expenditures to discuss two major uses of a regression model. These two uses are

1. Estimating the mean value of y for a given value of x. For instance, we can use our food expenditure regression model to estimate the mean food expenditure of all households with an income of $2000 per month.
2. Predicting a particular value of y for a given value of x. For instance, we can determine the expected food expenditure of a randomly selected household with a monthly income of $2000 using the food expenditure regression model.

12.8.1 USING THE REGRESSION MODEL FOR ESTIMATING THE MEAN VALUE OF y

Our population regression model is

$$y = A + Bx + \epsilon$$

As mentioned earlier in this chapter, the mean value of y for a given x is denoted by $\mu_{y|x}$ and read as "the mean value of y for a given value of x." Because of the assumption that the mean value of ϵ is zero, the mean value of y is given by

$$\mu_{y|x} = A + Bx$$

$\mu_{y|x}$ denotes the mean value of y for the population for a particular value of x. Our objective is to estimate this mean value. The value of $\hat{y}$, calculated from the sample regression line by substituting the value of x, is the *point estimate* of $\mu_{y|x}$.

For the example on incomes and food expenditures, the estimated sample regression line is (see Example 12-1)

$$\hat{y} = 1.91 + .20x$$

Suppose we want to estimate the mean food expenditure for all households with a monthly income of \$2000. We denote this population mean by $\mu_{y|20}$. Note that we have written $x = 20$ and not $x = 2000$ in $\mu_{y|20}$ because the units of measurement for the data used to estimate the regression line were in hundreds of dollars. Using the regression line, we find that the point estimate of $\mu_{y|20}$ is

$$\hat{y} = 1.91 + .20\,(20) = 5.91$$

Thus, based on the sample regression line, the point estimate for the mean food expenditure $\mu_{y|20}$ for all households with a monthly income of \$2000 is \$591 per month.

However, suppose we take a second sample of seven households from the same population and estimate the regression line for this sample. The point estimate of $\mu_{y|20}$ obtained from the regression line for the second sample is expected to be different. All possible samples of the same size taken from the same population will give different regression lines, as shown in Figure 12.22, and, hence, a different point estimate of $\mu_{y|x}$. Therefore, a confidence interval constructed for $\mu_{y|x}$ based on one sample will give a more reliable estimate of $\mu_{y|x}$ than will a point estimate.

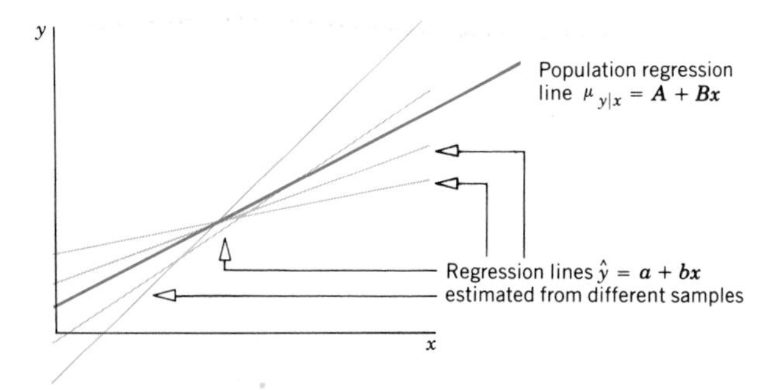

Figure 12.22 Population and sample regression lines.

To construct a confidence interval for $\mu_{y|x}$, we must know the mean, the standard deviation, and the shape of the sampling distribution of its point estimator $\hat{y}$.

The point estimator $\hat{y}$ of $\mu_{y|x}$ is normally distributed with a mean of $A + Bx$ and a standard deviation of

$$\sigma_{\hat{y}_m} = \sigma_\epsilon \sqrt{\frac{1}{n} + \frac{(x_0 - \bar{x})^2}{SS_{xx}}}$$

where $\sigma_{\hat{y}_m}$ is the standard deviation of $\hat{y}$ when it is used to estimate $\mu_{y|x}$, x_0 is the value of x for which we are estimating $\mu_{y|x}$, and σ_ϵ is the population standard deviation of ϵ.

However, as mentioned earlier, usually σ_ϵ is not known. Rather, it is estimated by the standard deviation s_e of sample errors. In this case, we replace σ_ϵ by s_e and $\sigma_{\hat{y}_m}$ by $s_{\hat{y}_m}$ in the foregoing expression. To make a confidence interval for $\mu_{y|x}$, we use the normal distribution when the sample size is large ($n \geq 30$) and the t distribution when the sample size is small ($n < 30$).

CONFIDENCE INTERVAL FOR $\mu_{y|x}$

The $(1 - \alpha)100\%$ confidence interval for $\mu_{y|x}$ for $x = x_0$ is

$$\hat{y} \pm t\, s_{\hat{y}_m}$$

where the value of t is obtained from the t distribution table for $\alpha/2$ area in the right tail and $df = n - 2$, and $s_{\hat{y}_m}$ is calculated as

$$s_{\hat{y}_m} = s_e \sqrt{\frac{1}{n} + \frac{(x_0 - \bar{x})^2}{SS_{xx}}}$$

Constructing a confidence interval for the mean value of y.

EXAMPLE 12-8 Refer to Example 12-1 on incomes and food expenditures. Find a 99% confidence interval for the mean food expenditure for all households with a monthly income of $2000.

Solution Using the regression line estimated in Example 12-1, the point estimate of the mean food expenditure for $x = 20$ is

$$\hat{y} = 1.91 + .20\,(20) = 5.91$$

The confidence level is 99%. Hence, the area in each tail is

$$\frac{\alpha}{2} = .5 - \left(\frac{.99}{2}\right) = .005$$

The degrees of freedom are

$$df = n - 2 = 7 - 2 = 5$$

From the t distribution table, the t value for .005 area in the right tail and 5 df is 4.032. From calculations in Examples 12-1 and 12-2 we know that

$$s_e = .842, \quad \bar{x} = 23.29, \quad \text{and} \quad SS_{xx} = 499.429$$

The standard deviation of $\hat{y}$ as an estimate of $\mu_{y|x}$ for $x = 20$ is calculated as follows.

$$s_{\hat{y}_m} = s_e \sqrt{\frac{1}{n} + \frac{(x_0 - \bar{x})^2}{SS_{xx}}} = (.842)\sqrt{\frac{1}{7} + \frac{(20 - 23.29)^2}{499.429}} = .342$$

Hence, the 99% confidence interval for $\mu_{y|20}$ is

$$\hat{y} \pm t\, s_{\hat{y}_m} = 5.91 \pm 4.032\,(.342) = 5.91 \pm 1.38 = 4.53 \text{ to } 7.29$$

Thus, with 99% confidence we can state that the mean food expenditure for all households with a monthly income of \$2000 is between \$453 and \$729. ■

12.8.2 USING THE REGRESSION MODEL FOR PREDICTING A PARTICULAR VALUE OF *y*

The second major use of a regression model is to predict a particular value of y for a given value of x, say x_0. For example, we may want to predict the food expenditure of a randomly selected household whose monthly income is \$2000. In this case, we are not interested in the mean food expenditure of all households with a monthly income of \$2000 but in the food expenditure of one particular household with a monthly income of \$2000. This predicted value of y is denoted by y_p. Again, to predict a single value of y for $x = x_0$ from the estimated sample regression line, we use the value of $\hat{y}$ as the point estimate of y_p. Using the estimated regression line, we find that $\hat{y}$ for $x = 20$ is

$$\hat{y} = 1.91 + .20\,(20) = 5.91$$

Thus, based on our regression line, the point estimate for the food expenditure of a given household with a monthly income of \$2000 is \$591 a month. Note that $\hat{y} = 5.91$ is the point estimate for the mean food expenditure for all households with $x = 20$ as well as for the predicted value of food expenditure of one household with $x = 20$.

Different regression lines estimated by using different samples of seven households each taken from the same population will give different values of the point estimator for the predicted value of y for $x = 20$. Hence, a confidence interval constructed for y_p based on one sample will give a more reliable estimate of y_p than will a point estimate. The confidence interval constructed for y_p is more commonly called a **prediction interval.**

The procedure used to construct a prediction interval for y_p is similar to constructing a confidence interval for $\mu_{y|x}$ except that the standard deviation of $\hat{y}$ is larger when we predict a single value of y than when we estimate $\mu_{y|x}$.

The point estimator $\hat{y}$ of y_p is normally distributed with a mean of $A + Bx$ and a standard deviation of

$$\sigma_{\hat{y}_p} = \sigma_\epsilon \sqrt{1 + \frac{1}{n} + \frac{(x_0 - \bar{x})^2}{SS_{xx}}}$$

where $\sigma_{\hat{y}_p}$ is the standard deviation of the predicted value of y, x_0 is the value of x for which we are predicting y, and σ_ϵ is the population standard deviation of ϵ.

However, usually σ_ϵ is not known. In such a case, we replace σ_ϵ by s_e and $\sigma_{\hat{y}_p}$ by $s_{\hat{y}_p}$ in the foregoing expression. To make a prediction interval for y_p, we use the normal distribution when the sample size is large ($n \geq 30$) and the t distribution when the sample size is small ($n < 30$).

PREDICTION INTERVAL FOR y_p

The $(1 - \alpha)100\%$ prediction interval for the predicted value of y, denoted by y_p, for $x = x_0$ is

$$\hat{y} \pm t\, s_{\hat{y}_p}$$

where the value of t is obtained from the t distribution table for $\alpha/2$ area in the right tail and $df = n - 2$. The value of $s_{\hat{y}_p}$ is calculated as follows.

$$s_{\hat{y}_p} = s_e \sqrt{1 + \frac{1}{n} + \frac{(x_0 - \bar{x})^2}{SS_{xx}}}$$

Making a prediction interval for a particular value of y.

EXAMPLE 12-9 Refer to Example 12-1 on incomes and food expenditures. Find a 99% prediction interval for the predicted food expenditure for a randomly selected household with a monthly income of $2000.

Solution Using the regression line estimated in Example 12-1, the point estimate of the predicted food expenditure for $x = 20$ is given by

$$\hat{y} = 1.91 + .20\,(20) = 5.91$$

The area in each tail for a 99% confidence level is

$$\frac{\alpha}{2} = .5 - \left(\frac{.99}{2}\right) = .005$$

The degrees of freedom are

$$df = n - 2 = 7 - 2 = 5$$

From the t distribution table, the t value for .005 area in the right tail and 5 df is 4.032. From calculations in Examples 12-1 and 12-2,

$$s_e = .842, \quad \bar{x} = 23.29, \quad \text{and} \quad SS_{xx} = 499.429$$

The standard deviation of $\hat{y}$ as an estimator of y_p for $x = 20$ is calculated as follows.

$$s_{\hat{y}_p} = s_e \sqrt{1 + \frac{1}{n} + \frac{(x_0 - \bar{x})^2}{SS_{xx}}}$$

$$= (.842) \sqrt{1 + \frac{1}{7} + \frac{(20 - 23.29)^2}{499.429}} = .909$$

Hence, the 99% prediction interval for y_p for $x = 20$ is

$$\hat{y} \pm t\, s_{\hat{y}_p} = 5.91 \pm 4.032\,(.909) = 5.91 \pm 3.67 = 2.24 \text{ to } 9.58$$

Thus, with 99% confidence we can state that the predicted food expenditure of a household with a monthly income of \$2000 is between \$224 and \$958. ■

As we can observe, this interval is much wider than the one for the mean value of y for $x = 20$ calculated in Example 12-8, which was \$453 to \$729. This is always true. The prediction interval for predicting a single value of y is always larger than the confidence interval for estimating the mean value of y for a certain value of x.

12.9 CAUTIONS IN USING REGRESSION

When carefully applied, regression is a very helpful technique for making predictions and estimations about one variable for a certain value of another variable. However, we need to be cautious while using the regression analysis, for it can give misleading results and predictions. Following are the two most important points to remember while using regression analysis.

EXTRAPOLATION

The regression line estimated for the sample data is valid only for the range of x values observed in the sample. For example, the values of x in the example on incomes and food expenditures vary from a minimum of 12 to a maximum of 37. Hence, the estimated regression line is only applicable for values of x between 12 and 37, that is, we should use this regression line to estimate the mean food expenditure or predict the food expenditure of a single household only for income levels between \$1200 and \$3700. If we estimate or predict y for a value of x either less than 12 or greater than 37, it is called *extrapolation*. This does not mean that we should never use the regression line for extrapolation. Instead, we should interpret such predictions cautiously and not attach much value to them.

Similarly, if the data used for a regression estimation are time-series data, the predicted values of y for periods outside the time interval used for the estimation of the regression line should be interpreted very cautiously. When using the estimated regression line for extrapolation, we are assuming that the same linear relationship between the two variables holds true for the values of x outside the given range. It is possible that the relationship between the two variables may not be linear outside the given range. Nonetheless, even if it is linear, adding a few more observations at either end will probably give a new estimation of the regression line.

CAUSALITY

The regression line does not prove causality between two variables. That is, it does not predict that a change in y is caused by a change in x. The information about causality is based on theory or common sense. A regression line only describes whether or not a significant quantitative relationship between x and y exists. Significant relationship means that we reject the null hypothesis H_0: $B = 0$ at a given significance level. The estimated regression line gives the change in y due to a change of one unit in x. Note that it does not indicate that the reason y has changed is because x has changed. In our example on incomes and food expenditures, it is economic theory and common sense, not the regression line, that tells us that food expenditure depends on income. The regression analysis simply helps to find if this dependence is significant.

EXERCISES

12.54 Briefly explain the difference between estimating the mean value of y and predicting a particular value of y using a regression model.

12.55 Construct a 99% confidence interval for the mean value of y and a 99% prediction interval for the predicted value of y for the following.

a. $\hat{y} = 3.25 + .80x$ for $x = 15$ given $s_e = .954$, $\bar{x} = 18.52$, $SS_{xx} = 144.65$, and $n = 10$

b. $\hat{y} = -27 + 7.67x$ for $x = 12$ given $s_e = 2.46$, $\bar{x} = 13.43$, $SS_{xx} = 369.77$, and $n = 10$

12.56 Construct a 95% confidence interval for the mean value of y and a 95% prediction interval for the predicted value of y for the following.

a. $\hat{y} = 13.40 + 2.58x$ for $x = 8$ given $s_e = 1.29$, $\bar{x} = 11.30$, $SS_{xx} = 210.45$, and $n = 12$

b. $\hat{y} = -8.6 + 3.72x$ for $x = 24$ given $s_e = 1.89$, $\bar{x} = 19.70$, $SS_{xx} = 315.40$, and $n = 12$

12.57 Refer to Exercise 12.24. Construct a 90% confidence interval for the mean salary of secretaries with 10 years of experience. Construct a 90% prediction interval for the salary of a randomly selected secretary with 10 years of experience.

12.58 Refer to data on hours studied and midterm scores of eight students given in Exercise 12.44. Construct a 99% confidence interval for $\mu_{y|x}$ for $x = 13$ and a 99% prediction interval for y_p for $x = 13$.

12.59 Using the data on midterm scores in math and economics for seven students given in Exercise 12.45, construct a 95% confidence interval for $\mu_{y|x}$ for $x = 80$ and a 95% prediction interval for y_p for $x = 80$.

12.60 Refer to data on yearly incomes and value of homes owned by nine homeowners given in Exercise12.46. Construct a 98% confidence interval for the mean value of homes for all homeowners with a yearly income of $35,000. Make a 98% prediction interval for the value of a home for a randomly selected homeowner with a yearly income of $35,000.

12.61 Refer to Exercise 12.47. Construct a 95% confidence interval for the mean yield of corn for all acres on which 90 pounds of fertilizer is used. Determine a 95% prediction interval for the yield of corn for a randomly selected acre on which 90 pounds of fertilizer is used.

12.62 Using the data on ages and cholesterol levels of 10 men given in Exercise 12.48, find a 99% confidence interval for the mean cholesterol level for all 53-year-old men. Make a 99% confidence interval for the cholesterol level for a randomly selected 53-year-old man.

12.63 Refer to Exercise 12.49. Construct a 99% confidence interval for the mean amount of charitable contributions made by all households with an income of $44,000 and make a 99% prediction interval for the charitable contributions made by a randomly selected household with an income of $44,000.

12.64 Refer to Exercise 12.50. Construct a 98% confidence interval for the mean starting salary of a college graduate with a GPA of 3.15. Construct a 98% prediction interval for the starting salary of a randomly selected college graduate with a GPA of 3.15.

GLOSSARY

Coefficient of determination A measure that gives the proportion of total variation in a dependent variable explained by an independent variable.

Degrees of freedom for a simple linear regression model Sample size minus 2, that is, $n - 2$.

Dependent variable The variable being predicted or explained.

Deterministic model A model in which the independent variable determines the dependent variable exactly.

Estimates of *A* and *B* The values of a and b that are calculated by using the sample data.

Independent variable The variable included in a regression model to explain the variation in the dependent variable.

Least squares method The method that fits a regression line through a scatter diagram such that the error sum of squares is minimum.

Linear correlation coefficient A measure of the strength of linear relationship between two variables.

Probabilistic model A model in which the independent variable does not determine the dependent variable exactly.

Random error term The difference between the actual value and the predicted value of y.

Scatter diagram A plot of the paired observations on x and y.

Simple linear regression A regression model with one dependent and one independent variable that assumes a straight-line relationship.

Slope The coefficient of x in a regression model that gives the change in y for a change of one unit in x.

SSE (error sum of squares) The sum of the squared differences between the actual and predicted values of y. It is that portion of SST that is not explained by the regression model.

SSR (regression sum of squares) That portion of SST that is explained by the regression model.

SST (total sum of squares) The sum of the squared differences between actual y values and $\bar{y}$.

Standard deviation of errors A measure of spread for the random errors.

***y*-intercept** The point at which a regression line intersects the vertical axis on which the dependent variable is marked. It is the value of y when x is zero.

KEY FORMULAS

1. **Least squares estimates of *A* and *B***

$$b = \frac{SS_{xy}}{SS_{xx}} \quad \text{and} \quad a = \bar{y} - b\bar{x}$$

where

$$SS_{xy} = \Sigma xy - \frac{(\Sigma x)(\Sigma y)}{n}$$

$$SS_{xx} = \Sigma x^2 - \frac{(\Sigma x)^2}{n}$$

2. **Standard deviation of the sample errors**

$$s_e = \sqrt{\frac{SS_{yy} - b\,SS_{xy}}{n - 2}}$$

where

$$SS_{yy} = \Sigma y^2 - \frac{(\Sigma y)^2}{n}$$

3. **Error sum of squares**

$$SSE = \Sigma e^2 = \Sigma(y - \hat{y})^2$$

4. **Total sum of squares**

$$SST = \Sigma y^2 - \frac{(\Sigma y)^2}{n}$$

5. **Regression sum of squares**

$$SSR = SST - SSE$$

6. **Coefficient of determination**

$$r^2 = \frac{b\,SS_{xy}}{SS_{yy}}$$

7. **Estimate of the standard deviation of *b***

$$s_b = \frac{s_e}{\sqrt{SS_{xx}}}$$

8. **The $(1 - \alpha)100\%$ confidence interval for *B***

$$b \pm t\,s_b$$

9. **Value of the test statistic *t* for *b***

$$t = \frac{b - B}{s_b}$$

10. Linear correlation coefficient

$$r = \frac{SS_{xy}}{\sqrt{SS_{xx}\, SS_{yy}}}$$

11. The $(1 - \alpha)100\%$ confidence interval for $\mu_{y|x}$

$$\hat{y} \pm t\, s_{\hat{y}_m}$$

where

$$s_{\hat{y}_m} = s_e \sqrt{\frac{1}{n} + \frac{(x_0 - \bar{x})^2}{SS_{xx}}}$$

12. The $(1 - \alpha)100\%$ prediction interval for y_p

$$\hat{y} \pm t\, s_{\hat{y}_p}$$

where

$$s_{\hat{y}_p} = s_e \sqrt{1 + \frac{1}{n} + \frac{(x_0 - \bar{x})^2}{SS_{xx}}}$$

SUPPLEMENTARY EXERCISES

12.65 The following table gives seven pairs of x and y values.

x	35	57	26	43	59	28	39
y	17	11	14	15	7	10	22

a. Compute SS_{xx}, SS_{yy}, and SS_{xy}.
b. Find the least squares regression line $\hat{y} = a + bx$.
c. Compute r and r^2.
d. Compute the standard deviation of errors.

12.66 The following table gives nine pairs of x and y values.

x	63	78	45	56	49	61	82	42	53
y	22	31	21	16	14	25	38	12	17

a. Compute SS_{xx}, SS_{yy}, and SS_{xy}.
b. Find the least squares regression line.
c. Compute r and r^2.
d. Compute the standard deviation of errors.

12.67 The following table gives eight pairs of x and y values.

x	13	21	15	18	9	25	11	27
y	34	48	41	38	22	56	35	52

a. Compute SS_{xx}, SS_{yy}, and SS_{xy}.
b. Find the least squares regression line.
c. Compute r and r^2.
d. Compute the standard deviation of errors.

12.68 The following table gives 10 pairs of x and y values.

x	18	22	38	26	49	16	29	35	46	57
y	13	9	5	5	2	11	10	3	2	1

a. Compute SS_{xx}, SS_{yy}, and SS_{xy}.
b. Find the least squares regression line.
c. Compute r and r^2.
d. Compute the standard deviation of errors.

12.69 The following table gives the total payroll (in millions of dollars) as of March 1989 and the percentage of games won during the 1988 season by six of the American League baseball teams.

Team	Total Payroll ($ millions)	Percentage of Games Won
Baltimore Orioles	9	34
California Angels	14	46
Kansas City Royals	18	52
Oakland Athletics	15	64
Seattle Mariners	10	42
Texas Rangers	11	44

Source: USA Today, March 30 and April 3, 1989. Copyright © 1989, *USA Today*. Adapted with permission.

a. Construct a scatter diagram for these data. Take total payroll as an independent variable and percentage of games won as a dependent variable. Does the scatter diagram exhibit a linear relationship between the two variables?
b. Find the least squares regression line $\hat{y} = a + bx$.
c. Give a brief interpretation of the values of a and b calculated in part b.
d. Plot the regression line on the scatter diagram and show the errors by drawing vertical lines between scatter points and the regression line.
e. Predict the percentage of games won for a team with a total payroll of $13 million.
f. Compute r and r^2.
g. Calculate the standard deviation of errors.
h. Make a 99% confidence interval for B.
i. Test at the 1% significance level if B is different from zero.

12.70 The following table gives the percentage of games won and the average attendance (in thousands) per home game for the 1988 season for seven of the National League baseball teams.

Team	Percentage of Games Won	Average Attendance per Home Game (thousands)
Atlanta Braves	34	11
Chicago Cubs	48	27
Los Angeles Dodgers	58	38
New York Mets	63	40
Philadelphia Phillies	40	26
St. Louis Cardinals	47	36
San Diego Padres	52	19

Source: USA Today, April 3, 1989. Copyright © 1989; *USA Today*. Adapted with permission.

a. Construct a scatter diagram for these data. Take percentage of games won as an independent variable and average attendance as a dependent variable. Does the scatter diagram exhibit a linear relationship between the two variables?
b. Find the least squares regression line $\hat{y} = a + bx$.
c. Give a brief interpretation of the values of y-intercept and slope.
d. Plot the regression line on the scatter diagram and show the errors by drawing vertical lines between scatter points and the regression line.
e. Predict the average attendance per home game for a team with 49% of games won.
f. Compute r and r^2.
g. Calculate the standard deviation of errors.
h. Make a 95% confidence interval for B.
i. Test at the 5% significance level if B is different from zero.

12.71 The following data give the information on ages (in years) and the number of breakdowns during the past month for a sample of seven machines of a large firm.

Age	12	7	2	8	13	9	4
Number of breakdowns	9	5	1	4	11	7	2

a. Taking age as an independent variable and the number of breakdowns as a dependent variable, what is your hypothesis about the sign of B in the regression line? (In other words, do you expect B to be positive or negative?)
b. Find the least squares regression line. Is the sign of b the same as you hypothesized for B in part a?
c. Give a brief interpretation of the values of a and b.
d. Compute r and r^2 and explain what they mean.
e. Compute the standard deviation of errors.
f. Construct a 98% confidence interval for B.
g. Test at the 2.5% significance level if B is positive.

12.72 The following table gives the information on the number of hours that eight bank loan officers slept the previous night and the number of loan applications that they processed the next day.

Number of hours	8	5	7	6	4	8	6	5
Number of applications	14	10	16	11	8	15	10	8

a. Taking the number of hours slept as an independent variable and the number of applications processed as a dependent variable, do you expect B to be positive or negative in the regression model $y = A + Bx + \epsilon$?
b. Find the least squares regression line. Is the sign of b the same as you hypothesized for B in part a?
c. Compute r and r^2 and explain what they mean.
d. Compute the standard deviation of errors.
e. Construct a 90% confidence interval for B.
f. Test at the 5% significance level if B is positive.

12.73 The following table gives information on the high school GPAs and the college GPAs of eight randomly selected students.

High school GPA	3.47	2.89	3.76	2.12	3.43	3.11	2.56	2.98
College GPA	3.23	3.47	2.56	1.95	3.86	3.56	2.78	2.78

a. Find the least squares regression line $\hat{y} = a + bx$. Take high school GPA as an independent variable and college GPA as a dependent variable.
b. Predict the college GPA of a student whose high school GPA is 3.20.
c. Compute r and r^2 and explain what they mean.
d. Compute the standard deviation of errors.
e. Construct a 95% confidence interval for B.
f. Test at the 2% significance level if B is different from zero.

12.74 The following table gives information on the temperature in a city and the volume of ice cream (in pounds) sold at an ice cream parlor for a random sample of eight days during the summer of 1990.

Temperature	93	86	77	89	98	102	87	79
Ice cream sold	187	169	123	198	232	267	158	117

a. Find the least squares regression line $\hat{y} = a + bx$. Take temperature as an independent variable and the volume of ice cream sold as a dependent variable.
b. Give a brief interpretation of the values of a and b.
c. Compute r and r^2 and explain what they mean.
d. Predict the amount of ice cream sold at this parlor on a day with a temperature of 95°F.
e. Compute the standard deviation of errors.
f. Construct a 99% confidence interval for B.
g. Test at the 1% significance level if B is different from zero.

12.75 The following table gives the heights (in inches) of nine randomly selected different seedlings at the end of a certain number of days after planting.

Days after planting	20	16	52	28	38	10	32	24	40
Height	21	19	38	27	32	7	27	28	37

a. With days after planting as an independent variable and the height of seedlings as a dependent variable, what do you expect the sign of B to be in the regression line $y = A + Bx + \epsilon$?
b. Find the least squares regression line $\hat{y} = a + bx$. Is the sign of b the same as you hypothesized for B in part a?
c. Give a brief interpretation of the values of a and b.
d. Compute r and r^2 and explain what they mean.
e. Predict the height of such a seedling at the end of 25 days after planting.
f. Compute the standard deviation of errors.
g. Construct a 98% confidence interval for B.
h. Test at the 1% significance level if B is positive.

12.76 The following table gives the average mortgage rate (in percent) and the median price of homes (in thousands of dollars) for the United States for the period 1981–1987. (Source: National Association of Realtors.)

Average mortgage rate	15.1	15.4	12.9	12.5	11.8	9.7	9.3
Median price	66.4	67.8	70.3	72.4	75.5	80.8	85.0

a. Find the least squares regression line with average mortgage rate as an independent variable and the median price of homes as a dependent variable.
b. What is the predicted median price of homes if the average mortgage rate is 13.5%?
c. Compute r and r^2 and explain what they mean.
d. Compute the standard deviation of errors.
e. Construct a 99% confidence interval for B.
f. Test at the 1% significance level if B is positive.

12.77 The following table gives the milk production (rounded to billions of pounds) for the years 1980 through 1988. (Source: U.S. Department of Agriculture.)

Year	Milk Production
1980	128
1981	133
1982	136
1983	140
1984	135
1985	143
1986	144
1987	143
1988	146

a. Assign a value of 1 to 1980, 2 to 1981, 3 to 1983, and so on. Call this new variable "time." Write a new table with variables *time* and *milk production*.
b. With time as an independent variable and milk production as a dependent variable, compute SS_{xx}, SS_{yy}, and SS_{xy}.
c. Construct a scatter diagram for these data. Does the scatter diagram exhibit a linear positive relationship between time and milk production?
d. Find the least squares regression line $\hat{y} = a + bx$.
e. Give a brief interpretation of the values of a and b. [*Hint:* The value of b will give an increase in the milk production (in billions of pounds) per year for the period 1980 through 1988 based on a linear relationship between the two variables.]
f. Compute the correlation coefficient r.
g. Predict the milk production for $x = 12$. Do you have any comment to make about this prediction?

12.78 The following table gives the gross national product (in hundreds of billions of dollars) for the years 1980 through 1988. (Source: U.S. Bureau of Economic Analysis.)

Year	Gross National Product
1980	27.32
1981	30.53
1982	31.66
1983	34.06
1984	37.72
1985	40.15
1986	42.32
1987	45.24
1988	48.81

a. Assign a value of 1 to 1980, 2 to 1981, 3 to 1983, and so on. Call this new variable "time." Write a new table with variables *time* and *gross national product*.
b. With time as an independent variable and gross national product as a dependent variable, compute SS_{xx}, SS_{yy}, and SS_{xy}.
c. Construct a scatter diagram for these data. Does the scatter diagram exhibit a linear positive relationship between time and gross national product?
d. Find the least squares regression line $\hat{y} = a + bx$.
e. Give a brief interpretation of the values of a and b. [See the hint in Exercise 12.77 part e.]
f. Compute the correlation coefficient r.
g. Predict the gross national product for $x = 13$. Do you have any comment to make about this prediction?

12.79 Refer to data on ages and the number of breakdowns for each of the seven machines given in Exercise 12.71. Construct a 99% confidence interval for the mean number of breakdowns for all machines with an age of 8 years. Find a 99% prediction interval for the number of breakdowns for a randomly selected machine with an age of 8 years.

12.80 Refer to data on high school and college GPAs of eight students given in Exercise 12.73. Determine a 95% confidence interval for the mean college GPA for all students with a high school GPA of 3.20. Make a 95% prediction interval for the college GPA of a randomly selected student with a high school GPA of 3.20.

12.81 Refer to data on temperature and the volume of ice cream sold at a parlor for a sample of eight days given in Exercise 12.74. Construct a 98% confidence interval for the mean volume of ice cream sold at this parlor on all days with a temperature of 95°F. Determine a 98% confidence interval for the volume of ice cream sold at this parlor on a randomly selected day with a temperature of 95°F.

12.82 Refer to data on heights of nine seedlings at the end of a certain number of days after planting given in Exercise 12.75. Make a 90% confidence interval for the mean height of all seedlings 25 days after planting. Construct a 90% prediction interval for the height of a randomly selected seedling 25 days after planting.

SELF-REVIEW TEST

1. A simple regression is a regression model that contains
   a. only one independent variable
   b. only one dependent variable
   c. more than one independent variable
   d. both a and b
2. A (simple) linear regression means that the relationship between the independent variable and the dependent variable is that of
   a. a straight line
   b. a curve
3. A deterministic regression model is a model that
   a. contains the random error term
   b. does not contain the random error term
   c. gives a nonlinear relationship
4. A probabilistic regression model is a model that
   a. contains the random error term
   b. does not contain the random error term
   c. shows an exact relationship
5. The least squares regression line minimizes the sum of
   a. errors **b.** squared errors **c.** predictions
6. The number of degrees of freedom for a simple regression model is
   a. $n - 1$ **b.** $n - 2$ **c.** $n - 5$
7. Indicate if the following statement is true or false:

   The coefficient of determination gives the proportion of total squared errors (SST) that is explained by the use of the regression model

8. Indicate if the following statement is true or false:

The simple linear correlation measures the strength of linear association between two variables

9. The value of the coefficient of determination is always in the range

a. 0 to 1 **b.** -1 to 1 **c.** -1 to 0

10. The value of the correlation coefficient is always in the range

a. 0 to 1 **b.** -1 to 1 **c.** -1 to 0

11. Explain why the random error term ϵ is added to the regression model.

12. Explain the difference between A and a and between B and b.

13. Briefly explain the assumptions of a regression model.

14. Briefly explain the difference between the population regression line and a sample regression line.

15. The following table gives the experience (in years) and the number of computers sold during the previous three months by seven salespersons.

Experience	4	12	9	6	10	16	7
Computers sold	19	42	28	31	39	35	21

a. Do you think experience depends on the number of computers sold or the computers sold depend on experience?

b. With experience as an independent variable and the number of computers sold as a dependent variable, what is your hypothesis about the sign of B in the regression model?

c. Construct a scatter diagram for these data. Does the scatter diagram exhibit a linear relationship between the two variables?

d. Find the least squares regression line. Is the sign of b the same as the one you hypothesized for B in part b?

e. Give a brief interpretation of the values of y-intercept and slope.

f. Compute r and r^2 and explain what they mean.

g. Predict the number of computers sold during the past three months by a salesperson with 11 years of experience.

h. Compute the standard deviation of errors.

i. Construct a 99% confidence interval for B.

j. Test at the 1% significance level if B is positive.

k. Construct a 95% confidence interval for the mean number of computers sold by all salespersons with an experience of 8 years.

l. Make a 95% prediction interval for the number of computers sold by a randomly selected salesperson with an experience of 8 years.

USING MINITAB

Following are the MINITAB commands that can be used to do the regression analysis discussed in this chapter.

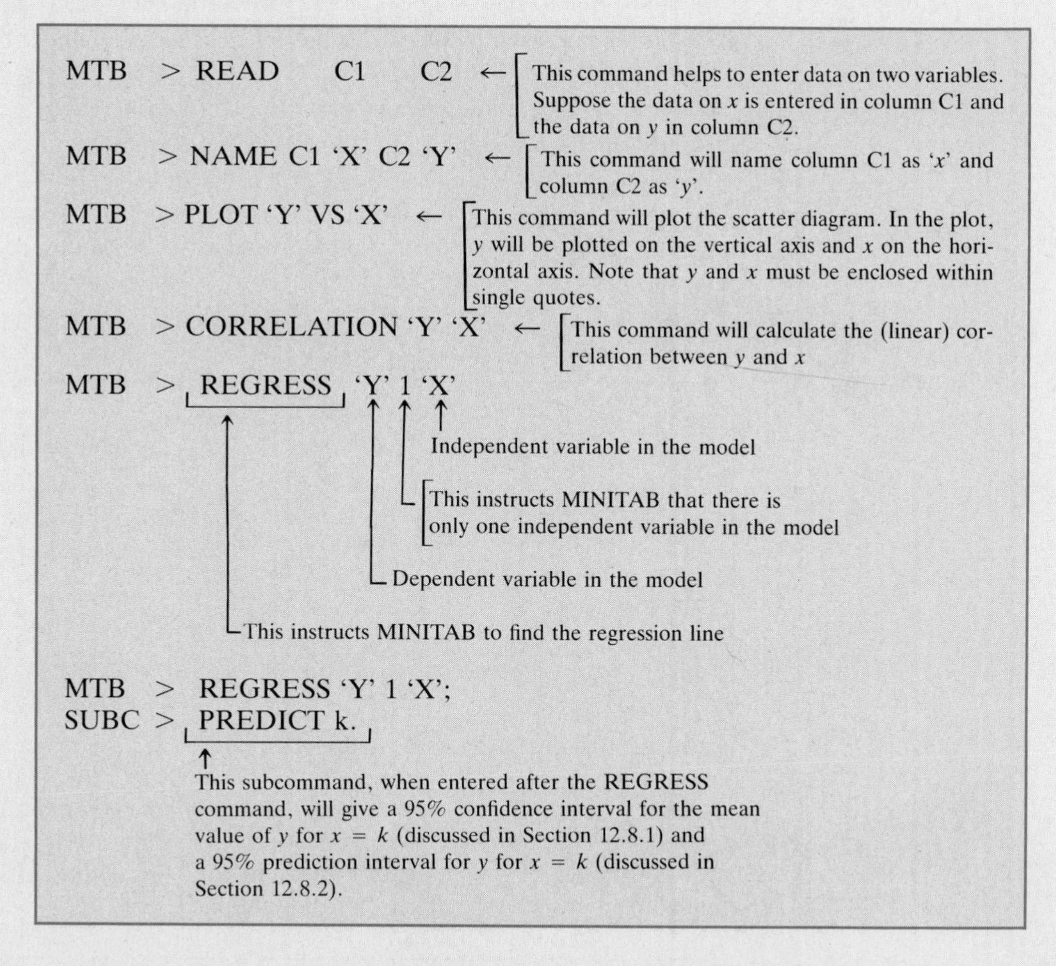

The following illustration describes the use of these MINITAB commands.

Illustration M12-1 Refer to data of Example 12-7 on the driving experiences (in years) and the monthly auto insurance premiums (in dollars) for eight auto drivers. That data set is reproduced below.

Driving Experience (years)	Monthly Auto Insurance Premium (dollars)
5	64
2	87
12	50
9	71
15	44
6	56
25	42
16	60

Using MINITAB,

(a) construct a scatter diagram for these data
(b) find the correlation between the two variables
(c) find the regression line with experience as an independent variable and premium as a dependent variable
(d) make a 90% confidence interval for B
(e) test at the 5% significance level if B is negative
(f) make a 95% confidence interval for the mean auto insurance premium for all drivers with a driving experience of 10 years and construct a 95% prediction interval for the auto insurance premium for a randomly selected driver with a driving experience of 10 years

Solution First we enter the data into MINITAB using READ command.

```
MTB   > READ C1 C2
DATA >  5  64
DATA >  2  87
DATA > 12  50
DATA >  9  71
DATA > 15  44
DATA >  6  56
DATA > 25  42
DATA > 16  60
DATA > END
```

In our example, experience is the independent variable and premium is the dependent variable. Hence, the data entered in column C1 represent the 'x' variable and those entered in column C2 represent the 'y' variable. Using the following command, we name column C1 as 'x' and column C2 as 'y'.

```
MTB > NAME C1 'X' C2 'Y'
```

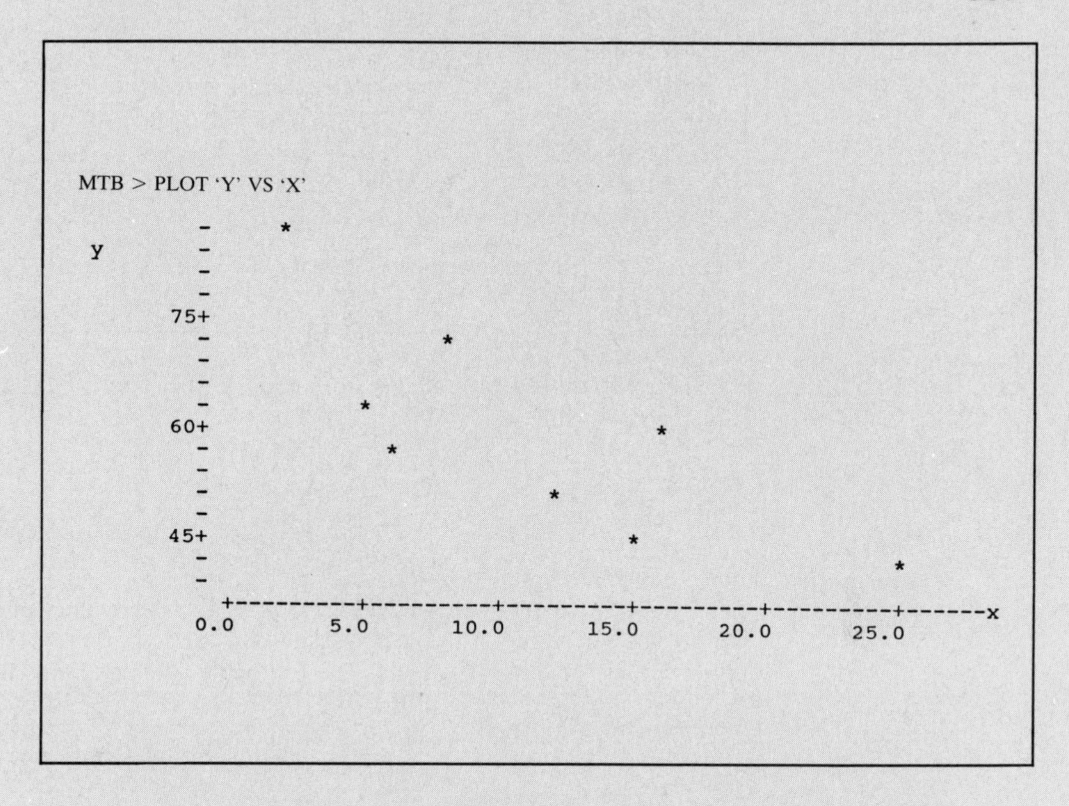

(a) The scatter diagram is constructed using the PLOT command as above.

(b) The CORRELATION command gives the linear correlation coefficient between x and y.

```
MTB > CORRELATION 'Y' 'X'
CORRELATION OF Y AND X = -0.768   <- [This is the linear correlation coefficient r
```

Hence, the linear correlation coefficient between x and y is

$$r = -.768$$

(c) The regression line is obtained by using the REGRESS command.

```
MTB > REGRESS 'Y' 1 'X'
THE REGRESSION EQUATION IS
Y = 76.7 - 1.55 X

PREDICTOR      COEF     STDEV    T-RATIO      P  ] <- This part
CONSTANT     76.660     6.961      11.01  0.000  ]    of the
X           -1.5476    0.5270      -2.94  0.026  ]    output is
                                                 ]    explained
S = 10.32    R-SQ = 59.0%    R-SQ(ADJ) = 52.1%   ]    below.
```

ANALYSIS OF VARIANCE

SOURCE	DF	SS	MS	F	P
REGRESSION	1	918.5	918.5	8.62	0.026
ERROR	6	639.0	106.5		
TOTAL	7	1557.5			

We have not discussed analysis of variance table for regression analysis in this chapter except for the column of SS

This MINITAB output has four main parts. The first part gives the regression equation, which is

$$\hat{Y} = 76.7 - 1.55X$$

Thus

$$a = 76.7 \quad \text{and} \quad b = -1.55$$

The second part of the MINITAB output gives the values of a and b, their standard deviations, the value of the test statistic t for the hypothesis tests about A and B, and the p-values for these two tests. The following chart explains this part of the output.

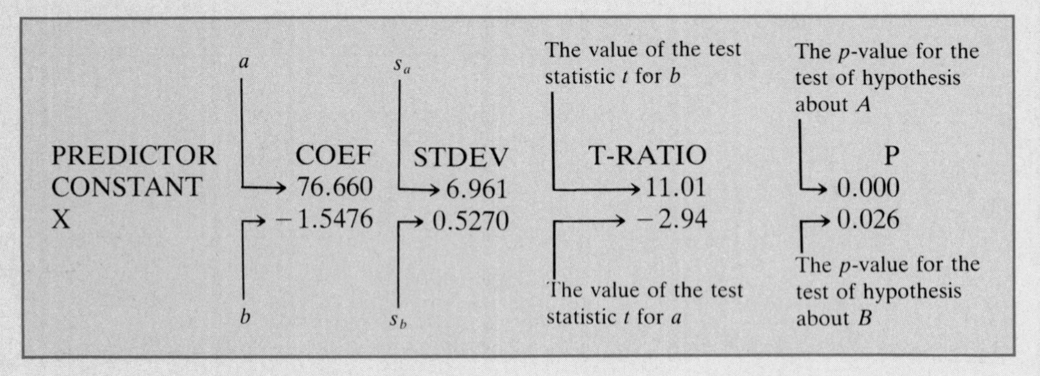

Note that we have not discussed the calculation of the standard deviation s_a of a and the test of hypothesis about a (and, hence, the value of the test statistic t for a and the p-value for the test of hypothesis about a) in this chapter. Therefore, in the row of CONSTANT in the above portion of the output, all the values except the coefficient value are irrelevant for this chapter. If we compare the various values in this chart with the ones calculated in Example 12-7, we notice a slight difference due to rounding.

The third part of the MINITAB output gives the standard deviation s_e of errors, the coefficient of determination r^2, and the adjusted r^2 (which is the value of r^2 adjusted for the degrees of freedom). However, we have not discussed the concept of adjusted r^2 in this chapter. The explanation of this part of the output is as follows.

S = 10.32 R − SQ = 59.0% R − SQ(ADJ) = 52.1%

s_e r^2 r^2 adjusted for df

The fourth part of the MINITAB output gives the analysis of variance table. We have not discussed such a table in this chapter except for the column of SS. The column of SS gives the values of SSR, SSE, and SST as follows.

SOURCE	SS	
REGRESSION	918.5	⟵ SSR
ERROR	639.0	⟵ SSE
TOTAL	1557.5	⟵ SST

(d) The 90% confidence interval for B is given by the formula

$$b \pm t\, s_b$$

From the second part of the MINITAB output explained in part (c), the row corresponding to x (or b) is

PREDICTOR	COEF	STDEV	T-RATIO	P
X	−1.5476	0.5270	−2.94	0.026

From this row of output,

$$b = -1.5476 \quad \text{and} \quad s_b = 0.5270$$

For a 90% confidence interval, $\alpha = .10$ and $\alpha/2 = .05$. The degrees of freedom are $n - 2 = 8 - 2 = 6$. Hence, from the t distribution table, the t value for .05 area in the right tail and 6 df is 1.943. The 90% confidence interval for B is

$$b \pm t\, s_b = -1.5476 \pm 1.943\,(.5270) = -1.5476 \pm 1.0240 = -2.57 \text{ to } -.52$$

(e) The null and alternative hypotheses are

$$H_0\text{: } B = 0 \quad (B \text{ is not negative})$$

$$H_1\text{: } B < 0 \quad (B \text{ is negative})$$

The significance level is .05. The degrees of freedom are 6. The test is left-tailed. From the t distribution table, the critical value of t for $\alpha = .05$ and $df = 6$ is -1.943.

We again use the information given in the row corresponding to x in what we called the second part of the MINITAB output. From that row of information,

$$\text{Value of the test statistic } t \text{ for } b = -2.94$$

Because the value of the test statistic $t = -2.94$ is less than the critical value of $t = -1.943$, it falls in the rejection region, which is in the left tail (see Figure 12.21 of Example 12-7). Consequently, we reject the null hypothesis and state that our sample information supports the hypothesis, at the 5% significance level, that B is negative.

(f) To make the 95% confidence interval for $\mu_{y|x}$ and prediction interval for y_p for $x = 10$, we use the subcommand PREDICT with the REGRESS command. This is illustrated below. Note that when we use the following command and subcommand, the MINITAB will give all the output that we listed earlier in response to the command REGRESS 'Y' 1 'X'. In addition, it will give the following output.

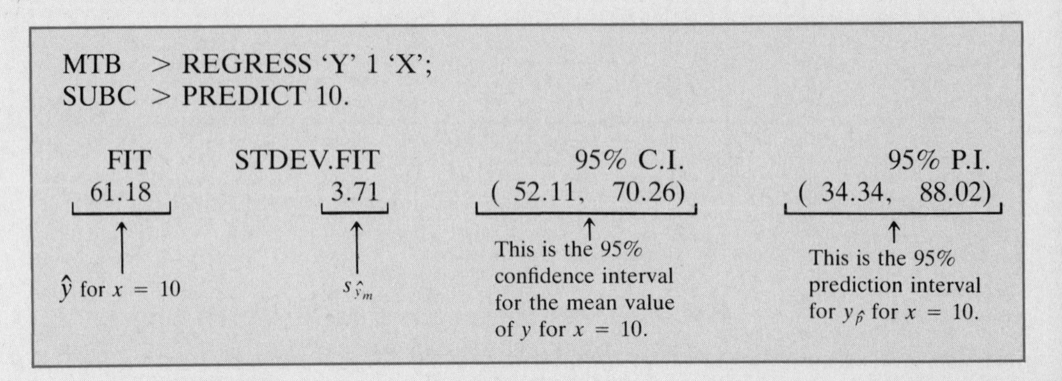

Thus, from this printout, the $\hat{y}$ for $x = 10$ is 61.18. This can be considered as a point estimate of the mean value of y as well as a point estimate of the predicted value of y for $x = 10$. The interval 52.11 to 70.26 in the printout gives a 95% confidence interval for the mean value $\mu_{y|10}$ of y for $x = 10$ as was discussed in section 12.8.1. The interval 34.34 to 88.02 gives a 95% prediction interval for y for $x = 10$ as was discussed in Section 12.8.2.

COMPUTER ASSIGNMENTS

M12.1 Professor Hamid Zangenehzadeh studied the relationship between student evaluations of a teacher and the expected grades of students in a course. The following table lists the (average of the) ratings of teachers by students and (the average of the) students' expected grades for 35 faculty members of the three departments of the School of Management at Widener University as reported in this study. (*Note:* The ratings of teachers reported here are what the author called unadjusted ratings of teachers.)

Ratings of Teachers	Students' Expected Grades	Ratings of Teachers	Students' Expected Grades
3.833	3.500	2.739	3.000
3.769	3.769	2.543	2.829
3.642	3.214	2.286	3.143
3.625	3.250	2.278	2.833
3.529	3.529	2.133	2.800
3.500	3.300	2.103	2.620
3.500	3.500	2.053	2.368
3.409	3.864	2.043	2.696
3.380	3.048	1.944	2.944
3.333	3.200	1.923	2.846
3.294	3.059	1.800	2.800
3.267	3.000	1.800	3.000
3.263	3.368	1.692	2.769

(*continued*)

Ratings of Teachers	Students' Expected Grades	Ratings of Teachers	Students' Expected Grades
3.120	3.440	1.692	3.462
3.045	2.909	1.688	3.125
3.000	3.500	1.667	4.000
3.000	3.500	1.625	2.375
2.923	2.538	1.333	3.555
2.826	3.086	0.521	2.652
2.778	3.111		

Source: Hamid Zangenehzadeh: "Grade Inflation: A Way Out," *Journal of Economic Education,* Summer 1988, 217–226. Reprinted with permission of Heldref Publications, Washington, D.C. Copyright © 1988 by the Heldref Publications.

Using MINITAB,

a. construct a scatter diagram for these data
b. find the correlation between the two variables
c. find the regression line with ratings of teachers as a dependent variable and students' expected grades as an independent variable
d. make a 95% confidence interval for B
e. test at the 1% significance level if B is positive

(*Note:* As the sample size is large, $n = 39$, the normal distribution can be applied to construct a confidence interval and to make a test of hypothesis about B in parts d and e if we desire so.)

M12.2 Refer to Data Set III on the heights and weights of NBA players given in Appendix B. Select a random sample of 25 players from that population and repeat all the parts of Illustration M12-1 for those data using MINITAB. Take weight as a dependent variable and height as an independent variable.

M12.3 Refer to data on the average mortgage rates (in percent) and the median price of homes (in thousands of dollars) given in Exercise 12.76. Using MINITAB,

a. construct a scatter diagram for these data
b. find the least squares regression line with average mortgage rate as an independent variable and the median price of homes as a dependent variable
c. compute the correlation coefficient
d. construct a 99% confidence interval for B
e. test at the 5% significance level if B is positive

13 ANALYSIS OF VARIANCE

13.1 THE *F* DISTRIBUTION
13.2 ONE-WAY ANALYSIS OF VARIANCE
SELF-REVIEW TEST
USING MINITAB

Chapter 10 described the procedures that are used to test hypotheses about the difference between two population means using the normal and t distributions. Also described in that chapter were the hypothesis testing procedures for the difference between two population proportions using the normal distribution. Then, Chapter 11 explained the procedures to test hypotheses about the equality of more than two population proportions using the chi-square distribution.

This chapter explains how to test the null hypothesis that the means of more than two populations are equal. For example, suppose that teachers at a school have devised three different methods to teach arithmetic. They want to find out if these three methods produce different mean scores. Let μ_1, μ_2, and μ_3 be the mean scores of all students who will be taught by methods I, II, and III, respectively. To test if the three teaching methods produce different means, we test the null hypothesis

$$H_0: \mu_1 = \mu_2 = \mu_3$$

against the alternative hypothesis

$$H_1\text{: All three means are not equal}$$

We use the *analysis of variance* procedure to perform this test of hypothesis.

The analysis of variance tests are made using the F distribution. First, the F distribution is described in Section 13.1. Then, Section 13.2 discusses the application of the one-way analysis of variance procedure to perform tests of hypotheses.

13.1 THE *F* DISTRIBUTION

Like the t and chi-square distributions, the shape of a particular ***F* distribution** (named after Sir Ronald Fisher) curve depends on the number of degrees of freedom. However, the F distribution has *two* numbers of degrees of freedom: *degrees of freedom for the numerator* and *degrees of freedom for the denominator*. These two numbers of degrees of freedom are the *parameters* of the F distribution. Each set of degrees of freedom for the numerator and for the denominator gives a different F distribution curve. The units of an F distribution are denoted by F, which assumes only nonnegative values. Like the normal, t, and chi-square distributions, the F distribution is also a continuous distribution. The shape of an F curve is skewed to the right, but the skewness decreases as the number of degrees of freedom increase.

THE *F* DISTRIBUTION

1. The F distribution is continuous and skewed to the right.
2. The F distribution has two numbers of degrees of freedom: *df* for the numerator and *df* for the denominator.
3. The units of an F distribution, denoted by F, are nonnegative.

Figure 13.1 gives three F distribution curves for three sets of degrees of freedom for the numerator and for the denominator. In the figure, the first number gives the degrees of freedom associated with the numerator and the second number gives the degrees of freedom associated with the denominator. We can observe from this figure that as the degrees of freedom increase the peak of the curve moves to the right, that is, the skewness decreases.

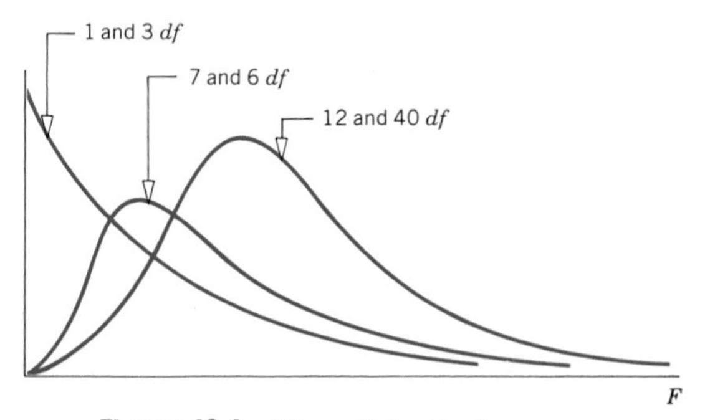

Figure 13.1 Three F distribution curves.

Table X in Appendix C (page 745) lists the values of F. To read Table X, we need to know three quantities: the degrees of freedom for the numerator, the degrees of freedom for the denominator, and an area in the right tail of an F curve. Note that the F distribution table (Table X) is read only for an area in the right tail of the F distribution curve. Also note that Table X has four parts. These four parts give the F values for an area of .01, .025, .05, and .10, respectively, in the right tail. Example 13-1 illustrates how to read Table X.

Reading the F distribution table.

EXAMPLE 13-1 Find the F value for 8 degrees of freedom for the numerator, 14 degrees of freedom for the denominator, and .05 area in the right tail.

Solution For an F distribution, degrees of freedom for the numerator and degrees of freedom for the denominator are usually written as follows.

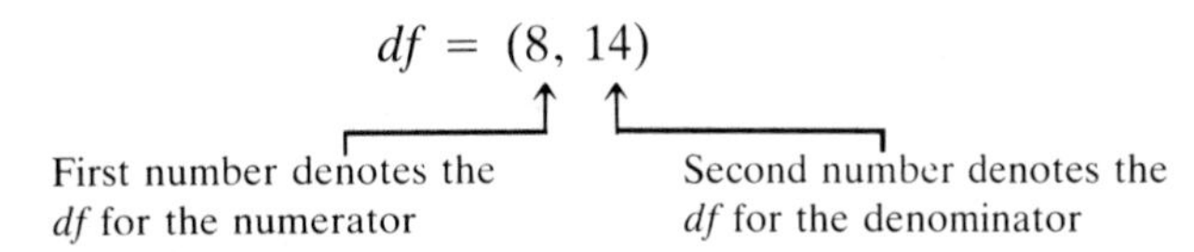

To find the required value of F, we consult the portion of Table X of Appendix C that corresponds to .05 area in the right tail of the F curve. The relevant portion of that table is shown as Table 13.1 below. To find the required F value, we locate 8 in the row for degrees of freedom for the numerator (at the top of Table X) and 14 in the column for degrees of freedom in the denominator (the first column on the left side in Table X). The entry where the column for 8 and the row for 14 intersect gives the required F value. This value of F is 2.70, as shown in Figure 13.2. The F value taken from this table for a test of hypothesis is called the *critical value* of F.

Table 13.1

Degrees of Freedom for Denominator	Degrees of Freedom for Numerator					
	1	2	...	8	...	100
1	161.5	199.5		238.9	...	253.0
2	18.51	19.00	...	19.37	...	19.49
.	...	...	...	...	...	...
.	...	...	...	...	...	...
14	4.60	3.74	...	2.70	...	2.19
.	...	...	...	...	...	...
.	...	...	...	...	...	...
100	3.94	3.09	...	2.03	...	1.39

The F value for 8 df for the numerator, 14 df for the denominator, and .05 area in the right tail.

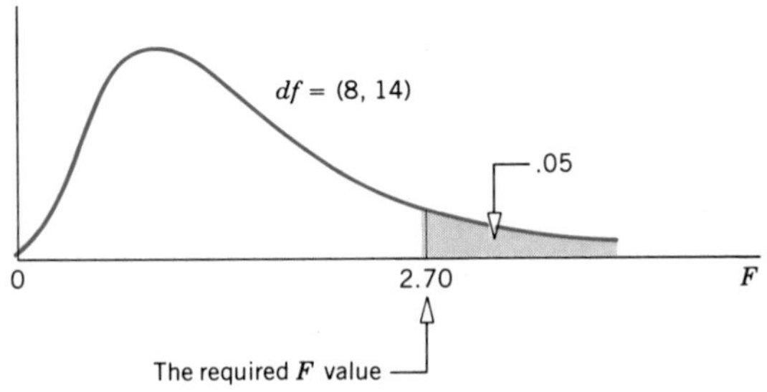

Figure 13.2 The critical value of F for 8 df for the numerator, 14 df for the denominator, and .05 area in the right tail.

■

EXERCISES

13.1 Find the critical value of F for the following.

a. $df = (5, 12)$ and area in the right tail = .05
b. $df = (4, 18)$ and area in the right tail = .05
c. $df = (12, 7)$ and area in the right tail = .05

13.2 Find the critical value of F for the following.

a. $df = (5, 12)$ and area in the right tail = .025
b. $df = (4, 18)$ and area in the right tail = .025
c. $df = (12, 7)$ and area in the right tail = .025

13.3 Determine the critical value of F for the following.

a. $df = (8, 11)$ and area in the right tail = .01
b. $df = (6, 12)$ and area in the right tail = .01
c. $df = (17, 5)$ and area in the right tail = .01

13.4 Determine the critical value of F for the following.

a. $df = (3, 14)$ and area in the right tail = .10
b. $df = (9, 10)$ and area in the right tail = .10
c. $df = (11, 4)$ and area in the right tail = .10

13.2 ONE-WAY ANALYSIS OF VARIANCE

As mentioned in the beginning of this chapter, the analysis of variance procedure is used to test the null hypothesis that the means of two or more populations are the same against the null hypothesis that all population means are not the same. Reconsider the example of teachers at a school who have devised three different methods to teach arithmetic. They want to find out if these three methods produce different mean scores. Let μ_1, μ_2, and μ_3 be the mean scores of all students who will be taught by methods I, II, and III, respectively. To test if the three teaching methods produce different means, we test the null hypothesis

$$H_0\text{: } \mu_1 = \mu_2 = \mu_3$$

against the alternative hypothesis

$$H_1\text{: All three means are not equal}$$

One method to test such a hypothesis is to test the three hypotheses $H_0: \mu_1 = \mu_2$, $H_0: \mu_1 = \mu_3$, and $H_0: \mu_2 = \mu_3$ separately using the procedure of Chapter 10. Besides being time consuming, such a procedure has other disadvantages. First, if we reject even one of these three hypotheses, then we must reject the null hypothesis H_0: $\mu_1 = \mu_2 = \mu_3$. Second, the type I error for the three tests (one for each test) together will give a very large type I error for the test H_0: $\mu_1 = \mu_2 = \mu_3$. Hence, we should prefer a procedure that can test the equality of three means in one test. The **ANOVA,** short for **analysis of variance,** provides us with such a procedure. It is used to compare two or more population means in a single test. (ANOVA can be used to compare two population means. However, the procedures learned in Chapter 10 are more efficient to perform tests of hypotheses about the difference between two population means.)

ANOVA

ANOVA is a procedure used to test the null hypothesis that the means of two or more populations are equal.

This section discusses the **one-way ANOVA** procedure to make tests comparing the means of several populations. By using a one-way ANOVA test, we analyze only one factor or variable. This procedure is called the analysis of variance because the test is based on the analysis of variation in the sample data. The application of ANOVA requires that the following assumptions hold true.

ASSUMPTIONS OF ANOVA

The following assumptions must hold true to use ANOVA.

1. The populations from which the samples are drawn are (approximately) normally distributed.
2. The populations from which the samples are drawn have the same variance (or standard deviation).
3. The samples drawn from different populations are random and independent.

For instance, in the example about three methods of teaching arithmetic, we first assume that the scores of all students taught by each method are (approximately) normally distributed. Second, the means of the distributions of scores for three teaching methods may or may not be the same, but all three distributions have the same variance σ^2. Third, when we take samples to make an ANOVA test, these samples are drawn independently and randomly from three different populations.

The ANOVA test is applied by calculating two estimates of the variance σ^2 of population distributions: the **variance between samples** and the **variance within samples.** The variance between samples is also called the **mean square between samples** or **MSB.** The variance within samples is also called the **mean square within samples** or **MSW.**

The variance between samples MSB gives an estimate of σ^2 based on the variation among the means of samples taken from different populations. For the example of three teaching methods, MSB will be based on the values of the mean scores of three samples of students taught by three different methods. The variance within samples MSW gives an estimate of σ^2 based on the variation within the data of samples. For the example of three teaching methods, MSW will be based on the scores of individual students included in the three samples taken from three populations.

The one-way ANOVA test is always right-tailed with the rejection region in the right tail. The hypothesis-testing procedure in ANOVA involves the same five steps that were used in earlier chapters. The next subsection explains how to calculate the value of the test statistic F for an ANOVA test.

13.2.1 CALCULATING THE VALUE OF THE TEST STATISTIC

The value of the test statistic F for a test of hypothesis using ANOVA is given by the ratio of two variances, the variance between samples (MSB) and the variance within samples (MSW).

TEST STATISTIC F FOR A ONE-WAY ANOVA TEST

The value of the test statistic F for an ANOVA test is calculated as

$$F = \frac{\text{Variance between samples}}{\text{Variance within samples}} \quad \text{or} \quad \frac{\text{MSB}}{\text{MSW}}$$

Example 13-2 describes the calculation of MSB, MSW, and the value of the test statistic F. Since the basic formulas are laborious to use, they are not presented here. We have used only the short-cut formulas to make calculations in this chapter.

Calculating the value of the test statistic F.

EXAMPLE 13-2 Fifteen fourth-grade students were randomly assigned to three groups in order to experiment with three different methods of teaching arithmetic. At the end of the semester, the same test was given to all 15 students. The following table gives the scores of students in three groups.

Method I	Method II	Method III
48	55	84
73	85	68
51	70	95
65	69	74
87	90	67

Calculate the value of the test statistic F. Assume that all the required assumptions mentioned in Section 13.2 hold true.

Solution In ANOVA terminology, the three methods used to teach arithmetic are called **treatments.** The table contains data on the scores of fourth-graders included in the three samples. Each sample of students is taught by a different method. Let

x = the score of a student

k = the number of different samples (or treatments)

n_i = the size of sample i

T_i = the sum of the values in sample i

N = the number of values in all samples $= n_1 + n_2 + n_3 + \cdots$

Σx = the sum of the values in all samples $= T_1 + T_2 + T_3 + \cdots$

Σx^2 = the sum of the squares of values in all samples

First we define three new concepts: SST, SSB, and SSW.

The total sum of squares, denoted by **SST,** is given by the formula

$$\text{SST} = \Sigma x^2 - \frac{(\Sigma x)^2}{N}$$

Note that this SST is the same as the one in regression analysis discussed in Chapter 12. The total sum of squares SST is separated into two portions: between-samples sum of squares, denoted by **SSB,** and within-samples sum of squares, denoted by **SSW.** The sum of SSB and SSW gives SST, that is,

$$\text{SST} = \text{SSB} + \text{SSW}$$

The SSB and SSW are calculated using the following formulas.

BETWEEN- AND WITHIN-SAMPLES SUM OF SQUARES

The between-samples sum of squares, denoted by SSB, is calculated as

$$\text{SSB} = \left(\frac{T_1^2}{n_1} + \frac{T_2^2}{n_2} + \frac{T_3^2}{n_3} + \cdots\right) - \frac{(\Sigma x)^2}{N}$$

The within-samples sum of squares, denoted by SSW, is calculated as

$$\text{SSW} = \Sigma x^2 - \left(\frac{T_1^2}{n_1} + \frac{T_2^2}{n_2} + \frac{T_3^2}{n_3} + \cdots\right)$$

Table 13.2 lists the scores of 15 students who were taught arithmetic by three different methods, the values of T_1, T_2, and T_3, and the values of n_1, n_2, and n_3.

Table 13.2

Method I	Method II	Method III
48	55	84
73	85	68
51	70	95
65	69	74
87	90	67
$T_1 = 324$	$T_2 = 369$	$T_3 = 388$
$n_1 = 5$	$n_2 = 5$	$n_3 = 5$

In Table 13.2, T_1 is obtained by adding the five scores of the first sample. Thus, $T_1 = 48 + 73 + 51 + 65 + 87 = 324$. Similarly, the sum of the values in the second and third samples give $T_2 = 369$ and $T_3 = 388$. Because there are five observations in each sample, $n_1 = n_2 = n_3 = 5$. The values of Σx and N are

$$\Sigma x = T_1 + T_2 + T_3 = 324 + 369 + 388 = 1081$$

$$N = n_1 + n_2 + n_3 = 5 + 5 + 5 = 15$$

To calculate Σx^2, we square all scores included in all three samples and then add them. Thus,

$$\Sigma x^2 = (48)^2 + (73)^2 + (51)^2 + (65)^2 + (87)^2 + (55)^2 + (85)^2 + (70)^2 + (69)^2 + (90)^2 + (84)^2 + (68)^2 + (95)^2 + (74)^2 + (67)^2$$
$$= 80{,}709$$

Substituting all the values in the formulas for SSB and SSW, we obtain the following values of SSB and SSW.

$$\text{SSB} = \left(\frac{(324)^2}{5} + \frac{(369)^2}{5} + \frac{(388)^2}{5}\right) - \frac{(1081)^2}{15} = 432.133$$

$$\text{SSW} = 80{,}709 - \left(\frac{(324)^2}{5} + \frac{(369)^2}{5} + \frac{(388)^2}{5}\right) = 2372.800$$

The value of SST is obtained by adding the values of SSB and SSW. Thus,

$$\text{SST} = 432.133 + 2372.800 = 2804.933$$

The variance between samples MSB and the variance within samples MSW are calculated using the following formulas.

MSB AND MSW

The MSB and MSW are calculated as

$$\text{MSB} = \frac{\text{SSB}}{k - 1} \quad \text{and} \quad \text{MSW} = \frac{\text{SSW}}{N - k}$$

where $k - 1$ and $N - k$ are respectively the *df* for the numerator and *df* for the denominator for the F distribution.

Consequently, the variance between samples is

$$\text{MSB} = \frac{\text{SSB}}{k - 1} = \frac{432.133}{3 - 1} = 216.067$$

The variance within samples is

$$\text{MSW} = \frac{\text{SSW}}{N - k} = \frac{2372.800}{15 - 3} = 197.733$$

The value of the test statistic F is given by the ratio of MSB and MSW. Therefore,

$$F = \frac{\text{MSB}}{\text{MSW}} = \frac{216.067}{197.733} = 1.09$$

For convenience, these calculations are often recorded in a table called the *ANOVA table*. Table 13.3 gives the general form of an ANOVA table.

Table 13.3 ANOVA Table

Source of Variation	Sum of Squares	Degrees of Freedom	Mean Square	Value of the Test Statistic
Between	SSB	$k - 1$	MSB	$F = \frac{\text{MSB}}{\text{MSW}}$
Within	SSW	$N - k$	MSW	
Total	SST	$N - 1$		

Substituting the values of various quantities in Table 13.3, we write an ANOVA table for our example as Table 13.4.

Table 13.4 ANOVA Table

Source of Variation	Sum of Squares	Degrees of Freedom	Mean Square	Value of the Test Statistic
Between	432.133	2	216.067	$F = \frac{216.067}{197.733} = 1.09$
Within	2372.800	12	197.733	
Total	2804.933	14		

■

13.2.2 ONE-WAY ANOVA TEST

Now suppose we want to test the null hypothesis that the mean scores are equal for all three groups of fourth-graders taught by the three different methods of Example 13-2 against the alternative hypothesis that the mean scores of all three groups are not equal. Note that in a one-way ANOVA test, the null hypothesis is that the means for all populations are equal. The alternative hypothesis is that all population means are not equal. In other words, the alternative hypothesis states that at least one of the population means is different from the others. Example 13-3 demonstrates how we can use one-way ANOVA to make this test.

Performing a one-way ANOVA test: all samples of the same size.

EXAMPLE 13-3 Reconsider Example 13-2 about the scores of 15 fourth-grade students who were randomly assigned to three groups in order to experiment with three different methods of teaching arithmetic. At the 1% significance level, can we reject the null hypothesis that the mean arithmetic score of all fourth-grade students taught by each of these three methods is the same? Assume that all the required assumptions mentioned in the beginning of Section 13.2 hold true.

Solution To make a test about the equality of means of three populations, we follow our standard procedure with five steps.

Step 1. *State the null and alternative hypotheses*

Let μ_1, μ_2, and μ_3 be the mean arithmetic scores of all fourth-grade students who are taught, respectively, by methods I, II, and III. The null and alternative hypotheses are

H_0: $\mu_1 = \mu_2 = \mu_3$ (the mean scores of three groups are equal)

H_1: All three means are not equal

Note that the alternative hypothesis states that at least one mean is different from the other two.

Step 2. *Select the distribution to use*

Because we are comparing three means for three normally distributed populations, we use the F distribution to make the test.

Step 3. *Determine the rejection and nonrejection regions*

The significance level is .01. Because a one-way ANOVA test is always right-tailed, the area in the right tail of the F curve is .01, which is the rejection region.

Next we need to know the *df* for the numerator and the denominator. In our example, the students were assigned to three different methods. As mentioned earlier, these methods are called treatments. The number of treatments is denoted by k. The total number of observations in all samples taken together is denoted by N. Then the number of degrees of freedom for the numerator is equal to $k - 1$ and the number of degrees of freedom for the denominator is equal to $N - k$. In our example, there are 3 treatments (methods of teaching) and 15 total observations (total number of students) in all three samples. Thus,

$$\text{Degrees of freedom for the numerator} = k - 1 = 3 - 1 = 2$$

$$\text{Degrees of freedom for the denominator} = N - k = 15 - 3 = 12$$

From Table X, we find the critical value of F for 2 *df* for the numerator, 12 *df* for the denominator, and .01 area in the right tail. This value is shown in Figure 13.3. The required value of F is 6.93.

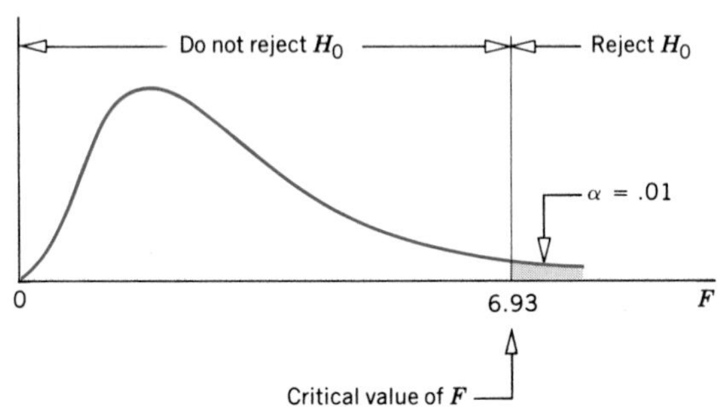

Figure 13.3 Critical value of F for $df = (2, 12)$ and $\alpha = .01$.

Thus, we will fail to reject H_0 if the value of the test statistic F is less than 6.93. We will reject H_0 if it is greater than 6.93.

Step 4. *Calculate the value of the test statistic*

We computed the value of the test statistic F for these data in Example 13.2. This value is

$$F = 1.09$$

Step 5. *Make a decision*

Because the value of the test statistic $F = 1.09$ is less than the critical value of $F = 6.93$, it falls in the nonrejection region. Hence, we fail to reject the null hypothesis, and conclude that the means of three populations are equal. In other words, the three different methods of teaching arithmetic do not seem to affect the mean scores of students. The difference in the three mean scores in the case of our three samples occurred only because of sampling error. ■

In Example 13-3, the sample sizes were the same for all treatments. Example 13-4 exhibits a case when the sample sizes are not the same for all treatments.

Performing a one-way ANOVA test: all samples not of the same size.

EXAMPLE 13-4 A bank manager wants to know if the mean number of customers served per hour by each of four tellers is the same. She observed each of the four tellers for a certain number of hours. The following table gives the number of customers served by the four tellers during each of those hours.

Teller A	Teller B	Teller C	Teller D
19	14	11	24
21	16	14	19
26	14	21	21
24	13	13	26
18	17	16	20
	13	18	

At the 5% significance level, test the null hypothesis that the mean number of customers served per hour by each of these four tellers is the same. Assume that all the required assumptions to apply the one-way ANOVA procedure hold true.

Solution

Step 1. *State the null and alternative hypotheses*

Let μ_1, μ_2, μ_3, and μ_4 be the mean number of customers served per hour by tellers A, B, C, and D, respectively. The null and alternative hypotheses are

H_0: $\mu_1 = \mu_2 = \mu_3 = \mu_4$ (the mean number of customers served per hour by each of four tellers is the same)

H_1: All four population means are not equal

Step 2. *Select the distribution to use*

Because we are testing for the equality of four means for four normally distributed populations, we use the F distribution to make the test.

Step 3. *Determine the rejection and nonrejection regions*

The significance level is .05. Hence, the area in the right tail of the F curve is .05.

In this example, there are 4 treatments (tellers) and 22 total observations in all four samples. Thus,

$$\text{Degrees of freedom for the numerator} = k - 1 = 4 - 1 = 3$$

$$\text{Degrees of freedom for the denominator} = N - k = 22 - 4 = 18$$

The critical value of F from Table X for 3 df for the numerator, 18 df for the denominator, and .05 area in the right tail is 3.16. This value is shown in Figure 13.4.

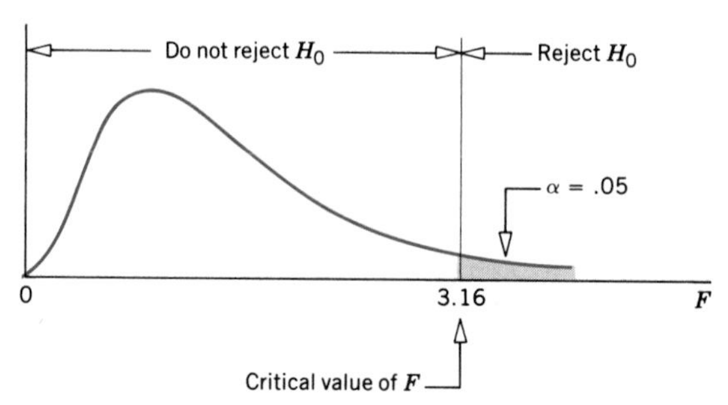

Figure 13.4 Critical value of F for $df = (3, 18)$ and $\alpha = .05$.

Step 4. *Calculate the value of the test statistic*

First we calculate SSB and SSW. Table 13.5 lists the number of customers served by the four tellers during the selected hours, the values of T_1, T_2, T_3, and T_4, and the values of n_1, n_2, n_3, and n_4.

Table 13.5

Teller A	Teller B	Teller C	Teller D
19	14	11	24
21	16	14	19
26	14	21	21
24	13	13	26
18	17	16	20
	13	18	
$T_1 = 108$	$T_2 = 87$	$T_3 = 93$	$T_4 = 110$
$n_1 = 5$	$n_2 = 6$	$n_3 = 6$	$n_4 = 5$

The values of Σx and N are

$$\Sigma x = T_1 + T_2 + T_3 + T_4 = 108 + 87 + 93 + 110 = 398$$

$$N = n_1 + n_2 + n_3 + n_4 = 5 + 6 + 6 + 5 = 22$$

The value of Σx^2 is calculated as follows.

$$\begin{aligned}\Sigma x^2 = {} & (19)^2 + (21)^2 + (26)^2 + (24)^2 + (18)^2 + (14)^2 + (16)^2 + (14)^2 \\ & + (13)^2 + (17)^2 + (13)^2 + (11)^2 + (14)^2 + (21)^2 + (13)^2 + (16)^2 \\ & + (18)^2 + (24)^2 + (19)^2 + (21)^2 + (26)^2 + (20)^2 \\ = {} & 7614\end{aligned}$$

Substituting all the values in the formulas for SSB and SSW, we obtain the following values of SSB and SSW.

$$\text{SSB} = \left(\frac{T_1^2}{n_1} + \frac{T_2^2}{n_2} + \frac{T_3^2}{n_3} + \frac{T_4^2}{n_4}\right) - \frac{(\Sigma x)^2}{N}$$

$$= \left(\frac{(108)^2}{5} + \frac{(87)^2}{6} + \frac{(93)^2}{6} + \frac{(110)^2}{5}\right) - \frac{(398)^2}{22} = 255.618$$

$$\text{SSW} = \Sigma x^2 - \left(\frac{T_1^2}{n_1} + \frac{T_2^2}{n_2} + \frac{T_3^2}{n_3} + \frac{T_4^2}{n_4}\right)$$

$$= 7614 - \left(\frac{(108)^2}{5} + \frac{(87)^2}{6} + \frac{(93)^2}{6} + \frac{(110)^2}{5}\right) = 158.200$$

Hence, the variance between samples MSB and the variance within samples MSW are

$$\text{MSB} = \frac{\text{SSB}}{k - 1} = \frac{255.618}{4 - 1} = 85.206$$

$$\text{MSW} = \frac{\text{SSW}}{N - k} = \frac{158.200}{22 - 4} = 8.789$$

The value of the test statistic F is given by the ratio of MSB and MSW, which is

$$F = \frac{\text{MSB}}{\text{MSW}} = \frac{85.206}{8.789} = 9.69$$

Writing the values of various quantities in the ANOVA table, we obtain Table 13.6.

Table 13.6 ANOVA Table

Source of Variation	Sum of Squares	Degrees of Freedom	Mean Square	Value of the Test Statistic
Between	255.618	3	85.206	
Within	158.200	18	8.789	$F = \frac{85.206}{8.789} = 9.69$
Total	413.818	21		

Step 5. *Make a decision*

Because the value of the test statistic $F = 9.69$ is greater than the critical value of $F = 3.16$, it falls in the rejection region. Consequently, we reject the null hypothesis and conclude that the mean number of customers served per hour by each of the four tellers is not the same. In other words, at least one of the four means is different from the other three. ■

EXERCISES

13.5 Describe the assumptions that must hold true to apply the one-way analysis of variance procedure to test hypotheses.

For the following exercises assume that all the required assumptions to apply the one-way ANOVA procedure hold true.

13.6 A farmer wants to check if three brands of fertilizer give the same average production of wheat per acre. He randomly assigned each fertilizer to eight one-acre tracts of land. The following table gives the production of wheat (in bushels) for each acre for three brands of fertilizer.

Fertilizer I	Fertilizer II	Fertilizer III
72	58	61
69	42	58
75	53	63
59	47	68
64	45	55
68	52	65
71	47	59
67	57	63

At the 5% significance level, test the null hypothesis that the mean yield of wheat for each of these three brands of fertilizer is the same.

13.7 Five seniors were randomly selected from each of three high schools. They were all given the same test in mathematics. The following table lists the scores of these seniors in that test.

School A	School B	School C
69	82	71
79	87	83
54	76	64
76	97	79
81	92	91

Using the 1% significance level, test the null hypothesis that the mean score of all high school seniors in mathematics at each of these three schools is the same.

13.8 Three brands of drugs were tested to find out if the mean time within which they provide relief from a headache is the same for these drugs. The first drug was administered to six randomly selected patients, the second to four randomly selected patients, and the third to five randomly selected patients. The following table gives the time taken (in minutes) by each patient to get relief from a headache after taking the medicine.

Drug I	Drug II	Drug III
25	15	44
38	21	39
40	19	52
65	23	58
47		73
52		

At the 2.5% significance level, test the null hypothesis that the mean time taken to provide relief from a headache is the same for all three drugs.

13.9 A company wanted to check if the average number of items manufactured per hour by each of three machines is the same. It observed each machine for six hours. The following table lists the number of items manufactured on each of the three machines during each hour.

Machine A	Machine B	Machine C
14	10	9
11	15	12
15	18	10
12	19	12
14	17	8
13	13	11

At the 5% significance level, will you reject the null hypothesis that the mean number of items manufactured per hour on each of these three machines is the same?

13.10 A dietitian wanted to test three different diets to find out if the mean weight loss for each of these diets is the same. She randomly selected 21 overweight persons, randomly divided them into three groups, and put each group on one of the three diets. The following table records the weights lost by these persons after being on these diets for 2 months.

Diet I	Diet II	Diet III
15	11	9
8	16	17
11	9	11
7	13	8
22	24	11
12	17	6
8	19	14

Using the 1% significance level, test the null hypothesis that the mean weight lost by all persons on each of the three diets is the same.

13.11 The following table gives the auto insurance premiums paid per month by randomly selected drivers insured with four different insurance companies.

Company A	Company B	Company C	Company D
65	48	57	62
73	69	61	53
54	88	89	45
43		77	51
		69	

Using the 1% significance level, test the null hypothesis that the mean auto insurance premium paid per month by all drivers insured by each of these four companies is the same.

13.12 The following table gives the number of classes missed during one semester by 25 randomly selected college students drawn from three different age groups.

Below 24	25 to 30	31 and Above
19	9	5
12	6	8
25	11	4
13	14	3
19	8	10
4	9	9
15	4	
12	13	
16	16	
9		

At the 2.5% significance level, test the null hypothesis that the mean number of classes missed during the semester by all students in each of these three age groups is the same.

13.13 A large company wants to check if the absentee rate is the same for different groups of women. The company randomly selected 19 women that belong to three groups: single, married with no children, and married with children. The number of days these women were absent during the last 6 months are recorded in the following table.

Single Women	Married Women with No Children	Married Women with Children
7	2	8
6	5	9
11	2	6
5	1	10
3	6	8
6	0	
4	2	

Using the 2.5% significance level, test the null hypothesis that the mean number of days on which women in each of these three age groups are absent is the same.

13.14 A large manufacturer of copying machines recently hired three new salespersons with degrees in marketing, mathematics, and sociology. The company wants to check if the fields of study have any effect on the mean number of sales made by these salespersons. The following table lists the number of sales made by these three salespersons during certain randomly selected days.

Salesperson with Marketing Degree	Salesperson with Mathematics Degree	Salesperson with Sociology Degree
9	2	4
10	1	1
3	3	1
7	2	3
4	5	6
12	3	8
8	1	1

Using the 5% significance level, can you reject the null hypothesis that the mean number of copying machines sold per day by these salespersons with degrees in different areas is the same?

GLOSSARY

Analysis of variance (ANOVA) A statistical technique used to test whether the means of two or more populations are equal.

***F* distribution** A continuous distribution that has two parameters: *df* for the numerator and *df* for the denominator.

Mean square between samples or MSB A measure of the variation among means of samples taken from different populations.

Mean square within samples or MSW A measure of the variation within data of all samples taken from different populations.

One-way ANOVA The analysis of variance technique that analyzes one variable only.

SSB The sum of squares between samples. Also called the sum of squares of the factor or treatment.

SST The total sum of squares given by the sum of SSB and SSW.

SSW The sum of squares within samples. Also called the sum of squares of errors.

KEY FORMULAS

Let

$$k = \text{the number of different samples (or treatments)}$$

$$n_i = \text{the size of sample } i$$

$$T_i = \text{the sum of the values in sample } i$$

$$N = \text{the number of values in all samples} = n_1 + n_2 + n_3 + \cdots$$

$$\Sigma x = \text{the sum of the values in all samples} = T_1 + T_2 + T_3 + \cdots$$

$$\Sigma x^2 = \text{the sum of the squares of values in all samples}$$

1. Degrees of freedom for the *F* distribution

$$\text{Degrees of freedom for the numerator} = k - 1$$

$$\text{Degrees of freedom for the denominator} = N - k$$

2. Between-Samples Sum of Squares

$$\text{SSB} = \left(\frac{T_1^2}{n_1} + \frac{T_2^2}{n_2} + \frac{T_3^2}{n_3} + \cdots\right) - \frac{(\Sigma x)^2}{N}$$

3. Within-Samples Sum of Squares

$$\text{SSW} = \Sigma x^2 - \left(\frac{T_1^2}{n_1} + \frac{T_2^2}{n_2} + \frac{T_3^2}{n_3} + \cdots\right)$$

4. Total Sum of Squares

$$SST = \Sigma x^2 - \frac{(\Sigma x)^2}{N}$$

Also

$$SST = SSB + SSW$$

5. Variance Between Samples

$$MSB = \frac{SSB}{k - 1}$$

6. Variance Within Samples

$$MSW = \frac{SSW}{N - k}$$

7. Value of the Test Statistic *F*

$$F = \frac{\text{Variance between samples}}{\text{Variance within samples}} \quad \text{or} \quad \frac{MSB}{MSW}$$

SUPPLEMENTARY EXERCISES

For the following exercises, assume that all the assumptions required to apply the one-way ANOVA hold true.

13.15 A professor wants to test for the equality of the mean scores of students enrolled in statistics classes at three different times: 8:00 AM, 1:00 PM, and 7:00 PM. The following table lists the midterm scores of students included in three random samples taken from classes meeting at these three times.

8:00 AM Classes	1:00 PM Classes	7:00 PM Classes
89	70	85
76	45	93
94	81	81
84	91	67
68	61	74
91	58	84
89	52	

Using the 5% significance level, test the null hypothesis that the mean scores of all students enrolled in statistics classes at three different times are equal.

13.16 A consumer agency wants to check if the mean lives of four brands of auto batteries, which sell for nearly the same price, are the same. The agency randomly selected a few batteries of each brand and tested them. The following table gives the lives of these batteries in thousands of hours.

Brand A	Brand B	Brand C	Brand D
62	53	57	56
68	67	61	53
51	78	81	45
49	69	72	67
55		61	

At the 5% significance level, will you reject the null hypothesis that the mean life of each of the four brands of batteries is the same?

13.17 The following table gives the starting salaries (in thousands of dollars) for 17 randomly selected college graduates who had four different majors.

Engineering	Business	Mathematics	Sociology
28.2	21.3	21.3	18.6
24.5	26.8	19.4	19.2
31.3	24.6	26.3	21.9
26.4	23.5	23.6	
29.1	25.8		

At the 1% significance level, test the null hypothesis that the mean starting salaries of all college graduates with these four majors are equal.

13.18 A farmer wants to test three brands of weight-gaining diets for chickens to determine if the mean weight gain for each of these brands is the same. He selected 15 chickens and randomly put each of them on one of these three brands of diets. The following table lists the weights gained (in pounds) by these chickens after a period of 1 month.

Brand A	Brand B	Brand C
.8	.6	1.2
1.3	1.1	.8
1.5	.6	.7
.9	.4	1.3
.6	.7	.9

At the 1% significance level, test the null hypothesis that the mean weight gained by all chickens is the same for each of these three diets.

13.19 The following table gives the response time (in minutes) of three fire companies in a city for certain randomly selected incidents after a fire was reported.

Company A	Company B	Company C
1.6	1.4	.8
.8	2.6	1.3
2.7	.9	1.7
1.2	3.5	.9
3.4	1.2	1.1
1.9	1.5	.7
4.3		2.1

At the 2.5% significance level, test the null hypothesis that the mean response time for each of these three fire companies for all fire incidents is the same.

13.20 The following table lists the number of violent crimes reported to police on randomly selected days for this year. The data are taken from three large cities of about the same size.

City A	City B	City C
5	2	8
9	4	12
10	1	5
3	11	3
5	4	9
7	6	11
13		

Using the 2.5% significance level, test the null hypothesis that the mean number of violent crimes reported per day is the same for each of these three cities.

13.21 The following table lists the prices of certain randomly selected college textbooks in statistics, psychology, economics, and history.

Statistics	Psychology	Economics	History
37	51	37	32
43	45	44	29
35	38	41	35
39	43	36	38
46		39	

Using the 5% significance level, test the null hypothesis that the mean prices of college textbooks in statistics, psychology, economics, and history are all equal.

13.22 A few samples of paints from each of three companies were tested for drying time. The following table records the drying times (in minutes) for these samples of paints.

Company A	Company B	Company C
42	57	45
53	63	49
43	61	51
47	54	58
42	51	44
51	60	41
56		47

Using the 1% significance level, test the null hypothesis that the mean drying times for paints of these three companies are equal.

13.23 Twenty cars of the same model were randomly divided into three groups to test for miles per gallon traveled on each of three different brands of gasoline. The following table lists the miles per gallon obtained by the cars from the use of these three brands of gasoline.

Brand A	Brand B	Brand C
26.2	24.6	27.3
28.5	23.6	29.4
25.0	22.7	25.7
26.7	25.1	27.3
25.8	26.0	24.8
27.3	22.2	26.1
28.4		24.6

At the 2.5% significance level, test the null hypothesis that the mean miles per gallon given for this model is the same for each of these three types of gasoline.

SELF-REVIEW TEST

1. The F distribution is

   a. continuous **b.** discrete

2. The F distribution is always

   a. symmetric **b.** skewed to the right **c.** skewed to the left

3. The units of the F distribution, denoted by F, are always

   a. nonpositive **b.** positive **c.** nonnegative

4. The one-way ANOVA test analyzes only one

   a. variable **b.** population **c.** sample

5. The one-way ANOVA test is always

   a. right-tailed **b.** left-tailed **c.** two-tailed

6. For a one-way ANOVA with k treatments and N observations in all samples taken together, the number of degrees of freedom for the numerator are

   a. $k - 1$ **b.** $N - k$ **c.** $N - 1$

7. For a one-way ANOVA with k treatments and N observations in all samples taken together, the number of degrees of freedom for the denominator are

   a. $k - 1$ **b.** $N - k$ **c.** $N - 1$

8. The ANOVA test can be applied to compare

   a. two or more population means
   b. more than four population means
   c. more than three population means

9. Briefly describe the assumptions of ANOVA as mentioned in this chapter.

10. The following table gives the hourly wage of computer programmers for samples taken from three cities.

New York	Boston	Los Angeles
$15.45	$23.50	$17.50
28.80	18.60	11.40
26.45	14.75	29.40
22.10	30.00	22.30
31.50	35.40	16.35
39.30	26.40	19.50
28.75		21.30

Using the 1% significance level, test the null hypothesis that the mean hourly wage for all computer programmers in each of these three cities is the same.

USING MINITAB

The first step in using MINITAB to solve a problem using the one-way analysis of variance procedure is to enter the given data in different columns. If all columns contain the same number of values (i.e., all samples are of the same size), we can use the READ command. However, if different columns contain different number of values (as in Illustration M13-1), we will use the SET command to enter data for each column one at a time. Suppose our example contains data on four samples. After entering data for these four samples in four columns C1, C2, C3, and C4, we use the following MINITAB command to perform one-way analysis of variance.

```
MTB > AOVONEWAY C1 C2 C3 C4
```

In the MINITAB command AOVONEWAY, AOV stands for analysis of variance and ONEWAY stands for one-way. Illustration M13-1 describes the use of MINITAB for a one-way analysis of variance.

Illustration M13-1 According to Example 13-4, a bank manager wants to know if the mean number of customers served per hour by each of four tellers is the same. She observed each of the four tellers for a certain number of hours. The following table gives the number of customers served by each of the four tellers during each of those hours.

Teller A	Teller B	Teller C	Teller D
19	14	11	24
21	16	14	19
26	14	21	21
24	13	13	26
18	17	16	20
	13	18	

At the 5% significance level, test the null hypothesis that the mean number of customers served per hour by each of these four tellers is the same. Assume that all the required assumptions to apply the one-way analysis of variance hold true.

Solution Let μ_1, μ_2, μ_3, and μ_4 be the mean number of customers served per hour by each of the four tellers, respectively. Then the null and alternative hypotheses are

$$H_0: \mu_1 = \mu_2 = \mu_3 = \mu_4 \quad \text{(all four population means are equal)}$$

$$H_1: \text{All four population means are not equal}$$

The number of values in each of the four samples is not the same. Hence, we use the SET command to enter data on four samples in four columns.

```
MTB  > NOTE: APPLICATION OF ONE-WAY ANOVA
MTB  > SET C1
DATA > 19  21  26  24  18
DATA > END
MTB  > SET C2
DATA > 14  16  14  13  17  13
DATA > END
MTB  > SET C3
DATA > 11  14  21  13  16  18
DATA > END
MTB  > SET C4
DATA > 24  19  21  26  20
DATA > END
MTB  > AOVONEWAY C1-C4 <--
```

This command instructs MINITAB to perform a one-way analysis of variance test for the data of columns C1, C2, C3, and C4

```
ANALYSIS OF VARIANCE
SOURCE     DF       SS       MS       F       P   <--
FACTOR      3   255.62    85.21    9.69   0.000
ERROR      18   158.20     8.79
TOTAL      21   413.82
```

Compare this table to the ANOVA Table 13.6 of Example 13-4

```
                                   INDIVIDUAL 95 PCT CI'S FOR
                                   MEAN BASED ON POOLED STDEV
                                   ------+---------+---------+---------+
LEVEL    N     MEAN    STDEV                            (-------*-------)
C1       5   21.600    3.362       (------*-------)
C2       6   14.500    1.643         (------*-------)
C3       6   15.500    3.619                            (-------*-------)
C4       5   22.000    2.915       ------+---------+---------+---------+
POOLED STDEV = 2.965                   14.0      17.5      21.0      24.5
```

Compare the analysis of variance table in the MINITAB solution with Table 13.6 of Example 13-4. The two sources of variation were called the variations between- and within-samples in the column labeled Source of Variation in Table 13.6. MINITAB calls these two sources respectively the *factor* and *error*. Also, MINITAB prints the *p*-value for the test.

The MINITAB solution also gives the following information.

1. The mean and standard deviation for the data of each column—C1, C2, C3, and C4. Thus, for example, the mean and standard deviation for the data of column C1 (for teller A) are 21.6 and 3.362, respectively.
2. The 95% confidence intervals for the means of populations corresponding to all samples, that is, the 95% confidence interval for the mean number of customers served per hour by each of the four tellers.
3. The pooled standard deviation, which is 2.965. This pooled standard deviation is nothing but the square root of what we called MSW in this chapter. The MSW calculated in Example 13-4 was 8.79. The squared root of 8.79 is 2.965, which is printed as pooled standard deviation in the MINITAB solution.

From the MINITAB printout, the value of the test statistic F is 9.69. The critical value of F from the F table for $\alpha = .05$, *df* for the numerator = 3, and *df* for the denominator = 18 is 3.16 (see Figure 13.4 of Example 13-4). The value of the test statistic $F = 9.69$ is larger than the critical value of $F = 3.16$ and it falls in the rejection region. Consequently, we reject the null hypothesis. Hence, the mean number of customers served per hour by each of the four tellers is not the same.

We can reach the same conclusion by considering the *p*-value. The *p*-value from the MINITAB solution is 0.000. Because this *p*-value is less than $\alpha = .05$, we reject the null hypothesis.

COMPUTER ASSIGNMENTS

M13.1 Refer to Exercise 13.6. Solve that exercise using MINITAB. (*Note:* Because the number of data values in each of the three columns in Exercise 13.6 is the same, the READ command can be used to enter data into MINITAB.)

M13.2 Refer to Exercise 13.21. Solve that exercise using MINITAB.

APPENDIX A DATA SOURCES, SAMPLING, AND THE USE OF RANDOM NUMBERS

A.1 SOURCES OF DATA

A.2 WHY SAMPLE?

A.3 A REPRESENTATIVE SAMPLE

A.4 SAMPLING AND NONSAMPLING ERRORS

A.5 RANDOM SAMPLING TECHNIQUES

USING MINITAB

In Chapter 1 we briefly explained a few terms and concepts such as sources of data, sample surveys, census, reasons for conducting a sample survey instead of a census, a representative sample, and simple random sample. In Chapter 7 we discussed sampling and nonsampling errors. However, nonsampling error is not discussed in detail in Chapter 7.

This appendix will explain these concepts in more detail. In addition to that, this appendix also discusses four sampling techniques and the use of a table of random numbers to select a sample.

A.1 SOURCES OF DATA

The availability of accurate data is essential for deriving reliable results and making accurate decisions. As with the truism "garbage in, garbage out" (GIGO), policy decisions adopted based on the results of poor data may prove to be disastrous.

Data sources can be divided into three categories: (1) internal sources, (2) external sources, and (3) surveys and experiments.

1. INTERNAL SOURCES

Many times data come from *internal sources,* such as a company's own personnel files or accounting records. A company that wants to forecast the future sales of its product might use data from its own records for past time periods. A police department might use data that exist in its own records to analyze changes in the nature of crimes over a period of time.

2. EXTERNAL SOURCES

All needed data may not be available from internal sources. Hence, to obtain data we may have to depend on sources outside the company. Such sources of data are called *external sources*. Data obtained from external sources may be primary or secondary data. Primary data are the data obtained from the organization which originally collected them. If we obtain data from the Bureau of Labor Statistics that were collected by this organization, then these are primary data. *Secondary data* are the data obtained from a source that did not originally collect them. For example, data originally collected by the Bureau of Labor Statistics and published in the *Statistical Abstract of the United States* are secondary data.

3. SURVEYS AND EXPERIMENTS

Sometimes the data we need may not be available from internal or external sources. In such cases, we may have to obtain data by conducting our own survey or experiment.

A. Surveys

In a survey we do not exercise any control over the factors when collecting information. For example, if we want to collect data on the money spent last month on clothes by various families, we will ask each of the families included in the survey how much it spent last month on clothes. Then we will record this information.

SURVEY

In a survey, data are collected from the members of a population or sample without exercising any particular control over the factors that may affect the characteristic of interest or the results of the survey.

A survey may be a census or a sample survey.

A Census

A **census** includes every member of the population of interest, which is called the *target population.*

CENSUS

A survey that includes every member of the population is called a census.

In practice, a census is rarely taken because it is very expensive and time consuming. Furthermore, in many cases it is impossible to identify each member of the target population. We will discuss these reasons in more detail in Section A.2.

A Sample Survey

Usually, to conduct research, we select a portion of the target population. This portion of the population is called a **sample.** Then we collect the required information from the elements included in the sample.

SAMPLE SURVEY

The technique of collecting information from a portion of the population is called a sample survey.

A survey can be conducted by personal interviews, by telephone, or by mail. The personal interview technique has the advantage of having a high response rate and a high quality of answers obtained. However, it is the most expensive and time-consuming technique. The telephone survey also gives a high response rate. Unlike personal interviews, it is less expensive and less time consuming. Nonetheless, a problem with this technique is that many people do not like to be called at home, and those who do not have a phone are left out of the survey. A survey conducted by mail is the least expensive method but the response rate is usually very low in such a survey. Many people included in such a survey do not return the questionnaires.

Conducting a survey with accurate and reliable results is not an easy task. To quote Warren Mitofsky, director of Elections and Surveys for CBS News, "Any damn fool with 10 phones and a typewriter thinks he can conduct a poll."† The preparation of a questionnaire is probably the most difficult part of a survey. The way a question

†"The Numbers Racket: How Polls and Statistics Lie," *U.S. News & World Report,* July 11, 1988.

is phrased can affect the results of the survey. The following is an excerpt from an article published in *Psychology Today*.

CASE STUDY A-1 IS IT A SIMPLE QUESTION?

Even the seemingly simplest of questions can yield complex answers. "Do you own a car?" asks Stanley Presser, a sociologist at the National Science Foundation in Washington, D.C. "That sounds like an awfully simple question. But is it really? What does 'you' mean? Suppose a wife is answering the poll, and the car is registered in her husband's name. How is she supposed to answer? What does 'own' mean? What if the car is on a long-term lease? What does 'car' mean? What if they have one of those new little vans, or a four-wheel-drive vehicle? My God, that sounds like a simple question! You can imagine how diverse the factors become in a more complicated one."

Suppose, however, that the question about car ownership had been preceded by a series of related questions: "Are you married? Does your spouse drive an automotive vehicle? Is it a car, a van or some other sort of vehicle? Is it leased, or does your spouse own it? Now about you—do you own a car?" Such a series of questions would serve to clarify the intended meaning of the one about car ownership.

Source: Rich Jaroslovsky, "What's on Your Mind, America?" *Psychology Today,* July–August 1988, pp. 54–59.

Even polls taken at the same time can produce dramatically different results depending on how a question is phrased. The following excerpts reproduced from *U.S. News & World Report* also show how the structure of the questions can change the results of a poll.

CASE STUDY A-2 HOW TO SKEW A POLL: LOADED QUESTIONS AND OTHER TRICKS

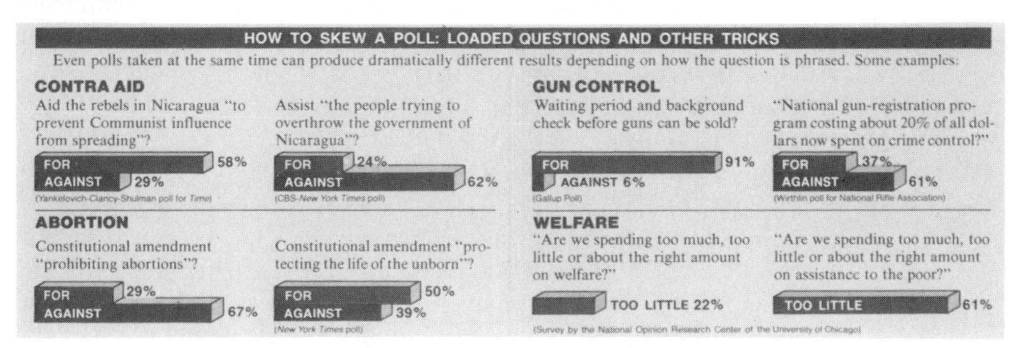
HOW TO SKEW A POLL: LOADED QUESTIONS AND OTHER TRICKS

Even polls taken at the same time can produce dramatically different results depending on how the question is phrased. Some examples:

CONTRA AID

Aid the rebels in Nicaragua "to prevent Communist influence from spreading"?
FOR 58%
AGAINST 29%
(Yankelovich-Clancy-Shulman poll for *Time*)

Assist "the people trying to overthrow the government of Nicaragua"?
FOR 24%
AGAINST 62%
(CBS-*New York Times* poll)

ABORTION

Constitutional amendment "prohibiting abortions"?
FOR 29%
AGAINST 67%

Constitutional amendment "protecting the life of the unborn"?
FOR 50%
AGAINST 39%
(*New York Times* poll)

GUN CONTROL

Waiting period and background check before guns can be sold?
FOR 91%
AGAINST 6%
(Gallup Poll)

"National gun-registration program costing about 20% of all dollars now spent on crime control?"
FOR 37%
AGAINST 61%
(Wirthlin poll for National Rifle Association)

WELFARE

"Are we spending too much, too little or about the right amount on welfare?"
TOO LITTLE 22%

"Are we spending too much, too little or about the right amount on assistance to the poor?"
TOO LITTLE 61%
(Survey by the National Opinion Research Center of the University of Chicago)

Source: *U.S. News & World Report,* July 11, 1988.

B. Experiments

In an experiment, we exercise control over some factors when collecting information.

AN EXPERIMENT

In an experiment, data are collected from members of a population or sample by exercising control over some or all of the factors that may affect the characteristic of interest or the results of the experiment.

For example, how is a new drug to be tested to find out whether or not it cures a disease? It is done by designing an experiment in which the patients under study are divided into two groups as follows.

1. The *treatment group*—the members of this group receive the actual drug.
2. The *control group*—the members of this group do not receive the actual drug but are given a substitute (called a placebo) that appears to be the actual drug.

The two groups are formed in such a way that the patients in one group are similar to the ones in the other group. Neither the doctors nor the patients know to which group a patient belongs. Such an experiment is called a *double-blind experiment.* Then, after a comparison of the percentage of patients cured in each of the two groups, a decision is made about the effectiveness or noneffectiveness of the new drug.

A.2 WHY SAMPLE?

As mentioned in the previous section, most of the time the surveys are conducted by using samples and not a census of the population. Some of the main reasons for conducting a sample survey instead of a census are as follows.

1. **Time.** In most cases, the size of the population is quite large. Consequently, conducting a census will take a long time whereas a sample survey can be conducted rapidly. It will be time consuming to interview or contact hundreds of thousands or even millions of members of a population. On the other hand, a survey of a sample of a few hundred elements may be completed in little time. In fact, because of the amount of time needed to conduct a census, by the time the census is completed the results may become obsolete.
2. **Cost.** The cost of collecting information from every member of a population may easily fall outside the limited budget of most, if not all, surveys. Hence, to stay within the available resources, conducting a sample survey may be the best approach.
3. **Impossible to take a census.** Sometimes it is impossible to take a census. First, it may not be possible to identify and access each member of the population. For example, if a team of biologists wants to conduct a survey about the mean length of a frog's jump or the mean weight of fish, it will not be possible to find each member of the population and include it in the survey. Second, sometimes conducting a survey means destroying the items included in the survey. For example, to estimate the mean life of light bulbs would necessitate burning out all the bulbs

included in the survey. The same is true about finding the average life of batteries. In such cases, only a portion of the population can be selected for the survey.

A.3 A REPRESENTATIVE SAMPLE

The results obtained from different samples will always differ from sample to sample. The results and decisions obtained based on a sample survey will be reliable and close to the results that would be obtained by conducting a population census only if the selected sample represents the characteristics of the population as closely as possible. Such a sample is called a **representative sample.**

REPRESENTATIVE SAMPLE

A sample that represents the characteristics of the population as closely as possible is called a representative sample.

Depending on how a sample is drawn, it may be a random sample or a nonrandom sample.

RANDOM SAMPLE

A random sample is a sample drawn in such a way that each member of the population has some chance for being selected in the sample.

A random sample is usually a representative sample. For a random sample, each member of the population may or may not have the same chance for being included in the sample. We will discuss four types of random samples in Section A.5 of this appendix.

Two types of nonrandom samples are a *convenience sample* and a *judgment sample*. In a **convenience sample,** the most accessible members of the population are selected to obtain the results quickly. For example, an opinion poll may be conducted in a few hours by collecting information from certain shoppers at a single shopping mall. In a **judgment sample,** the members are selected from the population based on the judgment and prior knowledge of an expert. Although such a sample may happen to be a representative sample, the chances of it being so are small. If the population is large, it is not an easy task to select a representative sample based on judgment.

The so called *pseudo polls* are examples of nonrepresentative samples. For instance, a survey conducted by a magazine that includes only its readers does not usually involve a representative sample. Similarly, a poll conducted by a television station giving two separate 900 telephone numbers for *yes* and *no* votes is not based on a representative sample. In these two examples, respondents will only be those people who read that magazine or watch that television station, who do not mind paying the postage and telephone charges, or who feel emotionally compelled to respond. To quote Larry King on this subject

> All over the board . . . The 900 telephone number is very popular these days, but viewers should be warned that in the case of political polling it has absolutely no basis

in fact. Poor people in the audience can't contribute to the survey, so it's faulty to begin with. . . . So next time you see a poll based on 900 numbers, treat it as some sort of middle-class amusement and forget about it. ("Larry King's People," *USA Today,* July 17, 1989. Copyright © 1989, *USA Today*. Reprinted with permission.)

A.4 SAMPLING AND NONSAMPLING ERRORS

The results obtained from a sample survey may contain two types of errors: the sampling and nonsampling errors.

A.4.1 SAMPLING ERROR

Usually, all samples taken from the same population will give different results because they contain different elements of the population. Moreover, the results obtained from any one sample will not be exactly the same as the ones obtained from the census. The difference between a sample result and the result we would have obtained by conducting a census is called the **sampling error,** assuming that the sample is random and no nonsampling error has been made.

> **SAMPLING ERROR**
>
> The sampling error is the difference between the result obtained from a sample survey and the result that would have been obtained if the whole population had been included in the survey.

The sampling error occurs because of chance and it cannot be avoided. A sampling error can occur only in a sample survey. It does not occur in a census. Sampling error is discussed in detail in Section 7.2 of Chapter 7 and an example of it is given there.

A.4.2 NONSAMPLING ERRORS

Nonsampling errors can occur both in a sample survey and in a census. Such errors occur because of human mistakes and not chance.

> **NONSAMPLING ERRORS**
>
> The errors that occur in the collection, recording, and tabulation of data are called nonsampling errors.

The following are the main reasons for the occurrence of nonsampling errors.

1. If a sample is nonrandom (and, hence, nonrepresentative), the sample results may be too different from the census results. The following quote from *U.S. News*

& World Report describes how even a randomly selected sample can become nonrandom if some of the members included in the sample cannot be contacted.

> A test poll conducted in the 1984 presidential election found that if the poll were halted after interviewing only those subjects who could be reached on the first try, Reagan showed a 3-percentage-point lead over Mondale. But when interviewers made a determined effort to reach everyone on their lists of randomly selected subjects—calling some as many as 30 times before finally reaching them—Reagan showed a 13 percent lead, much closer to the actual election result. As it turned out, people who were planning to vote Republican were simply less likely to be at home ("The Numbers Racket: How Polls and Statistics Lie," *U.S. News & World Report,* July 11, 1988.

2. The questions may be phrased in such a way that they are not fully understood by the members of the sample or population. As a result, the answers obtained are not accurate.
3. The respondents may intentionally give false information in response to some sensitive questions. For example, people may not tell the truth about drinking habits, incomes, or opinions about minorities. Sometimes the respondents may give wrong answers because of ignorance. For example, a person may not remember the exact amount he spent on clothes during the last year. If asked in a survey, he may give an inaccurate answer.
4. The poll-taker may make a mistake and enter a wrong number in the records or make an error while entering the data on a computer.

Nonsampling errors can be minimized if questions are prepared carefully and data are handled more cautiously.

A.5 RANDOM SAMPLING TECHNIQUES

There are many ways to select a random sample. Four of these techniques are discussed below.

1. SIMPLE RANDOM SAMPLING

A sample that assigns the same probability to each member of the population for being selected is called a **simple random sample.**

> **SIMPLE RANDOM SAMPLE**
>
> A simple random sample is a sample that is selected in such a way that each member of the population has the same chance of being included in the sample.

One way to select a simple random sample is by a lottery or drawing. For example, if we need to select five students from a class of 50, we write each of the 50 names on separate pieces of paper. Then, we place all 50 names in a box and mix them thoroughly. Next, we draw one name randomly from the box. We repeat this experiment four more times. The five drawn names comprise a simple random sample.

The second procedure used to select a simple random sample is by using a table of random numbers. Table I in Appendix C lists random numbers. These numbers are generated by a random process. Suppose we have a group of 400 persons and we need to select 30 persons randomly from this group. To select a simple random sample, we arrange the names of all 400 persons in alphabetic order and assign a three digit number, from 001 to 400, to each person.

Next, we use the random numbers table to select 30 persons. The random numbers in Table I are recorded in blocks of five digits. To use the table, we can start anywhere. One way to do so is to close your eyes and put a finger anywhere on the page and start at that point. From there, we can move in any direction. We need to pick three-digit numbers from the table because we have assigned three-digit numbers to the 400 persons in our population.

Suppose we start at the first block of the 31st row from the top of Table I. The five rows starting with the 31st row from that table are reproduced as Table A.1. The first block of five numbers in Table A.1 is 13049. We use the first three digits of this block to select the first person from the population. Hence, the first person selected is the one with a number 130. Suppose we move along the row to the right to make the next selection. The second block of five numbers in Table A.1 is 85293. The first three digits of this block give 852. However, we have only 400 persons in the population with assigned numbers of 001 to 400. Consequently, we cannot use 852 to select a person. Therefore, we move to the next block of five numbers without making a selection. The third block of numbers is 32747. The first three digits of this block give 327. Consequently, the second person selected is the one with a number 327. We continue this process until all 30 required persons are selected. This gives us a simple random sample of 30 persons.

Table A.1

13049	85293	32747	17728	50495	34617	73707	33976	86177
86544	52703	74990	98288	61833	48803	75258	83382	79099
77295	70694	97326	35430	53881	94007	70471	66815	73042
54637	32831	59063	72353	87365	15322	33156	40331	93942
50938	12004	18585	23896	62559	44470	27701	66780	56157

Although the table of random numbers given in Appendix C contains only 1485 blocks of five-digit numbers, we can easily construct a table of as many random numbers as we want using a computer software package such as MINITAB.

If we have access to a computer, we can use a statistical package, such as MINITAB, to select a simple random sample. The MINITAB sections at the end of Chapter 1 and this appendix explain and illustrate how we can draw such a sample by using MINITAB.

2. SYSTEMATIC RANDOM SAMPLING

The simple random sample procedure will become tedious if the size of the population is large. For example, if we need to select 150 households from a list of 45,000, it will be time consuming either to write the 45,000 names on pieces of papers and then select 150 households, or to assign a five-digit number to each of the 45,000 households and then select 150 households using the random number table. In such cases, it is more convenient to use systematic random sampling.

The procedure to select a systematic random sample is as follows. In the example just mentioned, we would arrange all 45,000 households alphabetically (or based on some other characteristic). Since, the sample size should equal 150, the ratio of population to sample size is 45,000/150 = 300. Using this ratio, we randomly select one household from the first 300 households in the arranged list either by using the lottery system or by using a random number table. Suppose, by using either of these methods, we select the 210th household. We will then select every 210th household from every 300 households in the list. In other words, our sample will include the households with numbers 210, 510, 810, 1110, 1410, 1710, and so on.

SYSTEMATIC RANDOM SAMPLING

In a systematic random sampling, we first randomly select one member from the first *k* units. Then every *k*th member, starting with the first selected member, is included in the sample.

Note that a systematic random sampling does not give a simple random sample because we cannot select two adjacent elements. Hence, every member of the population does not have the same probability of being selected.

3. STRATIFIED RANDOM SAMPLING

Suppose we need to select a sample from the population of a city and we want households with different income levels to be equally represented in the sample. In this case, instead of selecting a simple random sample or a systematic random sample, we may prefer to apply a different technique. First, we divide the whole population into different groups based on their income levels. For example, we may form three groups of low-, medium-, and high-income households. We will now have three *subpopulations,* which are usually called **strata.** We then select one sample from each subpopulation or stratum. The collection of all three samples selected from three strata gives us the required sample, called the **stratified random sample.** Usually, the sizes of the samples selected from different strata are proportionate to the sizes of the subpopulations in these strata. Note that the elements of each strata are identical with regard to the possession of a characteristic.

STRATIFIED RANDOM SAMPLING

In a stratified random sample, we first divide the population into subpopulations, which are called strata. Then, one sample is taken from each of the stratum. The collection of all samples from all strata gives the stratified random sample.

Thus, whenever we observe that a population differs widely in the possession of a characteristic, we may prefer to divide it into different strata and then select one sample from each stratum. We can divide the population on the basis of any characteristic such as income, expenditure, sex, education, race, employment, or family size.

4. CLUSTER SAMPLING

Sometimes the target population is scattered over a much wider geographical area. Consequently, if a simple random sample is taken, it may be costly to contact each member of the sample. In such a case, we divide the population into different geographical groups or clusters and take a random sample of certain clusters from all clusters. We then take a random sample of certain elements from each selected cluster. For example, suppose we are to conduct a survey of households in the state of New York. First, we divide the whole state of New York into, say, 40 regions, which will be called **clusters** or *primary units*. We make sure that all clusters are similar and, hence, representative of the population. We then select at random, say, 5 clusters from 40. Next, we randomly select certain households from each of these 5 clusters and conduct a survey of these selected households. This is called **cluster sampling**. Note that all clusters must be representative of the population.

CLUSTER SAMPLING

In cluster sampling, the whole population is first divided into (geographical) groups called clusters. Each cluster is representative of the population. Then a random sample of clusters is selected. Finally, a random sample of elements from each of the selected clusters is selected.

EXERCISES

A.1 Briefly describe the various sources of data.

A.2 Explain the difference between a sample survey and a census. Why is a sample survey usually preferred instead of a census?

A.3 What is a representative sample? Give one example of a nonrepresentative sample.

A.4 Explain briefly the meaning of sampling and nonsampling errors. Which of them occurs only in a sample survey and which one occurs both in a sample survey and a census?

A.5 Explain briefly the four sampling techniques discussed in this appendix.

A.6 Under what sampling technique do all elements of a population have the same chance of being selected in a sample?

GLOSSARY

Census A survey conducted by including every element of the population.

Cluster A subgroup (usually geographical) of the population that is representative of the population.

Cluster sampling The sampling technique under which the population is divided into clusters and a sample is chosen from one or a few clusters.

Convenience sample A sample that includes the most accessible members of the population.

Experiment Method of collecting data by controlling some or all factors.

Judgment sample A sample that includes the elements of the population selected based on the judgment and prior knowledge of an expert.

Nonsampling errors The errors that occur in the collection, recording, and tabulation of data.

Random sample A sample that assigns some chance to each member of the population for being selected in the sample.

Population The collection of all subjects of interest.

Representative sample A sample that contains the characteristics of the population as closely as possible.

Sample A portion of the population of interest.

Sampling error The difference between the result obtained from a sample survey and the result that would be obtained from the population census.

Simple random sample A sample chosen in such a way that each element of the population has the same probability of being included in the sample.

Stratified random sampling The sampling technique under which the population is divided into different strata and a sample is chosen from each stratum.

Stratum A subgroup of the population whose members are identical with regard to the possession of a characteristic.

Survey Collecting data from the elements of a population or sample.

Systematic random sampling Sampling method used to choose a sample by selecting every kth unit from the list.

USING MINITAB

We can use MINITAB to construct random number tables. For example, to construct a table of 75 three-digit random numbers, we will use the following MINITAB commands.

```
MTB  > RANDOM 75 OBSERVATIONS IN C1;
SUBC > INTEGERS BETWEEN 100 AND 999.
MTB  > PRINT C1
```

The first command instructs MINITAB to generate 75 random numbers and put them in column C1. Note the semicolon at the end of the first MINITAB command. The subcommand tells MINITAB to select integers that are between 100 and 999. Because these selected numbers will be between 100 and 999, they will automatically be three-digit numbers. Note that 100 is the smallest three-digit number and 999 is the largest three-digit number. Also, note the period at the end of the subcommand. The last MINITAB command will print the random numbers of column C1.

To generate two-digit random numbers, replace "100 AND 999" by "10 AND 99" in the subcommand, and to generate four-digit random numbers, replace "100 AND 999" by "1000 AND 9999" in the subcommand.

Illustration MA-1 Let us generate 16 three-digit random numbers. The following are the MINITAB commands and the MINITAB printout.

```
MTB  > NOTE: GENERATING 16 THREE-DIGIT RANDOM NUMBERS
MTB  > RANDOM 16 OBSERVATIONS IN C1;
SUBC > INTEGERS BETWEEN 100 AND 999.
MTB  > PRINT C1

C1
    465   838   628   524   115   176   699   322   967   539
    169   639   983   582   140   630
```

MINITAB can also be used to take a sample from a population. To do so, first enter the population data in MINITAB. Then use the MINITAB SAMPLE command to take a sample of any required size. Suppose we only have the population data on one variable. First, we enter these data in column C1 using the SET command. Then, we use the following MINITAB command to take a sample of 9 elements from the data of column C1. The sample data will be recorded in column C2. MINITAB command PRINT C2 will print the sample data.

```
MTB > SAMPLE 9 FROM C1 PUT IN C2
MTB > PRINT C2
```

Now suppose the population data are on four variables and we want to take a sample of size 12 from these data. First, we enter the population data in columns C1 to C4. Then, we use the following MINITAB command to take a sample of size 12 from the data of columns C1–C4. The sample data will be recorded in columns C5–C8. MINITAB command PRINT C5–C8 will print the sample data.

```
MTB > SAMPLE 12 FROM C1–C4 PUT IN C5–C8
MTB > PRINT C5-C8
```

The MINITAB section at the end of Chapter 1 illustrates how we can select a sample from a population by using MINITAB.

COMPUTER ASSIGNMENTS

MA.1 Using MINITAB, construct a table of 25 two-digit random numbers.
MA.2 Using MINITAB, construct a table of 60 four-digit random numbers.
MA.3 Using MINITAB, take a sample of 15 observations from Data Set III (NBA data) of Appendix B.

APPENDIX B

DATA SETS†

DATA SET I CITY DATA

DATA SET II DATA ON STATES

DATA SET III NBA DATA

DATA SET IV SAMPLE OF 500 OBSERVATIONS FROM MANCHESTER ROAD RACE DATA

†These data sets are available on a floppy disk in MINITAB and ASCII format. To obtain this disk, contact either John Wiley's College Division or your area representative of John Wiley. The disk contains the following files:

1. CITYDATA (This file contains Data Set I.)
2. STATDATA (This file contains Data Set II.)
3. NBA (This file contains Data Set III.)
4. RRACESAM (This file contains Data Set IV.)
5. ROADRACE (This file contains the population data for Data Set IV.)

DATA SET I CITY DATA†

Data on prices of selected products for selected cities across the country

Explanation of Columns

C1 Name of the city
C2 Price of half-gallon carton of whole milk
C3 Price of one dozen, Grade A, large eggs
C4 Price of 2-liter Coca Cola bottle, excluding any deposit
C5 Monthly rent of an unfurnished two-bedroom apartment (excluding all utilities except water), 1-½ or 2 baths, approximately 950 square feet
C6 Purchase price of 1800 square feet living area new house, on 8000 square feet lot, in urban area with all utilities
C7 Monthly telephone charges for a private residential line (customer owns instruments)
C8 Price of one gallon regular unleaded gasoline, national brand, including all taxes; cash price at self-service pump if available
C9 Price for woman's shampoo, trim, and blow dry
C10 Price of dry cleaning, man's two-piece suit
C11 Bowling price per line (game), evening rate
C12 Price of Paul Masson Chablis, 1.5-liter bottle

C1	C2	C3	C4	C5	C6	C7	C8	C9	C10	C11	C12
Alabama											
1. Dothan	1.40	1.01	1.55	284	93,750	18.88	1.031	15.70	4.67	1.45	8.18
2. Florence	1.52	1.00	1.41	312	83,500	24.86	1.039	15.30	4.51	1.75	6.78
3. Huntsville	1.35	1.01	1.37	422	103,500	25.81	1.091	15.29	5.20	1.72	6.62
4. Mobile	1.42	0.92	1.25	336	87,640	24.67	0.999	17.00	4.92	2.12	6.50
5. Montgomery	1.60	1.08	1.03	389	105,000	22.21	1.075	23.10	4.76	1.65	6.89
Alaska											
6. Anchorage	1.92	1.53	2.29	639	125,950	12.62	1.108	23.40	8.01	2.15	5.99
7. Fairbanks	2.05	1.65	2.43	580	100,000	14.42	1.265	20.50	8.99	2.70	5.59
8. Juneau	1.84	1.19	1.92	675	129,000	13.83	1.449	20.10	7.45	2.00	6.90
Arizona											
9. Flagstaff	1.24	1.12	1.59	540	103,600	14.87	1.143	16.22	6.45	1.85	4.00
10. Lake Havasu	1.31	1.13	1.55	462	109,667	21.80	1.128	18.00	6.00	1.75	4.22
11. Phoenix	1.26	1.04	1.17	512	96,225	15.65	1.061	18.81	7.02	2.20	4.41
12. Scottsdale	1.23	1.10	1.18	554	130,790	15.84	1.041	22.50	6.26	2.47	4.51
13. Tempe	1.19	1.13	1.19	489	101,700	17.06	1.063	14.79	6.76	2.28	4.75
14. Tucson	1.23	1.10	1.11	449	106,200	16.13	1.002	20.40	5.83	1.86	4.54
Arkansas											
15. Fayetteville	1.50	1.00	1.13	340	86,380	25.83	0.967	13.30	5.13	1.70	4.92
16. Fort Smith	1.39	1.24	1.19	348	77,810	21.22	0.959	18.60	4.60	1.85	5.14
17. Hot Springs	1.54	0.91	1.25	343	116,000	19.81	0.995	14.80	4.48	2.08	5.06
18. Jonesboro	1.46	1.09	1.06	383	85,667	20.42	0.952	13.33	5.65	1.75	5.40
19. Little Rock	1.55	1.08	1.18	400	91,720	22.79	0.943	25.20	5.44	2.00	5.10

†Data Set I is excerpted from *Cost of Living Index*, 23(2), Second Quarter 1990.

C1	C2	C3	C4	C5	C6	C7	C8	C9	C10	C11	C12
California											
20. Bakersfield	1.23	1.76	1.10	525	121,000	13.05	0.984	26.80	7.70	1.85	3.85
21. Blythe	1.31	1.32	1.04	474	120,750	19.65	1.173	20.60	5.50	1.50	3.52
22. Fresno	1.26	1.51	1.56	504	117,000	13.58	0.986	26.17	5.99	1.71	3.96
23. Los Angeles County	1.28	1.69	1.00	772	211,912	17.34	0.956	18.80	5.54	2.12	4.51
24. Orange County	1.28	1.69	1.00	710	242,985	16.75	0.966	27.00	5.55	2.50	4.57
25. Palm Springs	1.30	1.75	1.12	630	156,750	16.37	1.083	34.00	7.35	2.45	4.13
26. Riverside City	1.16	1.79	1.01	516	166,112	13.23	1.047	22.00	6.09	2.26	3.45
27. San Diego	1.32	1.71	1.14	899	232,800	14.26	1.106	23.48	5.98	2.22	3.83
28. Ventura County	1.28	1.69	1.00	675	245,114	16.70	1.015	19.30	6.24	2.25	4.51
29. Visalia	1.20	1.39	1.19	453	136,500	13.35	1.014	15.80	5.61	1.80	3.99
Colorado											
30. Boulder	1.44	0.97	1.15	501	120,680	17.49	0.999	15.00	7.26	1.75	5.37
31. Colorado Springs	1.40	1.06	1.36	362	92,400	16.08	0.989	16.49	4.63	1.59	4.40
32. Denver	1.38	0.94	1.12	471	105,186	17.24	1.000	20.18	5.76	1.88	4.69
33. Fort Collins	1.39	0.95	1.17	440	112,876	16.92	0.975	12.20	5.62	1.95	4.49
34. Gunnison	1.48	1.24	1.59	333	89,000	13.55	1.129	10.67	6.30	1.50	5.44
35. Longmont	1.39	0.98	1.11	410	104,900	19.40	0.969	15.30	6.40	1.77	4.59
Connecticut											
36. Meriden	1.70	1.13	1.18	565	212,500	16.43	1.049	18.60	6.62	2.50	5.99
37. New London	1.52	1.23	1.41	708	188,980	16.14	1.097	20.60	6.87	2.08	6.97
Delaware											
38. Dover	1.21	1.23	1.06	478	124,167	15.97	0.969	15.19	6.48	2.00	5.24
39. Wilmington	1.24	1.27	1.17	542	137,958	16.48	0.963	17.35	6.06	2.27	4.96
Florida											
40. Jacksonville	1.51	1.01	1.09	407	95,880	20.05	0.991	19.50	5.69	2.01	4.73
41. Miami/Dade County	1.55	1.12	1.03	617	122,376	18.00	1.008	27.60	6.05	2.31	4.85
42. Orlando	1.75	1.02	1.21	511	107,900	17.48	1.054	18.10	5.80	2.21	5.05
43. West Palm Beach	1.55	1.15	1.12	594	141,820	15.87	0.999	20.80	6.20	2.25	5.31
Georgia											
44. Americus	1.64	1.03	1.26	365	88,200	17.71	0.995	14.50	4.00	2.00	6.76
45. Athens	1.54	0.97	1.67	455	94,480	19.76	1.010	19.00	5.17	2.50	6.31
46. Atlanta	1.82	0.83	1.21	545	97,300	20.94	0.947	13.20	5.60	2.34	6.18
47. Augusta	1.46	0.95	1.07	360	94,600	18.95	0.917	22.75	5.01	2.12	6.77
48. Savannah	1.43	0.89	1.07	375	85,600	19.76	0.983	17.20	5.34	2.45	5.13
Hawaii											
49. Hilo	1.80	1.54	1.84	678	170,167	17.90	1.352	20.36	7.73	1.40	5.12
Idaho											
50. Boise	1.36	1.06	0.91	583	103,750	16.74	0.997	18.20	5.33	1.60	4.65
Illinois											
51. Champaign-Urbana	1.39	0.83	1.21	418	118,580	18.14	1.169	17.49	5.83	1.92	5.43
52. Decatur	1.25	0.85	1.40	399	84,500	18.10	1.079	14.80	5.43	1.56	4.25
53. Peoria	1.53	1.00	1.25	460	125,496	16.76	1.169	24.90	5.99	1.61	5.05
54. Quincy	1.30	0.92	1.40	384	103,500	17.04	1.064	15.50	5.68	1.00	4.00
55. Rockford	1.36	0.92	1.31	482	106,400	19.15	1.095	18.20	6.27	1.79	3.53
56. Springfield	1.42	0.87	1.19	410	107,533	17.26	1.136	19.10	5.47	1.65	4.21
Indiana											
57. Anderson	1.43	1.13	1.17	372	100,440	20.90	1.035	16.60	4.75	1.24	5.17
58. Bloomington	1.59	0.96	1.19	486	94,780	23.24	1.042	15.00	5.29	1.65	5.00
59. Evansville	1.50	0.84	1.49	378	92,200	22.77	1.052	14.38	5.63	1.64	5.19
60. Fort Wayne	1.32	0.65	1.04	440	93,067	24.09	1.051	19.40	5.83	1.82	4.93

C1	C2	C3	C4	C5	C6	C7	C8	C9	C10	C11	C12
61. Indianapolis	1.49	0.93	1.08	479	100,900	21.49	1.048	15.55	5.78	1.74	4.38
62. Muncie	1.35	1.03	1.27	475	111,828	19.06	1.037	15.40	5.52	1.48	6.16
63. South Bend	1.27	1.00	1.46	472	92,260	20.32	0.991	20.42	5.26	1.62	4.43
Iowa											
64. Ames	1.14	0.81	1.19	450	110,000	21.44	1.046	13.19	5.88	1.53	5.20
65. Cedar Rapids	1.16	0.92	1.30	402	106,705	22.09	1.107	16.34	6.25	1.59	4.90
66. Dubuque	1.17	0.83	1.18	550	108,800	23.25	1.061	13.89	5.35	1.33	4.89
67. Fort Dodge	1.11	0.96	1.56	425	95,400	12.11	1.059	16.00	5.48	1.62	5.24
68. Mason City	1.17	0.88	1.21	360	106,167	19.60	1.092	14.75	5.25	1.53	5.16
69. Sioux City	1.20	0.79	1.33	532	78,938	20.55	1.137	14.52	5.34	1.63	5.28
Kansas											
70. Garden City	1.33	0.89	1.29	378	76,675	15.98	1.097	15.39	4.92	1.85	5.24
71. Hays	1.37	0.82	1.09	312	83,960	15.87	0.995	12.67	5.82	1.65	4.74
72. Lawrence	1.36	1.10	1.28	453	97,850	16.97	0.987	16.00	5.65	1.52	4.99
73. Salina	1.29	0.89	1.18	309	82,625	16.76	0.983	12.60	5.91	1.60	4.99
74. Wichita	1.46	0.93	1.17	362	97,830	17.14	1.059	15.08	5.99	1.80	4.20
Kentucky											
75. Bowling Green	1.30	1.05	1.35	337	82,667	17.92	0.956	12.00	4.57	1.98	6.61
76. Lexington	1.50	0.98	1.07	504	105,984	26.10	0.983	20.25	5.14	1.71	5.32
77. Louisville	1.61	0.77	1.15	457	87,750	25.31	0.995	14.33	5.98	1.79	5.12
78. Owensboro	1.57	0.91	1.39	330	98,667	27.94	0.999	15.50	4.81	1.70	5.09
Louisiana											
79. Baton Rouge	1.55	1.02	1.25	326	87,800	21.54	0.985	17.10	6.07	2.24	3.81
80. Lake Charles	1.79	1.00	1.73	405	87,985	20.31	1.051	17.60	5.08	1.80	4.31
81. Monroe	1.52	1.04	1.05	381	84,917	21.50	1.090	14.75	6.03	2.00	4.63
82. New Orleans	1.48	1.19	0.97	339	89,600	22.37	1.055	22.60	5.03	1.60	4.10
Maryland											
83. Baltimore	1.39	1.20	1.43	524	123,700	23.19	1.047	16.20	6.26	2.38	5.65
84. Cumberland	1.07	1.07	1.24	357	96,900	20.62	1.117	15.10	4.99	1.98	5.06
85. Hagerstown	1.02	1.03	1.35	463	130,050	20.57	1.070	15.62	5.04	1.75	5.80
Massachusetts											
86. Springfield	1.37	1.48	1.32	602	170,470	14.09	0.970	18.71	7.46	2.01	4.93
87. Worcester	1.34	1.26	1.10	647	172,343	13.85	1.025	18.00	6.65	2.30	5.25
Michigan											
88. Ann Arbor	1.32	1.08	1.37	693	160,300	17.49	1.073	20.50	6.66	1.79	5.15
89. Grand Rapids	1.50	1.02	1.27	490	129,000	17.97	1.067	19.55	5.98	1.79	5.23
90. Kalamazoo	1.47	0.91	1.43	571	115,179	16.79	1.063	18.20	5.71	1.78	4.74
91. Marquette	1.41	0.90	1.34	453	102,659	15.33	1.092	16.00	5.35	1.48	5.10
Minnesota											
92. Minneapolis	1.24	0.82	1.27	542	113,660	19.24	1.093	11.80	6.18	1.41	4.19
93. Moorhead	1.24	0.95	1.39	440	97,108	20.27	1.129	13.67	5.65	1.46	4.48
94. Rochester	1.37	0.93	1.35	532	95,000	23.89	1.085	15.97	6.44	1.44	5.41
95. St. Cloud	1.28	0.74	1.31	418	82,580	22.13	1.101	14.90	5.88	1.59	4.29
Mississippi											
96. Laurel	1.59	1.07	1.09	298	77,200	27.25	1.079	17.50	4.90	1.75	5.02
Missouri											
97. Columbia	1.39	1.14	1.17	390	90,560	12.37	0.979	20.15	4.98	1.65	4.06
98. Joplin	1.51	1.06	1.21	356	86,143	16.73	1.013	13.05	5.21	1.59	4.29
99. Kansas City	1.24	1.03	1.21	470	98,000	21.00	1.011	14.35	5.52	1.81	4.25
100. Kirksville	1.23	1.01	1.24	413	90,200	17.96	0.999	10.17	4.78	1.60	4.50

C1	C2	C3	C4	C5	C6	C7	C8	C9	C10	C11	C12
Nebraska											
101. Grand Island	1.24	0.91	1.35	338	77,675	20.83	1.123	15.60	5.75	1.65	5.02
102. Hastings	1.28	0.92	1.34	339	77,723	16.39	1.119	16.39	6.32	1.38	4.60
103. Lincoln	1.30	0.95	1.14	429	90,200	16.46	1.027	16.80	6.04	1.74	4.04
104. Omaha	1.21	0.89	1.27	437	86,110	22.44	1.066	15.00	6.05	1.55	4.03
Nevada											
105. Carson City	1.16	1.06	1.30	478	135,345	14.37	1.041	18.40	7.06	1.70	3.61
106. Las Vegas	1.43	1.25	1.11	513	122,398	10.98	1.047	23.20	6.80	1.35	3.81
107. Reno-Sparks	1.13	1.01	1.27	501	122,917	13.50	1.036	13.69	5.81	1.48	4.31
New Hampshire											
108. Manchester	1.21	1.09	1.15	642	158,250	20.53	1.015	16.80	7.29	2.50	4.47
New Mexico											
109. Albuquerque	1.51	1.08	1.35	480	111,140	20.49	1.011	17.47	5.47	1.68	4.39
110. Farmington	1.46	1.01	1.16	390	101,500	22.73	1.075	14.75	4.78	1.25	4.65
111. Los Alamos	1.68	1.26	1.72	478	146,633	21.11	1.149	27.20	5.50	2.20	4.81
112. Roswell	1.48	1.16	1.39	382	98,440	23.27	1.103	16.00	4.93	1.75	5.16
113. Santa Fe	1.55	1.05	1.49	496	146,125	20.62	1.096	19.20	5.40	1.85	4.69
New York											
114. Albany	1.19	1.11	1.46	523	107,520	27.30	0.947	16.60	5.29	1.90	5.91
115. Buffalo	1.25	1.12	1.35	454	127,184	30.69	0.971	17.83	5.70	1.59	6.19
116. Elmira	1.13	1.11	1.87	398	126,500	23.58	0.972	14.81	5.77	1.57	4.89
117. Glens Falls	1.26	1.11	1.40	544	103,060	20.94	0.951	17.20	5.64	1.79	5.79
118. Nassau-Suffolk	1.48	1.28	1.15	905	235,000	34.40	1.157	20.70	5.93	2.48	6.65
119. Syracuse	1.33	1.25	1.62	411	109,950	26.99	0.931	11.60	5.59	1.60	5.27
North Carolina											
120. Burlington	1.47	1.11	1.23	372	113,650	16.64	1.004	16.20	5.51	2.00	3.79
121. Chapel Hill	1.53	1.12	1.29	472	128,700	16.66	1.076	19.80	6.05	2.25	5.01
122. Charlotte	1.47	1.04	1.25	426	112,800	17.98	1.015	20.60	5.52	2.10	4.03
123. Durham	1.49	1.11	1.25	454	111,390	18.01	1.059	22.60	5.43	2.25	5.49
124. Greenville	1.54	1.11	1.11	377	96,500	17.15	1.033	17.70	6.10	2.00	5.06
125. Hickory	1.48	1.05	1.05	448	117,475	21.07	1.055	17.50	5.23	1.90	4.20
126. Raleigh	1.43	1.00	1.25	471	113,900	17.98	1.014	19.60	6.20	1.96	5.49
127. Wilmington	1.39	1.01	1.17	419	115,880	16.63	1.055	16.80	5.00	2.18	5.09
128. Winston-Salem	1.46	1.05	1.16	407	118,467	19.19	1.059	21.33	5.85	2.42	3.69
North Dakota											
129. Fargo	1.24	1.04	1.27	422	104,900	18.72	1.129	14.00	5.31	1.58	5.74
Ohio											
130. Akron	1.17	0.90	1.09	449	86,700	22.45	1.027	12.40	4.65	1.68	5.15
131. Canton	1.19	0.98	1.05	383	86,333	21.86	0.985	13.30	5.21	1.33	4.99
132. Cincinnati	1.35	1.00	1.07	522	115,135	20.91	0.995	14.00	5.56	1.85	4.99
133. Cleveland	1.28	0.77	1.31	604	96,900	21.29	1.003	27.40	6.18	1.56	4.99
134. Columbus	1.34	1.09	1.46	485	112,420	21.29	0.997	20.00	6.50	1.90	4.99
135. Mt. Vernon	1.30	1.02	1.26	342	87,500	20.62	0.979	12.33	5.17	1.52	4.99
136. Toledo	1.32	0.92	1.17	473	131,600	20.52	0.973	22.30	5.80	1.81	5.20
137. Youngstown	1.20	1.00	1.01	349	89,000	21.17	0.999	14.40	5.17	1.58	5.29
Oklahoma											
138. Oklahoma City	1.33	1.07	1.49	390	81,261	20.49	0.997	20.10	5.54	1.91	5.31
139. Tulsa	1.27	1.16	1.29	467	76,996	19.49	0.993	17.80	5.27	1.83	4.89
Oregon											
140. Eugene	1.24	1.01	1.32	503	126,488	19.05	1.007	21.00	7.51	1.58	4.57
141. Klamath Falls	1.40	1.08	1.21	398	87,500	17.70	1.167	17.20	5.90	1.75	4.17
142. Salem	1.26	1.03	1.18	411	99,291	17.82	1.034	17.40	6.34	1.60	5.03

C1	C2	C3	C4	C5	C6	C7	C8	C9	C10	C11	C12
Pennsylvania											
143. Erie	1.15	1.04	1.37	447	118,800	20.58	0.967	13.40	5.73	1.48	5.59
144. Harrisburg	1.18	1.02	1.04	502	96,660	15.91	0.987	14.90	5.80	1.87	5.59
145. Lancaster	1.17	1.01	1.13	470	94,000	15.64	0.969	18.20	6.29	1.96	5.59
146. Mercer County	1.14	0.91	1.20	399	107,700	14.16	1.002	13.50	4.70	1.48	5.59
147. Philadelphia	1.09	1.08	1.43	708	143,000	13.90	1.039	24.80	6.10	1.77	5.59
148. Waynesboro	1.17	0.99	1.39	433	125,000	14.23	1.055	13.04	5.12	1.30	5.59
149. Williamsport	1.19	1.06	1.27	417	124,222	14.62	0.967	12.80	5.89	1.68	5.59
150. York	1.17	0.99	1.20	452	107,400	19.10	1.003	12.79	6.27	1.97	5.59
South Carolina											
151. Columbia	1.47	1.04	1.11	568	107,660	21.99	0.993	16.40	5.31	2.10	5.24
152. Florence	1.45	1.03	1.17	386	101,667	26.34	1.029	13.80	4.12	1.98	4.48
153. Greenville	1.63	1.02	1.09	373	102,000	25.87	0.967	22.00	5.74	2.09	5.21
154. Myrtle Beach	1.56	1.02	1.17	399	114,400	18.16	1.089	17.10	5.43	1.99	4.69
155. Spartanburg	1.51	0.97	1.01	398	85,980	22.41	0.975	19.00	5.62	2.17	5.21
South Dakota											
156. Rapid City	1.45	1.13	1.79	522	95,667	19.90	1.139	15.50	5.29	1.50	5.68
157. Sioux Falls	1.19	0.96	1.13	446	85,760	23.71	1.119	15.20	5.16	1.75	4.95
158. Vermillion	1.25	1.04	1.44	300	82,000	19.31	1.049	12.67	6.25	1.60	4.99
Tennessee											
159. Chattanooga	1.37	1.04	1.13	402	85,660	21.24	1.021	14.40	4.82	2.02	5.54
160. Dyersburg	1.45	0.72	1.39	312	79,856	15.56	1.144	14.60	5.47	1.65	8.32
161. Jackson	1.46	1.05	0.99	439	80,260	21.75	1.039	14.20	4.94	1.75	5.09
162. Knoxville	1.29	1.04	0.91	438	95,369	20.32	1.029	18.80	4.71	1.83	5.25
163. Memphis	1.45	1.04	1.28	380	89,950	20.19	0.965	15.19	5.33	1.76	4.89
164. Morristown	1.45	1.21	1.07	375	84,500	16.36	1.022	13.00	4.87	1.60	5.69
165. Nashville	1.32	0.97	1.36	519	111,450	19.46	0.989	15.00	5.40	1.70	4.96
Texas											
166. Amarillo	1.33	1.09	1.30	416	82,300	15.73	1.041	20.60	5.44	1.85	4.63
167. Austin	1.41	1.07	1.39	423	85,050	16.09	1.011	19.00	5.66	2.09	4.86
168. Corpus Christi	1.42	1.12	1.97	382	87,760	14.47	0.993	19.16	6.36	2.12	4.91
169. Dallas	1.60	1.07	1.52	532	95,404	18.37	0.958	24.20	7.03	2.39	4.45
170. El Paso	1.49	1.07	1.37	402	93,333	16.05	1.021	24.50	5.00	1.67	4.87
171. Houston	1.68	1.13	1.23	443	84,057	17.84	0.979	22.12	5.05	2.22	4.97
172. Midland	1.54	1.06	1.89	391	100,300	15.94	1.059	23.20	6.00	1.93	4.73
173. Odessa	1.52	1.08	1.89	313	94,660	15.82	1.049	17.60	4.85	1.54	4.82
174. San Antonio	1.34	1.11	1.80	389	92,200	15.27	0.955	18.62	5.18	2.16	4.71
175. Texarkana	1.41	1.10	1.22	331	78,300	15.25	1.029	17.60	5.98	1.57	5.54
176. Wichita Falls	1.26	0.84	1.05	383	85,660	14.31	1.001	18.00	4.96	1.75	4.75
Utah											
177. Cedar City	1.51	1.09	1.19	277	73,375	13.54	1.077	12.67	5.78	1.75	4.75
178. St. George	1.54	1.07	1.19	382	86,000	13.65	1.089	16.00	6.62	1.50	4.75
179. Salt Lake City	1.49	1.04	1.11	377	83,780	17.12	0.953	15.00	6.74	1.77	4.79
Vermont											
180. Montpelier-Barre	1.38	1.32	1.46	562	144,800	22.69	1.149	15.50	5.67	1.85	5.22
Virginia											
181. Hampton Roads	1.47	1.08	1.10	455	103,570	22.91	1.027	15.79	6.64	2.23	5.40
182. Lynchburg	1.55	1.11	1.05	385	110,000	17.19	1.006	18.62	6.41	2.05	5.20
183. Roanoke	1.52	1.01	1.13	407	113,411	20.13	1.004	15.00	5.54	1.83	5.72
Washington											
184. Seattle	1.36	1.04	1.37	608	137,183	18.50	0.986	14.38	7.12	2.00	5.51
185. Tacoma	1.29	1.04	1.19	458	103,000	16.57	1.009	20.00	7.01	1.89	4.53
186. Walla Walla	1.44	0.98	1.53	339	92,000	14.46	1.228	17.09	6.84	1.38	4.39

C1	C2	C3	C4	C5	C6	C7	C8	C9	C10	C11	C12
187. Wenatchee	1.36	1.12	1.40	455	87,500	16.78	1.174	17.62	5.83	2.20	3.14
188. Yakima	1.54	1.00	1.38	391	92,700	13.25	1.043	18.40	6.15	1.40	4.86
West Virginia											
189. Charleston	1.50	0.99	1.13	396	111,607	29.43	1.039	16.42	5.50	1.70	5.92
190. Wheeling	1.36	0.98	1.23	440	83,000	29.78	0.999	13.00	5.50	1.63	5.25
Wisconsin											
191. Eau Claire	1.37	0.92	1.37	470	94,120	21.50	1.079	15.20	6.18	1.62	4.81
192. Fond Du Lac	1.42	1.03	1.18	385	99,000	21.50	1.072	13.83	6.96	1.32	3.79
193. Green Bay	1.31	0.94	1.22	427	91,760	21.68	1.079	14.45	5.74	1.27	4.63
194. Janesville	1.34	0.94	1.17	445	103,167	20.90	1.079	13.47	6.08	1.81	4.52
195. Kenosha	1.45	0.88	1.02	496	124,500	25.75	1.123	11.70	5.17	1.35	4.38
196. La Crosse	1.23	0.97	1.11	387	89,900	15.57	1.076	13.40	6.20	1.28	3.63
197. Marshfield	1.46	0.86	1.13	409	92,560	22.46	1.119	12.50	5.28	1.70	5.19
198. Wausau	1.18	0.83	1.29	423	129,900	23.86	1.073	13.90	5.71	1.55	4.53
Wyoming											
199. Casper	1.45	0.97	1.71	311	96,500	20.32	0.959	18.25	6.04	1.50	5.09
200. Gillette	1.59	1.05	1.79	340	86,100	15.45	1.049	15.20	5.33	1.70	5.40

DATA SET II DATA ON STATES†

Information on different variables for 50 states

Explanation of Columns

C1 Name of the state
C2 The average salary of teachers, 1989 (Source: National Education Association)
C3 The current expenditure per pupil, 1988 (Source: National Education Association)
C4 Percentage of the population who received social security, 1988 (Source: U.S. Bureau of the Census)
C5 Percentage of the population who received food stamps, 1988 (Source: U.S. Department of Agriculture, Food and Nutrition Service)
C6 Female labor force participation rate, 1988 (Source: U.S. Bureau of Labor Statistics)
C7 Unemployment rate, 1988 (Source: U.S. Bureau of Labor Statistics)
C8 Per capita income, 1989 (Source: U.S. Bureau of Economic Analysis)
C9 Per capita bank deposits, 1988 (Source: Board of Governors of the Federal Reserve System)
C10 Per capita energy expenditure (in dollars), 1987 (Source: U.S. Energy Information Administration)
C11 Motor vehicle accident deaths per 100,000 population, 1987 (Source: U.S. National Center for Health Statistics)
C12 Birth rate (i.e., registered births per 1000 population), 1987 (Source: U.S. National Center for Health Statistics)
C13 Infant mortality rate (represents deaths of infants under 1 year old per 1000 live births), 1987 (Source: U.S. National Center for Health Statistics)

†Data Set II is reproduced from the *Statistical Abstract of the United States,* 1990, except for column labeled C3, which is reproduced from the *Statistical Abstract of the United States,* 1989.

	C1	C2	C3	C4	C5	C6	C7	C8	C9	C10	C11	C12	C13
1.	Alabama	25,190	2,752	16.7	10.3	50.8	7.2	11,040	6,539	1,794	29.0	14.6	12.2
2.	Alaska	41,754	7,038	5.9	4.8	64.1	9.3	16,357	7,138	2,639	17.1	22.2	10.4
3.	Arizona	28,684	3,265	15.8	7.0	56.4	6.3	13,017	6,718	1,583	27.2	18.7	9.5
4.	Arkansas	21,692	2,410	19.1	9.4	53.3	7.7	10,670	6,845	1,587	27.6	14.5	10.3
5.	California	35,285	3,994	12.5	6.0	57.0	5.3	16,035	7,500	1,363	20.9	18.2	9.0
6.	Colorado	29,558	4,359	12.0	6.1	61.6	6.4	14,110	6,408	1,404	18.7	16.3	9.8
7.	Connecticut	37,343	6,141	15.8	3.3	60.3	3.0	19,096	9,005	1,579	14.4	14.6	8.8
8.	Delaware	31,585	4,994	15.2	4.4	62.4	3.2	14,654	35,942	1,789	23.5	15.3	11.7
9.	Florida	26,974	4,389	20.3	5.1	55.2	5.0	14,338	8,045	1,401	23.3	14.6	10.6
10.	Georgia	28,038	2,939	13.3	7.3	59.7	5.8	12,886	7,057	1,645	26.0	16.5	12.7
11.	Hawaii	30,778	3,894	13.0	6.6	60.0	3.2	14,374	9,984	1,368	12.4	17.2	8.9
12.	Idaho	22,734	2,814	15.1	5.9	57.4	5.8	11,190	6,096	1,521	26.3	15.9	10.4
13.	Illinois	31,145	4,217	14.9	8.6	56.0	6.8	15,150	10,454	1,666	15.8	15.6	11.6
14.	Indiana	29,295	3,616	16.0	5.1	58.6	5.3	12,834	7,890	1,965	20.4	14.2	10.1
15.	Iowa	25,884	3,846	18.3	5.9	61.5	4.5	12,475	9,300	1,659	18.0	13.4	9.1
16.	Kansas	27,360	4,262	16.1	4.7	60.2	4.8	13,235	8,853	1,952	21.6	15.6	9.5
17.	Kentucky	24,932	3,355	16.9	12.3	50.9	7.9	11,081	7,955	1,717	23.7	13.8	9.7
18.	Louisiana	22,470	3,211	14.3	16.1	49.7	10.9	10,890	6,913	2,492	19.5	16.6	11.8
19.	Maine	24,938	4,276	17.3	6.6	57.0	3.8	12,955	5,333	1,636	18.3	14.2	8.3
20.	Maryland	33,700	4,871	12.8	5.3	61.8	4.5	16,397	7,919	1,494	18.0	16.0	11.5
21.	Massachusetts	31,909	5,396	16.1	5.6	60.0	3.3	17,456	12,244	1,486	12.8	14.4	7.2
22.	Michigan	34,419	4,122	15.7	9.4	56.2	7.6	14,094	7,435	1,582	18.8	15.3	10.7
23.	Minnesota	30,660	4,513	15.2	5.5	63.0	4.0	14,037	9,195	1,464	14.2	15.4	8.7
24.	Mississippi	22,579	2,760	16.9	18.6	50.8	8.4	9,612	6,302	1,605	30.9	15.7	13.7
25.	Missouri	25,981	3,566	17.4	7.6	58.4	5.7	13,340	8,993	1,569	21.2	14.7	10.2
26.	Montana	24,414	4,061	16.5	6.6	58.9	6.8	11,264	7,576	1,723	27.6	15.1	10.0
27.	Nebraska	23,845	3,641	16.5	5.7	60.5	3.6	12,773	9,403	1,639	18.6	14.9	8.6
28.	Nevada	28,840	3,829	13.9	3.5	66.1	5.2	14,799	6,177	1,673	24.3	16.6	9.6
29.	New Hampshire	26,702	3,990	14.3	1.8	63.3	2.4	17,049	7,840	1,552	16.5	16.1	7.8
30.	New Jersey	33,037	6,910	15.6	4.6	55.9	3.8	18,615	8,948	1,804	14.3	14.8	9.4
31.	New Mexico	25,139	3,880	13.6	9.3	53.5	7.8	10,752	5,724	1,636	35.1	18.2	8.1
32.	New York	36,654	6,864	15.6	8.3	51.3	4.2	16,036	14,736	1,292	14.0	15.3	10.7
33.	North Carolina	25,646	3,911	15.8	5.9	59.9	3.6	12,259	7,171	1,611	24.6	14.6	11.9
34.	North Dakota	22,249	3,353	16.6	5.4	60.2	4.8	11,388	9,161	2,131	16.8	15.3	8.7
35.	Ohio	29,671	4,019	16.2	9.7	54.5	6.0	13,326	7,293	1,721	16.9	14.6	9.3
36.	Oklahoma	22,000	3,051	16.0	8.1	54.6	6.7	10,875	7,122	1,610	20.5	14.6	9.6
37.	Oregon	29,390	4,574	17.1	7.4	59.0	5.8	12,776	5,528	1,492	23.6	14.2	10.4
38.	Pennsylvania	31,248	5,063	18.3	7.6	51.6	5.1	14,072	9,415	1,546	17.3	13.6	10.4
39.	Rhode Island	34,283	5,456	17.9	5.6	60.5	3.1	14,636	9,577	1,413	14.2	14.2	8.4
40.	South Carolina	25,498	3,075	14.9	7.3	56.5	4.5	11,102	4,421	1,681	31.7	15.4	12.7
41.	South Dakota	20,525	3,159	17.6	6.9	60.2	3.9	11,611	12,471	1,548	22.0	16.2	9.9
42.	Tennessee	25,619	3,189	16.3	10.0	53.9	5.8	12,212	7,317	1,799	26.7	14.0	11.7
43.	Texas	26,513	3,462	12.5	9.0	59.1	7.3	12,777	8,376	2,108	19.8	18.0	9.1
44.	Utah	22,828	2,658	10.6	5.3	59.4	4.9	10,564	5,124	1,419	18.3	21.0	8.8
45.	Vermont	26,861	4,949	15.2	5.7	62.8	2.8	12,941	8,657	1,554	20.5	14.9	8.5
46.	Virginia	29,056	4,145	13.3	5.4	60.9	3.9	15,050	7,997	1,586	18.2	15.3	10.2
47.	Washington	29,176	4,083	14.7	6.5	59.5	6.2	14,508	5,893	1,449	19.0	15.5	9.7
48.	West Virginia	21,904	3,895	19.5	13.5	41.1	9.9	10,306	7,389	1,646	26.5	11.8	9.8
49.	Wisconsin	30,779	4,991	17.0	6.1	61.8	4.3	13,296	7,334	1,472	17.4	14.8	8.6
50.	Wyoming	27,685	6,885	12.3	5.2	59.8	6.3	11,667	7,820	2,644	21.8	15.4	9.2

DATA SET III NBA DATA†

The heights and weights of NBA players who were on the rosters of National Basketball Association teams at the beginning of 1990–91 season

Explanation of Columns

C1 Name of the player
C2 Height of a player (in inches)
C3 Weight of a player (in pounds)

	C1	C2	C3
1.	Abdelnaby, Alaa	82	240
2.	Acres, Mark	83	225
3.	Adams, Michael	70	165
4.	Aguirre, Mark	78	232
5.	Ainge, Danny	77	185
6.	Alarie, Mark	80	225
7.	Alford, Steve	74	182
8.	Anderson, Greg	81	247
9.	Anderson, Nick	78	205
10.	Anderson, Ron	79	215
11.	Anderson, Willie	80	185
12.	Ansley, Michael	79	225
13.	Armstrong, B. J.	74	170
14.	Askins, Keith	80	197
15.	Babic, Milos	84	240
16.	Bagley, John	72	205
17.	Bailey, Thurl	83	232
18.	Bannister, Ken	81	260
19.	Barkley, Charles	78	250
20.	Barros, Dana	71	163
21.	Battle, John	74	175
22.	Battle, Kenny	78	211
23.	Bedford, William	85	235
24.	Benjamin, Benoit	84	260
25.	Bennett, Winston	79	210
26.	Bird, Larry	81	220
27.	Blackman, Rolando	78	206
28.	Blanks, Lance	76	195
29.	Blanton, Ricky	79	215
30.	Blaylock, Mookie	73	185
31.	Bol, Manute	91	225
32.	Bonner, Anthony	80	225
33.	Bogues, Tyrone	63	140
34.	Bowie, Sam	85	240
35.	Breuer, Randy	87	258
36.	Brickowski, Frank	82	240
37.	Brooks, Scott	71	165
38.	Brown, Chucky	80	214
39.	Brown, Dee	73	161
40.	Brown, Mike	81	260
41.	Brown, Tony	78	210
42.	Brundy, Stanley	79	215

†Data Set III, representing the 1990–91 National Basketball Association opening day team rosters is excerpted from the *USA Today*, November 2, 1990. The list includes the players on injury lists also.

	C1	C2	C3
43.	Bryant, Mark	81	245
44.	Buechler, Jud	78	220
45.	Bullard, Matt	82	225
46.	Burton, Willie	80	219
47.	Butler, Greg	83	240
48.	Cage, Michael	81	245
49.	Caldwell, Adrian	80	266
50.	Calloway, Rick	78	180
51.	Campbell, Elden	83	215
52.	Campbell, Tony	79	215
53.	Carr, Antoine	81	235
54.	Cartwright, Bill	85	245
55.	Catledge, Terry	80	230
56.	Causwell, Duane	84	240
57.	Ceballos, Cedric	78	210
58.	Chambers, Tom	82	230
59.	Chapman, Rex	76	195
60.	Cheeks, Maurice	73	180
61.	Chievous, Derrick	79	204
62.	Coffey, Richard	78	212
63.	Coleman, Derrick	82	230
64.	Coles, Bimbo	74	182
65.	Colter, Steve	75	165
66.	Conner, Lester	76	185
67.	Cook, Anthony	81	215
68.	Cooper, Wayne	82	220
69.	Corbin, Tyrone	78	222
70.	Corzine, Dave	83	265
71.	Cummings, Terry	81	240
72.	Curry, Dell	77	200
73.	Dailey, Quintin	74	180
74.	Daugherty, Brad	84	263
75.	Davis, Brad	75	183
76.	Davis, Terry	82	236
77.	Davis, Walter	78	200
78.	Dawkins, Johnny	74	170
79.	Divac, Vlade	85	248
80.	Donaldson, James	86	280
81.	Douglas, Sherman	73	180
82.	Dreiling, Greg	85	250
83.	Drew, Larry	74	190
84.	Drexler, Clyde	79	222
85.	Duckworth, Kevin	84	270
86.	Dudley, Chris	83	240
87.	Dumars, Joe	75	195
88.	Dunn, T. R.	76	192
89.	Eaton, Mark	88	290
90.	Edwards, Blue	77	200
91.	Edwards, James	85	252
92.	Edwards, Kevin	75	197
93.	Ehlo, Craig	79	205
94.	Elliott, Sean	80	205
95.	Ellis, Dale	79	215
96.	Ellison, Pervis	82	225
97.	English, Alex	79	190
98.	English, A. J.	75	180
99.	Ewing, Patrick	84	240
100.	Feiti, Dave	84	250

	C1	C2	C3
101.	Ferry, Danny	82	230
102.	Fleming, Vern	77	185
103.	Floyd, Sleepy	75	183
104.	Foster, Greg	83	240
105.	Gaines, Corey	76	195
106.	Gamble, Kevin	77	210
107.	Garland, Winston	74	175
108.	Garrick, Tom	74	195
109.	Gattison, Kenny	80	252
110.	George, Tate	77	190
111.	Gervin, Derrick	80	215
112.	Gill, Kendall	77	200
113.	Gilliam, Armon	81	245
114.	Glass, Gerald	78	221
115.	Gminski, Mike	83	260
116.	Grant, Gary	75	196
117.	Grant, Greg	67	145
118.	Grant, Harvey	81	215
119.	Grant, Horace	82	220
120.	Gray, Stuart	84	245
121.	Grayer, Jeff	76	213
122.	Green, A. C.	81	224
123.	Green, Rickey	72	172
124.	Green, Sidney	81	230
125.	Greenwood, David	81	225
126.	Griffith, Darrell	76	195
127.	Haley, Jack	82	240
128.	Hammonds, Tom	81	225
129.	Hansen, Bobby	78	195
130.	Hanzlik, Bill	79	200
131.	Hardaway, Tim	72	175
132.	Harper, Derek	76	200
133.	Harper, Ron	78	198
134.	Hastings, Scott	83	245
135.	Hawkins, Hersey	75	190
136.	Henson, Steve	71	177
137.	Higgins, Rod	79	205
138.	Higgins, Sean	81	195
139.	Hill, Tyrone	81	243
140.	Hinson, Roy	81	215
141.	Hodges, Craig	74	190
142.	Hoppen, Dave	83	240
143.	Hopson, Dennis	77	195
144.	Hornacek, Jeff	76	190
145.	Hughes, Mark	80	235
146.	Humphries, Jay	75	185
147.	Irvin, Byron	77	190
148.	Jackson, Chris	73	170
149.	Jackson, Mark	75	205
150.	Jamerson, Dave	77	192
151.	Jepsen, Les	84	237
152.	Johnson, Avery	71	174
153.	Johnson, Buck	79	206
154.	Johnson, Eddie	79	215
155.	Johnson, Kevin	73	190
156.	Johnson, Magic	81	220
157.	Johnson, Steve	82	240
158.	Johnson, Vinnie	74	200

	C1	C2	C3
159.	Jones, Charles	81	225
160.	Jordan, Michael	78	198
161.	Kemp, Shawn	82	240
162.	Kerr, Steve	75	180
163.	Kersey, Jerome	79	225
164.	Kessler, Alec	83	245
165.	Keys, Randolph	79	195
166.	Kimble, Bo	76	190
167.	King, Bernard	79	205
168.	King, Stacey	83	230
169.	Kite, Greg	83	260
170.	Kleine, Joe	84	271
171.	Knight, Negele	73	182
172.	Kofoed, Bart	76	210
173.	Koncak, Jon	84	260
174.	Kornet, Frank	81	231
175.	Krystkowiak, Larry	82	240
176.	Laimbeer, Bill	83	260
177.	Lane, Jerome	78	232
178.	Lang, Andrew	83	250
179.	Leckner, Eric	83	265
180.	Lee, Kurk	75	190
181.	Lever, Fat	75	175
182.	Levingston, Cliff	80	210
183.	Lewis, Reggie	79	195
184.	Liberty, Marcus	80	205
185.	Lichti, Todd	76	205
186.	Lister, Alton	84	240
187.	Lockhart, Ian	80	240
188.	Lohaus, Brad	83	235
189.	Long, Grant	81	230
190.	Mahorn, Rick	82	255
191.	Majerie, Dan	78	220
192.	Malone, Jeff	76	205
193.	Malone, Karl	81	256
194.	Malone, Moses	82	255
195.	Manning, Danny	82	230
196.	Marciulionis, Sarunas	77	200
197.	Martin, Jeff	77	195
198.	Massenburg, Tony	81	230
199.	Maxwell, Vernon	76	190
200.	Mays, Travis	74	190
201.	McCloud, George	80	215
202.	McCormick, Tim	84	240
203.	McCray, Rodney	80	235
204.	McDaniel, Xavier	79	205
205.	McHale, Kevin	82	225
206.	McKey, Derrick	81	210
207.	McMillan, Nate	77	197
208.	McNamara, Mark	83	235
209.	Meents, Scott	82	235
210.	Miller, Reggie	79	185
211.	Mitchell, Sam	79	210
212.	Moncrief, Sydney	75	181
213.	Morris, Chris	80	210
214.	Morton, John	75	183
215.	Mullin, Chris	79	215
216.	Munk, Chris	81	225
217.	Murphy, Tod	82	220
218.	Mustaf, Jerrod	82	244

	C1	C2	C3
219.	Nance, Larry	82	235
220.	Nealy, Ed	79	240
221.	Newman, Johnny	79	190
222.	Norman, Ken	80	219
223.	Oakley, Charles	81	245
224.	Ogg, Alan	86	235
225.	Olajuwon, Akeem	84	258
226.	Oliver, Brian	76	210
227.	Paddio, Gerald	79	205
228.	Palmer, Walter	85	215
229.	Parish, Robert	85.5	230
230.	Paxson, John	74	185
231.	Payne, Kenny	80	220
232.	Payton, Gary	76	190
233.	Perdue, Will	84	240
234.	Perkins, Sam	81	257
235.	Perry, Tim	81	220
236.	Person, Chuck	80	225
237.	Petersen, Jim	82	235
238.	Petrovic, Drazen	77	195
239.	Pierce, Ricky	77.5	225
240.	Pinckney, Ed	81	215
241.	Pippen, Scottie	79	210
242.	Polynice, Olden	84	244
243.	Popson, Dave	82	230
244.	Porter, Terry	75	195
245.	Pressey, Paul	77	203
246.	Price, Mark	72	178
247.	Pritchard, Kevin	75	185
248.	Quinnett, Brian	80	236
249.	Rambis, Kurt	80	213
250.	Rasmussen, Blair	84	260
251.	Reid, J. R.	81	256
252.	Reynolds, Jerry	80	206
253.	Rice, Glen	80	220
254.	Richardson, Pooh	73	180
255.	Richmond, Mitch	77	215
256.	Rivers, Glenn	76	185
257.	Roberts, Fred	82	245
258.	Robertson, Alvin	76	202
259.	Robinson, Cliff	82	225
260.	Robinson, David	85	235
261.	Robinson, Larry	77	185
262.	Rodman, Dennis	80	210
263.	Rollins, Tree	85	240
264.	Rudd, Delaney	74	195
265.	Salley, John	83	244
266.	Sampson, Ralph	88	235
267.	Sanders, Mike	78	215
268.	Schayes, Dan	83.5	268
269.	Scheffler, Steve	81	250
270.	Schintzius, Dwayne	85	260
271.	Schrempf, Detlef	82	230
272.	Scott, Byron	76	193
273.	Scott, Dennis	80	229
274.	Seikaly, Rony	83.5	252
275.	Shasky, John	83	240
276.	Shaw, Brian	78	190
277.	Sikma, Jack	84	262
278.	Simmons, Lionel	79	210

	C1	C2	C3
279.	Skiles, Scott	73	180
280.	Smith, Charles	73	160
281.	Smith, Charles	82	238
282.	Smith, Kenny	75	170
283.	Smith, Larry	80	251
284.	Smith, Michael	82	225
285.	Smith, Otis	77	210
286.	Smith, Tony	76	195
287.	Smits, Rik	88	265
288.	Smrek, Mike	84	260
289.	Sparrow, Rory	74	175
290.	Spencer, Felton	84	265
291.	Starks, John	77	180
292.	Stockton, John	73	175
293.	Strickland, Rod	75	175
294.	Sundvold, Jon	74	175
295.	Tarpley, Roy	84	250
296.	Teagle, Terry	77	195
297.	Theus, Reggie	79	213
298.	Thomas, Irving	80	225
299.	Thomas, Isiah	73	182
300.	Thomas, Jim	74	195
301.	Thompson, Billy	79	217
302.	Thompson, LaSalle	82	260
303.	Thompson, Mychal	82	235
304.	Thornton, Bob	82	225
305.	Thorpe, Otis	82	246
306.	Threatt, Sedale	74	177
307.	Tisdale, Wayman	81	240
308.	Tolbert, Tom	79	240
309.	Toolson, Andy	78	210
310.	Tripucka, Kelly	78	225
311.	Tucker, Trent	77	190
312.	Turner, Jeff	81	240
313.	Vandeweghe, Kiki	80	220
314.	Vaught, Loy	81	240
315.	Vincent, Sam	74	185
316.	Volkov, Alexander	82	218
317.	Vrankovic, Stojko	86	260
318.	Wagner, Milt	76.5	181
319.	Walker, Darrell	76	180
320.	Walker, Kenny	80	217
321.	Webb, Spud	67	135
322.	Wennington, Bill	84	260
323.	West, Doug	78	200
324.	West, Mark	82	246
325.	White, Randy	80	249
326.	Wiley, Morlon	76	192
327.	Wilkins, Dominique	80	200
328.	Wilkins, Eddie Lee	82.5	220
329.	Wilkins, Gerald	78	195
330.	Williams, Buck	80	225
331.	Williams, Herb	83	242
332.	Williams, Jayson	82	240
333.	Williams, John	83	238
334.	Williams, John	81	245
335.	Williams, Ken	81	205
336.	Williams, Michael	74	175
337.	Williams, Reggie	79	195

	C1	C2	C3
338.	Williams, Scott	82	230
339.	Willis, Kevin	84	235
340.	Wilson, Trevor	80	211
341.	Winchester, Kennard	77	212
342.	Wittman, Randy	78	210
343.	Wolf, Joe	83	230
344.	Wood, David	81	228
345.	Woodson, Mike	77	200
346.	Woolridge, Orlando	81	215
347.	Workman, Haywoode	75	180
348.	Worthy, James	81	225
349.	Wright, Howard	80	245
350.	Young, Danny	76	175

DATA SET IV SAMPLE OF 500 OBSERVATIONS FROM MANCHESTER ROAD RACE DATA

(This data set represents a random sample of 500 observations taken from the data on the Fifty-fourth Road Race held on Thanksgiving day, 22nd of November 1990, in Manchester (Connecticut). The total distance for that race is 4.748 miles. The following data set represents time (in minutes) taken by 500 participants to complete that race. The complete data set, which includes 7288 participants who completed that race, is available on a floppy disk, in MINITAB and ASCII format, from the publisher. The data were published in *The Hartford Courant* of November 26, 1990 and is reproduced here with the permission of *The Hartford Courant*.)

23.18	24.14	24.35	24.40	25.22	26.04	26.14	26.24	26.30	26.36	26.41	26.43	26.47
26.54	27.02	27.04	27.12	27.15	27.34	27.36	27.45	28.07	28.08	28.13	28.13	28.24
28.26	28.40	28.45	28.50	28.56	29.32	29.36	29.47	29.54	29.59	30.00	30.01	30.07
30.09	30.11	30.17	30.20	30.20	30.20	30.21	30.28	30.45	30.45	30.46	30.47	30.49
30.53	30.57	30.59	31.01	31.07	31.12	31.32	31.32	31.34	31.38	31.44	31.44	31.52
31.58	31.58	32.00	32.04	32.05	32.08	32.09	32.31	32.33	32.35	32.37	32.38	32.38
32.39	32.40	32.42	32.46	32.48	32.49	32.51	32.53	32.58	33.03	33.05	33.10	33.24
33.35	33.36	33.36	33.38	33.50	33.50	33.56	34.03	34.05	34.15	34.17	34.18	34.21
34.21	34.22	34.24	34.27	34.27	34.28	34.28	34.30	34.40	34.45	34.45	34.47	34.47
34.51	34.54	35.01	35.04	35.10	35.14	35.14	35.14	35.16	35.16	35.22	35.22	35.23
35.24	35.25	35.29	35.31	35.34	35.35	35.37	35.39	35.42	35.43	35.45	35.47	35.50
35.51	35.52	35.53	35.54	35.54	35.58	35.58	35.58	35.58	35.59	36.05	36.08	36.13
36.17	36.20	36.21	36.24	36.28	36.28	36.30	36.37	36.47	36.49	37.01	37.01	37.02
37.06	37.17	37.19	37.23	37.24	37.30	37.37	37.43	37.43	37.55	37.56	37.59	38.01
38.08	38.10	38.15	38.17	38.23	38.25	38.27	38.28	38.32	38.32	38.34	38.35	38.36
38.40	38.41	38.44	38.46	38.47	38.47	38.49	38.50	38.52	38.53	38.53	38.55	38.56
39.04	39.04	39.11	39.14	39.15	39.15	39.25	39.31	39.32	39.34	39.35	39.35	39.38
39.42	39.44	39.46	39.49	39.53	39.55	39.55	40.01	40.05	40.08	40.10	40.11	40.13
40.21	40.21	40.21	40.23	40.25	40.31	40.32	40.34	40.36	40.38	40.43	40.43	40.43
40.45	40.45	40.46	40.48	40.49	40.59	41.00	41.01	41.05	41.08	41.21	41.26	41.37
41.37	41.38	41.43	41.45	41.46	41.53	41.55	41.57	42.00	42.00	42.00	42.01	42.06
42.09	42.09	42.13	42.16	42.17	42.17	42.18	42.21	42.21	42.28	42.33	42.34	42.35
42.35	42.36	42.40	42.46	42.46	42.51	42.51	42.52	42.54	42.54	43.01	43.02	43.04
43.07	43.07	43.08	43.10	43.14	43.15	43.18	43.19	43.23	43.25	43.26	43.27	43.28
43.35	43.43	43.46	43.54	43.58	44.03	44.14	44.15	44.17	44.19	44.21	44.21	44.22
44.23	44.24	44.29	44.35	44.35	44.39	44.40	44.42	44.44	44.44	44.48	44.48	44.52
44.57	44.58	45.00	45.01	45.06	45.07	45.07	45.09	45.10	45.12	45.17	45.21	45.22
45.25	45.28	45.33	45.33	45.39	45.40	45.41	45.41	45.42	45.44	45.45	45.50	45.52
45.53	45.58	46.00	46.00	46.01	46.05	46.11	46.12	46.12	46.12	46.15	46.15	46.25
46.28	46.28	46.30	46.33	46.35	46.51	46.51	46.53	46.54	47.06	47.08	47.13	47.28
47.28	47.29	47.33	47.39	47.49	47.50	47.55	48.09	48.10	48.14	48.19	48.19	48.26
48.30	48.36	48.40	48.42	48.43	48.45	48.49	48.49	48.55	49.00	49.04	49.06	49.10
49.16	49.19	49.20	49.31	49.34	49.34	49.39	49.42	49.43	49.49	49.50	50.00	50.05
50.05	50.15	50.30	50.32	50.32	50.39	50.44	50.45	50.50	50.58	51.15	51.19	51.24
51.34	51.45	51.59	52.00	52.11	52.13	52.19	52.20	52.21	52.30	52.31	52.53	52.58
53.01	53.13	53.36	53.47	53.58	54.56	55.40	55.52	57.01	57.18	58.10	58.13	59.03
59.11	59.22	59.56	60.17	60.42	60.45	62.00	62.03	62.14	63.03	63.21	63.39	64.11
64.35	65.03	65.08	65.24	65.56	67.43	67.46	69.28	69.45	69.47	69.48	71.47	73.37
75.35	76.07	76.18	76.20	76.35	87.55							

APPENDIX C STATISTICAL TABLES†

TABLE I	RANDOM NUMBERS
TABLE II	FACTORIALS
TABLE III	VALUES OF $\binom{n}{x}$ (COMBINATION)
TABLE IV	TABLE OF BINOMIAL PROBABILITIES
TABLE V	VALUES OF $e^{-\lambda}$
TABLE VI	TABLE OF POISSON PROBABILITIES
TABLE VII	STANDARD NORMAL DISTRIBUTION TABLE
TABLE VIII	THE t DISTRIBUTION TABLE
TABLE IX	CHI-SQUARE DISTRIBUTION TABLE
TABLE X	THE F DISTRIBUTION TABLE

†All tables were prepared by the author. Tables I, IV, VI, and VII were constructed using MINITAB, and Tables VIII through X were made using SAS statistical software. Tables II, III, and V were prepared by using a calculator.

TABLE I RANDOM NUMBERS

57728	16308	27337	53884	60742	61693	39887	81779	36354
63962	45765	75060	46767	28844	32354	91463	25057	91907
51041	22252	38447	71567	95103	11124	34960	35710	91098
84048	53578	67379	42605	59122	39415	82869	86971	64817
17736	34458	67227	97041	77846	20338	52372	34645	56563
82238	83763	45464	18493	98489	72138	38942	97661	95788
28853	61793	44664	69427	68144	71949	57192	25592	49835
22251	73098	68108	87626	76724	56495	87357	83065	95316
66236	46591	69225	29867	60815	51931	40507	52568	47097
50006	91666	86406	92778	51232	38761	21861	98596	42673
68328	12840	61206	64298	27378	61452	13349	27223	79637
83039	25015	95983	82835	67268	23355	44647	25542	10536
53158	82329	81756	81429	54366	97530	51447	11324	49939
46802	61720	97508	73339	29277	17964	35421	39880	38180
25162	78468	44303	14425	42587	37212	58866	39008	91938
65957	15171	22417	95571	90679	54774	43979	71017	49647
10876	36062	91375	90128	14906	81447	49158	14703	89517
35354	66633	62311	58185	67310	95474	21878	89101	38299
70822	69983	23726	97422	46713	20340	42807	10859	26897
64299	12987	60370	70165	43306	14417	79261	53891	72816
74007	61658	86698	31571	75098	11676	35867	39764	47504
70909	68300	55074	42093	55745	80364	18488	47981	18702
67898	98830	97705	10723	82370	45586	19013	60915	84961
59386	25440	92441	14265	26123	85453	57326	72790	55243
71469	49833	95737	84195	78444	32104	89917	88361	35344
34064	12993	23818	28197	33755	96438	84223	10400	36797
86492	25367	65712	81581	89579	31759	56108	24476	47696
86914	87565	20344	39027	98338	95171	75562	54283	35342
88418	58064	13624	32978	90704	56218	84064	69990	45354
87948	83451	96217	40534	40775	74376	43157	74856	13950
13049	85293	32747	17728	50495	34617	73707	33976	86177
86544	52703	74990	98288	61833	48803	75258	83382	79099
77295	70694	97326	35430	53881	94007	70471	66815	73042
54637	32831	59063	72353	87365	15322	33156	40331	93942
50938	12004	18585	23896	62559	44470	27701	66780	56157
80999	49724	76745	25232	74291	74184	91055	58903	18172
71303	36255	77310	95847	30282	77207	34439	47763	99697
79264	16901	55814	89734	30255	87209	31629	19328	42532
30235	69368	38685	32790	58980	42159	88577	18427	73504
59110	69783	93713	29151	34933	95745	72271	38684	15426
28094	19560	27259	82736	49700	37876	52322	69562	75837
40341	20666	26662	16422	76351	70520	36890	86559	89160
30117	68850	28319	44992	68110	47007	22243	72813	60934
62287	44957	47690	79484	69449	27981	34770	34228	81686
96976	77830	61746	67846	15584	28070	79200	12663	63273
82584	34789	33494	55533	25040	84187	14479	26286	10665
35728	87881	70271	13115	35745	99145	92717	74357	16716
88458	63625	59577	92037	99012	40836	58817	30757	37934
49789	20873	53858	91356	11387	75208	33643	88210	42440
49131	34078	45396	56884	81416	46292	36012	30806	65220

TABLE I (*Continued*)

96256	82566	34796	88012	43066	35786	93715	15550	16690
43742	97487	68089	69887	23737	71136	21108	85204	60726
34527	87490	81183	95864	59430	19473	57978	39853	47877
58906	37390	88924	80917	58840	29907	99098	33761	50335
30438	12056	12104	61012	44674	49815	85298	94129	59542
79149	98261	48599	54336	71894	82889	51219	70291	60922
48703	25290	13835	35695	15440	52533	82849	25504	81623
96050	74505	18706	10572	66015	53509	48115	87578	86099
98859	48791	15048	73300	48045	33559	98939	98003	39453
32758	55597	11686	18385	31103	87621	39659	81413	68625
17238	73653	15557	79374	60965	75564	15872	34611	86497
79748	21687	94964	43348	26957	27085	81760	29099	23553
89199	75213	37815	99891	60990	37062	80331	54009	26812
87491	62544	51229	13028	81370	16309	28493	98555	24278
87338	17647	40018	48386	49992	44304	23330	38730	21601
48635	73063	37450	65403	65134	83119	16341	95766	83949
32197	94930	50586	88559	48025	97023	15372	18847	97168
83421	68819	69623	45088	54839	70855	86714	38202	98163
66167	84791	40631	33428	78200	41145	57816	86795	31646
26555	70521	69140	93495	38179	43253	78172	44239	60701
47786	37539	17452	88719	24423	59201	24979	51019	35458
47775	42564	15665	92454	98345	87963	81142	34356	41518
49414	83761	74309	82620	53677	34575	81871	76615	27653
75918	39825	60958	96584	26872	15379	84080	40371	35019
13440	85096	85668	24896	65261	83757	68388	29797	48376
39614	53926	97122	85279	15622	29329	59579	60250	73895
40067	48944	98882	39023	31677	41118	52818	29586	43848
65350	11148	63012	59418	54688	83692	95840	91627	84057
98902	62170	49281	29406	63143	43722	35838	98979	67024
10529	81048	29639	50740	93253	77339	80328	88580	30970
99123	48497	35247	33488	63781	19388	98534	27479	44269
65147	42913	50654	64220	13950	74293	53489	39014	86040
86886	15231	43834	88205	87159	30789	10959	81631	15575
17264	57846	52347	96649	69212	28053	62290	93328	98520
36110	87509	95913	66687	67149	81500	44107	27546	94868
58288	91109	66433	75388	80441	95720	64891	63049	68237
67834	18606	88840	39705	17329	15690	90382	35725	21362
67746	23016	87357	89427	98266	39452	58011	86665	70716
32196	36633	63350	73154	47699	15479	63905	81186	67181
33646	35175	41141	75793	58908	80681	88974	84611	46634
66973	98812	21094	45209	52503	51038	96306	75653	32482
74819	41419	91296	31736	99727	68791	93588	99566	98413
76495	37282	43051	17275	30370	76105	55926	98910	84767
39350	85262	59225	47343	63449	47004	94970	77067	16857
29657	66820	47420	37404	80296	94070	54249	60378	54670
98059	31868	86468	80389	66521	23304	99582	48791	74154
78131	47852	62735	79575	48757	25712	22468	66035	29237
36071	71312	33098	38558	57088	26162	32752	24827	95562
23742	21969	82378	28923	19944	91024	63237	36022	76979
48418	36759	92342	93571	86923	26627	46138	86343	21083

TABLE I *(Continued)*

74678	64188	39402	65189	30854	65086	43052	54042	79127
45693	86700	39667	53646	11663	86785	84727	83728	21758
68539	65113	68955	71627	38626	57160	63171	41707	51634
56807	61373	21941	71481	88523	72157	92088	41244	75735
29320	38387	89881	59789	50099	64811	31131	74334	65674
63399	73318	61578	28141	21655	65378	56261	69795	67096
33316	23627	55609	48463	92502	64287	99853	54497	85985
15403	81891	20190	72235	85636	16745	99483	43583	89137
76057	62447	71848	60035	76280	38017	26998	82690	16512
93885	29489	26222	77121	83244	74614	62527	36019	31265
66312	17182	96913	10736	52184	57082	58901	87749	34684
88960	52088	92432	18463	25562	20674	48988	41829	98681
38903	23457	87215	47089	38395	21929	94929	59489	79066
88281	90912	22965	16428	32289	99354	87068	55884	69518
98627	47123	32667	69196	70158	19828	42793	54593	53682
13183	37010	53184	53434	53631	96983	21201	10236	90134
69177	18284	72840	68433	88300	85396	10298	15680	13859
79501	74784	50483	37213	67077	59481	43976	35404	90683
52734	93419	47519	85203	27665	97179	47002	41258	39219
70256	58003	11565	67432	56505	18468	10293	46490	98191
47603	62142	37636	43374	19773	10538	27243	20800	50383
41507	43884	18253	81908	22803	84840	38968	70176	59393
56464	84865	65387	97484	95349	55548	54214	86814	20654
54523	35676	93542	20744	23942	22935	19794	53413	12979
86965	67669	42284	68532	24766	41411	97597	34998	70248
80707	35351	29958	12270	76227	31529	26105	94145	30469
57351	44045	67826	18191	75712	86420	36234	93377	20205
56860	89252	53362	82306	24114	91538	49114	67506	73489
87844	42168	83234	59134	17403	20418	65647	14702	59080
47157	27318	46686	59507	31598	16152	41184	95641	58835
63530	51286	46562	43739	51259	39836	72962	96998	89257
74595	35004	61728	28879	60412	81320	99003	20824	47086
26649	55512	58180	87954	91885	22660	31132	15752	11807
63596	44068	12648	91827	35448	59307	64466	68502	36292
66621	89136	46721	43322	78706	60249	90841	79917	18000
98128	99125	86432	87068	88376	65121	64402	55931	45748
85968	99264	64582	85694	29027	62883	53615	26692	73490
93011	71694	78514	63842	33754	84577	78698	38667	54673
36994	29619	36095	44782	85794	28498	25870	83655	97905
58857	32343	61392	65331	66939	51145	77060	85743	85278
97430	82854	28720	52153	37246	87152	95563	51769	79320
64642	69774	67582	95955	91433	95515	35211	39734	82631
56789	90056	28697	88922	39250	66008	55324	39129	63408
78707	36317	69939	63529	88044	66897	16846	67664	99997
20752	71605	38186	18221	79499	14660	86115	80339	34321
20794	82021	37432	97568	85812	97016	15655	40601	39475
84832	45347	60186	66673	62148	37683	79034	46572	69243
39960	63046	99657	28301	19953	84261	30215	52274	91374
51835	19676	40685	45677	57150	73208	59526	76240	24209
88213	83367	21935	72494	87548	43000	72275	81974	54718

TABLE I (*Continued*)

39746	38989	28721	38803	89668	57496	97127	59364	83335
95915	51291	14163	40972	33163	85169	66522	72010	53429
43937	78760	47672	69700	87058	19072	89435	13390	72315
28633	29330	20463	89033	16968	62815	65802	53006	70674
27415	32278	61924	61670	38880	13911	85037	93738	94913
10157	74513	43054	44601	35689	54559	91660	60035	83733
18041	40798	39274	72760	83644	48960	52193	95674	22516
22679	12792	60046	80515	12962	57351	36431	52277	50567
81468	99534	30455	17430	92600	85813	90223	15335	97102
50636	87932	25489	29395	87683	84579	10396	38276	33729
60635	13409	81824	77150	51472	65915	62520	33839	52209
68336	29892	94343	37822	55260	97321	20488	50172	45199
79273	96036	89979	78654	38959	36250	91126	90337	91381
71942	89335	75664	75278	40445	12818	24033	11809	44129
87842	60665	73523	55824	61257	22080	74425	54851	84786

TABLE II FACTORIALS

n	$n!$
0	1
1	1
2	2
3	6
4	24
5	120
6	720
7	5,040
8	40,320
9	362,880
10	3,628,800
11	39,916,800
12	479,001,600
13	6,227,020,800
14	87,178,291,200
15	1,307,674,368,000
16	20,922,789,888,000
17	355,687,428,096,000
18	6,402,373,705,728,000
19	121,645,100,408,832,000
20	2,432,902,008,176,640,000
21	51,090,942,171,709,440,000
22	1,124,000,727,777,607,680,000
23	25,852,016,738,884,976,640,000
24	620,448,401,733,239,439,360,000
25	15,511,210,043,330,985,984,000,000

TABLE III VALUES OF $\binom{n}{x}$ (COMBINATION)

n \ x	0	1	2	3	4	5	6	7	8	9	10	11	12	13	14	15	16	17	18	19	20
1	1	1																			
2	1	2	1																		
3	1	3	3	1																	
4	1	4	6	4	1																
5	1	5	10	10	5	1															
6	1	6	15	20	15	6	1														
7	1	7	21	35	35	21	7	1													
8	1	8	28	56	70	56	28	8	1												
9	1	9	36	84	126	126	84	36	9	1											
10	1	10	45	120	210	252	210	120	45	10	1										
11	1	11	55	165	330	462	462	330	165	55	11	1									
12	1	12	66	220	495	792	924	792	495	220	66	12	1								
13	1	13	78	286	715	1,287	1,716	1,716	1,287	715	286	78	13	1							
14	1	14	91	364	1,001	2,002	3,003	3,432	3,003	2,002	1,001	364	91	14	1						
15	1	15	105	455	1,365	3,003	5,005	6,435	6,435	5,005	3,003	1,365	455	105	15	1					
16	1	16	120	560	1,820	4,368	8,008	11,440	12,870	11,440	8,008	4,368	1,820	560	120	16	1				
17	1	17	136	680	2,380	6,188	12,376	19,448	24,310	24,310	19,448	12,376	6,188	2,380	680	136	17	1			
18	1	18	153	816	3,060	8,568	18,564	31,824	43,758	48,620	43,758	31,824	18,564	8,568	3,060	816	153	18	1		
19	1	19	171	969	3,876	11,628	27,132	50,388	75,582	92,378	92,378	75,582	50,388	27,132	11,628	3,876	969	171	19	1	
20	1	20	190	1,140	4,845	15,504	38,760	77,520	125,970	167,960	184,756	167,960	125,970	77,520	38,760	15,504	4,845	1,140	190	20	1
21	1	21	210	1,330	5,985	20,349	54,264	116,280	203,490	293,930	352,716	352,716	293,930	203,490	116,280	54,264	20,349	5,985	1,330	210	21
22	1	22	231	1,540	7,315	26,334	74,613	170,544	319,770	497,420	646,646	705,432	646,646	497,420	319,770	170,544	74,613	26,334	7,315	1,540	231
23	1	23	253	1,771	8,855	33,649	100,947	245,157	490,314	817,190	1,144,066	1,352,078	1,352,078	1,144,066	817,190	490,314	245,157	100,947	33,649	8,855	1,771
24	1	24	276	2,024	10,626	42,504	134,596	346,104	735,471	1,307,504	1,961,256	2,496,144	2,704,156	2,496,144	1,961,256	1,307,504	735,471	346,104	134,596	42,504	10,626
25	1	25	300	2,300	12,650	53,130	177,100	480,700	1,081,575	2,042,975	3,268,760	4,457,400	5,200,300	5,200,300	4,457,400	3,268,760	2,042,975	1,081,575	480,700	177,100	53,130

TABLE IV TABLE OF BINOMIAL PROBABILITIES

		p										
n	*x*	.05	.10	.20	.30	.40	.50	.60	.70	.80	.90	.95
1	0	.9500	.9000	.8000	.7000	.6000	.5000	.4000	.3000	.2000	.1000	.0500
	1	.0500	.1000	.2000	.3000	.4000	.5000	.6000	.7000	.8000	.9000	.9500
2	0	.9025	.8100	.6400	.4900	.3600	.2500	.1600	.0900	.0400	.0100	.0025
	1	.0950	.1800	.3200	.4200	.4800	.5000	.4800	.4200	.3200	.1800	.0950
	2	.0025	.0100	.0400	.0900	.1600	.2500	.3600	.4900	.6400	.8100	.9025
3	0	.8574	.7290	.5120	.3430	.2160	.1250	.0640	.0270	.0080	.0010	.0001
	1	.1354	.2430	.3840	.4410	.4320	.3750	.2880	.1890	.0960	.0270	.0071
	2	.0071	.0270	.0960	.1890	.2880	.3750	.4320	.4410	.3840	.2430	.1354
	3	.0001	.0010	.0080	.0270	.0640	.1250	.2160	.3430	.5120	.7290	.8574
4	0	.8145	.6561	.4096	.2401	.1296	.0625	.0256	.0081	.0016	.0001	.0000
	1	.1715	.2916	.4096	.4116	.3456	.2500	.1536	.0756	.0256	.0036	.0005
	2	.0135	.0486	.1536	.2646	.3456	.3750	.3456	.2646	.1536	.0486	.0135
	3	.0005	.0036	.0256	.0756	.1536	.2500	.3456	.4116	.4096	.2916	.1715
	4	.0000	.0001	.0016	.0081	.0256	.0625	.1296	.2401	.4096	.6561	.8145
5	0	.7738	.5905	.3277	.1681	.0778	.0312	.0102	.0024	.0003	.0000	.0000
	1	.2036	.3280	.4096	.3602	.2592	.1562	.0768	.0284	.0064	.0005	.0000
	2	.0214	.0729	.2048	.3087	.3456	.3125	.2304	.1323	.0512	.0081	.0011
	3	.0011	.0081	.0512	.1323	.2304	.3125	.3456	.3087	.2048	.0729	.0214
	4	.0000	.0004	.0064	.0283	.0768	.1562	.2592	.3601	.4096	.3281	.2036
	5	.0000	.0000	.0003	.0024	.0102	.0312	.0778	.1681	.3277	.5905	.7738
6	0	.7351	.5314	.2621	.1176	.0467	.0156	.0041	.0007	.0001	.0000	.0000
	1	.2321	.3543	.3932	.3025	.1866	.0937	.0369	.0102	.0015	.0001	.0000
	2	.0305	.0984	.2458	.3241	.3110	.2344	.1382	.0595	.0154	.0012	.0001
	3	.0021	.0146	.0819	.1852	.2765	.3125	.2765	.1852	.0819	.0146	.0021
	4	.0001	.0012	.0154	.0595	.1382	.2344	.3110	.3241	.2458	.0984	.0305
	5	.0000	.0001	.0015	.0102	.0369	.0937	.1866	.3025	.3932	.3543	.2321
	6	.0000	.0000	.0001	.0007	.0041	.0156	.0467	.1176	.2621	.5314	.7351
7	0	.6983	.4783	.2097	.0824	.0280	.0078	.0016	.0002	.0000	.0000	.0000
	1	.2573	.3720	.3670	.2471	.1306	.0547	.0172	.0036	.0004	.0000	.0000
	2	.0406	.1240	.2753	.3177	.2613	.1641	.0774	.0250	.0043	.0002	.0000
	3	.0036	.0230	.1147	.2269	.2903	.2734	.1935	.0972	.0287	.0026	.0002
	4	.0002	.0026	.0287	.0972	.1935	.2734	.2903	.2269	.1147	.0230	.0036
	5	.0000	.0002	.0043	.0250	.0774	.1641	.2613	.3177	.2753	.1240	.0406
	6	.0000	.0000	.0004	.0036	.0172	.0547	.1306	.2471	.3670	.3720	.2573
	7	.0000	.0000	.0000	.0002	.0016	.0078	.0280	.0824	.2097	.4783	.6983
8	0	.6634	.4305	.1678	.0576	.0168	.0039	.0007	.0001	.0000	.0000	.0000
	1	.2793	.3826	.3355	.1977	.0896	.0312	.0079	.0012	.0001	.0000	.0000
	2	.0515	.1488	.2936	.2965	.2090	.1094	.0413	.0100	.0011	.0000	.0000
	3	.0054	.0331	.1468	.2541	.2787	.2187	.1239	.0467	.0092	.0004	.0000
	4	.0004	.0046	.0459	.1361	.2322	.2734	.2322	.1361	.0459	.0046	.0004
	5	.0000	.0004	.0092	.0467	.1239	.2187	.2787	.2541	.1468	.0331	.0054

TABLE IV (*Continued*)

		p										
n	*x*	.05	.10	.20	.30	.40	.50	.60	.70	.80	.90	.95
	6	.0000	.0000	.0011	.0100	.0413	.1094	.2090	.2965	.2936	.1488	.0515
	7	.0000	.0000	.0001	.0012	.0079	.0312	.0896	.1977	.3355	.3826	.2793
	8	.0000	.0000	.0000	.0001	.0007	.0039	.0168	.0576	.1678	.4305	.6634
9	0	.6302	.3874	.1342	.0404	.0101	.0020	.0003	.0000	.0000	.0000	.0000
	1	.2985	.3874	.3020	.1556	.0605	.0176	.0035	.0004	.0000	.0000	.0000
	2	.0629	.1722	.3020	.2668	.1612	.0703	.0212	.0039	.0003	.0000	.0000
	3	.0077	.0446	.1762	.2668	.2508	.1641	.0743	.0210	.0028	.0001	.0000
	4	.0006	.0074	.0661	.1715	.2508	.2461	.1672	.0735	.0165	.0008	.0000
	5	.0000	.0008	.0165	.0735	.1672	.2461	.2508	.1715	.0661	.0074	.0006
	6	.0000	.0001	.0028	.0210	.0743	.1641	.2508	.2668	.1762	.0446	.0077
	7	.0000	.0000	.0003	.0039	.0212	.0703	.1612	.2668	.3020	.1722	.0629
	8	.0000	.0000	.0000	.0004	.0035	.0176	.0605	.1556	.3020	.3874	.2985
	9	.0000	.0000	.0000	.0000	.0003	.0020	.0101	.0404	.1342	.3874	.6302
10	0	.5987	.3487	.1074	.0282	.0060	.0010	.0001	.0000	.0000	.0000	.0000
	1	.3151	.3874	.2684	.1211	.0403	.0098	.0016	.0001	.0000	.0000	.0000
	2	.0746	.1937	.3020	.2335	.1209	.0439	.0106	.0014	.0001	.0000	.0000
	3	.0105	.0574	.2013	.2668	.2150	.1172	.0425	.0090	.0008	.0000	.0000
	4	.0010	.0112	.0881	.2001	.2508	.2051	.1115	.0368	.0055	.0001	.0000
	5	.0001	.0015	.0264	.1029	.2007	.2461	.2007	.1029	.0264	.0015	.0001
	6	.0000	.0001	.0055	.0368	.1115	.2051	.2508	.2001	.0881	.0112	.0010
	7	.0000	.0000	.0008	.0090	.0425	.1172	.2150	.2668	.2013	.0574	.0105
	8	.0000	.0000	.0001	.0014	.0106	.0439	.1209	.2335	.3020	.1937	.0746
	9	.0000	.0000	.0000	.0001	.0016	.0098	.0403	.1211	.2684	.3874	.3151
	10	.0000	.0000	.0000	.0000	.0001	.0010	.0060	.0282	.1074	.3487	.5987
11	0	.5688	.3138	.0859	.0198	.0036	.0005	.0000	.0000	.0000	.0000	.0000
	1	.3293	.3835	.2362	.0932	.0266	.0054	.0007	.0000	.0000	.0000	.0000
	2	.0867	.2131	.2953	.1998	.0887	.0269	.0052	.0005	.0000	.0000	.0000
	3	.0137	.0710	.2215	.2568	.1774	.0806	.0234	.0037	.0002	.0000	.0000
	4	.0014	.0158	.1107	.2201	.2365	.1611	.0701	.0173	.0017	.0000	.0000
	5	.0001	.0025	.0388	.1321	.2207	.2256	.1471	.0566	.0097	.0003	.0000
	6	.0000	.0003	.0097	.0566	.1471	.2256	.2207	.1321	.0388	.0025	.0001
	7	.0000	.0000	.0017	.0173	.0701	.1611	.2365	.2201	.1107	.0158	.0014
	8	.0000	.0000	.0002	.0037	.0234	.0806	.1774	.2568	.2215	.0710	.0137
	9	.0000	.0000	.0000	.0005	.0052	.0269	.0887	.1998	.2953	.2131	.0867
	10	.0000	.0000	.0000	.0000	.0007	.0054	.0266	.0932	.2362	.3835	.3293
	11	.0000	.0000	.0000	.0000	.0000	.0005	.0036	.0198	.0859	.3138	.5688
12	0	.5404	.2824	.0687	.0138	.0022	.0002	.0000	.0000	.0000	.0000	.0000
	1	.3413	.3766	.2062	.0712	.0174	.0029	.0003	.0000	.0000	.0000	.0000
	2	.0988	.2301	.2835	.1678	.0639	.0161	.0025	.0002	.0000	.0000	.0000
	3	.0173	.0852	.2362	.2397	.1419	.0537	.0125	.0015	.0001	.0000	.0000
	4	.0021	.0213	.1329	.2311	.2128	.1208	.0420	.0078	.0005	.0000	.0000
	5	.0002	.0038	.0532	.1585	.2270	.1934	.1009	.0291	.0033	.0000	.0000
	6	.0000	.0005	.0155	.0792	.1766	.2256	.1766	.0792	.0155	.0005	.0000
	7	.0000	.0000	.0033	.0291	.1009	.1934	.2270	.1585	.0532	.0038	.0002
	8	.0000	.0000	.0005	.0078	.0420	.1208	.2128	.2311	.1329	.0213	.0021
	9	.0000	.0000	.0001	.0015	.0125	.0537	.1419	.2397	.2362	.0852	.0173
	10	.0000	.0000	.0000	.0002	.0025	.0161	.0639	.1678	.2835	.2301	.0988

TABLE IV (*Continued*)

		p										
n	x	.05	.10	.20	.30	.40	.50	.60	.70	.80	.90	.95
	11	.0000	.0000	.0000	.0000	.0003	.0029	.0174	.0712	.2062	.3766	.3413
	12	.0000	.0000	.0000	.0000	.0000	.0002	.0022	.0138	.0687	.2824	.5404
13	0	.5133	.2542	.0550	.0097	.0013	.0001	.0000	.0000	.0000	.0000	.0000
	1	.3512	.3672	.1787	.0540	.0113	.0016	.0001	.0000	.0000	.0000	.0000
	2	.1109	.2448	.2680	.1388	.0453	.0095	.0012	.0001	.0000	.0000	.0000
	3	.0214	.0997	.2457	.2181	.1107	.0349	.0065	.0006	.0000	.0000	.0000
	4	.0028	.0277	.1535	.2337	.1845	.0873	.0243	.0034	.0001	.0000	.0000
	5	.0003	.0055	.0691	.1803	.2214	.1571	.0656	.0142	.0011	.0000	.0000
	6	.0000	.0008	.0230	.1030	.1968	.2095	.1312	.0442	.0058	.0001	.0000
	7	.0000	.0001	.0058	.0442	.1312	.2095	.1968	.1030	.0230	.0008	.0000
	8	.0000	.0000	.0011	.0142	.0656	.1571	.2214	.1803	.0691	.0055	.0003
	9	.0000	.0000	.0001	.0034	.0243	.0873	.1845	.2337	.1535	.0277	.0028
	10	.0000	.0000	.0000	.0006	.0065	.0349	.1107	.2181	.2457	.0997	.0214
	11	.0000	.0000	.0000	.0001	.0012	.0095	.0453	.1388	.2680	.2448	.1109
	12	.0000	.0000	.0000	.0000	.0001	.0016	.0113	.0540	.1787	.3672	.3512
	13	.0000	.0000	.0000	.0000	.0000	.0001	.0013	.0097	.0550	.2542	.5133
14	0	.4877	.2288	.0440	.0068	.0008	.0001	.0000	.0000	.0000	.0000	.0000
	1	.3593	.3559	.1539	.0407	.0073	.0009	.0001	.0000	.0000	.0000	.0000
	2	.1229	.2570	.2501	.1134	.0317	.0056	.0005	.0000	.0000	.0000	.0000
	3	.0259	.1142	.2501	.1943	.0845	.0222	.0033	.0002	.0000	.0000	.0000
	4	.0037	.0349	.1720	.2290	.1549	.0611	.0136	.0014	.0000	.0000	.0000
	5	.0004	.0078	.0860	.1963	.2066	.1222	.0408	.0066	.0003	.0000	.0000
	6	.0000	.0013	.0322	.1262	.2066	.1833	.0918	.0232	.0020	.0000	.0000
	7	.0000	.0002	.0092	.0618	.1574	.2095	.1574	.0618	.0092	.0002	.0000
	8	.0000	.0000	.0020	.0232	.0918	.1833	.2066	.1262	.0322	.0013	.0000
	9	.0000	.0000	.0003	.0066	.0408	.1222	.2066	.1963	.0860	.0078	.0004
	10	.0000	.0000	.0000	.0014	.0136	.0611	.1549	.2290	.1720	.0349	.0037
	11	.0000	.0000	.0000	.0002	.0033	.0222	.0845	.1943	.2501	.1142	.0259
	12	.0000	.0000	.0000	.0000	.0005	.0056	.0317	.1134	.2501	.2570	.1229
	13	.0000	.0000	.0000	.0000	.0001	.0009	.0073	.0407	.1539	.3559	.3593
	14	.0000	.0000	.0000	.0000	.0000	.0001	.0008	.0068	.0440	.2288	.4877
15	0	.4633	.2059	.0352	.0047	.0005	.0000	.0000	.0000	.0000	.0000	.0000
	1	.3658	.3432	.1319	.0305	.0047	.0005	.0000	.0000	.0000	.0000	.0000
	2	.1348	.2669	.2309	.0916	.0219	.0032	.0003	.0000	.0000	.0000	.0000
	3	.0307	.1285	.2501	.1700	.0634	.0139	.0016	.0001	.0000	.0000	.0000
	4	.0049	.0428	.1876	.2186	.1268	.0417	.0074	.0006	.0000	.0000	.0000
	5	.0006	.0105	.1032	.2061	.1859	.0916	.0245	.0030	.0001	.0000	.0000
	6	.0000	.0019	.0430	.1472	.2066	.1527	.0612	.0116	.0007	.0000	.0000
	7	.0000	.0003	.0138	.0811	.1771	.1964	.1181	.0348	.0035	.0000	.0000
	8	.0000	.0000	.0035	.0348	.1181	.1964	.1771	.0811	.0138	.0003	.0000
	9	.0000	.0000	.0007	.0116	.0612	.1527	.2066	.1472	.0430	.0019	.0000
	10	.0000	.0000	.0001	.0030	.0245	.0916	.1859	.2061	.1032	.0105	.0006
	11	.0000	.0000	.0000	.0006	.0074	.0417	.1268	.2186	.1876	.0428	.0049
	12	.0000	.0000	.0000	.0001	.0016	.0139	.0634	.1700	.2501	.1285	.0307
	13	.0000	.0000	.0000	.0000	.0003	.0032	.0219	.0916	.2309	.2669	.1348
	14	.0000	.0000	.0000	.0000	.0000	.0005	.0047	.0305	.1319	.3432	.3658
	15	.0000	.0000	.0000	.0000	.0000	.0000	.0005	.0047	.0352	.2059	.4633

TABLE IV (*Continued*)

		p										
n	x	.05	.10	.20	.30	.40	.50	.60	.70	.80	.90	.95
16	0	.4401	.1853	.0281	.0033	.0003	.0000	.0000	.0000	.0000	.0000	.0000
	1	.3706	.3294	.1126	.0228	.0030	.0002	.0000	.0000	.0000	.0000	.0000
	2	.1463	.2745	.2111	.0732	.0150	.0018	.0001	.0000	.0000	.0000	.0000
	3	.0359	.1423	.2463	.1465	.0468	.0085	.0008	.0000	.0000	.0000	.0000
	4	.0061	.0514	.2001	.2040	.1014	.0278	.0040	.0002	.0000	.0000	.0000
	5	.0008	.0137	.1201	.2099	.1623	.0667	.0142	.0013	.0000	.0000	.0000
	6	.0001	.0028	.0550	.1649	.1983	.1222	.0392	.0056	.0002	.0000	.0000
	7	.0000	.0004	.0197	.1010	.1889	.1746	.0840	.0185	.0012	.0000	.0000
	8	.0000	.0001	.0055	.0487	.1417	.1964	.1417	.0487	.0055	.0001	.0000
	9	.0000	.0000	.0012	.0185	.0840	.1746	.1889	.1010	.0197	.0004	.0000
	10	.0000	.0000	.0002	.0056	.0392	.1222	.1983	.1649	.0550	.0028	.0001
	11	.0000	.0000	.0000	.0013	.0142	.0666	.1623	.2099	.1201	.0137	.0008
	12	.0000	.0000	.0000	.0002	.0040	.0278	.1014	.2040	.2001	.0514	.0061
	13	.0000	.0000	.0000	.0000	.0008	.0085	.0468	.1465	.2463	.1423	.0359
	14	.0000	.0000	.0000	.0000	.0001	.0018	.0150	.0732	.2111	.2745	.1463
	15	.0000	.0000	.0000	.0000	.0000	.0002	.0030	.0228	.1126	.3294	.3706
	16	.0000	.0000	.0000	.0000	.0000	.0000	.0003	.0033	.0281	.1853	.4401
17	0	.4181	.1668	.0225	.0023	.0002	.0000	.0000	.0000	.0000	.0000	.0000
	1	.3741	.3150	.0957	.0169	.0019	.0001	.0000	.0000	.0000	.0000	.0000
	2	.1575	.2800	.1914	.0581	.0102	.0010	.0001	.0000	.0000	.0000	.0000
	3	.0415	.1556	.2393	.1245	.0341	.0052	.0004	.0000	.0000	.0000	.0000
	4	.0076	.0605	.2093	.1868	.0796	.0182	.0021	.0001	.0000	.0000	.0000
	5	.0010	.0175	.1361	.2081	.1379	.0472	.0081	.0006	.0000	.0000	.0000
	6	.0001	.0039	.0680	.1784	.1839	.0944	.0242	.0026	.0001	.0000	.0000
	7	.0000	.0007	.0267	.1201	.1927	.1484	.0571	.0095	.0004	.0000	.0000
	8	.0000	.0001	.0084	.0644	.1606	.1855	.1070	.0276	.0021	.0000	.0000
	9	.0000	.0000	.0021	.0276	.1070	.1855	.1606	.0644	.0084	.0001	.0000
	10	.0000	.0000	.0004	.0095	.0571	.1484	.1927	.1201	.0267	.0007	.0000
	11	.0000	.0000	.0001	.0026	.0242	.0944	.1839	.1784	.0680	.0039	.0001
	12	.0000	.0000	.0000	.0006	.0081	.0472	.1379	.2081	.1361	.0175	.0010
	13	.0000	.0000	.0000	.0001	.0021	.0182	.0796	.1868	.2093	.0605	.0076
	14	.0000	.0000	.0000	.0000	.0004	.0052	.0341	.1245	.2393	.1556	.0415
	15	.0000	.0000	.0000	.0000	.0001	.0010	.0102	.0581	.1914	.2800	.1575
	16	.0000	.0000	.0000	.0000	.0000	.0001	.0019	.0169	.0957	.3150	.3741
	17	.0000	.0000	.0000	.0000	.0000	.0000	.0002	.0023	.0225	.1668	.4181
18	0	.3972	.1501	.0180	.0016	.0001	.0000	.0000	.0000	.0000	.0000	.0000
	1	.3763	.3002	.0811	.0126	.0012	.0001	.0000	.0000	.0000	.0000	.0000
	2	.1683	.2835	.1723	.0458	.0069	.0006	.0000	.0000	.0000	.0000	.0000
	3	.0473	.1680	.2297	.1046	.0246	.0031	.0002	.0000	.0000	.0000	.0000
	4	.0093	.0700	.2153	.1681	.0614	.0117	.0011	.0000	.0000	.0000	.0000
	5	.0014	.0218	.1507	.2017	.1146	.0327	.0045	.0002	.0000	.0000	.0000
	6	.0002	.0052	.0816	.1873	.1655	.0708	.0145	.0012	.0000	.0000	.0000
	7	.0000	.0010	.0350	.1376	.1892	.1214	.0374	.0046	.0001	.0000	.0000
	8	.0000	.0002	.0120	.0811	.1734	.1669	.0771	.0149	.0008	.0000	.0000
	9	.0000	.0000	.0033	.0386	.1284	.1855	.1284	.0386	.0033	.0000	.0000
	10	.0000	.0000	.0008	.0149	.0771	.1669	.1734	.0811	.0120	.0002	.0000
	11	.0000	.0000	.0001	.0046	.0374	.1214	.1892	.1376	.0350	.0010	.0000
	12	.0000	.0000	.0000	.0012	.0145	.0708	.1655	.1873	.0816	.0052	.0002
	13	.0000	.0000	.0000	.0002	.0045	.0327	.1146	.2017	.1507	.0218	.0014

TABLE IV (*Continued*)

		p										
n	*x*	.05	.10	.20	.30	.40	.50	.60	.70	.80	.90	.95
	14	.0000	.0000	.0000	.0000	.0011	.0117	.0614	.1681	.2153	.0700	.0093
	15	.0000	.0000	.0000	.0000	.0002	.0031	.0246	.1046	.2297	.1680	.0473
	16	.0000	.0000	.0000	.0000	.0000	.0006	.0069	.0458	.1723	.2835	.1683
	17	.0000	.0000	.0000	.0000	.0000	.0001	.0012	.0126	.0811	.3002	.3763
	18	.0000	.0000	.0000	.0000	.0000	.0000	.0001	.0016	.0180	.1501	.3972
19	0	.3774	.1351	.0144	.0011	.0001	.0000	.0000	.0000	.0000	.0000	.0000
	1	.3774	.2852	.0685	.0093	.0008	.0000	.0000	.0000	.0000	.0000	.0000
	2	.1787	.2852	.1540	.0358	.0046	.0003	.0000	.0000	.0000	.0000	.0000
	3	.0533	.1796	.2182	.0869	.0175	.0018	.0001	.0000	.0000	.0000	.0000
	4	.0112	.0798	.2182	.1491	.0467	.0074	.0005	.0000	.0000	.0000	.0000
	5	.0018	.0266	.1636	.1916	.0933	.0222	.0024	.0001	.0000	.0000	.0000
	6	.0002	.0069	.0955	.1916	.1451	.0518	.0085	.0005	.0000	.0000	.0000
	7	.0000	.0014	.0443	.1525	.1797	.0961	.0237	.0022	.0000	.0000	.0000
	8	.0000	.0002	.0166	.0981	.1797	.1442	.0532	.0077	.0003	.0000	.0000
	9	.0000	.0000	.0051	.0514	.1464	.1762	.0976	.0220	.0013	.0000	.0000
	10	.0000	.0000	.0013	.0220	.0976	.1762	.1464	.0514	.0051	.0000	.0000
	11	.0000	.0000	.0003	.0077	.0532	.1442	.1797	.0981	.0166	.0002	.0000
	12	.0000	.0000	.0000	.0022	.0237	.0961	.1797	.1525	.0443	.0014	.0000
	13	.0000	.0000	.0000	.0005	.0085	.0518	.1451	.1916	.0955	.0069	.0002
	14	.0000	.0000	.0000	.0001	.0024	.0222	.0933	.1916	.1636	.0266	.0018
	15	.0000	.0000	.0000	.0000	.0005	.0074	.0467	.1491	.2182	.0798	.0112
	16	.0000	.0000	.0000	.0000	.0001	.0018	.0175	.0869	.2182	.1796	.0533
	17	.0000	.0000	.0000	.0000	.0000	.0003	.0046	.0358	.1540	.2852	.1787
	18	.0000	.0000	.0000	.0000	.0000	.0000	.0008	.0093	.0685	.2852	.3774
	19	.0000	.0000	.0000	.0000	.0000	.0000	.0001	.0011	.0144	.1351	.3774
20	0	.3585	.1216	.0115	.0008	.0000	.0000	.0000	.0000	.0000	.0000	.0000
	1	.3774	.2702	.0576	.0068	.0005	.0000	.0000	.0000	.0000	.0000	.0000
	2	.1887	.2852	.1369	.0278	.0031	.0002	.0000	.0000	.0000	.0000	.0000
	3	.0596	.1901	.2054	.0716	.0123	.0011	.0000	.0000	.0000	.0000	.0000
	4	.0133	.0898	.2182	.1304	.0350	.0046	.0003	.0000	.0000	.0000	.0000
	5	.0022	.0319	.1746	.1789	.0746	.0148	.0013	.0000	.0000	.0000	.0000
	6	.0003	.0089	.1091	.1916	.1244	.0370	.0049	.0002	.0000	.0000	.0000
	7	.0000	.0020	.0545	.1643	.1659	.0739	.0146	.0010	.0000	.0000	.0000
	8	.0000	.0004	.0222	.1144	.1797	.1201	.0355	.0039	.0001	.0000	.0000
	9	.0000	.0001	.0074	.0654	.1597	.1602	.0710	.0120	.0005	.0000	.0000
	10	.0000	.0000	.0020	.0308	.1171	.1762	.1171	.0308	.0020	.0000	.0000
	11	.0000	.0000	.0005	.0120	.0710	.1602	.1597	.0654	.0074	.0001	.0000
	12	.0000	.0000	.0001	.0039	.0355	.1201	.1797	.1144	.0222	.0004	.0000
	13	.0000	.0000	.0000	.0010	.0146	.0739	.1659	.1643	.0545	.0020	.0000
	14	.0000	.0000	.0000	.0002	.0049	.0370	.1244	.1916	.1091	.0089	.0003
	15	.0000	.0000	.0000	.0000	.0013	.0148	.0746	.1789	.1746	.0319	.0022
	16	.0000	.0000	.0000	.0000	.0003	.0046	.0350	.1304	.2182	.0898	.0133
	17	.0000	.0000	.0000	.0000	.0000	.0011	.0123	.0716	.2054	.1901	.0596
	18	.0000	.0000	.0000	.0000	.0000	.0002	.0031	.0278	.1369	.2852	.1887
	19	.0000	.0000	.0000	.0000	.0000	.0000	.0005	.0068	.0576	.2702	.3774
	20	.0000	.0000	.0000	.0000	.0000	.0000	.0000	.0008	.0115	.1216	.3585
21	0	.3406	.1094	.0092	.0006	.0000	.0000	.0000	.0000	.0000	.0000	.0000
	1	.3764	.2553	.0484	.0050	.0003	.0000	.0000	.0000	.0000	.0000	.0000

TABLE IV (*Continued*)

						p						
n	x	.05	.10	.20	.30	.40	.50	.60	.70	.80	.90	.95
	2	.1981	.2837	.1211	.0215	.0020	.0001	.0000	.0000	.0000	.0000	.0000
	3	.0660	.1996	.1917	.0585	.0086	.0006	.0000	.0000	.0000	.0000	.0000
	4	.0156	.0998	.2156	.1128	.0259	.0029	.0001	.0000	.0000	.0000	.0000
	5	.0028	.0377	.1833	.1643	.0588	.0097	.0007	.0000	.0000	.0000	.0000
	6	.0004	.0112	.1222	.1878	.1045	.0259	.0027	.0001	.0000	.0000	.0000
	7	.0000	.0027	.0655	.1725	.1493	.0554	.0087	.0005	.0000	.0000	.0000
	8	.0000	.0005	.0286	.1294	.1742	.0970	.0229	.0019	.0000	.0000	.0000
	9	.0000	.0001	.0103	.0801	.1677	.1402	.0497	.0063	.0002	.0000	.0000
	10	.0000	.0000	.0031	.0412	.1342	.1682	.0895	.0176	.0008	.0000	.0000
	11	.0000	.0000	.0008	.0176	.0895	.1682	.1342	.0412	.0031	.0000	.0000
	12	.0000	.0000	.0002	.0063	.0497	.1402	.1677	.0801	.0103	.0001	.0000
	13	.0000	.0000	.0000	.0019	.0229	.0970	.1742	.1294	.0286	.0005	.0000
	14	.0000	.0000	.0000	.0005	.0087	.0554	.1493	.1725	.0655	.0027	.0000
	15	.0000	.0000	.0000	.0001	.0027	.0259	.1045	.1878	.1222	.0112	.0004
	16	.0000	.0000	.0000	.0000	.0007	.0097	.0588	.1643	.1833	.0377	.0028
	17	.0000	.0000	.0000	.0000	.0001	.0029	.0259	.1128	.2156	.0998	.0156
	18	.0000	.0000	.0000	.0000	.0000	.0006	.0086	.0585	.1917	.1996	.0660
	19	.0000	.0000	.0000	.0000	.0000	.0001	.0020	.0215	.1211	.2837	.1981
	20	.0000	.0000	.0000	.0000	.0000	.0000	.0003	.0050	.0484	.2553	.3764
	21	.0000	.0000	.0000	.0000	.0000	.0000	.0000	.0006	.0092	.1094	.3406
22	0	.3235	.0985	.0074	.0004	.0000	.0000	.0000	.0000	.0000	.0000	.0000
	1	.3746	.2407	.0406	.0037	.0002	.0000	.0000	.0000	.0000	.0000	.0000
	2	.2070	.2808	.1065	.0166	.0014	.0001	.0000	.0000	.0000	.0000	.0000
	3	.0726	.2080	.1775	.0474	.0060	.0004	.0000	.0000	.0000	.0000	.0000
	4	.0182	.1098	.2108	.0965	.0190	.0017	.0001	.0000	.0000	.0000	.0000
	5	.0034	.0439	.1898	.1489	.0456	.0063	.0004	.0000	.0000	.0000	.0000
	6	.0005	.0138	.1344	.1808	.0862	.0178	.0015	.0000	.0000	.0000	.0000
	7	.0001	.0035	.0768	.1771	.1314	.0407	.0051	.0002	.0000	.0000	.0000
	8	.0000	.0007	.0360	.1423	.1642	.0762	.0144	.0009	.0000	.0000	.0000
	9	.0000	.0001	.0140	.0949	.1703	.1186	.0336	.0032	.0001	.0000	.0000
	10	.0000	.0000	.0046	.0529	.1476	.1542	.0656	.0097	.0003	.0000	.0000
	11	.0000	.0000	.0012	.0247	.1073	.1682	.1073	.0247	.0012	.0000	.0000
	12	.0000	.0000	.0003	.0097	.0656	.1542	.1476	.0529	.0046	.0000	.0000
	13	.0000	.0000	.0001	.0032	.0336	.1186	.1703	.0949	.0140	.0001	.0000
	14	.0000	.0000	.0000	.0009	.0144	.0762	.1642	.1423	.0360	.0007	.0000
	15	.0000	.0000	.0000	.0002	.0051	.0407	.1314	.1771	.0768	.0035	.0001
	16	.0000	.0000	.0000	.0000	.0015	.0178	.0862	.1808	.1344	.0138	.0005
	17	.0000	.0000	.0000	.0000	.0004	.0063	.0456	.1489	.1898	.0439	.0034
	18	.0000	.0000	.0000	.0000	.0001	.0017	.0190	.0965	.2108	.1098	.0182
	19	.0000	.0000	.0000	.0000	.0000	.0004	.0060	.0474	.1775	.2080	.0726
	20	.0000	.0000	.0000	.0000	.0000	.0001	.0014	.0166	.1065	.2808	.2070
	21	.0000	.0000	.0000	.0000	.0000	.0000	.0002	.0037	.0406	.2407	.3746
	22	.0000	.0000	.0000	.0000	.0000	.0000	.0000	.0004	.0074	.0985	.3235
23	0	.3074	.0886	.0059	.0003	.0000	.0000	.0000	.0000	.0000	.0000	.0000
	1	.3721	.2265	.0339	.0027	.0001	.0000	.0000	.0000	.0000	.0000	.0000
	2	.2154	.2768	.0933	.0127	.0009	.0000	.0000	.0000	.0000	.0000	.0000
	3	.0794	.2153	.1633	.0382	.0041	.0002	.0000	.0000	.0000	.0000	.0000
	4	.0209	.1196	.2042	.0818	.0138	.0011	.0000	.0000	.0000	.0000	.0000
	5	.0042	.0505	.1940	.1332	.0350	.0040	.0002	.0000	.0000	.0000	.0000

TABLE IV (*Continued*)

n	x	p = .05	.10	.20	.30	.40	.50	.60	.70	.80	.90	.95
	6	.0007	.0168	.1455	.1712	.0700	.0120	.0008	.0000	.0000	.0000	.0000
	7	.0001	.0045	.0883	.1782	.1133	.0292	.0029	.0001	.0000	.0000	.0000
	8	.0000	.0010	.0442	.1527	.1511	.0584	.0088	.0004	.0000	.0000	.0000
	9	.0000	.0002	.0184	.1091	.1679	.0974	.0221	.0016	.0000	.0000	.0000
	10	.0000	.0000	.0064	.0655	.1567	.1364	.0464	.0052	.0001	.0000	.0000
	11	.0000	.0000	.0019	.0332	.1234	.1612	.0823	.0142	.0005	.0000	.0000
	12	.0000	.0000	.0005	.0142	.0823	.1612	.1234	.0332	.0019	.0000	.0000
	13	.0000	.0000	.0001	.0052	.0464	.1364	.1567	.0655	.0064	.0000	.0000
	14	.0000	.0000	.0000	.0016	.0221	.0974	.1679	.1091	.0184	.0002	.0000
	15	.0000	.0000	.0000	.0004	.0088	.0584	.1511	.1527	.0442	.0010	.0000
	16	.0000	.0000	.0000	.0001	.0029	.0292	.1133	.1782	.0883	.0045	.0001
	17	.0000	.0000	.0000	.0000	.0008	.0120	.0700	.1712	.1455	.0168	.0007
	18	.0000	.0000	.0000	.0000	.0002	.0040	.0350	.1332	.1940	.0505	.0042
	19	.0000	.0000	.0000	.0000	.0000	.0011	.0138	.0818	.2042	.1196	.0209
	20	.0000	.0000	.0000	.0000	.0000	.0002	.0041	.0382	.1633	.2153	.0794
	21	.0000	.0000	.0000	.0000	.0000	.0000	.0009	.0127	.0933	.2768	.2154
	22	.0000	.0000	.0000	.0000	.0000	.0000	.0001	.0027	.0339	.2265	.3721
	23	.0000	.0000	.0000	.0000	.0000	.0000	.0000	.0003	.0059	.0886	.3074
24	0	.2920	.0798	.0047	.0002	.0000	.0000	.0000	.0000	.0000	.0000	.0000
	1	.3688	.2127	.0283	.0020	.0001	.0000	.0000	.0000	.0000	.0000	.0000
	2	.2232	.2718	.0815	.0097	.0006	.0000	.0000	.0000	.0000	.0000	.0000
	3	.0862	.2215	.1493	.0305	.0028	.0001	.0000	.0000	.0000	.0000	.0000
	4	.0238	.1292	.1960	.0687	.0099	.0006	.0000	.0000	.0000	.0000	.0000
	5	.0050	.0574	.1960	.1177	.0265	.0025	.0001	.0000	.0000	.0000	.0000
	6	.0008	.0202	.1552	.1598	.0560	.0080	.0004	.0000	.0000	.0000	.0000
	7	.0001	.0058	.0998	.1761	.0960	.0206	.0017	.0000	.0000	.0000	.0000
	8	.0000	.0014	.0530	.1604	.1360	.0438	.0053	.0002	.0000	.0000	.0000
	9	.0000	.0003	.0236	.1222	.1612	.0779	.0141	.0008	.0000	.0000	.0000
	10	.0000	.0000	.0088	.0785	.1612	.1169	.0318	.0026	.0000	.0000	.0000
	11	.0000	.0000	.0028	.0428	.1367	.1488	.0608	.0079	.0002	.0000	.0000
	12	.0000	.0000	.0008	.0199	.0988	.1612	.0988	.0199	.0008	.0000	.0000
	13	.0000	.0000	.0002	.0079	.0608	.1488	.1367	.0428	.0028	.0000	.0000
	14	.0000	.0000	.0000	.0026	.0318	.1169	.1612	.0785	.0088	.0000	.0000
	15	.0000	.0000	.0000	.0008	.0141	.0779	.1612	.1222	.0236	.0003	.0000
	16	.0000	.0000	.0000	.0002	.0053	.0438	.1360	.1604	.0530	.0014	.0000
	17	.0000	.0000	.0000	.0000	.0017	.0206	.0960	.1761	.0998	.0058	.0001
	18	.0000	.0000	.0000	.0000	.0004	.0080	.0560	.1598	.1552	.0202	.0008
	19	.0000	.0000	.0000	.0000	.0001	.0025	.0265	.1177	.1960	.0574	.0050
	20	.0000	.0000	.0000	.0000	.0000	.0006	.0099	.0687	.1960	.1292	.0238
	21	.0000	.0000	.0000	.0000	.0000	.0001	.0028	.0305	.1493	.2215	.0862
	22	.0000	.0000	.0000	.0000	.0000	.0000	.0006	.0097	.0815	.2718	.2232
	23	.0000	.0000	.0000	.0000	.0000	.0000	.0001	.0020	.0283	.2127	.3688
	24	.0000	.0000	.0000	.0000	.0000	.0000	.0000	.0002	.0047	.0798	.2920
25	0	.2774	.0718	.0038	.0001	.0000	.0000	.0000	.0000	.0000	.0000	.0000
	1	.3650	.1994	.0236	.0014	.0000	.0000	.0000	.0000	.0000	.0000	.0000
	2	.2305	.2659	.0708	.0074	.0004	.0000	.0000	.0000	.0000	.0000	.0000
	3	.0930	.2265	.1358	.0243	.0019	.0001	.0000	.0000	.0000	.0000	.0000
	4	.0269	.1384	.1867	.0572	.0071	.0004	.0000	.0000	.0000	.0000	.0000
	5	.0060	.0646	.1960	.1030	.0199	.0016	.0000	.0000	.0000	.0000	.0000

TABLE IV *(Continued)*

n	x	p = .05	.10	.20	.30	.40	.50	.60	.70	.80	.90	.95
	6	.0010	.0239	.1633	.1472	.0442	.0053	.0002	.0000	.0000	.0000	.0000
	7	.0001	.0072	.1108	.1712	.0800	.0143	.0009	.0000	.0000	.0000	.0000
	8	.0000	.0018	.0623	.1651	.1200	.0322	.0031	.0001	.0000	.0000	.0000
	9	.0000	.0004	.0294	.1336	.1511	.0609	.0088	.0004	.0000	.0000	.0000
	10	.0000	.0001	.0118	.0916	.1612	.0974	.0212	.0013	.0000	.0000	.0000
	11	.0000	.0000	.0040	.0536	.1465	.1328	.0434	.0042	.0001	.0000	.0000
	12	.0000	.0000	.0012	.0268	.1140	.1550	.0760	.0115	.0003	.0000	.0000
	13	.0000	.0000	.0003	.0115	.0760	.1550	.1140	.0268	.0012	.0000	.0000
	14	.0000	.0000	.0001	.0042	.0434	.1328	.1465	.0536	.0040	.0000	.0000
	15	.0000	.0000	.0000	.0013	.0212	.0974	.1612	.0916	.0118	.0001	.0000
	16	.0000	.0000	.0000	.0004	.0088	.0609	.1511	.1336	.0294	.0004	.0000
	17	.0000	.0000	.0000	.0001	.0031	.0322	.1200	.1651	.0623	.0018	.0000
	18	.0000	.0000	.0000	.0000	.0009	.0143	.0800	.1712	.1108	.0072	.0001
	19	.0000	.0000	.0000	.0000	.0002	.0053	.0442	.1472	.1633	.0239	.0010
	20	.0000	.0000	.0000	.0000	.0000	.0016	.0199	.1030	.1960	.0646	.0060
	21	.0000	.0000	.0000	.0000	.0000	.0004	.0071	.0572	.1867	.1384	.0269
	22	.0000	.0000	.0000	.0000	.0000	.0001	.0019	.0243	.1358	.2265	.0930
	23	.0000	.0000	.0000	.0000	.0000	.0000	.0004	.0074	.0708	.2659	.2305
	24	.0000	.0000	.0000	.0000	.0000	.0000	.0000	.0014	.0236	.1994	.3650
	25	.0000	.0000	.0000	.0000	.0000	.0000	.0000	.0001	.0038	.0718	.2774

TABLE V VALUES OF $e^{-\lambda}$

λ	$e^{-\lambda}$	λ	$e^{-\lambda}$
0.0	1.000000	5.5	.004087
0.1	.904837	5.6	.003698
0.2	.818731	5.7	.003346
0.3	.740818	5.8	.003028
0.4	.670320	5.9	.002739
0.5	.606531	6.0	.002479
0.6	.548812	6.1	.002243
0.7	.496585	6.2	.002029
0.8	.449329	6.3	.001836
0.9	.406570	6.4	.001662
1.0	.367879	6.5	.001503
1.1	.332871	6.6	.001360
1.2	.301194	6.7	.001231
1.3	.272532	6.8	.001114
1.4	.246597	6.9	.001008
1.5	.223130	7.0	.000912
1.6	.201897	7.1	.000825
1.7	.182684	7.2	.000747
1.8	.165299	7.3	.000676
1.9	.149569	7.4	.000611
2.0	.135335	7.5	.000553
2.1	.122456	7.6	.000500
2.2	.110803	7.7	.000453
2.3	.100259	7.8	.000410
2.4	.090718	7.9	.000371
2.5	.082085	8.0	.000335
2.6	.074274	8.1	.000304
2.7	.067206	8.2	.000275
2.8	.060810	8.3	.000249
2.9	.055023	8.4	.000225
3.0	.049787	8.5	.000203
3.1	.045049	8.6	.000184
3.2	.040762	8.7	.000167
3.3	.036883	8.8	.000151
3.4	.033373	8.9	.000136
3.5	.030197	9.0	.000123
3.6	.027324	9.1	.000112
3.7	.024724	9.2	.000101
3.8	.022371	9.3	.000091
3.9	.020242	9.4	.000083
4.0	.018316	9.5	.000075
4.1	.016573	9.6	.000068
4.2	.014996	9.7	.000061
4.3	.013569	9.8	.000055
4.4	.012277	9.9	.000050
4.5	.011109	10.0	.0000454
4.6	.010052	11.0	.0000167
4.7	.009095	12.0	.00000614
4.8	.008230	13.0	.00000226
4.9	.007447	14.0	.00000083
5.0	.006738	15.0	.00000031
5.1	.006097	16.0	.00000011
5.2	.005517	17.0	.00000004
5.3	.004992	18.0	.00000001
5.4	.004517	19.0	.000000006
		20.0	.000000002

TABLE VI TABLE OF POISSON PROBABILITIES

					λ					
x	**0.1**	**0.2**	**0.3**	**0.4**	**0.5**	**0.6**	**0.7**	**0.8**	**0.9**	**1.0**
0	.9048	.8187	.7408	.6703	.6065	.5488	.4966	.4493	.4066	.3679
1	.0905	.1637	.2222	.2681	.3033	.3293	.3476	.3595	.3659	.3679
2	.0045	.0164	.0333	.0536	.0758	.0988	.1217	.1438	.1647	.1839
3	.0002	.0011	.0033	.0072	.0126	.0198	.0284	.0383	.0494	.0613
4	.0000	.0001	.0003	.0007	.0016	.0030	.0050	.0077	.0111	.0153
5	.0000	.0000	.0000	.0001	.0002	.0004	.0007	.0012	.0020	.0031
6	.0000	.0000	.0000	.0000	.0000	.0000	.0001	.0002	.0003	.0005
7	.0000	.0000	.0000	.0000	.0000	.0000	.0000	.0000	.0000	.0001

					λ					
x	**1.1**	**1.2**	**1.3**	**1.4**	**1.5**	**1.6**	**1.7**	**1.8**	**1.9**	**2.0**
0	.3329	.3012	.2725	.2466	.2231	.2019	.1827	.1653	.1496	.1353
1	.3662	.3614	.3543	.3452	.3347	.3230	.3106	.2975	.2842	.2707
2	.2014	.2169	.2303	.2417	.2510	.2584	.2640	.2678	.2700	.2707
3	.0738	.0867	.0998	.1128	.1255	.1378	.1496	.1607	.1710	.1804
4	.0203	.0260	.0324	.0395	.0471	.0551	.0636	.0723	.0812	.0902
5	.0045	.0062	.0084	.0111	.0141	.0176	.0216	.0260	.0309	.0361
6	.0008	.0012	.0018	.0026	.0035	.0047	.0061	.0078	.0098	.0120
7	.0001	.0002	.0003	.0005	.0008	.0011	.0015	.0020	.0027	.0034
8	.0000	.0000	.0001	.0001	.0001	.0002	.0003	.0005	.0006	.0009
9	.0000	.0000	.0000	.0000	.0000	.0000	.0001	.0001	.0001	.0002

					λ					
x	**2.1**	**2.2**	**2.3**	**2.4**	**2.5**	**2.6**	**2.7**	**2.8**	**2.9**	**3.0**
0	.1225	.1108	.1003	.0907	.0821	.0743	.0672	.0608	.0550	.0498
1	.2572	.2438	.2306	.2177	.2052	.1931	.1815	.1703	.1596	.1494
2	.2700	.2681	.2652	.2613	.2565	.2510	.2450	.2384	.2314	.2240
3	.1890	.1966	.2033	.2090	.2138	.2176	.2205	.2225	.2237	.2240
4	.0992	.1082	.1169	.1254	.1336	.1414	.1488	.1557	.1622	.1680
5	.0417	.0476	.0538	.0602	.0668	.0735	.0804	.0872	.0940	.1008
6	.0146	.0174	.0206	.0241	.0278	.0319	.0362	.0407	.0455	.0504
7	.0044	.0055	.0068	.0083	.0099	.0118	.0139	.0163	.0188	.0216
8	.0011	.0015	.0019	.0025	.0031	.0038	.0047	.0057	.0068	.0081
9	.0003	.0004	.0005	.0007	.0009	.0011	.0014	.0018	.0022	.0027
10	.0001	.0001	.0001	.0002	.0002	.0003	.0004	.0005	.0006	.0008
11	.0000	.0000	.0000	.0000	.0000	.0001	.0001	.0001	.0002	.0002
12	.0000	.0000	.0000	.0000	.0000	.0000	.0000	.0000	.0000	.0001

TABLE VI (*Continued*)

x	λ 3.1	3.2	3.3	3.4	3.5	3.6	3.7	3.8	3.9	4.0
0	.0450	.0408	.0369	.0334	.0302	.0273	.0247	.0224	.0202	.0183
1	.1397	.1304	.1217	.1135	.1057	.0984	.0915	.0850	.0789	.0733
2	.2165	.2087	.2008	.1929	.1850	.1771	.1692	.1615	.1539	.1465
3	.2237	.2226	.2209	.2186	.2158	.2125	.2087	.2046	.2001	.1954
4	.1733	.1781	.1823	.1858	.1888	.1912	.1931	.1944	.1951	.1954
5	.1075	.1140	.1203	.1264	.1322	.1377	.1429	.1477	.1522	.1563
6	.0555	.0608	.0662	.0716	.0771	.0826	.0881	.0936	.0989	.1042
7	.0246	.0278	.0312	.0348	.0385	.0425	.0466	.0508	.0551	.0595
8	.0095	.0111	.0129	.0148	.0169	.0191	.0215	.0241	.0269	.0298
9	.0033	.0040	.0047	.0056	.0066	.0076	.0089	.0102	.0116	.0132
10	.0010	.0013	.0016	.0019	.0023	.0028	.0033	.0039	.0045	.0053
11	.0003	.0004	.0005	.0006	.0007	.0009	.0011	.0013	.0016	.0019
12	.0001	.0001	.0001	.0002	.0002	.0003	.0003	.0004	.0005	.0006
13	.0000	.0000	.0000	.0000	.0001	.0001	.0001	.0001	.0002	.0002
14	.0000	.0000	.0000	.0000	.0000	.0000	.0000	.0000	.0000	.0001

x	λ 4.1	4.2	4.3	4.4	4.5	4.6	4.7	4.8	4.9	5.0
0	.0166	.0150	.0136	.0123	.0111	.0101	.0091	.0082	.0074	.0067
1	.0679	.0630	.0583	.0540	.0500	.0462	.0427	.0395	.0365	.0337
2	.1393	.1323	.1254	.1188	.1125	.1063	.1005	.0948	.0894	.0842
3	.1904	.1852	.1798	.1743	.1687	.1631	.1574	.1517	.1460	.1404
4	.1951	.1944	.1933	.1917	.1898	.1875	.1849	.1820	.1789	.1755
5	.1600	.1633	.1662	.1687	.1708	.1725	.1738	.1747	.1753	.1755
6	.1093	.1143	.1191	.1237	.1281	.1323	.1362	.1398	.1432	.1462
7	.0640	.0686	.0732	.0778	.0824	.0869	.0914	.0959	.1002	.1044
8	.0328	.0360	.0393	.0428	.0463	.0500	.0537	.0575	.0614	.0653
9	.0150	.0168	.0188	.0209	.0232	.0255	.0281	.0307	.0334	.0363
10	.0061	.0071	.0081	.0092	.0104	.0118	.0132	.0147	.0164	.0181
11	.0023	.0027	.0032	.0037	.0043	.0049	.0056	.0064	.0073	.0082
12	.0008	.0009	.0011	.0014	.0016	.0019	.0022	.0026	.0030	.0034
13	.0002	.0003	.0004	.0005	.0006	.0007	.0008	.0009	.0011	.0013
14	.0001	.0001	.0001	.0001	.0002	.0002	.0003	.0003	.0004	.0005
15	.0000	.0000	.0000	.0000	.0001	.0001	.0001	.0001	.0001	.0002

x	λ 5.1	5.2	5.3	5.4	5.5	5.6	5.7	5.8	5.9	6.0
0	.0061	.0055	.0050	.0045	.0041	.0037	.0033	.0030	.0027	.0025
1	.0311	.0287	.0265	.0244	.0225	.0207	.0191	.0176	.0162	.0149
2	.0793	.0746	.0701	.0659	.0618	.0580	.0544	.0509	.0477	.0446
3	.1348	.1293	.1239	.1185	.1133	.1082	.1033	.0985	.0938	.0892
4	.1719	.1681	.1641	.1600	.1558	.1515	.1472	.1428	.1383	.1339

TABLE VI (*Continued*)

x	λ 5.1	5.2	5.3	5.4	5.5	5.6	5.7	5.8	5.9	6.0
5	.1753	.1748	.1740	.1728	.1714	.1697	.1678	.1656	.1632	.1606
6	.1490	.1515	.1537	.1555	.1571	.1584	.1594	.1601	.1605	.1606
7	.1086	.1125	.1163	.1200	.1234	.1267	.1298	.1326	.1353	.1377
8	.0692	.0731	.0771	.0810	.0849	.0887	.0925	.0962	.0998	.1033
9	.0392	.0423	.0454	.0486	.0519	.0552	.0586	.0620	.0654	.0688
10	.0200	.0220	.0241	.0262	.0285	.0309	.0334	.0359	.0386	.0413
11	.0093	.0104	.0116	.0129	.0143	.0157	.0173	.0190	.0207	.0225
12	.0039	.0045	.0051	.0058	.0065	.0073	.0082	.0092	.0102	.0113
13	.0015	.0018	.0021	.0024	.0028	.0032	.0036	.0041	.0046	.0052
14	.0006	.0007	.0008	.0009	.0011	.0013	.0015	.0017	.0019	.0022
15	.0002	.0002	.0003	.0003	.0004	.0005	.0006	.0007	.0008	.0009
16	.0001	.0001	.0001	.0001	.0001	.0002	.0002	.0002	.0003	.0003
17	.0000	.0000	.0000	.0000	.0000	.0001	.0001	.0001	.0001	.0001

x	λ 6.1	6.2	6.3	6.4	6.5	6.6	6.7	6.8	6.9	7.0
0	.0022	.0020	.0018	.0017	.0015	.0014	.0012	.0011	.0010	.0009
1	.0137	.0126	.0116	.0106	.0098	.0090	.0082	.0076	.0070	.0064
2	.0417	.0390	.0364	.0340	.0318	.0296	.0276	.0258	.0240	.0223
3	.0848	.0806	.0765	.0726	.0688	.0652	.0617	.0584	.0552	.0521
4	.1294	.1249	.1205	.1162	.1118	.1076	.1034	.0992	.0952	.0912
5	.1579	.1549	.1519	.1487	.1454	.1420	.1385	.1349	.1314	.1277
6	.1605	.1601	.1595	.1586	.1575	.1562	.1546	.1529	.1511	.1490
7	.1399	.1418	.1435	.1450	.1462	.1472	.1480	.1486	.1489	.1490
8	.1066	.1099	.1130	.1160	.1188	.1215	.1240	.1263	.1284	.1304
9	.0723	.0757	.0791	.0825	.0858	.0891	.0923	.0954	.0985	.1014
10	.0441	.0469	.0498	.0528	.0558	.0588	.0618	.0649	.0679	.0710
11	.0244	.0265	.0285	.0307	.0330	.0353	.0377	.0401	.0426	.0452
12	.0124	.0137	.0150	.0164	.0179	.0194	.0210	.0227	.0245	.0263
13	.0058	.0065	.0073	.0081	.0089	.0099	.0108	.0119	.0130	.0142
14	.0025	.0029	.0033	.0037	.0041	.0046	.0052	.0058	.0064	.0071
15	.0010	.0012	.0014	.0016	.0018	.0020	.0023	.0026	.0029	.0033
16	.0004	.0005	.0005	.0006	.0007	.0008	.0010	.0011	.0013	.0014
17	.0001	.0002	.0002	.0002	.0003	.0003	.0004	.0004	.0005	.0006
18	.0000	.0001	.0001	.0001	.0001	.0001	.0001	.0002	.0002	.0002
19	.0000	.0000	.0000	.0000	.0000	.0000	.0001	.0001	.0001	.0001

x	λ 7.1	7.2	7.3	7.4	7.5	7.6	7.7	7.8	7.9	8.0
0	.0008	.0007	.0007	.0006	.0006	.0005	.0005	.0004	.0004	.0003
1	.0059	.0054	.0049	.0045	.0041	.0038	.0035	.0032	.0029	.0027
2	.0208	.0194	.0180	.0167	.0156	.0145	.0134	.0125	.0116	.0107
3	.0492	.0464	.0438	.0413	.0389	.0366	.0345	.0324	.0305	.0286
4	.0874	.0836	.0799	.0764	.0729	.0696	.0663	.0632	.0602	.0573

TABLE VI (*Continued*)

					λ					
x	**7.1**	**7.2**	**7.3**	**7.4**	**7.5**	**7.6**	**7.7**	**7.8**	**7.9**	**8.0**
5	.1241	.1204	.1167	.1130	.1094	.1057	.1021	.0986	.0951	.0916
6	.1468	.1445	.1420	.1394	.1367	.1339	.1311	.1282	.1252	.1221
7	.1489	.1486	.1481	.1474	.1465	.1454	.1442	.1428	.1413	.1396
8	.1321	.1337	.1351	.1363	.1373	.1381	.1388	.1392	.1395	.1396
9	.1042	.1070	.1096	.1121	.1144	.1167	.1187	.1207	.1224	.1241
10	.0740	.0770	.0800	.0829	.0858	.0887	.0914	.0941	.0967	.0993
11	.0478	.0504	.0531	.0558	.0585	.0613	.0640	.0667	.0695	.0722
12	.0283	.0303	.0323	.0344	.0366	.0388	.0411	.0434	.0457	.0481
13	.0154	.0168	.0181	.0196	.0211	.0227	.0243	.0260	.0278	.0296
14	.0078	.0086	.0095	.0104	.0113	.0123	.0134	.0145	.0157	.0169
15	.0037	.0041	.0046	.0051	.0057	.0062	.0069	.0075	.0083	.0090
16	.0016	.0019	.0021	.0024	.0026	.0030	.0033	.0037	.0041	.0045
17	.0007	.0008	.0009	.0010	.0012	.0013	.0015	.0017	.0019	.0021
18	.0003	.0003	.0004	.0004	.0005	.0006	.0006	.0007	.0008	.0009
19	.0001	.0001	.0001	.0002	.0002	.0002	.0003	.0003	.0003	.0004
20	.0000	.0000	.0001	.0001	.0001	.0001	.0001	.0001	.0001	.0002
21	.0000	.0000	.0000	.0000	.0000	.0000	.0000	.0000	.0001	.0001

					λ					
x	**8.1**	**8.2**	**8.3**	**8.4**	**8.5**	**8.6**	**8.7**	**8.8**	**8.9**	**9.0**
0	.0003	.0003	.0002	.0002	.0002	.0002	.0002	.0002	.0001	.0001
1	.0025	.0023	.0021	.0019	.0017	.0016	.0014	.0013	.0012	.0011
2	.0100	.0092	.0086	.0079	.0074	.0068	.0063	.0058	.0054	.0050
3	.0269	.0252	.0237	.0222	.0208	.0195	.0183	.0171	.0160	.0150
4	.0544	.0517	.0491	.0466	.0443	.0420	.0398	.0377	.0357	.0337
5	.0882	.0849	.0816	.0784	.0752	.0722	.0692	.0663	.0635	.0607
6	.1191	.1160	.1128	.1097	.1066	.1034	.1003	.0972	.0941	.0911
7	.1378	.1358	.1338	.1317	.1294	.1271	.1247	.1222	.1197	.1171
8	.1395	.1392	.1388	.1382	.1375	.1366	.1356	.1344	.1332	.1318
9	.1255	.1269	.1280	.1290	.1299	.1306	.1311	.1315	.1317	.1318
10	.1017	.1040	.1063	.1084	.1104	.1123	.1140	.1157	.1172	.1186
11	.0749	.0775	.0802	.0828	.0853	.0878	.0902	.0925	.0948	.0970
12	.0505	.0530	.0555	.0579	.0604	.0629	.0654	.0679	.0703	.0728
13	.0315	.0334	.0354	.0374	.0395	.0416	.0438	.0459	.0481	.0504
14	.0182	.0196	.0210	.0225	.0240	.0256	.0272	.0289	.0306	.0324
15	.0098	.0107	.0116	.0126	.0136	.0147	.0158	.0169	.0182	.0194
16	.0050	.0055	.0060	.0066	.0072	.0079	.0086	.0093	.0101	.0109
17	.0024	.0026	.0029	.0033	.0036	.0040	.0044	.0048	.0053	.0058
18	.0011	.0012	.0014	.0015	.0017	.0019	.0021	.0024	.0026	.0029
19	.0005	.0005	.0006	.0007	.0008	.0009	.0010	.0011	.0012	.0014
20	.0002	.0002	.0002	.0003	.0003	.0004	.0004	.0005	.0005	.0006
21	.0001	.0001	.0001	.0001	.0001	.0002	.0002	.0002	.0002	.0003
22	.0000	.0000	.0000	.0000	.0001	.0001	.0001	.0001	.0001	.0001

TABLE VI *(Continued)*

	λ									
x	9.1	9.2	9.3	9.4	9.5	9.6	9.7	9.8	9.9	10
0	.0001	.0001	.0001	.0001	.0001	.0001	.0001	.0001	.0001	.0000
1	.0010	.0009	.0009	.0008	.0007	.0007	.0006	.0005	.0005	.0005
2	.0046	.0043	.0040	.0037	.0034	.0031	.0029	.0027	.0025	.0023
3	.0140	.0131	.0123	.0115	.0107	.0100	.0093	.0087	.0081	.0076
4	.0319	.0302	.0285	.0269	.0254	.0240	.0226	.0213	.0201	.0189
5	.0581	.0555	.0530	.0506	.0483	.0460	.0439	.0418	.0398	.0378
6	.0881	.0851	.0822	.0793	.0764	.0736	.0709	.0682	.0656	.0631
7	.1145	.1118	.1091	.1064	.1037	.1010	.0982	.0955	.0928	.0901
8	.1302	.1286	.1269	.1251	.1232	.1212	.1191	.1170	.1148	.1126
9	.1317	.1315	.1311	.1306	.1300	.1293	.1284	.1274	.1263	.1251
10	.1198	.1209	.1219	.1228	.1235	.1241	.1245	.1249	.1250	.1251
11	.0991	.1012	.1031	.1049	.1067	.1083	.1098	.1112	.1125	.1137
12	.0752	.0776	.0799	.0822	.0844	.0866	.0888	.0908	.0928	.0948
13	.0526	.0549	.0572	.0594	.0617	.0640	.0662	.0685	.0707	.0729
14	.0342	.0361	.0380	.0399	.0419	.0439	.0459	.0479	.0500	.0521
15	.0208	.0221	.0235	.0250	.0265	.0281	.0297	.0313	.0330	.0347
16	.0118	.0127	.0137	.0147	.0157	.0168	.0180	.0192	.0204	.0217
17	.0063	.0069	.0075	.0081	.0088	.0095	.0103	.0111	.0119	.0128
18	.0032	.0035	.0039	.0042	.0046	.0051	.0055	.0060	.0065	.0071
19	.0015	.0017	.0019	.0021	.0023	.0026	.0028	.0031	.0034	.0037
20	.0007	.0008	.0009	.0010	.0011	.0012	.0014	.0015	.0017	.0019
21	.0003	.0003	.0004	.0004	.0005	.0006	.0006	.0007	.0008	.0009
22	.0001	.0001	.0002	.0002	.0002	.0002	.0003	.0003	.0004	.0004
23	.0000	.0001	.0001	.0001	.0001	.0001	.0001	.0001	.0002	.0002
24	.0000	.0000	.0000	.0000	.0000	.0000	.0000	.0001	.0001	.0001

	λ									
x	11	12	13	14	15	16	17	18	19	20
0	.0000	.0000	.0000	.0000	.0000	.0000	.0000	.0000	.0000	.0000
1	.0002	.0001	.0000	.0000	.0000	.0000	.0000	.0000	.0000	.0000
2	.0010	.0004	.0002	.0001	.0000	.0000	.0000	.0000	.0000	.0000
3	.0037	.0018	.0008	.0004	.0002	.0001	.0000	.0000	.0000	.0000
4	.0102	.0053	.0027	.0013	.0006	.0003	.0001	.0001	.0000	.0000
5	.0224	.0127	.0070	.0037	.0019	.0010	.0005	.0002	.0001	.0001
6	.0411	.0255	.0152	.0087	.0048	.0026	.0014	.0007	.0004	.0002
7	.0646	.0437	.0281	.0174	.0104	.0060	.0034	.0019	.0010	.0005
8	.0888	.0655	.0457	.0304	.0194	.0120	.0072	.0042	.0024	.0013
9	.1085	.0874	.0661	.0473	.0324	.0213	.0135	.0083	.0050	.0029
10	.1194	.1048	.0859	.0663	.0486	.0341	.0230	.0150	.0095	.0058
11	.1194	.1144	.1015	.0844	.0663	.0496	.0355	.0245	.0164	.0106
12	.1094	.1144	.1099	.0984	.0829	.0661	.0504	.0368	.0259	.0176
13	.0926	.1056	.1099	.1060	.0956	.0814	.0658	.0509	.0378	.0271
14	.0728	.0905	.1021	.1060	.1024	.0930	.0800	.0655	.0514	.0387

TABLE VI (*Continued*)

					λ					
x	**11**	**12**	**13**	**14**	**15**	**16**	**17**	**18**	**19**	**20**
15	.0534	.0724	.0885	.0989	.1024	.0992	.0906	.0786	.0650	.0516
16	.0367	.0543	.0719	.0866	.0960	.0992	.0963	.0884	.0772	.0646
17	.0237	.0383	.0550	.0713	.0847	.0934	.0963	.0936	.0863	.0760
18	.0145	.0255	.0397	.0554	.0706	.0830	.0909	.0936	.0911	.0844
19	.0084	.0161	.0272	.0409	.0557	.0699	.0814	.0887	.0911	.0888
20	.0046	.0097	.0177	.0286	.0418	.0559	.0692	.0798	.0866	.0888
21	.0024	.0055	.0109	.0191	.0299	.0426	.0560	.0684	.0783	.0846
22	.0012	.0030	.0065	.0121	.0204	.0310	.0433	.0560	.0676	.0769
23	.0006	.0016	.0037	.0074	.0133	.0216	.0320	.0438	.0559	.0669
24	.0003	.0008	.0020	.0043	.0083	.0144	.0226	.0328	.0442	.0557
25	.0001	.0004	.0010	.0024	.0050	.0092	.0154	.0237	.0336	.0446
26	.0000	.0002	.0005	.0013	.0029	.0057	.0101	.0164	.0246	.0343
27	.0000	.0001	.0002	.0007	.0016	.0034	.0063	.0109	.0173	.0254
28	.0000	.0000	.0001	.0003	.0009	.0019	.0038	.0070	.0117	.0181
29	.0000	.0000	.0001	.0002	.0004	.0011	.0023	.0044	.0077	.0125
30	.0000	.0000	.0000	.0001	.0002	.0006	.0013	.0026	.0049	.0083
31	.0000	.0000	.0000	.0000	.0001	.0003	.0007	.0015	.0030	.0054
32	.0000	.0000	.0000	.0000	.0001	.0001	.0004	.0009	.0018	.0034
33	.0000	.0000	.0000	.0000	.0000	.0001	.0002	.0005	.0010	.0020
34	.0000	.0000	.0000	.0000	.0000	.0000	.0001	.0002	.0006	.0012
35	.0000	.0000	.0000	.0000	.0000	.0000	.0000	.0001	.0003	.0007
36	.0000	.0000	.0000	.0000	.0000	.0000	.0000	.0001	.0002	.0004
37	.0000	.0000	.0000	.0000	.0000	.0000	.0000	.0000	.0001	.0002
38	.0000	.0000	.0000	.0000	.0000	.0000	.0000	.0000	.0000	.0001
39	.0000	.0000	.0000	.0000	.0000	.0000	.0000	.0000	.0000	.0001

TABLE VII STANDARD NORMAL DISTRIBUTION TABLE

The entries in the table give the areas under the standard normal curve from 0 to z.

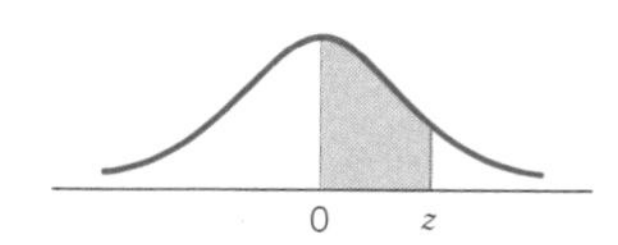

z	.00	.01	.02	.03	.04	.05	.06	.07	.08	.09
0.0	.0000	.0040	.0080	.0120	.0160	.0199	.0239	.0279	.0319	.0359
0.1	.0398	.0438	.0478	.0517	.0557	.0596	.0636	.0675	.0714	.0753
0.2	.0793	.0832	.0871	.0910	.0948	.0987	.1026	.1064	.1103	.1141
0.3	.1179	.1217	.1255	.1293	.1331	.1368	.1406	.1443	.1480	.1517
0.4	.1554	.1591	.1628	.1664	.1700	.1736	.1772	.1808	.1844	.1879
0.5	.1915	.1950	.1985	.2019	.2054	.2088	.2123	.2157	.2190	.2224
0.6	.2257	.2291	.2324	.2357	.2389	.2422	.2454	.2486	.2517	.2549
0.7	.2580	.2611	.2642	.2673	.2704	.2734	.2764	.2794	.2823	.2852
0.8	.2881	.2910	.2939	.2967	.2995	.3023	.3051	.3078	.3106	.3133
0.9	.3159	.3186	.3212	.3238	.3264	.3289	.3315	.3340	.3365	.3389
1.0	.3413	.3438	.3461	.3485	.3508	.3531	.3554	.3577	.3599	.3621
1.1	.3643	.3665	.3686	.3708	.3729	.3749	.3770	.3790	.3810	.3830
1.2	.3849	.3869	.3888	.3907	.3925	.3944	.3962	.3980	.3997	.4015
1.3	.4032	.4049	.4066	.4082	.4099	.4115	.4131	.4147	.4162	.4177
1.4	.4192	.4207	.4222	.4236	.4251	.4265	.4279	.4292	.4306	.4319
1.5	.4332	.4345	.4357	.4370	.4382	.4394	.4406	.4418	.4429	.4441
1.6	.4452	.4463	.4474	.4484	.4495	.4505	.4515	.4525	.4535	.4545
1.7	.4554	.4564	.4573	.4582	.4591	.4599	.4608	.4616	.4625	.4633
1.8	.4641	.4649	.4656	.4664	.4671	.4678	.4686	.4693	.4699	.4706
1.9	.4713	.4719	.4726	.4732	.4738	.4744	.4750	.4756	.4761	.4767
2.0	.4772	.4778	.4783	.4788	.4793	.4798	.4803	.4808	.4812	.4817
2.1	.4821	.4826	.4830	.4834	.4838	.4842	.4846	.4850	.4854	.4857
2.2	.4861	.4864	.4868	.4871	.4875	.4878	.4881	.4884	.4887	.4890
2.3	.4893	.4896	.4898	.4901	.4904	.4906	.4909	.4911	.4913	.4916
2.4	.4918	.4920	.4922	.4925	.4927	.4929	.4931	.4932	.4934	.4936
2.5	.4938	.4940	.4941	.4943	.4945	.4946	.4948	.4949	.4951	.4952
2.6	.4953	.4955	.4956	.4957	.4959	.4960	.4961	.4962	.4963	.4964
2.7	.4965	.4966	.4967	.4968	.4969	.4970	.4971	.4972	.4973	.4974
2.8	.4974	.4975	.4976	.4977	.4977	.4978	.4979	.4979	.4980	.4981
2.9	.4981	.4982	.4982	.4983	.4984	.4984	.4985	.4985	.4986	.4986
3.0	.4987	.4987	.4987	.4988	.4988	.4989	.4989	.4989	.4990	.4990

TABLE VIII THE t DISTRIBUTION TABLE

The entries in the table give the critical values of t for the specified number of degrees of freedom and areas in the right tail.

	Area in the Right Tail under the t Distribution Curve					
df	.10	.05	.025	.01	.005	.001
1	3.078	6.314	12.706	31.821	63.657	318.309
2	1.886	2.920	4.303	6.965	9.925	22.327
3	1.638	2.353	3.182	4.541	5.841	10.215
4	1.533	2.132	2.776	3.747	4.604	7.173
5	1.476	2.015	2.571	3.365	4.032	5.893
6	1.440	1.943	2.447	3.143	3.707	5.208
7	1.415	1.895	2.365	2.998	3.499	4.785
8	1.397	1.860	2.306	2.896	3.355	4.501
9	1.383	1.833	2.262	2.821	3.250	4.297
10	1.372	1.812	2.228	2.764	3.169	4.144
11	1.363	1.796	2.201	2.718	3.106	4.025
12	1.356	1.782	2.179	2.681	3.055	3.930
13	1.350	1.771	2.160	2.650	3.012	3.852
14	1.345	1.761	2.145	2.624	2.977	3.787
15	1.341	1.753	2.131	2.602	2.947	3.733
16	1.337	1.746	2.120	2.583	2.921	3.686
17	1.333	1.740	2.110	2.567	2.898	3.646
18	1.330	1.734	2.101	2.552	2.878	3.610
19	1.328	1.729	2.093	2.539	2.861	3.579
20	1.325	1.725	2.086	2.528	2.845	3.552
21	1.323	1.721	2.080	2.518	2.831	3.527
22	1.321	1.717	2.074	2.508	2.819	3.505
23	1.319	1.714	2.069	2.500	2.807	3.485
24	1.318	1.711	2.064	2.492	2.797	3.467
25	1.316	1.708	2.060	2.485	2.787	3.450
26	1.315	1.706	2.056	2.479	2.779	3.435
27	1.314	1.703	2.052	2.473	2.771	3.421
28	1.313	1.701	2.048	2.467	2.763	3.408
29	1.311	1.699	2.045	2.462	2.756	3.396
30	1.310	1.697	2.042	2.457	2.750	3.385
31	1.309	1.696	2.040	2.453	2.744	3.375
32	1.309	1.694	2.037	2.449	2.738	3.365
33	1.308	1.692	2.035	2.445	2.733	3.356
34	1.307	1.691	2.032	2.441	2.728	3.348
35	1.306	1.690	2.030	2.438	2.724	3.340
36	1.306	1.688	2.028	2.434	2.719	3.333
37	1.305	1.687	2.026	2.431	2.715	3.326
38	1.304	1.686	2.024	2.429	2.712	3.319

TABLE VIII (*Continued*)

	Area in the Right Tail under the *t* Distribution Curve					
df	.10	.05	.025	.01	.005	.001
39	1.304	1.685	2.023	2.426	2.708	3.313
40	1.303	1.684	2.021	2.423	2.704	3.307
41	1.303	1.683	2.020	2.421	2.701	3.301
42	1.302	1.682	2.018	2.418	2.698	3.296
43	1.302	1.681	2.017	2.416	2.695	3.291
44	1.301	1.680	2.015	2.414	2.692	3.286
45	1.301	1.679	2.014	2.412	2.690	3.281
46	1.300	1.679	2.013	2.410	2.687	3.277
47	1.300	1.678	2.012	2.408	2.685	3.273
48	1.299	1.677	2.011	2.407	2.682	3.269
49	1.299	1.677	2.010	2.405	2.680	3.265
50	1.299	1.676	2.009	2.403	2.678	3.261
51	1.298	1.675	2.008	2.402	2.676	3.258
52	1.298	1.675	2.007	2.400	2.674	3.255
53	1.298	1.674	2.006	2.399	2.672	3.251
54	1.297	1.674	2.005	2.397	2.670	3.248
55	1.297	1.673	2.004	2.396	2.668	3.245
56	1.297	1.673	2.003	2.395	2.667	3.242
57	1.297	1.672	2.002	2.394	2.665	3.239
58	1.296	1.672	2.002	2.392	2.663	3.237
59	1.296	1.671	2.001	2.391	2.662	3.234
60	1.296	1.671	2.000	2.390	2.660	3.232
61	1.296	1.670	2.000	2.389	2.659	3.229
62	1.295	1.670	1.999	2.388	2.657	3.227
63	1.295	1.669	1.998	2.387	2.656	3.225
64	1.295	1.669	1.998	2.386	2.655	3.223
65	1.295	1.669	1.997	2.385	2.654	3.220
66	1.295	1.668	1.997	2.384	2.652	3.218
67	1.294	1.668	1.996	2.383	2.651	3.216
68	1.294	1.668	1.995	2.382	2.650	3.214
69	1.294	1.667	1.995	2.382	2.649	3.213
70	1.294	1.667	1.994	2.381	2.648	3.211
71	1.294	1.667	1.994	2.380	2.647	3.209
72	1.293	1.666	1.993	2.379	2.646	3.207
73	1.293	1.666	1.993	2.379	2.645	3.206
74	1.293	1.666	1.993	2.378	2.644	3.204
75	1.293	1.665	1.992	2.377	2.643	3.202
∞	1.282	1.645	1.960	2.326	2.576	3.090

TABLE IX CHI-SQUARE DISTRIBUTION TABLE

The entries in the table give the critical values of χ^2 for the specified number of degrees of freedom and areas in the right tail.

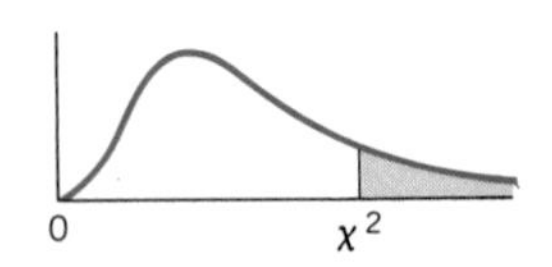

df	Area in the Right Tail under the Chi-square Distribution Curve									
	.995	.990	.975	.950	.900	.100	.050	.025	.010	.005
1	0.000	0.000	0.001	0.004	0.016	2.706	3.841	5.024	6.635	7.879
2	0.010	0.020	0.051	0.103	0.211	4.605	5.991	7.378	9.210	10.597
3	0.072	0.115	0.216	0.352	0.584	6.251	7.815	9.348	11.345	12.838
4	0.207	0.297	0.484	0.711	1.064	7.779	9.488	11.143	13.277	14.860
5	0.412	0.554	0.831	1.145	1.610	9.236	11.070	12.833	15.086	16.750
6	0.676	0.872	1.237	1.635	2.204	10.645	12.592	14.449	16.812	18.548
7	0.989	1.239	1.690	2.167	2.833	12.017	14.067	16.013	18.475	20.278
8	1.344	1.646	2.180	2.733	3.490	13.362	15.507	17.535	20.090	21.955
9	1.735	2.088	2.700	3.325	4.168	14.684	16.919	19.023	21.666	23.589
10	2.156	2.558	3.247	3.940	4.865	15.987	18.307	20.483	23.209	25.188
11	2.603	3.053	3.816	4.575	5.578	17.275	19.675	21.920	24.725	26.757
12	3.074	3.571	4.404	5.226	6.304	18.549	21.026	23.337	26.217	28.300
13	3.565	4.107	5.009	5.892	7.042	19.812	22.362	24.736	27.688	29.819
14	4.075	4.660	5.629	6.571	7.790	21.064	23.685	26.119	29.141	31.319
15	4.601	5.229	6.262	7.261	8.547	22.307	24.996	27.488	30.578	32.801
16	5.142	5.812	6.908	7.962	9.312	23.542	26.296	28.845	32.000	34.267
17	5.697	6.408	7.564	8.672	10.085	24.769	27.587	30.191	33.409	35.718
18	6.265	7.015	8.231	9.390	10.865	25.989	28.869	31.526	34.805	37.156
19	6.844	7.633	8.907	10.117	11.651	27.204	30.144	32.852	36.191	38.582
20	7.434	8.260	9.591	10.851	12.443	28.412	31.410	34.170	37.566	39.997
21	8.034	8.897	10.283	11.591	13.240	29.615	32.671	35.479	38.932	41.401
22	8.643	9.542	10.982	12.338	14.041	30.813	33.924	36.781	40.289	42.796
23	9.260	10.196	11.689	13.091	14.848	32.007	35.172	38.076	41.638	44.181
24	9.886	10.856	12.401	13.848	15.659	33.196	36.415	39.364	42.980	45.559
25	10.520	11.524	13.120	14.611	16.473	34.382	37.652	40.646	44.314	46.928
26	11.160	12.198	13.844	15.379	17.292	35.563	38.885	41.923	45.642	48.290
27	11.808	12.879	14.573	16.151	18.114	36.741	40.113	43.195	46.963	49.645
28	12.461	13.565	15.308	16.928	18.939	37.916	41.337	44.461	48.278	50.993
29	13.121	14.256	16.047	17.708	19.768	39.087	42.557	45.722	49.588	52.336
30	13.787	14.953	16.791	18.493	20.599	40.256	43.773	46.979	50.892	53.672
40	20.707	22.164	24.433	26.509	29.051	51.805	55.758	59.342	63.691	66.766
50	27.991	29.707	32.357	34.764	37.689	63.167	67.505	71.420	76.154	79.490
60	35.534	37.485	40.482	43.188	46.459	74.397	79.082	83.298	88.379	91.952
70	43.275	45.442	48.758	51.739	55.329	85.527	90.531	95.023	100.425	104.215
80	51.172	53.540	57.153	60.391	64.278	96.578	101.879	106.629	112.329	116.321
90	59.196	61.754	65.647	69.126	73.291	107.565	113.145	118.136	124.116	128.299
100	67.328	70.065	74.222	77.929	82.358	118.498	124.342	129.561	135.807	140.169

TABLE X THE *F* DISTRIBUTION TABLE

Area in the Right Tail under the Curve = .01

Degrees of Freedom for Numerator (columns); Degrees of Freedom for Denominator (rows)

	1	2	3	4	5	6	7	8	9	10	11	12	15	20	25	30	40	50	100
1	4052	5000	5403	5625	5764	5859	5928	5981	6022	6056	6083	6106	6157	6209	6240	6261	6287	6303	6334
2	98.50	99.00	99.17	99.25	99.30	99.33	99.36	99.37	99.39	99.40	99.41	99.42	99.43	99.45	99.46	99.47	99.47	99.48	99.49
3	34.12	30.82	29.46	28.71	28.24	27.91	27.67	27.49	27.35	27.23	27.13	27.05	26.87	26.69	26.58	26.50	26.41	26.35	26.24
4	21.20	18.00	16.69	15.98	15.52	15.21	14.98	14.80	14.66	14.55	14.45	14.37	14.20	14.02	13.91	13.84	13.75	13.69	13.58
5	16.26	13.27	12.06	11.39	10.97	10.67	10.46	10.29	10.16	10.05	9.96	9.89	9.72	9.55	9.45	9.38	9.29	9.24	9.13
6	13.75	10.92	9.78	9.15	8.75	8.47	8.26	8.10	7.98	7.87	7.79	7.72	7.56	7.40	7.30	7.23	7.14	7.09	6.99
7	12.25	9.55	8.45	7.85	7.46	7.19	6.99	6.84	6.72	6.62	6.54	6.47	6.31	6.16	6.06	5.99	5.91	5.86	5.75
8	11.26	8.65	7.59	7.01	6.63	6.37	6.18	6.03	5.91	5.81	5.73	5.67	5.52	5.36	5.26	5.20	5.12	5.07	4.96
9	10.56	8.02	6.99	6.42	6.06	5.80	5.61	5.47	5.35	5.26	5.18	5.11	4.96	4.81	4.71	4.65	4.57	4.52	4.41
10	10.04	7.56	6.55	5.99	5.64	5.39	5.20	5.06	4.94	4.85	4.77	4.71	4.56	4.41	4.31	4.25	4.17	4.12	4.01
11	9.65	7.21	6.22	5.67	5.32	5.07	4.89	4.74	4.63	4.54	4.46	4.40	4.25	4.10	4.01	3.94	3.86	3.81	3.71
12	9.33	6.93	5.95	5.41	5.06	4.82	4.64	4.50	4.39	4.30	4.22	4.16	4.01	3.86	3.76	3.70	3.62	3.57	3.47
13	9.07	6.70	5.74	5.21	4.86	4.62	4.44	4.30	4.19	4.10	4.02	3.96	3.82	3.66	3.57	3.51	3.43	3.38	3.27
14	8.86	6.51	5.56	5.04	4.69	4.46	4.28	4.14	4.03	3.94	3.86	3.80	3.66	3.51	3.41	3.35	3.27	3.22	3.11
15	8.68	6.36	5.42	4.89	4.56	4.32	4.14	4.00	3.89	3.80	3.73	3.67	3.52	3.37	3.28	3.21	3.13	3.08	2.98
16	8.53	6.23	5.29	4.77	4.44	4.20	4.03	3.89	3.78	3.69	3.62	3.55	3.41	3.26	3.16	3.10	3.02	2.97	2.86
17	8.40	6.11	5.18	4.67	4.34	4.10	3.93	3.79	3.68	3.59	3.52	3.46	3.31	3.16	3.07	3.00	2.92	2.87	2.76
18	8.29	6.01	5.09	4.58	4.25	4.01	3.84	3.71	3.60	3.51	3.43	3.37	3.23	3.08	2.98	2.92	2.84	2.78	2.68
19	8.18	5.93	5.01	4.50	4.17	3.94	3.77	3.63	3.52	3.43	3.36	3.30	3.15	3.00	2.91	2.84	2.76	2.71	2.60
20	8.10	5.85	4.94	4.43	4.10	3.87	3.70	3.56	3.46	3.37	3.29	3.23	3.09	2.94	2.84	2.78	2.69	2.64	2.54
21	8.02	5.78	4.87	4.37	4.04	3.81	3.64	3.51	3.40	3.31	3.24	3.17	3.03	2.88	2.79	2.72	2.64	2.58	2.48
22	7.95	5.72	4.82	4.31	3.99	3.76	3.59	3.45	3.35	3.26	3.18	3.12	2.98	2.83	2.73	2.67	2.58	2.53	2.42
23	7.88	5.66	4.76	4.26	3.94	3.71	3.54	3.41	3.30	3.21	3.14	3.07	2.93	2.78	2.69	2.62	2.54	2.48	2.37
24	7.82	5.61	4.72	4.22	3.90	3.67	3.50	3.36	3.26	3.17	3.09	3.03	2.89	2.74	2.64	2.58	2.49	2.44	2.33
25	7.77	5.57	4.68	4.18	3.85	3.63	3.46	3.32	3.22	3.13	3.06	2.99	2.85	2.70	2.60	2.54	2.45	2.40	2.29
30	7.56	5.39	4.51	4.02	3.70	3.47	3.30	3.17	3.07	2.98	2.91	2.84	2.70	2.55	2.45	2.39	2.30	2.25	2.13
40	7.31	5.18	4.31	3.83	3.51	3.29	3.12	2.99	2.89	2.80	2.73	2.66	2.52	2.37	2.27	2.20	2.11	2.06	1.94
50	7.17	5.06	4.20	3.72	3.41	3.19	3.02	2.89	2.78	2.70	2.63	2.56	2.42	2.27	2.17	2.10	2.01	1.95	1.82
100	6.90	4.82	3.98	3.51	3.21	2.99	2.82	2.69	2.59	2.50	2.43	2.37	2.22	2.07	1.97	1.89	1.80	1.74	1.60

TABLE X *(Continued)*

Area in the Right Tail under the Curve = .025

.025

0 F

Degrees of Freedom for Denominator

	Degrees of Freedom for Numerator																	
	1	2	3	4	5	6	7	8	9	10	11	12	15	20	25	30	40	50	100
1	647.8	799.5	864.2	899.6	921.8	937.1	948.2	956.7	963.3	968.6	973.0	976.7	984.9	993.1	998.1	1001	1006	1008	1013
2	38.51	39.00	39.17	39.25	39.30	39.33	39.36	39.37	39.39	39.40	39.41	39.41	39.43	39.45	39.46	39.46	39.47	39.48	39.49
3	17.44	16.04	15.44	15.10	14.88	14.73	14.62	14.54	14.47	14.42	14.37	14.34	14.25	14.17	14.12	14.08	14.04	14.01	13.96
4	12.22	10.65	9.98	9.61	9.36	9.20	9.07	8.98	8.90	8.84	8.79	8.75	8.66	8.56	8.50	8.46	8.41	8.38	8.32
5	10.01	8.43	7.76	7.39	7.15	6.98	6.85	6.76	6.68	6.62	6.57	6.52	6.43	6.33	6.27	6.23	6.18	6.14	6.08
6	8.81	7.26	6.60	6.23	5.99	5.82	5.70	5.60	5.52	5.46	5.41	5.37	5.27	5.17	5.11	5.07	5.01	4.98	4.92
7	8.07	6.54	5.89	5.52	5.29	5.12	4.99	4.90	4.82	4.76	4.71	4.67	4.57	4.47	4.40	4.36	4.31	4.28	4.21
8	7.57	6.06	5.42	5.05	4.82	4.65	4.53	4.43	4.36	4.30	4.24	4.20	4.10	4.00	3.94	3.89	3.84	3.81	3.74
9	7.21	5.72	5.08	4.72	4.48	4.32	4.20	4.10	4.03	3.96	3.91	3.87	3.77	3.67	3.60	3.56	3.51	3.47	3.40
10	6.94	5.46	4.83	4.47	4.24	4.07	3.95	3.85	3.78	3.72	3.66	3.62	3.52	3.42	3.35	3.31	3.26	3.22	3.15
11	6.72	5.26	4.63	4.28	4.04	3.88	3.76	3.66	3.59	3.53	3.47	3.43	3.33	3.23	3.16	3.12	3.06	3.03	2.96
12	6.55	5.10	4.47	4.12	3.89	3.73	3.61	3.51	3.44	3.37	3.32	3.28	3.18	3.07	3.01	2.96	2.91	2.87	2.80
13	6.41	4.97	4.35	4.00	3.77	3.60	3.48	3.39	3.31	3.25	3.20	3.15	3.05	2.95	2.88	2.84	2.78	2.74	2.67
14	6.30	4.86	4.24	3.89	3.66	3.50	3.38	3.29	3.21	3.15	3.09	3.05	2.95	2.84	2.78	2.73	2.67	2.64	2.56
15	6.20	4.77	4.15	3.80	3.58	3.41	3.29	3.20	3.12	3.06	3.01	2.96	2.86	2.76	2.69	2.64	2.59	2.55	2.47
16	6.12	4.69	4.08	3.73	3.50	3.34	3.22	3.12	3.05	2.99	2.93	2.89	2.79	2.68	2.61	2.57	2.51	2.47	2.40
17	6.04	4.62	4.01	3.66	3.44	3.28	3.16	3.06	2.98	2.92	2.87	2.82	2.72	2.62	2.55	2.50	2.44	2.41	2.33
18	5.98	4.56	3.95	3.61	3.38	3.22	3.10	3.01	2.93	2.87	2.81	2.77	2.67	2.56	2.49	2.44	2.38	2.35	2.27
19	5.92	4.51	3.90	3.56	3.33	3.17	3.05	2.96	2.88	2.82	2.76	2.72	2.62	2.51	2.44	2.39	2.33	2.30	2.22
20	5.87	4.46	3.86	3.51	3.29	3.13	3.01	2.91	2.84	2.77	2.72	2.68	2.57	2.46	2.40	2.35	2.29	2.25	2.17
21	5.83	4.42	3.82	3.48	3.25	3.09	2.97	2.87	2.80	2.73	2.68	2.64	2.53	2.42	2.36	2.31	2.25	2.21	2.13
22	5.79	4.38	3.78	3.44	3.22	3.05	2.93	2.84	2.76	2.70	2.65	2.60	2.50	2.39	2.32	2.27	2.21	2.17	2.09
23	5.75	4.35	3.75	3.41	3.18	3.02	2.90	2.81	2.73	2.67	2.62	2.57	2.47	2.36	2.29	2.24	2.18	2.14	2.06
24	5.72	4.32	3.72	3.38	3.15	2.99	2.87	2.78	2.70	2.64	2.59	2.54	2.44	2.33	2.26	2.21	2.15	2.11	2.02
25	5.69	4.29	3.69	3.35	3.13	2.97	2.85	2.75	2.68	2.61	2.56	2.51	2.41	2.30	2.23	2.18	2.12	2.08	2.00
30	5.57	4.18	3.59	3.25	3.03	2.87	2.75	2.65	2.57	2.51	2.46	2.41	2.31	2.20	2.12	2.07	2.01	1.97	1.88
40	5.42	4.05	3.46	3.13	2.90	2.74	2.62	2.53	2.45	2.39	2.33	2.29	2.18	2.07	1.99	1.94	1.88	1.83	1.74
50	5.34	3.97	3.39	3.05	2.83	2.67	2.55	2.46	2.38	2.32	2.26	2.22	2.11	1.99	1.92	1.87	1.80	1.75	1.66
100	5.18	3.83	3.25	2.92	2.70	2.54	2.42	2.32	2.24	2.18	2.12	2.08	1.97	1.85	1.77	1.71	1.64	1.59	1.48

TABLE X (*Continued*)

Area in the Right Tail under the Curve = .05

Degrees of Freedom for Denominator	Degrees of Freedom for Numerator																		
	1	2	3	4	5	6	7	8	9	10	11	12	15	20	25	30	40	50	100
1	161.5	199.5	215.7	224.6	230.2	234.0	236.8	238.9	240.5	241.9	243.0	243.9	246.0	248.0	249.3	250.1	251.1	251.8	253.0
2	18.51	19.00	19.16	19.25	19.30	19.33	19.35	19.37	19.38	19.40	19.40	19.41	19.43	19.45	19.46	19.46	19.47	19.48	19.49
3	10.13	9.55	9.28	9.12	9.01	8.94	8.89	8.85	8.81	8.79	8.76	8.74	8.70	8.66	8.63	8.62	8.59	8.58	8.55
4	7.71	6.94	6.59	6.39	6.26	6.16	6.09	6.04	6.00	5.96	5.94	5.91	5.86	5.80	5.77	5.75	5.72	5.70	5.66
5	6.61	5.79	5.41	5.19	5.05	4.95	4.88	4.82	4.77	4.74	4.70	4.68	4.62	4.56	4.52	4.50	4.46	4.44	4.41
6	5.99	5.14	4.76	4.53	4.39	4.28	4.21	4.15	4.10	4.06	4.03	4.00	3.94	3.87	3.83	3.81	3.77	3.75	3.71
7	5.59	4.74	4.35	4.12	3.97	3.87	3.79	3.73	3.68	3.64	3.60	3.57	3.51	3.44	3.40	3.38	3.34	3.32	3.27
8	5.32	4.46	4.07	3.84	3.69	3.58	3.50	3.44	3.39	3.35	3.31	3.28	3.22	3.15	3.11	3.08	3.04	3.02	2.97
9	5.12	4.26	3.86	3.63	3.48	3.37	3.29	3.23	3.18	3.14	3.10	3.07	3.01	2.94	2.89	2.86	2.83	2.80	2.76
10	4.96	4.10	3.71	3.48	3.33	3.22	3.14	3.07	3.02	2.98	2.94	2.91	2.85	2.77	2.73	2.70	2.66	2.64	2.59
11	4.84	3.98	3.59	3.36	3.20	3.09	3.01	2.95	2.90	2.85	2.82	2.79	2.72	2.65	2.60	2.57	2.53	2.51	2.46
12	4.75	3.89	3.49	3.26	3.11	3.00	2.91	2.85	2.80	2.75	2.72	2.69	2.62	2.54	2.50	2.47	2.43	2.40	2.35
13	4.67	3.81	3.41	3.18	3.03	2.92	2.83	2.77	2.71	2.67	2.63	2.60	2.53	2.46	2.41	2.38	2.34	2.31	2.26
14	4.60	3.74	3.34	3.11	2.96	2.85	2.76	2.70	2.65	2.60	2.57	2.53	2.46	2.39	2.34	2.31	2.27	2.24	2.19
15	4.54	3.68	3.29	3.06	2.90	2.79	2.71	2.64	2.59	2.54	2.51	2.48	2.40	2.33	2.28	2.25	2.20	2.18	2.12
16	4.49	3.63	3.24	3.01	2.85	2.74	2.66	2.59	2.54	2.49	2.46	2.42	2.35	2.28	2.23	2.19	2.15	2.12	2.07
17	4.45	3.59	3.20	2.96	2.81	2.70	2.61	2.55	2.49	2.45	2.41	2.38	2.31	2.23	2.18	2.15	2.10	2.08	2.02
18	4.41	3.55	3.16	2.93	2.77	2.66	2.58	2.51	2.46	2.41	2.37	2.34	2.27	2.19	2.14	2.11	2.06	2.04	1.98
19	4.38	3.52	3.13	2.90	2.74	2.63	2.54	2.48	2.42	2.38	2.34	2.31	2.23	2.16	2.11	2.07	2.03	2.00	1.94
20	4.35	3.49	3.10	2.87	2.71	2.60	2.51	2.45	2.39	2.35	2.31	2.28	2.20	2.12	2.07	2.04	1.99	1.97	1.91
21	4.32	3.47	3.07	2.84	2.68	2.57	2.49	2.42	2.37	2.32	2.28	2.25	2.18	2.10	2.05	2.01	1.96	1.94	1.88
22	4.30	3.44	3.05	2.82	2.66	2.55	2.46	2.40	2.34	2.30	2.26	2.23	2.15	2.07	2.02	1.97	1.94	1.91	1.85
23	4.28	3.42	3.03	2.80	2.64	2.53	2.44	2.37	2.32	2.27	2.24	2.20	2.13	2.05	2.00	1.96	1.91	1.88	1.82
24	4.26	3.40	3.01	2.78	2.62	2.51	2.42	2.36	2.30	2.25	2.22	2.18	2.16	2.03	1.97	1.94	1.89	1.86	1.80
25	4.24	3.39	2.99	2.76	2.60	2.49	2.40	2.34	2.28	2.24	2.20	2.16	2.09	2.01	1.96	1.92	1.87	1.84	1.78
30	4.17	3.32	2.92	2.69	2.53	2.42	2.33	2.27	2.21	2.16	2.13	2.09	2.01	1.93	1.88	1.84	1.79	1.76	1.70
40	4.08	3.23	2.84	2.61	2.45	2.34	2.25	2.18	2.12	2.08	2.04	2.00	1.92	1.84	1.78	1.74	1.69	1.66	1.59
50	4.03	3.18	2.79	2.56	2.40	2.29	2.20	2.13	2.07	2.03	1.99	1.95	1.87	1.78	1.73	1.69	1.63	1.60	1.52
100	3.94	3.09	2.70	2.46	2.31	2.19	2.10	2.03	1.97	1.93	1.89	1.85	1.77	1.68	1.62	1.57	1.52	1.48	1.39

TABLE X *(Continued)*

Area in the Right Tail under the Curve = .10

Degrees of Freedom for Denominator	Degrees of Freedom for Numerator																		
	1	2	3	4	5	6	7	8	9	10	11	12	15	20	25	30	40	50	100
1	39.86	49.50	53.59	55.83	57.24	58.20	58.91	59.44	59.86	60.19	60.47	60.71	61.22	61.74	62.05	62.26	62.53	62.69	63.01
2	8.53	9.00	9.16	9.24	9.29	9.33	9.35	9.37	9.38	9.39	9.40	9.41	9.42	9.44	9.45	9.46	9.47	9.47	9.48
3	5.54	5.46	5.39	5.34	5.31	5.28	5.27	5.25	5.24	5.23	5.22	5.22	5.20	5.18	5.17	5.17	5.16	5.15	5.14
4	4.54	4.32	4.19	4.11	4.05	4.01	3.98	3.95	3.94	3.92	3.91	3.90	3.87	3.84	3.83	3.82	3.80	3.80	3.78
5	4.06	3.78	3.62	3.52	3.45	3.40	3.37	3.34	3.32	3.30	3.28	3.27	3.24	3.21	3.19	3.17	3.16	3.15	3.13
6	3.78	3.46	3.29	3.18	3.11	3.05	3.01	2.98	2.96	2.94	2.92	2.90	2.87	2.84	2.81	2.80	2.78	2.77	2.75
7	3.59	3.26	3.07	2.96	2.88	2.83	2.78	2.75	2.72	2.70	2.68	2.67	2.63	2.59	2.57	2.56	2.54	2.52	2.50
8	3.46	3.11	2.92	2.81	2.73	2.67	2.62	2.59	2.56	2.54	2.52	2.50	2.46	2.42	2.40	2.38	2.36	2.35	2.32
9	3.36	3.01	2.81	2.69	2.61	2.55	2.51	2.47	2.44	2.42	2.40	2.38	2.34	2.30	2.27	2.25	2.23	2.22	2.19
10	3.29	2.92	2.73	2.61	2.52	2.46	2.41	2.38	2.35	2.32	2.30	2.28	2.24	2.20	2.17	2.16	2.13	2.12	2.09
11	3.23	2.86	2.66	2.54	2.45	2.39	2.34	2.30	2.27	2.25	2.23	2.21	2.17	2.12	2.10	2.08	2.05	2.04	2.01
12	3.18	2.81	2.61	2.48	2.39	2.33	2.28	2.24	2.21	2.19	2.17	2.15	2.10	2.06	2.03	2.01	1.99	1.97	1.94
13	3.14	2.76	2.56	2.43	2.35	2.28	2.23	2.20	2.16	2.14	2.12	2.10	2.05	2.01	1.98	1.96	1.93	1.92	1.88
14	3.10	2.73	2.52	2.39	2.31	2.24	2.19	2.15	2.12	2.10	2.07	2.05	2.01	1.96	1.93	1.91	1.89	1.87	1.83
15	3.07	2.70	2.49	2.36	2.27	2.21	2.16	2.12	2.09	2.06	2.04	2.02	1.97	1.92	1.89	1.87	1.85	1.83	1.79
16	3.05	2.67	2.46	2.33	2.24	2.18	2.13	2.09	2.06	2.03	2.01	1.99	1.94	1.89	1.86	1.84	1.81	1.79	1.76
17	3.03	2.64	2.44	2.31	2.22	2.15	2.10	2.06	2.03	2.00	1.98	1.96	1.91	1.86	1.83	1.81	1.78	1.76	1.73
18	3.01	2.62	2.42	2.29	2.20	2.13	2.08	2.04	2.00	1.98	1.95	1.93	1.89	1.84	1.80	1.78	1.75	1.74	1.70
19	2.99	2.61	2.40	2.27	2.18	2.11	2.06	2.02	1.98	1.96	1.93	1.91	1.86	1.81	1.78	1.76	1.73	1.71	1.67
20	2.97	2.59	2.38	2.25	2.16	2.09	2.04	2.00	1.96	1.94	1.91	1.89	1.84	1.79	1.76	1.74	1.71	1.69	1.65
21	2.96	2.57	2.36	2.23	2.14	2.08	2.02	1.98	1.95	1.92	1.90	1.87	1.83	1.78	1.74	1.72	1.69	1.67	1.63
22	2.95	2.56	2.35	2.22	2.13	2.06	2.01	1.97	1.93	1.90	1.88	1.86	1.81	1.76	1.73	1.70	1.67	1.65	1.61
23	2.94	2.55	2.34	2.21	2.11	2.05	1.99	1.95	1.92	1.89	1.87	1.84	1.80	1.74	1.71	1.69	1.66	1.64	1.59
24	2.93	2.54	2.33	2.19	2.10	2.04	1.98	1.94	1.91	1.88	1.85	1.83	1.78	1.73	1.70	1.67	1.64	1.62	1.58
25	2.92	2.53	2.32	2.18	2.09	2.02	1.97	1.93	1.89	1.87	1.84	1.82	1.77	1.72	1.68	1.66	1.63	1.61	1.56
30	2.88	2.49	2.28	2.14	2.05	1.98	1.93	1.88	1.85	1.82	1.79	1.77	1.72	1.67	1.63	1.61	1.57	1.55	1.51
40	2.84	2.44	2.23	2.09	2.00	1.93	1.87	1.83	1.79	1.76	1.74	1.71	1.66	1.61	1.57	1.54	1.51	1.48	1.43
50	2.81	2.41	2.20	2.06	1.97	1.90	1.84	1.80	1.76	1.73	1.70	1.68	1.63	1.57	1.53	1.50	1.46	1.44	1.39
100	2.76	2.36	2.14	2.00	1.91	1.83	1.78	1.73	1.69	1.66	1.64	1.61	1.56	1.49	1.45	1.42	1.38	1.35	1.29

ANSWERS TO ODD-NUMBERED EXERCISES AND SELF-REVIEW TESTS

(Note: Due to differences in rounding, the answers obtained by readers may differ slightly from the ones given below.)

CHAPTER 1

1.11 a. quantitative b. quantitative c. qualitative
d. quantitative e. quantitative f. qualitative

1.17 a. $\Sigma x = 13$ b. $\Sigma(x - 1) = 7$ c. $(\Sigma x)^2 = 169$ d. $\Sigma x^2 = 35$

1.19 a. $\Sigma x = 61$ b. $\Sigma(x - 4) = 33$ c. $(\Sigma x)^2 = 3721$ d. $\Sigma x^2 = 555$

1.21 a. $\Sigma f = 59$ b. $\Sigma m^2 = 819$ c. $\Sigma mf = 576$ d. $\Sigma m^2 f = 7614$ e. $\Sigma(m - 10)^2 f = 1994$

1.23 a. $\Sigma x = 88$ b. $\Sigma y = 58$ c. $\Sigma xy = 855$ d. $\Sigma x^2 = 1590$ e. $\Sigma y^2 = 622$

1.27 a. sample b. population for that week c. sample d. population e. population

1.29 a. $\Sigma x = 56$ b. $\Sigma(x - 6) = 20$ c. $(\Sigma x)^2 = 3136$ d. $\Sigma x^2 = 586$

Self-Review Test: Chapter 1

1. b 2. c

3. a. sampling without replacement b. representative sample c. sampling with replacement

4. a. qualitative variable b. quantitative (discrete) variable c. quantitative (continuous) variable

6. a. cross-section data b. time-series data c. cross-section data

7. a. $\Sigma x = 36$ b. $\Sigma x^2 = 232$ c. $(\Sigma x)^2 = 1296$ d. $\Sigma(x - 2) = 24$ e. $\Sigma(x - 2)^2 = 112$

8. a. $\Sigma m = 45$ b. $\Sigma f = 112$ c. $\Sigma mf = 975$ d. $\Sigma m^2 f = 9855$

CHAPTER 2

2.1 c. 52% 2.3 c. 46.7%

2.9 b. Yes, each class has a width equal to 5.

2.11 a. 1–100, 101–200, 201–300, 301–400, 401–500, 501–600, 601–700

2.13 c. skewed d. 4%

2.17 c. 68% 2.25 c. skewed to the right

2.27 c. 76.6% 2.29 c. 57.9%

2.37 80% received food stamps worth $2400 or less.

2.57 d. 53.3% 2.59 c. 27.1%

2.61 d. Boundaries of the third class are 2000.5 and 3000.5; width = 1000

2.63 c. 5.8% 2.65 c. Class width is equal to 15 for each class.

Self-Review Test: Chapter 2

2. a. 5 b. 10 c. 54.5 d. 39.5 e. 49 f. 200 g. .105

4. a. & b.

Class	f	Relative Frequency	Percentage
B	9	.45	45
F	4	.20	20
M	6	.30	30
S	1	.05	5

c. 30% of the children live with mother only.

5. a. & b.

Class	f	Relative Frequency	Percentage
1–4	3	.125	12.5
5–8	6	.250	25.0
9–12	7	.292	29.2
13–16	6	.250	25.0
17–20	2	.083	8.3

c. 12.5 + 25.0 = 37.5% of the instructors have eight or fewer years of teaching experience.

6.

Class	f	Cumulative Relative Frequency	Cumulative Percentage
1–4	3	.125	12.5
1–8	9	.375	37.5
1–12	16	.667	66.7
1–16	22	.917	91.7
1–20	24	1.000	100.0

Approximately 59% of the instructors have a teaching experience of 11 or fewer years.

7.

0	47
1	112578
2	689
3	11289
4	137
5	1
6	27

8. 30 33 37 42 44 46 47 49 51 53 53 56 60 67 67 71 79

CHAPTER 3

3.1 $\bar{x}$ = \$490.75; median = \$475.50

3.3 $\bar{x}$ = \$18.34; median = \$17.44; these data have no mode.

3.5 $\bar{x}$ = \$4.97; median = \$4.85

3.7 $\bar{x}$ = 9.75; median = 11.50; these data have no mode.

3.9 $\bar{x}$ = 166.10 days; median = 165 days; mode = 154 and 163 days

3.11 a. $\bar{x}$ = \$10,728.11; median = \$1500
b. Yes, \$85,000 is an outlier. $\bar{x}$ = \$1444.13; median = \$1200.
Mean is affected more than median. c. median

3.13 mean = 6.67 cars; median = 6.50 cars; mode = 3, 6, and 7 cars

3.15 combined mean = \$41.89 3.17 age of the sixth person = 48 years

3.19 mean for data set I = 9.40; mean for data set II = 18.80.
The mean of the second data set is twice that of the first data set.

3.23 mean = \$86. Deviations of values from the mean are −4, 30, −21, −11, 6. The sum of these deviations is zero.

3.25 range = 81; s^2 = 557.268; s = 23.61

3.27 range = 26 cars; s^2 = 72.278; s = 8.50 cars

3.29 range = 16 miles per hour; s^2 = 22.667; s = 4.76 miles per hour

3.31 range = \$29 hundred; s^2 = 82.933; s = \$9.11 hundred

3.33 range = \$44 thousand; s^2 = 206.473; s = \$14.37 thousand

3.35 s for data set I = 4.04; s for data set II = 8.07
The standard deviation of the second data set is twice the standard deviation of the first data set.

3.37 $\bar{x}$ = 10.30; s^2 = 24.163; s = 4.92

3.39 $\bar{x}$ = 9.40; s^2 = 37.711; s = 6.14

3.41 $\bar{x}$ = \$27.20; s^2 = 176.694; s = \$13.29

3.43 $\bar{x}$ = 74.40 hours; s^2 = 506.092; s = 22.50 hours

3.45 $\bar{x}$ = 11.83 patients; s^2 = 36.751; s = 6.06 patients

3.47 $\bar{x}$ = 36.80 minutes; s^2 = 597.714; s = 24.45 minutes

3.49 $\bar{x}$ = 1.57 cars; s^2 = .975; s = .99 cars

3.51 **a.** at least 75% **b.** at least 89% **c.** at least 84%

3.53 **a.** at least 75% **b.** at least 89%

3.55 **a.** 68% **b.** 95% **c.** 99.7%

3.57 **a.** 68% **b.** 95% **c.** 99.7%

3.59 **a.** Q_1 = 77; Q_2 = 87.5; Q_3 = 93 **b.** P_{30} = 77 **c.** 57.14

3.61 **a.** Q_1 = 35; Q_2 = 40; Q_3 = 43 **b.** P_{79} = 45 **c.** 36.67

3.63 **a.** Q_1 = 4; Q_2 = 6.5; Q_3 = 8.5 **b.** 66.67

3.65 **a.** Q_1 = 8; Q_2 = 10; Q_3 = 14 **b.** P_{82} = 14 **c.** 40

3.75 **a.** $\bar{x}$ = \$531.44; median = \$88; mode = \$25 and \$100
b. Yes. After dropping the outlier 6000: $\bar{x}$ = 166.87; median = 76; mode = 25 and 100. Mean changes by a larger amount when we drop the outlier.

3.77 **a.** $\bar{x}$ = \$27.33 thousand; median = \$22.70 thousand; no mode
b. range = \$50 thousand, s^2 = 301.582, s = \$17.37 thousand

3.79 **a.** $\bar{x}$ = 7.33 citations; median = 7.50 citations; mode = 4, 7, and 8 citations
b. range = 14 citations; s^2 = 17.152; s = 4.14 citations

3.81 **a.** $\bar{x}$ = 5.16 inches; s^2 = 6.994; s = 2.64 inches

3.83 $\bar{x}$ = \$126; s^2 = 3180.808; s = \$56.40. These summary measures are the sample statistics because data are for 100 randomly selected persons only.

3.85 μ = 4.75 days; σ^2 = 13.388; σ = 3.66 days. These summary measures are the population parameters because data include all 40 employees of the company.

3.87 **a.** at least 75% **b.** at least 56% **c.** at least 84%

3.89 **a.** 68% **b.** 95% **c.** 99.7%

3.91 **a.** Q_1 = 3.5; Q_2 = 7; Q_3 = 10.5
b. P_{79} = 9 **c.** percentile rank of 7 = 44.44

Self-Review Test: Chapter 3

1. b **2.** a and d **3.** c **4.** c **5.** b **6.** b

7. a **8.** a **9.** b **10.** a **11.** b **12.** a

13. $\bar{x}$ = 9.8; median = 8; mode = 6; range = 20; s^2 = 42.178; s = 6.49

16. The value of the standard deviation for a data set is zero when all the values in the data set are the same.

17. **a.** The frequency column gives the number of days for which the items sold were in the corresponding class.
b. $\bar{x}$ = 37.88 items; s^2 = 40.369; s = 6.35 items

18. **a.** at least 75% **b.** at least 89%

19. **a.** 95% **b.** 99.7%

20. Q_1 = 10; Q_2 = 12; Q_3 = 18 **21.** P_{68} = 16

22. percentile rank of 15 = 56.25

24. $\bar{x}$ = \$466.43 **25.** 3.17 **26.** 20% trimmed mean = 85.17

27. a. mean for data set I = 19.75; mean for data set II = 16.75. The mean of the second data set is equal to the mean of the first data set minus 3.
b. s for data set I = 11.32; s for data set II = 11.32. The standard deviations of the two data sets are equal.

CHAPTER 4

4.5 Possible outcomes: CC, CW, WC, WW

4.7 Possible outcomes: FFF, FFA, FAF, FAA, AFF, AFA, AAF, AAA

4.9 a. RG and GR. A compound event.
b. RR, RG, and GR. A compound event.
c. RR, GR, and RG. A compound event.
d. GR. A simple event.

4.11 a. GG, GD, DG. A compound event.
b. GD and DG. A compound event.
c. GD. A simple event.
d. GD, DG, and DD. A compound event.

4.15 −.35, 1.56, 5/3, −2/7

4.17 .267 4.19 .540 4.21 .510

4.23 a. .333 b. .667

4.25 a. .539 b. .461

4.27 a. .304 b. .217 c. .268 d. .211

4.29 .885 4.31 a. .304 b. .696

4.33 .858 4.35 .200

4.37 a. .512 b. .771 c. .512 d. .519

4.39 a. i. .530 ii. .846 iii. .859 iv. .719
b. Events "scientist" and "engineer" are mutually exclusive.
Events "engineer" and "male" are not mutually exclusive.
c. Events "female" and "engineer" are dependent.

4.41 a. i. .558 ii. .443 iii. .564 iv. .458
b. Events "male" and "in favor" are not mutually exclusive.
Events "in favor" and "against" are mutually exclusive.
c. Events "female" and "in favor" are dependent.

4.43 dependent events

4.45 a. mutually exclusive and dependent events
b. yes; $P(B) = .250$; $P(A) = .750$

4.47 $P(A) = .333$; $P(\bar{A}) = .667$

4.49 $P(\bar{A}) = .850$

4.51 a. .128 b. .234 c. .443

4.53 a. .165 b. .285 c. .370

4.55 a. .014 b. .039

4.57 .800

4.59 a. .056 b. .175

4.61 a. i. .417 ii. .083 b. .000

4.63 .333 4.65 .238 4.67 .058 4.69 .913

4.71 a. .312 b. .182

4.73 .908 4.75 .507 4.77 1296

4.79 792 4.81 960

4.83 a. .440 b. .860

4.85 a. .420 b. .470

4.87 a. .843 b. .653

4.89 a. .927 b. .480

4.91 .810 4.93 .830 4.95 .846

4.97 .880 4.99 b. .538

4.101 a. .818 b. .688

4.103 Possible outcomes: *DD*, *DR*, *RD*, *RR*

4.105 Possible outcomes: *PP*, *PF*, *FP*, *FF*

4.107 a. *GG*, *GB*, *BG*. A compound event.
b. *GB*, *BG*, *BB*. A compound event.
c. *GB*, *BG*. A compound event.

4.109 a. .257 b. .143

4.111 a. i. .395 ii. .574 iii. .760 iv. .835 v. .300 vi. .700
b. dependent and mutually nonexclusive events

4.113 a. i. .563 ii. .275 iii. .409 iv. .711 v. .163 vi. .838
b. dependent and mutually nonexclusive events

4.115 a. i. .750 ii. .700 iii. .225 iv. .775
b. dependent and mutually nonexclusive events

4.117 i. .436 ii. .884

4.119 a. .470 b. .320 c. .360 d. .830

Self-Review Test: Chapter 4

1. a 2. b 3. c 4. a
5. a 6. b 7. c 8. b
9. b 10. c 11. b 12. .500

13. a. dependent and mutually nonexclusive events
b. i. .400 ii. .679

14. .780 15. .160 16. .423

17. a. i. .360 ii. .275 iii. .350 iv. .650

18. dependent and mutually nonexclusive events

CHAPTER 5

5.3 a. discrete random variable b. continuous random variable
c. continuous random variable d. discrete random variable
e. discrete random variable f. continuous random variable

5.7 a. not a valid probability distribution
b. not a valid probability distribution
c. a valid probability distribution

5.9 b. i. .14 ii. .32 iii. .50 iv. .44

5.11 a.

x	2	3	4	5	6	7	8
$P(x)$	.0667	.1600	.2133	.2933	.1733	.0667	.0267

b. i. .2933 ii. .7733 iii. .8399 iv. .7333

5.13

x	0	1	2
$P(x)$	.1444	.4712	.3844

5.15

x	0	1	2
$P(x)$	.7056	.2688	.0256

5.17

x	0	1	2
$P(x)$	.1474	.5052	.3474

5.19 **a.** $\mu = 5.10$; $\sigma = 1.145$ **b.** $\mu = .92$; $\sigma = .977$

5.21 $\mu = 2.43$ members; $\sigma = 1.202$ members

5.23 $\mu = 6.41$ pigs; $\sigma = 1.866$ pigs

5.25 $\mu = 2.34$ students; $\sigma = 1.306$ students

5.27 $\mu = 4.65$ shoe pairs; $\sigma = 1.434$ shoe pairs

5.29 $\mu = 1.24$ persons; $\sigma = .686$ persons

5.31 $\mu = -\$.40$; $\sigma = \$54.780$

5.33 $\mu = 1.20$ males; $\sigma = .674$ males

5.35 720; 39,916,800; 120; 40,320; 28; 1; 1; 15; 330

5.37 66 **5.39** 220 **5.41** 15,504 **5.43** n and p

5.45 **a.** a binomial experiment **b.** not a binomial experiment **c.** a binomial experiment

5.47 **a.** .5905 **b.** .1147 **c.** .0425 **5.49** .0991

5.51 **a.** .9998 **b.** .4401 **c.** .0429

5.53 .1102 **5.55** .1641

5.57 **a.** .0835 **b.** .1582 **c.** .5817 **5.59** .1829

5.61 **a.** .9888 **b.** .8761 **c.** .3232

5.63 **a.** .0600 **b.** .3311 **c.** .2782 **5.65** .2362

5.67

x	0	1	2	3	4
$P(x)$	.0081	.0756	.2646	.4116	.2401

$\mu = 2.80$ households; $\sigma = .917$ households

5.69

x	0	1	2	3	4	5
$P(x)$	.4783	.3720	.1240	.0230	.0026	.0002

$\mu = .70$ freshmen; $\sigma = .794$ freshmen

5.73 **a.** .1992 **b.** .0771

5.75 **a.**

x	0	1	2	3	4	5
$P(x)$	.5488	.3293	.0988	.0198	.0030	.0004

b.

x	0	1	2	3	4	5	6	7	8	9
$P(x)$	.1653	.2975	.2678	.1607	.0723	.0260	.0078	.0020	.0005	.0001

5.77 .0829 **5.79** .0978

5.81 **a.** .2311 **b.** **i.** .8741 **ii.** .0244 **iii.** .6920

5.83 **a.** .0985 **b.** **i.** .1700 **ii.** .0068

5.85 **a.** .2231

b.

x	0	1	2	3	4	5	6	7	8
$P(x)$	.2231	.3347	.2510	.1255	.0471	.0141	.0035	.0008	.0001

5.87 **a.** .3614

b.

x	0	1	2	3	4	5	6	7
$P(x)$	.3012	.3614	.2169	.0867	.0260	.0062	.0012	.0002

5.89 **a.** .08 **b.** .88 **c.** .45 **d.** .96

5.91

x	0	1	2
$P(x)$	.1156	.4488	.4356

5.93 μ = 4.16 cars; σ = 1.037 cars **5.95** μ = 1.01 accidents; σ = 1.127 accidents

5.97 720; 40,320; 3,628,800; 1; 120; 28

5.99 1820 **5.101** .2780 **5.103** .1652 **5.105** .3336

5.107 **a.** .0003 **b.** .2499 **c.** .8924

5.109 **a.** .5011 **b.** .0162

5.111 **a.** .0915 **b.** **i.** .0037 **ii.** .0996

5.113 **a.** .0383

b.

x	0	1	2	3	4	5	6
$P(x)$	.4493	.3595	.1438	.0383	.0077	.0012	.0002

Self-Review Test: Chapter 5

2. probability distribution table

3. a **4.** b **5.** a **7.** b

8. a **9.** **a.** ii **b.** i **c.** iii **10.** a

12.

x	0	1	2
$P(x)$	.5625	.3750	.0625

13. μ = .50 households; σ = .612 households

14.

x	0	1	2
$P(x)$	.5625	.3750	.0625

15. **a.** .1678 **b.** .8821 **c.** .0017

16. **a.** **i.** .0361 **ii.** .9473 **iii.** .0526

b.

x	0	1	2	3	4	5	6	7	8	9
$P(x)$	.1353	.2707	.2707	.1804	.0902	.0361	.0120	.0034	.0009	.0002

CHAPTER 6

6.3 **a.** .4713 **b.** .4599 **c.** .0967 **d.** .0603 **e.** .9429

6.5 **a.** .0594 **b.** .0244 **c.** .9798 **d.** .9686

6.7 **a.** .5 approximately **b.** .5 approximately **c.** .00 approximately
d. .00 approximately

6.9 **a.** .9626 **b.** .4830 **c.** .4706 **d.** .0838

6.11 **a.** .0207 **b.** .2430 **c.** .1841 **d.** .9564

6.13 **a.** .7823 **b.** .8553 **c.** .5 approximately **d.** .5 approximately
e. .00 approximately **f.** .00 approximately

6.15 **a.** 1.40 **b.** −2.20 **c.** −1.40 **d.** 2.80

6.17 **a.** .4599 **b.** .1210 **c.** .2223

6.19 **a.** .3336 **b.** .9564 **c.** .9564 **d.** .00

6.21 **a.** .2178 **b.** .5997

6.23 a. .7967 b. .3372 c. .0475 d. .7734

6.25 a. .1867 b. .4036

6.27 a. .7291 b. .1573

6.29 a. .0571 b. .9370

6.31 a. 8.59% b. 57.85%

6.33 a. 13.14% b. 65%

6.35 4.75%

6.37 a. .1711 b. .2267

6.39 a. 81.59% b. 84.13%

6.41 2.64%

6.43 a. .51 approximately b. −.75 approximately c. −.82 approximately d. 1.25 approximately

6.45 a. 1.96 b. −1.65 approximately c. −3.08 d. 2.33 approximately

6.47 a. .0764 b. .6793 c. .8413 d. .8238

6.49 .7398

6.51 .0382

6.53 .0934

6.55 .2776

6.57 .2461

6.59 a. .0778 b. .2514

6.61 a. .0779 b. .0392

6.63 a. .2351 b. .0823

6.65 a. .0207 b. .2705

6.67 a. .7333 b. .0228

6.69 a. 5.82% b. 3.01%

6.71 a. 66.78% b. 4.01%

6.73 a. 18.10% b. 79.67%

6.75 .0471

6.77 .6361

6.79 .0195

6.81 .7553

Self-Review Test: Chapter 6

1. a
2. a
3. d
4. b
5. a
6. c
7. b
8. b
9. a. .1823 b. .9264 c. .1170 d. .7611
10. a. −1.28 b. .61 c. 1.65 d. −1.07
11. a. .7894 b. .0336 c. .6141 d. .0817
12. a. .2725 b. .0192 c. .0474

CHAPTER 7

7.7 a. $\mu_{\bar{x}} = 90$; $\sigma_{\bar{x}} = 5.06$ b. $\mu_{\bar{x}} = 90$; $\sigma_{\bar{x}} = 2.70$

7.9 $\mu_{\bar{x}} = 461$ acres; $\sigma_{\bar{x}} = 6.26$ acres

7.11 $\mu_{\bar{x}} = 172.2$ pounds; $\sigma_{\bar{x}} = 2.39$ pounds

7.15 $\mu_{\bar{x}} = 68$ miles per hour; $\sigma_{\bar{x}} = .67$ miles per hour; the normal distribution

7.17 $\mu_{\bar{x}} = 3.02$; $\sigma_{\bar{x}} = .04$; approximately normal distribution

7.19 $\mu_{\bar{x}} = \$70$; $\sigma_{\bar{x}} = \$2.64$; approximately normal distribution

7.21 $\mu_{\bar{x}} = \$36{,}011$; $\sigma_{\bar{x}} = \$626.10$; approximately normal distribution

7.23 a. .1530 b. .7611

7.25 a. .1940 b. .8749
7.27 a. .0014 b. .0681
7.29 a. .00 b. .9505
7.31 a. .9782 b. .0048
7.33 a. .8664 b. .0668
7.35 a. .60 b. 10
7.37 $\mu_{\hat{p}} = p = .69$; $\sigma_{\hat{p}} = .033$; approximately normal distribution
7.39 $\mu_{\hat{p}} = p = .591$; $\sigma_{\hat{p}} = .055$; approximately normal distribution
7.41 a. .1983 b. .7088
7.43 a. .7620 b. .7224
7.45 a. .8469 b. .8944
7.47 $\mu_{\bar{x}} = 7.6$ days; $\sigma_{\bar{x}} = .098$ days
7.49 $\mu_{\bar{x}} = 80$ minutes; $\sigma_{\bar{x}} = 1.668$ minutes; the normal distribution
7.51 $\mu_{\bar{x}} = 70$ pounds; $\sigma_{\bar{x}} = .949$ pounds; approximately normal distribution
7.53 a. .1229 b. .0516
7.55 a. .1446 b. .1276
7.57 a. .1401 b. .1367
7.59 a. .7763 b. .7852
7.61 $\mu_{\hat{p}} = .82$; $\sigma_{\hat{p}} = .019$; approximately normal distribution
7.63 a. .2515 b. .2119
7.65 a. .9201 b. .9015

Self-Review Test: Chapter 7

1. b 2. b 3. a 4. a 5. b
6. b 7. c 8. a 9. b 10. a
11. c 12. a
14. a. $\mu_{\bar{x}} = 145$ pounds; $\sigma_{\bar{x}} = 3.6$ pounds; approximately normal distribution
b. $\mu_{\bar{x}} = 145$ pounds; $\sigma_{\bar{x}} = 1.8$ pounds; approximately normal distribution
15. $\mu_{\bar{x}} = 47$ minutes; $\sigma_{\bar{x}} = 1.188$ minutes; approximately normal distribution
16. a. .2793 b. .1292 c. .6628 d. .6046
17. a. $\mu_{\hat{p}} = .52$; $\sigma_{\hat{p}} = .050$; approximately normal distribution
b. $\mu_{\hat{p}} = .52$; $\sigma_{\hat{p}} = .025$; approximately normal distribution
18. a. .9306 b. .2164 c. .1335 d. .8984

CHAPTER 8

8.3 a. 14.53 to 17.47 b. 14.76 to 17.24 c. 14.96 to 17.04
8.5 $787.65 to $810.35 8.7 $1736.59 to $1811.41
8.9 8.13 to 8.27 hours 8.11 3.53 to 3.67 kilograms
8.13 $38,797.26 to $39,148.74 8.15 a. $349.42 to $390.58
8.21 a. 60.92 to 70.08 b. 61.73 to 69.27 c. 61.95 to 69.05
8.23 $32,157.58 to $33,842.42 8.25 2.29 to 3.05
8.27 36.84 to 39.16 bushels 8.29 31.75 to 32.21 ounces
8.31 a. 65.19 to 70.81 minutes 8.33 14.52 to 32.98 hours

8.37 a. .73 to .79 b. .74 to .78 c. .72 to .80
8.39 .32 to .36 **8.41** .24 to .28 **8.43** .41 to .47
8.45 .30 to .36 **8.47** .65 to .77 **8.49** .29 to .37
8.51 .45 to .51 **8.53** a. $n = 299$ b. $n = 126$ c. $n = 61$
8.55 $n = 84$ **8.57** $n = 221$
8.59 a. $n = 1849$ b. $n = 601$ c. $n = 6807$
8.61 $n = 4161$ **8.63** $n = 1417$ **8.65** 22.20 to 22.76 hours
8.67 \$6.59 to \$6.91 **8.69** \$2004.57 to \$2255.43
8.71 25.15 to 32.65 hours **8.73** \$314.41 to \$365.59
8.75 8.65 to 10.35 hours **8.77** .32 to .38
8.79 .54 to .58 **8.81** .08 to .14
8.83 $n = 60$ **8.85** $n = 1068$

Self-Review Test: Chapter 8

1. a. Estimation means assigning values to a *population parameter* based on the value of a *sample statistic*.
   b. An estimator is the *sample statistic* used to estimate a *population parameter*.
   c. The value of a *sample statistic* is called the point estimate of the corresponding *population parameter*.

2. b 3. a 4. a 5. d 6. b
7. \$2.12 to \$2.58 8. \$255,509.69 to \$339,218.31
9. .34 to .40 10. $n = 41$ 11. $n = 1849$ 12. $n = 782$

CHAPTER 9

9.5 a. H_0: $\mu = 9.5$ hours; H_1: $\mu \neq 9.5$ hours; a two-tailed test
b. H_0: $\mu = 2.9$; H_1: $\mu < 2.9$; a left-tailed test
c. H_0: $\mu = 175$; H_1: $\mu > 175$; a right-tailed test
d. H_0: $\mu = 5$ hours; H_1: $\mu \neq 5$ hours; a two-tailed test

9.7 a. Critical values: $z = -1.65$ and 1.65; test statistic: $z = -1.53$; do not reject H_0
b. Critical value: $z = -2.33$; test statistic: $z = -5.85$; reject H_0
c. Critical value: $z = 1.65$; test statistic: $z = 7.91$; reject H_0

9.9 H_0: $\mu = 45$ months; H_1: $\mu < 45$ months; critical value: $z = -2.33$; test statistic: $z = -1.87$; do not reject H_0

9.11 H_0: $\mu = \$120{,}000$; H_1: $\mu > \$120{,}000$; critical value: $z = 2.33$; test statistic: $z = 5.06$; reject H_0

9.13 H_0: $\mu = 28$ hours; H_1: $\mu < 28$ hours; critical value: $z = -2.33$; test statistic: $z = -2.05$; do not reject H_0

9.15 H_0: $\mu = \$9.66$; H_1: $\mu > \$9.66$; critical value: $z = 1.96$; test statistic: $z = 19.26$; reject H_0

9.17 H_0: $\mu = 16.6$; H_1: $\mu \neq 16.6$; critical values: $z = -1.96$ and 1.96; test statistic: $z = 1.45$; do not reject H_0

9.19 H_0: $\mu = 18$; H_1: $\mu \neq 18$; critical values: $z = -2.58$ and 2.58; test statistic: $z = 7.50$; reject H_0

9.23 a. p-value = .0046 b. p-value = .0017 c. p-value = .0162

9.25 H_0: $\mu = \$720$; H_1: $\mu > \$720$; z for $\bar{x} = 751$ is 2.31; p-value = .0104

9.27 H_0: $\mu = 7.6$ years; H_1: $\mu < 7.6$ years; z for $\bar{x} = 6.9$ is -2.36; p-value = .0091

9.29 a. Critical values: $t = -2.797$ and 2.797; test statistic: $t = 4.082$; reject H_0
b. Critical value: $t = -2.131$; test statistic: $t = -1.818$; do not reject H_0
c. Critical value: $t = 1.328$; test statistic: $t = 2.236$; reject H_0

9.31 H_0: $\mu \le 7$; H_1: $\mu > 7$; critical value: $t = 1.729$; test statistic: $t = 7.198$; reject H_0

9.33 H_0: $\mu = \$1045$; H_1: $\mu > \$1045$; critical value: $t = 2.473$; test statistic: $t = 4.134$; reject H_0

9.35 H_0: $\mu = 292$; H_1: $\mu \neq 292$; critical values: $t = -2.093$ and 2.093; test statistic: $t = 1.830$; do not reject H_0

9.37 H_0: $\mu = \$6800$; H_1: $\mu > \$6800$; critical value: $t = 2.492$; test statistic: $t = 5.183$; reject H_0

9.39 H_0: $\mu = \$119{,}500$; H_1: $\mu > \$119{,}500$; critical value: $t = 2.485$; test statistic: $t = 4.103$; reject H_0

9.41 H_0: $\mu \ge 1200$; H_1: $\mu < 1200$; critical value: $t = -2.492$; test statistic: $t = -4.118$; reject H_0

9.43 **a.** Critical values: $z = -1.65$ and 1.65; test statistic: $z = .60$; do not reject H_0
b. Critical value: $z = -1.65$; test statistic: $z = -4.12$; reject H_0
c. Critical value: $z = 2.33$; test statistic: $z = 1.25$; do not reject H_0

9.45 H_0: $p = .25$; H_1: $p < .25$; critical value: $z = -2.33$; test statistic: $z = -1.16$; do not reject H_0

9.47 H_0: $p = .19$; H_1: $p > .19$; critical value: $z = 1.96$; test statistic: $z = 1.56$; do not reject H_0

9.49 H_0: $p = .211$; H_1: $p \neq .211$; critical values: $z = -1.96$ and 1.96; test statistic: $z = 1.06$; do not reject H_0

9.51 H_0: $p = .40$; H_1: $p > .40$; critical value: $z = 1.65$; test statistic: $z = 4.67$; reject H_0

9.53 H_0: $p = .64$; H_1: $p < .64$; critical value: $z = -1.96$; test statistic: $z = -3.33$; reject H_0

9.55 H_0: $p = .05$; H_1: $p \neq .05$; critical values: $z = -2.58$ and 2.58; test statistic: $z = 2.27$; do not reject H_0

9.57 H_0: $\mu = 2500$; H_1: $\mu < 2500$; critical value: $z = -1.96$; test statistic: $z = -1.77$; do not reject H_0

9.59 H_0: $\mu = 27$; H_1: $\mu \neq 27$; critical values: $z = -1.65$ and 1.65; test statistic: $z = 10.62$; reject H_0

9.61 H_0: $\mu = \$38{,}973$; H_1: $\mu > \$38{,}973$; critical value: $z = 2.33$; test statistic: $z = 16.13$; reject H_0

9.63 H_0: $\mu = \$322.26$; H_1: $\mu > \$322.26$; z for $\bar{x} = 329.50$ is 2.01; p-value $= .0222$

9.65 H_0: $\mu = \$1035$; H_1: $\mu \neq \$1035$; critical values: $t = -2.779$ and 2.779; test statistic: $t = 6.906$; reject H_0

9.67 H_0: $\mu \le 25$ minutes; H_1: $\mu > 25$ minutes; critical value: $t = 2.602$; test statistic: $t = 2.083$; do not reject H_0

9.69 H_0: $\mu = 7.6$ pounds; H_1: $\mu < 7.6$ pounds; critical value: $t = -2.508$; test statistic: $t = -4.192$; reject H_0

9.71 H_0: $p = .413$; H_1: $p > .413$; critical value: $z = 2.33$; test statistic: $z = 1.88$; do not reject H_0

9.73 H_0: $p = .22$; H_1: $p < .22$; critical value: $z = -1.65$; test statistic: $z = -.77$; do not reject H_0

9.75 H_0: $p = .59$; H_1: $p \neq .59$; critical values: $z = -1.96$ and 1.96; test statistic: $z = 1.82$; do not reject H_0

Self-Review Test: Chapter 9

1. a **2.** b **3.** a **4.** b **5.** a

6. a **7.** a **8.** c **9.** b **10.** d

11. c **12.** a **13.** b

14. H_0: $\mu = \$1365$; H_1: $\mu \neq \$1365$; critical values: $z = -2.58$ and 2.58; test statistic: $z = 4.46$; reject H_0

15. H_0: $\mu \geq 36$ months; H_1: $\mu < 36$ months; critical value: $t = -1.753$; test statistic: $t = -3.478$; reject H_0
16. H_0: $p = .73$; H_1: $p > .73$; critical value: $z = 2.33$; test statistic: $z = 1.82$; do not reject H_0
17. H_0: $\mu = 60$ minutes; H_1: $\mu \neq 60$ minutes; z for $\bar{x} = 62.6$ is 2.36; p-value $= .0182$

CHAPTER 10

10.3 -4.29 to -2.71

10.5 H_0: $\mu_1 - \mu_2 = 0$; H_1: $\mu_1 - \mu_2 \neq 0$; critical values: $z = -2.58$ and 2.58; test statistic: $z = -8.73$; reject H_0

10.7 H_0: $\mu_1 - \mu_2 = 0$; H_1: $\mu_1 - \mu_2 < 0$; critical value: $z = -1.65$; test statistic: $z = -8.73$; reject H_0

10.9 a. -1.37 to .77
b. H_0: $\mu_1 - \mu_2 = 0$; H_1: $\mu_1 - \mu_2 \neq 0$; critical values: $z = -2.58$ and 2.58; test statistic: $z = -.55$; do not reject H_0

10.11 a. \$2.04 to \$2.26
b. H_0: $\mu_1 - \mu_2 = 0$; H_1: $\mu_1 - \mu_2 > 0$; critical value: $z = 1.65$; test statistic: $z = 32.09$; reject H_0

10.13 a. -390.51 to -169.49 grams
b. H_0: $\mu_1 - \mu_2 = 0$; H_1: $\mu_1 - \mu_2 < 0$; critical value: $z = -1.65$; test statistic: $z = -5.50$; reject H_0

10.15 a. 4.92 to 84.88 minutes
b. H_0: $\mu_1 - \mu_2 = 0$; H_1: $\mu_1 - \mu_2 > 0$; critical value: $z = 2.05$; test statistic: $z = 2.62$; reject H_0

10.17 a. $-$\$436.36 to \$832.36
b. H_0: $\mu_1 - \mu_2 = 0$; H_1: $\mu_1 - \mu_2 \neq 0$; critical values: $z = -1.96$ and 1.96; test statistic: $z = .61$; do not reject H_0

10.19 a. -13.62 to .82
b. H_0: $\mu_1 - \mu_2 = 0$; H_1: $\mu_1 - \mu_2 < 0$; critical value: $z = -1.65$; test statistic: $z = -2.29$; reject H_0

10.21 1.21 to 9.29

10.23 H_0: $\mu_1 - \mu_2 = 0$; H_1: $\mu_1 - \mu_2 \neq 0$; critical values: $t = -2.701$ and 2.701; test statistic: $t = 3.514$; reject H_0

10.25 H_0: $\mu_1 - \mu_2 = 0$; H_1: $\mu_1 - \mu_2 > 0$; critical value: $t = 1.683$; test statistic: $t = 3.514$; reject H_0

10.27 a. 1.50 to 16.50 items
b. H_0: $\mu_1 - \mu_2 = 0$; H_1: $\mu_1 - \mu_2 > 0$; critical value: $t = 1.746$; test statistic: $t = 2.542$; reject H_0

10.29 a. 1.57 to 4.43 miles per hour
b. H_0: $\mu_1 - \mu_2 = 0$; H_1: $\mu_1 - \mu_2 > 0$; critical value: $t = 2.416$; test statistic: $t = 4.243$; reject H_0

10.31 a. -14.52 to 4.52 minutes
b. H_0: $\mu_1 - \mu_2 = 0$; H_1: $\mu_1 - \mu_2 < 0$; critical value: $t = -2.412$; test statistic: $t = -1.413$; do not reject H_0

10.33 a. \$4909.87 to \$7610.13
b. H_0: $\mu_1 - \mu_2 = 0$; H_1: $\mu_1 - \mu_2 > 0$; critical value: $t = 2.400$; test statistic: $t = 7.766$; reject H_0

10.35 4.49 to 13.23

10.37 H_0: $\mu_1 - \mu_2 = 0$; H_1: $\mu_1 - \mu_2 \neq 0$; critical values: $t = -2.756$ and 2.756; test statistic: $t = 5.586$; reject H_0

10.39 H_0: $\mu_1 - \mu_2 = 0$; H_1: $\mu_1 - \mu_2 > 0$; critical value: $t = 1.699$; test statistic: $t = 5.586$; reject H_0

10.41 a. .85 to 17.15 items
b. H_0: $\mu_1 - \mu_2 = 0$; H_1: $\mu_1 - \mu_2 > 0$; critical value: $t = 1.796$; test statistic: $t = 2.429$; reject H_0

10.43 a. 1.52 to 4.48 miles per hour
b. H_0: $\mu_1 - \mu_2 = 0$; H_1: $\mu_1 - \mu_2 > 0$; critical value: $t = 2.445$; test statistic: $t = 4.132$; reject H_0

10.45 a. −8.40 to 3.80
b. H_0: $\mu_1 - \mu_2 = 0$; H_1: $\mu_1 - \mu_2 < 0$; critical value: $t = -1.688$; test statistic: $t = -1.025$; do not reject H_0

10.47 a. 11.03 to 23.97 b. 49.69 to 62.11 c. 25.37 to 33.23

10.49 a. Critical values: $t = -2.060$ and 2.060; test statistic: $t = 12.549$; reject H_0
b. Critical value: $t = 2.624$; test statistic: $t = 2.306$; do not reject H_0
c. Critical value: $t = -1.328$; test statistic: $t = -14.397$; reject H_0

10.51 a. −15.24 to 17.82
b. H_0: $\mu_d = 0$; H_1: $\mu_d \neq 0$; critical values: $t = -2.447$ and 2.447; test statistic: $t = .289$; do not reject H_0

10.53 a. −16.63 to −3.37
b. H_0: $\mu_d = 0$; H_1: $\mu_d < 0$; critical value: $t = -1.895$; test statistic: $t = -2.857$; reject H_0

10.55 a. −3.22 to −.22 miles per gallon
b. H_0: $\mu_d = 0$; H_1: $\mu_d < 0$; critical value: $t = -3.365$; test statistic: $t = -2.950$; do not reject H_0

10.57 −.052 to .152

10.59 H_0: $p_1 - p_2 = 0$; H_1: $p_1 - p_2 \neq 0$; critical values: $z = -1.96$ and 1.96; test statistic: $z = .94$; do not reject H_0

10.61 H_0: $p_1 - p_2 = 0$; H_1: $p_1 - p_2 > 0$; critical value: $z = 2.05$; test statistic: $z = .96$; do not reject H_0

10.63 a. .056 to .124
b. H_0: $p_1 - p_2 = 0$; H_1: $p_1 - p_2 > 0$; critical value: $z = 2.33$; test statistic: $z = 6.92$; reject H_0

10.65 a. .074 to .188
b. H_0: $p_1 - p_2 = 0$; H_1: $p_1 - p_2 > 0$; critical value: $z = 2.33$; test statistic: $z = 5.95$; reject H_0

10.67 a. .035 to .105
b. H_0: $p_1 - p_2 = 0$; H_1: $p_1 - p_2 > 0$; critical value: $z = 2.33$; test statistic: $z = 3.89$; reject H_0

10.69 a. −.194 to −.124
b. H_0: $p_1 - p_2 = 0$; H_1: $p_1 - p_2 < 0$; critical value: $z = -2.33$; test statistic: $z = -10.60$; reject H_0

10.71 −5.26 to −2.84

10.73 H_0: $\mu_1 - \mu_2 = 0$; H_1: $\mu_1 - \mu_2 < 0$; critical value: $z = -1.65$; test statistic: $z = -8.65$; reject H_0

10.75 H_0: $\mu_1 - \mu_2 = 0$; H_1: $\mu_1 - \mu_2 \neq 0$; critical values: $t = -2.024$ and 2.024; test statistic: $t = 3.589$; reject H_0

10.77 a. 16.14 to 51.26 b. 8.46 to 10.74 c. 14.59 to 17.81

10.79 −.04 to .16

10.81 H_0: $p_1 - p_2 = 0$; H_1: $p_1 - p_2 > 0$; critical value: $z = 2.05$; test statistic: $z = 1.54$; do not reject H_0

10.83 **a.** \$.42 to \$.72
b. H_0: $\mu_1 - \mu_2 = 0$; H_1: $\mu_1 - \mu_2 > 0$; critical value: $z = 2.33$; test statistic: $z = 9.05$; reject H_0

10.85 **a.** −\$14.41 to −\$9.59
b. H_0: $\mu_1 - \mu_2 = 0$; H_1: $\mu_1 - \mu_2 < 0$; critical value: $z = -2.33$; test statistic: $z = -9.77$; reject H_0

10.87 **a.** −10.62 to −3.38 bushels
b. H_0: $\mu_1 - \mu_2 = 0$; H_1: $\mu_1 - \mu_2 < 0$; critical value: $t = -1.675$; test statistic: $t = -3.885$; reject H_0

10.89 **a.** −1.51 to −.69
b. H_0: $\mu_1 - \mu_2 = 0$; H_1: $\mu_1 - \mu_2 < 0$; critical value: $t = -1.675$; test statistic: $t = -7.237$; reject H_0

10.91 **a.** −10.59 to −3.41 bushels
b. H_0: $\mu_1 - \mu_2 = 0$; H_1: $\mu_1 - \mu_2 < 0$; critical value: $t = -1.676$; test statistic: $t = -3.919$; reject H_0

10.93 **a.** −1.50 to −.70
b. H_0: $\mu_1 - \mu_2 = 0$; H_1: $\mu_1 - \mu_2 < 0$; critical value: $t = -1.677$; test statistic: $t = -7.333$; reject H_0

10.95 **a.** −9.54 to −.24
b. H_0: $\mu_d = 0$; H_1: $\mu_d < 0$; critical value: $t = -2.896$; test statistic: $t = -2.424$; do not reject H_0

10.97 **a.** .095 to .165
b. H_0: $p_1 - p_2 = 0$; H_1: $p_1 - p_2 > 0$; critical value: $z = 2.05$; test statistic: $z = 8.13$; reject H_0

Self-Review Test: Chapter 10

1. a

3. **a.** 1.62 to 2.78
b. H_0: $\mu_1 - \mu_2 = 0$; H_1: $\mu_1 - \mu_2 > 0$; critical value: $z = 1.65$; test statistic: $z = 9.87$; reject H_0

4. **a.** −2.72 to −1.88
b. H_0: $\mu_1 - \mu_2 = 0$; H_1: $\mu_1 - \mu_2 < 0$; critical value: $t = -2.416$; test statistic: $t = -11.005$; reject H_0

5. **a.** −2.70 to −1.90 hours
b. H_0: $\mu_1 - \mu_2 = 0$; H_1: $\mu_1 - \mu_2 < 0$; critical value: $t = -2.421$; test statistic: $t = -11.500$; reject H_0

6. **a.** −3.43 to 1.43 items
b. H_0: $\mu_d = 0$; H_1: $\mu_d \neq 0$; critical values: $t = -2.447$ and 2.447; test statistic: $t = -1.527$; do not reject H_0

7. **a.** −.052 to .092
b. H_0: $p_1 - p_2 = 0$; H_1: $p_1 - p_2 \neq 0$; critical values: $z = -2.58$ and 2.58; test statistic: $z = .61$; do not reject H_0

CHAPTER 11

11.1 $\chi^2 = 20.483$ **11.3** $\chi^2 = 8.547$ **11.5** $\chi^2 = 14.860$

11.9 Critical value: $\chi^2 = 15.086$; test statistic: $\chi^2 = 5.972$; do not reject H_0

11.11 Critical value: $\chi^2 = 9.488$; test statistic: $\chi^2 = 11.800$; reject H_0

11.13 Critical value: $\chi^2 = 9.488$; test statistic: $\chi^2 = 3.193$; do not reject H_0

11.15 Critical value: $\chi^2 = 6.251$; test statistic: $\chi^2 = 8.721$; reject H_0

11.17 Critical value: $\chi^2 = 3.841$; test statistic: $\chi^2 = 6.250$; reject H_0

11.19 Critical value: $\chi^2 = 3.841$; test statistic: $\chi^2 = 2.819$; do not reject H_0

11.21 Critical value: $\chi^2 = 5.991$; test statistic: $\chi^2 = 68.375$; reject H_0

11.23 Critical value: $\chi^2 = 4.605$; test statistic: $\chi^2 = 1.410$; do not reject H_0

11.25 Critical value: $\chi^2 = 12.592$; test statistic: $\chi^2 = 21.586$; reject H_0

11.27 Critical value: $\chi^2 = 3.841$; test statistic: $\chi^2 = 16.850$; reject H_0

11.29 a. .9482 to 4.3698; .974 to 2.090
b. 1.5695 to 13.1276; 1.253 to 3.623

11.31 H_0: $\sigma^2 = .50$; H_1: $\sigma^2 > .50$; critical value: $\chi^2 = 24.996$; test statistic: $\chi^2 = 27.000$; reject H_0

11.33 H_0: $\sigma^2 = 1.7$; H_1: $\sigma^2 \neq 1.7$; critical values: $\chi^2 = 5.009$ and 24.736; test statistic: $\chi^2 = 27.529$; reject H_0

11.35 a. .0069 to .0389; .083 to .197
b. H_0: $\sigma^2 = .01$; H_1: $\sigma^2 > .01$; critical value: $\chi^2 = 30.144$; test statistic: $\chi^2 = 26.600$; do not reject H_0

11.37 a. H_0: $\sigma^2 \leq .003$; H_1: $\sigma^2 > .003$; critical value: $\chi^2 = 48.278$; test statistic: $\chi^2 = 54.133$; reject H_0
b. .0034 to .0120; .058 to .110

11.39 Critical value: $\chi^2 = 9.488$; test statistic: $\chi^2 = 10.281$; reject H_0

11.41 Critical value: $\chi^2 = 11.345$; test statistic: $\chi^2 = 2.768$; do not reject H_0

11.43 Critical value: $\chi^2 = 11.345$; test statistic: $\chi^2 = 12.229$; reject H_0

11.45 Critical value: $\chi^2 = 6.635$; test statistic: $\chi^2 = 10.203$; reject H_0

11.47 Critical value: $\chi^2 = 4.605$; test statistic: $\chi^2 = 13.529$; reject H_0

11.49 Critical value: $\chi^2 = 16.812$; test statistic: $\chi^2 = 10.181$; do not reject H_0

11.51 a. 3.8577 to 16.5902; 1.964 to 4.073
b. 8.9161 to 27.4121; 2.986 to 5.236

11.53 H_0: $\sigma^2 = .7$; H_1: $\sigma^2 > .7$; critical value: $\chi^2 = 28.845$; test statistic: $\chi^2 = 32.000$; reject H_0

11.55 H_0: $\sigma^2 = 10.0$; H_1: $\sigma^2 \neq 10.0$; critical values: $\chi^2 = 8.231$ and 31.526; test statistic: $\chi^2 = 25.740$; do not reject H_0

11.57 a. H_0: $\sigma^2 = 5000$; H_1: $\sigma^2 < 5000$; critical value: $\chi^2 = 8.907$; test statistic: $\chi^2 = 12.160$; do not reject H_0
b. 1679.9757 to 7965.4133; 40.988 to 89.249

Self-Review Test: Chapter 11

1. b 2. a 3. c 4. a 5. b
6. b 7. c 8. b 9. a

10. Critical value: $\chi^2 = 9.210$; test statistic: $\chi^2 = 111.005$; reject H_0

11. Critical value: $\chi^2 = 11.345$; test statistic: $\chi^2 = 28.435$; reject H_0

12. Critical value: $\chi^2 = 3.841$; test statistic: $\chi^2 = 4.888$; reject H_0

13. a. .0316 to .1457; .178 to .382
b. H_0: $\sigma^2 = .03$; H_1: $\sigma^2 > .03$; critical value: $\chi^2 = 42.980$; test statistic: $\chi^2 = 48.000$; reject H_0

CHAPTER 12

12.3 a. y-intercept = −60; slope = 8 b. y-intercept = 44; slope = 4
c. y-intercept = 75; slope = −5

12.7 a. SS_{xx} = 153.714; SS_{xy} = 100.143
b. $\hat{y} = .19 + .65x$

12.9 b. $\hat{y} = 107.97 - 11.35x$ e. $2852

12.11 b. $\hat{y} = 210.35 + 24.56x$ e. $824.35

12.13 a. $\mu_{yx} = 29.48 + 1.48x$ b. population regression line; the values give A and B
d. 49.31%

12.17 a. SS_{xx} = 363.875; SS_{yy} = 67.500; SS_{xy} = 143.250
b. s_e = 1.392
c. SST = 67.500; SSE = 11.111; SSR = 56.389
d. r^2 = .83

12.19 a. SS_{xx} = 1230.400; SS_{yy} = 9185.600; SS_{xy} = 2690.800
b. s_e = 20.288
c. SST = 9185.600; SSE = 3301.018; SSR = 5884.582
d. r^2 = .64

12.21 a. s_e = 6.564 b. r^2 = .76

12.23 a. s_e = 60.410 b. r^2 = .88

12.25 a. $\hat{y} = -1.28 + .05x$ b. .00 to .10
c. H_0: $B = 0$; H_1: $B > 0$; critical value: t = 2.306; test statistic: t = 3.571; reject H_0

12.27 a. $\hat{y} = 106.82 - 5.05x$ b. −8.36 to −1.74
c. H_0: $B = 0$; H_1: $B < 0$; critical value: t = −2.571; test statistic: t = −3.927; reject H_0

12.29 a. $\hat{y} = -125.61 + 8.76x$ b. 2.81 to 14.71
c. H_0: $B = 0$; H_1: $B > 0$; critical value: t = 3.747; test statistic: t = 5.513; reject H_0

12.31 a. $\hat{y} = 2.22 + 2.19x$ b. .52 to 3.86
c. H_0: $B = 1.50$; H_1: $B > 1.50$; critical value: t = 2.896; test statistic: t = 1.194;
do not reject H_0

12.33 a. positive b. positive c. positive d. negative e. zero

12.35 a. close to zero b. r = .11

12.37 a. positively related b. close to 1 c. r = .97

12.39 −.87; yes

12.41 .80

12.43 a. positively related b. .74

12.45 a. SS_{xx} = 885.429; SS_{yy} = 751.429; SS_{xy} = 750.429
b. $\hat{y} = 12.81 + .85x$ e. r = .92; r^2 = .85
f. s_e = 4.766 g. 89.31 h. .20 to 1.50
i. H_0: $B = 0$; H_1: $B > 0$; critical value: t = 2.015; test statistic: t = 5.313; reject H_0

12.47 a. SS_{xx} = 2104.857; SS_{yy} = 1488.857; SS_{xy} = 1726.857
b. $\hat{y} = 43.53 + .82x$ d. r = .98; r^2 = .95
e. s_e = 3.817 f. 129.63 g. .54 to 1.10
h. H_0: $B = 0$; H_1: $B \neq 0$; critical values: t = −2.571 and 2.571; test statistic: t = 9.880;
reject H_0

12.49 a. SS_{xx} = 1510.400; SS_{yy} = 500.400; SS_{xy} = 770.400
b. $\hat{y} = -7.63 + .51x$ d. r = .89; r^2 = .79
e. s_e = 3.666 f. .19 to .83
g. H_0: $B = 0$; H_1: $B > 0$; critical value: t = 2.896; test statistic: t = 5.426; reject H_0

12.51 a. $SS_{xx} = 82.000$; $SS_{yy} = 7944.875$; $SS_{xy} = 781.500$
b. $\hat{y} = 5.40 + 9.53x$ d. $r = .97$; $r^2 = .94$
e. $s_e = 9.103$ f. 7.07 to 11.99
g. H_0: $B = 0$; H_1: $B \neq 0$; critical values: $t = -3.707$ and 3.707; test statistic: $t = 9.483$; reject H_0

12.53 a. $SS_{xx} = 1812$; $SS_{yy} = 138$; $SS_{xy} = -481$
b. $\hat{y} = 98.34 - .27x$ d. $r = -.96$; $r^2 = .94$
e. $s_e = 1.275$ f. $-.37$ to $-.17$
g. H_0: $B = 0$; H_1: $B < 0$; critical value: $t = -3.365$; test statistic: $t = -9.000$; reject H_0

12.55 a. 13.87 to 16.63; 11.76 to 18.74 b. 62.36 to 67.72; 56.36 to 73.72

12.57 18.12 to 20.26; 15.86 to 22.52

12.59 76.15 to 85.47; 67.70 to 93.92

12.61 113.60 to 121.06; 106.83 to 127.83

12.63 9.67 to 19.95; 1.48 to 28.14

12.65 a. $SS_{xx} = 1018.000$; $SS_{yy} = 147.429$; $SS_{xy} = -154.000$
b. $\hat{y} = 19.86 - .15x$ c. $r = -.40$; $r^2 = .16$ d. $s_e = 4.987$

12.67 a. $SS_{xx} = 299.875$; $SS_{yy} = 849.500$; $SS_{xy} = 471.750$
b. $\hat{y} = 13.46 + 1.57x$ c. $r = .93$; $r^2 = .87$ d. $s_e = 4.259$

12.69 b. $\hat{y} = 18.39 + 2.23x$ e. 47.38% f. $r = .75$; $r^2 = .56$
g. $s_e = 7.514$ h. -2.28 to 6.74
i. H_0: $B = 0$; H_1: $B \neq 0$; critical values: $t = -4.604$ and 4.604; test statistic: $t = 2.276$; do not reject H_0

12.71 a. positive b. $\hat{y} = -1.43 + .89x$; yes d. $r = .97$; $r^2 = .94$
e. $s_e = .943$ f. .56 to 1.22
g. H_0: $B = 0$; H_1: $B > 0$; critical value: $t = 2.571$; test statistic: $t = 9.175$; reject H_0

12.73 a. $\hat{y} = 1.20 + .60x$ b. 3.12 c. $r = .51$; $r^2 = .26$
d. $s_e = .579$ e. $-.41$ to 1.61
f. H_0: $B = 0$; H_1: $B \neq 0$; critical values: $t = -3.143$ and 3.143; test statistic: $t = 1.453$; do not reject H_0

12.75 a. positive b. $\hat{y} = 6.57 + .68x$; yes d. $r = .93$; $r^2 = .86$
e. 23.57 f. $s_e = 3.821$ g. .37 to .99
h. H_0: $B = 0$; H_1: $B > 0$; critical value: $t = 2.998$; test statistic: $t = 6.602$; reject H_0

12.77 b. $SS_{xx} = 60$; $SS_{yy} = 288$; $SS_{xy} = 121$
d. $\hat{y} = 128.57 + 2.02x$ f. $r = .92$ g. 152.81 billion pounds

12.79 4.25 to 7.13; 1.63 to 9.75

12.81 198.41 to 234.97; 171.60 to 261.78

Self-Review Test: Chapter 12

1. d 2. a 3. b 4. a 5. b
6. b 7. true 8. true 9. a 10. b

15. a. The number of computers sold depends on experience.
b. positive d. $\hat{y} = 16.63 + 1.54x$; yes
f. $r = .71$; $r^2 = .51$ g. 33.57 or 34 approximately
h. $s_e = 6.686$ i. -1.20 to 4.28
j. H_0: $B = 0$; H_1: $B > 0$; critical value: $t = 3.365$; test statistic: $t = 2.268$; do not reject H_0
k. 22.15 to 35.75 l. 10.47 to 47.43

CHAPTER 13

13.1 a. 3.11 b. 2.93 c. 3.57

13.3 a. 4.74 b. 4.82 c. 9.72

13.7 Critical value: $F = 6.93$; test statistic: $F = 2.89$; do not reject H_0

13.9 Critical value: $F = 3.68$; test statistic: $F = 6.94$; reject H_0

13.11 Critical value: $F = 5.95$; test statistic: $F = 1.63$; do not reject H_0

13.13 Critical value: $F = 4.69$; test statistic: $F = 10.20$; reject H_0

13.15 Critical value: $F = 3.59$; test statistic: $F = 4.73$; reject H_0

13.17 Critical value: $F = 5.74$; test statistic: $F = 7.44$; reject H_0

13.19 Critical value: $F = 4.62$; test statistic: $F = 2.05$; do not reject H_0

13.21 Critical value: $F = 3.34$; test statistic: $F = 4.34$; reject H_0

13.23 Critical value: $F = 4.62$; test statistic: $F = 6.56$; reject H_0

Self-Review Test: Chapter 13

1. a 2. b 3. c 4. a

5. a 6. a 7. b 8. a

10. Critical value: $F = 6.11$; test statistic: $F = 2.31$; do not reject H_0

INDEX

Addition rule of probability, 206
  for mutually exclusive events, 208–209
Alpha (α), 381
  level of significance, 381, 429
  probability of type I error, 429
Alternative hypothesis, 426–427
Analysis of variance (ANOVA), one-way, 666–667, 671
  alternative hypothesis, 666
  assumptions of, 667
  between-samples sum of squares (SSB), 669
  null hypothesis, 666
  table, 671
  test statistic, 668
  total sum of squares (SST), 669
  variance (or mean square):
    between samples (MSB), 667, 670
    within samples (MSW), 667, 670
  within-samples sum of squares (SSW), 669
Arithmetic mean, 96
Average, 96

Bar chart, 41
Bar graph, 41–42, 58, 75–76
Bell-shaped distribution/curve, 128
Bernoulli trials, 249
β (Beta), probability of type II error, 429
Bimodal distribution, 104
Binomial distribution, 248, 250
  mean of, 260–261
  normal approximation to, 321
  probability histogram, 255
  probability of failure for, 249
  probability of success for, 249
  standard deviation of, 260–261
  table, 726
  using the table of, 257
Binomial experiment, 249
  conditions of, 249
Binomial formula, 251
Binomial parameters, 251
Binomial random variable, 250
BMDP, 26
Boundaries, class, 48
Box-and-whisker plot, 136

Categorical variable, 15
Cell, 176, 558
Census, 7, 690
Central limit theorem, 353, 364
Central tendency, measures of, 96
Chance experiment, 224
Chance variable, 224
Chebyshev's theorem, 126
Chi-square distribution, 546
  degrees of freedom, 546
  goodness-of-fit test, 548–551
  table of, 744
  test of homogeneity, 566
  test of independence, 558–559
  using the table of, 547–548
Class, 47
  boundaries, 48
  frequency, 39
  limits, 48
  midpoint/mark, 49
  width/size, 48, 49–50
Classical probability, 168
  concept, 168
  rule, 168
Cluster, 698
Cluster sampling, 698
Coefficient of determination, 614, 617
Coefficient of linear correlation, 626
Combinations, 244–245
Combined mean, 108
Complementary events, 186
Composite events, 163
Compound events, 162–163
Conditional probability, 177–178, 195
Confidence coefficient, 381
Confidence interval, 381
  for the difference between two means:
    large and independent samples, 476–477
    small and independent samples, 489–490, 498
  for the difference between two proportions, large and independent samples, 515
  for the mean:
    large sample, 381–382
    small sample, 390–393
  for the mean of paired differences, 506
  for the mean value of y for a given x, 640
  for the predicted value of y for a given x, 642
  for the proportion, large sample, 400–401
  for the slope of regression line, 620
  for the standard deviation, 572
  for the variance, 572

Confidence interval:
lower limit of, 380
maximum error of, 382
upper limit of, 380
Confidence level, 381
Constant, 12
Contingency table, 176, 558
Continuous probability distribution, 286–287
Continuous random variable, 225
Continuous variable, 15
Control group, 692
Convenience sample, 693
Correction factor, finite population, 346, 363
Correction for continuity, 323
Correlation coefficient, 626
population, 625
sample, 625–626
Counting rule, 200
Critical value, 427
Cross-section data, 16
Cumulative frequency distribution, 66
Cumulative percentage distribution, 67
Cumulative relative frequency distribution, 67

Data, 13
cross-section, 16
grouped, 48
primary, 689
qualitative, 15
quantitative, 14
ranked, 100–101
raw, 38
secondary, 689
sources of, 17, 689
time-series, 17
ungrouped, 38
Data set, 5, 13
Degrees of freedom, defined, 391–392
for the chi-square distribution, 546
for the *F* distribution, 664
for the *t* distribution, 392
Dependent events, 184
Dependent samples, 474
Dependent variable in regression, 594
Descriptive statistics, 4–5
Determining the sample size:
for estimating the mean, 407
for estimating the proportion, 409
Deterministic regression model, 597
Deviation from the mean, 111
Difference between two population means:
confidence interval for, 476–477, 489–490, 498
tests of hypotheses about, 479–480, 491, 499
Difference between two population proportions:
confidence interval for, 515
tests of hypotheses about, 516–517
Difference between two sample means:
mean of, 475–476
sampling distribution of, 475–476
standard deviation of, 475–476
Difference between two sample proportions:
mean of, 514–515
sampling distribution of, 514–515
standard deviation of, 514–515
Discrete random variable, 224
expected value of, 233
mean of, 233–234
probability distribution of, 226
standard deviation of, 238
variance of, 238
Discrete variable, 14
Dispersion, measures of, 109
Distribution:
bell-shaped, 128
binomial, 248, 250
chi-square, 546
cumulative frequency, 66
cumulative percentage, 67
cumulative relative frequency, 67
F, 664
frequency, 39, 47–48
normal, 291–292
percentage, 40, 52
Poisson, 264–266
probability, 226, 286–287
relative frequency, 40, 52
sampling, 341
standard normal, 294–295
t, 391–392

Element, 5, 12
Elementary event, 162
Empirical rule, 128–129
Equally likely outcomes or events, 168
Equation of a regression model, 597
Error:
nonsampling, 343, 694
random, 597
sampling, 343, 694
standard, 345
sum of squares (SSE), 600
type I, 428–429
type II, 428–429
Estimate, 379
interval, 380
point, 379
Estimated regression line, 598
Estimates of regression coefficients, 598
Estimation, 378
Estimator, 379
Event(s), 162
complement of, 186
complementary, 186
compound or composite, 162–163
dependent, 184
equally likely, 168
impossible, 167
independent, 183
intersection of, 190
mutually exclusive, 182

simple or elementary, 162
sure, 167
union of, 204
Expected frequencies:
for a goodness-of-fit test, 549–550
for a test of independence or homogeneity, 559–560
Expected value of a discrete random variable, 233
Experiment, 160, 692
binomial, 249
double-blind, 692
multinomial, 548–549
random or chance, 224
statistical, 160
Extrapolation, 643
Extreme values, 98

Factorial formula, 243
Factorials, 242–243
table of, 724
Failure, probability of, 249
F distribution, 664
degrees of freedom for, 664
table of, 745
Finite population correction factor, 346, 363
First quartile, 130–131
Fisher, Sir Ronald, 664
Frequency, 39, 47–48
cumulative, 66
cumulative relative, 67
expected, 549–550, 559–560
histogram, 53
joint, 558
observed, 549–550, 559
polygon, 55
relative, 40, 52
Frequency density, relative, 286
Frequency distribution:
curve, 55
for qualitative data, 39
for quantitative data, 47–48

Galton, Sir Francis, 594
Goodness-of-fit test, 548–551
degrees of freedom for, 550
expected frequencies for, 549–550
observed frequencies for, 549–550
test statistic for, 551
Gossett, W. S., 391
Grouped data, 48

Hinge:
lower, 136
upper, 136
Histogram(s), 53
frequency, 53
percentage, 53
probability, 228
relative frequency, 53
shapes of, 58
skewed to the left, 58–59
skewed to the right, 58–59
symmetric, 58–59
uniform/rectangular, 58–59
Homogeneity, chi-square test of, 566
Hypothesis:
alternative, 426–427
null, 426–427
Hypothesis test, *see* Test of hypothesis

Impossible event, 167
Independence, chi-square test of, 558–559
degrees of freedom for, 559
Independent events, 183
Independent samples, 474
Independent variable in a regression model, 594
Inductive statistics, 6
Inferential statistics, 5–6
Intersection of events, 190
Interval estimate, 380

Joint frequency, 558
Joint probability, 191
of independent events, 195–196
of mutually exclusive events, 199
Judgment sample, 693

Law of large numbers, 171
Leaf/leaves, 70–71
Least squares, method of, 599–601
Least squares regression line, 599–601
Left-tailed test, 431–432
Level of confidence, 381
Level of significance, 381, 429
Linear correlation, 625–627
Linear correlation coefficient, 626
Linear regression, 594–595
Linear relationship, 595
Line graph, 75
Lower class boundary, 48
Lower class limit, 48
Lower hinge, 136

Marginal probability, 176–177
Matched samples, 474, 503–504
Maximum error:
of estimating the mean, 382
of estimating the proportion, 409
Mean, 96–97, 117–118
arithmetic, 96
of the binomial distribution, 260–261
combined, 108
confidence interval for, 381–382, 390–393
deviation from, 111
of the discrete random variable, 233–234
for grouped data, 117–118
of the normal distribution, 292
population, 97
sample, 97

Mean *(Continued)*
of the sampling distribution of the difference between:
two sample means, 475–476
two sample proportions, 415
of the sampling distribution of the mean of paired differences, 505
sampling distribution of the sample, 341
of the sampling distribution of sample mean, 345–346
of the sampling distribution of the slope of a regression line, 620
shape of the sampling distribution of sample, 349–350, 352–354
standard deviation of the sample, 345–347
tests of hypotheses about, 434–435, 446–447
trimmed, 109
for ungrouped data, 96–97
Mean deviation, 111
Measurement, 13
Measures of central tendency for ungrouped data, 96
combined mean, 108
mean, 96–97
median, 100–101
mode, 103
trimmed mean, 109
Measures of dispersion for ungrouped data, 109
range, 110
standard deviation, 110–112
variance, 110–112
Measures of position, 130
percentile rank, 133–134
percentiles, 132–133
quartiles, 130–131
Measures of spread, 109
Median for ungrouped data, 100–101
Method of least squares, 599–601
MINITAB:
environment, 27
input, 27
introduction, 26
output, 27
worksheet, 29, 32
MINITAB commands:
ADD, 32
ALTERNATIVE, 466
AOVONEWAY, 685
BINOMIAL, 279
BOXPLOT, 153
CDF, 334
CHISQUARE, 588–589
COPY, 31
CORRELATION, 655
DELETE, 31
DESCRIBE, 153
DIVIDE, 31–32
DOTPLOT, 90
END, 28
ERASE, 31
HELP COMMANDS, 31
HELP HELP, 31
HELP OVERVIEW, 31
HISTOGRAM, 88
INFORMATION, 29
INSERT, 31
LET, 31
MAXIMUM, 153
MEAN, 153
MEDIAN, 153
MINIMUM, 153
MULTIPLY, 31–32
NAME, 28, 655
NOPAPER, 27
NORMAL, 334
PAPER, 27
PDF, 279
PLOT, 655
POISSON, 281
POOLED, 534
PREDICT, 655
PRINT, 28
RANDOM, 700
RANGE, 153
READ, 27, 30
REGRESS, 655
RETRIEVE, 29
SAMPLE, 29–30, 701
SAVE, 29
SET, 27–28
SORT, 32
STDEV, 153
STEM-AND-LEAF, 90
STOP, 27
SUBTRACT, 32
SUM, 153
TINTERVAL, 420, 540
TTEST, 468, 540
TWOSAMPLE, 534
ZINTERVAL, 418
ZTEST, 466
MINITAB prompt, 27
DATA 〉, 28
MTB 〉, 27
SUBCOM 〉, 89
Mode for ungrouped data, 103
Multimodal distribution, 104
Multinomial experiment, 548–549
Multiplication rule of probability, 191
for independent events, 195–196
Mutually exclusive events, 182

Negative correlation, 625–626
Nonlinear regression, 594
Nonlinear relationship, 595
Nonrejection region, 427–428
Nonsampling errors, 343, 694

Normal distribution, 291–292
  as an approximation to binomial distribution, 321
  characteristics of, 292–293
  equation of, 293
  mean of, 292
  parameters of, 293
  standard, 294–295
  standard deviation of, 292
  standardizing, 303–304
  table of the standard, 741
  using the table of the standard, 295–296
Normal random variable, 292
Null hypothesis, 426–427

Observation, 13
Observed frequencies, 549
  for a goodness-of-fit test, 549–550
  for a test of independence or homogeneity, 559
Odds, 173–174
Ogive, 67–68
One-tailed test, 430
One-way analysis of variance, 666–667, 671
  assumptions of, 667
One-way ANOVA table, 671
Outcome, 160
Outliers, 98

Paired differences, 504
  confidence interval for the mean of, 506
  mean of the population, 504
  mean of the sample, 504–505
  sampling distribution of the mean of the sample, 505
  standard deviation of the population, 504
  standard deviation of the sample, 504–505
  test of hypothesis about the mean of, 508
Paired samples, 474, 503–504
Parameters, 115
Percentage distribution:
  cumulative, 67
  for qualitative data, 40
  for quantitative data, 52
Percentile rank, 133–134
Percentiles, 132–133
Pie chart, 43
Point estimate, 379
Poisson, Simeon D., 264
Poisson distribution, 264–266
  conditions to apply, 265
  formula, 266
  parameter of, 266
  table, 735
  using the table of, 268–269
Polygon, 55
  frequency, 55
  percentage, 55
  relative frequency, 55
Pooled sample proportion, 517
Pooled standard deviation for two samples, 489
Population, 5, 7
  distribution, 340
  mean, 97
  proportion, 360–361
  size, 51
  standard deviation, 111–112, 120
  variance, 111–112, 120
Population correlation coefficient, 625
Population parameters, 115
Population regression line, 597–598
Position, measures of, 130
Positive correlation, 625–626
Prediction interval for y for a given x, 642
Probabilistic regression model, 597
Probability, 6, 167
  as an area under the curve, 287
  concept of classical, 168
  conditional, 177–178, 195
  histogram, 228
  of intersection of events, 191
  joint, 191
  marginal, 176–177
  of mutually exclusive events, 199
  properties of, 167
  relative frequency concept of, 170
  subjective, 172
  of type I error, 429
  of type II error, 429
  of union of events, 206
Probability distribution, 226, 286–287
  binomial, 248, 250
  characteristics of, 227
  normal, 291–292
  of a discrete random variable, 226
  of a continuous random variable, 286–287
  Poisson, 264–266
Proportion:
  population, 360–361
    confidence interval for, 400–401
    maximum error of estimate for, 409
    tests of hypotheses about, 453
  sample, 360–361
    mean of the sampling distribution of, 363
    sampling distribution of, 361
    standard deviation of, 363–364
Pseudo polls, 693
p-value:
  for a one-tailed test, 442–443
  for a two-tailed test, 443
p-value approach to hypothesis testing, 442–443

Qualitative data, 15
  frequency distribution for, 39
  percentage distribution for, 40
  relative frequency for, 40

Qualitative variable, 15
Quantitative data, 14
  cumulative frequency distribution for, 66
  cumulative percentage distribution for, 67
  cumulative relative frequency distribution for, 67
  frequency distribution for, 47–48
  percentage distribution for, 52
  relative frequency distribution for, 52
Quantitative variable, 14
Quartiles, 130–131
  first, 130–131
  second, 130–131
  third, 130–131

Random error, 597
Random experiment, 224
Random numbers:
  table of, 720
  using the table of, 696–697
Random sample, 10, 693
  simple, 10, 695
Random variable, 224
  continuous, 225
  discrete, 224
Range, 110
Rank, percentile, 133–134
Ranked data, 100–101
Raw data, 38
Real class limits, 48
Rectangular histogram, 58–59
Regression:
  analysis, 596
  assumptions of, 607–609
  coefficient of determination, 614, 617
  confidence interval for the mean value of y for a given x, 640
  confidence interval for the slope, 620
  dependent variable, 594
  deterministic model, 597
  error sum of squares (SSE), 600
  independent variable, 594
  least squares line, 599–601
  least squares method, 599–601
  linear, 594–595
  model, 594–596
  nonlinear, 594
  observed or actual value of y, 599
  predicted value of y, 598–599
  prediction interval for y for a given x, 642
  probabilistic model, 597
  regression sum of squares (SSR), 616
  scatter diagram, 598–599
  simple, 594
  standard deviation of errors, 612–613
  tests of hypotheses about the slope, 621
  total sum of squares (SST), 615
Regression line/model
  estimated, 598
  population, 597–598
  random error of, 597
  sample, 598
  slope of the, 595
  true, 597–598
  y-intercept of, 595
Regression model, equation of, 597
Regression sum of squares (SSR), 616
Rejection region, 427–428
Relative frequency, 40, 52
  cumulative, 67
  histogram, 53
  polygon, 55
  for qualitative data, 40
  for quantitative data, 52
Relative frequency concept of probability, 170
Relative frequency density, 286
Relative frequency distribution, *see* Relative frequency
Representative sample, 9, 693
Residual, 599
Right-tailed test, 432–433

Sample, 5, 7, 690
  cluster, 698
  convenience, 693
  judgment, 693
  mean, 97
  proportion, 360–361
  random, 10, 693
  representative, 9, 693
  simple random, 10, 695–696
  standard deviation, 111–112, 120
  stratified, 697–698
  systematic, 696–697
  variance, 111–112, 120
Sample points, 160
Samples:
  dependent, 474
  independent, 474
  paired, 474, 503–504
  matched, 474, 503–504
Sample size, 51
  determination of:
    for estimating the mean, 407
    for estimating the proportion, 409
Sample space, 160
Sample statistic, 115
Sample survey, 8, 690
Sampling:
  cluster, 698
  error, 343, 694
  with replacement, 11
  without replacement, 11
  stratified, 697–698
  systematic, 696–697
Sampling distribution, 341
  of the difference between two sample means, large and independent samples, 475–476

of the difference between two sample proportions, large and independent samples, 514–515
of the mean of sample paired differences, 505
of the sample mean, 341
of the sample proportion, 361
of the slope of a regression line, 619–620
SAS, 26
Scatter diagram, 598–599
Scattergram, 598–599
Second quartile, 130–131
Significance level, 381, 429
Simple event, 162
Simple linear regression, 594–596
Simple random sample, 10, 695–696
Single-valued classes, 57
Size:
population, 51
sample, 51
Skewed histogram, 58–59
to the left, 58–59
to the right, 58–59
Slope of the regression line, 595
confidence interval for, 620
tests of hypotheses about, 621
Sources of data, 17, 689
external, 17, 689
internal, 17, 689
SPSS, 26
Standard deviation:
of the binomial distribution, 260–261
confidence interval for, 572
of the difference between two sample means, 475–476
of the difference between two sample proportions, 514–515
of the discrete random variable, 238
of the errors for a regression model, 612–613
for grouped data, 120
of the mean of paired differences, 505
of the normal distribution, 292
population, 111–112, 120
sample, 111–112, 120
of the sample mean, 346–347
of the sample proportion, 363–364
of the slope of a regression line, 620
of the standard normal distribution, 294
for ungrouped data, 110–112
use of, 126
Standard error, 345
Standardizing a normal distribution, 303–304
Standard normal distribution, 294–295
table of, 741
using the table of, 295–296
Standard score, 294
Standard units, 294
Statistic(s), 4, 115
descriptive, 4–5
inferential, 5–6
Statistically not significant, 440
Statistically significant, 440
Stem, 70–71
Stem-and-leaf display, 70
grouped, 72
ranked, 71
Strata, 697
Stratified sampling, 697–698
Subjective probability, 172
Success, probability of, 249
Summation notation, 18
Sum of squares:
error (SSE), 600
regression (SSR), 616
total (SST), 615
Sure event, 167
Survey, 7, 689–690
Symmetric histogram, 58–59
Systematic sampling, 696–697

t distribution, 391–392
table of, 742
degrees of freedom of, 391–392
Tables:
binomial probabilities, 726
chi-square distribution, 744
combinations, 725
factorials, 724
Poisson probabilities, 735
random numbers, 720
the F distribution, 745
the standard normal distribution, 741
the t distribution, 742
Tallies, 40
Target population, 7, 690
Test of hypothesis:
ANOVA, 671
about the difference between two means:
large and independent samples, 479–480
small and independent samples, 491, 499
about the difference between two proportions, large and independent samples, 516–517
goodness-of-fit, 548–551
homogeneity, 566
independence, 558–559
left-tailed, 431–432
about the mean:
large sample, 434–435
small sample, 446–447
about the mean of paired differences, 508
nonrejection region for, 427–428
one-tailed, 430
about the proportion, large sample, 453
p-value approach to make, 442–443
rejection region for, 427–428
right-tailed, 432–433
significance level for, 381, 429
about the slope of regression line, 621
steps of, 434

Test of hypothesis *(Continued)*
two-tailed, 430–431
type I error of, 428–429
type II error of, 428–429
about the variance, 574
Test statistic, tests of hypotheses:
ANOVA, 668
difference between two means:
large and independent samples, 480
small and independent samples, 491, 499
difference between two proportions, large and independent samples, 517
goodness-of-fit, 551
independence or homogeneity, 559
mean:
large samples, 435
small samples, 447
mean of paired differences, 508
proportion, large samples, 453
slope of regression line, 621
variance, 574
Third quartile, 130–131
Time-series data, 17
graphs for, 74–76
Total sum of squares:
ANOVA, 669
regression, 615
Treatment, 668
Treatment group, 692
Tree diagram, 160
Trial, 249
Trimmed mean, 109
Truncation, 53
Two-tailed test of hypothesis, 430–431
Type I error, 428–429
probability of, 429
Type II error, 428–429
probability of, 429
Typical value, 96

Ungrouped data, 38
Uniform histogram, 58–59
Unimodal distribution, 104
Union of events, 204
probability of, 206
Upper class boundary, 48
Upper class limit, 48
Upper hinge, 136

Variable, 12, 14–15
categorical, 15
continuous, 15
dependent, 594
discrete, 14
independent, 594
qualitative, 15
quantitative, 14
random, 224
Variance:
analysis of, 666–667
confidence interval for, 572
of a discrete random variable, 238
for grouped data, 120
population, 111–112, 120
sample, 111–112, 120
between samples, 667, 670
within samples, 667, 670
tests of hypotheses about, 574
for ungrouped data, 110–112
Variation, measures of, 109
Venn diagram, 160

Whiskers, 136–137

y-intercept in a regression model, 595

z score, 294–295
z value, 294–295

TABLE VII STANDARD NORMAL DISTRIBUTION TABLE

The entries in the table give the areas under the standard normal curve from 0 to z.

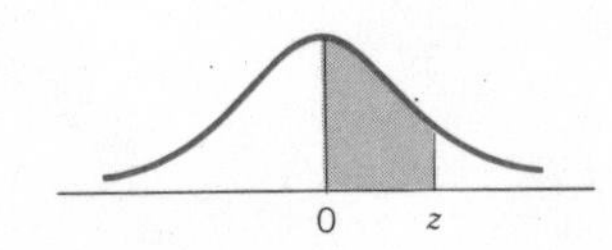

z	.00	.01	.02	.03	.04	.05	.06	.07	.08	.09
0.0	.0000	.0040	.0080	.0120	.0160	.0199	.0239	.0279	.0319	.0359
0.1	.0398	.0438	.0478	.0517	.0557	.0596	.0636	.0675	.0714	.0753
0.2	.0793	.0832	.0871	.0910	.0948	.0987	.1026	.1064	.1103	.1141
0.3	.1179	.1217	.1255	.1293	.1331	.1368	.1406	.1443	.1480	.1517
0.4	.1554	.1591	.1628	.1664	.1700	.1736	.1772	.1808	.1844	.1879
0.5	.1915	.1950	.1985	.2019	.2054	.2088	.2123	.2157	.2190	.2224
0.6	.2257	.2291	.2324	.2357	.2389	.2422	.2454	.2486	.2517	.2549
0.7	.2580	.2611	.2642	.2673	.2704	.2734	.2764	.2794	.2823	.2852
0.8	.2881	.2910	.2939	.2967	.2995	.3023	.3051	.3078	.3106	.3133
0.9	.3159	.3186	.3212	.3238	.3264	.3289	.3315	.3340	.3365	.3389
1.0	.3413	.3438	.3461	.3485	.3508	.3531	.3554	.3577	.3599	.3621
1.1	.3643	.3665	.3686	.3708	.3729	.3749	.3770	.3790	.3810	.3830
1.2	.3849	.3869	.3888	.3907	.3925	.3944	.3962	.3980	.3997	.4015
1.3	.4032	.4049	.4066	.4082	.4099	.4115	.4131	.4147	.4162	.4177
1.4	.4192	.4207	.4222	.4236	.4251	.4265	.4279	.4292	.4306	.4319
1.5	.4332	.4345	.4357	.4370	.4382	.4394	.4406	.4418	.4429	.4441
1.6	.4452	.4463	.4474	.4484	.4495	.4505	.4515	.4525	.4535	.4545
1.7	.4554	.4564	.4573	.4582	.4591	.4599	.4608	.4616	.4625	.4633
1.8	.4641	.4649	.4656	.4664	.4671	.4678	.4686	.4693	.4699	.4706
1.9	.4713	.4719	.4726	.4732	.4738	.4744	.4750	.4756	.4761	.4767
2.0	.4772	.4778	.4783	.4788	.4793	.4798	.4803	.4808	.4812	.4817
2.1	.4821	.4826	.4830	.4834	.4838	.4842	.4846	.4850	.4854	.4857
2.2	.4861	.4864	.4868	.4871	.4875	.4878	.4881	.4884	.4887	.4890
2.3	.4893	.4896	.4898	.4901	.4904	.4906	.4909	.4911	.4913	.4916
2.4	.4918	.4920	.4922	.4925	.4927	.4929	.4931	.4932	.4934	.4936
2.5	.4938	.4940	.4941	.4943	.4945	.4946	.4948	.4949	.4951	.4952
2.6	.4953	.4955	.4956	.4957	.4959	.4960	.4961	.4962	.4963	.4964
2.7	.4965	.4966	.4967	.4968	.4969	.4970	.4971	.4972	.4973	.4974
2.8	.4974	.4975	.4976	.4977	.4977	.4978	.4979	.4979	.4980	.4981
2.9	.4981	.4982	.4982	.4983	.4984	.4984	.4985	.4985	.4986	.4986
3.0	.4987	.4987	.4987	.4988	.4988	.4989	.4989	.4989	.4990	.4990